4. Answers to Problems-
provide answers to all end-of-chapter questions and problems.

s referring to a decoder, an en-

~~~~~~~~~~~~coder, MUX

(b) Uses SELECT inputs.   MUX, DEMUX
(c) Can be used in parallel-to-serial conversion.   MUX
(d) Produces a binary code at its output.   encoder
(e) Only one of its outputs can be active at one time.   decoder, DEMUX
(f) Can be used to route an input signal to one of several possible outputs.   DEMUX
(g) Can be used to generate arbitrary logic functions.   MUX

9-38. Use D as the data input; use A, B, C as the select inputs.

9-38. Show how the 7442 decoder can be used as 1-to-8 demultiplexer. (*Hint:* See Problem 9-7.)

9-39. Each DEMUX output goes LOW one at a time in sequence.

9-39. Apply the waveforms of Figure 9-54 to the inputs of the 74LS138 DE-MUX of Figure 9-31(a) as follows:

$$D \rightarrow A_2, \quad C \rightarrow A_1, \quad B \rightarrow A_0, \quad A \rightarrow \overline{E}_1$$

Draw the waveforms at the DEMUX outputs.

---

5. Icons- descriptive images

Acetates-
indicate the number of the overhead projector acetate that is related to this material.

3 📽

Several methods exist for designing counters that follow arbitrary sequences. We will present the details for one common method that uses J-K flip-flops in a synchronous counter configuration. The same method can be used in designs with D flip-flops. The technique is one of several design procedures that are part of an area of digital circuit design called **sequential circuit design** which is normally part of an advanced course.

Troubleshooting Problems-
indicate problems that will exercise the student's troubleshooting reasoning skills.

9-45. A test of the security monitoring system of Figure 9-33 produces the results recorded below.

Condition	LEDs
All doors closed	All LEDs off
Door 0 open	LED 0 flashing
Door 1 open	LED 2 flashing
Door 2 open	LED 1 flashing
Door 3 open	LED 3 flashing
Door 4 open	LED 4 flashing
Door 5 open	LED 6 flashing
Door 6 open	LED 5 flashing
Door 7 open	LED 7 flashing

What are the possible faults that could cause this operation?

More Difficult Problems-
denote problems that require more than the normal effort and understanding to solve. Often a design problem.

9-41. Five lines going to the remote panel.

🔆 **9-41.** Modify the system of Figure 9-33 to handle 16 doors. Use a 74150 sixteen-input MUX and two 74LS138 DEMUXes. How many lines are going to the remote monitoring panel?

9-42. Draw the waveforms at Z, O_0, O_1, O_2, and O_3 in Figure 9-34 for the following register data: $[A] = 0011$, $[B] = 0110$, $[C] = 1001$, $[D] = 0111$.   See Solutions Manual.

DIGITAL SYSTEMS

Principles and Applications

FIFTH EDITION

Ronald J. Tocci

Monroe Community College

PRENTICE HALL
Englewood Cliffs, New Jersey 07632

Library of Congress Cataloging-in-Publication Data

Tocci, Ronald, J.
    Digital systems : principles and applications / Ronald J. Tocci. –
– Annotated instructor's ed.
        p.   cm.
    Includes index.
    ISBN 0-13-213224-9
    1. Digital electronics.   I. Title.
TK7868.D5T62   1991   90-7767
621.381--dc20   CIP

Editorial/production supervision: Ed Jones
Interior and cover design: Suzanne Behnke
Cover photo: The Image Bank
   photographer: Larry Keenan
Manufacturing buyers: Dave Dickey,
   Mary McCartney, and Ed O'Dougherty
Page layout: Thomas Nery

Printed in the United States of America

10 9 8 7 6 5 4 3 2

ISBN 0-13-213224-9

PRENTICE-HALL INTERNATIONAL (UK) LIMITED, *London*
PRENTICE-HALL OF AUSTRALIA PTY. LIMITED, *Sydney*
PRENTICE-HALL CANADA INC., *Toronto*
PRENTICE-HALL HISPANOAMERICANA, S.A., *Mexico*
PRENTICE-HALL OF INDIA PRIVATE LIMITED, *New Delhi*
PRENTICE-HALL OF JAPAN, INC., *Tokyo*
SIMON & SCHUSTER ASIA PTE. LTD., *Singapore*
EDITORA PRENTICE-HALL DO BRASIL, LTDA., *Rio de Janeiro*

CONTENTS

ABOUT THE ANNOTATED INSTRUCTOR'S EDITION

We are offering, for the first time ever in a digital systems textbook, a unique instructional aid that we hope you will find very helpful: the *Annotated Instructor's Edition*. This volume contains the complete student text along with additional features especially for the instructor. To enhance your classroom presentation the margins contain four different types of notes:

Teaching Tips suggest alternate ways of presenting material, additional examples, and points to stress.

Discussion Questions are questions or exercises that can be incorporated into your classroom teaching to evoke discussion and provide immediate feedback on student understanding.

Common Student Mistakes describe the most likely student mistakes or misconceptions and help you to correct erroneous thinking before it goes too far.

Answers to Problems provide answers to all end-of-chapter questions and problems in the margins located as close to the question as possible.

The margins also employ three different icons or symbols that denote special characteristics of the associated material. The three icons are used to denote:

1. Material for which there is an acetate (overhead transparency).
2. Troubleshooting questions
3. More difficult problems

These instructor aids have been designed to help you supplement the text material in your lecture and recitation classes so that these classes become more than a simple rehash of the book. Of course, many of you do this already with examples and insights from your own experiences. If you would like to share these with other instructors, please send them to me for incorporation in the next edition. My address is

Ronald J. Tocci
Monroe Community College
1000 East Henrietta Road
Rochester, New York 14623

Be sure to look at the inside front cover to see descriptions and examples of each type of instructor note and icon.

PREFACE

This book is a comprehensive up-to-date study of the principles and techniques of modern digital systems. It is intended for use in vocational, two-year, and four-year programs in technology, engineering, and computer science. Although a background in basic electronic devices can be helpful, a major portion of the material requires no electronics training. Those portions of the text that utilize electronic concepts can be skipped over without adversely affecting the comprehension of the logic principles.

lems have been added to each chapter so that there are now about 450 of them. Some of these are applications that show how the logic devices or circuits presented in the chapter are used in a typical microcomputer system. The more difficult problems have been flagged as such for the instructor's convenience. The "Answers to Selected Problems" at the end of the text now includes answers to almost all of the end-of-chapter questions and problems. Finally, a third color has been added to the book's design to help highlight certain elements of the material.

GENERAL IMPROVEMENTS

This fifth edition contains several general improvements to the previous edition. In addition to being updated, some of the material has been rewritten for greater clarity and thoroughness. The chapter glossaries have been combined into one comprehensive glossary located in Appendix I for more easy reference. The answers to *all* section review questions have now been provided and placed at the end of the chapter. Examples have been added to further illustrate new principles or procedures, bringing the total number of examples to over 200. Short-answer drill questions are now included with the end-of-chapter problems to provide a quick check on the reader's understanding of the basic concepts. Prob-

SPECIFIC CHANGES

In addition to these pedagogical improvements, there are several changes in the topical coverage. The major ones are:

Chapter 4: New section on parity generator and checker circuits.

Chapter 5: Introduction of state transition diagrams. New section on microcomputer application.

Chapter 7: New optional section on design of synchronous counters. Extensive use of state transition diagrams. Introduction to and problems using a bidirectional shift register (74194).

Chapter 8: Material on different IC packages.

Chapter 9: New section on code converter circuits.

Chapter 10: Added flowchart and timing diagram for SAC operation. New section on other ADC methods including tracking ADC and dual-slope ADC. New section on digital storage oscilloscope.

Chapter 11: Added function generator application. Expanded material on programmable logic devices to include PROM, PLA, PAL, and EPLDs with examples of how to determine and symbolize fuse conditions for a desired truth table. Added section on NVRAM. Decreased material on magnetic core and added coverage of optical disk and CD ROM.

RETAINED FEATURES

This edition retains all of the features that made the previous editions so widely accepted. It utilizes a block diagram approach to teach the basic logic operations without confusing the reader with the details of internal operation. All but the most basic electrical characteristics of the logic ICs are withheld until the reader has a firm understanding of logic principles. In Chapter 8 the reader is introduced to the internal IC circuitry. At that point, the reader can interpret a logic block's input and output characteristics and "fit" it properly into a complete system.

The treatment of each new topic or device typically follows these steps: the principle of operation is introduced; thoroughly explained examples and applications are presented often using actual ICs; short review questions are posed at the end of the section; and finally, in-depth problems are available at the end of the chapter. Ranging from simple to complex, these problems provide instructors with a wide choice of student assignments. These problems are often intended to reinforce the material without simple repetition of the principles. They require the student to demonstrate comprehension of the principles by applying them to different situations. This also helps the student develop confidence and expand his/her knowledge of the material.

The IEEE/ANSI standard for logic symbols is introduced and discussed with minimum disruption of the topic flow and, if desired, can be omitted completely or in part. The extensive troubleshooting coverage is spread over Chapters 4 through 11 and includes presentation of troubleshooting principles and techniques, case studies, 25 troubleshooting examples, and 60 *real* troubleshooting problems. When supplemented with hands-on lab exercises, this material can help foster the development of good troubleshooting skills.

SEQUENCING

It is a rare instructor who uses the chapters of a textbook in the sequence in which they are presented. In fact, I must admit that, for many different reasons, I do not use my own books in that way. This book was written so that, for the most part, each chapter builds on previous material, but it is possible to alter the chapter sequence somewhat. The first part of Chapter 6 (arithmetic operations) can be covered right after Chapter 2 (number systems), although this would produce a long interval before the arithmetic circuits of Chapter 6 are encountered. Much of the material in Chapter 8 (IC characteristics) can be covered earlier (e.g., after Chapter 4 or 5) without causing any serious problems.

This book can be used in either a one-term course or in a two-term sequence. When used in one term, it may be necessary, depending on available class hours, to omit some topics. Here is a list of sections and chapters that can be deleted with minimum disruption. Obviously, the choice of deletions will depend on factors such as program/course objectives and student background.

1. *Chapter 1:* all
2. *Chapter 2:* Sections 6 and 7
3. *Chapter 4:* Sections 7 and 8; Sections 10–13 if troubleshooting is not to be covered
4. *Chapter 5:* Sections 3 and 25
5. *Chapter 6:* Sections 5, 7, 11, 13, 16–18, 20
6. *Chapter 7:* Sections 10, 14–23
7. *Chapter 8:* Sections 9, 17–21
8. *Chapter 9:* Sections 6, 10, 11
9. *Chapter 10:* Sections 6, 13–16
10. *Chapter 11:* Sections 9, 18–22
11. *Chapter 12:* all

SUPPLEMENTS

An extensive package of teaching and learning tools has been developed to supplement this textbook. Each component of this package provides a unique function that can be used independently. The complete set of supplements provides one of the most comprehensive educational programs available on the subject matter.

Laboratory Manuals   Two lab manuals with different approaches have been developed to accompany this textbook.

The first is written by *Jim C. DeLoach* of DeVry Institute of Technology and *Frank J. Ambrosio* of Monroe Community College. This manual contains over forty experiments that are coordinated with the textbook topic coverage to provide practical laboratory experience with the emphasis on analysis and troubleshooting. Several design projects are also included. There is enough material to use in a two-semester laboratory sequence if desired. Each experiment ends with review questions to help measure the student's understanding of the circuits, concepts, and techniques covered in the experimental procedure.

The manual presents information on breadboarding techniques, logic probes, logic pulsers, and logic analyzers. It provides a list of all necessary ICs, components, lab equipment, and digital trainer requirements.

Gregory L. Moss of Purdue University has prepared a lab manual that emphasizes simulation and design. Its 25 major topical units each contain several laboratory projects so that there are enough projects for a two-semester sequence. Alternatively, an instructor can select appropriate projects to fit particular course needs.

In addition to extensive applications of standard logic ICs, the manual provides experience with computer tools used by industry in logic circuit design and development. *Logic circuit simulation* is introduced to students with an easy-to-use but powerful text-based simulator. Custom implementation of logic circuits using *programmable logic devices* is included in several lab projects throughout the manual. Both of these areas may be treated as optional for digital courses that do not currently utilize the computer as a design tool; these courses can use the numerous standard design projects included in the manual.

Instructor Solutions Manual   This manual contains detailed solutions to all end-of-chapter problems from the text. It also includes answers and expected results for the experiments from the two laboratory manuals just described.

Software   There are two excellent software packages that accompany this book and can be used to greatly enhance the text material: *DrillMaker* and *P/C-SILOS (educational version)*. Both are available to instructors free upon adoption of this book.

DrillMaker by John Eubanks, Coastal Computer Company, is a $5\frac{1}{4}$-inch diskette that will run on IBM PCs and compatibles. This program contains interactive drills on logic fundamentals such as

number systems, codes, Boolean algebra, binary and hexadecimal arithmetic, and Karnaugh maps. It allows drilling on a single subject or a variety of subjects and gives complete worked-out solutions for each question.

The program also permits the user to obtain a K-map solution for any inputted truth table up to four variables and allows "don't care" entries. The computer screen shows exactly how the map is looped and gives the minimum Boolean expression.

A second, more comprehensive program called *LogicMaker* is also available from Coastal Computer Company. (This is not free upon adoption.) It provides an extensive set of activities that can be used to reinforce the learning of digital electronics starting from basic logic gates on up to MSI chips. Its many options include:

(1) Over 100 exercises in which a logic circuit using actual ICs is drawn on the screen and the student is asked to determine the logic levels and/or timing diagrams. An oscilloscope display is presented on the screen for waveform displays, and a waveform editor allows the student to input waveforms

(2) Circuit simulation whereby the student can build and test a logic circuit on the computer screen using any of a wide variety of TTL ICs from a stored IC directory. Once the circuit has been constructed on screen, the student can supply input waveforms and logic levels with which the circuit operation will be tested

(3) Over 100 design projects that ask the student to use given ICs to implement both combinatorial and sequential circuits to satisfy given input/output relationships

(4) Interactive troubleshooting exercises with simulated logic probe and oscilloscope measurements

LogicMaker also provides extensive HELP messages to guide the new user through the many features of the program operation. It even allows the instructor to change existing exercises or to add new exercises to suit his/her requirements.

This software has many more features too numerous to mention here. It has been thoroughly class-tested and has received enthusiastic endorsements from students and instructors who have used it in their courses. For more information on *LogicMaker*, contact

> John M. Eubanks
> Coastal Computer Company
> 1609 Country Club Drive
> Rocky Mountain, NC 27804
> Phone: 919 442-7436

P/C-SILOS, the second software package that is free upon adoption, performs powerful, interactive logic simulation on a PC. P/C-SILOS is used just like an operating system. You issue commands to read data files, start the simulation, and output the results. It can handle up to a maximum of 200 gates and requires 320KB of memory. Upon adoption of this book, Prentice-Hall will provide the instructor with one educational version package, consisting of one disk and one user manual. Contact your local Prentice Hall College Sales representative or write to:

> Ben T. Colt
> Marketing Department
> Prentice Hall
> Englewood Cliffs, NJ 07632

Transparency Masters   These are 135 high-quality, enlarged copies of diagrams from the text that would be difficult and/or time-consuming to redraw in the classroom. Transparencies can be made from these masters and used to enhance your classroom presentation.

Acetates (Transparencies)   These are a set of 25 two-color transparencies that provide the instructor with examples and applications that supplement the text material. They can be used in lectures as examples that the student has not previously

seen, or they can be worked out in problem sessions, or they can even be given as a quiz.

Test Item File   This is a hardcopy set of 750 questions that can be used for tests and quizzes. It is also available on the Prentice Hall Data Manager microcomputer testing services for the IBM PC and clones.

Student Study Guide   Prepared by Frank J. Ambrosio, this provides over 200 questions and problems covering all the text material. Answers and detailed solutions are given for each exercise. This guide provides an opportunity for the student to develop greater proficiency and subject mastery, and can be valuable in examination preparation.

ACKNOWLEDGMENTS

Prior to starting work on this edition, we sent an extensive questionnaire to many users and former users of previous editions. I am grateful to all of those who responded with their comments, critiques, and suggestions. Their input was invaluable as I went through the process of deciding what changes to incorporate in the new edition. My thanks go out to:

Robert Arndt
Belleville Area College

Forrest Barker
Los Angeles City College

Charles Barnette
Oklahoma State University of Technology

Bruce E. Bothwell
Chemeketa Community College

Dave Brown
IVTC Northeast

Keith Brown
St. Clair College

Jerry Bylander
Grayson Community College

Pat Chambers
Triton College

James Charbonneau
Indiana Vocational Technical College

Rabin Chatterjee
Centennial College

Alan H. Czarapata
Montgomery College

Herb Daugherty
IVTC Muncie

Louis Deutsch
Microcomputer Technology Institute

Receil Fielding
Northwestern Electronics Institute

William Franken
Maple Wood Community College

Louis Gallegos
ITT Technical Institute

Gary Grizzle
ITT Technical Schools

Gary F. House
DeVry Atlanta

David B. Kilborn
Northwestern Electronics Institute

Fred Kirk
Columbia State Community College

David Lewick
Minneapolis Technical Institute

Kenneth Markowitz
New York City Technical College

Tony McRae
Collin County Community College

Dennis Meyer
Mid-State Technical College

Jim L. Pannell
DeVry Institute of Technology

Wesley Prescod
Fanshawe College

Dave Prophet
Frederick Community College

Robert N. Reaves
Durham Technical Institute

Bruce Reed
Howard Community College

Gordon Rehmeyer
Essex Community College

Carl Tucker
Palomar College

Dr. James H. Yu
San Jose State University

O. S. Zemlak
Algonquin College

I am particularly grateful to Gregory L. Moss of Purdue University for his many excellent suggestions, most of which are part of this edition.

As always, the people at Prentice Hall were a pleasure to work with. I was especially fortunate to again have Ed Jones as my production editor. (I

hope he has forgiven me for my unintentional failure to acknowledge his major contribution to the last edition.) His thoroughness and attention to detail far exceeds that of any of the numerous production editors who worked with me on other projects. Once again, my series editor Alice Barr provided the combination of open-minded leadership, gentle encouragement, and infectious enthusiasm that I needed to keep me going. Alice's assistant Audrey Hellman not only served as a skillful substitute on those many occasions when Alice was on the road,

but our phone conversations never failed to brighten my day.

Finally, another special thank you to my colleague and friend, Frank Ambrosio, for his excellent work on the glossary, index, and Instructor's Resource Manual. The timely and dependable way in which he completed his tasks made it easy for me to stay on schedule.

Ronald J. Tocci
Monroe Community College

TROUBLESHOOTING

It is generally agreed that one of the most important skills needed by a technician is the ability to troubleshoot. There is no general agreement among educators, however, as to whether it is possible to effectively teach newcomers to the field of electronics the skills needed to become good troubleshooters. Some people believe that troubleshooting can only be learned on the job under real-world conditions. Others feel that students must have a special aptitude for troubleshooting or they can never become efficient troubleshooters. After twenty-five years of experience in industry and education, I am convinced that technician-level programs can imbue students with an understanding of troubleshooting fundamentals and a knowledge of how to apply a systematic approach while providing them with the opportunity to develop and exercise their logical reasoning skills.

Electronic and computer technology programs can incorporate troubleshooting into both the theory and lab portions of their courses. I suggest that the following guidelines can be helpful in developing the troubleshooting component of your instructional programs.

1. Case studies, troubleshooting examples, and troubleshooting problems can help students develop the analytical skills that are the most essential part of effective troubleshooting. The abundant number of these provided in this text should be integrated into lecture, recitation, problem sessions, and homework assignments.

2. After a new circuit has been analyzed during lecture, you can pose a "what if" question to the class asking what would happen to the operation if a certain fault were placed in the circuit. This type of repetitive activity gradually builds up the students' repertoire of causes and effects that can be recognized during real troubleshooting.

3. A major part of real-world troubleshooting involves schematic reading. Wherever possible, you should expose students to schematics of real equipment, preferably equipment they have used or will be using, like power supplies, meters, oscilloscopes, computers, and terminals. Initially, you can assign small sections of a schematic to be analyzed. Then you can develop some troubleshooting problems or case studies on those sections. As the students get more and more circuits and principles under their belts, you can increase the scope of the assignments to include more of the schematic and even have them trace signals from one schematic to another. I will say more about this in a later section.

4. A substantial portion of the lab activity should be dedicated to exercises that involve actual hands-on troubleshooting of defective or faulty circuits or systems. *No circuit is too simple to troubleshoot*. It is a good idea to introduce the student to troubleshooting with relatively simple circuits using a wide variety of faults; this will help to develop a sense of accomplishment and confidence. The circuits can gradually be made more complex as the

student gains competence at each level of complexity.

5. The approach that has worked best for me in the lab is first to have the students analyze and test a working circuit or system so that they become familiar with all of the operating modes. You can then insert a fault into the circuit or give them a prebuilt bugged circuit to test and troubleshoot. Later in the training program you may want to have the students troubleshoot a circuit they have never seen before to give them a taste of "cold turkey" troubleshooting.

6. I suggest that you require students to record test results in a table or in some other orderly form. Encourage them to examine these recorded results in order to detect any pattern that may be used to help hone in on the problem area. This step is especially important in the more complex sequential circuits, and it is one that students seem to resist even though it often leads to quicker fault detection.

7. Work closely with each student or lab team. Let them watch you as you troubleshoot some piece of faulty lab equipment and explain what you are doing at each step in the process. Invite them to tell you what they think you should do next, or what they think the possible faults might be. I have found this activity to be very valuable in giving students a troubleshooting approach to emulate.

8. Electronics instructors have come up with a number of clever ways to bug a working circuit. Here are some of the common ones.

(a) Bad ICs. Keep all bad ICs that you come across. Put a sticker under each IC with a description of its fault, or with a number that is referenced to a list of IC faults.

(b) Often you don't have an IC with the type of fault you want the student to experience. Sometimes you can create faulty ICs by purposely zapping them with voltages that are either too large or of the wrong polarity. This is a fun project to involve students in. If this is not feasible or does not produce the desired fault, you can try soldering a very fine wire between IC pins under the IC to emulate an internal short.

(c) Defective components (resistors, capacitors, diodes, transistors). Inevitably, first year basic electricity and electronics labs are great sources for these.

(d) Insert wrong-valued components into a circuit. I once spent an hour troubleshooting a dense PC board where the problem turned out to be a 10K resistor where a 1K was supposed to be.

(e) Breaking or bending IC pins before inserting them into sockets or patching boards.

(f) If wire-wrap is being used, you can clip a wire from its post right where the insulation begins. Then rewrap the insulated end around the post. It will look just like the original wrap.

(g) If patching wire is being used, you can make up some wires with opens in the middle by stripping both ends, sliding the insulation back toward one end, clipping the bare wire at the other end, reinserting it, and sliding the insulation back over it to cover the break.

I am sure that many of you have other clever bugging tricks that have proven to be successful. If you would be willing to share them, send them on to me and I will put them in the next edition. My address is Monroe Community College, 1000 E. Henrietta Road, Rochester, New York, 14623.

NONTRADITIONAL TESTING METHODS

I am taking this opportunity to offer some observations on an important and, in my opinion, most difficult part of our jobs as teachers—the evaluation of student learning. I will also describe two nontraditional methods of testing that I have used over the years and have found to have many advantages (as well as a few disadvantages) compared to the traditional methods.

In my twenty-five years of teaching I have had many discussions with students and have had countless opportunities to observe the effects of testing on students, and I am convinced that the traditional methods of testing reveal only the most superficial aspects of a student's understanding. Where do we get the notion that a test passer has learned anything other than how to pass the test? How often do you hear a student lament, after taking a test, "I thought I really knew that stuff, but did I ever choke," or "If I only had more time I could've done a lot better," or "I blew it because nothing I studied was on the test"?

Too much pressure is put on the student to get the answers right the first time and within a restricted time interval. A lot of students have extreme difficulty functioning under these conditions, and who are we to say that if they can't, then they must not be capable of functioning under more realistic circumstances?

We tend to place too much emphasis on problem *doing* rather than problem *solving*. Traditional tests often contain cookbook type problems that follow a step-by-step procedure which, if done correctly, will lead to the predetermined answers. The type of student who does well on this type of examination is the one who will practice at a particular type of problem until s/he has mastered the procedure. Unfortunately, the student concentrates almost solely on the process and is rarely motivated to understand the concepts and principles underlying the process.

The purpose of testing has always been to measure the degree of student learning. I believe that testing can also serve as an additional and effective learning experience. Unfortunately, under traditional testing, students rarely realize any significant educational benefits. Rather, they look upon the tests as being punitive.

Two nontraditional testing schemes that I believe overcome many of the drawbacks of conventional methods are *graded problem sessions* and *takehome exams*.

GRADED PROBLEM SESSIONS Seldom, if ever, are the student and instructor seen as partners in the process of evaluating the student's learning. Using a properly designed problem session format, the student and instructor can *jointly* participate in the evaluation process, with the major goal being to improve the student's performance in weak areas. A graded problem session combines the evaluation and learning functions into one activity.

I will now describe the problem session format I have used with gratifying albeit hard-earned results.

I have found that for maximum efficiency the number of students should be limited to twelve, assuming one instructor and a normal mix of student abilities. At least two hours (three is preferable) should be devoted to the problem session to allow each student an adequate share of the instructor's attention.

The students are given three or four problems to work on at their own pace. They are allowed about 15–20 minutes before I begin circulating among the group to check the progress of each student. If I find a student having difficulty with a particular problem, I will consult with him individually to determine where the difficulty lies. Then, depending on how "lost" he is, I will do one or more of the following: clarify or restate the problem, give him a hint or a push to get started, help him find the flaws in his reasoning, or refer him to his textbook or notes. I leave him to complete the problem and I go on to work with other students.

As each student completes a problem, she lets me check it over unless I am working with another student, in which case she can go on to the next problem until I am available. While checking a problem solution, I typically ask the student to explain one or more steps in her procedure. This helps in developing the student's ability to communicate technical concepts and ideas verbally as well as on paper—a skill that is noticeably deficient in our graduates. If I am satisfied with the student's understanding of the principles and concepts, she is allowed to resume working on the other problems. If not, I will modify the problem and have her repeat it until she demonstrates an understanding.

It is my responsibility to try to see that each student eventually finishes each type of problem correctly without assistance during the problem session time. Sometimes this is not possible, and I will allow him to finish it at home or during one of my office hours. It is necessary that each student complete all the problems so that good analytical and problem-solving skills can to be developed.

My problem sessions are graded because I believe that the students need that as motivation to prepare for them. It should be no surprise that fair grading is difficult under the conditions I just described. It seems reasonable that a student who comes in and does the problems correctly with little or no instructor assistance should be evaluated higher than one who requires assistance and several iterations before achieving success. My grading scheme employs point deductions based on the amount of instructor assistance and the number of times the problem had to be repeated. Sometimes a student is obviously unprepared for the problem session and apathetic about the problem session process. In these cases a very low grade (even zero) is justified. That will usually convince the student to change his tune for the next problem session. While I feel the grading is necessary, it is more important that the students leave the problem session feeling confident in their understanding of the material covered; as such, I tend to be reasonably lenient in grading—anyone who completes the work receives at least a grade of C.

Conducting problem sessions in a manner similar to what I have described is a lot more work than traditional lectures or recitation classes. In addition to having to engage in more subjective grading, the instructor expends a great deal of energy that cannot be compared to any other teaching activity except, perhaps, some laboratory teaching situations. It is not unusual for the instructor to emerge from a problem session both mentally and physically exhausted. But I contend that the educational and psychological benefits to the student outweigh these drawbacks. It is my experience that the problem sessions significantly enhance student learning as demonstrated by student performance on follow-up exams. They enjoy the problem sessions because they are in a "learning partnership" with the instructor in which they can concentrate on clearing up their misconceptions and improving their understanding in an essentially nonpunitive atmosphere.

TAKEHOME EXAMS   In our two courses on Digital Electronics and Digital Systems/Microprocessors, we evaluate student competence in the theory parts of the courses through periodic half-hour quizzes, a midterm takehome exam, and a final

takehome exam. The quizzes usually cover one chapter, sometimes two. We have found it preferable to have a larger number of short quizzes than to have three or four unit exams that require students to study three or four chapters at a time. For one thing, students have more opportunities to make up for "having a bad day (or two)." For another, the students are required to master the material in smaller chunks at more frequent intervals instead of cramming every four weeks.

The midterm and final takehome exams are comprehensive exams that are distributed to the students three or four weeks before they are due to be handed in. This relatively long time frame is necessary if most of the following benefits of the takehome exam are to be realized.

(1) *A wide variety of problems can be used.* They can be real world, multi-faceted, open-ended problems that require in-depth analysis. Some can require research from resources other than the course textbooks. Others can require circuit design.

(2) *The exam can utilize schematics of actual equipment.* This makes the problems more realistic—less contrived—and fosters an attitude among the students that this is the kind of work they may be called upon to do on the job. By using schematics from different equipment and different manufacturers, the students learn that schematic reading is not as straightforward as it might have appeared in textbooks or labs.

For example, we have used schematics from a Hewlett-Packard oscilloscope and have asked the students to analyze the horizontal sweep circuit and its control logic. As another example, we have used the complete set of schematics for both a KIM-1 and a TRS-80 microcomputer and have had the students analyze almost all of the major functions. We are currently beginning to develop takehome exams built around schematics for a 68000-based microcomputer. Realistic troubleshooting problems have also been devised around these schematics.

(3) *The takehome exam demands better quality work.* We expect the submitted exam to be typed, well-organized, and neat, and to employ the standard principles of good technical writing. Most of the exam questions require that an explanation or defense accompany the technical presentation. These requirements help to develop and improve technical communication skills.

(4) *Test anxiety is virtually eliminated.* The conditions and requirements the students are working under are more like those they will experience on the job.

(5) *Student confidence and pride are generated.* This is an intangible benefit that cannot be underestimated. The exams are challenging and realistic and require a lot of time and effort. Many students experience a great sense of accomplishment.

(6) *Learning takes place.* Typical student comments are: "Boy, it was a lot of work, but I learned a lot" and "I feel like I know so much more than before I started work on the exam." A good part of the learning is achieved as a result of the discussions and interchange of ideas that take place among the students. We teachers hate to admit it, but very often students learn better from other students than from us.

Of course the takehome exam approach has some drawbacks. The exams are more difficult to devise than the traditional ones if they are to realize the benefits I have just cited. The grading of the exams is more time-consuming because much more of it is subjective and involves reading verbal explanations and descriptions. The takehome exam process by its very nature is subject to abuse by some students who manage to take inordinate advantage of the work and ideas of others. This can be minimized somewhat by a strong statement to the effect that discussion and idea interchange among students is expected and, indeed, encouraged, but that the submitted exam must be the individual's own work, and any evidence of outright plagiarism will be severely penalized. In my experience this has not been a big problem because most students take the exam seriously and view it as an important part of the educational process.

Here is an example of the type of question that can be part of a well-designed takehome exam. Notice how the student is required to verbally support all answers.

Refer to sheet 2 of the TRS-80 schematic and locate the dynamic RAM chips. Answer the following. Show all steps, and explain when necessary.

(a) How did you tell that they were DRAM and not SRAM?

(b) Determine the total DRAM capacity.
(c) Describe the function of IC Z43.
(d) Show how to determine the total DRAM address range.
(e) Modify the circuit to expand the capacity to 32K × 8. Describe your modification.
(f) Find the area where the RAS, CAS, and MUX signals are generated. Describe how the circuit generates these signals and, starting with the MASTER CLOCK signal, draw the timing diagram showing each of these signals.

DIGITAL SYSTEMS

Principles and Applications

DIGITAL SYSTEMS

Principles and Applications

FIFTH EDITION

Ronald J. Tocci

Monroe Community College

PRENTICE HALL
Englewood Cliffs, New Jersey 07632

Library of Congress Cataloging-in-Publication Data

Tocci, Ronald J.
  Digital systems : principles and applications / Ronald J. Tocci. -
- 5th ed.
        p.   cm.
  Includes index.
  ISBN 0-13-213133-1
  1. Digital electronics.   I. Title.
TK7868.D5T62   1991b
  621.381--dc20

                                                    90-42716
                                                    CIP

Editorial/production supervision: Ed Jones
Interior design: Suzanne Behnke
Cover design: Suzanne Behnke
Cover photo: Image Bank
  Photographer: Larry Keenan
Manufacturing buyers: Dave Dickey,
  Mary McCartney, and Ed O'Dougherty
Page layout: Thomas Nery

© 1991, 1988, 1985, 1980, 1977 by Prentice-Hall, Inc.
A Division of Simon & Schuster
Englewood Cliffs, New Jersey 07632

Printed in the United States of America
10   9   8   7   6   5   4   3

ISBN   0-13-213133-1

PRENTICE-HALL INTERNATIONAL (UK) LIMITED, *London*
PRENTICE-HALL OF AUSTRALIA PTY. LIMITED, *Sydney*
PRENTICE-HALL CANADA INC., *Toronto*
PRENTICE-HALL HISPANOAMERICANA, S.A., *Mexico*
PRENTICE-HALL OF INDIA PRIVATE LIMITED, *New Delhi*
PRENTICE-HALL OF JAPAN, INC., *Tokyo*
SIMON & SCHUSTER ASIA PTE. LTD., *Singapore*
EDITORA PRENTICE-HALL DO BRASIL, LTDA., *Rio de Janeiro*

CONTENTS

5 FLIP-FLOPS AND RELATED DEVICES 150

6 DIGITAL ARITHMETIC: OPERATIONS
AND CIRCUITS 229

8   INTEGRATED-CIRCUIT LOGIC FAMILIES   366

9   MSI LOGIC CIRCUITS   451

PREFACE

This book is a comprehensive up-to-date study of the principles and techniques of modern digital systems. It is intended for use in two-year and four-year programs in technology, engineering, and computer science. Although a background in basic electronic devices is helpful, a major portion of the material requires no electronics training. Those portions of the text that utilize electronic concepts can be skipped over without adversely affecting the comprehension of the logic principles.

GENERAL IMPROVEMENTS

This fifth edition contains several general improvements to the previous edition. In addition to being updated, some of the material has been rewritten for greater clarity and thoroughness. The chapter glossaries have been combined into one comprehensive glossary located in Appendix I for more easy reference. The answers to *all* section review questions have now been provided and placed at the end of the chapter. Examples were added where it was felt they were needed to further illustrate the new principles or procedures, bringing the total number of examples to over 200. Short-answer drill questions are now included with the end-of-chapter problems to provide a quick check on the reader's understanding of the basic concepts. Problems have been added to each chapter so that there are now about 450 of them. Some of these are applications that show how the logic devices or circuits presented in the chapter are used in a typical microcomputer system. The more difficult problems have been flagged as such for the instructor's convenience. The Answers to Selected Problems at the end of the text now include answers to almost all of the end-of-chapter questions and problems. Finally, a third color has been added to the book's design to help highlight certain elements of the material.

SPECIFIC CHANGES

In addition to these pedagogical improvements, there are several changes in the topical coverage. The major ones are:

Chapter 4: New section on parity generator and checker circuits.

Chapter 5: Introduction of state transition diagrams. New section on microcomputer application.

Chapter 7: New optional section on design of synchronous counters. Extensive use of state transition diagrams. Introduction to and problems using a bidirectional shift register (the 74194).

Chapter 8: Material on different IC packages.

Chapter 9: New section on code converter circuits.

Chapter 10: Added flowchart and timing diagram for SAC operation. New section on other ADC methods including tracking ADC and dual-slope ADC. New section on digital storage oscilloscope.

Chapter 11: Added function generator application. Expanded material on programmable logic devices to include PROM, PLA, PAL, and EPLDs with examples of how to determine and symbolize fuse conditions for a desired truth table. Added section on NVRAM. Decreased material on magnetic core and added coverage of optical disk and CD ROM.

RETAINED FEATURES

This edition retains all of the features that made the previous editions so widely accepted. It utilizes a block diagram approach to teach the basic logic operations without confusing the reader with the details of internal operation. All but the most basic electrical characteristics of the logic ICs are withheld until the reader has a firm understanding of logic principles. In Chapter 8 the reader is introduced to the internal IC circuitry. At that point, the reader can interpret a logic block's input and output characteristics and "fit" it properly into a complete system.

The treatment of each new topic or device typically follows these steps: the principle of operation is introduced; thoroughly explained examples and applications are presented often using actual ICs; short review questions are posed at the end of the section; and finally, in-depth problems are available at the end of the chapter. Ranging from simple to complex, these problems provide instructors with a wide choice of student assignments. These problems are often intended to reinforce the material without simple repetition of the principles. They require the student to demonstrate comprehension of the principles by applying them to different situations. This also helps the student develop confidence and expand his or her knowledge of the material.

The IEEE/ANSI standard for logic symbols is introduced and discussed with minimum disruption of the topic flow, and, if desired, can be omitted completely or in part. The extensive troubleshooting coverage is spread over Chapters 4 through 11 and includes presentation of troubleshooting principles and techniques, case studies, 25 troubleshooting examples, and 60 *real* troubleshooting problems. When supplemented with hands-on lab exercises, this material can help foster the development of good troubleshooting skills.

SEQUENCING

It is a rare instructor who uses the chapters of a textbook in the sequence in which they are presented. In fact, I must admit that, for many different reasons, I do not use my own books in that way. This book was written so that, for the most part, each chapter builds on previous material, but it is possible to alter the chapter sequence somewhat. The first part of Chapter 6 (arithmetic operations) can be covered right after Chapter 2 (number systems), although this would produce a long interval before the arithmetic circuits of Chapter 6 are encountered. Much of the material in Chapter 8 (IC characteristics) can be covered earlier (e.g., after Chapter 4 or 5) without causing any serious problems.

This book can be used in either a one-term course or in a two-term sequence. When used in one term, it may be necessary, depending on available class hours, to omit some topics. Here is a list of sections and chapters that can be deleted with minimum disruption. Obviously, the choice of deletions will depend on factors such as program or course objectives and student background.

1. *Chapter 1:* all
2. *Chapter 2:* Sections 6 and 7
3. *Chapter 4:* Sections 7 and 8; Sections 10–13 if troubleshooting is not to be covered
4. *Chapter 5:* Sections 3 and 25
5. *Chapter 6:* Sections 5, 7, 11, 13, 16, 18, 20
6. *Chapter 7:* Sections 10, 14–23
7. *Chapter 8:* Sections 9, 17–21
8. *Chapter 9:* Sections 6, 10, 11
9. *Chapter 10:* Sections 6, 13–16
10. *Chapter 11:* Sections 9, 18–22
11. *Chapter 12:* all

ACKNOWLEDGMENTS

Prior to starting work on this edition, we sent an extensive questionnaire to many users and former users of previous editions. I am grateful to all of those who responded with their comments, critiques, and suggestions. Their input was invaluable as I went through the process of deciding what changes to incorporate in the new edition. I am particularly grateful to Gregory L. Moss of Purdue University for his many excellent suggestions. All of them were carefully considered, and most of them are part of this edition.

As always, the people at Prentice-Hall were a pleasure to work with. I was especially fortunate to have Ed Jones as my production editor again (I hope he has forgiven me for my unintentional failure to acknowledge his major contribution to the last edition). His thoroughness and attention to detail far exceeds that of any of the numerous production editors who have worked with me on previous writing projects. Once again, Alice Barr provided the combination of open-minded leadership, gentle encouragement, and infectious enthusiasm that I needed to keep me going.

Finally, another special thank you to my colleague and friend Frank Ambrosio, for his work on the glossary, index, and Instructor's Manual. The timely and dependable way in which he completed his tasks made it easy for me to stay on schedule.

Ron Tocci
Monroe Community College

1

INTRODUCTORY CONCEPTS

Objectives

Upon completion of this chapter, you will be able to:

- Distinguish between analog and digital representations.
- Name the advantages, disadvantages, and major differences among analog, digital, and hybrid systems.
- Understand the need for analog-to-digital converters (ADCs) and digital-to-analog converters (DACs).
- Convert between decimal and binary numbers.
- Identify typical digital signals.
- Cite several integrated-circuit fabrication technologies.
- Identify a timing diagram.
- State the differences between parallel and serial transmission.
- Describe the property of memory.
- Describe the major parts of a digital computer and understand their functions.

Introduction

When most of us hear the term "digital," we immediately think of "digital calculator" or "digital computer." This can probably be attributed to the dramatic way that low-cost, powerful calculators and computers have become accessible to the average person. It is important to realize that calculators and computers represent only one of the many applications of digital circuits and principles. Digital circuits are used in electronic products such as video games, microwave ovens, and automobile control systems, and in test equipment such as meters, generators, and oscilloscopes. Digital techniques have also replaced a lot of the older "analog circuits" used in consumer products such as radios, TV sets, and high-fidelity sound recording and playback equipment.

In this book we are going to study the principles and techniques that are common to all digital systems from the simplest on/off switch to the most complex computer. If this book is successful, you should gain a deep understanding

of how all digital systems work, and you should be able to apply this understanding to the analysis and troubleshooting of any digital system.

We start by introducing some underlying concepts that are a vital part of digital technology; these concepts will be expanded on as they are needed later in the text. We will also introduce some of the terminology that is necessary when embarking on a new field of study, and will add to it in every chapter. A complete glossary of terminology is presented in Appendix I.

1-1   NUMERICAL REPRESENTATIONS

In science, technology, business, and, in fact, most other fields of endeavor, we are constantly dealing with *quantities*. Quantities are measured, monitored, recorded, manipulated arithmetically, observed, or in some other way utilized in most physical systems. It is important when dealing with various quantities that we be able to represent their values efficiently and accurately. There are basically two ways of representing the numerical value of quantities: **analog** and **digital**.

Analog Representations   In *analog representation* one quantity is represented by another which is proportional to the first. An example is an automobile speedometer, in which the deflection of the needle is proportional to the speed of the auto. The angular position of the needle represents the value of the auto's speed, and the needle follows any changes that occur as the auto speeds up or slows down.

Another example is the common room thermostat, in which the bending of the bimetallic strip is proportional to the room temperature. As the temperature changes gradually, the curvature of the strip changes proportionally.

Still another example of an analog quantity is found in the familiar audio microphone. In this device an output voltage is generated in proportion to the amplitude of the sound waves that impinge on the microphone. The variations in the output voltage follow the same variations as the input sound.

Analog quantities such as those cited above have an important characteristic: *they can vary over a continuous range of values*. The automobile speed can have any value between zero and, say, 100 mph. Similarly, the microphone output might be anywhere within a range of zero to 10 mV (for example, 1 mV, 2.3724 mV, 9.9999 mV).

Digital Representations   In *digital representation* the quantities are represented not by proportional quantities but by symbols called *digits*. As an example, consider the digital watch, which provides the time of day in the form of decimal digits which represent hours and minutes (and sometimes seconds). As we know, the time of day changes continuously, but the digital watch reading does not change continuously; rather, it changes in steps of one per minute (or per second). In other words, this digital representation of the time of day changes in *discrete* steps, as compared with the representation of time provided by an analog watch, where the dial reading changes continuously.

The major difference between analog and digital quantities, then, can be simply stated as follows:

$$\text{analog} \equiv \text{continuous}$$
$$\text{digital} \equiv \text{discrete (step by step)}$$

Because of the discrete nature of digital representations, there is no ambiguity when reading the value of a digital quantity, whereas the value of an analog quantity is often open to interpretation.

___ EXAMPLE 1-1 _____

Which of the following involve analog quantities and which involve digital quantities?

(a) Ten-position switch

(b) Current meter

(c) Temperature

(d) Sand grains on the beach

(e) Radio volume control

Solution

(a) Digital

(b) Analog

(c) Analog

(d) Digital, since the number of grains can be only certain discrete (integer) values and not any value over a continuous range

(e) Analog

___ REVIEW QUESTION* _____

1. Concisely describe the major difference between analog and digital quantities.

1-2 DIGITAL AND ANALOG SYSTEMS

A **digital system** is a combination of devices designed to manipulate physical quantities or information that are represented in digital form; that is, they can take on only discrete values. These devices are most often electronic, but they can also be mechanical, magnetic, or pneumatic. Some of the more familiar digital systems include digital computers and calculators, digital watches, traffic-signal controllers, and typewriters.

An **analog system** contains devices that manipulate physical quantities that are represented in analog form. In an analog system, the quantities can vary over a continuous range of values. For example, the amplitude of the output signal to the speaker in a radio receiver can have any value between zero and its maximum limit. Other common analog systems are the telephone system, magnetic tape recording and playback equipment, and the automobile odometer.

*Answers to review questions are found at the end of the chapter in which they occur.

Advantages of Digital Techniques   An increasing majority of applications in electronics, as well as in most other technologies, use digital techniques to perform operations that were once performed using analog methods. The chief reasons, for the shift to digital technology are:

1. *Digital systems are easier to design.* This is because the circuits that are used are *switching circuits,* where *exact* values of voltage or current are not important, only the range (HIGH or LOW) in which they fall.

2. *Information storage is easy.* This is accomplished by special switching circuits that can latch onto information and hold it for as long as necessary.

3. *Accuracy and precision are greater.* Digital systems can handle as many digits of precision as you need simply by adding more switching circuits. In analog systems, precision is usually limited to three or four digits because the values of voltage and current are directly dependent on the circuit component values.

4. *Operation can be programmed.* It is fairly easy to design digital systems whose operation is controlled by a set of stored instructions called a *program.* Analog systems can also be *programmed,* but the variety and complexity of the available operations is severely limited.

5. *Digital circuits are less affected by noise.* Spurious fluctuations in voltage (noise) are not as critical in digital systems because the exact value of a voltage is not important, as long as the noise is not large enough to prevent us from distinguishing a HIGH from a LOW.

6. *More digital circuitry can be fabricated on IC chips.* It is true that analog circuitry has also benefited from the tremendous development of IC technology, but its relative complexity, and its use of devices that cannot be economically integrated (high-value capacitors, precision resistors, inductors, transformers), have prevented analog systems from achieving the same high degree of integration.

Limitations of Digital Techniques   There is really only one major drawback when using digital techniques:

The real world is mainly analog

Most physical quantities are analog in nature, and it is these quantities that are often the inputs and outputs that are being monitored, operated on, and controlled by a system. Some examples are temperature, pressure, position, velocity, liquid level, flow rate, and so on. We are in the habit of expressing these quantities *digitally,* such as when we say that the temperature is 64° (63.8° when we want to be more accurate); but we are really making a digital approximation to an inherently analog quantity.

To take advantage of digital techniques when dealing with analog inputs and outputs, three steps must be followed:

1. Convert the "real-world" analog inputs to digital form.
2. Process (operate on) the digital information.
3. Convert the digital outputs back to real-world analog form.

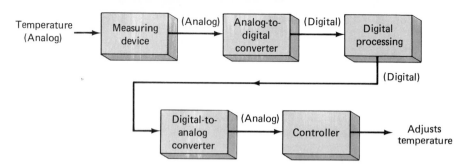

FIGURE 1-1   Block diagram of a temperature control system that requires analog/digital conversions in order to allow the use of digital processing techniques.

Figure 1-1 shows a block diagram of this for a typical temperature control system. As the diagram shows, the analog temperature is measured and the measured value is then converted to a digital quantity by an **analog-to-digital converter (ADC).** The digital quantity is then processed by the digital circuitry, which may or may not include a digital computer. Its digital output is converted back to an analog quantity by a **digital-to-analog converter (DAC).** This analog output is fed to a controller which takes some kind of action to adjust the temperature.

The need for conversion between analog and digital forms of information can be considered a drawback because of the added complexity and expense. Another factor that is often important is the extra time required to perform these conversions. In many applications, these factors are outweighed by the numerous advantages of using digital techniques, and so the conversion between analog and digital quantities has become quite commonplace in the current technology.

There are situations, however, where using only analog techniques is simpler and more economical. For example, the process of signal amplification is most easily accomplished using analog circuitry.

It is becoming more and more common to see both digital and analog techniques employed within the same system in order to profit from the advantages of each. In these *hybrid* systems, one of the most important parts of the design phase involves determining what parts of the system are to be analog and what parts are to be digital.

Finally, it is safe to predict that as time goes on, we should see a greater and greater shift to digital techniques as the economic benefits of integration become even more overriding than they are now.

___ REVIEW QUESTIONS ___

1. What are the advantages of digital techniques over analog?
2. What is the chief limitation to the use of digital techniques?

1-3   DIGITAL NUMBER SYSTEMS

Many number systems are in use in digital technology. The most common are the decimal, binary, octal, and hexadecimal systems. The decimal system is clearly the most familiar to us because it is a tool that we use every day. Examining some of its characteristics will help us to better understand the other systems.

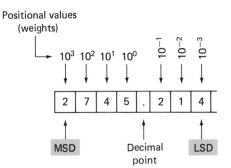

FIGURE 1-2   Decimal position values as powers of 10.

Decimal System

The *decimal system* is composed of *10* numerals or symbols. These 10 symbols are 0, 1, 2, 3, 4, 5, 6, 7, 8, 9; using these symbols as *digits* of a number, we can express any quantity. The decimal system, also called the *base-10* system because it has 10 digits, has evolved naturally as a result of the fact that man has 10 fingers. In fact, the word "digit" is derived from the Latin word for "finger."

The decimal system is a *positional-value system* in which the value of a digit depends on its position. For example, consider the decimal number 453. We now that the digit 4 actually represents 4 *hundreds,* the 5 represents 5 *tens,* and the 3 represents 3 *units*. In essence, the 4 carries the most weight of the three digits; it is referred to as the *most significant digit (MSD)*. The 3 carries the least weight and is called the *least significant digit (LSD)*.

Consider another example, 27.35. This number is actually equal to 2 tens plus 7 units plus 3 tenths plus 5 hundredths, or $2 \times 10 + 7 \times 1 + 3 \times 0.1 + 5 \times 0.01$. The decimal point is used to separate the integer and fractional parts of the number.

More rigorously, the various positions relative to the decimal point carry weights that can be expressed as powers of 10. This is illustrated in Figure 1-2, where the number 2745.214 is represented. The decimal point separates the positive powers of 10 from the negative powers. The number 2745.214 is thus equal to

$$(2 \times 10^{+3}) + (7 \times 10^{+2}) + (4 \times 10^{1}) + (5 \times 10^{0})$$
$$+ (2 \times 10^{-1}) + (1 \times 10^{-2}) + (4 \times 10^{-3})$$

In general, any number is simply the sum of the products of each digit value and its positional value.

Decimal Counting

When counting in the decimal system, we start with 0 in the units position and take each symbol (digit) in progression until we reach 9. Then we add a 1 to the next higher position and start over with zero in the first position (see Figure 1-3). This process continues until the count of 99 is reached. Then we add a 1 to the third position and start over with zeros in the first two positions. The same pattern is followed continuously as high as we wish to count.

It is important to note that in decimal counting the units position (LSD) changes upward with each step in the count, the tens position changes upward every 10 steps in the count, the hundreds position changes upward every 100 steps in the count, and so on.

Another characteristic of the decimal system is that using only two decimal places we can count through $10^2 = 100$ different numbers (0 to 99).* With three

---

*Zero is counted as a number.

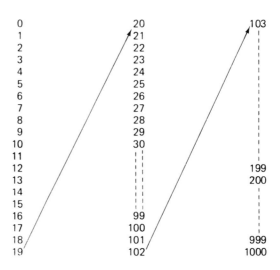

0	20	103
1	21	⋮
2	22	
3	23	
4	24	
5	25	
6	26	
7	27	
8	28	
9	29	
10	30	
11	⋮	
12		199
13		200
14		
15		
16	99	
17	100	
18	101	999
19	102	1000

FIGURE 1-3   Decimal counting.

places we can count through 1000 numbers (0 to 999); and so on. In general, with N places or digits we can count through 10^N different numbers, starting with and including zero. The largest number will always be $10^N - 1$.

Teaching Tip: Students need to be constantly reminded that zero is a number in the counting sequence so that there is a difference of 1 between the number of counts and the maximum count for any given number of digits.

Binary System   Unfortunately, the decimal number system does not lend itself to convenient implementation in digital systems. For example, it is very difficult to design electronic equipment so that it can work with 10 different voltage levels (each one representing one decimal character, 0 through 9). On the other hand, it is very easy to design simple, accurate electronic circuits that operate with only two voltage levels. For this reason, almost every digital system uses the binary number system (base 2) as the basic number system of its operations, although other systems are often used in conjunction with binary.

In the *binary system* there are only two symbols or possible digit values, 0 and 1. Even so, this base-2 system can be used to represent any quantity that can be represented in decimal or other number systems. In general though, it will take a greater number of binary digits to express a given quantity.

All the statements made earlier concerning the decimal system are equally applicable to the binary system. The binary system is also a positional-value system, wherein each binary digit has its own value or weight expressed as a power of 2. This is illustrated in Figure 1-4. Here, places to the left of the *binary point* (counterpart of the decimal point) are positive powers of 2 and places to the right are nega-

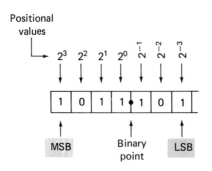

Positional values

FIGURE 1-4   Binary position values as powers of 2.

tive powers of 2. The number 1011.101 is shown represented in the figure. To find its equivalent in the decimal system we simply take the sum of the products of each digit value (0 or 1) and its positional value:

$$1011.101_2 = (1 \times 2^3) + (0 \times 2^2) + (1 \times 2^1) + (1 \times 2^0)$$
$$+ (1 \times 2^{-1}) + (0 \times 2^{-2}) + (1 \times 2^{-3})$$
$$= 8 + 0 + 2 + 1 + 0.5 + 0 + 0.125$$
$$= 11.625_{10}$$

Notice in the preceding operation that subscripts (2 and 10) were used to indicate the base in which the particular number is expressed. This convention is used to avoid confusion whenever more than one number system is being employed.

In the binary system, the term *binary digit* is often abbreviated to the term **bit,** which we will use henceforth. Thus, in the number expressed in Figure 1-4 there are 4 bits to the left of the binary point, representing the integer part of the number, and 3 bits to the right of the binary point, representing the fractional part. The most significant bit (MSB) is the leftmost bit (largest weight). The least significant bit (LSB) is the rightmost bit (smallest weight). These are indicated in Figure 1-4.

Discussion Question: What is the weight of the LSB of every binary integer?

Binary Counting   When we deal with binary numbers, we will usually be restricted to a specific number of bits. This restriction is based on the circuitry that is used to represent these binary numbers. Let's use 4-bit binary numbers to illustrate the method for counting in binary.

The sequence (shown in Figure 1-5) begins with all bits at 0; this is called the *zero count*. For each successive count, the units (2^0) position *toggles;* that is, it changes from one binary value to the other. Each time the units bit changes from a 1 to a 0, the twos (2^1) position will toggle. Each time the twos position changes from 1 to 0, the fours (2^2) position will toggle (change states). Likewise, each time the fours position goes from 1 to 0, the eights (2^3) position toggles. This same process would be continued for the higher-order bit positions if the binary number had more than 4 bits.

FIGURE 1-5   Binary counting sequence.

Weights →	$2^3 = 8$	$2^2 = 4$	$2^1 = 2$	$2^0 = 1$	Decimal equivalent
	0	0	0	0	0
	0	0	0	1	1
	0	0	1	0	2
	0	0	1	1	3
	0	1	0	0	4
	0	1	0	1	5
	0	1	1	0	6
	0	1	1	1	7
	1	0	0	0	8
	1	0	0	1	9
	1	0	1	0	10
	1	0	1	1	11
	1	1	0	0	12
	1	1	0	1	13
	1	1	1	0	14
	1	1	1	1	15

LSB

The binary counting sequence has an important characteristic, as shown in Figure 1-5. The units bit (LSB) changes either from 0 to 1 or 1 to 0 with *each* count. The second bit (twos position) stays at 0 for two counts, then at 1 for two counts, then at 0 for two counts, and so on. The third bit (fours position) stays at 0 for four counts, then at 1 for four counts, and so on. The fourth bit (eights position) stays at 0 for eight counts, then at 1 for eight counts. If we wanted to count further we would add more places, and this pattern would continue with 0s and 1s alternating in groups of 2^{N-1}. For example, using a fifth binary place, the fifth bit would alternate sixteen 0s, then sixteen 1s, and so on.

As we saw for the decimal system, it is also true for the binary system that by using N bits or places we can go through 2^N counts. For example, with 2 bits we can go through $2^2 = 4$ counts (00_2 through 11_2); with 4 bits we can go through $2^4 = 16$ counts (0000_2 through 1111_2); and so on. The last count will always be all 1s and is equal to $2^N - 1$ in the decimal system. For example, using 4 bits, the last count is $1111_2 = 2^4 - 1 = 15_{10}$.

___ EXAMPLE 1-2 ___

What is the largest number that can be represented using 8 bits?

Solution $2^N - 1 = 2^8 - 1 = 255_{10} = 11111111_2$.

___ REVIEW QUESTIONS ___

1. What is the decimal equivalent of 1101011_2?
2. What is the next binary number following 10111_2 in the counting sequence?
3. What is the largest decimal value that can be represented using 12 bits?

1-4   REPRESENTING BINARY QUANTITIES

In digital systems the information that is being processed is usually present in binary form. Binary quantities can be represented by any device that has only two operating states or possible conditions. For example, a switch has only two states: open or closed. We can arbitrarily let an open switch represent binary 0 and a closed switch represent binary 1. With this assignment we can now represent any binary number as illustrated in Figure 1-6(a), where the states of the various switches represent 10010_2.

Another example is shown in Figure 1-6(b), where holes punched in paper are used to represent binary numbers. A punched hole is a binary 1 and absence of a hole is a binary 0.

There are numerous other devices which have only two operating states or can be operated in two extreme conditions. Among these are: light bulb (bright or dark), diode (conducting or nonconducting), relay (energized or deenergized), transistor (cut off or saturated), photocell (illuminated or dark), thermostat (open or closed), mechanical clutch (engaged or disengaged), and magnetic tape (magnetized or demagnetized).

In electronic digital systems, binary information is represented by voltages (or currents) that are present at the inputs and outputs of the various circuits. Typically,

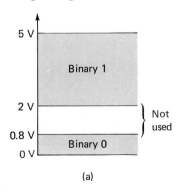

(a)

(b)

FIGURE 1-6 Using (a) switches and (b) punched paper tape to represent binary numbers.

the binary 0 and 1 are represented by two nominal voltage levels. For example, zero volts (0 V) might represent binary 0, and +5 V might represent binary 1. In actuality, because of circuit variations, the 0 and 1 would be represented by voltage ranges. This is illustrated in Figure 1-7(a), where any voltage between 0 and 0.8 V represents a 0 and any voltage between 2 and 5 V represents a 1. All input and output signals will normally fall within one of these ranges except during transitions from one level to another. Figure 1-7(b) shows a typical digital signal as it sequences through the binary value 01010.

We can now see another significant difference between digital and analog systems. In digital systems, the exact value of a voltage *is not* important; for example, a voltage of 3.6 V means the same as a voltage of 4.3 V. In analog systems, the exact value of a voltage *is* important. For instance, if the analog voltage is proportional to the temperature measured by a transducer, the 3.6 V would represent a different temperature than would 4.3 V. In other words, the voltage value carries significant information. This characteristic means that the design of accurate analog circuitry is generally more difficult than that of digital circuitry because of the way in which exact voltage values will fluctuate with variations in component values, temperature, and noise.

Teaching Tip: You can mention here that when a voltage does not fall in either of the prescribed ranges, it is usually an indication of a circuit malfunction or a loading problem.

1-5 DIGITAL CIRCUITS

As explained in Section 1-4, digital circuits are designed to produce output voltages that fall within the prescribed 0 and 1 voltage ranges such as those defined in Figure 1-7. Likewise, digital circuits are designed to respond predictably to input voltages

FIGURE 1-7 (a) Typical voltage assignments in digital system; (b) typical digital signal.

(a)

(b)

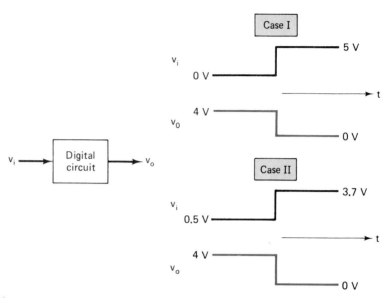

FIGURE 1-8   A digital circuit responds to an input's binary level (0 or 1) and not to its actual voltage.

that are within the defined 0 and 1 ranges. What this means is that a digital circuit will respond in the same way to all input voltages that fall within the allowed 0 range; similarly, it will not distinguish between input voltages that lie within the allowed 1 range.

To illustrate, Figure 1-8 represents a typical digital circuit with input v_i and output v_o. The output is shown for two different input signal waveforms. Note that v_o is the same for both cases because the two input waveforms, while they differ in their exact voltage levels, are at the same binary levels.

Logic Circuits

The manner in which a digital circuit responds to an input is referred to as the circuit's *logic*. Each type of digital circuit obeys a certain set of logic rules. For this reason, digital circuits are also called *logic circuits*. We will use both terms interchangeably throughout the text. In Chapter 3 we will see more clearly what is meant by a circuit's "logic."

We will be studying all the types of logic circuits that are currently used in digital systems. Initially, our attention will be focused only on the logical operation which these circuits perform—that is, the relationship between the circuit inputs and outputs. We will defer any discussion of the internal circuit operation of these logic circuits until after we have developed an understanding of their logical operation.

Digital Integrated Circuits

Almost all the digital circuits used in modern digital systems are integrated circuits (ICs). The wide variety of available logic ICs has made it possible to construct complex digital systems that are smaller and more reliable than the discrete-component counterparts.

Several integrated-circuit fabrication technologies are used to produce digital ICs, the most common being TTL, CMOS, NMOS, and ECL. Each differs in the type of circuitry used to provide the desired logic operation. For example, TTL (transistor-transistor logic) uses the bipolar transistor as its main circuit element,

while CMOS (complementary metal-oxide-semiconductor) uses the enhancement MOSFET as its principal circuit element. We will learn about the various IC technologies, their characteristics, and their relative advantages and disadvantages after we master the basic logic circuit types.

——— REVIEW QUESTIONS ———

1. *True or false:* The exact value of an input voltage is critical for a digital circuit.
2. Can a digital circuit produce the same output voltage for different input voltage values?
3. A digital circuit is also referred to as a _____ circuit.

1-6 PARALLEL AND SERIAL TRANSMISSION

One of the most common operations that occur in any digital system is the transmission of information from one place to another. The information can be transmitted over a distance as small as a fraction of an inch on the same circuit board, or over a distance of many miles when an operator at a computer terminal is communicating with a computer in another city. The information that is transmitted is in binary form and is generally represented as voltages at the outputs of a sending circuit that are connected to the inputs of a receiving circuit. Figure 1-9 illustrates the two basic methods for digital information transmission: *parallel* and *serial*.

Figure 1-9(a) shows how the binary number 10110 is transmitted from circuit A to circuit B using parallel transmission. Each bit of the binary number is represented by one of the circuit A outputs, where output A_4 is the MSB and A_0 is the

FIGURE 1-9 (a) Parallel transmission uses one connecting line per bit, and all bits are transmitted simultaneously;

Parallel Transmission

(a)

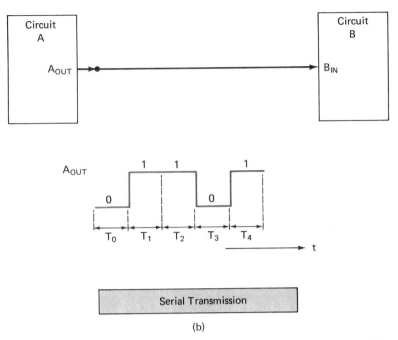

FIGURE 1-9 (*cont.*)   (b) serial transmission uses only one signal line, and the individual bits are transmitted serially (one at a time).

LSB. Each of the circuit *A* outputs is connected to a corresponding input of circuit *B* so that all 5 bits of information are transmitted simultaneously (in parallel).

In Figure 1-9(b) there is only one connection from circuit *A* to circuit *B* when serial transmission is used. Here the output of circuit *A* will produce a digital signal whose voltage level will change at regular intervals in accordance with the binary number being transmitted. In this way the information is being transmitted a bit at a time (serially) over the one signal line. The **timing diagram** in Figure 1-9(b) shows how the signal level varies with time. During the first time interval, T_0, the signal is at the 0 level; during interval T_1 the signal is at the 1 level; and so forth.

The principal trade-off between parallel and serial representations is one of speed versus circuit simplicity. The transmission of binary data from one part of a digital system to another can be done more quickly using parallel representation because all the bits are transmitted simultaneously, while serial representation transmits one bit at a time. On the other hand, parallel requires more signal lines connected between the sender and receiver of the binary data than does serial. In other words, parallel is faster and serial requires fewer signal lines. This comparison between parallel and serial methods for representing binary information will be encountered many times in discussions throughout the text.

Teaching Tip: A good example to cite in which serial transmission is more economical is the common telephone transmission system.

—— REVIEW QUESTION ——————————————————

1. Describe the relative advantages of parallel and serial transmission of binary data.

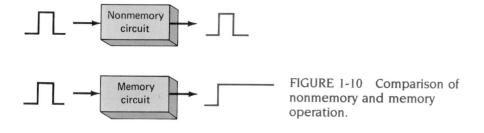

FIGURE 1-10 Comparison of nonmemory and memory operation.

1-7 MEMORY

When an input signal is applied to most devices or circuits, the output somehow changes in response to the input, and when the input signal is removed, the output returns to its original state. These circuits do not exhibit the property of *memory*, since their outputs revert back to normal. In digital circuitry certain types of devices and circuits do have memory. When an input is applied to such a circuit, the output will change its state, but it will remain in the new state even after the input is removed. This property of retaining its response to a momentary input is called **memory**. Figure 1-10 illustrates nonmemory and memory operations.

Memory devices and circuits play an important role in digital systems because they provide means for storing binary numbers either temporarily or permanently, with the ability to change the stored information at any time. As we shall see, the various memory elements include magnetic types and those which utilize electronic latching circuits (called *latches* and *flip-flops*).

1-8 DIGITAL COMPUTERS

Digital techniques have found their way into innumerable areas of technology, but the area of automatic **digital computers** is by far the most notable and most extensive. Although digital computers affect some part of all of our lives, it is doubtful that many of us know exactly what a computer does. In simplest terms, *a computer is a system of hardware that performs arithmetic operations, manipulates data (usually in binary form), and makes decisions.*

For the most part, human beings can do whatever computers can do, but computers can do it with much greater speed and accuracy. This is in spite of the fact that computers perform all their calculations and operations one step at a time. For example, a human being can take a list of 10 numbers and find their sum all in one operation by listing the numbers one over the other and adding them column by column. A computer, on the other hand, can add numbers only two at a time, so that adding this same list of numbers will take nine actual addition steps. Of course, the fact that the computer requires only a microsecond or less per step makes up for this apparent inefficiency.

A computer is faster and more accurate than people are, but unlike most people it has to be given a complete set of instructions that tell it *exactly* what to do at each step of its operation. This step of instructions, called a **program**, is prepared by one or more persons for each job the computer is to do. Programs are placed in the computer's memory unit in binary-coded form, with each instruction having a unique

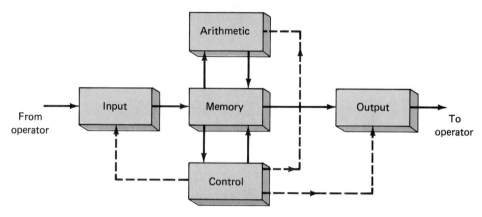

FIGURE 1-11   Functional diagram of a digital computer.

code. The computer takes these instruction codes from memory *one at a time* and performs the operation called for by the code. Much more will be said about this later.

Major Parts of a Computer   There are several types of computer systems, but each can be broken down into the same functional units. Each unit performs specific functions, and all units function together to carry out the instructions given in the program. Figure 1-11 shows the five major functional parts of a digital computer and their interaction. The solid lines with arrows represent the flow of information. The dashed lines with arrows represent the flow of timing and control signals.

The major functions of each unit are:

1. **Input unit.** Through this unit a complete set of instructions and data is fed into the computer system and into the memory unit, to be stored until needed. The information typically enters the input unit by means of punched cards, magnetic tape, magnetic disk, or a keyboard.
2. **Memory unit.** The memory stores the instructions and data received from the input unit. It stores the results of arithmetic operations received from the arithmetic unit. It also supplies information to the output unit.
3. **Control unit.** This unit takes instructions from the memory unit one at a time and interprets them. It then sends appropriate signals to all the other units to cause the specific instruction to be executed.
4. **Arithmetic/logic unit.** All arithmetic calculations and logical decisions are performed in this unit, which can then send results to the memory unit to be stored.
5. **Output unit.** This unit takes data from the memory unit and prints out, displays, or otherwise presents the information to the operator (or process, in the case of a process control computer).

How Many Types of Computers Are There?   The number of computer types depends on the criteria used to classify them. Computers will differ in their physical size, operating speed, memory capacity, and processing capability, as well as other characteristics. The most common way to classify computers is by their physical size, which usually, but not always, is indicative of their relative capabilities. The three basic classifications are **microcomputer, minicomputer,** and **mainframe.**

The microcomputer is the smallest type of computer. It generally consists of several IC chips including a **microprocessor** chip, memory chips, and input/output interface chips along with input/output devices such as a keyboard and video display. Microcomputers resulted from the tremendous advances in IC fabrication technology that make it possible to pack more and more circuitry into a small space. For example, the microprocessor chip contains the circuitry for both the arithmetic/logic unit and the control unit.

Minicomputers are larger than microcomputers and have prices that go into the tens of thousands of dollars (including input/output peripheral equipment). "Minis" are widely used in industrial control systems, scientific applications for schools and research laboratories, and business applications for small businesses. Although more expensive than microcomputers, minicomputers continue to be widely used because they are generally faster and possess more capabilities. These differences in speed and capability, however, are rapidly shrinking.

Mainframes, the largest computers, are found in corporations, banks, universities, and scientific laboratories. These "maxicomputers" can cost as much as several million dollars and include complete systems of peripheral equipment such as magnetic tape units, magnetic disk units, card punchers and readers, keyboards, printers, and many more. Applications of mainframes range from computationally oriented science and engineering problem solving to the data-oriented business applications, where emphasis is on maintaining and updating large quantities of data and information.

REVIEW QUESTIONS

1. Explain how a digital circuit that has memory differs from one that does not.
2. Name the five major functional units of a computer.
3. Name the three classes of computers based on physical size.

PROBLEMS

SECTION 1-2

1-1. Which of the following are analog quantities, and which are digital?
   (a) Number of atoms in a sample of material   digital
   (b) Altitude of an aircraft   analog
   (c) Pressure in a bicycle tire   analog
   (d) Current through a speaker   analog
   (e) Timer setting on a microwave oven   digital

SECTION 1-3

1-2. Convert the following binary numbers to their equivalent decimal values.

(a) $11001_2 =$ _____ 25 _____ $_{10}$

(b) 1001.1001 9.5625

(c) 10011011001.10110 1241.6875

1-3. Using 6 bits, show the binary counting sequence from 000000 to 111111.

1-4. What is the maximum number that we can count up to using 10 bits?   1023

1-5. How many bits are needed to count up to a maximum of 511?   nine

SECTION 1-6

1-6. Suppose that the decimal integer values from 0 to 15 are to be transmitted.

(a) How many lines will be needed if parallel representation is used?   four

(b) How many will be needed if serial representation is used?   one

ANSWERS TO SECTION REVIEW QUESTIONS

SECTION 1-1

1. Analog quantities can take on *any* value over a continuous range; digital quantities can take on only *discrete* values.

SECTION 1-2

1. Easier to design; easier to store information; greater accuracy and precision; programmability; less affected by noise; higher degree of integration.   **2.** Real-world physical quantities are analog.

SECTION 1-3

1. 107_{10} **2.** 11000_2 **3.** 4095_{10}

SECTION 1-5

1. False   **2.** Yes, provided that the two input voltages are within the same logic level range.
3. logic

SECTION 1-6

1. Parallel is faster; serial requires only one signal line.

SECTION 1-8

1. One that has memory will have its output changed and *remain* changed in response to a momentary change in the input signal.
2. Input, output, memory, arithmetic/logic, control.   **3.** Microcomputer, minicomputer, maxicomputer (mainframe).

2

NUMBER SYSTEMS AND CODES

Outline

Upon completion of this chapter, you will be able to:

- Use two different methods to perform decimal-to-binary conversions.
- Cite several advantages of the octal and hexadecimal number systems.
- Convert from the hexadecimal or octal number system to either the decimal or binary number system.
- Express decimal numbers using the BCD code.
- Understand the difference between the BCD code and the straight binary code.
- Cite the major differences between the Gray code and the binary code.
- Understand the need for alphanumeric codes, especially the ASCII code.
- Describe the parity method for error detection.
- Determine parity (odd or even) of digital data.

Introduction

The binary number system is the most important one in digital systems, but several others are also important. The decimal system is important because it is universally used to represent quantities outside a digital system. This means that there will be situations where decimal values have to be converted to binary values before they are entered into the digital system. For example, when you punch a decimal number into your hand calculator (or computer), the circuitry inside the device converts the decimal number to a binary value.

Likewise, there will be situations where the binary values at the outputs of a digital circuit have to be converted to decimal values for presentation to the outside world. For example, your calculator (or computer) uses binary numbers to calculate answers to a problem, then converts the answers to a decimal value before displaying them.

In addition to binary and decimal, two other number systems find wide-spread applications in digital systems. The *octal* (base-8) and *hexadecimal (base-16)* number systems are both used for the same purpose—to provide an efficient means for representing large binary numbers. As we shall see, both these number systems have the advantage that they can be easily converted to and from binary.

In a digital system, three or four of these number systems may be in use at the same time, so that an understanding of the system operation requires the ability to convert from one number system to another. This chapter will show you how to perform these conversions. Although some of them will not be of immediate use in our study of digital systems, you will need them when you begin to study microprocessors.

This chapter will also introduce some of the *binary codes* that are used to represent various kinds of information. These binary codes will use 1s and 0s, but in a way that differs somewhat from the binary number system.

2-1   BINARY-TO-DECIMAL CONVERSIONS

As explained in Chapter 1, the binary number system is a positional system where each binary digit (bit) carries a certain weight based on its position relative to the LSB. Any binary number can be converted to its decimal equivalent simply by summing together the weights of the various positions in the binary number which contain a 1. To illustrate:

$$1 \quad 1 \quad 0 \quad 1 \quad 1_2 \quad (binary)$$
$$2^4 + 2^3 + 0 + 2^1 + 2^0 = 16 + 8 + 2 + 1$$
$$= 27_{10} \quad (decimal)$$

Let's try another example with a greater number of bits:

$$1 \quad 0 \quad 1 \quad 1 \quad 0 \quad 1 \quad 0 \quad 1_2 =$$
$$2^7 + 0 + 2^5 + 2^4 + 0 + 2^2 + 0 + 2^0 = 181_{10}$$

Note that the procedure is to find the weights (i.e., powers of 2) for each bit position that contains a 1, and then to add them up. Also note that the MSB has a weight of 2^7 even though it is the eighth bit; this is because the LSB is the first bit and has a weight of 2^0.

—— REVIEW QUESTIONS ——

1. Convert 100011011011_2 to its decimal equivalent.
2. What is the weight of the MSB of a 16-bit number?

2-2   DECIMAL-TO-BINARY CONVERSIONS

There are two ways to convert a decimal *whole* number to its equivalent binary-system representation. The first method is the reverse of the process described in Section 2.1. The decimal number is simply expressed as a sum of powers of 2 and

then 1s and 0s are written in the appropriate bit positions. To illustrate:

$$45_{10} = 32 + 8 + 4 + 1 = 2^5 + 0 + 2^3 + 2^2 + 0 + 2^0$$
$$= 1 \quad 0 \quad 1 \quad 1 \quad 0 \quad 1_2$$

Common Student Mistake: Forgetting to include the 2^0 position.

Note that a 0 is placed in the 2^1 and 2^4 positions, since all positions must be accounted for. Another example is the following:

$$76_{10} = 64 + 8 + 4 = 2^6 + 0 + 0 + 2^3 + 2^2 + 0 + 0$$
$$= 1 \quad 0 \quad 0 \quad 1 \quad 1 \quad 0 \quad 0_2$$

Repeated Division   Another method uses repeated division by 2. The conversion, illustrated below for 25_{10}, requires repeatedly dividing the decimal number by 2 and writing down the remainder after each division until a quotient of 0 is obtained. Note that the binary result is obtained by writing the first remainder as the LSB and the last remainder as the MSB.

Common Student Mistake: Getting the order of the bits reversed. Advise converting back to decimal to check results.

$$\frac{25}{2} = 12 + \text{remainder of } 1$$
$$\frac{12}{2} = 6 + \text{remainder of } 0$$
$$\frac{6}{2} = 3 + \text{remainder of } 0$$
$$\frac{3}{2} = 1 + \text{remainder of } 1$$
$$\frac{1}{2} = 0 + \text{remainder of } 1$$

$$\text{MSB} \qquad \text{LSB}$$
$$25_{10} = 1 \quad 1 \quad 0 \quad 0 \quad 1_2$$

If a calculator is being used to perform the divisions by 2, the remainders can be determined by noting whether the quotient has a fractional part or not. For example, the calculator would produce $25/2 = 12.5$. The .5 indicates that there is a remainder of 1. The calculator would also produce $12/2 = 6.0$, which indicates a remainder of 0. Another example is shown below as it would occur using a calculator.

$$\frac{37}{2} = 18.5 \longrightarrow \text{remainder of } 1 \text{ (LSB)}$$
$$\frac{18}{2} = 9.0 \longrightarrow \qquad\qquad 0$$
$$\frac{9}{2} = 4.5 \longrightarrow \qquad\qquad 1$$
$$\frac{4}{2} = 2.0 \longrightarrow \qquad\qquad 0$$
$$\frac{2}{2} = 1.0 \longrightarrow \qquad\qquad 0$$
$$\frac{1}{2} = 0.5 \longrightarrow \qquad\qquad 1 \text{ (MSB)}$$

Thus, $37_{10} = 100101_2$.

1. Convert 83_{10} to binary using both methods.
2. Convert 729_{10} to binary using both methods. Check your answer by converting back to decimal.

2-3 OCTAL NUMBER SYSTEM

The octal number system is very important in digital computer work. The octal number system has a base of *eight*, meaning that it has eight possible digits: 0, 1, 2, 3, 4, 5, 6, and 7. Thus, each digit of an octal number can have any value from 0 to 7. The digit positions in an octal number have weights as follows:

$$\text{----} \boxed{8^4 \mid 8^3 \mid 8^2 \mid 8^1 \mid 8^0 \mid 8^{-1} \mid 8^{-2} \mid 8^{-3} \mid 8^{-4} \mid 8^{-5}} \text{-----}$$

octal point

Octal-to-Decimal Conversion An octal number, then, can be easily converted to its decimal equivalent by multiplying each octal digit by its positional weight. For example,

$$372_8 = 3 \times (8^2) + 7 \times (8^1) + 2 \times (8^0)$$
$$= 3 \times 64 + 7 \times 8 + 2 \times 1$$
$$= 250_{10}$$

Another example:

$$24.6_8 = 2 \times (8^1) + 4 \times (8^0) + 6 \times (8^{-1})$$
$$= 20.75_{10}$$

Decimal-to-Octal Conversion A decimal integer can be converted to octal by using the same repeated-division method that we used in the decimal-to-binary conversion, but with a division factor of 8 instead of 2. An example is shown below.

$$\frac{266}{8} = 33 + \text{remainder of } 2$$
$$\frac{33}{8} = 4 + \text{remainder of } 1$$
$$\frac{4}{8} = 0 + \text{remainder of } 4$$
$$266_{10} = \boxed{412_8}$$

Note that the first remainder becomes the least significant digit (LSD) of the octal number, and the last remainder becomes the most significant digit (MSD).

If a calculator is used to perform the divisions in the above process, the result will include a decimal fraction instead of a remainder. The remainder can be ob-

tained, however, by multiplying the decimal fraction by 8. For example, 266/8 produces 33.25. The remainder becomes $0.25 \times 8 = 2$. Similarly, 33/8 will be 4.125 and the remainder becomes $0.125 \times 8 = 1$.

Octal-to-Binary Conversion   The primary advantage of the octal number system is the ease with which conversion can be made between binary and octal numbers. The conversion from octal to binary is performed by converting *each* octal digit to its 3-bit binary equivalent. The eight possible digits are converted as indicated in Table 2-1.

TABLE 2-1

Octal Digit	0	1	2	3	4	5	6	7
Binary Equivalent	000	001	010	011	100	101	110	111

Using these conversions, any octal number is converted to binary by individually converting each digit. For example, we can convert 472_8 to binary as follows:

$$
\begin{array}{ccc}
4 & 7 & 2 \\
\downarrow & \downarrow & \downarrow \\
100 & 111 & 010
\end{array}
$$

Discussion Question: How can you prove that 472_8 is equivalent to 100111010_2?

Hence, octal 472 is equivalent to binary 100111010. As another example, consider converting 5431_8 to binary:

$$
\begin{array}{cccc}
5 & 4 & 3 & 1 \\
\downarrow & \downarrow & \downarrow & \downarrow \\
101 & 100 & 011 & 001
\end{array}
$$

Thus, $5431_8 = 101100011001_2$.

Binary-to-Octal Conversion   Converting from binary integers to octal integers is simply the reverse of the foregoing process. The bits of the binary number are grouped into groups of *three* bits starting at the LSB. Then each group is converted to its octal equivalent (Table 2-1). To illustrate, consider the conversion of 100111010_2 to octal.

$$
\begin{array}{ccc}
\underbrace{1\ 0\ 0} & \underbrace{1\ 1\ 1} & \underbrace{0\ 1\ 0} \\
\downarrow & \downarrow & \downarrow \\
4 & 7 & 2_8
\end{array}
$$

Sometimes the binary number will not have even groups of 3 bits. For those cases, we can add one or two 0s to the left of the MSB of the binary number to fill out the last group. This is illustrated below for the binary number 11010110.

$$
\begin{array}{ccc}
\underbrace{0\ 1\ 1} & \underbrace{0\ 1\ 0} & \underbrace{1\ 1\ 0} \\
\downarrow & \downarrow & \downarrow \\
3 & 2 & 6_8
\end{array}
$$

Note that a 0 was placed to the left of the MSB in order to produce even groups of 3 bits.

Counting in Octal

The largest octal digit is 7, so that in counting in octal, a digit is incremented upward from 0 to 7. Once it reaches 7, it recycles to 0 on the next count and causes the next higher digit to be incremented. This is illustrated in the following sequences of octal counting: (a) 65, 66, 67, 70, 71; (b) 275, 276, 277, 300.

With N octal digits, we can count from zero up to $8^N - 1$, for a total of 8^N different counts. For example, with three octal digits we can count from 000_8 to 777_8, which is a total of $8^3 = 512_{10}$ different octal numbers.

Usefulness of Octal System

The ease with which conversions can be made between octal and binary makes the octal system attractive as a "shorthand" means of expressing large binary numbers. In computer work, binary numbers with up to 64 bits are not uncommon. These binary numbers, as we shall see, do not always represent a numerical quantity but are often some type of code which conveys non-numerical information. In computers, binary numbers might represent (1) actual numerical data; (2) numbers corresponding to a location (address) in memory; (3) an instruction code; (4) a code representing alphabetic and other nonnumerical characters; or (5) a group of bits representing the status of devices internal or external to the computer.

When dealing with a large quantity of binary numbers of many bits, it is convenient and more efficient for us to write the numbers in octal rather than binary. Keep in mind, however, that the digital circuits and systems work strictly in binary; we are using octal only as a convenience for the operators of the system.

Teaching Tip: Another way to look at the octal counting sequence is to take the decimal counting sequence and eliminate all counts that contain an 8 or a 9 digit. For example, where the decimal sequence beginning with 256 is 256, 257, 258, 259, 260, 261, . . . the octal sequence beginning with 256_8 is 256, 257, 260, 261. . . .

_____ EXAMPLE 2-1 _____

Convert 177_{10} to its 8-bit binary equivalent by first converting to octal.

Solution

$$\frac{177}{8} = 22 + \text{remainder of } 1$$

$$\frac{22}{8} = 2 + \text{remainder of } 6$$

$$\frac{2}{8} = 0 + \text{remainder of } 2$$

Thus, $177_{10} = 261_8$. Now we can convert this octal number to its binary equivalent 010110001_2, so that we finally have

$$177_{10} = 10110001_2$$

Note that we chop off the leading 0 to express the result as 8 bits.

This method of decimal-to-octal-to-binary conversion is often quicker than going directly from decimal to binary, especially for large numbers. Likewise, it is often quicker to convert binary to decimal by first converting to octal.

_____ REVIEW QUESTIONS _____

1. Convert 614_8 to decimal.
2. Convert 146_{10} to octal, then from octal to binary.
3. Convert 10011101_2 to octal.

4. Write the next three numbers in this octal counting sequence: 624, 625, 626, ——, ——, ——.
5. Convert 975_{10} to binary by first converting to octal.
6. Convert binary 1010111011 to decimal by first converting to octal.

2-4 HEXADECIMAL NUMBER SYSTEM

The hexadecimal system uses base 16. Thus, it has 16 possible digit symbols. It uses the digits 0 through 9 plus the letters A, B, C, D, E, and F as the 16 digit symbols. Table 2-2 shows the relationships between hexadecimal, decimal, and binary. Note that each hexadecimal digit represents a group of four binary digits. It is important to remember that hex (abbreviation for hexadecimal) digits A through F are equivalent to the decimal values 10 through 15.

Hex-to-Decimal Conversion   A hex number can be converted to its decimal equivalent by using the fact that each hex digit position has a weight that is a power of 16. The LSD has a weight of $16^0 = 1$, the next higher digit has a weight of $16^1 = 16$, the next higher digit has a weight of $16^2 = 256$, and so on. The conversion process is demonstrated in the examples below:

$$356_{16} = 3 \times 16^2 + 5 \times 16^1 + 6 \times 16^0$$
$$= 768 + 80 + 6$$
$$= 854_{10}$$
$$2AF_{16} = 2 \times 16^2 + 10 \times 16^1 + 15 \times 16^0$$
$$= 512 + 160 + 15$$
$$= 687_{10}$$

Note that in the second example the value 10 was substituted for A and the value 15 for F in the conversion to decimal.

For practice, verify that $1BC2_{16}$ is equal to 7106_{10}.

TABLE 2-2

Hexadecimal	Decimal	Binary
0	0	0000
1	1	0001
2	2	0010
3	3	0011
4	4	0100
5	5	0101
6	6	0110
7	7	0111
8	8	1000
9	9	1001
A	10	1010
B	11	1011
C	12	1100
D	13	1101
E	14	1110
F	15	1111

Decimal-to-Hex Conversion Recall that we did decimal-to-binary conversion using repeated division by 2, and decimal-to-octal using repeated division by 8. Likewise, decimal-to-hex conversion can be done using repeated division by 16. The following two examples will illustrate.

___ EXAMPLE 2-2 _____

Convert 423_{10} to hex.

Solution

$$\frac{423}{16} = 26 + \text{remainder of } 7$$

$$\frac{26}{16} = 1 + \text{remainder of } 10$$

$$\frac{1}{16} = 0 + \text{remainder of } 1$$

$$423_{10} = 1A7_{16}$$

___ EXAMPLE 2-3 _____

Convert 214_{10} to hex.

Solution

$$\frac{214}{16} = 13 + \text{remainder of } 6$$

$$\frac{13}{16} = 0 + \text{remainder of } 13$$

$$214_{10} = D6_{16}$$

Again note how the remainders of the division processes form the digits of the hex number. Also note that any remainders that are greater than 9 are represented by the letters A through F.

If a calculator is being used to perform the divisions in the conversion process, the results will include a decimal fraction instead of a remainder. The remainder can be obtained by multiplying the fraction by 16. To illustrate, in Example 2-3 the calculator would have produced

$$\frac{214}{16} = 13.375$$

The remainder becomes $(0.375) \times 16 = 6$.

Hex-to-Binary Conversion Like the octal number system, the hexadecimal number system is used primarily as a "shorthand" method for representing binary

Discussion Question:
Which would you rather
have as your weekly
salary—76_8, 76_{10}, or
76_{16}?

numbers. It is a relatively simple matter to convert a hex number to binary. *Each* hex digit is converted to its 4-bit binary equivalent (Table 2-2). This is illustrated below for $9F2_{16}$.

$$9F2_{16} = \quad 9 \qquad\qquad F \qquad\qquad 2$$
$$\downarrow \qquad\qquad \downarrow \qquad\qquad \downarrow$$
$$1\ 0\ 0\ 1 \quad 1\ 1\ 1\ 1 \quad 0\ 0\ 1\ 0$$
$$= 100111110010_2$$

For practice, verify that $BA6_{16} = 101110100110_2$.

Binary-to-Hex Conversion   Conversion from binary to hex is just the reverse of the process above. The binary number is grouped into groups of *four* bits, and each group is converted to its equivalent hex digit. Zeros are added, as needed, to complete a 4-bit group (shown shaded).

$$1\ 1\ 1\ 0\ 1\ 0\ 0\ 1\ 1\ 0_2 = \quad 0\ 0\ \ 1\ 1 \quad 1\ 0\ 1\ 0 \quad 0\ 1\ 1\ 0$$
$$3 \qquad\quad A \qquad\quad 6$$
$$= 3A6_{16}$$

In order to perform these conversions between hex and binary, it is necessary to know the 4-bit binary numbers (0000 through 1111) and their equivalent hex digits. Once these are mastered, the conversions can be performed quickly without the need for any calculations. This is why hex (and octal) are so useful in representing large binary numbers.

For practice, verify that $101011111_2 = 15F_{16}$.

Counting in Hexadecimal   When counting in hex, each digit position can be incremented (increased by 1) from 0 to F. Once a digit position reaches the value F, it is reset to 0 and the next digit position is incremented. This is illustrated in the following hex counting sequences:

(a) 38, 39, 3A 3B, 3C, 3D, 3E, 3F, 40, 41, 42

(b) 6F8, 6F9, 6FA, 6FB, 6FC, 6FD, 6FE, 6FF, 700

Note that when there is a 9 in a digit position, it becomes an A when it is incremented.

──── EXAMPLE 2-4 ────────────────────────────

Convert decimal 378 to a 16-bit binary number by first converting to hexadecimal.

Solution

$$\frac{378}{16} = 23 + \text{remainder of } 10$$
$$\frac{23}{16} = 1 + \text{remainder of } 7$$
$$\frac{1}{16} = 0 + \text{remainder of } 1$$

Thus, $378_{10} = 17A_{16}$. This hex value can easily be converted to binary

000101111010. Finally, we can express 378_{10} as a 16-bit number by adding four leading 0s:

$$378_{10} = 0000 \quad 0001 \quad 0111 \quad 1010_2$$

___

___ EXAMPLE 2-5 ___

Convert $B2F_{16}$ to octal.

Solution   It's easiest to first convert hex to binary, then to octal.

$$
\begin{aligned}
B2F_{16} &= 1011 \quad 0010 \quad 1111 && \{\text{convert to binary}\} \\
&= 101 \quad 100 \quad 101 \quad 111 && \{\text{group bits by 3s}\} \\
&= 5 \quad 4 \quad 5 \quad 7_8 && \{\text{convert to octal}\}
\end{aligned}
$$

___

___ REVIEW QUESTIONS ___

1. Convert $24CE_{16}$ to decimal.
2. Convert 3117_{10} to hex, then from hex to binary.
3. Convert 10010111101101101_2 to hex.
4. Write the next four numbers in this hex counting sequence: E9A, E9B, E9C, E9D, ——, ——, ——.
5. Convert 3527_8 to hex.

2-5   BCD CODE

When numbers, letters, or words are represented by a special group of symbols, we say that they are being encoded, and the group of symbols is called a **code**. Probably one of the most familiar codes is the Morse code, where series of dots and dashes represent letters of the alphabet.

We have seen that any decimal number can be represented by an equivalent binary number. The group of 0s and 1s in the binary number can be thought of as a code representing the decimal number. When a decimal number is represented by its equivalent binary number, we call it **straight binary coding**.

Digital systems all use some form of binary numbers for their internal operation, but the external world is decimal in nature. This means that conversions between the decimal and binary systems are being performed often. We have seen that the conversions between decimal and binary can become long and complicated for large numbers. For this reason, a means of encoding decimal numbers that combines some features of both the decimal and binary systems is used in certain situations.

Binary-Coded-Decimal Code   If *each* digit of a decimal number is represented by its binary equivalent, the result is a code called **binary-coded-decimal** (hereafter abbreviated BCD). Since a decimal digit can be as large as 9, four bits are required to code each digit (the binary code for 9 is 1001).

To illustrate the BCD code, take a decimal number such as 874. Each *digit* is changed to its binary equivalent as follows:

$$\begin{array}{ccc} 8 & 7 & 4 \quad \text{(decimal)} \\ \downarrow & \downarrow & \downarrow \\ 1000 & 0111 & 0100 \quad \text{(BCD)} \end{array}$$

As another example, let us change 943 to its BCD-code representation:

$$\begin{array}{ccc} 9 & 4 & 3 \quad \text{(decimal)} \\ \downarrow & \downarrow & \downarrow \\ 1001 & 0100 & 0011 \quad \text{(BCD)} \end{array}$$

Once again, each decimal digit is changed to its straight binary equivalent. Note that 4 bits are *always* used for each digit.

The BCD code, then, represents each digit of the decimal number by a 4-bit binary number. Clearly only the 4-bit binary numbers from 0000 through 1001 are used. The BCD code does not use the numbers 1010, 1011, 1100, 1101, 1110, and 1111. In other words, only 10 of the 16 possible 4-bit binary code groups are used. If any of the "forbidden" 4-bit numbers ever occurs in a machine using the BCD code, it is usually an indication that an error has occurred.

—— EXAMPLE 2-6 ——————————————

Convert 0110100000111001 (BCD) to its decimal equivalent.

Solution   Divide the BCD number into 4-bit groups and convert each to decimal.

$$\begin{array}{cccc} \underline{0110} & \underline{1000} & \underline{0011} & \underline{1001} \\ 6 & 8 & 3 & 9 \end{array}$$

—— EXAMPLE 2-7 ——————————————

Convert the BCD number 011111000001 to its decimal equivalent.

Solution

$$\begin{array}{ccc} \underline{0111} & \underline{1100} & \underline{0001} \\ 7 & \downarrow & 1 \end{array}$$

forbidden code group indicates error in BCD number

Comparison of BCD and Binary   It is important to realize that BCD is not another number system like binary, octal, decimal, and hexadecimal. It is, in fact, the decimal system with each digit encoded in its binary equivalent. It is also important to understand that a BCD number is *not* the same as a straight binary number. A straight binary code takes the *complete* decimal number and represents it in binary; the BCD code converts *each* decimal *digit* to binary individually. To illustrate, take the number 137 and compare its straight binary and BCD codes:

$$137_{10} = 10001001_2 \qquad \text{(binary)}$$
$$137_{10} = 0001 \quad 0011 \quad 0111 \qquad \text{(BCD)}$$

The BCD code requires 12 bits while the straight binary code requires only 8 bits to represent 137. BCD requires more bits than straight binary to represent decimal numbers of more than one digit. This is because BCD does not use all possible 4-bit groups, as pointed out earlier, and is therefore somewhat inefficient.

The main advantage of the BCD code is the relative ease of converting to and from decimal. Only the 4-bit code groups for the decimal digits 0 through 9 need to be remembered. This ease of conversion is especially important from a hardware standpoint because in a digital system it is the logic circuits that perform the conversions to and from decimal.

Teaching Tip: Several common examples of the use of BCD are found in calculators, DVMs, frequency counters, digital clocks. You may want to discuss the reasons it is used in these devices by giving an advance glimpse of the encoding/decoding functions. It might be worth mentioning why most computers do not use BCD instead of binary for their internal operations (because it has more complex arithmetic operations, and it requires more bits to represent a decimal value).

REVIEW QUESTIONS

1. Represent the decimal value 178 by its straight binary equivalent. Then encode the same decimal number using BCD.
2. How many bits are required to represent an eight-digit decimal number in BCD?
3. What is an advantage of encoding a decimal number in BCD as compared with straight binary? What is a disadvantage?

2-6 EXCESS-3 CODE

The *excess-3 code* is related to the BCD code and is sometimes used instead because it possesses advantages in certain arithmetic operations. The excess-3 code for a decimal number is performed in the same manner as BCD except that 3 is added to *each* decimal digit before encoding it in binary. For example, to encode the decimal digit 4 into excess-3 code, we must first add 3 to obtain 7. Then the 7 is encoded in its equivalent 4-bit binary code 0111.

As another example, let us convert 48 into its excess-3 code representation:

$$
\begin{array}{cc}
4 & 8 \\
+\ \underline{3} & +\ \underline{3} \qquad \text{add 3 to each digit} \\
7 & 11 \\
\downarrow & \downarrow \\
0111 & 1011 \qquad \text{convert to a 4-bit binary code}
\end{array}
$$

Table 2-3 lists the BCD and excess-3 code representations for the decimal digits. Note that both codes use only 10 of the 16 possible 4-bit code groups. The excess-3 code, however, does not use the same code groups. For excess-3, the invalid code groups are 0000, 0001, 0010, 1101, 1110, and 1111.

TABLE 2-3

Decimal	BCD	Excess-3
0	0000	0011
1	0001	0100
2	0010	0101
3	0011	0110
4	0100	0111
5	0101	1000
6	0110	1001
7	0111	1010
8	1000	1011
9	1001	1100

TABLE 2-4

Decimal	Binary Code	Gray Code	Decimal	Binary Code	Gray Code
0	0000	0000	8	1000	1100
1	0001	0001	9	1001	1101
2	0010	0011	10	1010	1111
3	0011	0010	11	1011	1110
4	0100	0110	12	1100	1010
5	0101	0111	13	1101	1011
6	0110	0101	14	1110	1001
7	0111	0100	15	1111	1000

2-7 GRAY CODE

The *Gray code* belongs to a class of codes called *minimum-change codes,* in which only *one* bit in the code group changes when going from one step to the next. The Gray code is an *unweighted* code, meaning that the bit positions in the code groups do not have any specific weight assigned to them. Because of this, the Gray code is not suited for arithmetic operations but finds application in input/output devices and some types of analog-to-digital converters.

Table 2-4 shows the Gray-code representation for the decimal numbers 0 through 15, together with the straight binary code. If we examine the Gray-code groups for each decimal number, it can be seen that in going from *any* one decimal number to the next, only *one* bit of the Gray code changes. For example, in going from 3 to 4, the Gray code changes from 0010 to 0110, with only the second bit from the left changing. Going from 14 to 15 the Gray-code bits change from 1001 to 1000, with only the last bit changing. This is the principal characteristic of the Gray code. Compare this with the binary code, where anywhere from one to all of the bits change in going from one step to the next.

The Gray code is often used in situations where other codes, such as binary, might produce erroneous or ambiguous results during those transitions in which more than one bit of the code is changing. For instance, using binary code and going from 0111 to 1000 requires that all four bits change simultaneously. Depending on the device or circuit that is generating the bits, there may be a significant difference in the transition times of the different bits. If so, the transition from 0111 to 1000 could produce one or more intermediate states. For example, if the most significant bit changes faster than the rest, the following transitions will occur:

The occurrence of 1111 is only momentary but it could conceivably produce erroneous operation of the elements that are being controlled by the bits. Obviously, using the Gray code would eliminate this problem, since only one bit change occurs per transition and no "race" between bits can occur.

1. What is the principal advantage of the Gray code over straight binary?

2-8 ALPHANUMERIC CODES

In addition to numerical data, a computer must be able to handle nonnumerical information. In other words, a computer should recognize codes that represent letters of the alphabet, punctuation marks, and other special characters as well as numbers. These codes are called *alphanumeric codes*. A complete alphanumeric code would include the 26 lowercase letters, 26 uppercase letters, 10 numeric digits, 7 punctuation marks, and anywhere from 20 to 40 other characters, such as +, /, #, %, *, and so on. We can say that an alphanumeric code represents all of the various characters and functions that are found on a standard typewriter (or computer) keyboard.

ASCII Code   The most widely used alphanumeric code, the American Standard Code for Information Interchange (ASCII), is used in most microcomputers and

TABLE 2-5 Partial Listing of ASCII Code

Character	7-Bit ASCII	Octal	Hex	Character	7-Bit ASCII	Octal	Hex
A	100 0001	101	41	Y	101 1001	131	59
B	100 0010	102	42	Z	101 1010	132	5A
C	100 0011	103	43	0	011 0000	060	30
D	100 0100	104	44	1	011 0001	061	31
E	100 0101	105	45	2	011 0010	062	32
F	100 0110	106	46	3	011 0011	063	33
G	100 0111	107	47				
H	100 1000	110	48	4	011 0100	064	34
I	100 1001	111	49	5	011 0101	065	35
J	100 1010	112	4A	6	011 0110	066	36
K	100 1011	113	4B	7	011 0111	067	37
L	100 1100	114	4C	8	011 1000	070	38
				9	011 1001	071	39
M	100 1101	115	4D	blank	010 0000	040	20
N	100 1110	116	4E	.	010 1110	056	2E
O	100 1111	117	4F	(	010 1000	050	28
P	101 0000	120	50	+	010 1011	053	2B
Q	101 0001	121	51	$	010 0100	044	24
R	101 0010	122	52	*	010 1010	052	2A
				)	010 1001	051	29
S	101 0011	123	53	—	010 1101	055	2D
T	101 0100	124	54	/	010 1111	057	2F
U	101 0101	125	55	,	010 1100	054	2C
V	101 0110	126	56	=	011 1101	075	3D
W	101 0111	127	57	⟨RETURN⟩	000 1101	015	0D
X	101 1000	130	58	⟨LINEFEED⟩	000 1010	012	0A

minicomputers, and in many mainframes. The ASCII code (pronounced "askee") is a 7-bit code, and so it has $2^7 = 128$ possible code groups. This is more than enough to represent all of the standard keyboard characters as well as control functions such as the ⟨RETURN⟩ and ⟨LINEFEED⟩ functions. Table 2-5 shows a partial listing of the ASCII code. In addition to the binary code group for each character, the table gives the octal and hexadecimal equivalents.

___ EXAMPLE 2-8 _____

The following is a message encoded in ASCII code. What is the message?

<div align="center">1001000   1000101   1001100   1010000</div>

Solution   Convert each 7-bit code to its hex equivalent. The results are

<div align="center">48   45   4C   50</div>

Now locate these hex values in Table 2-5 and determine the character represented by each. The results are

<div align="center">H   E   L   P</div>

The ASCII code is used for the transfer of alphanumeric information between a computer and input/output devices such as video terminals or printers. A computer also uses it internally to store the information that an operator types in at the computer's keyboard. The following example illustrates this.

___ EXAMPLE 2-9 _____

An operator is typing in a BASIC program at the keyboard of a certain microcomputer. The computer converts each keystroke into its ASCII code and stores the code in memory. Determine the codes that will be entered into memory when the operator types in the following BASIC statement:

<div align="center">GOTO 25</div>

Solution   Locate each character (including the space) in Table 2.5 and record its ASCII code.

G	1000111
O	1001111
T	1010100
O	1001111
(space)	0100000
2	0110010
5	0110101

___ REVIEW QUESTIONS _____

1. Encode the following message in ASCII code using the hex representation: "COST = $72."
2. The following ASCII-coded message is stored in successive memory locations in a computer:

<div align="center">1010011   1010100   1001111,   1010000</div>

What is the message?

2-9 PARITY METHOD FOR ERROR DETECTION

The movement of binary data and codes from one location to another is the most frequent operation performed in digital systems. Here are some examples:

- Transmission of digitized voice over a microwave link
- The storage and retrieval of data from external memory devices such as magnetic tape and disk
- The transmission of information from a computer to a remote user terminal or another computer

Whenever information is transmitted from one device (the transmitter) to another device (the receiver), there is a possibility that errors can occur such that the receiver does not receive the identical information that was sent by the transmitter. The major cause of any transmission errors is **electrical noise,** which consists of spurious fluctuations in voltage or current that are present in all electronic systems to varying degrees. Figure 2-1 is a simple illustration of a type of transmission error.

The transmitter sends a relative noise-free serial digital signal over a signal line to a receiver. However, by the time the signal reaches the receiver, it contains a certain degree of noise superimposed on the original signal. Occasionally, the noise is large enough in amplitude that it will alter the logic level of the signal as it does at point x. When this occurs, the receiver may incorrectly interpret that bit as a logic 1, which is not what the transmitter has sent.

Most modern digital equipment is designed to be relatively error-free, and the probability of errors such as that shown in Figure 2-1 is very low. However we must realize that digital systems often transmit thousands, even millions of bits per second, so that even a very low rate of occurrence of errors can produce an occasional error that might prove to be bothersome, if not disastrous. For this reason, many digital systems employ some method for detection (and sometimes correction) of errors. One of the simplest and most widely used schemes for error detection is the **parity method**.

Parity Bit   A *parity bit* is an extra bit that is attached to a code group that is being transferred from one location to another. The parity bit is made either 0 or 1, depending on the number of 1s that are contained in the code group. Two different methods are used.

In the *even-parity* method, the value of the parity bit is chosen so that the total number of 1s in the code group (including the parity bit) is an *even* number. For example, suppose that the group is 1000011. This is the ASCII character C. The code

FIGURE 2-1   Example of noise causing an error in the transmission of digital data.

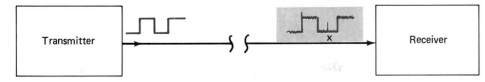

group has *three* 1s. Therefore, we will add a parity bit of 1 to make the total number of 1s an even number. The *new* code group, including the parity bit, thus becomes

$$1 \quad 1 \quad 0 \quad 0 \quad 0 \quad 0 \quad 1 \quad 1$$

↑————————————added parity bit*

If the code group contains an even number of 1s to begin with, the parity bit is given a value of 0. For example, if the code group were 1000001 (the ASCII code for "A"), the assigned parity bit would be 0, so that the new code, including the parity bit, would be 01000001.

The *odd-parity* method is used in exactly the same way except that the parity bit is chosen so the total number of 1s (including the parity bit) is an *odd* number. For example, for the code group 1000001, the assigned parity bit would be a 1. For the code group 1000011, the parity bit would be a 0.

Regardless of whether even parity or odd parity is used, the parity bit becomes an actual part of the code word. For example, adding a parity bit to the 7-bit ASCII code produces an 8-bit code. Thus, the parity bit is treated just like any other bit in the code.

Common Student Mistake: Students often make the mistake of assuming that the parity bit is wrong whenever the receiver determines that there is a parity error, when in fact the error can be with any of the bits including the parity bit.

The parity bit is used to detect any *single-bit* errors that occur during the transmission of a code from one location to another (e.g., from a computer to a video terminal). For example, suppose the character "A" is being transmitted and *odd* parity is being used. The transmitted code would be

$$1 \quad 1 \quad 0 \quad 0 \quad 0 \quad 0 \quad 0 \quad 1$$

When the receiver circuit receives this code, it will check that the code contains an odd number of 1s (including the parity bit). If so, the receiver will assume that the code has been correctly received. Now, suppose that because of some noise or malfunction the receiver actually receives the following code:

$$1 \quad 1 \quad 0 \quad 0 \quad 0 \quad 0 \quad 0 \quad 0$$

The receiver will find that this code has an *even* number of 1s. This tells the receiver that there must be an error in the code, since presumably the transmitter and receiver have agreed to use odd parity. There is no way, however, that the receiver can tell which bit is in error, since it does not know what the code is supposed to be.

It should be apparent that this parity method would not work if *two* bits were in error, because two errors would not change the "oddness" or "evenness" of the number of 1s in the code. In practice, the parity method is used only in situations where the probability of a single error is very low and the probability of double errors is essentially zero.

When the parity method is being used, the transmitter and receiver must have agreement, in advance, as to whether odd or even parity is being used. There is no advantage of one over the other, although even parity seems to be used more often. The transmitter must attach an appropriate parity bit to each unit of information that it transmits. For example, if the transmitter is sending ASCII-coded data, it will attach a parity bit to each 7-bit ASCII code group. When the receiver examines the data that it has received from the transmitter, it checks each code group to see that the total number of 1s (including the parity bit) is consistent with the agreed-upon

---

*The parity bit can be placed at either end of the code group but is usually placed to the left of the MSB.

type of parity. This is often called *checking the parity* of the data. In the event that it detects an error, the receiver may send a message to the transmitter asking it to retransmit the last set of data. The exact procedure that is followed when an error is detected will depend on the particular system design.

Teaching Tip: You may wish to give a few examples of the types of actions taken when a receiver detects a parity error, such as setting a flag in a special register, sending a message back to the transmitter, printing a special character on the screen or paper, or turning on a parity-error indicator LED.

—— EXAMPLE 2-10 ——

A transmitter is sending ASCII-coded data to a receiver with an even-parity bit. Show the actual codes when the transmitter is sending the message "HELLO."

Solution   First look up the ASCII codes for each character in the message. Then for each code, count the number of 1s. If it is an even number, attach a 0 as the MSB. If it is an odd number, attach a 1. Thus the resulting 8-bit codes will all have an even number of 1s (including parity).

```
          ┌── attached even-parity bits
          ↓
H-    0   1  0  0  1  0  0  0
E-    1   1  0  0  0  1  0  1
L-    1   1  0  0  1  1  0  0
L-    1   1  0  0  1  1  0  0
O-    1   1  0  0  1  1  1  1
```

—— REVIEW QUESTIONS ——

1. Attach an odd-parity bit to the ASCII code for the $ symbol, and express the result in hexadecimal.
2. Attach an even-parity bit to the BCD code for decimal 69.
3. Why can't the parity method detect a double error in transmitted data?

PROBLEMS

SECTIONS 2-1 and 2-2

2-1. Convert these binary numbers to decimal:
    (a) 10110   22
    (b) 10001101   141
    (c) 100100001001   2313
    (d) 1111010111   983
    (e) 10111111   191

2-2. Convert the following decimal values to binary:
    (a) 37   100101
    (b) 14   1110
    (c) 189   10111101
    (d) 205   11001101
    (e) 2313   100100001001
    (f) 511   111111111

2-3. What is the largest decimal value that can be represented by an 8-bit binary number? A 16-bit number?   25; 65, 535

2-4. Convert each octal number to its decimal equivalent:
(a) 743 483   (d) 257 175
(b) 36 30   (e) 1204 644
(c) 3777 2047

2-5. Convert each of the following decimal numbers to octal:
(a) 59 73   (d) 65,535 200000
(b) 372 564   (e) 255 377
(c) 919 1627

2-6. (a) 111100011 (b) 011110
(c) 011/11111111
(d) 010101111
(e) 001010000100
2-7. (a) 26 (b) 215 (c) 4411
(d) 1727 (e) 277
2-8. 165, 166, 167, 170, 171,
172, 173, 174, 175, 176, 177,
200

2-6. Convert each of the octal values from Problem 2-4 to binary.

2-7. Convert the binary numbers in Problem 2-1 to octal.

2-8. List the octal numbers in sequence from 165_8 to 200_8.

2-9. When a large decimal number is to be converted to binary, it is some-
times easier to convert it first to octal, and then from octal to binary.
Try this procedure for 2313_{10} and compare it with the procedure used
in Problem 2-2(e). 4411

SECTION 2-4

2-10. Convert these hex values to decimal:
(a) 92 146   (d) 2C0 704
(b) 1A6 422   (e) 7FF 2047
(c) 37FD 14,333

2-11. Convert these decimal values to hex:
(a) 75 4B   (d) 25,619 6413
(b) 314 13A   (e) 4095 FFF
(c) 2048 800

2-12. (a) 16 (b) 8D (c) 909
(d) 3D7 (e) BF
2-13. (a) 10010010
(b) 000110100110
(c) 0011011111111101
(d) 001011000000
(e) 011111111111
2-14. (b) 000 to FFF

2-12. Convert the binary numbers in Problem 2-1 to hexadecimal.

2-13. Convert the hex values in Problem 2-10 to binary.

2-14. In most microcomputers the addresses of memory locations are
specified in hexadecimal. These addresses are sequential numbers that
identify each memory circuit.
(a) A particular microcomputer can store an 8-bit number in each
memory location. If the memory addresses range from 0000_{16} to
$FFFF_{16}$, how many memory locations are there? 65,536
(b) Another microcomputer is specified to have 4096 memory loca-
tions. What range of hex addresses does this computer use?

2-15. 280, 281, 282, 283, 284,
285, 286, 287, 288, 289, 28A,
28B, 28C, 28D, 28E, 28F, 290,
291, 292, 293, 294, 295, 296,
297, 298, 299, 29A, 29B, 29C,
29D, 29E, 29F, 2A0

2-15. List the hex numbers in sequence from 280 to 2A0.

SECTION 2-5

2-16. Encode these decimal numbers in BCD:
(a) 47 01000111   (d) 42,689.627 0100001001101000100101100010 0111
(b) 962 100101100010 (e) 1204 0001001000000100
(c) 187 000110000111

2-17. ten; twelve

2-17. How many bits are required to represent the decimal numbers in the
range from 0 to 999 using straight binary code? Using BCD code?

2-18. The following numbers are in BCD. Convert them to decimal.
(a) 1001011101010010 9752   (c) 0111011101110101 7775
(b) 000110000100 184   (d) 010010010010 492

2-19. Represent the statement "X = 25/Y" in ASCII code (excluding quotes). Attach an even parity bit.

2-19. X: 1011000; =: 0111101; 2: 0110010; 5: 0111101; /: 0101111; Y: 1011001

2-20. Attach an *even* parity bit to each of the ASCII codes for Problem 2-19 and give the results in hex.

2-21. The following code groups are being transmitted. Attach an *even*-parity bit to each group.
   (a) 10110110   110110110
   (b) 00101000   000101000
   (c) 11110111   111110111

SECTION 2-9

2-22. Convert the following decimal numbers to BCD code and then attach an *odd*-parity bit:
   (a) 74   101110100
   (b) 38   000111000
   (c) 165   0000101100101
   (d) 9201   11001001000000001

2-23. In a certain digital system, the decimal numbers from 000 through 999 are represented in BCD code. An *odd*-parity bit is also included at the end of each code group. Examine each of the code groups below and assume that each one has just been transferred from one location to another. Some of the groups contain errors. Assume that *no more than* two errors have occurred for each group. Determine which of the code groups have a single error and which of them *definitely* have a double error. (*Hint:* Remember that this is a BCD code.)
   (a) 1001010110000   no single-bit error
   └─parity bit
   (b) 0100011101100   single error
   (c) 0111110000011   double error
   (d) 1000011000101   no single errors

2-24. Suppose that the receiver received the following data from the transmitter of Example 2-10:

$$\begin{array}{cccccccc} 0 & 1 & 0 & 0 & 1 & 0 & 0 & 0 \\ 1 & 1 & 0 & 0 & 0 & 1 & 0 & 1 \\ 1 & 1 & 0 & 0 & 1 & 1 & 0 & 0 \\ 1 & 1 & 0 & 0 & 1 & 0 & 0 & 0 \\ 1 & 1 & 0 & 0 & 1 & 1 & 0 & 0 \end{array}$$

What errors can the receiver determine in this received data?

2-24. First three code groups are OK. Fourth code has a single error. Last code looks OK to receiver even though it has a double error.

DRILL QUESTIONS

2-25. Perform each of the following conversions. For some of them, you may want to try several methods to see which one works best for you. For example, a binary-to-decimal conversion may be done directly, or it may be done as a binary-to-octal conversion followed by an octal-to-decimal conversion.
   (a) $1417_{10} = $ ___10110001001___ $_2$
   (b) $255_{10} = $ ___11111111___ $_2$
   (c) 11010001_2 ___209___ $_{10}$
   (j) $3E1C_{16} = $ ___15,900___ $_{10}$
   (k) $1600_{10} = $ ___640___ $_{16}$
   (l) $38,187_{10} = $ ___952B___ $_{16}$

(d) $1110101000100111_2 =$ <u>59,943</u>$_{10}$

(e) $2497_{10} =$ <u>4701</u>$_8$

(f) $511_{10} =$ <u>777</u>$_8$

(g) $235_8 =$ <u>157</u>$_{10}$

(h) $4316_8 =$ <u>2254</u>$_{10}$

(i) $7A9_{16} =$ <u>1961</u>$_{10}$

(m) $865_{10} =$ <u>100001100101</u> (BCD)

(n) 100101000111 (BCD) $=$ <u>947</u>$_{10}$

(o) $465_8 =$ <u>135</u>$_{16}$

(p) $B34_{16} =$ <u>5464</u>$_8$

(q) 01110100(BCD) $=$ <u>1001010</u>$_2$

(r) $111010_2 =$ <u>01011000</u> (BCD)

2-26. (a) 100101   (b) 00110111
(c) 01101010
(d) 0110011 0110111
(e) 45   (f) 25

2-26. Represent the decimal value 37 in each of the following ways:
(a) straight binary, (b) BCD, (c) excess-3, (d) ASCII (i.e., treat each digit as a character), (e) octal, (f) hex.

2-27. Fill in the blanks with the correct word or words.
(a) Conversion from decimal to <u>octal</u> requires repeated division by 8.
(b) Conversion from decimal to hex requires repeated division by <u>16</u>.
(c) In the BCD code, each <u>digit</u> is converted to its 4-bit binary equivalent.
(d) The <u>Gray</u> code has the characteristic that only one bit changes in going from one step to the next.
(e) A transmitter attaches a <u>parity</u> to a code group to allow the receiver to detect <u>single-bit errors</u>.
(f) The <u>ASCII</u> code is the most common alphanumeric code used in computer systems.
(g) <u>octal</u> and <u>hexadecimal</u> are often used as a convenient way to represent large binary numbers.

ANSWERS TO SECTION REVIEW QUESTIONS

SECTION 2-1

1. 2267   **2.** 32768

SECTION 2-2

1. 1010011   **2.** 1011011001

SECTION 2-3

1. 396   **2.** 222; 010010010   **3.** 235
4. 627, 630, 631   **5.** 1111001111   **6.** 699

SECTION 2-4

1 9422   **2.** C2D; 110000101101
3. 97B5   **4.** E9E, E9F, EA0, EA1   **5.** 757

SECTION 2-5

1. 101100102_2; 000101111000 (BCD)
2. 32   **3.** Advantage—ease of conversion; disadvantage—BCD requires more bits.

SECTION 2-7

1. No intermediate states occur during transitions.

SECTION 2-8

1. 43, 4F, 53, 54, 20, 3D, 20, 24, 37, 32
2. STOP

SECTION 2-9

1. A4   **2.** 001101001   **3.** Two errors in the data would not change the oddness or evenness of the number of 1s in the data.

3

LOGIC GATES AND BOOLEAN ALGEBRA

Outline

Objectives

Upon completion of this chapter, you will be able to:

- Analyze the INVERTER circuit.
- Describe the operation of and construct the truth tables for the AND, NAND, OR, and NOR gates.
- Draw timing diagrams for the various logic-circuit gates.
- Simplify complex logic circuits by applying the various Boolean algebra laws and rules.
- Simplify intricate Boolean equations by applying DeMorgan's theorems.
- Use either of the universal gates (NAND or NOR) to implement the circuit represented by a Boolean expression.
- Explain the advantages of constructing a logic circuit diagram using the alternate gate symbols, versus the standard logic-gate symbols.
- Describe the concept of active-LOW and active-HIGH logic signals.
- Draw and interpret logic circuits that use the new IEEE/ANSI standard symbols.

Introduction

As pointed out in Chapter 1, digital (logic) circuits operate in the binary mode where each input and output voltage is either a 0 or 1; the 0 and 1 designations represent predefined voltage ranges. This characteristic of logic circuits allows us to use **Boolean algebra** as a tool for the analysis and design of digital systems. In this chapter we will study *logic* gates, which are the most fundamental logic circuits, and we will see how their operation can be described using Boolean algebra. We will also see how logic gates can be combined to produce logic circuits, and how these circuits can be described and analyzed using Boolean algebra.

3-1 BOOLEAN CONSTANTS AND VARIABLES

Boolean algebra differs in a major way from ordinary algebra in that Boolean constants and variables are allowed to have only two possible values, 0 or 1. A Boolean variable is a quantity that may, at different times, be equal to either 0 or 1. The Boolean variables are often used to represent the voltage level present on a wire or at the input/output terminals of a circuit. For example, in a certain digital system the Boolean value of 0 might be assigned to any voltage in the range from 0 to 0.8 V while the Boolean value of 1 might be assigned to any voltage in the range 2 to 5 V*.

Thus, Boolean 0 and 1 do not represent actual numbers but instead represent the state of a voltage variable, or what is called its **logic level.** A voltage in a digital circuit is said to be at the logic level 0 or the logic level 1, depending on its actual numerical value. In the digital logic field several other terms are used synonymously with 0 and 1. Some of the more common ones are shown in Table 3-1. We will use the 0/1 and LOW/HIGH designations most of the time.

Boolean algebra is used to express the effects that various digital circuits have on logic inputs, and to manipulate logic variables for the purpose of determining the best method for performing a given circuit function. In all our work to follow we shall use letter symbols to represent logic variables. For example, A might represent a certain digital circuit input or output, and at any time we must have either $A = 0$ or $A = 1$: if not one, then the other.

Because only two values are possible, Boolean algebra is relatively easy to work with as compared with ordinary algebra. In Boolean algebra there are no fractions, decimals, negative numbers, square roots, cube roots, logarithms, imaginary numbers, and so on. In fact, in Boolean algebra there are only *three* basic operations:

1. *Logical addition,* also called *OR addition* or simply the *OR operation.* The common symbol for this operation is the plus sign $(+)$.

2. *Logical multiplication,* also called *AND multiplication* or simply the *AND operation.* The common symbol for this operation is the multiplication sign $(\cdot)$.

3. *Logical complementation or inversion,* also called the *NOT operation.* The common symbol for this operation is the overbar $(^-)$.

TABLE 3-1

LOGIC 0	LOGIC 1
False	True
Off	On
Low	High
No	Yes
Open switch	Closed switch

*Voltages between 0.8 and 2 V are undefined (neither 0 nor 1) and under normal circumstances should not occur.

Inputs		Output

A	B	x
0	0	?
0	1	?
1	0	?
1	1	?

(a)

A	B	C	x
0	0	0	?
0	0	1	?
0	1	0	?
0	1	1	?
1	0	0	?
1	0	1	?
1	1	0	?
1	1	1	?

(b)

A	B	C	D	x
0	0	0	0	?
0	0	0	1	?
0	0	1	0	?
0	0	1	1	?
0	1	0	0	?
0	1	0	1	?
0	1	1	0	?
0	1	1	1	?
1	0	0	0	?
1	0	0	1	?
1	0	1	0	?
1	0	1	1	?
1	1	0	0	?
1	1	0	1	?
1	1	1	0	?
1	1	1	1	?

(c)

FIGURE 3-1   Truth tables for (a) two-input, (b) three-input, and (c) four-input cases.

3-2   TRUTH TABLES

Many logic circuits have more than one input and only one output. A *truth table* shows how the logic circuit's output responds to the various combinations of logic levels at the inputs. The format for two-, three-, and four-input truth tables is shown in Figure 3-1.

In each truth table the possible combinations of 0 and 1 logic levels for the inputs (A, B, C, D) are listed on the left, and the resultant logic level for output x is listed on the right. The entries for x are shown as "?" for now, because these values will be different for each type of logic circuit.

Note that there are 4 table entries for the two-input truth table, 8 entries for a three-input truth table, and 16 entries for the four-input truth table. The number of input combinations will equal 2^N for an N-input truth table. Also note that the list of all possible input combinations follows the binary counting sequence, and so it is an easy matter to write down all the combinations without missing any.

3-3   OR OPERATION

Let A and B represent two independent logic variables. When A and B are combined using the OR operation, the result, x, can be expressed as

$$x = A + B$$

In this expression the + sign does not stand for ordinary addition; it stands for the OR operation, whose rules are given in the truth table shown in Figure 3-2(a).

It should be apparent from the truth table that except for the case where $A = B = 1$, the OR operation is the same as ordinary addition. However, for $A = B = 1$ the OR sum is 1 (not 2 as in ordinary addition). This is easy to remember if we recall that only 0 and 1 are possible values in Boolean algebra, so that the

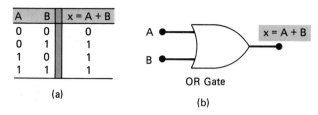

A	B	x = A + B
0	0	0
0	1	1
1	0	1
1	1	1

(a)

$x = A + B$

OR Gate

(b)

FIGURE 3-2   (a) Truth table defining the OR operation; (b) circuit symbol for a two-input OR gate.

largest value we can get is 1. This same result is true if we have $x = A + B + C$, for the case where $A = B = C = 1$. That is,

$$x = 1 + 1 + 1 = 1$$

We can therefore say that the OR operation result will be 1 if any one *or* more variables is a 1. This is also clear in the table in Figure 3-2(a).

The expression $x = A + B$ is read as "x equals A OR B." The key thing to remember is that the $+$ sign stands for the OR operation as defined by the truth table in Figure 3.2(a), and not for ordinary addition.

OR Gate   In digital circuitry an *OR gate* is a circuit that has two or more inputs and whose output is equal to the OR sum of the inputs. Figure 3-2(b) shows the symbol for a two-input OR gate. The inputs A and B are logic voltage levels and the output x is a logic voltage level whose value is the result of the OR operation on A and B; that is, $x = A + B$. In other words, the OR gate operates in such a way that its output is HIGH (logic 1) if either input A *or* B *or both* are at a logic 1 level. The OR gate output will be LOW (logic 0) only if all its inputs are at logic 0.

Teaching Tip: Another way to characterize an OR gate is as a "one or more" gate where the output will go HIGH when one or more inputs are HIGH.

This same idea can be extended to more than two inputs. Figure 3-3 shows a three-input OR gate and its truth table. Examination of this truth table shows again that the output will be 1 for every case where one or more inputs are 1. This general principle is the same for OR gates with any number of inputs.

Using the language of Boolean algebra, the output x can be expressed as $x = A + B + C$, where again it must be emphasized that the $+$ represents the OR operation. The output of any OR gate, then, can be expressed as the OR of its various inputs. We will put this to use when we subsequently analyze logic circuits.

Summary of the OR Operation   The important points to remember concerning the OR operation and OR gates are:

FIGURE 3-3   Symbol and truth table for a three-input OR gate.

A	B	C	x = A + B + C
0	0	0	0
0	0	1	1
0	1	0	1
0	1	1	1
1	0	0	1
1	0	1	1
1	1	0	1
1	1	1	1

1. The OR operation produces a result of 1 when *any* of the input variables is 1.
2. The OR operation produces a result of 0 *only* when all the input variables are 0.
3. With the OR operation, $1 + 1 = 1$, $1 + 1 + 1 = 1$, and so on.

—— EXAMPLE 3-1 ——————————————————————

In many industrial control systems it is required to activate an output function whenever any one of several inputs is activated. For example, in a chemical process it may be desired that an alarm be activated whenever the process temperature exceeds a maximum value *or* whenever the pressure goes above a certain limit. Figure 3-4 is a block diagram of this situation. The temperature transducer circuit produces

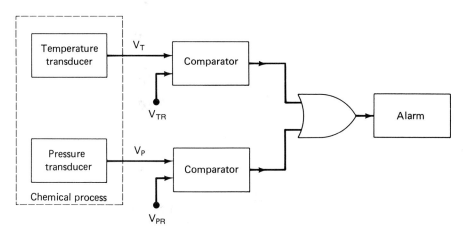

FIGURE 3-4   Example of the use of OR gate in alarm system.

an output voltage proportional to the process temperature. This voltage, V_T, is compared with a temperature reference voltage, V_{TR}, in a voltage comparator circuit. The comparator output is normally a low voltage (logic 0), but it switches to a high voltage (logic 1) when V_T exceeds V_{TR}, indicating that process temperature is excessive. A similar arrangement is used for the pressure measurement, so that its associated comparator output goes from low to high when the pressure is excessive.

Since we want the alarm to be activated when either temperature *or* pressure is too high, it should be apparent that the two comparator outputs can be fed to a two-input OR gate. The OR gate output thus goes HIGH (1) for either alarm condition and will activate the alarm. This same idea can obviously be extended to situations with more than two process variables.

—— EXAMPLE 3-2 ——————————————————————

Determine the OR gate output in Figure 3-5. The OR gate inputs A and B are varying according to the timing diagrams shown. For example, A starts out LOW at time t_0, goes HIGH at t_1, back LOW at t_3, and so on.

Solution   The OR gate output is determined by realizing that it will be HIGH whenever *any* of the inputs are at the high level. When A goes HIGH at t_1, OUTPUT will go HIGH. OUTPUT will stay HIGH until t_4, when both inputs are LOW. Note that the changes in the input levels that occur at t_2 and t_3 have no effect on OUTPUT, since one of the inputs remains at the HIGH level while the other is changing. As

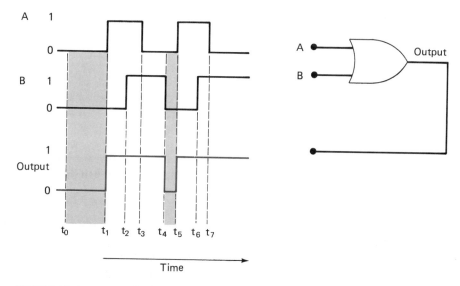

FIGURE 3-5   Example 3-2.

long as one input to an OR gate is HIGH, the output will stay HIGH regardless of what is happening at the other inputs. The same reasoning is used to determine the rest of the timing diagram for OUTPUT.

___ EXAMPLE 3-3A ___

For the situation depicted in Figure 3-6, determine the waveform at the OR gate output.

Solution   The three OR gate inputs A, B, and C are varying, as shown by their waveform diagrams. The OR gate output is determined by realizing that it will be high whenever *any* of the three inputs is at a high level. Using this reasoning, the OR output waveform is as shown in the figure. Particular attention should be paid to what occurs at time t_1. The diagram shows that at that instant of time, input A is going from high to low while input B is going from low to high. Since these inputs are

FIGURE 3-6   Example 3-3.

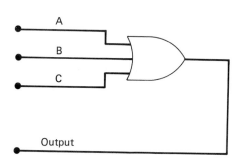

making their transitions at approximately the same time and since these transitions take a certain amount of time, there is a short interval when these OR gate inputs are both in the undefined range between 0 and 1. When this occurs, the OR gate output also becomes a value in this range, as evidenced by the glitch or spike on the output waveform at t_1. The occurrence of this glitch and its size (amplitude and width) depend on the speed with which the input transitions occur.

Discussion Question: What will appear at the OR gate output if the *C* input is kept constantly HIGH?

_____ EXAMPLE 3-3B _____

What would happen to the glitch in the output if input *C* sat in the HIGH state while *A* and *B* were changing at time t_1?

Solution   With the *C* input HIGH at t_1, the OR gate output will remain HIGH regardless of what is occurring at the other inputs, because any HIGH input will keep an OR gate output HIGH. Therefore, the glitch will not appear in the output.

_____ REVIEW QUESTIONS _____

1. What is the only set of input conditions that will produce a LOW output for any OR gate?
2. Write the Boolean expression for a six-input OR gate.
3. If the *A* input in Figure 3-6 is permanently kept at the 1 level, what will the resultant output waveform be?

3-4   AND OPERATION

If two logic variables *A* and *B* are combined using AND multiplication, the result, *x*, can be expressed as

$$x = A \cdot B$$

In this expression the · sign stands for the Boolean operation of AND multiplication, whose rules are given in the truth table shown in Figure 3-7(a).

It should be apparent from the table that AND multiplication is *exactly* the same as ordinary multiplication. Whenever *A* or *B* is 0, their product is zero; when both *A and B* are 1, their product is 1. We can therefore say that in the AND operation the result will be 1 *only if* all the inputs are 1; for all other cases the result is 0.

The expression $x = A \cdot B$ is read "*x* equals *A* AND *B*." The multiplication

FIGURE 3-7   (a) Truth table for the AND operation; (b) AND gate symbol.

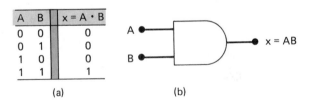

A	B	$x = A \cdot B$
0	0	0
0	1	0
1	0	0
1	1	1

(a)   (b)

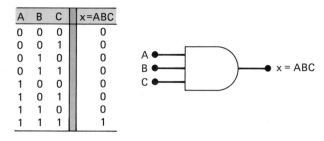

A	B	C	x=ABC
0	0	0	0
0	0	1	0
0	1	0	0
0	1	1	0
1	0	0	0
1	0	1	0
1	1	0	0
1	1	1	1

FIGURE 3-8   Truth table and symbol for a three-input AND gate.

sign is generally omitted as in ordinary algebra, so that the expression becomes $x = AB$. The key thing to remember is that the AND operation is the same as ordinary multiplication, where the variables can be either 0 or 1.

AND Gate   A two-input AND gate is shown symbolically in Figure 3-7(b). The AND gate output is equal to the AND product of the logic inputs; that is, $x = AB$. In other words, the AND gate is a circuit that operates in such a way that its output is high only when all its inputs are high. For all other cases the AND gate output is low.

Teaching Tip: An AND gate can also be characterized as an "all" gate since its output will go HIGH only when all its inputs are HIGH.

This same operation is characteristic of AND gates with more than two inputs. For example, a three-input AND gate and its accompanying truth table are shown in Figure 3-8. Once again, note that the gate output is 1 only for the case where $A = B = C = 1$. The expression for the output is $x = ABC$. For a four-input AND gate, the output is $x = ABCD$, and so on.

Note the difference between the symbols for the AND gate and the OR gate. Whenever you see the AND symbol on a logic-circuit diagram, it tells you that the output will go HIGH *only* when *all* inputs are HIGH. Whenever you see the OR symbol, it means that the output will go HIGH when *any* input is HIGH.

Summary of the AND Operation

1. The AND operation is performed exactly like ordinary multiplication of 1s and 0s.
2. An output equal to 1 occurs only for the single case where all inputs are 1.
3. The output is 0 for any case where one or more inputs are 0.

—— EXAMPLE 3-4 ——

Determine the output x from the AND gate in Figure 3-9 for the given input waveforms.

Solution   The output of an AND gate is determined by realizing that it will be high only when all inputs are high at the same time. For the input waveforms given, this condition is met only during intervals t_2—t_3 and t_6—t_7. At all other times, one or more of the inputs is a 0, thereby producing a low output. Note that input level changes that occur while the other input is low have no effect on the output.

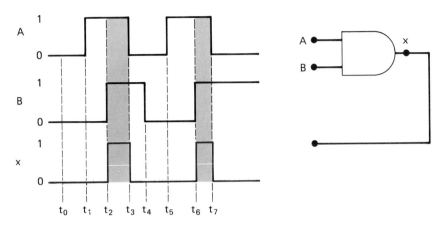

FIGURE 3-9   Example 3-4.

___ EXAMPLE 3-5A _____

Determine the output waveform for the AND gate shown in Figure 3-10.

FIGURE 3-10   Example 3-5.

Solution   The output x will be at 1 only when A and B are both high at the same time. Using this fact, the x waveform can be determined as shown in the figure.

Notice that the x waveform is 0 whenever B is 0, regardless of the signal at A. Also notice that whenever B is 1 the x waveform is the same as A. Thus we can think of the B input as a *control* input whose logic level determines whether the A waveform gets through to the x output or not. In this situation, the AND gate is used as an *inhibit circuit*. We can say that $B = 0$ is the inhibit condition producing a 0 output. Conversely, $B = 1$ is the *enable* condition, which enables A to reach the output. This inhibit operation is an important application of AND gates which will be encountered later.

___ EXAMPLE 3-5B _____

Discussion Question: Ask what will appear at x if the B input is kept at the 1 level.

What will happen to the x output waveform in Figure 3-10 if the B input is kept at the 0 level?

Solution   With B kept LOW, the x output will also stay LOW. This can be reasoned in two different ways. First, with $B = 0$ we have $x = A \cdot B = A \cdot 0 = 0$, since anything multiplied (ANDed) by 0 will be 0. Another way to look at it is that an AND gate requires that all inputs be HIGH in order for the output to be HIGH and this cannot happen if B is kept LOW.

1. What is the only input combination that will produce a HIGH at the output of a five-input AND gate?
2. What logic level should be applied to the second input of a two-input AND gate if the logic signal at the first input is to be inhibited (prevented) from reaching the output?
3. *True or false:* An AND gate output will always differ from an OR gate output for the same input conditions.

3-5   NOT OPERATION

The NOT operation is unlike the OR and AND operations in that it can be performed on a single input variable. For example, if the variable A is subjected to the NOT operation, the result x can be expressed as

$$x = \overline{A}$$

where the overbar represents the NOT operation. This expression is read "x equals NOT A" or "x equals the *inverse* of A" or "x equals the *complement* of A." Each of these is in common usage and all indicate that the logic value of $x = \overline{A}$ *is opposite to* the logic value of A. The truth table in Figure 3-11(a) clarifies this for the two cases $A = 0$ and $A = 1$. That is,

$$\overline{1} = 0 \qquad \text{because NOT 1 is 0}$$

and

$$\overline{0} = 1 \qquad \text{because NOT 0 is 1}$$

The NOT operation is also referred to as **inversion** or **complementation,** and these terms will be used interchangeably throughout the text. Although we will always use the overbar indicator to represent inversion, it is important to mention that another indicator for inversion is the prime symbol ('). That is,

$$A' = \overline{A}$$

Both should be recognized as indicating the inversion operation.

NOT Circuit (INVERTER)   Figure 3-11(b) shows the symbol for a NOT circuit, which is more commonly called an INVERTER. This circuit *always* has only a

FIGURE 3-11   (a) Truth table; (b) symbol for the NOT circuit; (c) corresponding waveforms.

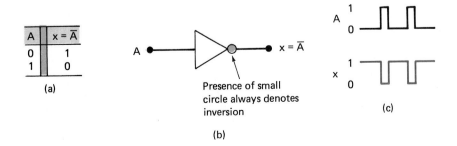

single input, and its output logic level is always opposite to the logic level of this input. Figure 3-11(c) shows how the INVERTER affects an input signal. It inverts (complements) the input signal at all points on the waveform.

Summary of Boolean Operations   The rules for the OR, AND, and NOT operations may be summarized as follows:

OR	AND	NOT
$0 + 0 = 0$	$0 \cdot 0 = 0$	$\bar{0} = 1$
$0 + 1 = 1$	$0 \cdot 1 = 0$	$\bar{1} = 0$
$1 + 0 = 1$	$1 \cdot 0 = 0$	
$1 + 1 = 1$	$1 \cdot 1 = 1$	

___ REVIEW QUESTIONS ___

1. The output of the INVERTER of Figure 3-11 is connected to the input of a second INVERTER. Determine the output level of the second INVERTER for each level of input A.
2. The output of the AND gate in Figure 3-7 is connected to the input of an INVERTER. Write the truth table showing the INVERTER output, y, for each combination of inputs A and B.

3-6   DESCRIBING LOGIC CIRCUITS ALGEBRAICALLY

Any logic circuit, no matter how complex, may be completely described using the Boolean operations previously defined, because the OR gate, AND gate, and NOT circuit are the basic building blocks of digital systems. For example, consider the circuit in Figure 3-12. This circuit has three inputs, A, B, and C, and a single output, x. Utilizing the Boolean expression for each gate, we can easily determine the expression for the output.

The expression for the AND gate output is written $A \cdot B$. This AND output is connected as an input to the OR gate along with C, another input. The OR gate operates on its inputs so that its output is the OR sum of the inputs. Thus, we can express the OR output as $x = A \cdot B + C$. (This final expression could also be written as $x = C + A \cdot B$, since it does not matter which term of the OR sum is written first.)

FIGURE 3-12   Logic circuit with its Boolean expression.

$$x = A \cdot B + C$$

FIGURE 3-13 Logic circuit whose expression requires parentheses.

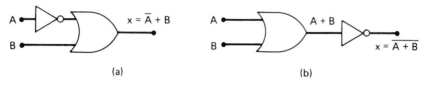

(a)   (b)

FIGURE 3-14 Circuits using INVERTERs.

Occasionally, there may be confusion as to which operation in an expression is performed first. The expression $A \cdot B + C$ can be interpreted in two different ways: (1) $A \cdot B$ is ORed with C, or (2) A is ANDed with the term $B + C$. To avoid this confusion, it will be understood that if an expression contains both AND and OR operations, the AND operations are performed first, unless there are *parentheses* in the expression, in which case the operation inside the parentheses is to be performed first. This is the same rule that is used in ordinary algebra to determine the order of operations.

To illustrate further, consider the circuit in Figure 3-13. The expression for the OR-gate output is simply $A + B$. This output serves as an input to the AND gate along with another input, C. Thus, we express the output of the AND gate as $x = (A + B) \cdot C$. Note the use of parentheses here to indicate that A and B are ORed *first*, before their OR sum is ANDed with C. Without the parentheses it would be interpreted *incorrectly*, since $A + B \cdot C$ means A is ORed with the product $B \cdot C$.

Circuits Containing INVERTERs   Whenever an INVERTER is present in a logic-circuit diagram, its output expression is simply equal to the input expression with a bar over it. Figure 3-14 shows two examples using INVERTERs. In Figure 3-14(a) the input A is fed through an INVERTER, whose output is therefore $\overline{A}$. The INVERTER output is fed to an OR gate together with B, so that the *OR* output is equal to $\overline{A} + B$. Note that the bar is over the A alone, indicating that A is first inverted and then ORed with B.

In Figure 3-14(b) the output of the OR gate is equal to $A + B$ and is fed through an INVERTER. The INVERTER output is therefore equal to $(\overline{A + B})$, since it inverts the *complete* input expression. Note that the bar covers the entire expression $(A + B)$. This is important because, as will be shown later, the expressions $(\overline{A + B})$ and $(\overline{A} + \overline{B})$ are *not* equivalent. The expression $(\overline{A + B})$ means that A is ORed with B and then their OR sum is inverted, whereas the expression $(\overline{A} + \overline{B})$ indicates that A is inverted and B is inverted and the results are then ORed together.

Teaching Tip: An analogous mathematical principle is that $(A + B)^2$ is not the same as $A^2 + B^2$.

Figure 3-15 shows two more examples, which should be studied carefully. Note especially the use of *two* separate sets of parentheses in Figure 3-15(b). Also

(a)

(b)

FIGURE 3-15   More examples.

notice in Figure 3-15(a) that the input variable *A* is connected as input to two different gates.

—— REVIEW QUESTION ——

1. In Figure 3-15(a) change each AND gate to an OR gate, and change the OR gate to an AND gate. Then write the expression for output *x*.

3-7   EVALUATING LOGIC-CIRCUIT OUTPUTS

Once the Boolean expression for a circuit output is obtained, the logic level of the output can be determined for any values of circuit inputs. For example, suppose that we want to know the logic level of the output *x* for the circuit in Figure 3-15(a) for the case where $A = 0$, $B = 1$, $C = 1$, and $D = 1$. As in ordinary algebra, the value of *x* can be found by "plugging" the values of the variables into the expression and performing the indicated operations as follows:

$$x = \overline{A}BC\,(\overline{A + D})$$
$$= \overline{0} \cdot 1 \cdot 1 \cdot (\overline{0 + 1})$$
$$= 1 \cdot 1 \cdot 1 \cdot (\overline{0 + 1})$$
$$= 1 \cdot 1 \cdot 1 \cdot (\overline{1})$$
$$= 1 \cdot 1 \cdot 1 \cdot 0$$
$$= 0$$

As another illustration, let us evaluate the output of the circuit in Figure 3-15(b) for $A = 0$, $B = 0$, $C = 1$, $D = 1$, and $E = 1$.

$$
\begin{aligned}
x &= [D + \overline{(A + B)C}] \cdot E \\
&= [1 + \overline{(0 + 0) \cdot 1}] \cdot 1 \\
&= [1 + \overline{0 \cdot 1}] \cdot 1 \\
&= [1 + \overline{0}] \cdot 1 \\
&= [1 + 1] \cdot 1 \\
&= 1 \cdot 1 \\
&= 1
\end{aligned}
$$

In general, the following rules must always be followed when evaluating a Boolean expression:

1. First, perform all inversions of single terms; that is, $\overline{0} = 1$ or $\overline{1} = 0$.
2. Then perform all operations within parentheses.
3. Perform an AND operation before an OR operation unless parentheses indicate otherwise.
4. If an expression has a bar over it, perform the operations of the expression first and then invert the result.

Common Student Mistake: A common error at this stage is to invert each individual value under a large inversion sign before combining values: $\overline{1.0 + 1} = \overline{1} + \overline{1} = \overline{1} = 0$. It is important to stress that the operations under the large inversion sign must be performed first.

Determining Output Level from a Diagram The output logic level for given input levels can also be determined directly from the circuit diagram *without* using the Boolean expression. This technique is often used by technicians during the troubleshooting or testing of a logic system since it also tells them what each gate output is supposed to be as well as the final output. To illustrate, the circuit of Figure 3-15(a) is redrawn in Figure 3-16 with the input levels $A = 0$, $B = 1$, $C = 1$, and $D = 1$. The procedure is to start from the inputs and to proceed through each INVERTER and gate, writing down each of their outputs in the process until the final output is reached.

In Figure 3-16, AND gate 1 has *all* three inputs at the 1 level because the INVERTER changes the $A = 0$ to a 1. This condition produces a 1 at the AND output since $1 \cdot 1 \cdot 1 = 1$. The OR gate has inputs of 1 and 0, which produces a 1 output since $1 + 0 = 1$. This 1 is inverted to 0 and applied to AND gate 2 along with the 1 from the first AND output. The 0 and 1 inputs to AND gate 2 produce an output of 0, because $0 \cdot 1 = 0$.

FIGURE 3-16 Determining output level from circuit diagram.

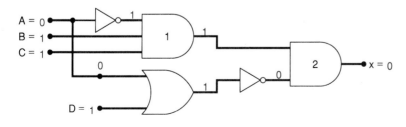

EXAMPLE 3-6

Determine the output in Figure 3-16 for the condition where all inputs are LOW.

Solution   With $A = B = C = D = 0$, the output of AND gate 1 will be LOW. This LOW places a LOW at the input of AND gate 2, which automatically produces $x = 0$, regardless of the levels at other points in the circuit. This example shows that it is not always necessary to determine the logic level at every point in order to determine the output level.

___ REVIEW QUESTIONS ___

1. Use the expression for x to determine the output of the circuit in Figure 3-15(a) for the conditions $A = 0$, $B = 1$, $C = 1$, and $D = 0$.
2. Use the expression for x to determine the output of the circuit in Figure 3-15(b) for the conditions $A = B = E = 1$, $C = D = 0$.
3. Determine the answers to questions 1 and 2 by finding the logic levels present at each gate input and output as was done in Figure 3-16.

3-8 IMPLEMENTING CIRCUITS FROM BOOLEAN EXPRESSIONS

If the operation of a circuit is defined by a Boolean expression, a logic-circuit diagram can be implemented directly from that expression. For example, if we needed a circuit that was defined by $x = A \cdot B \cdot C$, we would immediately know that all that was needed was a three-input AND gate. If we needed a circuit that was defined by $x = A + \overline{B}$, we would use a two-input OR gate with an INVERTER on one of the inputs. The same reasoning used for these simple cases can be extended to more complex circuits.

Suppose that we wanted to construct a circuit whose output is $y = AC + B\overline{C} + \overline{A}BC$. This Boolean expression contains three terms (AC, $B\overline{C}$, $\overline{A}BC$), which are ORed together. This tells us that a three-input OR gate is required with inputs that are equal to AC, $B\overline{C}$, and $\overline{A}BC$, respectively. This is illustrated in Figure 3-17(a), where a three-input OR gate is drawn with inputs labeled as AC, $B\overline{C}$, and $\overline{A}BC$.

Each OR-gate input is an AND product term, which means that an AND gate with appropriate inputs can be used to generate each of these terms. This is shown in Figure 3-17(b), which is the final circuit diagram. Note the use of INVERTERs to produce the $\overline{A}$ and $\overline{C}$ terms required in the expression.

This same general approach can always be followed, although we shall find that there are some clever, more efficient techniques that can be employed. For now, however, this straightforward method will be used to minimize the number of new things that are to be learned.

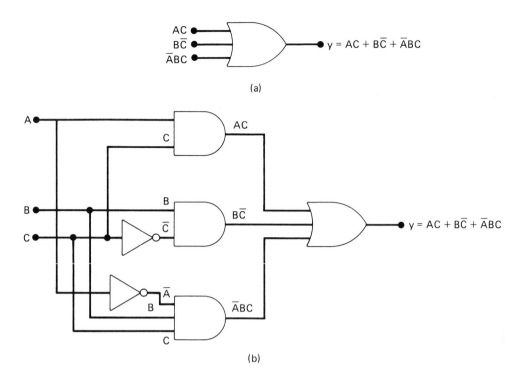

$$y = AC + B\bar{C} + \bar{A}BC$$

(a)

$$y = AC + B\bar{C} + \bar{A}BC$$

(b)

FIGURE 3-17   Constructing a logic circuit from a Boolean expression.

— EXAMPLE 3-7 —————————————————————————

Draw the circuit diagram that implements the expression $x = AB + \bar{B}C$.

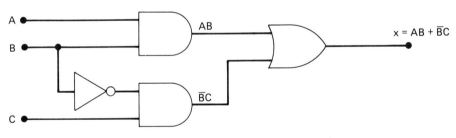

FIGURE 3-18   Example 3-7.

Solution   This expression indicates that the terms AB and $\bar{B}C$ are inputs to an OR gate, and each of these two terms is generated from a separate AND gate. The result is shown in Figure 3-18.

— REVIEW QUESTION —————————————————————

1. Draw the circuit diagram that implements the expression $x = \bar{A}BC\,(\overline{A + D})$ using gates with no more than three inputs.

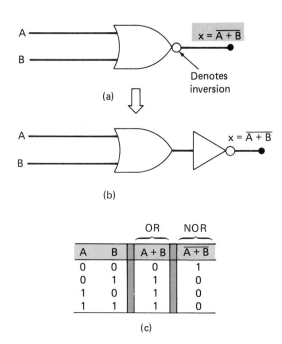

FIGURE 3-19   (a) NOR symbol; (b) equivalent circuit; (c) truth table.

3-9   NOR GATES AND NAND GATES

Two other types of logic gates, NOR gates and NAND gates, are used extensively in digital circuitry. These gates actually combine the basic operations AND, OR, and NOT, which make it relatively easy to describe them using the Boolean algebra operations learned previously.

NOR Gate   The symbol for a two-input NOR gate is shown in Figure 3-19(a). It is the same as the OR gate symbol except that it has a small circle on the output. This small circle represents the inversion operation. Thus, the NOR gate operates like an OR gate followed by an INVERTER, so that the circuits in Figure 3-19(a) and (b) are equivalent, and the output expression for the NOR gate is $x = \overline{A + B}$.

The truth table in Figure 3-19(c) shows that the NOR gate output is the exact inverse of the OR-gate output for all possible input conditions. Whereas an OR-gate output goes HIGH when any input is HIGH, the NOR gate output goes LOW when any input is HIGH. This same operation can be extended to NOR gates with more than two inputs.

———— EXAMPLE 3-8 ————

Discussion Question: What will be the output waveform if A is held at a constant HIGH level? At a constant LOW level?

Determine the waveform at the output of a NOR gate for the input waveforms shown in Figure 3-20.

Solution   There are several ways to determine the NOR output waveform. One way is to first find the OR output waveform and then invert it (change all 1s to 0s, and vice versa). Another way utilizes the fact that a NOR-gate output will be HIGH

FIGURE 3-20   Example 3-8.

only when all inputs are LOW. Thus you can examine the input waveforms, find those time intervals where they are all LOW, and make the NOR output HIGH for those intervals. The NOR output will be LOW for all other time intervals. The resultant output waveform is shown in the figure.

—— EXAMPLE 3-9 ——————————————————————————

Determine the Boolean expression for a three-input NOR gate followed by an INVERTER.

Solution   Refer to Figure 3.21, where the circuit diagram is shown. The expression at the NOR output is $(\overline{A + B + C})$, which is then fed through an INVERTER to produce

$$x = \overline{(\overline{A + B + C})}$$

A
B $\overline{A+B+C}$ $x = \overline{\overline{A+B+C}} = A+B+C$
C

FIGURE 3-21   Example 3-9.

The presence of the double inversion signs indicates that the quantity $(A + B + C)$ has been inverted and then inverted again. It should be clear that this simply results in the expression $(A + B + C)$ being unchanged. That is,

$$x = \overline{(\overline{A + B + C})} = A + B + C$$

Whenever two inversion bars are over the same variable or quantity, they cancel each other out, as in the example above. However, in cases such as $\overline{A} + \overline{B}$ the inversion bars do not cancel. This is because the smaller inversion bars invert the single variables A and B, respectively, while the wide bar inverts the quantity $(\overline{A} + \overline{B})$. Thus, $\overline{\overline{A} + \overline{B}} \neq A + B$. Similarly, $\overline{\overline{A}\,\overline{B}} \neq AB$.

Discussion Question: How can we prove that $\overline{\overline{A} + \overline{B}} \neq A + B$?

NAND Gate   The symbol for a two-input NAND gate is shown in Figure 3-22(a). It is the same as the AND-gate symbol except for the small circle on its output. Once again this small circle denotes the inversion operation. Thus the NAND operates like an AND gate followed by an INVERTER, so that the circuits of Figure 3-22(a) and (b) are equivalent, and the output expression for the NAND gate is $x = \overline{AB}$.

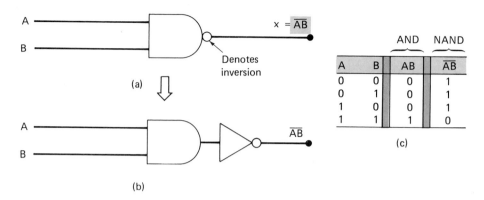

(a)

(b)

A	B	AB	$\overline{AB}$
0	0	0	1
0	1	0	1
1	0	0	1
1	1	1	0

(c)

FIGURE 3-22   (a) NAND symbol; (b) equivalent circuit; (c) truth table.

The truth table in Figure 3-22(c) shows that the NAND-gate output is the exact inverse of the AND gate for all possible input conditions. The AND output goes HIGH only when all inputs are HIGH, while the NAND output goes LOW only when all inputs are HIGH. This same characteristic is true of NAND gates having more than two inputs.

—— EXAMPLE 3-10 ——

Determine the output waveform of a NAND gate having the inputs shown in Figure 3-23.

FIGURE 3-23   Example 3-10.

Discussion Question: What will be the output waveform if *A* is held at a constant LOW level? At a constant HIGH level?

Solution   The output can be determined in several ways. One way is to first draw the output for an AND gate, and then invert it. Another way utilizes the fact that a NAND output will be LOW only when all inputs are HIGH. Thus, you can find those time intervals during which the inputs are all HIGH, and make the NAND output LOW for those intervals. The output will be HIGH at all other times.

—— EXAMPLE 3-11 ——

Implement the logic circuit that has the expression $x = \overline{AB \cdot (\overline{C + D})}$ using only NOR and NAND gates.

Solution   The $(\overline{C + D})$ term is the expression for the output of a NOR gate. This term is ANDed with *A* and *B* and the result is inverted; this, of course, is the NAND operation. Thus, the circuit is implemented as shown in Figure 3-24. Note that the

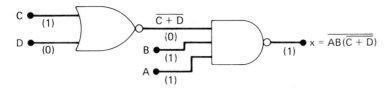

FIGURE 3-24   Example 3-11.

NAND gate first ANDs the A, B and $(\overline{C + D})$ terms, and then inverts the *complete* result.

____ EXAMPLE 3-12 ____

Determine the output level in Figure 3-24 for $A = B = C = 1$ and $D = 0$.

Solution   In the first method we use the expression for x.

$$x = \overline{AB(\overline{C + D})}$$
$$= \overline{1 \cdot 1 \cdot (\overline{1 + 0})}$$
$$= \overline{1 \cdot 1 \cdot (\overline{1})}$$
$$= \overline{1 \cdot 1 \cdot 0}$$
$$= \overline{0} = 1$$

In the second method we write down the input logic levels on the circuit diagram (shown in parentheses) and follow these levels through each gate to the final output. The NOR gate has inputs of 1 and 0 to produce an output of 0 (an OR would have produced an output of 1). The NAND gate thus has input levels of 0, 1, and 1 to produce an output of 1 (an AND would have produced an output of 0).

Discussion Question: Is an AND gate with inverters on its inputs the same as a NAND gate?

____ REVIEW QUESTIONS ____

1. What is the only set of input conditions that will produce a HIGH output from a three-input NOR gate?

2. What type of gate is equivalent to a NAND gate followed by an IN-VERTER?
3. Change the NOR gate of Figure 3-24 to a NAND gate, and change the NAND to a NOR. What is the new expression for x?

3-10   BOOLEAN THEOREMS

We have seen how Boolean algebra can be used to help analyze a logic circuit and express its operation mathematically. We will continue our study of Boolean algebra by investigating the various Boolean theorems (rules) that can help us to simplify logic expressions and logic circuits. The first group of theorems is given in Figure 3-25. In each theorem, x is a logic variable that can be either a 0 or a 1. Each theorem is accompanied by a logic-circuit diagram that demonstrates its validity.

Theorem (1) states that if any variable is ANDed with 0, the result has to be 0.

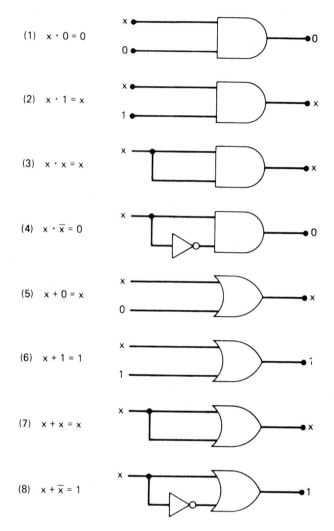

(1) $x \cdot 0 = 0$	
(2) $x \cdot 1 = x$	
(3) $x \cdot x = x$	
(4) $x \cdot \bar{x} = 0$	
(5) $x + 0 = x$	
(6) $x + 1 = 1$	
(7) $x + x = x$	
(8) $x + \bar{x} = 1$	

FIGURE 3-25   Single-variable theorems.

This is easy to remember because the AND operation is just like ordinary multiplication, where we know that anything multiplied by 0 is 0. We also know that the output of an AND gate will be 0 whenever any input is 0, regardless of the level on the other input.

Theorem (2) is also obvious by comparison with ordinary multiplication.

Theorem (3) can be proved by trying each case. If $x = 0$, then $0 \cdot 0 = 0$; if $x = 1$, then $1 \cdot 1 = 1$. Thus, $x \cdot x = x$.

Theorem (4) can be proved in the same manner. However, it can also be reasoned that at any time either x or its inverse $\bar{x}$ has to be at the 0 level, and so their AND product always has to be 0.

Theorem (5) is straightforward, since 0 *added* to anything does not affect its value, either in regular addition or in OR addition.

Theorem (6) states that if any variable is ORed with 1, the result will always be 1. Checking this for both values of x: $0 + 1 = 1$ and $1 + 1 = 1$. Equivalently, we can remember that an OR-gate output will be 1 when *any* input is 1, regardless of the value of the other input.

Theorem (7) can be proved by checking for both values of x: $0 + 0 = 0$ and $1 + 1 = 1$.

Theorem (8) can be proved similarly, or we can just reason that at any time either x or $\bar{x}$ has to be at the 1 level so that we are always ORing a 0 and a 1, which always results in 1.

Before introducing any more theorems, it should be pointed out that in applying theorems (1) through (8) the variable x may actually represent an expression containing more than one variable. For example, if we have $A\bar{B}\,(\overline{A\bar{B}})$, we can invoke theorem (4) by letting $x = A\bar{B}$. Thus, we can say that $A\bar{B}\,(\overline{A\bar{B}}) = 0$. The same idea can be applied to the use of any of these theorems.

Multivariable Theorems   The theorems presented below involve more than one variable:

(9) $\quad x + y = y + x$

(10) $\quad x \cdot y = y \cdot x$

(11) $\quad x + (y + z) = (x + y) + z = x + y + z$

(12) $\quad x(yz) = (xy)z = xyz$

(13a) $\quad x(y + z) = xy + xz$

(13b) $\quad (w + x)(y + z) = wy + xy + wz + xz$

(14) $\quad x + xy = x$

(15) $\quad x + \bar{x}y = x + y$

Theorems (9) and (10) are called the *commutative laws*. These laws indicate that the order in which we OR or AND two variables is unimportant; the result is the same.

Theorems (11) and (12) are the *associative laws,* which state that we can group the variables in an AND expression or OR expression any way we want.

Theorem (13) is the *distributive law,* which states that an expression can be expanded by multiplying term by term just the same as in ordinary algebra. This theorem also indicates that we can factor an expression. That is, if we have a sum of two (or more) terms, each of which contains a common variable, the common variable can be factored out just like in ordinary algebra. For example, if we have the expression $ABC + A\bar{B}\bar{C}$, we can factor out the $\bar{B}$ variable:

$$A\bar{B}C + \bar{A}\,\bar{B}\,\bar{C} = \bar{B}(AC + \bar{A}\,\bar{C})$$

As another example, consider the expression $ABC + ABD$. Here the two terms have the variables A an B in common, and so $A \cdot B$ can be factored out of both terms. That is,

$$ABC + ABD = AB(C + D)$$

Theorems (9) to (13) are easy to remember and use since they are identical to those of ordinary algebra. Theorems (14) and (15), on the other hand, do not have any counterparts in ordinary algebra. Each can be proved by trying all possible cases for x and y. This is illustrated for theorem (14) as follows:

Case 1. For $x = 0$, $y = 0$,

$$x + xy = x$$
$$0 + 0 \cdot 0 = 0$$
$$0 = 0$$

Case 2. For $x = 0$, $y = 1$,

$$x + xy = x$$
$$0 + 0 \cdot 1 = 0$$
$$0 + 0 = 0$$
$$0 = 0$$

Case 3. For $x = 1$, $y = 0$,

$$x + xy = x$$
$$1 + 1 \cdot 0 = 1$$
$$1 + 0 = 1$$
$$1 = 1$$

Case 4. For $x = 1$, $y = 1$,

$$x + xy = x$$
$$1 + 1 \cdot 1 = 1$$
$$1 + 1 = 1$$
$$1 = 1$$

Theorem (14) can also be proved by factoring and using theorems (6) and (2) as follows:

$$x + xy = x(1 + y)$$
$$= x \cdot 1 \quad \text{[using theorem (6)]}$$
$$= x \quad \text{[using theorem (2)]}$$

Teaching Tip: It may be instructive to show how theorem (15) can be proven by using theorem (14) to substitute $x + xy$ for x into the left side of theorem (15).

All of these Boolean theorems can be useful in simplifying a logic expression—that is, in reducing the number of terms in the expression. When this is done, the reduced expression will produce a circuit that is less complex than the one which the original expression would have produced. A good portion of the next chapter will be devoted to the process of circuit simplification. For now, the following examples will serve to illustrate how the Boolean theorems can be applied.

—— EXAMPLE 3-13 ——————————————————

Simplify the expression $y = A\bar{B}D + A\bar{B}\bar{D}$.

Solution   Factor out the common variables $A\bar{B}$ using theorem (13):
$$y = A\bar{B}(D + \bar{D})$$
Using theorem (8), the term in parentheses is equivalent to 1. Thus,
$$y = A\bar{B} \cdot 1$$
$$y = A\bar{B} \quad \text{[using theorem (2)]}$$

—— EXAMPLE 3-14 ——————————————————

Simplify $z = (\bar{A} + B)(A + B)$.

Solution   The expression can be expanded by multiplying out the terms [theorem (13)]:
$$z = \bar{A} \cdot A + \bar{A} \cdot B + B \cdot A + B \cdot B$$
Invoking theorem (4), the term $\bar{A} \cdot A = 0$. Also, $B \cdot B = B$ [theorem (3)]:
$$z = 0 + \bar{A} \cdot B + B \cdot A + B = \bar{A}B + AB + B$$

Factoring out the variable B [theorem (13)], we have

$$z = B(\overline{A} + A + 1)$$

Finally, using theorems (2) and (6),

$$z = B$$

———— EXAMPLE 3-15 ————————————————————

Simplify $x = ACD + A\overline{B}CD$.

Solution Factoring out the common variables CD, we have

$$x = CD(A + \overline{A}B)$$

Utilizing theorem (15), we can replace $A + \overline{A}B$ by $A + B$, so

$$x = CD(A + B)$$
$$= ACD + BCD$$

———— REVIEW QUESTIONS ————————————————————

1. Use theorems (13) and (14) to simplify $y = A\overline{C} + AB\overline{C}$.
2. Use theorems (13) and (8) to simplify $y = \overline{A}\,\overline{B}C\overline{D} + \overline{A}B\overline{C}\overline{D}$.

3-11   DeMORGAN'S THEOREMS

Two of the most important theorems of Boolean algebra were contributed by a great mathematician named DeMorgan. *DeMorgan's theorems* are extremely useful in simplifying expressions in which a product or sum of variables is inverted. The two theorems are:

$$(16)\ \overline{(x + y)} = \overline{x} \cdot \overline{y}$$

$$(17)\ \overline{(x \cdot y)} = \overline{x} + \overline{y}$$

Theorem (16) says that when the OR sum of two variables is inverted, this is the same as inverting each variable individually and then ANDing these inverted variables. Theorem (17) says that when the AND product of two variables is inverted, this is the same as inverting each variable individually and then ORing them. Each of DeMorgan's theorems can be readily proven by checking for all possible combinations of x and y. This will be left as an end-of-chapter exercise.

Although these theorems have been stated in terms of single variables x and y, they are equally valid for situations where x and/or y are expressions that contain more than one variable. For example, let's apply theorem (6) to the expression $\overline{(A\overline{B} + C)}$ as shown below:

$$\overline{(A\overline{B} + C)} = \overline{(A\overline{B})} \cdot \overline{C}$$

Note that here we treated $A\overline{B}$ as x and C as y. The result can be further simplified since we have a product $A\overline{B}$ that is inverted. Using theorem (17), the expression becomes

$$\overline{A\overline{B}} \cdot \overline{C} = (\overline{A} + \overline{\overline{B}}) \cdot \overline{C}$$

Notice that we can replace $\bar{\bar{B}}$ by B, so that we finally have
$$(\overline{A} + B) \cdot \overline{C} = \overline{A}\,\overline{C} + B\overline{C}$$
This final result contains only inverter signs that invert a single variable.

——— EXAMPLE 3-16 ———————————————————

Simplify the expression $z = \overline{(\overline{A} + C) \cdot (B + \overline{D})}$ to one having only single variables inverted.

Solution   Using theorem (17), we can rewrite this as
$$z = \overline{(\overline{A} + C)} + \overline{(B + \overline{D})}$$
We can think of this as breaking the large inverter sign down the middle and changing the AND sign ($\cdot$) to an OR sign ($+$). Now the term $\overline{(\overline{A} + C)}$ can be simplified by applying theorem (16). Likewise, $\overline{(B + \overline{D})}$ can be simplified:
$$z = \overline{(\overline{A} + C)} + \overline{(B + \overline{D})}$$
$$= (\overline{\bar{A}} \cdot \overline{C}) + \overline{B} \cdot \overline{\overline{D}}$$
Here we have broken the larger inverter signs down the middle and replaced the ($+$) with a ($\cdot$). Canceling out the double inversions, we have finally
$$z = A\overline{C} + \overline{B}D$$

Common Student Mistake: Notice the use of parentheses in example I to keep the *B* and *C* variables together. Students very often fail to do this. Show them what happens if parentheses are not used in this example.

Example 3-16 points out that when using DeMorgan's theorems to reduce an expression, we may break an inverter sign at any point in the expression and change the operator at that point in the expression to its opposite ($+$ is changed to $\cdot$, and vice versa). This procedure is continued until the expression is reduced to one in which only single variables are inverted. Two more examples are given below.

1. $z = \overline{A + \overline{B} \cdot C}$
$\quad = \overline{A} \cdot \overline{(\overline{B} \cdot C)}$
$\quad = \overline{A} \cdot (\overline{\overline{B}} + \overline{C})$
$\quad = \overline{A} \cdot (B + \overline{C})$

2. $q = \overline{(A + BC) \cdot (D + EF)}$
$\quad = \overline{(A + BC)} + \overline{(D + EF)}$
$\quad = (\overline{A} \cdot \overline{BC}) + (\overline{D} \cdot \overline{EF})$
$\quad = [\overline{A} \cdot (\overline{B} + \overline{C})] + [\overline{D} \cdot (\overline{E} + \overline{F})]$
$\quad = \overline{A}\overline{B} + \overline{A}\overline{C} + \overline{D}\overline{E} + \overline{D}\overline{F}$

Discussion Question: How can DeMorgan's theorems be used to show that $\overline{A + B} = A + B$?

DeMorgan's theorems are easily extended to more than two variables. For example, it can be proved that
$$\overline{x + y + z} = \overline{x} \cdot \overline{y} \cdot \overline{z}$$
$$\overline{x \cdot y \cdot z} = \overline{x} + \overline{y} + \overline{z}$$
and so on for more variables. Again, realize that any one of these variables can be an expression rather than a single variable.

Implications of DeMorgan's Theorems   Let us examine these theorems (16) and (17) from the standpoint of logic circuits. First, consider theorem (16),
$$\overline{x + y} = \overline{x} \cdot \overline{y}$$
The left-hand side of the equation can be viewed as the output of a NOR gate whose inputs are x and y. The right-hand side of the equation, on the other hand, is the result of first inverting both x and y and then putting them through an AND gate. These two representations are equivalent and are illustrated in Figure 3-26(a). What this means is than an AND gate with INVERTERs on each of its inputs is equivalent

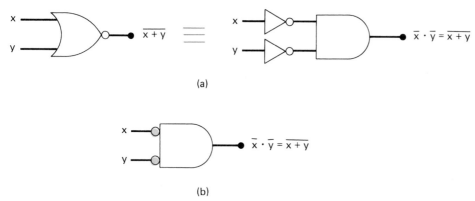

(a)

$$\bar{x} \cdot \bar{y} = \overline{x + y}$$

(b)

FIGURE 3-26   (a) Equivalent circuits implied by theorem (16); (b) alternative symbol for the NOR function.

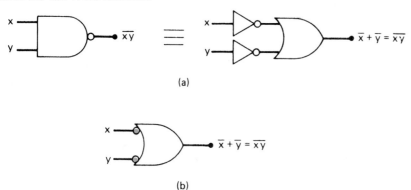

(a)

$$\bar{x} + \bar{y} = \overline{xy}$$

(b)

FIGURE 3-27   (a) Equivalent circuits implied by theorem (17); (b) alternative symbol for the NAND function.

to a NOR gate. In fact, both representations are used to represent the NOR function. When the AND gate with inverted inputs is used to represent the NOR function, it is usually drawn as shown in Figure 3-26(b), where the small circles on the inputs represent the inversion operation.

Now consider theorem (17),

$$\overline{x \cdot y} = \bar{x} + \bar{y}$$

The left side of the equation can be implemented by a NAND gate with inputs x and y. The right side can be implemented by first inverting inputs x and y and then putting them through an OR gate. These two equivalent representations are shown in Figure 3-27(a). The OR gate with INVERTERs on each of its inputs is equivalent to the NAND gate. In fact, both representations are used to represent the NAND function. When the OR gate with inverted inputs is used to represent the NAND function, it is usually drawn as shown in Figure 3-27(b), where the circles again represent inversion.

_____ EXAMPLE 3-17 _____

Implement a circuit having the output expression $z = \bar{A} + \bar{B} + C$ using a NAND gate and an INVERTER.

Solution   Invert the right side of the expression twice. This is legal, since a double inversion does not change the logic value.

$$z = \overline{\overline{\bar{A} + \bar{B} + C}}$$

$$z = \overline{A \cdot B \cdot \overline{C}} = \overline{\overline{A}} + \overline{\overline{B}} + \overline{\overline{C}} = \overline{A} + \overline{B} + C$$

FIGURE 3-28   Example 3-17.

Now use DeMorgan's theorem to break the lower inversion bar:

$$z = \overline{\overline{A}} \cdot \overline{\overline{B}} \cdot \overline{C}$$

Cancel the double inversions over A and B to obtain

$$z = A \cdot B \cdot \overline{C}$$

In this new form, it is easy to see how a NAND gate can be used to produce z. The circuit is drawn in Figure 3-28.

——— REVIEW QUESTIONS ———

1. Use DeMorgan's theorems to convert the expression $z = \overline{(A + B) \cdot \overline{C}}$ to one that has only single-variable inversions.
2. Repeat question 1 for the expression $y = \overline{\overline{RST} + \overline{Q}}$.
3. Implement a circuit having output expression $z = \overline{A} \, \overline{B} C$ using only a NOR gate and an INVERTER.

3-12   UNIVERSALITY OF NAND GATES AND NOR GATES

All Boolean expressions consist of various combinations of the basic operations of OR, AND, and INVERT. Therefore, any expression can be implemented using OR gates, AND gates, and INVERTERs. It is possible, however, to implement any logic expression using *only* NAND gates and no other type of gate. This is because NAND gates, in the proper combination, can be used to perform each of the Boolean operations OR, AND, and INVERT. This is demonstrated in Figure 3-29.

First, in Figure 3-29(a) we have a two-input NAND gate whose inputs are purposely connected together so that the variable A is applied to both. In this configuration, the NAND simply acts as INVERTER, since its output is $x = \overline{A \cdot A} = \overline{A}$.

In Figure 3-29(b) we have two NAND gates connected so that the AND operation is performed. NAND gate 2 is used as an INVERTER to change $\overline{AB}$ to $\overline{\overline{AB}} = AB$, which is the desired AND function.

The OR operation can be implemented using NAND gates connected as shown in Figure 3-29(c). Here NAND gates 1 and 2 are used as INVERTERs to invert the inputs, so that the final output is $x = \overline{\overline{A} \cdot \overline{B}}$, which can be simplified *to* $x = A + B$ using DeMorgan's theorem.

In a similar manner, it can be shown that NOR gates can be arranged to implement any of the Boolean operations. This is illustrated in Figure 3-30. Part (a) shows that a NOR gate with its inputs connected together behaves as an INVERTER, since the output is $x = \overline{A + A} = \overline{A}$.

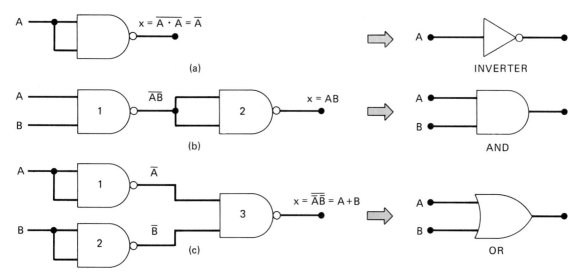

FIGURE 3-29   NAND gates can be used to implement any Boolean function.

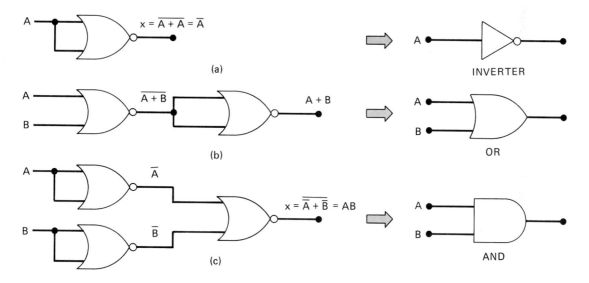

FIGURE 3-30   NOR gates can be used to implement any Boolean operation.

In Figure 3-30(b) two NOR gates are arranged so that the OR operation is performed. NOR gate 2 is used as an INVERTER to change $\overline{A + B}$ to $\overline{\overline{A + B}} = A + B$, which is the desired OR function.

The AND operation can be implemented with NOR gates as shown in Figure 3-30(c). Here NOR gates 1 and 2 are used as INVERTERs to invert the inputs, so that the final output is $x = \overline{\overline{A} + \overline{B}}$, which can be simplified to $x = A \cdot B$ by use of DeMorgan's theorem.

Since any of the Boolean operations can be implemented using only NAND gates, any logic circuit can be constructed using only NAND gates. The same is true for NOR gates. This characteristic of NAND and NOR gates can be very useful in logic-circuit design, as the following example illustrates.

EXAMPLE 3-18

A logic-circuit designer has to implement a circuit that satisfies the expression $x = AB + CD$ using the minimum number of ICs. He has the TTL integrated circuits shown in Figure 3-31 available for his use. Each of these ICs is a *quad*, which means that it contains *four* identical two-input gates on one chip.

Solution   The straightforward method for implementing the given expression uses two AND gates and an OR gate as shown in Figure 3-32(a). This implementa-

FIGURE 3-31   ICs available for Example 3-18.

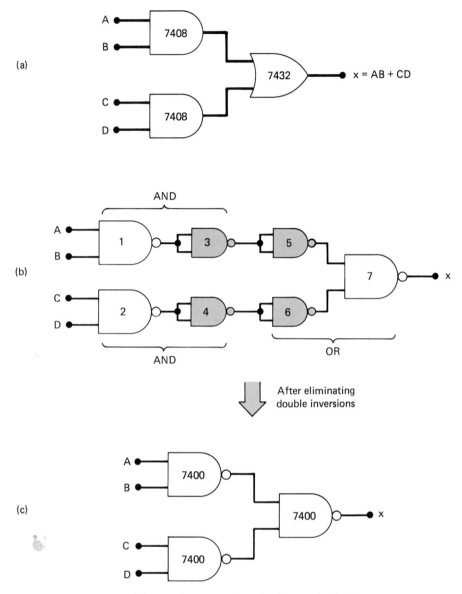

FIGURE 3-32   Possible implementations for Example 3-18.

tion requires the 7408 and 7432 ICs, and some of the gates on each IC are wasted since they are not used.

Another implementation can be accomplished by taking the circuit of Figure 3-32(a) and replacing each AND gate and OR gate by its equivalent NAND-gate implementation from Figure 3.29. The result is shown in Figure 3-32(b).

At first glance this new circuit looks as if it requires seven NAND gates. However, NAND gates 3 and 5 are connected as INVERTERs in series and can be eliminated from the circuit since they perform a double inversion of the signal out of NAND gate 1. Similarly, NAND gates 4 and 6 can be eliminated. The final circuit, after eliminating the double INVERTERs, is drawn in Figure 3-32(c).

This final circuit is more efficient than the one in Figure 3-32(a) because it uses three two-input NAND gates that can be implemented from one IC, the 7400.

Teaching Tip: Show what happens if we try to implement the expression $x = AB + CD$ using all NOR gates. Is it more efficient than using ANDs and ORs?

1. How many different ways do we now have to implement the inversion operation in a logic circuit?
2. Implement the expression $x = (A + B)(C + D)$ using OR and AND gates. Then implement the expression using only NOR gates by converting each OR and AND gate to its NOR implementation from Figure 3-30. Which circuit is more efficient?
3. Write the output expression for the circuit of Figure 3-32(c) and use DeMorgan's theorems to show that it is equivalent to the expression for the circuit of Figure 3-32(a).

3-13 ALTERNATE LOGIC-GATE REPRESENTATIONS

We have introduced the five basic logic gates (AND, OR, INVERTER, NAND, and NOR) and the standard symbols used to represent them on logic-circuit diagrams. Although you may find that many circuit diagrams still use these standard symbols exclusively, it has become increasingly more common to find circuit diagrams that utilize an alternate set of symbols *in addition* to the standard symbols.

FIGURE 3-33 Standard and alternate symbols for various logic gates and inverter.

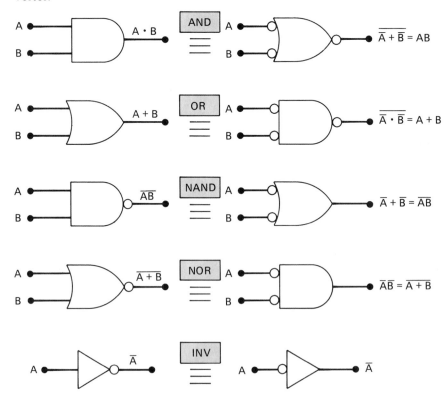

Before discussing the reasons for using an alternate symbol for a logic gate, we will present the alternate symbols for each gate and show that they are equivalent to the standard symbols. Refer to Figure 3-33; the left side of the illustration shows the standard symbol for each logic gate, and the right side shows the alternate symbol. The alternate symbol for each gate is obtained from the standard symbol by doing the following:

1. Invert each input and output of the standard symbol. This is done by adding bubbles (small circles) on input and output lines that do not have bubbles, and by removing bubbles that are already there.

2. Change the operation symbol from AND to OR, or from OR to AND. (In the special case of the INVERTER, the operation symbol is not changed.)

For example, the standard NAND symbol is an AND symbol with a bubble on its output. Following the steps outlined above, remove the bubble from the output, and add a bubble to each input. Then change the AND symbol to an OR symbol. The result is an OR symbol with bubbles on its inputs.

We can easily prove that this alternate symbol is equivalent to the standard symbol by using DeMorgan's theorems and recalling that the bubble represents an inversion operation. The output expression from the standard NAND symbol is $\overline{AB} = \overline{A} + \overline{B}$, which is the same as the output expression for the alternate symbol. This same procedure can be followed for each pair of symbols in Figure 3-33.

Several points should be stressed regarding the logic symbol equivalences:

1. The equivalences are valid for gates with *any* number of inputs.

2. None of the standard symbols have bubbles on their inputs, and all the alternate symbols do.

3. The standard and alternate symbols for each gate represent the same physical circuit; there is no difference in the circuits represented by the two symbols.

4. NAND and NOR gates are inverting gates, and so both the standard and alternate symbols for each will have a bubble on *either* the input or the output. AND and OR gates are *noninverting* gates, and so the alternate symbols for each will have bubbles on *both* inputs and output.

Logic-Symbol Interpretation

Each of the logic-gate symbols of Figure 3-33 provides a unique interpretation of how the gate operates. Before we can demonstrate these interpretations, we must first establish the concept of *active* logic levels.

When an input or output line on a logic gate symbol has *no bubble* on it, that line is said to be active-HIGH. When an input or output line *does* have a *bubble* on it, that line is said to be active-LOW. The presence or absence of a bubble, then, determines the active-HIGH/active-LOW status of a gate's inputs and output, and is used to interpret the gate operation.

To illustrate, Figure 3-34(a) shows the standard symbol for a NAND gate. The standard symbol has a bubble on its output and no bubbles on its inputs. Thus, it has an active-LOW output and active-HIGH inputs. The logic operation represented by this symbol can therefore be interpreted as follows:

The output goes LOW only when all the inputs are HIGH

Note that this says that the output will go to its active state only when *all* the inputs are in their active states. The word "all" is used because of the AND symbol.

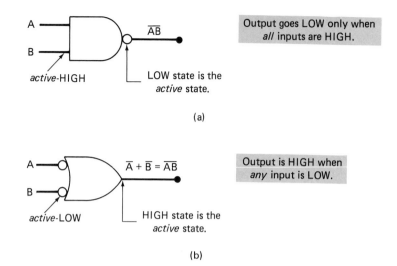

(a)

(b)

FIGURE 3-34   Interpretation of the two NAND gate symbols.

The alternate symbol for a NAND gate shown in Figure 3-34(b) has an active-HIGH output and active-LOW inputs, and so its operation can be stated as

The output goes HIGH when any input is LOW

Again, this says that the output will be in its active state whenever *any* of the inputs are in their active states. The word "any" is used because of the OR symbol.

With a little thought, it can be seen that the two interpretations for the NAND symbols in Figure 3-34 are different ways of saying the same thing.

Summary   At this point you are probably wondering why there is a need to have two different symbols and interpretations for each logic gate. Hopefully, the reasons will become clear after reading the next section. For now, let us summarize the important points concerning the logic-gate representations.

1. To obtain the alternate symbol for a logic gate, take the standard symbol and change its operation symbol (OR to AND, or AND to OR) and change the bubbles on both inputs and output (that is, delete bubbles that are present, and add bubbles where there are none).

2. To interpret the logic-gate operation, first note which logic state, 0 or 1, is the active state for the inputs, and which is the active state for the output. Then, realize that the output's active state is produced by having *all* the inputs in their active state (if an AND symbol is used), or by having any of the inputs in their active state (if an OR symbol is used).

———— EXAMPLE 3-19 ————————————————————————————

Give the interpretation of the two OR gate symbols.

Solution   The results are shown in Figure 3-35. Note how the word *any* is used when the gate symbol includes an OR symbol and the word *all* is used when it includes an AND symbol.

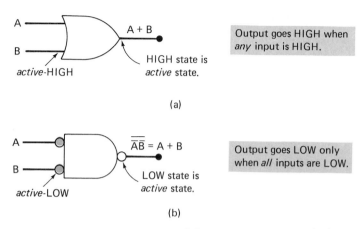

FIGURE 3-35   Interpretation of the two OR gate symbols.

1. Write the interpretation of the operation performed by the standard NOR gate symbol in Figure 3-33.
2. Repeat question 1 for the alternate NOR gate symbol.
3. Repeat question 1 for the alternate AND gate symbol.
4. Repeat question 1 for the standard AND gate symbol.

3-14   WHICH GATE REPRESENTATION TO USE

Some logic-circuit designers and many textbooks use only the standard logic-gate symbols in their circuit schematics. While this practice is not incorrect, it does nothing to make the circuit operation easier to follow. Proper use of the alternate gate symbols in the circuit diagram can make the circuit operation much clearer. This can be illustrated by considering the example shown in Figure 3-36.

The circuit in Figure 3-36(a) contains three NAND gates connected to produce an output Z that depends on inputs A, B, C, D. The circuit diagram uses the standard symbol for each of the NAND gates. While this diagram is logically correct, it does not facilitate any understanding of how the circuit functions. The improved circuit representations given in Figure 3-36(b) and (c), however, can be analyzed more easily to determine the circuit operation.

The representation of Figure 3-36(b) is obtained from the original circuit diagram by replacing NAND gate 3 with its alternate symbol. In this diagram, output Z is taken from a NAND gate symbol that has an active-HIGH output. Thus, we can say that Z will go HIGH when either X or Y is LOW. Now, since X and Y each appear at the output of NAND symbols having active-LOW outputs, we can say that X will go LOW only if $A = B = 1$, and Y will go LOW only if $C = D = 1$. Putting this all together, we can describe the circuit operation as follows:

Output Z will go HIGH whenever either $A = B = 1$ or $C = D = 1$

This description can be translated to truth-table form by setting $Z = 1$ for those cases where $A = B = 1$, and for those cases where $C = D = 1$. For all other cases, Z is made a 0. The resultant truth table is shown in Figure 3-36(d).

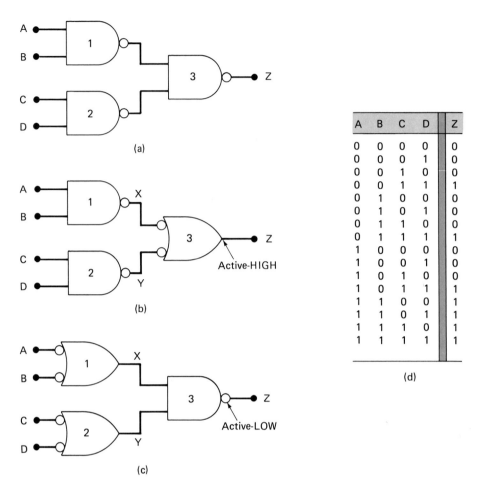

A	B	C	D	Z
0	0	0	0	0
0	0	0	1	0
0	0	1	0	0
0	0	1	1	1
0	1	0	0	0
0	1	0	1	0
0	1	1	0	0
0	1	1	1	1
1	0	0	0	0
1	0	0	1	0
1	0	1	0	0
1	0	1	1	1
1	1	0	0	1
1	1	0	1	1
1	1	1	0	1
1	1	1	1	1

(d)

FIGURE 3-36   (a) Original circuit using standard NAND symbols; (b) equivalent representation where output Z is active-HIGH; (c) equivalent representation where output Z is active-LOW; (d) truth table.

The representation of Figure 3-36(c) is obtained from the original circuit diagram by replacing NAND gates 1 and 2 by their alternate symbols. In this equivalent representation the Z output is taken from a NAND gate that has an active-LOW output. Thus, we can say that Z will go LOW only when $X = Y = 1$. Since X and Y are active-HIGH outputs, we can say that X will be HIGH when either A or B is LOW, and Y will be HIGH when either C or D is LOW. Putting this all together, we can describe the circuit operation as follows:

Output Z will go LOW only when A or B is LOW and C or D is LOW

This description can be translated to truth-table form by making $Z = 0$ for all cases where at least one of the A or B inputs is LOW at the same time that at least one of the C or D inputs is LOW. For all other cases, Z is made a 1. The resultant truth table is the same as that obtained for the circuit diagram of Figure 3-36(b).

Which Circuit Diagram Should Be Used?

The answer to this question depends on the particular function being performed by the circuit output. If the circuit is being used to cause some action (e.g., turn on a device, activate another logic

circuit) when output Z goes to the 1 state, then we say that Z is active-HIGH, and the circuit diagram of Figure 3-36(b) should be used. On the other hand, if the circuit is being used to cause some action when Z goes to the 0 state, then Z is active-LOW and the diagram of Figure 3-36(c) should be used.

Of course, there will be situations where *both* output states are used to produce different actions and either one can be considered to be the active state. For these cases, either circuit representation can be used.

Bubble Placement   Refer to the circuit representation of Figure 3-36(b) and note that the symbols for NAND gates 1 and 2 were chosen to have active-LOW outputs to match the active-LOW inputs of NAND gate 3. Refer to the circuit representation of Figure 3-36(c) and note that the symbols for NAND gates 1 and 2 were chosen to have active-HIGH outputs to match the active-HIGH inputs of NAND gate 3. This leads to the following general rule for preparing logic-circuit schematics:

> ***Whenever possible, connect bubble outputs to bubble inputs, and nonbubble outputs to nonbubble inputs***

The following examples will show how this rule can be applied.

———— EXAMPLE 3-20 ————————————————————

The logic circuit in Figure 3-37(a) is being used to activate an alarm when its output Z goes HIGH. Modify the circuit diagram so that it more effectively represents the circuit operation.

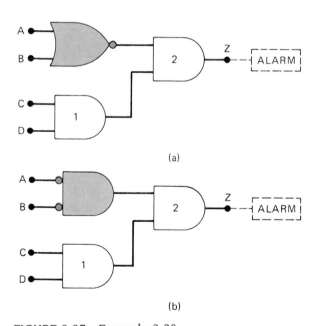

FIGURE 3-37   Example 3-20.

Solution   Since $Z = 1$ will activate the alarm, Z is active-HIGH. Thus, the AND gate 2 symbol does not have to be changed. The NOR gate symbol should be changed to the alternate symbol with a nonbubble (active-HIGH) output to match the nonbubble input of AND gate 2. The result is shown in Figure 3-37(b).

EXAMPLE 3-21

When the output of the logic circuit in Figure 3-38(a) goes LOW, it activates another logic circuit. Modify the circuit diagram to more effectively represent the circuit operation.

FIGURE 3-38   Example 3-21.

Solution   Since Z is active-LOW, the symbol for OR gate 2 has to be changed to its alternate symbol as shown in Figure 3-38(b). The new OR-gate 2 symbol has bubble inputs, and so the AND gate and OR gate 1 symbols have to be changed to bubbled outputs as shown in Figure 3-38(b). The INVERTER already has a bubble output.

Teaching Tip: It cannot be overemphasized that the three circuit diagrams in Figure 3-36 represent the same physical circuit, the same chips connected in the same way. It should also be stressed that not all circuit schematics, commercial or otherwise, use the alternate gate symbols, so it is often necessary to analyze the circuit using the conventional techniques.

Analyzing Circuits   When a logic-circuit schematic is drawn using the rules we followed in these examples, it is much easier for an engineer or technician (or student) to follow the signal flow through the circuit and to determine the input conditions that are needed to activate the output. This will be illustrated in the following examples—which, incidentally, use circuit diagrams taken from the logic schematics of an actual microcomputer.

EXAMPLE 3-22

The logic circuit in Figure 3-39 generates an output, *MEM*, that is used to activate the memory ICs in a particular microcomputer. Determine the input conditions necessary to activate *MEM*.

FIGURE 3-39   Example 3-22.

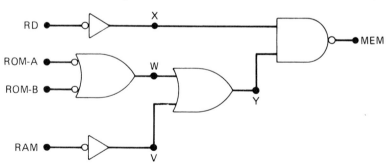

Solution   One way to do this would be to write the expression for *MEM* in terms of the inputs *RD*, *ROM-A*, *ROM-B*, and *RAM*, and to evaluate it for the 16 possible combinations of these inputs. While this method would work, it would require a lot more work than is necessary.

A more efficient method is to interpret the circuit diagram using the ideas we have been developing in the last two sections. These are the steps:

1. *MEM* is active-LOW, and it will go LOW only when *X* and *Y* are HIGH.
2. *X* will be HIGH only when *RD* = 0.
3. *Y* will be HIGH when either *W* or *V* is HIGH.
4. *V* will be HIGH when *RAM* = 0.
5. *W* will be HIGH when either *ROM-A* or *ROM-B* = 0.
6. Putting this all together, *MEM* will go LOW only when *RD* = 0 *and* at least one of the three inputs *ROM-A*, *ROM-B*, or *RAM* is LOW.

—— EXAMPLE 3-23 ——————————————————

The logic circuit in Figure 3-40 is used to control the motor on a cassette recorder when the microcomputer is sending data to or receiving data from the recorder. The circuit will turn on the motor when *CASS* = 1. Determine the input conditions necessary to turn on the motor.

Solution   Once again we will interpret the diagram in a step-by-step fashion:

1. *CASS* is active-HIGH, and it will go HIGH only when $X = Y = 0$.
2. *X* will be LOW when either *IN* or *OUT* is HIGH.
3. *Y* will be LOW only when $W = 0$ and $A_0 = 0$.
4. *W* will be LOW only when A_1 through A_7 are all HIGH.
5. Putting this all together, *CASS* will be HIGH when $A_1 = A_2 = A_3 = A_4 = A_5 = A_6 = A_7 = 1$ and $A_0 = 0$, and either *IN* or *OUT* or both are 1.

FIGURE 3-40   Example 3-23.

Note the strange symbol for the 8-input CMOS NAND gate (74C00); also note that the signal A_7 is connected to two of the NAND inputs.

Asserted Levels

We have been describing logic signals as being active-LOW or active-HIGH. For example, the output *MEM* in Figure 3.39 is active-LOW, and the output *CASS* in Figure 3-40 is active-HIGH, since these are the output states that cause something to happen. Likewise, Figure 3-40 has active HIGH inputs A_1 to A_7, and active-LOW input A_0.

When a logic signal is in its active state, it can be said to be **asserted.** For example, when we say that input A_0 is asserted, we are saying that it is in its active-LOW state. When a logic signal is not in its active state, it is said to be **unasserted.** Thus, when we say that *CASS* is unasserted, we mean that it is in its inactive state (LOW).

Clearly, the terms "asserted" and "unasserted" are synonymous with "active" and "inactive," respectively:

$$\textbf{asserted} = \textbf{active}$$

$$\textbf{unasserted} = \textbf{inactive}$$

Both sets of terms are in common use in the digital field, and so you should recognize both ways of describing a logic signal's active state.

Labeling Active-LOW Logic Signals

It has become common practice to use an overbar to label active-LOW signals. The overbar serves as another indication that the signal is active-LOW; of course, the absence of an overbar means that the signal is active-HIGH.

Teaching Tip: Other circuit schematic conventions that are used to denote that a signal is active-LOW are: MEM*, MEM L, and MEM′.

To illustrate, all of the signals in Figure 3-39 are active-LOW, and so they can be labeled as follows:

$$\overline{RD}, \quad \overline{ROM\text{-}A}, \quad \overline{ROM\text{-}B}, \quad \overline{RAM}, \quad \overline{MEM}$$

Remember, the overbar is simply a way to emphasize that these are active-LOW signals. We will employ this convention for labeling logic signals whenever appropriate.

▬ REVIEW QUESTIONS ▬

1. Use the method of Examples 3-22 and 3-23 to determine the input conditions needed to activate the output of the circuit in Figure 3-37(b).
2. Repeat question 1 for the circuit of Figure 3-38(b).
3. How many NAND gates are in Figure 3-39?
4. How many NOR gates are in Figure 3-40?
5. What will be the output level in Figure 3-38(b) when all the inputs are asserted?
6. What inputs are required to assert the alarm output in Figure 3-37(b)?

3-15   IEEE/ANSI STANDARD LOGIC SYMBOLS

The logic symbols that we have used throughout this chapter are well-known standard symbols that have been widely used in the digital industry for many years. These symbols work well for the basic logic gates because each gate symbol has a

distinctive shape, and each gate input has the same function. They do not provide enough useful information, however, for the more complex logic devices such as flip-flops, counters, decoders, multiplexers, memories, and microprocessor interface ICs. These complex ICs often have several inputs and outputs with different functions and modes of operation.

In order to provide more useful information about these complex logic devices, a new set of standard symbols was introduced in 1984 under IEEE/ANSI Standard 91–1984. These newer symbols are gradually being accepted by more and more electronics companies and IC manufacturers, and have begun to appear in their published literature. In addition, military contracts in the United States now require the use of these new symbols. It is therefore important to become familiar with the new standard symbols since eventually they will completely replace the traditional ones, although this process may take a long time.

The principal difference in the new standard is that it uses rectangular symbols for all devices instead of different symbol shapes for each device. A special notation system is used to indicate how the inputs and outputs are related. Figure 3-41 shows the traditional symbols with the newer rectangular symbols for the basic logic gates. Study them carefully and note the following points:

1. The newer symbols use a small right triangle ◁ in place of the small bubble of the traditional symbols. Like the bubble, this triangle indicates an inversion of the logic level. The presence or absence of the triangle also signifies whether an input or output is active-LOW or active-HIGH.

Teaching Tip: A complete, easy to follow description of the new IEEE logic symbology standard may be found in a Texas Instrument publication entitled *Using Functional Logic Symbols—Application of IEEE Std 91-1984.*

FIGURE 3-41   Standard logic symbols: (a) traditional; (b) rectangular.

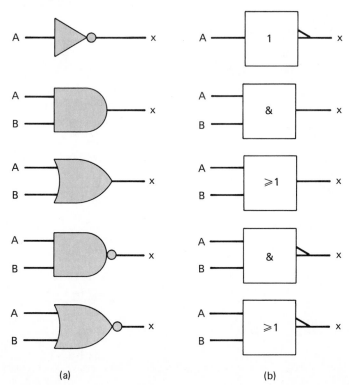

(a)   (b)

2. A special notation inside each rectangular symbol describes the logic relation between inputs and output. The "1" inside the INVERTER symbol denotes a device with only *one* input; the triangle on the output indicates that the output will go to its active-LOW state when that one input is in its active-HIGH state.

   The "&" inside the AND symbol means that the output will go to its active-HIGH state when all the inputs are in their active-HIGH state. The " ≥ " inside the OR gate means that the output will go to its active state (HIGH) whenever *one or more* inputs are in their active state (HIGH).

3. The rectangular symbols for the NAND and NOR are the same as those for the AND and OR, respectively, with the addition of the small inversion triangle on the output.

IEEE/ANSI Symbols for Logic-Gate ICs   The rectangular symbols can also be used to represent the complete logic of an IC package that contains a number of independent gates. This is illustrated in Figure 3-42 for the 7404 TTL hex* INVERTER IC, and in Figure 3-43 for the 7420 dual four-input NAND IC. Each logic gate is represented as a separate rectangular block. Note how the rectangular symbols indicate the logic operation notation only in the top block; it is understood to apply to all the other blocks representing the other gates on the chip.

It is important to understand the difference between the two alternate ways to represent a logic gate in a circuit, and the two different standards for logic-gate symbols. You choose which set of standard logic symbols to use—either the traditional

FIGURE 3-42   7404 hex INVERTER IC: (a) traditional logic symbol; (b) rectangular logic symbol. The notation "1" appears only in the top rectangle, but applies to all blocks below.

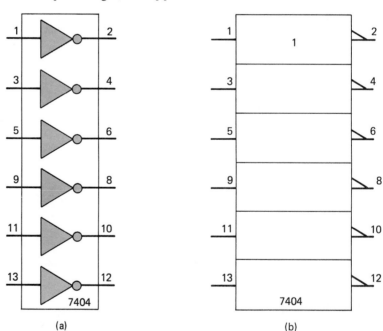

(a)   (b)

*The term "hex" when used to describe an IC indicates that the IC contains *six* circuits of the same type.

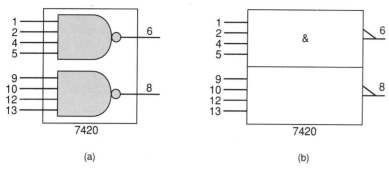

(a)   (b)

FIGURE 3-43   7420 dual four-input NAND IC: (a) traditional symbol; (b) rectangular symbol.

symbols (distinctive shapes for each type of gate) or the new standard rectangular symbols. Regardless of which symbols you choose to use, there are two possible ways to represent a gate in a circuit diagram depending on its active output state. This is illustrated in Example 3-24.

EXAMPLE 3-24

Figure 3-44(a) shows the two representations for a NOR gate using the traditional logic symbols. Remember, the choice of which representation to use in a circuit dia-

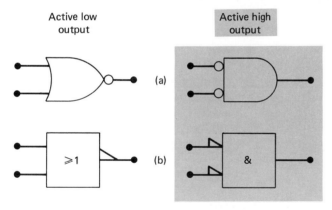

FIGURE 3-44   Both representations of a NOR gate using the two types of symbols: (a) traditional; (b) rectangular.

Teaching Tip: It should be made very clear that the IEEE symbols and the traditional symbols are just different conventions for representing the logic gates in a circuit schematic. Once a symbol convention is chosen, a logic gate will have two alternate representations (within the convention being used). This concept can be reinforced by repeating Example 3-24 for one or more of the other gates.

gram is determined by the desired active state of the output. Redraw the two representations using the new IEEE/ANSI symbols.

Solution   Figure 3.44(b) shows the results.

IEEE/ANSI Symbols for Complex ICs

There would be no real advantage to the new symbols if all we had to deal with were the basic logic gates. For the more complex logic devices, however, the new standard symbols with their *dependency notation* will specify the complete logic operation of the device. This makes it virtually unnecessary to refer to the manufacturer's data manual to find out how a particular logic IC is functioning in a circuit. We will see examples of this when we encounter the more complex logic circuits in subsequent chapters.

Throughout this text we will continue to use the traditional logic symbols in our circuit diagrams. There are two reasons for this: (1) the logic circuit schematics for most of the digital equipment already in the field will still use the traditional symbology, and (2) it will be some time before the many thousands of engineers and technicians make the complete switch to the new standard. We will also present and describe the new IEEE/ANSI symbols for most of the logic devices that are covered in the text. In this way, you can become familiar with the various aspects of the dependency notation that is the major benefit provided by the new standard.

REVIEW QUESTIONS

1. What is the major advantage of the new IEEE/ANSI symbols?
2. Draw all of the basic logic gates using both the traditional symbols and the IEEE standard symbols.
3. Repeat question 2 for the alternate representation of each gate.

PROBLEMS

SECTION 3-3

3-1. x is LOW for all those intervals when all inputs are LOW.

3-1. Draw the output waveform for the circuit of Figure 3-45.

FIGURE 3-45

3-2. x is LOW for all those intervals when B and C are LOW.

3-2. Suppose the A input in Figure 3-45 is unintentionally shorted to ground (i.e., $A = 0$). Draw the resulting output waveform.

3-3. Output will be a constant HIGH.

3-3. Suppose that the A input in Figure 3-45 is unintentionally shorted to the +5 V supply line (i.e., $A = 1$). Draw the resulting output waveform.

3-4. (a) It may be that the second input is never HIGH while the first input is LOW.   (b) It may be that there is no interval when either one input or the other is HIGH.

3-4. Read the statements below concerning an OR gate. At first they may appear to be valid, but after some thought you should realize that neither one is *always* true. Prove this by showing a specific example to refute each statement.
   (a) If the output waveform from an OR gate is the same as the waveform at one of its inputs, then the other input is being held permanently LOW.

(b) If the output waveform from an OR gate is always HIGH, then one of its inputs is being held permanently HIGH.

3.5. How many different sets of input conditions will produce a HIGH output from a five-input OR gate?   Thirty-one.

SECTION 3-4

3-6. Change the OR gate in Figure 3-45 to an AND gate.
  (a) Draw the output waveform.
  (b) Draw the output waveform if the *A* input is permanently shorted to ground.
  (c) Draw the output waveform if *A* is permanently shorted to +5 V.

3-7. Refer to Figure 3-4. Modify the circuit so that the alarm is to be activated only when the pressure and temperature exceed their maximum limits at the same time.   Change the OR to an AND.

3-8. Change the OR gate in Figure 3-6 to an AND gate and draw the output waveform.   Output is always LOW.

3-6. (a) Output is HIGH only when all inputs are HIGH.
(b) Output at a constant LOW.
(c) Output is HIGH for those intervals when *B* and *C* are both HIGH.

SECTIONS 3-5 to 3-7

3-9. Add an INVERTER to the output of the OR gate from Figure 3-45. Draw the waveform at the INVERTER output.

3-10. Write the Boolean expression for the output *x* in Figure 3-46(a). Determine the value of *x* for all possible input conditions and list the values in a truth table.

3-9. Output is inverse of that from Problem 3-1.

3-10. $x = (\overline{A} + \overline{B})BC$
Truth-table shows $x = 1$ only when $A = B = C$.

FIGURE 3-46

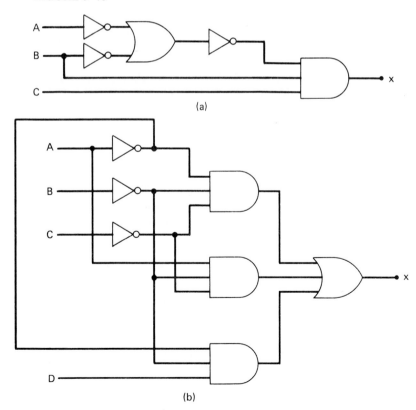

(a)

(b)

3-11. Repeat Problem 3-10 for the circuit in Figure 3-46(b).

3-12. Change each OR to an AND, and each AND to an OR in Figure 3-15(b). Then write the expression for the output.

3-13. Determine the complete truth table for the circuit of Figure 3-16 by finding the logic levels present at each gate output for each of the 16 possible combinations of input levels.

SECTION 3-8

3-14. For each of the following expressions, construct the corresponding logic circuit, using AND and OR gates and INVERTERs.
(a) $x = \overline{AB(C + D)}$
(b) $z = \overline{(A + B + \overline{CD}\overline{E})} + \overline{B}C\overline{D}$
(c) $y = \overline{(\overline{M + N} + \overline{P}Q)}$
(d) $x = \overline{W + P\overline{Q}}$
(e) $z = MN(P + \overline{N})$

SECTION 3-9

3-15. Apply the input waveforms of Figure 3-47 to a NOR gate and draw the output waveform.

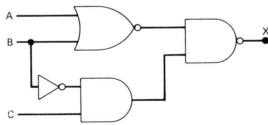

FIGURE 3-47

3-16. Repeat Problem 3-15 with the C input held permanently LOW. Then repeat for C held permanently HIGH.

3-17. Repeat Problem 3-15 for a NAND gate.

3-18. Repeat Problem 3-16 for a NAND gate.

3-19. Write the output expression for the circuit of Figure 3-48.
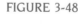

3-20. Determine the complete truth table for the circuit of Figure 3-48.

3-21. Modify the circuits that were constructed in Problem 3-14 so that NAND gates and NOR gates are used wherever appropriate.

FIGURE 3-48

3-22. DRILL QUESTION

Complete each expression.

(a) $A + 1 =$ 1

(b) $A \cdot A =$ A

(c) $B \cdot \bar{B} =$ 0

(d) $C + C =$ C

(e) $x \cdot 0 =$ 0

(f) $D \cdot 1 =$ D

(g) $D + 0 =$ D

(h) $C + \bar{C} =$ 1

(i) $G + GF =$ G

(j) $y + \overline{w}y =$ y

3-23. Prove theorem (15) by trying all possible cases.   Every case works!

3-24. (a) Simplify the following expression using theorems (13b), (3), and (4): $x = M\bar{N}P + \bar{M}N\bar{P}$

$$x = (M + N)(\bar{M} + P)(\bar{N} + \bar{P})$$

(b) Simplify the following expression using theorems (13a), (8), and (6): $z = B\bar{C}$

$$z = \bar{A}B\bar{C} + AB\bar{C} + B\bar{C}D$$

SECTIONS 3-11 and 3-12

3-25. Prove DeMorgan's theorems by trying all possible cases.

3-25. All cases work!

3-26. Simplify each of the following expressions using DeMorgan's theorems:

(a) $\overline{\overline{ABC}}$ $A + \bar{B} + C$

(b) $\overline{\bar{A} + \bar{B}C}$ $A(B + \bar{C})$

(c) $\overline{AB\overline{CD}}$ $\bar{A} + \bar{B} + CD$

(d) $\overline{A(B + \bar{C})D}$ $\bar{A} + B + \bar{C} + \bar{D}$

(e) $\overline{(M + \bar{N})(\bar{M} + N)}$ $\bar{M}N + M\bar{N}$

(f) $\overline{\overline{\overline{ABCD}}}$ $(\bar{A} + \bar{B})C + \bar{D}$

3-27. Use DeMorgan's theorems to simplify the expression for the output of Figure 3-48. $x = A + B + \bar{C}$

3-28. Convert the circuit of Figure 3-46(b) to one using only NAND gates. Then write the output expression for the new circuit, simplify it using DeMorgan's theorems, and compare it with the expression for the original circuit. $x = \bar{A}\bar{B}\bar{C} + AB\bar{C} + \bar{A}\bar{B}D$

3-29. Convert the circuit of Figure 3-46(a) to one using only NOR gates. Then write the expression for the new circuit, simplify it using DeMorgan's theorems, and compare it with the expression for the original circuit. $x = ABC$

3-30. Show how a two-input NAND gate can be constructed from two-input NOR gates.   Put an INVERTER on each input and on the output.

3-31. Show how a two-input NOR gate can be constructed from two-input NAND gates.   Put an INVERTER on each input and on the output.

SECTIONS 3-13 and 3-14

3-32. Draw the standard representations for each of the basic logic gates. Then draw the alternate representations.

3-32. See Figure 3-33.

3-33. For each statement below, draw the appropriate logic-gate representation and indicate the type of gate:

(a) A HIGH output occurs only when all three inputs are LOW.

(b) A LOW output occurs when any of the four inputs is LOW.

(c) A LOW output occurs only when all eight inputs are HIGH.

3-33. (a) NOR gate alternate representation. (b) AND gate alternate representation. (c) NAND gate alternate representation.

3-34. (a) Change the NOR and INVERTER to their alternate representations. (b) x is LOW for $ABC = 001$.

3-34. The output of the circuit of Figure 3-48 is supposed to turn on an indicator lamp when it goes LOW.
    (a) Modify the circuit diagram so that it more effectively represents the circuit operation.
    (b) Use the new circuit diagram to determine the input conditions that will activate the output. Do this by working back from the output using the information given by the gate symbols as was done in Examples 3-22 and 3-23. Compare the results with the truth table obtained in Problem 3-20.

3-35. (a) Z is HIGH only when $ABCD = 0011$. (b) Change NOR to its standard representation. Change both ANDs to alternate representations. Z will be LOW when A or B is HIGH, or when C or D is LOW.

3-35. (a) Determine the input conditions needed to activate output Z in Figure 3-37(b). Do this by working back from the output as was done in Examples 3-22 and 3-23.
    (b) Assume that it is the LOW state of Z that is to activate the alarm. Change the circuit diagram to reflect this, and then use the revised diagram to determine the input conditions needed to activate the alarm.

3-36. Connect an INVERTER between A_1 and the 74C30. Use the alternate representation.

3-36. Modify the circuit of Figure 3-40 so that $A_1 = 0$ is needed to produce $CASS = 1$ instead of $A_1 = 1$.

3-37. X will go HIGH when $E = 1$, or $D = 0$, or $B = C = 0$, or when $B = 1$ and $A = 0$.

3-37. Determine the input conditions needed to cause the output in Figure 3-49 to go to its active state.

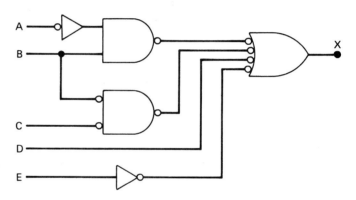

FIGURE 3-49

3-38. X is LOW for $EDCBA = $ 01011, 01100, 01101, and 01111.

3-38. Use the results of Problem 3-37 to obtain the complete truth table for the circuit of Figure 3-49.

FIGURE 3-50

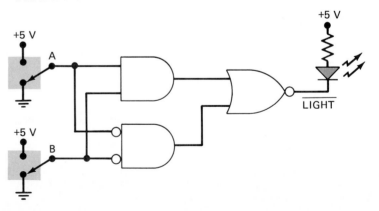

3-39. What is the asserted state for the output of Figure 3-49? For the output of Figure 3-36(c)? (a) HIGH (b) LOW

3-40. Figure 3-50 shows an application of logic gates that simulates a two-way switch like the ones used in our homes to turn a light on or off from two different switches. Here the light is an LED which will be ON (conducting) when the NOR gate output is LOW. Note that this output is labeled $\overline{LIGHT}$ to indicate that it is active-LOW. Determine the input conditions needed to turn on the LED. Then verify that the circuit operates as a two-way switch using switches A and B. In Chapter 4 you will learn how to design circuits like this one to produce a given relationship between inputs and outputs.

SECTION 3-15

3-41. Redraw the circuit of Figure 3-49 using the IEEE/ANSI symbology.

3-42. Determine the Boolean expression for the output Z in Figure 3-51.

***3-43.** The output of the circuit of Figure 3-51 is supposed to be active-LOW. Redraw it to more effectively represent the circuit operation.

3-44. Use the redrawn version of the circuit of Figure 3-51 and do each of the following:

(a) Determine the various input conditions that will produce the active-LOW output state. Do this using only the circuit diagram without writing the expression for Z and without generating the complete truth table. The results should be

A	B	C	D	E	F
1	1	1	1	1	- 1
1	1	1	1	0	1
1	1	1	0	1	1

(b) Verify that the simplified expression for output Z is given by:
$$Z = \overline{ABCF\,(D + E)}$$

3-40. $\overline{LIGHT} = 0$ when $A = B = 0$ or $A = B = 1$.

3-41. Substitute the appropriate ANSI symbols.

3-42. $Z = \overline{(\overline{AB} + \overline{C})F + \overline{C}} + \overline{(\overline{D + E})}$

3-43. Change output OR gate to its alternate representation. Change upper NOR gate to its alternate representation.

3-44. Everything checks out as expected.

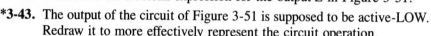

FIGURE 3-51

*The light bulb to the left indicates a more challenging problem throughout the text.

(c) Plug each of the sets of conditions from (a) into the expression you determined in (b), and verify that each one produces $Z = 0$.

MICROCOMPUTER APPLICATION

 3-45. Refer to Example 3-23. The inputs A_7 through A_0 are *address* inputs that are supplied to this circuit from outputs of the microprocessor chip in a microcomputer. The 8-bit address code A_7 to A_0 selects which device the microprocessor wants to activate. In Example 3-23, the required address code to activate the cassette recorder was A_7 through $A_0 = 11111110_2 = FE_{16}$.

Modify the circuit so that the microprocessor has to supply an address code of $4A_{16}$ to activate the recorder.

3-45. Put INVERTERs on the A_7, A_5, A_4, and A_2 inputs to the 74C30.

ANSWERS TO SECTION REVIEW QUESTIONS

SECTION 3-3

1. All inputs LOW.
2. $x = A + B + C + D + E + F$.
3. Constant HIGH.

SECTION 3-4

1. All five inputs $= 1$.
2. A LOW input will keep output LOW.
3. False. See truth table of each gate.

SECTION 3-5

1. Output of second INVERTER will be same as input A.   **2.** y will be LOW only for $A = B = 1$.

SECTION 3-6

1. $x = \overline{A} + B + C + \overline{AD}$

SECTION 3-7

1. $x = 1$ **2.** $x = 1$

SECTION 3-8

1. See Figure 3-15(a).

SECTION 3-9

1. All inputs LOW.   **2.** AND gate.
3. $x = A + B + \overline{CD}$.

SECTION 3-10

1. $y = A\overline{C}$ **2.** $y = \overline{A}\,\overline{B}\,\overline{D}$

SECTION 3-11

1. $z = \overline{A}\,\overline{B} + C$ **2.** $y = (\overline{R} + S + \overline{T})Q$
3. Same as Figure 3-28 except NAND is replaced by NOR.

SECTION 3-12

1. Three   **2.** NOR circuit is more efficient because it can be implemented with one 7402 IC.

SECTION 3-13

1. Output goes LOW when any input is HIGH.   **2.** Output goes HIGH only when all inputs are LOW.   **3.** Output goes LOW when any input is LOW.   **4.** Output goes HIGH only when all inputs are HIGH.

SECTION 3-14

1. Z will go HIGH when $A = B = 0$ and $C = D = 1$.   **2.** Z will go LOW when $A = B = 0$, $E = 1$, and either C or D or both are 0.   **3.** Two.   **4.** Two.   **5.** LOW.
6. $A = B = 0$, $C = D = 1$.

SECTION 3-15

1. The IEEE symbols with their dependency notation specify the complete operation of the logic device.   **2.** See Figure 3-41.   **3.** See Figure 3-44.

4

COMBINATORIAL LOGIC CIRCUITS

Outline

Objectives

Upon completion of this chapter, you will be able to:

- Convert a logic expression into a sum-of-products expression.
- Perform the necessary steps to derive a sum-of-products expression in order to design a combinatorial logic circuit in its simplest form.
- Use the Karnaugh map as a tool to simplify and design logic circuits.
- Explain the operation of both exclusive-OR and exclusive-NOR circuits.
- Design logic circuits with and without the help of a truth table.
- Identify and understand inhibit circuits.
- Cite the basic characteristics of digital ICs.
- Understand the inherent operative differences between TTL and CMOS.
- Use the basic troubleshooting rules of digital systems.
- Deduce from measured results the faults of malfunctioning combinatorial logic circuits.

Introduction

In Chapter 3 we studied the operation of all the basic logic gates, and we used Boolean algebra to describe and analyze circuits that were made up of combinations of logic gates. These circuits can be classified as *combinatorial* logic circuits because, at any time, the logic level at the output depends on the combination of logic levels present at the inputs. A combinatorial circuit has no *memory* characteristic, and so its output depends *only* on the current value of its inputs.

In this chapter we will continue our study of combinatorial logic circuits. To start, we will go further into the simplification of logic circuits. Two methods will be used: one will use Boolean algebra theorems, the other a *mapping* technique. In addition, we will study simple techniques for designing logic circuits to satisfy a given set of requirements. A complete study of logic-circuit design is not one of our objectives, but the methods we introduce are more than sufficient for the types of design situations that a technician will encounter.

The last portion of the chapter is devoted to the troubleshooting of combinatorial circuits. This first exposure to troubleshooting should begin to develop the type of analytical skills needed for sucessful troubleshooting. To make this material as practical as possible, we will first present some of the basic characteristics of logic-gate ICs in the TTL and CMOS logic families along with a description of the most common types of faults encountered in digital IC circuits.

4-1 SUM-OF-PRODUCTS FORM

The methods of logic-circuit simplification and design that we will study require the logic expression to be in a *sum-of-products* form. Some examples of this form are

1. $ABC + \overline{A}B\overline{C}$
2. $AB + \overline{A}B\overline{C} + \overline{CD} + D$
3. $\overline{A}B + C\overline{D} + EF + GK + H\overline{L}$

Each of these sum-of-products expressions consists of two or more AND terms (products) that are ORed together. Each AND term consists of one or more variables appearing in either complemented or uncomplemented form. For example, in the sum-of-products expression $ABC + \overline{A}B\overline{C}$, the first AND product contains the variables A, B, and C in their uncomplemented (not inverted) form. The second AND term contains A and C in their complemented (inverted) form. Note that in a sum-of-products expression, one inversion sign *cannot* cover more than one variable in a term (e.g., we cannot have $\overline{ABC}$ or $\overline{RST}$).

4-2 SIMPLIFYING LOGIC CIRCUITS

Once the expression for a logic circuit has been obtained, we may be able to reduce it to a simpler form containing fewer terms or fewer variables in one or more terms. The new expression can then be used to implement a circuit that is equivalent to the original circuit but that contains fewer gates and connections.

To illustrate, the circuit of Figure 4-1(a) can be simplified to produce the cir-

FIGURE 4-1 It is often possible to simplify a logic circuit such as that in part (a) to produce a more efficient implementation (b).

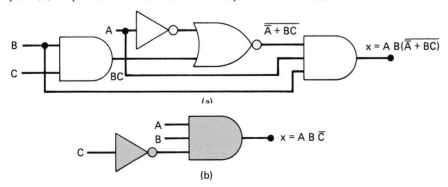

cuit of Figure 4-1(b). Since both circuits perform the same logic, it should be obvious that the simpler circuit is more desirable because it contains fewer gates and will therefore be smaller and cheaper than the original. Furthermore, the circuit reliability will improve because there are fewer interconnections that can be potential circuit faults.

In subsequent sections we will study two methods for simplifying logic circuits. One method will utilize the Boolean algebra theorems and, as we shall see, is greatly dependent on inspiration and experience. The other method (Karnaugh mapping) is a systematic, cookbook approach. Some instructors may wish to skip over this latter method because it is somewhat mechanical and probably does not contribute to a better understanding of Boolean algebra. This can be done without affecting the continuity or clarity of the rest of the text.

4-3 ALGEBRAIC SIMPLIFICATION

The Boolean algebra theorems that we studied in Chapter 3 can be used to help us simplify the expression for a logic circuit. Unfortunately, it is not always obvious which theorems should be applied in order to produce the simplest result. Furthermore, there is no easy way to tell whether the simplified expression is in its simplest form or whether it could have been simplified further. Thus, algebraic simplification often becomes a process of trial and error. With experience, however, one can become adept at obtaining reasonably good results.

The examples that follow will illustrate many of the ways in which the Boolean theorems can be applied in trying to simplify an expression. You should notice that these examples contain two essential steps:

1. The original expression is put into the sum-of-products form by repeated application of DeMorgan's theorems and multiplication of terms.

2. Once it is in this form, the product terms are checked for common factors, and factoring is performed wherever possible. Hopefully, the factoring results in the elimination of one or more terms.

Teaching Tip: Some students will want to factor the *A* variable from each term. Show them that this will lead to the same final result but will take more steps. Stress that there may be several possible ways to start the factoring. With experience, one can become adept at choosing the initial step that leads to the fewest total steps. The important thing is that proper application of the Boolean theorems will always lead to a simplified expression, regardless of the initial step.

___ EXAMPLE 4-1 ___

Simplify the logic circuit shown in Figure 4-2(a).

Solution   The first step is to determine the expression for the output. The result is

$$z = ABC + A\bar{B}\cdot(\overline{\overline{A}\,\overline{C}})$$

Once the expression is determined, it is usually a good idea to break down all large inverter signs using DeMorgan's theorems and then multiply out all terms.

$$\begin{aligned}
z &= ABC + A\bar{B}(\overline{\overline{A}} + \overline{\overline{C}}) &&\text{[theorem (17)]}\\
&= ABC + A\bar{B}(A + C) &&\text{[cancel double inversions]}\\
&= ABC + A\bar{B}A + A\bar{B}C &&\text{[multiply out]}\\
&= ABC + A\bar{B} + A\bar{B}C &&\text{[}A\cdot A = A\text{]}
\end{aligned}$$

With the expression in sum-of-products form, we should look for common variables among the various terms with the intention of factoring. The first and third terms above have *AC* in common, which can be factored out:

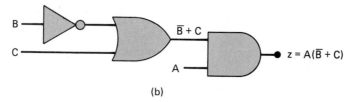

FIGURE 4-2   Example 4-1.

$$z = AC(B + \bar{B}) + A\bar{B}$$

Since $B + \bar{B} = 1$, then

$$z = AC(1) + A\bar{B}$$
$$= AC + A\bar{B}$$

We can now factor out A, which results in

$$z = A(C + \bar{B})$$

This result can be simplified no further. Its circuit implementation is shown in Figure 4-2(b). It is obvious that the circuit in (b) is a great deal simpler than the original circuit in (a).

EXAMPLE 4-2

Simplify the expression $z = ABC + AB\bar{C} + A\bar{B}C$.

Solution   We will look at two different ways to arrive at the same result.

Method 1: The first two terms in the expression have the variables AB in common. Thus,

$$z = AB(C + \bar{C}) + A\bar{B}C$$
$$= AB(1) + A\bar{B}C$$
$$= AB + A\bar{B}C$$

We can factor the variable A from both terms:

$$z = A(B + \bar{B}C)$$

Invoking theorem (15),

$$z = A(B + C)$$

Method 2: The original expression is $z = ABC + AB\bar{C} + A\bar{B}C$. The first two terms have the variables AB in common. The first and last terms have the variables AC in common. How do we know whether to factor AB from the first two terms or AC from the two end terms? Actually, we can do both by using the ABC term *twice*. In other words, we can rewrite the expression as

$$z = ABC + AB\bar{C} + A\bar{B}C + ABC$$

where we have added an extra term ABC. This is valid and will not change the value of the expression, since $ABC + ABC = ABC$ [theorem (7)]. Now we can factor AB from the first two terms and AC from the last two terms:

$$z = AB(C + \overline{C}) + AC(\overline{B} + B)$$
$$= AB \cdot 1 + AC \cdot 1$$
$$= AB + AC = A(B + C)$$

This is, of course, the same result as method 1. This trick of using the same term twice can always be used. In fact, the same term can be used more than twice if necessary.

▬ EXAMPLE 4-3 ▬

Simplify $z = \overline{A}C(\overline{ABD}) + \overline{ABCD} + A\overline{BC}$.

Solution   First, use DeMorgan's theorem on the first term:
$$z = \overline{A}C(A + \overline{B} + \overline{D}) + \overline{ABCD} + A\overline{BC} \qquad \text{(step 1)}$$

Multiplying out,
$$z = \overline{A}CA + \overline{A}C\overline{B} + \overline{A}C\overline{D} + \overline{ABCD} + A\overline{BC} \qquad (2)$$

Since $\overline{A} \cdot A = 0$, the first term is eliminated:
$$z = \overline{AB}C + \overline{A}C\overline{D} + \overline{ABCD} + A\overline{BC} \qquad (3)$$

This is the desired sum-of-products form. Now we have to look for common factors among the various product terms. The idea is to check for the largest common factor between any two or more product terms. For example, the first and last terms have the common factor $\overline{BC}$, and the second and third terms share the common factor $\overline{AD}$. We can factor these out as follows:
$$z = \overline{B}C(\overline{A} + A) + \overline{AD}(C + B\overline{C}) \qquad (4)$$

Now, since $\overline{A} + A = 1$, and $C + B\overline{C} = C + B$ [theorem (15)], we have
$$z = \overline{B}C + \overline{AD}(B + C) \qquad (5)$$

5 This same result could have been reached with other choices for the factoring. For example, we could have factored C from the first, second, and fourth product terms in step 3 to obtain
$$z = C(\overline{AB} + \overline{AD} + A\overline{B}) + \overline{ABCD}$$

The expression inside the parentheses can be factored further:
$$z = C(\overline{B}[\overline{A} + A] + \overline{AD}) + \overline{ABCD}$$

Since $\overline{A} + A = 1$, this becomes
$$z = C(\overline{B} + \overline{AD}) + \overline{ABCD}$$

Multiplying out yields
$$z = \overline{B}C + \overline{A}C\overline{D} + \overline{ABCD}$$

Now we can factor $\overline{AD}$ from the second and third terms to get
$$z = \overline{B}C + \overline{AD}(C + B\overline{C})$$

Using theorem (15) the expression in parentheses becomes $B + C$. Thus we finally have
$$z = \overline{B}C + \overline{AD}(B + C)$$

This is the same result as we obtained earlier, but it took us many more steps. This illustrates why you should look for the largest common factors: it will generally lead to the final expression in the fewest steps.

EXAMPLE 4-4

Simplify the expression $x = (\overline{A} + B)(A + B + D)\overline{D}$.

Solution   The expression can be put into sum-of-products form by multiplying out all the terms. The result is

$$x = \overline{A}A\overline{D} + \overline{A}B\overline{D} + \overline{A}D\overline{D} + BA\overline{D} + BB\overline{D} + BD\overline{D}$$

The first term can be eliminated, since $\overline{A}A = 0$. Likewise, the third and sixth terms can be eliminated, since $D\overline{D} = 0$. The fifth term can be simplified to $B\overline{D}$, since $BB = B$. This gives us

$$x = \overline{A}B\overline{D} + AB\overline{D} + B\overline{D}$$

We can factor $B\overline{D}$ from each term to obtain

$$x = B\overline{D}(\overline{A} + A + 1)$$

Clearly, the term inside the parentheses is always 1, and so we finally have

$$x = B\overline{D}$$

Teaching Tip: Some students may wonder why the *A* variable does not appear in the final expression. It may seem that the original circuit was poorly designed. That may be so, or it may be that the circuit for *x* utilized outputs from other logic circuits that were being used for other purposes. In cases like this, it is a good idea to generate truth tables for the original and final expressions to verify that they are the same.

EXAMPLE 4-5

Simplify the circuit of Figure 4-3(a).

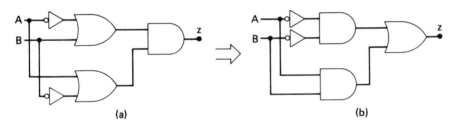

FIGURE 4-3   Example 4-5.

Solution   The expression for output z is

$$z = (\overline{A} + B)(A + \overline{B})$$

Multiplying out to get the sum-of-products form, we obtain

$$z = \overline{A}A + \overline{A}\overline{B} + BA + B\overline{B}$$

We can eliminate $\overline{A}A = 0$ and $B\overline{B} = 0$ to end up with

$$z = \overline{A}\overline{B} + AB$$

This expression is implemented in Figure 4-3(b), and if we compare it with the original circuit, we see that both circuits contain the same number of gates and connections. In this case the simplification process produced an equivalent, but not simpler, circuit.

Common Student Mistake: A common error here is to try to apply theorem 8 ($x + \overline{x} = 1$) and make $AB + \overline{A}\,\overline{B} = 1$.

EXAMPLE 4-6

Simplify $x = A\overline{B}C + \overline{A}BD + \overline{CD}$.

Solution   You can try, but you will not be able to simplify this expression any further.

1. State which of the following expressions are *not* in the sum-of-products form: (a) $RS\overline{T} + \overline{R}S\overline{T} + \overline{T}$, (b) $AD\overline{C} + \overline{A}DC$, (c) $MN\overline{P} + (M + \overline{N})P$, (d) $AB + A\overline{B}\overline{C} + A\overline{B}CD$.
2. Simplify the circuit in Figure 4-1(a) to arrive at the circuit of Figure 4-1(b).
3. Change each AND gate in Figure 4-1(a) to a NAND gate. Determine the new expression for x and simplify it.

4-4 DESIGNING COMBINATORIAL LOGIC CIRCUITS

When the desired output level of a logic circuit is given for all possible input conditions, the results can be conveniently displayed in a truth table. The Boolean expression for the required circuit can then be derived from the truth table. For example, consider Figure 4-4(a), where a truth table is shown for a circuit that has two inputs, A and B, and output x. The table shows that output x is to be at the 1 level *only* for the case where $A = 0$ and $B = 1$. It now remains to determine what logic circuit will produce this desired operation. It should be apparent that one possible solution is that shown in Figure 4-4(b). Here an AND gate is used with inputs $\overline{A}$ and B, so that $x = \overline{A} \cdot B$. Obviously x will be 1 *only if* both inputs to the AND gate are 1, namely $\overline{A} = 1$ (which means $A = 0$) and $B = 1$. For all other values of A and B, the output x has to be 0.

A similar approach can be used for the other input conditions. For instance, if x were to be high only for the $A = 1$, $B = 0$ condition, the resulting circuit would be an AND gate with inputs A and $\overline{B}$. In other words, for any of the four possible input conditions we can generate a high x output by using an AND gate with appropriate inputs. The four different cases are shown in Figure 4-5. Each of the AND gates shown generates an output that is 1 *only* for the one given input condition and is 0 for all other conditions. It should be noted that the AND inputs are inverted or not inverted depending on the values that the variables have for the given condition. If the variable is 0 for the given condition, it is inverted before entering the AND gate.

Let us now consider the case shown in Figure 4-6(a), where we have a truth table which indicates that the output x is to be 1 for two different cases: $A = 0$, $B = 1$ and $A = 1$, $B = 0$. How can this be implemented? We know that the AND term $\overline{A} \cdot B$ will generate a 1 only for the $A = 0$, $B = 1$ condition, and the AND term $A \cdot \overline{B}$ will generate a 1 for the $A = 1$, $B = 0$ condition. Since x has to be

FIGURE 4-4

A	B	x
0	0	0
0	1	1
1	0	0
1	1	0

(a)   (b)

$x = \overline{A}B$

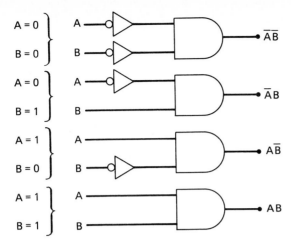

FIGURE 4-5   AND gates used to generate
high outputs for each of the possible input
conditions.

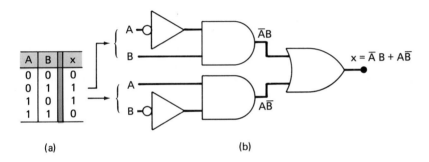

A	B	x
0	0	0
0	1	1
1	0	1
1	1	0

(a)   (b)

FIGURE 4-6   Each set of input conditions that is to produce a
HIGH output is implemented by a separate AND gate. The AND
outputs are ORed to produce final output.

HIGH for *either* condition, it should be clear that these terms should be ORed to-
gether to produce the desired output, x. This implementation is shown in Figure
4-6(b), where the resulting expression for the output is $x = \overline{A}B + A\overline{B}$.

In this example, an AND term is generated for each case in the table where the
output x is to be a 1. The AND-gate outputs are then ORed together to produce the
total output x, which will be 1 when either AND term is 1. This same procedure can
be extended to examples with more than two inputs. Consider the following truth
table for a three-input circuit:

A	B	C	x	
0	0	0	0	
0	0	1	0	
0	1	0	1	$\rightarrow \quad \overline{A}B\overline{C}$
0	1	1	1	$\rightarrow \quad \overline{A}BC$
1	0	0	0	
1	0	1	0	
1	1	0	0	
1	1	1	1	$\rightarrow \quad ABC$

Here there are three cases where the output x is to be 1. The required AND term for each of these cases is shown. Again, note that for each case where a variable is 0, it appears complemented in the AND term. The final expression for x is obtained by ORing the three AND terms. Thus

$$x = \overline{A}B\overline{C} + \overline{A}BC + ABC$$

This expression can be implemented with three AND gates feeding an OR gate.

Discussion Question: In the circuit implementation for x, how many OR-gate inputs can be in the HIGH state at one time?

In all the preceding examples the expression for the output x was derived from the truth table in sum-of-products form. The general procedure for obtaining the output expression from a truth table can be summarized as follows:

1. Write an AND term for each case in the table where the output is to be a 1.
2. Each AND term contains each input variable in either inverted or noninverted form. If the variable is 0 for that particular case in the table, it is inverted in the AND term; otherwise it is not.
3. All the AND terms are then ORed together to produce the final expression for the output.

Complete Design Procedure   Once the output expression has been determined from the truth table in sum-of-products form, it can easily be implemented using AND and OR gates and INVERTERs. There will be one AND gate for each term in the expression and one OR gate, which is fed by the outputs of each AND gate. Usually, however, the expression can be simplified, thereby resulting in a more efficient circuit. The following example illustrates the complete design procedure.

▬ EXAMPLE 4-7

Design a logic circuit that has three inputs, A, B, and C, and whose output will be high only when a majority of the inputs is high.

Solution   The first step is to set up the truth table based on the problem statement. The eight possible input cases are shown in Figure 4-7(a). On the basis of the problem statement, the output x should be a 1 whenever two or more inputs are 1. For all other cases the output should be 0. The next step is to write the AND terms for each case where $x = 1$, as shown in the figure. The expression for x can then be written as

$$x = \overline{A}BC + A\overline{B}C + AB\overline{C} + ABC$$

This expression can be simplified in several ways. Perhaps the quickest way is to realize that the last term ABC has two variables in common with each of the other terms. Thus, we can use the ABC term to factor with each of the other terms. The expression is rewritten with the ABC term occurring three times (recall that this is legal in Boolean algebra):

$$x = \overline{A}BC + ABC + A\overline{B}C + ABC + AB\overline{C} + ABC$$

Factoring the appropriate pairs of terms, we have

$$x = BC(\overline{A} + A) + AC(\overline{B} + B) + AB(\overline{C} + C)$$

Since each term in parentheses is equal to 1, we have

$$x = BC + AC + AB$$

This expression is implemented in Figure 4-7(b). Since the expression is in sum-of-products form, the circuit consists of a group of AND gates working into a single OR gate.

A	B	C	x	
0	0	0	0	
0	0	1	0	
0	1	0	0	
0	1	1	1	$\overline{A}BC$
1	0	0	0	
1	0	1	1	$A\overline{B}C$
1	1	0	1	$AB\overline{C}$
1	1	1	1	ABC

(a)

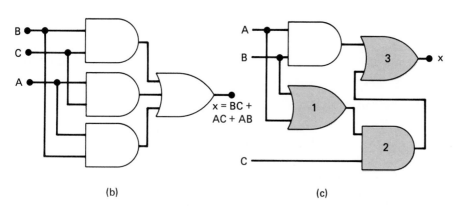

(b)   (c)

FIGURE 4-7   Example 4-7.

This expression can be factored further, but it will not result in the elimination of any terms. It might, however, produce a slightly simpler circuit. For example, let's factor C from the first two terms to obtain

$$x = C(B + A) + AB$$

This expression is implemented in Figure 4-7(c). The resulting circuit requires the same number of gates as the circuit in Figure 4-7(b), but it requires one less connection because the final OR gate has only two inputs. In practice, a logic-circuit designer may still select the implementation in Figure 4-7(b) because of some other considerations, one of which is the fact that the A and B input signals in Figure 4-7(c) have to propagate through three gates (1, 2, and 3) before reaching the output. This might be an important consideration in a high-speed digital system.

___

_____ EXAMPLE 4-8 _____

Refer to Figure 4-8(a), where four logic-signal lines A, B, C, D are being used to represent a 4-bit binary number with A as the MSB and D as the LSB. The binary inputs are fed to a logic circuit that produces a HIGH output only when the binary number is greater than $0110_2 = 6_{10}$. Design this circuit.

Solution   The truth table is shown in Figure 4-8(b). For each case in the truth table we have indicated the decimal equivalent of the binary number represented by the *ABCD* combination.

The output z is set equal to 1 for all those cases where the binary number is greater than 0110. For all other cases, z is set equal to 0. This truth table gives us

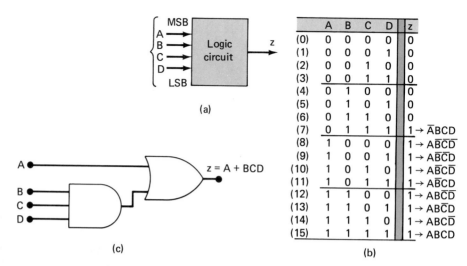

FIGURE 4-8   Example 4-8.

Teaching Tip: Again, some students may be tempted to factor A from the second through ninth terms. Show how this will lead to the same final expression.

the following sum-of-products expression:

$$z = \bar{A}BCD + A\bar{B}\bar{C}\bar{D} + A\bar{B}\bar{C}D + A\bar{B}C\bar{D} + A\bar{B}CD + AB\bar{C}\bar{D} + AB\bar{C}D + ABC\bar{D} + ABCD$$

Simplification of this expression will be a formidable task, but with a little care it can be accomplished. The step-by-step process involves factoring and eliminating terms of the form $A + \bar{A}$:

$$z = \bar{A}BCD + A\bar{B}\bar{C}(\bar{D} + D) + A\bar{B}C(\bar{D} + D) + AB\bar{C}(\bar{D} + D) + ABC(\bar{D} + D)$$

$$= \bar{A}BCD + A\bar{B}\bar{C} + A\bar{B}C + AB\bar{C} + ABC$$

$$= \bar{A}BCD + A\bar{B}(\bar{C} + C) + AB(\bar{C} + C)$$

$$= \bar{A}BCD + A\bar{B} + AB$$

$$= \bar{A}BCD + A(\bar{B} + B)$$

$$= \bar{A}BCD + A$$

This can be reduced further by invoking theorem (15), which says that $x + \bar{x}y = x + y$. In this case $x = A$ and $y = BCD$. Thus,

$$z = \bar{A}BCD + A = BCD + A$$

This final expression is implemented in Figure 4-8(c).

As this example demonstrates, the algebraic simplification method can be quite lengthy when the original expression contains a large number of terms. This is a limitation that is not shared by the Karnaugh mapping method, as we will see later.

Discussion Question: Now that we have the final expression, it may be instructive to ask the students to look at the truth table to see if the expression could have been obtained by inspection. Tell them to look for easily identifiable patterns such as a 1 output any time a variable or combination of variables is in a specific state. Stress that the ability to come up with an expression by inspection may develop as one gets more experience with logic, but is something that not everyone will be good at. Of course, the step-by-step process can always be relied on to obtain the simplified expression.

Implementing the Final Design

In the design examples we have shown, the final circuit was implemented using AND and OR gates. In fact, the sum-of-products form always yields a circuit containing one or more AND gates driving a single OR gate. One of the reasons for using the sum-of-products form is that it can be implemented using all NAND gates with little, if any, increase in circuit complexity over the AND/OR implementation. Since NAND gates are the most readily available logic gates in the TTL logic family, this is an important characteristic.

To illustrate, Figure 4-9 shows the equivalent NAND implementation for the circuits of Figures 4-7(b) and 4-8(c). You may wish to work out these conversions yourself as a review of the procedure covered in Chapter 3.

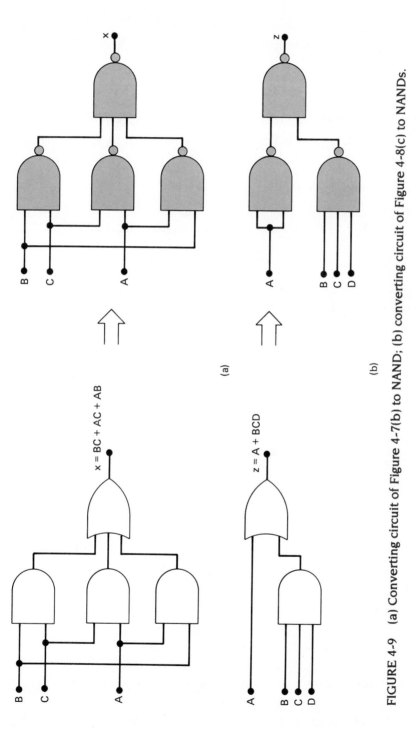

FIGURE 4-9 (a) Converting circuit of Figure 4-7(b) to NAND; (b) converting circuit of Figure 4-8(c) to NANDs.

Comparing the NAND implementation with the circuit in Figure 4-9(a), we see that they are identical in structure; that is, *each* gate of the original circuit has been replaced by a single NAND gate. This characteristic is true only if the original circuit is in sum-of-products form. The only exception to this is when the sum-of-products form contains a single-variable term such as $z = A + BCD$ in Figure 4-9(b). Here the NAND implementation requires an extra NAND gate used as an INVERTER on the A input.

We can streamline the process of converting a *sum-of-products* circuit from AND/OR to NAND gates as follows:

1. Replace each AND gate, OR gate, and INVERTER by a *single* NAND gate.
2. Use a NAND gate to invert any single variable that is feeding the final OR gate.

Check out this process for the circuits in Figure 4-9.

_____ EXAMPLE 4-9A _____

Design a logic circuit with inputs P, Q, R so that output S is HIGH whenever P is 0 or whenever $Q = R = 1$.

Solution   The truth table is shown below. There are five different input conditions that are to produce a HIGH output.

P	Q	R	S	
0	0	0	1	$\overline{P}\overline{Q}\overline{R}$
0	0	1	1	$\overline{P}\overline{Q}R$
0	1	0	1	$\overline{P}Q\overline{R}$
0	1	1	1	$\overline{P}QR$
1	0	0	0	
1	0	1	0	
1	1	0	0	
1	1	1	1	PQR

The sum-of-products expression becomes
$$S = \overline{P}\overline{Q}\overline{R} + \overline{P}\overline{Q}R + \overline{P}Q\overline{R} + \overline{P}QR + PQR$$
We can begin the simplification process by factoring out $\overline{P}\overline{Q}$ from terms 1 and 2, and factoring out $\overline{P}Q$ from terms 3 and 4:
$$S = \overline{P}\overline{Q}(\overline{R} + R) + \overline{P}Q(\overline{R} + R) + PQR$$
Now we can eliminate the $\overline{R} + R$ terms, since they equal 1:
$$S = \overline{P}\overline{Q} + \overline{P}Q + PQR$$
Factoring $\overline{P}$ from terms 1 and 2 allows us to eliminate Q from these terms:
$$S = \overline{P} + PQR$$
Here we can apply theorem (15) $(x + \overline{x}y = x + y)$ to obtain
$$S = \overline{P} + QR$$
The AND/OR implementation for this circuit is shown in Figure 4-10(a).

FIGURE 4-10 (a) AND/OR implementation for Example 4-9A; (b) NAND implementation of same circuit; (c) simplification achieved by eliminating double INVERTERs.

___ EXAMPLE 4-9B ___

Implement the design of Example 4-9A using all NAND gates.

Solution   Replace each OR and AND gate in Figure 4-10(a) with a NAND gate, and replace the INVERTER by a NAND gate INVERTER (labeled 1). In addition, since the top input to the OR gate is a single variable ($\overline{P}$), a NAND gate INVERTER (labeled 2) has to be placed at that input. The resulting circuit is shown in Figure 4-10(b). Clearly, we can eliminate INVERTERS 1 and 2 (since they form a double inversion of input P) to produce the final circuit of Figure 4-10(c).

Teaching Tip: Some students may be confused as to why the NAND INVERTER 2 is added to the circuit when there is already an INVERTER on the P input. Stress that INVERTER 2 is part of the process of converting the circuit to NAND gates and has nothing to do with the fact that P is inverted in the original circuit.

___ REVIEW QUESTIONS ___

1. Write the sum-of-products expression for a circuit with four inputs and an ouput that is to be HIGH only when input A is LOW at the same time that exactly two other inputs are LOW.
2. Implement the expression of question 1 using all NAND gates. How many are required?

4-5   KARNAUGH MAP METHOD*

 6

The Karnaugh map is a graphical device used to simplify a logic equation or to convert a truth table to its corresponding logic circuit in a simple, orderly process. Although a Karnaugh map (henceforth abbreviated *K map*) can be used for problems involving any number of input variables, its practical usefulness is limited to six

___

*This topic may be omitted without affecting the continuity of the remainder of the text.

variables. The following discussion will be limited to problems with up to four inputs, since even five- and six-input problems are too involved and are best done by a computer program.

Karnaugh Map Format   The K map, like a truth table, is a means for showing the relationship between logic inputs and the desired output. Figure 4-11 shows three examples of K maps for two, three, and four variables, together with the corresponding truth tables. These examples illustrate the following important points:

1. The truth table gives the value of output X for each combination of input values. The K map gives the same information in a different format. Each case in the truth table corresponds to a square in the K map. For example, in Figure 4-11(a), the $A = 0$, $B = 0$ condition in the truth table corresponds to the $\overline{AB}$ square in the K map. Since the truth table shows $X = 1$ for this case, a 1 is placed in the $\overline{AB}$ square in the K map. Similarly, the $A = 1$, $B = 1$ condition

FIGURE 4-11   Karnaugh maps and truth tables for (a) two, (b) three, and (c) four variables.

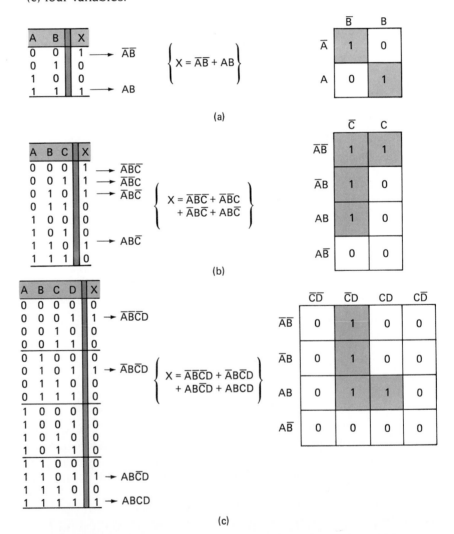

in the truth table corresponds to the AB square of the K map. Since $X = 1$ for this case, a 1 is placed in the AB square. All other squares are filled with 0s. This same idea is used in the three- and four-variable maps shown in the figure.

2. The K-map squares are labeled so that horizontally adjacent squares differ only in one variable. For example, the upper left-hand square in the four-variable map is $\overline{ABCD}$, while the square immediately to its right is $\overline{ABC}D$ (only the D variable is different). Similarly, vertically adjacent squares differ only in one variable. For example, the upper left-hand square is $\overline{ABCD}$ while the square directly below it is $\overline{A}B\overline{CD}$ (only the B variable is different).

   Note that each square in the top row is considered to be adjacent to a corresponding square in the bottom row. For example, the $\overline{ABCD}$ square in the top row is adjacent to the $A\overline{BCD}$ square in the bottom row, since they differ only in the A variable. You can think of the top of the map as being wrapped around to touch the bottom of the map. Likewise, squares in the leftmost column are adjacent to corresponding squares in the rightmost column.

3. In order for vertically and horizontally adjacent squares to differ in only one variable, the top-to-bottom labeling must be done in the order shown—$\overline{AB}$, $\overline{A}B$, AB, $A\overline{B}$. The same is true of the left-to-right labeling.

4. Once a K map has been filled with 0s and 1s, the sum-of-products expression for the output X can be obtained by ORing together those squares that contain a 1. In the three-variable map of Figure 4-11(b), the $\overline{AB}\overline{C}$, $\overline{A}BC$, $\overline{A}B\overline{C}$, and $AB\overline{C}$ squares contain a 1, so that $X = \overline{AB}\overline{C} + \overline{A}BC + \overline{A}B\overline{C} + AB\overline{C}$.

Looping   The expression for output X can be simplified by properly combining those squares in the K map which contain 1s. The process for combining these 1s is called *looping*.

Looping Groups of Two (Pairs)   Figure 4-12(a) is the K map for a particular three-variable truth table. This map contains a pair of 1s that are vertically adjacent to each other; the first represents $\overline{A}B\overline{C}$ and the second represents $AB\overline{C}$. Note that in these two terms only the A variable appears in both normal and complemented form (B and $\overline{C}$ remain unchanged). These two terms can be looped (combined) to give a resultant that eliminates the A variable since it appears in both uncomplemented and complemented forms. This is easily proved as follows:

$$X = \overline{A}B\overline{C} + AB\overline{C}$$
$$= B\overline{C}(\overline{A} + A)$$
$$= B\overline{C}(1) = B\overline{C}$$

This same principle holds true for any pair of vertically or horizontally adjacent 1s. Figure 4-12(b) shows an example of two horizontally adjacent 1s. These two can be looped and the C variable eliminated since it appears in both its uncomplemented and complemented forms to give a resultant of $X = \overline{A}B$.

Another example is shown in Figure 4-12(c). In a K map the top row and bottom row of squares are considered to be adjacent. Thus, the two 1s in this map can be looped to provide a resultant of $\overline{ABC} + A\overline{B}C = \overline{B}C$.

Figure 4-12(d) shows a K map that has two pairs of 1s which can be looped. The two 1s in the top row are horizontally adjacent. The two 1s in the bottom row are also adjacent, since in a K map the leftmost column and rightmost column of squares are considered to be adjacent. When the top pair of 1s is looped, the D vari-

Discussion Question: Why doesn't the looping of two diagonally adjacent 1s result in the elimination of a variable?

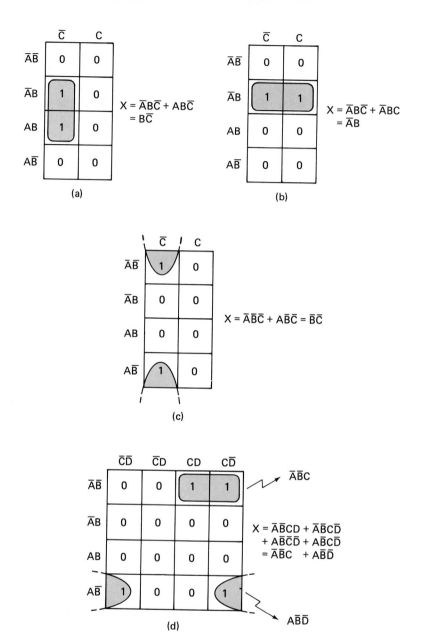

FIGURE 4-12   Examples of looping pairs of adjacent 1s.

able is eliminated (since it appears as both D and $\overline{D}$) to give the term $\overline{A}\overline{B}C$. Looping the bottom pair eliminates the C variable to give the term $A\overline{B}\overline{D}$. These two terms are ORed to give the final result for X.

To summarize:

> ***Looping a pair of adjacent 1s in a K map eliminates the variable that appears in complemented and uncomplemented form.***

Looping Groups of Four (Quads)   A K map may contain a group of four 1s that are adjacent to each other. This group is called a *quad*. Figure 4-13 shows

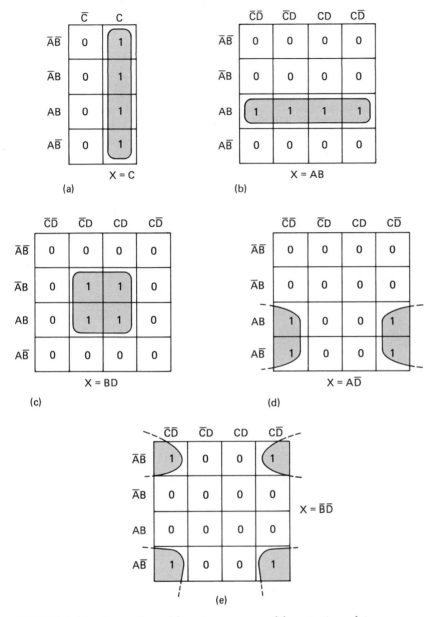

FIGURE 4-13 Examples of looping groups of four 1s (quads).

several examples of quads. In part (a) the four 1s are vertically adjacent and in part (b) they are horizontally adjacent. The K map in Figure 4-13(c) contains four 1s in a square, and they are considered adjacent to each other. The four 1s in Figure 4-13(d) are also adjacent, as are those in Figure 4-13(e) because, as pointed out earlier, the top and bottom rows and the leftmost and rightmost columns are considered to be adjacent to each other.

When a quad is looped, the resultant term will contain only the variables that do not change form for all the squares in the quad. For example, in Figure 4-13(a) the four squares which contain a 1 are $\overline{A}\overline{B}C$, $\overline{A}BC$, ABC, and $A\overline{B}C$. Examination of these terms reveals that only the variable C remains unchanged (both A and B appear in complemented and uncomplemented form). Thus, the resultant expression for X is

simply $X = C$. This can be proved as follows:

$$X = \overline{A}\overline{B}C + \overline{A}BC + ABC + A\overline{B}C$$
$$= \overline{A}C(\overline{B} + B) + AC(B + \overline{B})$$
$$= \overline{A}C + AC$$
$$= C(\overline{A} + A) = C$$

As another example, consider Figure 4-13(d), where the four squares containing 1s are $AB\overline{C}\overline{D}$, $A\overline{B}C\overline{D}$, $ABC\overline{D}$, and $A\overline{B}C\overline{D}$. Examination of these terms indicates that only the variables A and $\overline{D}$ remain unchanged, so that the simplified expression for X is

$$X = A\overline{D}$$

This can be proved in the same manner that was used above.

The reader should check each of the other cases in Figure 4-13 to verify the indicated expressions for X. To summarize:

> ***Looping a quad of 1s eliminates the two variables that appear in both complemented and uncomplemented form.***

Looping Groups of Eight (Octets)   A group of eight 1s that are adjacent to one another is called an *octet*. Several examples of octets are shown in Figure 4-14. When an octet is looped in a four-variable map, three of the four variables are eliminated because only one variable remains unchanged. For example, examination of

FIGURE 4-14   Examples of looping groups of eight 1s (octets).

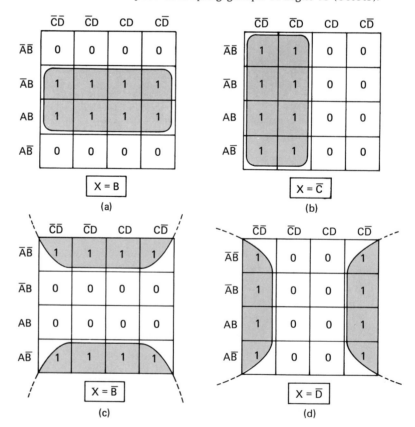

the eight looped squares in Figure 4-14(a) shows that only the variable B is in the same form for all eight squares; the other variables appear in complemented and uncomplemented form. Thus, for this map, $X = B$. The reader can verify the results for the other examples in Figure 4-14.

To summarize:

Looping an octet of 1s eliminates the three variables that appear in both complemented and uncomplemented form.

Complete Simplification Process   We have seen how looping of pairs, quads, and octets on a K map can be used to obtain a simplified expression. We can summarize the rule for loops of *any* size: *When a variable appears in both complemented and uncomplemented form within a loop, that variable is eliminated from the expression. Variables that are the same for all squares of the loop must appear in the final expression.*

It should be clear that a larger loop of 1s eliminates more variables. To be exact, a loop of two eliminates one variable, a loop of four eliminates two, and a loop of eight eliminates three. This principle will now be used to obtain a simplified logic expression from a K map that contains any combination of 1s and 0s.

The procedure will first be outlined and then applied to several examples. The steps below are followed in using the K-map method for simplifying a Boolean expression:

1. Construct the K map and place 1s in those squares corresponding to the 1s in the truth table. Place 0s in the other squares.
2. Examine the map for adjacent 1s and loop those 1s which are *not* adjacent to any other 1s. These are called *isolated* 1s.
3. Next, look for those 1s which are adjacent to only one other 1. Loop *any* pair containing such a 1.
4. Loop any octet even if some of the 1s have already been looped.
5. Loop any quad that contains one or more 1s which have not already been looped.
6. Loop any pairs necessary to include any 1s that have not yet been looped, making sure to use the minimum number of loops.
7. Form the OR sum of all the terms generated by each loop.

These steps will be followed exactly and referred to in the following examples. In each case, the resulting logic expression will be in its simplest sum-of-products form.

━━ EXAMPLE 4-10 ━━━━━━━━━━━━━━━━━━━━━━━━━━━━━━

Figure 4-15(a) shows the K map for a four-variable problem. We will assume that the map was obtained from the problem truth table (step 1). The squares are numbered for convenience in identifying each loop.

Step 2. Square 4 is the only square containing a 1 that is not adjacent to any other 1. It is looped and is referred to as loop 4.

Step 3. Square 15 is adjacent *only* to square 11. This pair is looped and referred to as loop 11, 15.

Step 4. There are no octets.

(a)

(b)

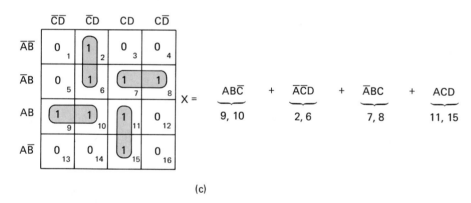

(c)

FIGURE 4-15 Examples 4-10 to 4-12.

Step 5. Squares 6, 7, 10, and 11 form a quad. This quad is looped (loop 6, 7, 10, 11). Note that square 11 is used again, even though it was part of loop 11, 15.

Step 6. All 1s have already been looped.

Step 7. Each loop generates a term in the expression for X. Loop 4 is simply $\overline{A}\,\overline{B}C\overline{D}$. Loop 11, 15 is ACD (the B variable is eliminated). Loop 6, 7, 10, 11 is BD (A and C are eliminated).

— EXAMPLE 4-11 —

Consider the K map in Figure 4-15(b). Once again we can assume that step 1 has already been performed.

Step 2. There are no isolated 1s.

Step 3. The 1 in square 3 is adjacent *only* to the 1 in square 7. Looping this pair (loop 3, 7) produces the term $\overline{A}CD$.

Step 4. There are no octets.

Step 5. There are two quads. Squares 5, 6, 7, and 8 form one quad. Looping this quad produces the term $\overline{A}B$. The second quad is made up of squares 5, 6, 9, and 10. This quad is looped because it contains two squares that have not been looped previously. Looping this quad produces $B\overline{C}$.

Step 6. All 1s have already been looped.

Step 7. The terms generated by the three loops are ORed together to obtain the expression for X.

▬▬ EXAMPLE 4-12 ▬▬▬▬▬▬▬▬▬▬▬▬▬▬▬▬▬▬▬▬▬

Consider the K map in Figure 4-15(c).

Step 2. There are no isolated 1s.

Step 3. The 1 in square 2 is adjacent only to the 1 in square 6. This pair is looped to produce $\overline{A}\,\overline{C}D$. Similarly, square 9 is adjacent only to square 10. Looping this pair produces $A\overline{B}\overline{C}$. Likewise, loop 7, 8 and loop 11,15 produce the terms $\overline{A}BC$ and ACD, respectively.

Teaching Tip: This example illustrates why isolated pairs should be looped before any quads are looped. It may be instructive to show what expression would result if the quad (6, 7, 10, 11) were looped first.

Step 4. There are no octets.

Step 5. There is one quad formed by squares 6, 7, 10, and 11. This quad, however, is *not* looped, because all the 1s in the quad have been included in other loops.

Step 6. All 1s have already been looped.

Step 7. The expression for X is shown in the figure.

▬▬ EXAMPLE 4-13 ▬▬▬▬▬▬▬▬▬▬▬▬▬▬▬▬▬▬▬▬▬

Consider the K map in Figure 4-16(a).

Step 2. There are no isolated 1s.

FIGURE 4-16 The same K map with two equally good solutions.

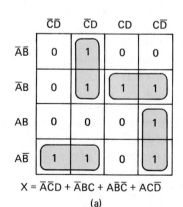

$$X = \overline{A}\,\overline{C}D + \overline{A}BC + A\overline{B}\overline{C} + AC\overline{D}$$

(a)

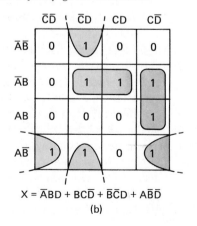

$$X = \overline{A}BD + BC\overline{D} + \overline{B}\overline{C}D + A\overline{B}\overline{D}$$

(b)

Teaching Tip: It may be disconcerting to some students that the two possible loopings in this example produce completely different expressions. This means that a given truth table doesn't necessarily have a unique simplified expression. This is analogous in mathematics to a graph of a function of two variables that can be expressed by two different equations—which we know is not possible. This kind of occurrence is a property of Boolean algebra only.

Step 3. There are no 1s that are adjacent to only one other 1.

Step 4. There are no octets.

Step 5. There are no quads.

Steps 6 and 7. There are many possible pairs. The looping must use the minimum number of loops to account for all the 1s. For this map there are *two* possible loopings, which require only four looped pairs. Figure 4-16(a) shows one solution and its resultant expression. Figure 4-16(b) shows the other. Note that both expressions are of the same complexity, and so neither is better than the other.

___

___ EXAMPLE 4-14 ___

Use the K map to simplify the expression $x = \overline{ABC} + \overline{B}C + \overline{A}B$.

Solution   In this problem we are not given the truth table from which to fill in the K map. Instead, we must fill in the K map by taking each of the product terms in the expression and placing 1s in the corresponding squares.

The first term, $\overline{ABC}$, tells us to enter a 1 in the $\overline{ABC}$ square of the map (see Figure 4-17). The second term, $\overline{B}C$, tells us to enter a 1 in each square that contains

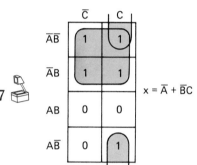

FIGURE 4-17   Example 4-14.

$x = \overline{A} + \overline{B}C$

a $\overline{B}C$ in its label. In Figure 4-17 this would be the $A\overline{B}C$ and $\overline{A}\,\overline{B}C$ squares. Likewise, the $\overline{A}B$ term tells us to place a 1 in the $\overline{A}B\overline{C}$ and $\overline{A}BC$ squares. All other squares will be filled with 0s.

Now the K map can be looped for simplification. The result is $x = \overline{A} + \overline{B}C$, as shown in the figure.

___

"Don't Care" Conditions   Some logic circuits can be designed so that there are certain input conditions for which there are no specified output levels, usually because these input conditions will never occur. In other words, there will be certain combinations of input levels where we "don't care" whether the output is HIGH or LOW. This is illustrated in the truth table of Figure 4-18(a).

Here the output z is not specified as either 0 or 1 for the conditions A, B, $C = 1, 0, 0$ and A, B, $C = 0, 1, 1$. Instead, an x is shown for these conditions. The x represents the "don't care" condition. A "don't care" condition can come about for several reasons, the most common being that in some situations certain input combinations can never occur, and so there is no specified output for these conditions.

A circuit designer is free to make the output for any "don't care" condition either a 0 or a 1 in order to produce the simplest output expression. For example, the K map for this truth table is shown in Figure 4-18(b) with an x placed in the $A\overline{B}\overline{C}$ and

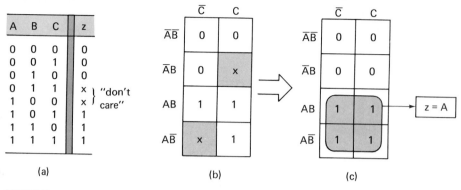

A	B	C	z
0	0	0	0
0	0	1	0
0	1	0	0
0	1	1	x
1	0	0	x
1	0	1	1
1	1	0	1
1	1	1	1

(a)   (b)   (c)

FIGURE 4-18 "Don't care" conditions should be changed to 0 or 1 to produce K-map looping that yields the simplest expression.

$\overline{A}BC$ squares. The designer here would be wise to change the x in the $A\overline{B}\,\overline{C}$ square to a 1 and the x in the $\overline{A}BC$ square to a 0, since this would produce a quad that can be looped to produce $z = A$, as shown in Figure 4-18(c).

Whenever "don't care" conditions occur, we have to decide which ones to change to 0 and which to 1 to produce the best K-map looping (i.e., the simplest expression). This decision is not always an easy one. Several end-of-chapter problems will provide practice in dealing with "don't care" cases.

Summary   The K-map process has several advantages over the algebraic method. K mapping is a more orderly process with well-defined steps as compared with the trial-and-error process sometimes used in algebraic simplification. K mapping usually requires fewer steps, especially for expressions containing many terms, and it always produces a minimum expression.

Nevertheless, some instructors prefer the algebraic method because it requires a thorough knowledge of Boolean algebra and is not simply a mechanical procedure. Each method has its advantages, and though most logic designers are adept at both, being proficient in one method is all that is necessary to produce acceptable results.

___ REVIEW QUESTIONS ___

1. Use K mapping to simplify the expression of Example 4-7.
2. Use K mapping to simplify the expression of Example 4-8. This should emphasize the advantage of K mapping for expressions containing many terms.
3. What is meant by a "don't care" condition?

4-6   EXCLUSIVE-OR AND EXCLUSIVE-NOR CIRCUITS

Two special logic circuits that occur quite often in digital systems are the *exclusive-OR* and *exclusive-NOR* circuits.

Exclusive-OR   Consider the logic circuit of Figure 4-19(a). The output expres-

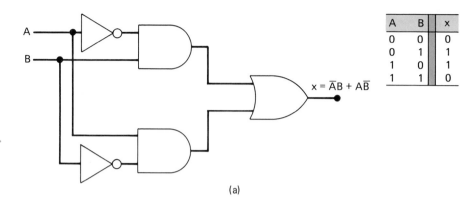

A	B	x
0	0	0
0	1	1
1	0	1
1	1	0

$x = \overline{A}B + A\overline{B}$

(a)

(b)

(c)

FIGURE 4-19 (a) Exclusive-OR circuit and truth table: (b) traditional EX-OR gate symbol; (c) IEEE/ANSI symbol for EX-OR gate.

sion of this circuit is

$$x = \overline{A}B + A\overline{B}$$

The accompanying truth table shows that $x = 1$ for two cases: $A = 0$, $B = 1$ (the $\overline{A}B$ term) and $A = 1$, $B = 0$ (the $A\overline{B}$ term). In other words, **this circuit produces a HIGH output whenever the two inputs are at opposite levels.** This is the exclusive-OR circuit, which will hereafter be abbreviated EX-OR.

This particular combination of logic gates occurs quite often and is very useful in certain applications. In fact, the EX-OR circuit has been given a symbol of its own, shown in Figure 4-19(b). This symbol is assumed to contain all the logic contained in the EX-OR circuit and therefore has the same logic expression and truth table. This EX-OR circuit is commonly referred to as an EX-OR *gate*, and we consider it as another type of logic gate. The new IEEE/ANSI symbol for an EX-OR gate is shown in Figure 4-19(c). It indicates that the output will be active-HIGH *only* when one input *alone* is HIGH.

An EX-OR gate has only *two* inputs; there are no three-input or four-input EX-OR gates. The two inputs are combined so that $x = \overline{A}B + A\overline{B}$. A shorthand way that is sometimes used to indicate the EX-OR output expression is

$$x = A \oplus B$$

where the symbol $\oplus$ represents the EX-OR gate operation.

The characteristics of an EX-OR gate are summarized as follows:

1. It has only two inputs and its output is
$$x = \overline{A}B + A\overline{B} = A \oplus B$$

2. Its output is *HIGH* only when the two inputs are at *different* levels.

Several ICs are available that contain EX-OR gates. Those listed below are *quad* EX-OR chips containing four EX-OR gates.

Teaching Tip: It may be instructive to point out here that an OR gate is often referred to as an *inclusive*-OR gate since it includes the case where both inputs are HIGH, as compared with the exclusive-OR gate, which excludes that case.

- **7486**   Quad EX-OR (TTL family)
- **74C86**   Quad EX-OR (CMOS family)
- **74HC86**   Quad EX-OR (high-speed CMOS)

Exclusive-NOR   The exclusive-NOR circuit (abbreviated EX-NOR) operates completely opposite to the EX-OR circuit. Figure 4-20(a) shows an EX-NOR circuit and its accompanying truth table. The output expression is

$$x = AB + \overline{A}\,\overline{B}$$

which indicates along with the truth table that x will be 1 for two cases: $A = B = 1$ (the AB term) and $A = B = 0$ (the $\overline{A}\,\overline{B}$ term). In other words, **this circuit produces a high output whenever the two inputs are at the same level.**

It should be apparent that the output of the EX-NOR circuit is the exact inverse of the output of the EX-OR circuit. The traditional symbol for an EX-NOR gate is obtained by simply adding a small circle at the output of the EX-OR symbol [Figure 4-20(b)]. The IEEE/ANSI symbol adds the small triangle on the output of the EX-OR symbol. Both symbols indicate an output that goes to its active-LOW state when *only* one input is HIGH.

The EX-NOR gate also has only *two* inputs, and it combines them so that its output is

$$x = AB + \overline{A}\,\overline{B}$$

A shorthand way to indicate the output expression of the EX-NOR is

$$x = \overline{A \oplus B}$$

Discussion Question: Use Boolean algebra to prove that the EX-NOR expression is the inverse of the EX-OR expression.

FIGURE 4-20   (a) Exclusive-NOR circuit; (b) traditional symbol for EX-NOR gate; (c) IEEE/ANSI symbol.

(a)

(b)   (c)

which is simply the inverse of the EX-OR operation. The EX-NOR gate is summarized as follows:

1. It has only two inputs and its output is
$$x = AB + \overline{A}\overline{B} = \overline{A \oplus B}$$

2. Its output is HIGH only when the two inputs are at the *same* level.

Several ICs are available that contain EX-NOR gates. Those listed below are quad EX-NOR chips containing four EX-NOR gates.

- **74LS266**   Quad EX-NOR (TTL family)
- **74C266**   Quad EX-NOR (CMOS)
- **74HC266**   Quad EX-NOR (high-speed CMOS)

Each of these EX-NOR chips, however, has special output circuitry that limits its use to special types of applications. Very often, a logic designer will obtain the EX-NOR function simply by connecting the output of an EX-OR to an INVERTER.

_____ EXAMPLE 4-15 _____

Determine the output waveform for the input waveforms given in Figure 4-21.

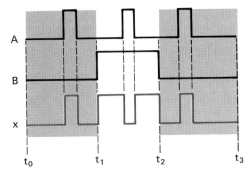

FIGURE 4-21   Example 4-15.

Discussion Question: How can an EX-NOR be used as a controlled INVERTER?

Solution   The output waveform is obtained using the fact that the EX-OR output will go HIGH only when its inputs are at different levels. The resulting output waveform reveals several interesting points:

1. The x waveform matches the A input waveform during those time intervals when $B = 0$. This occurs during the time intervals t_0 to t_1 and t_2 to t_3.

2. The x waveform is the *inverse* of the A input waveform during those time intervals when $B = 1$. This occurs during the interval t_1 to t_2.

3. These observations show that an EX-OR gate can be used as a *controlled INVERTER;* that is, one of its inputs can be used to control whether the signal at the other input will be inverted or not. This property will be useful in certain applications.

_____ EXAMPLE 4-16 _____

$x_1 x_0$ represents a 2-bit binary number that can have any value (00, 01, 10, or 11); for example, when $x_1 = 1$ and $x_0 = 0$, the binary number is 10, and so on. Similarly,

y_1y_0 represents another 2-bit binary number. Design a logic circuit, using x_1, x_0, y_1, and y_0 inputs, whose output will be HIGH only when the two binary numbers x_1x_0 and y_1y_0 are *equal*.

Solution   The first step is to construct a truth table for the 16 input conditions.

x_1	x_0	y_1	y_0	x (OUTPUT)
0	0	0	0	1
0	0	0	1	0
0	0	1	0	0
0	0	1	1	0
0	1	0	0	0
0	1	0	1	1
0	1	1	0	0
0	1	1	1	0
1	0	0	0	0
1	0	0	1	0
1	0	1	0	1
1	0	1	1	0
1	1	0	0	0
1	1	0	1	0
1	1	1	0	0
1	1	1	1	1

The output z has to be high whenever the x_1x_0 values match the y_1y_0 values; that is, whenever $x_1 = y_1$ and $x_0 = y_0$. The table shows that there are four such cases. We could now continue with the normal procedure, which would be to obtain a sum-of-products expression for z, attempt to simplify it, and then implement the result. However, the nature of this problem makes it ideally suited for implementation using EX-NOR gates, and a little thought will produce a simple solution with minimum work. Refer to Figure 4-22; in this logic diagram x_1 and y_1 are fed to one EX-NOR gate and x_0 and y_0 are fed to another EX-NOR gate. The output of each EX-NOR will be HIGH only when its inputs are equal. Thus, for $x_0 = y_0$ and $x_1 = y_1$ both EX-NOR outputs will be high. This is the condition we are looking for, because

Discussion Question: Show how to detect when two 8-bit numbers are the same.

FIGURE 4-22   Circuit for detecting equality of two 2-bit binary numbers.

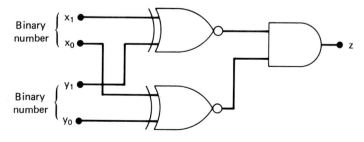

it means that the two 2-bit numbers are equal. The AND gate output will be HIGH only for this case, thereby producing the desired output.

___ EXAMPLE 4-17 ___

When simplifying the expression for the output of a combinatorial logic circuit, you may encounter the EX-OR or EX-NOR operations as you are factoring. This will often lead to the use of EX-OR or EX-NOR gates in the implementation of the final circuit. To illustrate, simplify the circuit of Figure 4-23(a).

Solution   The unsimplified expression for the circuit is obtained as
$$Z = ABCD + A\overline{B}\overline{C}D + \overline{A}\overline{D}$$
We can factor AD from the first two terms:
$$Z = AD(BC + \overline{B}\,\overline{C}) + \overline{A}\overline{D}$$

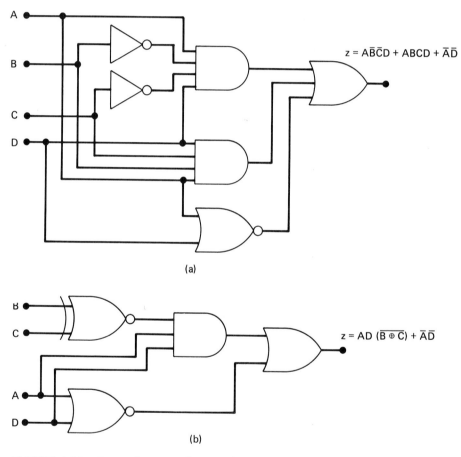

(a)

(b)

FIGURE 4-23   Example 4-17, showing how an EX-NOR gate may be used to simplify circuit implementation.

At first glance you might think that the expression in parentheses can be replaced by 1. But that would be true only if it were $BC + \overline{BC}$. You should recognize the expression in parentheses as the EX-NOR combination of B and C. This fact can be used to reimplement the circuit as shown in Figure 4-23(b). This circuit is much

simpler than the original, since it uses gates with fewer inputs, and two INVERTERs have been eliminated.

REVIEW QUESTIONS

1. Use Boolean algebra to prove that the EX-NOR output expression is the exact inverse of the EX-OR output expression.
2. What is the output of an EX-NOR gate when a logic signal and its exact inverse are connected to its inputs?
3. A logic designer needs an INVERTER, and all that is available is one EX-OR gate from a 7486 chip. Does he need another chip?

4-7 PARITY GENERATOR AND CHECKER

In Chapter 2 we saw that a transmitter can attach a parity bit to a set of data bits before transmitting the data bits to a receiver. We also saw how this allows the receiver to detect any single-bit errors that may have occurred during the transmission. Figure 4-24 shows an example of one type of logic circuitry that is used for parity

FIGURE 4-24 EX-OR gates used to implement parity generator and parity checker for even-parity system.

(a)

(b)

generation and parity checking. This particular example uses a group of 4 bits as the data to be transmitted, and uses an even parity bit. It can readily be adapted to use odd parity and any number of bits.

In Figure 4-24(a), the set of data to be transmitted is applied to the parity-generator circuit, which produces the even-parity bit, P, at its output. This parity bit is transmitted to the receiver along with the original data bits, making a total of 5 bits. In Figure 4-24(b), these 5 bits (data + parity) enter the receiver's parity-checker circuit, which produces an error output, E, that indicates whether or not a single-bit error has occurred.

It should not be too surprising that both these circuits employ EX-OR gates, when we consider that a single EX-OR gate operates in such a way that it produces a 1 output if an odd number of its inputs are 1, and a 0 output if an even number of its inputs are 1.

Teaching Tip: It should be pointed out here that the parity generator/checker circuits of Figure 4-24 can suffer from prohibitive propagation delays because of the many levels of gating. If the total delay is excessive, an alternate circuit design must be used.

━━━ EXAMPLE 4-18 ━━

Determine the parity generator's output for each of the following sets of input data, $D_3 D_2 D_1 D_0$; (a) 0111; (b) 1001; (c) 0000; (d) 0100.

Solution   For each case, apply the data levels to the parity-generator inputs and trace them through each gate to the P output. The results are: (a) 1; (b) 0; (c) 0; and (d) 1. Note that P is a 1 only when the original data contain an odd number of 1s. Thus, the total number of 1s sent to the receiver (data + parity) will be even.

━━━ EXAMPLE 4-19 ━━

Determine the parity checker's output for each of the following sets of data from the transmitter:

	P	D_3	D_2	D_1	D_0
(a)	0	1	0	1	0
(b)	1	1	1	1	0
(c)	1	1	1	1	1
(d)	1	0	0	0	0

Solution   For each case, apply these levels to the parity-checker inputs and trace them through to the E output. The results are (a) 0; (b) 0; (c) 1; (d) 1. Note that a 1 is produced at E only when an *odd* number of 1s appear in the inputs to the parity checker. This indicates that an error has occurred, since even parity is being used.

━━━ EXAMPLE 4-20 ━━

Does the parity-checker circuit have any way of "knowing" which input bit is in error?

Solution   No. The parity checker doesn't know what the state of each input bit should be; it knows only that an even number of 1s should be present. Regardless of which bit is wrong, a single-bit error will change the total number of 1s from even to odd (either by removing a 1 or by adding an extra 1) and cause E to go HIGH.

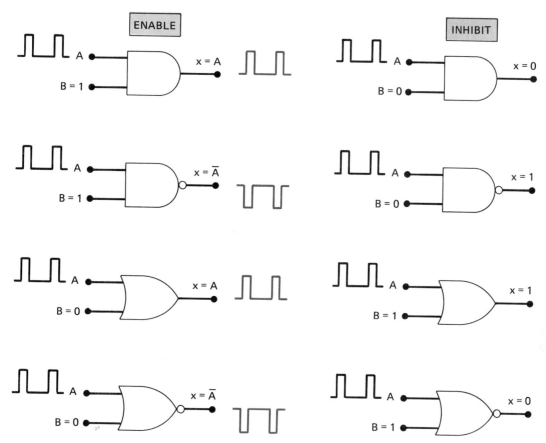

FIGURE 4-25 Four basic gates can either enable or inhibit the passage of an input signal, A, under control of the logic level at control input B.

4-8 INHIBIT CIRCUITS

Each of the basic logic gates can be used to control the passage of an input logic signal through to the output. This is depicted in Figure 4-25, where a logic signal, A, is applied to one input of each of the basic logic gates. The other input of each gate is the control input, B. The logic level at this control input will determine whether the input signal is **enabled** to reach the output, or **inhibited** from reaching the output.

Examine Figure 4-25 and you should notice that when the noninverting gates (AND, OR) are enabled, the output will follow the A signal exactly. Conversely, when the inverting gates (NAND, NOR) are enabled, the output will be the exact inverse of the A signal.

Also notice that AND and NOR gates produce a constant LOW output when they are in the inhibited condition. Conversely, the NAND and OR gates produce a constant HIGH output in the inhibited condition.

There will be many situations in digital-circuit design where the passage of a logic signal is to be enabled or inhibited, depending on conditions present at one or more control inputs. Several are shown in the following examples.

EXAMPLE 4-21

Discussion Question: Modify the inhibit circuit of Example 4-21 so that the output will be HIGH when the signal is inhibited. This will require changing the AND to a NAND, and inverting the *A* signal.

EXAMPLE 4-21

Design a logic circuit that will allow a signal to pass to the output only when control inputs *B* and *C* are both HIGH; otherwise, the output will stay LOW.

Solution   An AND gate should be used because the signal is to be passed without inversion, and the inhibit output condition is a LOW. Since the enable condition has to occur only when $B = C = 1$, a three-input AND gate is used, as shown in Figure 4-26(a).

EXAMPLE 4-22

Design a logic circuit that allows a signal to pass to the output only when one, but not both, of the control inputs are HIGH; otherwise, the output will stay HIGH.

Solution   The result is drawn in Figure 4-26(b). An OR gate is used because we

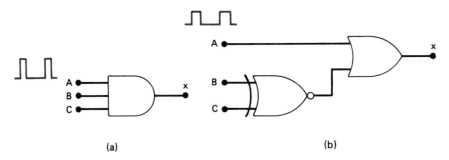

(a)   (b)

FIGURE 4-26   Examples 4-21 and 4-22.

want the output inhibit condition to be a HIGH, and we do not want to invert the signal. Control inputs *B* and *C* are combined in an EX-NOR gate. When *B* and *C* are different, the EX-NOR sends a LOW to enable the OR gate. When *B* and *C* are the same, the EX-NOR sends a HIGH to inhibit the OR gate.

EXAMPLE 4-23

Design a logic circuit with input signal *A*, control input *B*, and outputs *X* and *Y* to operate as follows:

FIGURE 4-27   Example 4-23.

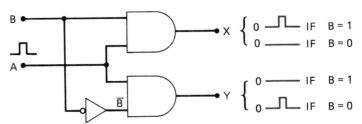

1. When $B = 1$, output X will follow input A, and output Y will be 0.

2. When $B = 0$, output X will be 0, and output Y will follow input A.

Solution   The two outputs will be 0 when inhibited and will follow the input signal when enabled. Thus, an AND gate should be used for each output. Since X is to be enabled when $B = 1$, its AND gate must be controlled by B, as shown in Figure 4-27. Since Y is to be enabled when $B = 0$, its AND gate is controlled by $\overline{B}$.

This circuit is called a *pulse-steering* circuit because it steers the input pulse to one output or the other depending on B.

_____ REVIEW QUESTIONS _____

1. Design a logic circuit with three inputs A, B, C and an output that goes LOW only when A is HIGH while B and C are different.
2. Which logic gates produce a 1 output in the inhibit state?
3. Which logic gates pass the inverse of the input signal when they are enabled?

4-9   BASIC CHARACTERISTICS OF DIGITAL ICs

Digital ICs are a collection of resistors, diodes, and transistors fabricated on a single piece of semiconductor material (usually silicon) called a *substrate* which is commonly referred to as a "chip." The chip is enclosed in a protective plastic or ceramic package from which pins extend for connecting the IC to other devices. The most common type of package is a **dual-in-line package (DIP),** shown in Figure 4-28. It is so called because it consists of two parallel rows of pins. The pins are numbered counterclockwise when viewed from the top of the package with respect to an identifying notch or dot at one end of the chip. The DIP shown is a 14-pin package; 16-, 20-, 24-, 28-, 40-, and 64-pin packages are also used.

Digital ICs are often categorized according to their circuit complexity as measured by the number of equivalent logic gates on the substrate. There are currently five standard levels of complexity that are defined as follows:

COMPLEXITY	NUMBER OF GATES
Small-scale integration (SSI)	Fewer than 12
Medium-scale integration (MSI)	12 to 99
Large-scale integration (LSI)	100 to 9999
Very large-scale integration (VLSI)	10,000 to 99,999
Ultra large-scale integration (ULSI)	100,000 or more

All of the specific ICs referred to in Chapter 3 and this chapter are SSI chips having a small number of gates. In modern digital systems, medium-scale integration (MSI) and large-scale integration (LSI) devices perform most of the functions that once required several circuit boards full of SSI devices. However, SSI chips are still used as the "interface" between these more complex chips. Typically, small combinations of discrete gates are used to connect the larger ICs to each other or to exter-

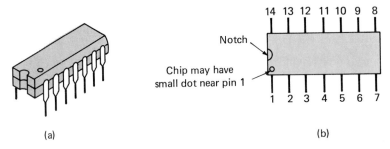

(a)   (b)

FIGURE 4-28 (a) Dual-in-line package; (b) top view showing pin numbers.

nal devices. Thus, it is necessary to know how to analyze, design, test, and troubleshoot simple combinatorial circuits.

Bipolar and Unipolar Digital ICs   Digital ICs can also be categorized according to the principal type of electronic component used in their circuitry. *Bipolar ICs* are those that are made using the bipolar junction transistor (NPN and PNP) as their main circuit element. *Unipolar ICs* are those that use the unipolar field-effect transistors (P-channel and N-channel MOSFETs) as their main element.

The most widely used family of bipolar digital ICs is the TTL (transistor-transistor logic) family. Figure 4-29(a) shows a standard TTL INVERTER circuit. Notice that it contains several bipolar transistors. The TTL family is especially prominent in small-scale and medium-scale integration (SSI and MSI) and has been a front-runner in these categories for quite some time. Its leading position in SSI and

FIGURE 4-29 (a) TTL INVERTER circuit; (b) CMOS INVERTER circuit. Pin numbers are given in parentheses.

(a)

MSI, however, is being challenged by the CMOS (complementary metal-oxide-semiconductor) logic family, which belongs to the category of unipolar digital ICs. Figure 4-29(b) shows a standard CMOS INVERTER circuit that uses enhancement-type MOSFETs as its main circuit element. Later in the text we will take a detailed look at these TTL and CMOS circuits.

TTL Family   The TTL logic family actually consists of several subfamilies or series. Table 4-1 lists the name of each TTL series together with the prefix designation used to identify different ICs as being part of that series. For example, ICs that are part of the standard TTL series have an identification number that starts with 74. The 7402, 7438, and 74123 are all ICs in this series. Likewise ICs that are part of the low-power Schottky TTL series will have an identification number that starts with 74LS. The 74LS02, 74LS38, and 74LS123 are examples of devices in the 74LS series.

The differences between the various TTL series are in their electrical characteristics, such as power dissipation, delay times, and switching speed. They do not differ in the pin layout or logic operations performed by the internal circuitry. For example, the 7402, 74H02, 74L02, 74S02, 74LS02, 74ALS02, and 74AS02 are all quad two-input NOR gates. We will compare the electrical characteristics of the different TTL series in Chapter 8.

Teaching Tip: It can be mentioned that it is not always wise to replace an IC from one series with one of a different series, because of the different electrical characteristics. Much more on this in Chapter 8.

CMOS Family   There are several CMOS series available. They are listed in Table 4-2. The 4000 and 14000 series are the oldest CMOS series. These series contain many of the same logic functions as the TTL family, but they were not designed

Teaching Tip: Even though it will be covered in detail in Chapter 8, you may want to briefly discuss why there are so many different logic subfamilies.

TABLE 4-1   Various Series within the TTL Logic Family

TTL Series	Prefix	Example IC
Standard TTL	74	7404 (hex INVERTER)
High-speed TTL	74H	74H04 (hex INVERTER)
Low-power TTL	74L	74L04 (hex INVERTER)
Schottky TTL	74S	74S04 (hex INVERTER)
Low-power Schottky TTL	74LS	74LS04 (hex INVERTER)
Advanced Schottky TTL	74AS	74AS04 (hex INVERTER)
Advanced low-power Schottky TTL	74ALS	74ALS04 (hex INVERTER)

TABLE 4-2   Various Series within CMOS Logic Family

CMOS Series	Prefix	Example IC
Metal-gate CMOS	40 or 140	4001 or 14001 (quad NOR gates)
Metal-gate, pin-compatible with TTL	74C	74C02 (quad NOR gates)
Silicon-gate, pin-compatible with TTL, high-speed	74HC	74HC02 (quad NOR gates)
Silicon-gate, high-speed, electrically compatible with TTL	74HCT	74HCT02 (quad NOR gates)

to be pin-compatible with TTL devices. For example, the 4001 quad NOR chip contains four two-input NOR gates, as does the TTL 7402 chip, but the gate inputs and outputs on the CMOS chip will not have the same pin numbers as the corresponding signals on the TTL chip.

The 74C, 74HC, 74HCT, 74AC and 74ACT series are newer CMOS series, the latter two being the most recent. All of them are pin-compatible with correspondingly numbered TTL devices. For example, the 74C02, 74HC02, and 74HCT02 are all pin-compatible with the 7402, 74H02, 74LS02, and so on. The 74HC and 74HCT series operate at a higher speed than the 74C devices. The 74HCT series is designed to be electrically compatible with TTL devices. This means that 74HCT integrated circuits, unlike 74C and 74HC integrated circuits, can be connected directly to TTL devices without any interfacing circuitry.

Power and Ground   To use digital ICs, it is necessary to make the proper connections to the IC pins. The most important connections are *dc power* and *ground*. These are required for the circuits on the chip to operate correctly. Referring to Figure 4-29, you can see that both the TTL and CMOS circuits have a dc power supply voltage connected to one of their pins, and ground to another. The power supply pin is labeled V_{CC} for the TTL circuit, and V_{DD} for the CMOS circuit. Many of the newer CMOS integrated circuits that are designed to be compatible with TTL integrated circuits also use V_{CC} as their power pin.

If either the power or ground connection is not made to the IC, the logic gates on the chip will not respond properly to the logic inputs and will not produce the expected output logic levels.

Logic-Level Voltage Ranges   For TTL devices, V_{CC} is nominally +5 V. For CMOS integrated circuits, V_{DD} can range from +3 to +18 V, although +5 V is most often used when CMOS integrated circuits are used in the same circuit with TTL integrated circuits.

For standard TTL devices the acceptable voltage ranges for the logic 0 and logic 1 levels are defined as shown in Figure 4-30(a). A logic 0 is any voltage in the range 0 to 0.8 V; a logic 1 is any voltage between 2 V and 5 V. Voltages that are not in either of these ranges are said to be **indeterminate** and should not be used as inputs to any TTL device. The IC manufacturers cannot guarantee how a TTL circuit

Common Student Mistake:
Students often confuse the V_{CC} input voltage with the logic input voltage. They need to be reminded that V_{CC} is the voltage needed by the IC to operate properly, and it must be held reasonably constant. The logic inputs to the individual gates on the IC can be any voltage within the allowed 0 or 1 ranges, and can change at any time.

FIGURE 4-30   Logic-level voltage ranges for TTL and CMOS digital ICs.

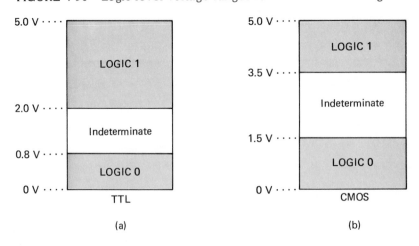

(a)

(b)

will respond to input levels that are in the indeterminate range (0.8 to 2.0 V).

The logic level ranges for CMOS integrated circuits operating with $V_{DD} = +5$ V are shown in Figure 4-30(b). Voltages between 0 and 1.5 V are defined as a logic 0, and voltages from 3.5 to 5 V as a logic 1. The indeterminate range includes voltages between 1.5 and 3.5 V.

Unconnected (Floating) Inputs   What happens when the input to a digital IC is left unconnected? An unconnected input is often called a "floating" input. The answer to this question will be different for TTL and CMOS.

A floating TTL input acts just like a logic 1. In other words, the IC will respond as if the input had a logic HIGH level applied to it. This characteristic is often used when testing a TTL circuit. A lazy technician might leave certain inputs unconnected instead of connecting them to a logic HIGH. Although this is logically correct, it is not a recommended practice, especially in final circuit designs, since the floating TTL input is extremely susceptible to picking up noise signals that can adversely affect the device's operation.

A floating TTL input will measure a dc level of between 1.4 and 1.8 V when checked with a VOM or oscilloscope. Even though this is in the indeterminate range for TTL, it will produce the same response as a logic 1. Being aware of this characteristic of a floating TTL input can be valuable when troubleshooting TTL circuits.

If a CMOS input is left floating, it may have disastrous results. The IC may become overheated and eventually destroy itself. For this reason all inputs to a CMOS integrated circuit must be connected to a LOW or a HIGH level or to the output of another IC. A floating CMOS input will not measure as a specific dc voltage, but will fluctuate randomly as it picks up noise. Thus, it does not act as logic 1 or logic 0, and so its effect on the output is unpredictable. Sometimes the output will oscillate as a result of the noise picked up by the floating input.

Logic-Circuit Connection Diagrams   A connection diagram shows *all* electrical connections, pin numbers, IC numbers, component values, signal names, and supply voltages. Figure 4-31 shows a typical connection diagram for a simple logic circuit. Examine it carefully and note the following important points:

1. The circuit uses logic gates from two different ICs. The two INVERTERs are

FIGURE 4-31   Typical logic-circuit connection diagram.

Ic	Type
Z1	7404 hex inverter
Z2	7400 quad nand

part of a 7404 chip which has been given the designation Z1. The 7404 contains six INVERTERs; two of them are used in this circuit and each is labeled as being part of chip Z1. Likewise, the two NAND gates are part of a 7400 chip that contains four NAND gates. All of the gates on this chip are designated with the label Z2. By numbering each gate as Z1, Z2, Z3, and so on, we can keep track of which gate is part of which chip. This is especially valuable in more complex circuits containing many ICs with several gates per chip.

2. Each gate input and output pin number is indicated on the diagram. These pin numbers and the IC labels are used to easily reference any point in the circuit. For example, Z1 pin 2 prefers to the output pin of the top INVERTER. Likewise, we can say that Z1 pin 4 is connected to Z2 pin 9.

3. The power and ground connections to each IC are shown on the diagram. For example, Z1 pin 14 is connected to +5 V and Z1 pin 7 is connected to ground. These connections provide power to *all* of the six INVERTERs that are part of Z1.

Manufacturers of electronic equipment generally supply detailed schematics that use a format similar to that in Figure 4-31. These connection diagrams are a virtual necessity when troubleshooting a faulty circuit.

—— REVIEW QUESTIONS ——

1. What is the most common type of digital IC package?
2. Name the five common categories of digital ICs according to complexity.
3. *True or false:* A 7474 chip will contain the same logic and pin layout as the 74LS74.
4. *True or false:* A 74HC74 chip will contain the same logic and pin layout as the 7474.
5. Which CMOS series is not pin-compatible with TTL?
6. What is the acceptable voltage range of a logic 0 for TTL? What is it for a logic 1?
7. Repeat question 6 for CMOS operating at $V_{DD} = 5$ V.
8. How does a TTL integrated circuit respond to a floating input?
9. How does a CMOS integrated circuit respond to a floating input?

4-10 TROUBLESHOOTING DIGITAL SYSTEMS

There are three basic steps in fixing a digital circuit or system that has a fault (failure):

1. **Fault detection.** Observe the circuit/system operation and compare it with the expected correct operation.
2. **Fault isolation.** Perform tests and make measurements to isolate the fault.
3. **Fault correction.** Replace the faulty component, repair faulty connection, remove short, and so on.

Although these steps may seem relatively apparent and straightforward, the ac-

tual troubleshooting procedure that is followed is highly dependent on the type and complexity of the circuitry, and on the kinds of troubleshooting tools and documentation that are available.

Good troubleshooting techniques can only be learned in a laboratory environment through experimentation and actual troubleshooting of faulty circuits and systems. There is absolutely no better way to become an effective troubleshooter than to do as much troubleshooting as possible, and no amount of textbook reading can provide that kind of experience. We can, however, help you to develop the analytical skills that are the most essential part of effective troubleshooting. We will describe the types of faults that are common to systems that are made primarily from digital ICs and tell you how to recognize them. We will then present typical case studies to illustrate the analytical processes involved in troubleshooting. In addition, there will be end-of-chapter problems to provide you with the opportunity to go through these analytical processes to reach conclusions about faulty digital circuits.

For all of the troubleshooting discussions throughout the text, we will assume that a technician always has a *logic probe* and an *oscilloscope* available, since these are standard tools in any digital lab. We will further assume that the logic probe has one or more indicator lights that can indicate the following different conditions at the probe tip:

1. Constant LOW level
2. Constant HIGH level
3. Repetitive pulse train (pulsing)
4. A LOW level that is momentarily pulsed HIGH
5. A HIGH level that is momentarily pulsed LOW
6. An open circuit or an indeterminate logic level

In the troubleshooting examples we will often state the logic probe indication at various points in the circuit.

4-11 INTERNAL DIGITAL IC FAULTS

The most common internal failures of digital ICs are:

1. Malfunction in the internal circuitry
2. Inputs or outputs shorted to ground or V_{CC}
3. Inputs or outputs open-circuited
4. Short between two pins (other than ground or V_{CC})

We will now describe each of these types of failure.

Malfunction in Internal Circuitry   This is usually caused by one of the internal components failing completely or operating outside its specifications. When this happens the IC outputs do not respond properly to its inputs. There is no way to predict what the outputs will do, because it depends on what internal component has failed. Examples of this type of failure would be a base-emitter short in transistor Q_4 or an extremely large resistance value for R_2 in the TTL INVERTER of Figure 4-29(a). This type of internal IC failure is not as common as the other three.

FIGURE 4-32 (a) IC input internally shorted to ground; (b) IC input internally shorted to supply voltage. This type of failure forces input signal at shorted pin to stay in one state.

FIGURE 4-33 (a) IC output internally shorted to ground; (b) output internally shorted to supply voltage. This type of failure does not affect signals at the IC inputs.

Teaching Tip: Students often have difficulty realizing how a shorted input pin will affect the output of the logic circuit that drives that input. Until they learn more about the internal circuitry of an IC output, it is best to just assume that such a situation will cause the output to be stuck at 0 V or +5 V.

Teaching Tip: It should be pointed out how an ohmmeter or continuity checker can be used to help detect internal shorts to ground or V_{CC}. Of course, this must be done with power removed from the circuit.

Input Internally Shorted to Ground or Supply This type of internal failure will cause the input to be stuck in the LOW or HIGH state. Figure 4-32(a) shows input pin 2 of a NAND gate shorted to ground within the IC. This will cause pin 2 always to be in the LOW state. If this input pin is being driven by a logic signal B, it will effectively short B to ground. Thus, this type of fault will affect the output of the device that is generating the B signal.

Similarly, an IC input pin could be internally shorted to +5 V as in Figure 4-32(b). This would keep that pin stuck in the HIGH state. If this input pin is being driven by a logic signal A, it would effectively short A to +5 V.

Output Internally Shorted to Ground or Supply This type of internal failure will cause the output pin to be stuck in the LOW or HIGH state. Figure 4-33(a) shows pin 3 of the NAND gate shorted to ground within the IC. This output is stuck LOW, and it will not respond to the conditions applied to input pins 1 and 2; in other words, logic inputs A and B will have no effect on output X.

An IC output pin can also be shorted to +5 V within the IC as shown in Figure 4-33(b). This forces the output pin 3 to be stuck HIGH regardless of the state of the signals at the input pins. Note that this type of failure has no effect on the logic signals at the IC inputs.

_____ EXAMPLE 4-24 _____

Refer to the circuit of Figure 4-34. A technician uses a logic probe to determine the conditions at the various IC pins. The results are recorded in the figure. Examine

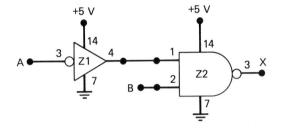

Pin	Condition
Z1-3	Pulsing
Z1-4	LOW
Z2-1	LOW
Z2-2	HIGH
Z2-3	HIGH

FIGURE 4-34   Example 4-24.

these results and determine if the circuit is working properly. If not, suggest some of the possible faults.

Solution   Output pin 4 of the INVERTER should be pulsing, since its input is pulsing. The recorded results, however, show that pin 4 is stuck LOW. Since this is connected to Z2 pin 1, this keeps the NAND output HIGH. From our preceding discussion, we can list three possible faults that could produce this operation.

First, there could be an internal component failure in the INVERTER that prevents it from responding properly to its input. Second, pin 4 of the INVERTER could be internally shorted to ground, thereby keeping it stuck LOW. Third, pin 1 of Z2 could be shorted to ground internal to Z2. This would prevent the INVERTER output pin from changing.

Discussion Question: How do we know that the problem can't be a broken connection between Z1-4 and Z2-1?

Open-Circuited Input or Output
Sometimes the very fine conducting wire that connects an IC pin to the IC's internal circuitry will break, producing an open circuit. Figure 4-35(a) shows this for an input pin, and Figure 4-35(b) for an output pin.

If an input to an IC is internally open, the logic signal applied to that input will have no effect on the output. The open ciruit leaves the input in the floating state. As we stated earlier, TTL and CMOS will respond differently to an open input. TTL integrated circuits will respond as if the open input is a logic 1; CMOS integrated circuits will respond erratically and may even become damaged from overheating.

When the output from an IC is internally open there will be no voltage present

Discussion Question: Can an ohmmeter or continuity checker be used to detect this type of fault?

FIGURE 4-35   (a) IC with internally open input will not respond to signals applied to that input pin; (b) an internally open output pin will produce no output voltage.

(a)

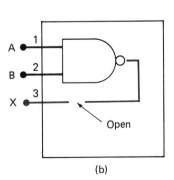

(b)

at the output pin regardless of the applied input conditions. If this output is connected to the input of one or more other ICs, this produces an open or floating input to those ICs.

_____ EXAMPLE 4-25 _____

Refer to the circuit of Figure 4-36 and the recorded logic probe indications. What are some of the possible faults that could produce the recorded results? Assume that the ICs are TTL.

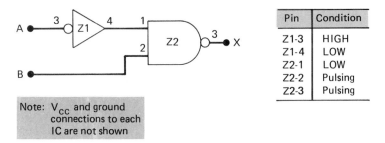

Pin	Condition
Z1-3	HIGH
Z1-4	LOW
Z2-1	LOW
Z2-2	Pulsing
Z2-3	Pulsing

Note: V_{CC} and ground connections to each IC are not shown

FIGURE 4-36   Example 4-25.

Solution   Examination of the recorded results indicates that the INVERTER appears to be working properly, but the NAND output is inconsistent with its inputs. The NAND output should be HIGH, since its input pin 1 is LOW. This LOW should prevent the NAND gate from responding to the pulses at pin 2. It is probable that this LOW is not reaching the internal NAND gate circuitry because of an internal open. Because the IC is TTL, this open circuit would produce the same effect as a logic HIGH at pin 1. If the IC had been CMOS, the internal open circuit at pin 1 might have produced an indeterminate output and possible overheating and destruction of the chip.

From our earlier statement regarding open TTL inputs, you might have expected that the voltage of pin 1 of Z2 would be 1.4 to 1.8 V and should have been registered as indeterminate by the logic probe. This would have been true if the open circuit had been *external* to the NAND chip. There is no open circuit between Z1 pin 4 and Z2 pin 1, and so the voltage at Z1 pin 4 is reaching Z2 pin 1, but it becomes disconnected *inside* the NAND chip.

_____

Short between Two Pins   An internal short between two pins of an IC will force the logic signals at those pins always to be identical. Whenever two signals that are supposed to be different show the same logic-level variations, there is a good possibility that the signals are shorted together.

Discussion Question: What waveform can you expect at the X output in Figure 4-37?

Consider the circuit in Figure 4-37, where pins 5 and 6 of the NOR gate are internally shorted together. The short causes the two INVERTER output pins to be connected together so that the signals at Z1 pin 2 and Z1 pin 4 have to be identical even when the two INVERTER input signals are trying to produce different outputs. To illustrate, consider the input waveforms shown in the diagram. Even though these input waveforms are different, the waveforms at outputs Z1-2 and Z1-4 are the same.

During the interval t_1 to t_2, both INVERTERs have a HIGH input and both are trying to produce a LOW output, so that their being shorted together makes no dif-

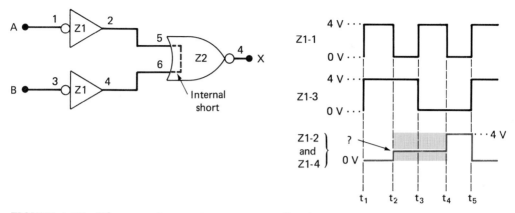

FIGURE 4-37 When two input pins are internally shorted, the signals driving these pins are forced to be identical, and usually a signal with three distinct levels results.

ference. During the interval t_4 to t_5, both INVERTERs have a LOW input and are trying to produce a HIGH output, so that again their being shorted has no effect. However, during the intervals t_2 to t_3 and t_3 to t_4, one INVERTER is trying to produce a HIGH output while the other is trying to produce a LOW output. For this situation the actual voltage level that appears at the shorted outputs will depend on the internal IC circuitry. For TTL devices it will usually be a voltage in the high end of the logic 0 range (i.e., close to 0.8 V), although it may also be in the indeterminate range. For CMOS devices it will often be a voltage in the indeterminate range.

Whenever you see a waveform like the Z1-2, Z1-4 signal in Figure 4-37 with three different levels, you should suspect that two output signals may be shorted together.

REVIEW QUESTIONS

1. List the different internal digital IC faults.
2. Which internal IC fault can produce signals that show three different voltage levels?

4-12 EXTERNAL FAULTS

We have seen how to recognize the effects of various faults internal to digital ICs. There are many more things that can go wrong external to the ICs; we will describe the most common ones in this section.

Open Signal Lines   This category includes any fault that produces a break or discontinuity in the conducting path such that a voltage level or signal is prevented from going from one point to another. Some of the causes of open signal lines are:

1. Broken wire
2. Poor solder connection; loose wire-wrap connection

3. Crack or cut trace on a printed circuit board (some of these are hairline cracks that are difficult to see without a magnifying glass)
4. Bent or broken pin on an IC
5. Faulty IC socket such that the IC pin does not make good contact with the socket

This type of circuit fault is easily verified by disconnecting power from the circuit and checking with an ohmmeter between the two points in question.

_____ EXAMPLE 4-26 _____

Consider the CMOS circuit of Figure 4-38 and the accompanying logic probe indications. What is the most probable circuit fault?

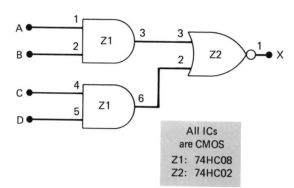

Pin	Condition
Z1-1	Pulsing
Z1-2	HIGH
Z1-3	Pulsing
Z1-4	LOW
Z1-5	Pulsing
Z1-6	LOW
Z2-3	Pulsing
Z2-2	Indeterminate
Z2-1	Indeterminate

All ICs are CMOS
Z1: 74HC08
Z2: 74HC02

FIGURE 4-38   Example 4-26.

Discussion Question: Why can't the fault be an internal open circuit in Z2?

Solution   The indeterminate level at the NOR gate output is probably due to the indeterminate input at pin 2. Since there is a LOW at Z1-6, this LOW should also be at Z2-2. Clearly, the LOW from Z1-6 is not reaching Z2-2, and there must be an open circuit in the signal path between these two points. The location of this open circuit can be determined by starting at Z1-6 with the logic probe and tracing the LOW level toward Z2-2 until it changes into an indeterminate level.

Shorted Signal Lines   This type of fault has the same effect as an internal short between IC pins. It causes two signals to be exactly the same. The main causes are:

1. **Sloppy wiring.** An example of this is stripping too much insulation from ends of wires that are in close proximity.
2. **Solder bridges.** These are splashes of solder that short two or more points together. They commonly occur between points that are very close together, such as adjacent pins on a chip.
3. **Incomplete etching.** The copper between adjacent conducting paths on a printed circuit board is not completely etched away.

An ohmmeter check can also be used to verify that two signal lines are shorted together.

Faulty Power Supply   All digital systems have one or more dc power supplies that supply the V_{CC} and V_{DD} voltages required by the chips. A faulty power supply or one that is overloaded (supplying more than its rated amount of current) will provide poorly regulated supply voltages to the ICs, and the ICs either will not operate, or will operate erratically.

A power supply may go out of regulation because of a fault in its internal circuitry, or because the circuits that it is powering are drawing more current than the supply is designed for. This can happen if a chip or a component has a fault that causes it to draw much more current than normal.

It is a good troubleshooting practice to check the voltage levels at each power supply in the system to see that they are within their specified ranges. It is also a good idea to check them on an oscilloscope to verify that there is no significant amount of ac ripple on the dc levels, and that the voltage levels stay regulated during the system operation.

One of the most common signs of a faulty power supply is one or more chips operating erratically or not at all. Some ICs are more tolerant of power supply variations and may operate properly, while others do not. You should always check the power and ground levels at each IC that appears to be operating incorrectly.

—— REVIEW QUESTIONS ——

1. What are the most common types of external faults?
2. List some of the causes of signal path open circuits.
3. What symptoms are caused by a faulty power supply?

4-13   TROUBLESHOOTING CASE STUDY

The following example will illustrate the analytical processes involved in troubleshooting digital circuits. Although the example is a fairly simple combinatorial logic circuit, the reasoning and troubleshooting procedures used can be applied to the more complex digital circuits that we encounter in subsequent chapters.

—— EXAMPLE 4-27 ——

Consider the circuit of Figure 4-39. The output Y is supposed to go HIGH for either of the following conditions:

1. $A = 1$, $B = 0$ regardless of level on C
2. $A = 0$, $B = 1$, $C = 1$

You may wish to verify this for yourself.

When the circuit is tested, the technician observes that the output Y goes HIGH whenever A is HIGH or C is HIGH regardless of the level at B. She takes logic probe measurements for the condition where $A = B = 0$, $C = 1$ and comes up with the indications recorded in Figure 4-39.

Examine the recorded levels and list the possible causes for the malfunction. Then develop a step-by-step procedure to determine the exact fault.

Pin	Condition
Z1-1	LOW
Z1-2	LOW
Z1-3	HIGH
Z2-4	LOW
Z2-5	HIGH
Z2-6, 10	HIGH
Z2-13	HIGH
Z2-12	HIGH
Z2-9, 11	LOW
Z2-8	HIGH

FIGURE 4-39   Example 4-27.

Solution   All of the NAND gate outputs are correct for the levels present at their inputs. The EX-OR gate, however, should be producing a LOW at output pin 3, since both of its inputs are at the same LOW level. It appears that Z1-3 is stuck HIGH even though its inputs should produce a LOW. There are several possible causes for this:

Teaching Tip: Why couldn't the fault be an internal open circuit at Z1-3?

1. An internal component failure in Z1 that prevents its output from going LOW

2. An external short to V_{CC} from any point along the conductors connected to node X (shaded in diagram)

3. Pin 3 of Z1 internally shorted to V_{CC}

4. Pin 5 of Z2 internally shorted to V_{CC}

5. Pin 13 of Z2 internally shorted to V_{CC}

All of these possibilities except for the first one will short node X (and every IC pin connected to it) directly to V_{CC}.

The following procedure can be used to isolate the fault. This procedure is not the only approach that can be used, and as we stated earlier, the actual troubleshooting procedure that a technician uses is very dependent on what test equipment is available.

1. Check the V_{CC} and ground levels at the appropriate pins of Z1. Although it is unlikely that the absence of either of these might cause Z1-3 to stay HIGH, it is a good idea to make this check on any IC that is producing an incorrect output.

2. Turn off power to the circuit and use an ohmmeter to check for a short (resistance less than 1 Ω) between node X and any point connected to V_{CC} (such as Z1-14 or Z2-14). If no short is indicated, the last four possibilities in our list can be eliminated. This means that it is very likely that Z1 has an internal failure and should be replaced.

3. If step 2 shows that there is a short from node X to V_{CC}, perform a thorough visual examination of the circuit board and look for solder bridges, unetched copper slivers, uninsulated wires touching each other, and any other possible cause of an external short to V_{CC}. A likely spot for a solder bridge would be between adjacent pins 13 and 14 of Z2. Pin 14 is connected to V_{CC} and pin 13 to node X. If an external short is found, remove it and perform an ohmmeter check to verify that node X is no longer shorted to V_{CC}.

4. If step 3 does not reveal an external short, the three possibilities that remain are internal shorts to V_{CC} at Z1-3, Z2-13, or Z2-5. One of these is shorting node X to V_{CC}.

To determine which of these IC pins is the culprit, we should disconnect each of them from node X *one at a time* and recheck for a short to V_{CC} after each disconnection. When the pin that is internally shorted to V_{CC} is disconnected, node X will no longer be shorted to V_{CC}.

The process of disconnecting each suspected pin from node X can be easy or difficult depending on how the circuit is constructed. If the ICs are in sockets, all you need to do is to pull the IC from its socket, bend out the suspected pin, and reinsert the IC into its socket. If the ICs are soldered into a printed circuit board, you will have to cut the trace that is connected to the pin (or cut the pin) and repair the cut trace (or pin) when you are finished.

There is a troubleshooting technique that makes it unnecessary to bend pins or cut traces when trying to isolate a short. It involves using a tool called a *current tracer* to trace the flow of current through the short circuit as the node is being pulsed. The current tracer senses the changing magnetic field around the conductor through which the current is being shorted. We will examine this in Chapter 8.

Example 4-27, although fairly simple, shows you the kinds of thinking that a troubleshooter must employ in order to isolate a fault. You will have the opportunity to begin developing your own troubleshooting skills by working on Problems 4-34 to 4-44.

PROBLEMS

SECTIONS 4-2 and 4-3

4-1. Simplify the following expressions using Boolean algebra.
(a) $x = ABC + \overline{A}C$ $C(\overline{A} + B)$
(b) $y = (Q + R)(\overline{Q} + \overline{R})$ $\overline{Q}R + Q\overline{R}$
(c) $w = \overline{ABC} + A\overline{B}C + \overline{A}$ $C + \overline{A}$
(d) $q = \overline{RST}\,(\overline{R} + S + \overline{T})$ $\overline{R}S\overline{T}$
(e) $x = \overline{A}B\overline{C} + \overline{A}BC + ABC + A\overline{B}\overline{C} + A\overline{B}C$ $BC + \overline{B}(A + \overline{C})$
(f) $z = (B + \overline{C})(\overline{B} + C) + \overline{A}+B+\overline{C}$ $BC + \overline{BC} + A\overline{B}$
(g) $y = (\overline{C + D}) + \overline{A}C\overline{D} + A\overline{B}\overline{C} + \overline{A}\overline{B}CD + AC\overline{D}$ $\overline{D} + A\overline{B}\overline{C} + \overline{A}\overline{B}C$

4-2. Simplify the circuit of Figure 4-40 using Boolean algebra.   **4.2.** $x = Q(M + N)$

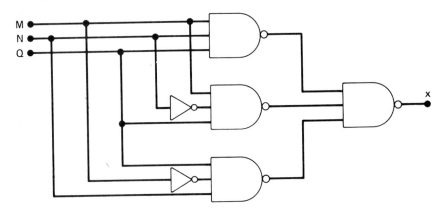

M
N
Q

x

FIGURE 4-40

4-3. Change each gate in Figure 4-40 to a NOR gate and simplify the circuit using Boolean algebra. $x = MN + Q$

4-4. Design the logic circuit corresponding to the following truth table. $x = \overline{A}B + \overline{B}\overline{C} + BC$

A	B	C	x
0	0	0	1
0	0	1	0
0	1	0	1
0	1	1	1
1	0	0	1
1	0	1	0
1	1	0	0
1	1	1	1

4-5. Design a logic circuit whose output is HIGH *only* when a majority of inputs A, B, and C are LOW. $x = \overline{A}\overline{B} + \overline{A}\overline{C} + \overline{B}\overline{C}$

4-6. Three photocells are being illuminated by three different flashing lights. The lights are supposed to be flashing in sequence so that at no time should all three lights be on at the same time or off at the same time. Each photocell is used to monitor one of the lights, and each photocell is in a circuit that produces a LOW output voltage when the photocell is dark and a HIGH output voltage when the photocell is illuminated. Design a logic circuit that has as its inputs the photocell-circuit outputs and which produces a HIGH output whenever the three lights are *all* on or *all* off at the same time. $X = \overline{A}\overline{B}\overline{C} + ABC$

4-7. A 4-bit binary number is represented as $A_3A_2A_1A_0$, where A_3, A_2, A_1, and A_0 represent the individual bits with A_0 equal to the LSB. Design a logic circuit that will produce a HIGH output whenever the binary number is greater than 0010 and less than 1000. $x = \overline{A}_3(A_2 + A_1A_0)$

4-8. Figure 4-41 shows a diagram for an automobile alarm circuit used to detect certain undesirable conditions. The three switches are used to

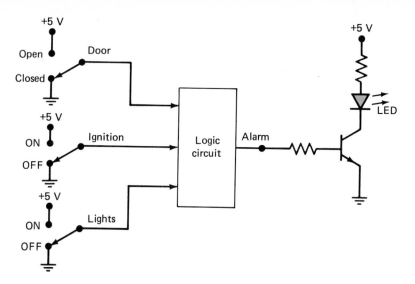

FIGURE 4-41

indicate the status of the door by the driver's seat, the ignition, and the headlights, respectively. Design the logic circuit with these three switches as inputs so that the alarm will be activated whenever either of the following conditions exists:

- The headlights are on while the ignition is off.
- The door is open while the ignition is on. $X = ID + \bar{I}L$

4-9. Implement the circuit of Problem 4-4 using all NAND gates.

4-10. Implement the circuit of Problem 4-5 using all NAND gates.

4-11. Implement the expression $z = \bar{D} + ABC + \overline{AC}$ using ANDs, ORs, and INVERTERs; then convert to all NAND gates. See Solutions Manual.

4-9. Change each gate to its NAND equivalent, and then cancel all double INVERTERs.

4-10. Change each gate to its NAND equivalent, and then cancel all double INVERTERs.

SECTION 4-5

4-12. Simplify the expression of Problem 4-1(e) using the K map.

4-13. Simplify the expression of Problem 4-1(g) using the K map.

4-14. Simplify the expression from Problem 4-7 using a K map.

4-15. Determine the minimum expression for each K map in Figure 4-42.

4-12. $x = BC + \bar{B}\bar{C} + A\bar{B}$ or $x = BC + \bar{B}\bar{C} + AC$
4-13. $y = \bar{D} + \bar{A}\bar{B}C + A\bar{B}\bar{C}$
4-14. $x = \bar{A}_3 A_2 + \bar{A}_3 A_1 A_0$
4-15. (a) $x = \bar{A}B + \bar{A}C + \bar{B}D + ABC\bar{D}$
(b) $x = \overline{AD} + \bar{B}C + \bar{B}D$
(c) $y = \bar{B} + AC$

FIGURE 4-42

	$\bar{C}\bar{D}$	$\bar{C}D$	CD	$C\bar{D}$
$\bar{A}\bar{B}$	1	1	1	1
$\bar{A}B$	1	1	0	0
AB	0	0	0	1
$A\bar{B}$	0	1	1	0

(a)

$f = \bar{A}\bar{C} + \bar{B}D + \bar{A}\bar{B} + ABC\bar{D}$

	$\bar{C}\bar{D}$	$\bar{C}D$	CD	$C\bar{D}$
$\bar{A}\bar{B}$	1	0	1	1
$\bar{A}B$	1	0	0	1
AB	0	0	0	0
$A\bar{B}$	1	0	1	1

(b)

$f = \bar{A}\bar{D} + \bar{B}C + \bar{B}D$

	$\bar{C}$	C
$\bar{A}\bar{B}$	1	1
$\bar{A}B$	0	0
AB	1	0
$A\bar{B}$	1	X

(c)

$f = \bar{B} + AC$

FIGURE 4-43

4-16. $X = B\overline{C} + AD$; the don't cares in squares 6, 10, 11 and 15 are made 1s.

4-16. Figure 4-43 shows a *BCD counter* that produces a 4-bit output representing the BCD code for the number of pulses that have been applied to the counter input. For example, after four pulses have occurred, the counter outputs are $DCBA = 0100_2 = 4_{10}$. The counter resets to 0000 on the tenth pulse and starts counting over again. In other words, the $DCBA$ outputs will never represent a number greater than $1001_2 = 9_{10}$. Design the logic circuit that produces a HIGH output whenever the count is 2, 3, or 9. Use K mapping and take advantage of the "don't care" conditions.

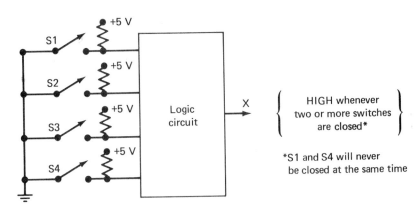

FIGURE 4-44

4-17. $X = \overline{S}_1\overline{S}_2 + \overline{S}_1\overline{S}_3 + \overline{S}_3\overline{S}_4 + \overline{S}_2\overline{S}_3 + \overline{S}_2\overline{S}_4$

4-17. Figure 4-44 shows four switches that are part of the control circuitry in a copy machine. The switches are at various points along the path of the copy paper as the paper passes through the machine. Each switch is normally open, and as the paper passes over a switch, the switch closes. It is impossible for switches S_1 and S_4 to be closed at the same time. Design the logic circuit to produce a HIGH output whenever *two or more* switches are closed at the same time. Use K mapping and take advantage of the "don't care" conditions.

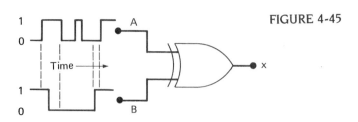

FIGURE 4-45

4-18. (a) Determine the output waveform for the circuit of Figure 4-45.   **4.18a.** $x = 1$ when A and B are
   (b) Repeat with the B input held LOW. $x = A$ at different levels.
   (c) Repeat with B held HIGH. $x = $ inverse of A

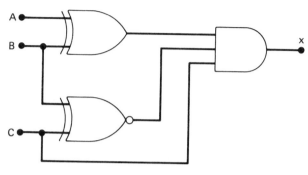

FIGURE 4-46

4-19. Determine the input conditions needed to produce $x = 1$ in Figure
   4-46. $ABC = 011$

4-20. Redo Problem 4-6 using EX-OR or EX-NOR gates and one other gate.   **4-20.** Put A and B into an EX-
OR, put B and C into an EX-OR,
and feed both EX-ORs into a
NOR gate.

FIGURE 4-47

 4-21. Figure 4-47 represents a *relative-magnitude detector* that takes two 3-
   bit binary numbers $x_2x_1x_0$ and $y_2y_1y_0$ and determines whether they are
   equal and, if not, which one is larger. There are three outputs, defined
   as follows:
   1. $M = 1$ only if the two input numbers are equal.
   2. $N = 1$ only if $x_2x_1x_0$ is greater than $y_2y_1y_0$.
   3. $P = 1$ only if $y_2y_1y_0$ is greater than $x_2x_1x_0$.

   Design the logic circuitry for the comparator. The circuit has *six* in-
   puts and *three* outputs and is therefore much too complex to handle us-
   ing the truth-table approach. Refer to Example 4-16 as a hint to how
   you might start to solve this problem.   See Solutions Manual.

MORE DESIGN PROBLEMS

 4-22. Figure 4-48 represents a multiplier circuit that takes 2-bit binary num-
   bers x_1x_0 and y_1y_0 and produces an output binary number $z_3z_2z_1z_0$ that
   is equal to the arithmetic product of the two input numbers. Design
   the logic circuit for the multiplier. (*Hint:* The logic circuit will have
   four inputs and four outputs.)

4-22. $z_3 = y_1y_0x_1x_0$;
$z_2 = y_1x_1(\bar{y}_0 + \bar{x}_0)$;
$z_1 = y_0x_1(\bar{y}_1 + \bar{x}_0) + y_1x_0(\bar{y}_0 + \bar{x}_1)$; $z_0 = y_0x_0$

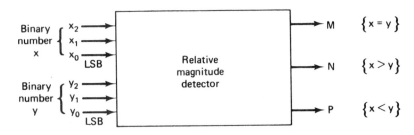

FIGURE 4-48

4-23. A BCD code is being transmitted to a remote receiver. The bits are A_3, A_2, A_1, A_0 with A_3 as the MSB. The receiver circuitry includes a *BCD error-detector* circuit that examines the received code to see if it is a legal BCD code (i.e., ≤ 1001). Design this circuit to produce a HIGH for any error condition. $x = A_3 A_2 + A_3 A_1$

4-24. Design a logic circuit whose output is HIGH whenever A and B are both HIGH as long as C and D are either both LOW or both HIGH. Try to do this without using a truth table. Then check your result by constructing a truth table for your circuit to see if it agrees with the problem statement. $AB \cdot (\overline{C \oplus D})$

4-25. Four large tanks at a chemical plant contain different liquids being heated. Liquid-level sensors are being used to detect whenever the level in tank A or tank B rises above a predetermined level. Temperature sensors in tanks C and D detect when the temperature in either of these tanks drops below a prescribed temperature limit. Assume that the liquid-level sensor outputs A and B are LOW when the level is satisfactory and HIGH when the level is too high. Also, the temperature-sensor outputs C and D are LOW when the temperature is satisfactory and HIGH when the temperature is too low. Design a logic circuit that will detect whenever the level in tank A or tank B is too high at the same time that the temperature in either tank C or tank D is too low. $x = (A + B)(C + D)$

4-26. N-S (green) = $\overline{CD}(A + B) + AB(\overline{C} + \overline{D})$

4-26. Figure 4-49 shows the intersection of a main highway with a secondary access road. Vehicle-detection sensors are placed along lanes C and D (main road) and lanes A and B (access road). These sensor outputs are LOW (0) when no vehicle is present and HIGH (1) when a vehicle is present. The intersection traffic light is to be controlled according to the following logic.

  1. The east-west (E-W) traffic light will be green whenever *both* lanes C and D are occupied.

  2. The E-W light will be green whenever *either* C or D is occupied but lanes A and B are not *both* occupied.

  3. The north-south (N-S) light will be green whenever *both* lanes A and B are occupied but C and D are not *both* occupied.

  4. The N-S light will also be green when *either* A or B is occupied while C and D are *both* vacant.

  5. The E-W light will be green when *no* vehicles are present.

  Using the sensor outputs A, B, C, and D as inputs, design a logic circuit to control the traffic light. There should be two outputs, N-S and

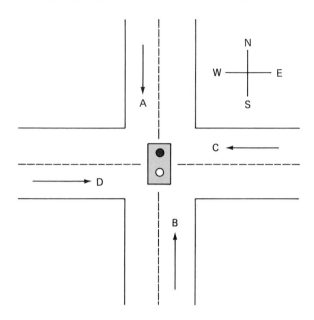

FIGURE 4-49

E-W, which go HIGH when the corresponding light is to be *green*. Simplify the circuit as much as possible and show *all* steps.

SECTION 4-7

4-27. Redesign the parity generator and checker of Figure 4-24 to operate using odd parity. (*Hint:* what is the relationship between an odd-parity bit and an even-parity bit for the same set of data bits?)

4-27. Change the last EX-OR to an EX-NOR.

4-28. Redesign the parity generator and checker of Figure 4-24 to operate on 8 data bits. For both circuits add four EX-ORs in the same pattern.

SECTION 4-8

4-29. Design a logic circuit that will allow input signal A to pass through to the output only when control input B is LOW while control input C is HIGH; otherwise, the output is LOW. $x = A\bar{B}C$

4-30. Design a circuit that will *inhibit* the passage of an input signal only when control inputs B, C, and D are all HIGH; the output is to be HIGH in the inhibited condition. $x = A + BCD$

4-31. Design a logic circuit that controls the passage of a signal A according to the following requirements:
1. Output X will equal A when control inputs B and C are the same.
2. X will remain HIGH when B and C are different.

4-31. $X = A + (B \oplus C)$

4-32. Design a logic circuit that has two signal inputs A_1 and A_0 and a control input S so that it functions according to the requirements given in Figure 4-50. This type of circuit is called a *multiplexer* (covered in Chapter 9). $z = A_0 \cdot \bar{S} + A_1 \cdot S$

4-33. Use K mapping to design a circuit to meet the requirements of Example 4-16. Compare this circuit with the solution in Figure 4-22. This points out that the K-map method cannot take advantage of the EX-OR and EX-NOR gate logic. The designer has to be able to determine when these gates are applicable.

4-33. $z = \bar{x}_1\bar{x}_0\bar{y}_1\bar{y}_0 + \bar{x}_1x_0\bar{y}_1y_0 + x_1\bar{x}_0y_1\bar{y}_0 + x_1x_0y_1y_0$

FIGURE 4-50

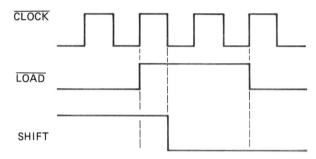

FIGURE 4-51

SECTIONS 4-9 to 4-13*

4-34. The signals shown in Figure 4-51 are applied to the inputs of the circuit of Figure 4-31. Suppose there is an internal open circuit at Z1-4.
   (a) What will a logic probe indicate at Z1-4? indeterminate
   (b) What dc voltage reading would you expect a VOM to register at Z1-4? (Remember that the ICs are TTL.) 1.4 V to 1.8 V
   (c) Sketch what you think the $\overline{CLKOUT}$ and $\overline{SHIFTOUT}$ signals will look like. See Solutions Manual

4-35. Since the ICs are TTL, the short will probably cause the signal at Z2-9 and Z2-10 to go LOW when either SHIFT or Z1-4 tries to go LOW (i.e., the circuit is a wired-AND).

4-35. The signals of Figure 4-51 are applied to the inputs of the circuit of Figure 4-31. Suppose there is an internal short between pins 9 and 10 of Z2. Sketch the probable signal at Z2-10, $\overline{CLKOUT}$, and $\overline{SHIFTOUT}$.

4-36. Assume that the ICs in Figure 4-31 are CMOS. Describe how the circuit operation would be affected by an open circuit in the conductor connecting Z2-2 and Z2-10. Voltages at Z2-2 and Z2-3 are unpredictable.

4-37. In Example 4-24 we listed three possible faults for the situation of Figure 4-34. What procedure would you follow to determine which of the faults is the actual one? First isolate Z1-4 from Z2-1 by cutting trace.

4-38. Faulty V_{CC} or ground; Z2-1 or Z2-2 or Z2-3 internally open

4-38. Refer to the circuit of Figure 4-36. Assume that the devices are CMOS. Also assume that the logic probe indication at Z2-3 is "indeterminate" rather than "pulsing." List the possible faults and write a procedure to follow to determine the actual fault.

4-39. yes: c, e, f; no: a, b, d, g

4-39. Refer to the logic circuit of Figure 4-39. Recall that output Y is supposed to be HIGH for either of the following conditions:
   1. $A = 1, B = 0$, regardless of C
   2. $A = 0, B = 1, C = 1$

*The logic probe to the left indicates a troubleshooting problem throughout the text.

When testing the circuit, the technician observes that Y goes HIGH only for the first condition, but stays LOW for all other input conditions. Consider the following list of possible faults. For each one indicate "yes" or "no" as to whether or not it could be the actual fault. Explain your reasoning for each "no" response.

(a) An internal short to ground at Z2-13.
(b) An open circuit in the connection to Z2-13.
(c) An internal short to V_{CC} at Z2-11.
(d) An open circuit in the V_{CC} connection to Z2.
(e) An internal open circuit at Z2-9.
(f) An open in the connection from Z2-11 to Z2-9.
(g) A solder bridge between pins 6 and 7 of Z2.

4-40. Develop a procedure for isolating the fault that is causing the malfunction described in Problem 4-39.

4-41. Assume that the gates in Figure 4-39 are all CMOS. When the technician tests the circuit he finds that it operates correctly except for the following conditions:

1. $A = 1, B = 0, C = 0$
2. $A = 0, B = 1, C = 1$

For these conditions, the logic probe indicates indeterminate levels at Z2-6, Z2-11, and Z2-8. What do you think is the probable fault in the circuit? Explain your reasoning.

4-42. Figure 4-52 is a combinatorial logic circuit that operates an alarm in a car whenever the driver and/or passenger seats are occupied and the seat belts are not fastened when the car is started. The active-HIGH signals *DRIV* and *PASS* indicate the presence of the driver and passenger, respectively, and are taken from pressure-actuated switches in the seats. The signal *IGN* is active-HIGH when the ignition switch is on. The signal $\overline{BELTD}$ is active-LOW and indicates that the driver's seat belt is *unfastened;* $\overline{BELTP}$ is the corresponding signal for the passenger seat belt. The alarm will be activated (LOW) whenever the car is started and either of the front seats is occupied and its seat belt is not fastened.

(a) Verify that the circuit will function as described.

4-40. Apply the $ABC = 011$ condition to the inputs and trace logic levels.
4-41. Likely fault is a short between Z2-6 and Z2-11.

4-42. (b) circuit would not detect unfastened passenger seat belt.
(c) circuit acts as if a passenger is always present with a fastened seat belt.

 11

FIGURE 4-52

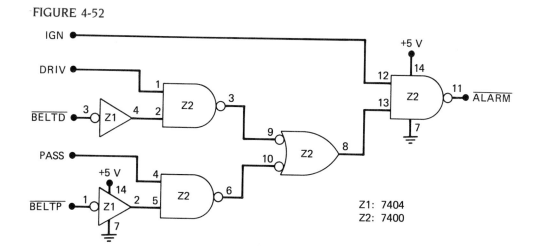

Z1: 7404
Z2: 7400

(b) Describe how this alarm system would operate if Z1-2 were internally shorted to ground.

(c) Describe how it would operate if there were an open connection from Z2-6 to Z2-10.

 4-43. Suppose the system of Figure 4-52 is functioning in such a way that the alarm is activated as soon as the driver and/or passenger are seated and the car is started, regardless of the status of the seat belts. What are the possible faults? What procedure would you follow to find the actual fault?   See Solutions Manual.

4-44. Suppose the alarm system of Figure 4-52 is operating in such a way that the alarm goes on continuously as soon as the car is started regardless of the state of the other inputs. List the possible faults and write a procedure to isolate the fault.   See Solutions Manual.

DRILL QUESTION

4-45. Define each of the following terms:
(a) Karnaugh map
(b) sum-of-products form
(c) parity generator
(d) octet
(e) inhibit circuit
(f) don't care state
(g) floating input
(h) indeterminate voltage level   see Glossary

MICROCOMPUTER APPLICATIONS

4-46a. 00_{16} to EF_{16}
4-46c. F8 to FF_{16}

4-46. In a microcomputer, the microprocessor unit (MPU) is always communicating with one of the following: (1) random-access memory (RAM), which stores programs and data that can be readily changed; (2) read-only memory (ROM), which stores programs and data that never change; (3) external input/output devices (I/O) such as keyboard, video displays, printers, and disk drives. As it is executing a program, the MPU will generate an address code that selects which type of device (RAM, ROM, or I/O) it wants to communicate with. Figure 4-53 shows a typical arrangement where the MPU outputs an

FIGURE 4-53

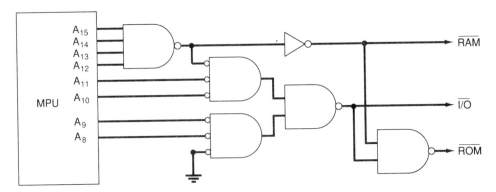

8-bit address code A_{15} through A_8. Actually, the MPU outputs a 16-bit address code, but the low-order bits A_7 through A_0 are not used in the device selection process. The address code is applied to a logic circuit which uses it to generate the device select signals—$\overline{\text{RAM}}$, $\overline{\text{ROM}}$, and $\overline{\text{I/O}}$.

Analyze this circuit and determine the following:
(a) the range of addresses A_{15} through A_8 that will activate $\overline{\text{RAM}}$.
(b) the range of addresses that activate $\overline{\text{I/O}}$.   F0$_{16}$
(c) the range of addresses that activate $\overline{\text{ROM}}$.

Express the addresses in binary and hexadecimal. For example, the answer to (a) is A_{15} to $A_8 = 00000000_2$ to $11101111_2 = 00_{16}$ to EF_{16}.

4-47. In some microcomputers the MPU can be *disabled* for short periods of time while another device controls the RAM, ROM, and I/O. During these intervals a special control signal ($\overline{\text{DMA}}$) is activated by the MPU and is used to disable (deactivate) the device select logic so that the $\overline{\text{RAM}}$, $\overline{\text{ROM}}$, and $\overline{\text{I/O}}$ are all in their inactive state. Modify the circuit of Figure 4-53 so that $\overline{\text{RAM}}$, $\overline{\text{ROM}}$, and $\overline{\text{I/O}}$ will be deactivated whenever the signal $\overline{\text{DMA}}$ is active, regardless of the state of the address code.

4-47. Connect each output ($\overline{\text{RAM}}$, $\overline{\text{I/O}}$, $\overline{\text{ROM}}$) to a separate 2-input OR gate. Invert $\overline{\text{DMA}}$ and connect it to the other input of each OR gate.

ANSWERS TO SECTION REVIEW QUESTIONS

SECTION 4-3

1. Expression (b) is not in sum-of-products form, because of the inversion sign over both the C and D variables (i.e., the $\overline{ACD}$ term). Expression (c) is not in sum-of-products form, because of the $(M + \overline{N})P$ term.   **3.** $x = \overline{A} + \overline{B} + \overline{C}$.

SECTION 4-4

1. $x = \overline{A}\overline{B}\overline{C}D + \overline{A}B\overline{C}\overline{D} + \overline{A}B\overline{C}D$ **2.** eight

SECTION 4-5

1. $x = AB + AC + BC$.   **2.** $x = A + BCD$.
3. An input condition for which there is no specific required output condition.

SECTION 4-6

2. A constant LOW.   **3.** No, the available EX-OR gate can be used as an INVERTER by connecting one of its inputs to a constant HIGH (see Example 4-15).

SECTION 4-8

1. $x = \overline{A(B \oplus C)}$ **2.** OR, NAND
3. NAND, NOR

SECTION 4-9

1. DIP.   **2.** SSI, MSI, LSI, VLSI, ULSI.
3. True.   **4.** True.   **5.** 40 or 140 series.
6. 0 V to 0.8 V; 2.0 V to 5.0 V.   **7.** 0 V to 1.5 V; 3.5 V to 5.0 V.   **8.** As if the input were HIGH.   **9.** Unpredictably; may overheat and be destroyed.

SECTION 4-11

1. Open inputs or outputs; inputs or outputs shorted to V_{CC}; inputs or outputs shorted to ground; pins shorted together; internal circuit failures.   **2.** Pins shorted together.

SECTION 4-12

1. Open signal lines; shorted signal lines; faulty power supply.   **2.** Broken wires; pool solder connections; cracks or cuts in PC board; bent or broken IC pins; faulty IC sockets.   **3.** ICs operating erratically or not at all.

5

FLIP-FLOPS AND RELATED DEVICES

Outline

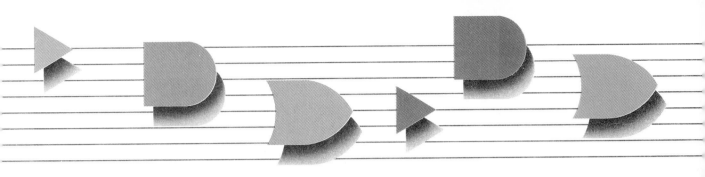

Objectives

Upon completion of this chapter, you will be able to:

- Construct and analyze the operation of a latch flip-flop made from NAND or NOR gates.
- Debounce a mechanical switch by using a latch circuit.
- Describe the difference between synchronous and asynchronous systems.
- Understand several types of edge-triggered flip-flops, such as the J-K, D-type, and S-C.
- Analyze and apply the various flip-flop timing parameters specified by the manufacturers.
- Describe a pulse-steering and an edge-detector circuit.
- Understand the major differences between parallel and serial data transfers.
- Draw the output timing waveforms of several types of flip-flops in response to a set of input signals.
- Analyze the various IEEE/ANSI flip-flop symbols.
- Use state transition diagrams to describe counter operation.
- Cite various flip-flop applications.
- Use flip-flops in synchronization circuits.
- Connect shift registers as data transfer circuits.
- Employ flip-flops as frequency-division and counting circuits.
- Understand the typical characteristics of Schmitt triggers.
- Apply two different types of one-shots in circuit design.
- Design a free-running oscillator using a 555 timer.
- Recognize and predict the effects of clock skew on synchronous circuits.
- Troubleshoot various types of flip-flop circuits.

Introduction

The logic circuits considered thus far have been combinatorial circuits whose output levels at any instant of time are dependent on the levels present at the inputs at that time. Any prior input-level conditions have no effect on the present outputs because combinatorial logic circuits have no memory. Most digital sys-

151

Combinatorial outputs

Memory outputs

Combinatorial logic gates

Memory elements

External inputs

FIGURE 5-1   General digital system diagram.

tems are made up of both combinatorial circuits and memory elements.

Figure 5-1 shows a block diagram of a general digital system that combines combinatorial logic gates with memory devices. The combinatorial portion accepts logic signals from external inputs and from the outputs of the memory elements. The combinatorial circuit operates on these inputs to produce various outputs, some of which are used to determine the binary values to be stored in the memory elements. The outputs of some of the memory elements, in turn, go to the inputs of logic gates in the combinatorial circuits. This process indicates that the external outputs of a digital system are a function of both its external inputs and the information stored in its memory elements.

The most important memory element is the *flip-flop*, which is made up of an assembly of logic gates. Even though a logic gate, by itself, has no storage capability, several can be connected together in ways that permit information to be stored. There are several different gate arrangements that are used to produce these flip-flops (abbreviated FF).

Figure 5-2(a) is the general type of symbol used for a flip-flop. It shows two outputs, labeled Q and $\overline{Q}$, that are the inverse of each other. Actually, any letter can be used, but Q is the one most often used. The Q output is called the *normal* FF output, and $\overline{Q}$ is the *inverted* FF output. Whenever we refer to the state of a FF, we are referring to the state of its normal (Q) output; it is understood that its in-

FIGURE 5-2   General flip-flop symbol and definition of two possible output states.

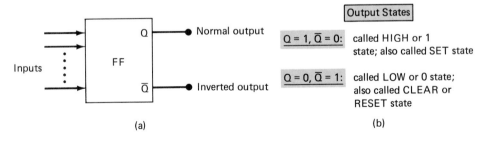

Output States
$Q = 1, \overline{Q} = 0$: called HIGH or 1 state; also called SET state
$Q = 0, \overline{Q} = 1$: called LOW or 0 state; also called CLEAR or RESET state

(a)

(b)

verted output ($\overline{Q}$) is in the opposite state. For example, if we say that a FF is in the HIGH (1) state, we mean that $Q = 1$; if we say that a FF is in the LOW (0) state, we mean that $Q = 0$. Of course, the $\overline{Q}$ state will always be the inverse of Q.

A FF, then, has the two allowed operating states indicated in Figure 5-2(b). Note the different ways that are used to refer to these two states. You should become familiar with each of these since they are all in common use.

As the symbol in Figure 5-2(a) implies, a FF can have one or more inputs. These inputs are used to cause the FF to switch back and forth ("flip-flop") between its possible output states. As we shall see, a FF input only has to be pulsed momentarily to cause a change in the FF output state, and the output will remain in that new state even after the input pulse is over. This is the FF's *memory* characteristic.

The flip-flop is known by other names, including *latch* and *bistable multivibrator*. The term "latch" is used for certain types of flip-flops that we will describe. The term "bistable multivibrator" is the more technical name for a flip-flop, but it is too much of a mouthful to be used regularly.

5-1 NAND GATE LATCH

The most basic FF circuit can be constructed from either two NAND gates or two NOR gates. The NAND gate version, called a *NAND gate latch* or simply a *latch,* is shown in Figure 5-3(a). The two NAND gates are cross-coupled so that the output of NAND-1 is connected to one of the inputs of NAND-2, and vice versa. The gate outputs, labeled Q and $\overline{Q}$, respectively, are the latch outputs. Under normal conditions, these outputs will always be the inverse of each other. The latch inputs are labeled SET and CLEAR, for reasons that will become clear.

The SET and CLEAR inputs are normally resting in the HIGH state and one of them will be pulsed LOW whenever we want to change the latch outputs. We begin our analysis by showing that there are two equally likely output states when SET = CLEAR = 1. One possibility is shown in Figure 5-3(a), where we have $Q = 0$ and $\overline{Q} = 1$. With $Q = 0$, the inputs to NAND-2 are 0 and 1, which produce $\overline{Q} = 1$. The 1 from $\overline{Q}$ causes NAND-1 to have a 1 at both inputs to produce a 0 out-

FIGURE 5-3   NAND latch has two possible resting states when SET = CLEAR = 1.

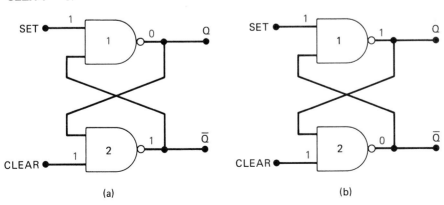

(a)   (b)

put at Q. In effect, what we have is the LOW at the NAND-1 output producing a HIGH at the NAND-2 output, which, in turn, keeps the NAND-1 output LOW.

The second possibility is shown in Figure 5-3(b), where $Q = 1$ and $\overline{Q} = 0$. The HIGH from NAND-1 produces a LOW at the NAND-2 output, which, in turn, keeps the NAND-1 output HIGH. Thus, there are two possible output states when SET = CLEAR = 1; as we shall soon see, the one that actually exists will depend on what has occurred previously at the inputs.

Setting the Latch (FF)

Now let's investigate what happens when the SET input is momentarily pulsed LOW while CLEAR is kept HIGH. Figure 5-4(a) shows what happens when $Q = 0$ prior to the occurrence of the pulse. As SET is pulsed LOW at time t_0, Q will go HIGH, and this HIGH will force $\overline{Q}$ to go LOW so that NAND-1 now has two LOW inputs. Thus, when SET returns to the 1 state at t_1, the NAND-1 output *remains* HIGH, which, in turn, keeps the NAND-2 output LOW.

Figure 5-4(b) shows what happens when $Q = 1$ and $\overline{Q} = 0$ prior to the application of the SET pulse. Since $\overline{Q} = 0$ is already keeping the NAND-1 output HIGH, the LOW pulse at SET will not change anything. Thus, when SET returns HIGH,

Discussion Question: What determines the minimum pulse duration that can be applied to the SET input in Figure 5-4(a) and still cause Q to latch to the HIGH state?

FIGURE 5-4 Pulsing the SET input to the 0 state will always produce the $Q = 1$, $\overline{Q} = 0$: output state (a) $Q = 0$ prior to SET pulse; (b) $Q = 1$ prior to SET pulse.

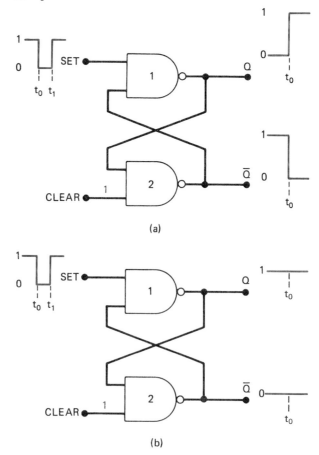

(a)

(b)

the latch outputs are still in the $Q = 1$, $\overline{Q} = 0$ state.

We can summarize Figure 5-4 by stating that a LOW pulse on the SET input will always cause the latch to end up in the $Q = 1$ state. This operation is called **setting** the latch or FF. In fact, the $Q = 1$ state is also called the **set state**.

Clearing the Latch (FF)   Now let's consider what occurs when the CLEAR input is pulsed LOW while SET is kept HIGH. Figure 5-5(a) shows what happens when $Q = 0$ and $\overline{Q} = 1$ prior to the application of the pulse. Since $Q = 0$ is already keeping the NAND-2 output HIGH, the LOW pulse at CLEAR will not have any effect. When CLEAR returns HIGH, the latch outputs are still $Q = 0$ and $\overline{Q} = 1$.

Figure 5-5(b) shows the situation where $Q = 1$ prior to the occurrence of the CLEAR pulse. As CLEAR is pulsed LOW at t_0, $\overline{Q}$ will go HIGH, and this HIGH forces Q to go LOW so that NAND-2 now has two LOW inputs. Thus, when CLEAR returns HIGH at t_1, the NAND-2 output *remains* HIGH, which, in turn, keeps the NAND-1 output LOW.

Figure 5-5 can be summarized by stating that a LOW pulse on the CLEAR input will always cause the latch to end up in the $Q = 0$ state. This operation is called **clearing** or **resetting** the latch, and the $Q = 0$ state is also called the **cleared** or **reset** state.

FIGURE 5-5   Pulsing the CLEAR input to the LOW state will always produce $Q = 0$, $\overline{Q} = 1$: (a) $Q = 0$ prior to CLEAR pulse; (b) $Q = 1$ prior to CLEAR pulse.

(a)

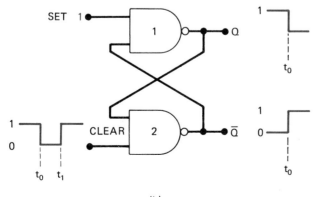

(b)

Simultaneous Setting and Clearing   The last case to consider is the case where the SET and CLEAR inputs are simultaneously pulsed LOW. This will produce HIGH levels at both NAND outputs so that $Q = \overline{Q} = 1$. Clearly, this is an undesired condition, since the two outputs are supposed to be inverses of each other. Furthermore, when the SET and CLEAR inputs return HIGH, the resulting output state will depend on which input returns HIGH first. Simultaneous transitions back to the 1 state will produce unpredictable results. For these reasons the SET = CLEAR = 0 condition is normally not used for the NAND latch.

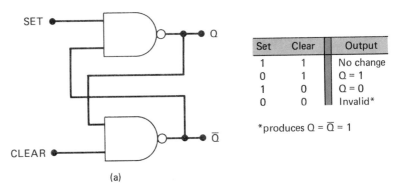

Set	Clear	Output
1	1	No change
0	1	Q = 1
1	0	Q = 0
0	0	Invalid*

*produces $Q = \overline{Q} = 1$

(a)

FIGURE 5-6   (a) NAND latch; (b) truth table.

Summary of NAND Latch   The operation described above can be conveniently placed in a truth table (Figure 5-6) and is summarized as follows:

Teaching Tip: You may want to qualify the statement about the $S = C = 1$ condition having no effect on the output state. This would not be true if S and C were both at 0 and then were simultaneously switched to the 1 state. Of course, this situation is not normally used.

1. **SET = CLEAR = 1.** This condition is the normal resting state and it has no effect on the output state. The Q and $\overline{Q}$ outputs will remain in whatever state they were prior to this input condition.

2. **SET = 0, CLEAR = 1.** This will always cause the output to go to the $Q = 1$ state, where it will remain even after SET returns HIGH. This is called *setting* the latch.

3. **SET = 1, CLEAR = 0.** This will always produce the $Q = 0$ state, where the output will remain even after CLEAR returns HIGH. This is called *clearing* or *resetting* the latch.

4. **SET = CLEAR = 0.** This condition tries to set and clear the latch at the same time and can produce ambiguous results. It should not be used.

Alternate Representations   From the description of the NAND latch operation, it should be clear that the SET and CLEAR inputs are active-LOW. The SET input will set $Q = 1$ when SET goes LOW; the CLEAR input will clear $Q = 0$ when CLEAR goes LOW. For this reason, the NAND latch is often drawn using the alternate representation for each NAND gate, as shown in Figure 5-7(a). The bubbles on the SET and CLEAR inputs emphasize the fact that these inputs are active-LOW.

Figure 5-7(b) shows a simplified block representation that we will sometimes use. The S and C labels represent the SET and CLEAR inputs, and the bubbles indicate the active-LOW nature of these inputs. Whenever we use this block symbol, it represents a NAND latch.

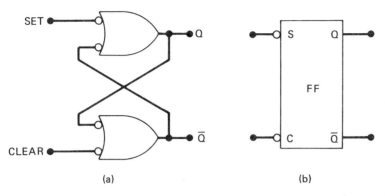

FIGURE 5-7   (a) NAND latch equivalent representation; (b) simplified block symbol.

Terminology   The action of *clearing* a FF or a latch is also called *resetting*, and both terms are used interchangeably in the digital field. In fact, a CLEAR input can also be called a RESET input, and a SET-CLEAR latch can be called a SET-RESET latch.

___ EXAMPLE 5-1 ___

The waveforms of Figure 5-8 are applied to the inputs of a NAND gate latch. Assume that initially $Q = 0$, and determine the Q waveform.

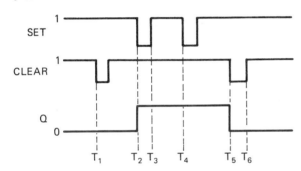

FIGURE 5-8   Example 5.1

Solution   Initially, SET = CLEAR = 1 so that Q will remain in the 0 state. The LOW pulse that occurs on the CLEAR input at time T_1 will have no effect, since Q is already in the cleared (0) state.

The only way that Q can go to the 1 state is by a LOW pulse on the SET input. This occurs at time T_2 when SET first goes LOW. When SET returns HIGH at T_3, Q will remain in its new HIGH state.

At time T_4 when SET goes LOW again, there will be no effect on Q because Q is already set to the 1 state.

The only way to bring Q back to the 0 state is by a LOW pulse on the CLEAR input. This occurs at time T_5. When CLEAR returns to 1 at time T_6, Q remains in the LOW state.

This example shows that the latch output "remembers" the last input that was activated and will not change states until the opposite input is activated.

Common Student Mistake: Students tend to want something to happen when the pulse at T_4 occurs. Remind them of how the situation in Figure 5-4(b) works; the LOW at $\overline{Q}$ is already keeping the NAND-1 output HIGH, so that the LOW at SET has no new effect.

EXAMPLE 5-2

It is virtually impossible to obtain a "clean" voltage transition from a mechanical switch, because of the phenomenon of **contact bounce**. This is illustrated in Figure 5-9(a), where the action of moving the switch from contact position 1 to 2 produces several output voltage transitions as the switch bounces (makes and breaks contact with contact 2 several times) before coming to rest on contact 2.

The multiple transitions on the output signal generally last no longer than a few milliseconds, but they would be unacceptable in many applications. A NAND latch can be used to prevent the presence of switch bounce from affecting the output. Describe the operation of the "switch debouncing" circuit in Figure 5-9(b).

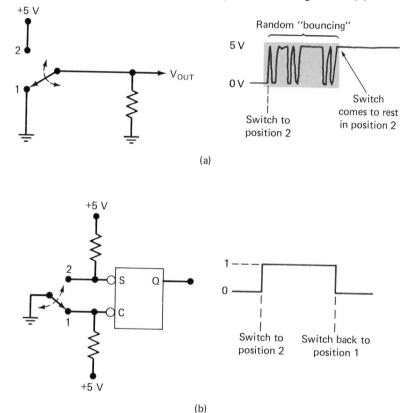

FIGURE 5-9 (a) Mechanical contact bounce will produce multiple transitions; (b) NAND latch used to debounce a mechanical switch.

Solution   Assume that the switch is resting in position 1 so that the CLEAR input is LOW and $Q = 0$. When the switch is moved to position 2, CLEAR will go HIGH, and a LOW will appear on the SET input as the switch first makes contact. This will set $Q = 1$ within a matter of a few nanoseconds (the response time of the NAND gate). Now if the switch bounces off contact 2, SET and CLEAR will both be HIGH, and Q will not be affected; it will stay HIGH. Thus, nothing will happen at Q as the switch bounces on and off contact 2 before finally coming to rest in position 2.

Likewise, when the switch is moved from position 2 back to position 1, it will place a LOW on the CLEAR input as it first makes contact. This clears Q to the LOW state, where it will remain even if the switch bounces on and off contact 1 several times before coming to rest.

Thus, the output at Q will consist of a single transition each time the switch is moved from one position to the other.

5-2 NOR GATE LATCH

Two cross-coupled NOR gates can be used as a NOR gate latch. The arrangement, shown in Figure 5-10(a), is similar to the NAND latch except that the Q and $\overline{Q}$ outputs have reversed positions.

The analysis of the operation of the NOR latch can be performed in exactly the same manner as for the NAND latch. The results are given in the truth table in Figure 5-10(b) and are summarized as follows:

1. **SET = CLEAR = 0.** This is the normal resting state for the NOR latch and it has no effect on the output state. Q and $\overline{Q}$ will remain in whatever state they were prior to the occurrence of this input condition.
2. **SET = 1, CLEAR = 0.** This will always set $Q = 1$, where it will remain even after SET returns to 0.

Teaching Tip: The $S = C = 0$ condition would have an effect if S and C were returning from the 1 state simultaneously. Again, this situation is not normally used.

FIGURE 5-10 (a) NOR gate latch; (b) truth table; (c) simplified block symbol.

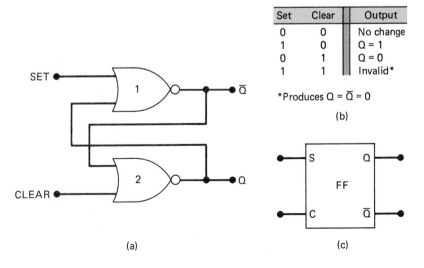

Set	Clear	Output
0	0	No change
1	0	Q = 1
0	1	Q = 0
1	1	Invalid*

*Produces $Q = \overline{Q} = 0$

(b)

(a)   (c)

3. **SET = 0, CLEAR = 1.** This will always clear $Q = 0$, where it will remain even after CLEAR returns to 0.

4. **SET = 1, CLEAR = 1.** This condition tries to set and clear the latch at the same time, and it produces $Q = \overline{Q} = 0$. If the inputs are returned to 0 simultaneously, the resulting output state is unpredictable. This input condition should not be used.

The NOR gate latch operates exactly like the NAND latch except that the SET and CLEAR inputs are active-HIGH rather than active-LOW, and the normal resting state is SET = CLEAR = 0. Q will be set HIGH by a HIGH pulse on the SET input, and it will be cleared LOW by a HIGH pulse on the CLEAR input. The simplified block symbol for the NOR latch in Figure 5-10(c) is shown with no bubbles on the S and C inputs; this indicates that these inputs are active-HIGH.

—— EXAMPLE 5-3 ——————————————————————————

Assume that $Q = 0$ initially, and determine the Q waveform for the NOR latch inputs of Figure 5-11.

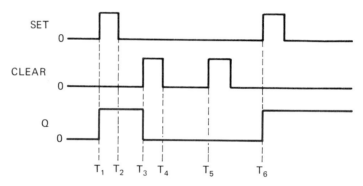

FIGURE 5-11   Example 5-3.

Solution   Initially, SET = CLEAR = 0, which has no effect in Q, and Q stays LOW. When SET goes HIGH at time T_1, Q will be set to 1 and will remain there even after SET returns to 0 at T_2.

At T_3 the CLEAR input goes HIGH and clears Q to the 0 state, where it remains even after CLEAR returns LOW at T_4.

The CLEAR pulse at T_5 has no effect on Q, since Q is already LOW. The SET pulse at T_6 again sets Q back to 1, where it will stay.

This example shows that the FF "remembers" the last input that was activated, and it will not change states until the opposite input is activated.

—— EXAMPLE 5-4 ——————————————————————————

Figure 5-12 shows a simple circuit that can be used to detect the interruption of a light beam. The light is focused on a phototransistor that is connected in the common-emitter configuration to operate as a switch. Assume that the latch has previously been cleared to the 0 state by momentarily opening switch S_1 and describe what happens if the light beam is momentarily interrupted.

Solution   With light on the phototransistor, we can assume that it is fully conducting so that the resistance between the collector and emitter is very small. Thus,

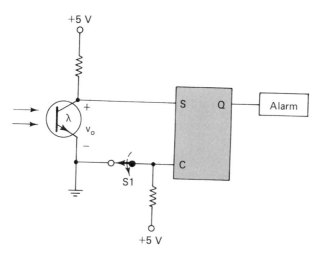

FIGURE 5-12   Example 5-4.

v_0 will be close to 0 V. This places a LOW on the SET input of the latch so that SET = CLEAR = 0.

When the light beam is interrupted, the phototransistor turns off and its collector-emitter resistance becomes very high (i.e., essentially an open circuit). This causes v_0 to rise to approximately 5 V; this activates the SET input and turns on the alarm.

Now, the alarm will remain on because Q will remain HIGH even if the light beam was only momentarily interrupted, and v_0 goes back to 0 V. The alarm can be turned off only by momentarily opening S_1 to produce a HIGH on the CLEAR input.

In this application, the latch's memory characteristic is used to convert a momentary occurrence (beam interruption) into a constant output.

Discussion Question: What would happen if the light beam is permanently broken, and S_1 is momentarily opened to try to turn off the alarm?

Flip-Flop State on Power-Up   When power is applied to a circuit, it is not possible to predict the starting state of a flip-flop's output if its SET and CLEAR inputs are in their inactive state (e.g., $S = C = 1$ for a NAND latch, $S = C = 0$ for a NOR latch). There is just as much chance that the starting state will be at $Q = 0$ as $Q = 1$. It will depend on things like internal propagation delays, parasitic capacitance, and external loading. If a latch or FF has to start off in a particular state to ensure the proper operation of a circuit, then it has to be placed in that state by momentarily activating the SET or CLEAR input at the start of the circuit's operation. This is often achieved by application of a pulse to the appropriate input.

― REVIEW QUESTIONS ―

1. What is the normal resting state of the NOR latch inputs? What is the active state?
2. When a FF is set, what are the states of Q and $\overline{Q}$?
3. What is the only way to cause the Q output of a NOR latch to change from 1 to 0?
4. If the NOR latch in Figure 5-12 were replaced by a NAND latch, why wouldn't the circuit work properly?

5-3 TROUBLESHOOTING CASE STUDY

The following two examples will present an illustration of the kinds of reasoning used in troubleshooting a circuit containing a latch.

___ EXAMPLE 5-5 ___

Analyze and describe the operation of the circuit in Figure 5-13.

Switch position		X_A	X_B
A		Pulses	LOW
B		LOW	Pulses

FIGURE 5-13   Examples 5-5 and 5-6.

Solution   The switch is used to set or clear the NAND latch to produce clean bounce-free signals at Q and $\overline{Q}$. These latch outputs control the passage of the 1-kHz pulse signal through to the AND outputs of X_A and X_B.

When the switch moves to position A, the latch is set to $Q = 1$. This enables the 1-kHz pulses to pass through to X_A, while the LOW at $\overline{Q}$ keeps $X_B = 0$. When the switch moves to position B, the latch is cleared to $Q = 0$, which keeps $X_A = 0$, while the HIGH at $\overline{Q}$ enables the pulses to pass through to X_B.

___ EXAMPLE 5-6 ___

A technician tests the circuit of Figure 5-13 and records the following observations:

Switch Position	Set (Z1-1)	Clear (Z1-5)	Q (Z1-3)	$\overline{Q}$ (Z1-6)	X_A (Z2-3)	X_B (Z2-6)
A	LOW	HIGH	LOW	HIGH	LOW	Pulses
B	HIGH	LOW	LOW	HIGH	LOW	Pulses

He notices that when the switch is in position B the circuit functions correctly, but in position A the latch does not set to the $Q = 1$ state. What are the possible faults that could produce this malfunction?

Solution   There are several possibilities:

1. An internal open connection at Z1-1. This would prevent Q from responding to the SET input.
2. An internal component failure in NAND gate Z1 that prevents it from responding properly.
3. The Q output stuck LOW. This could be caused by:
   (a) Z1-3 internally shorted to ground.
   (b) Z1-4 internally shorted to ground.
   (c) Z2-2 internally shorted to ground.
   (d) The Q node externally shorted to ground.
   An ohmmeter check from Q to ground will determine if any of these conditions are present. A visual check should reveal any external short.

   What about $\overline{Q}$ internally or externally shorted to V_{CC}? A little thought will lead to the conclusion that this could not be the fault. If $\overline{Q}$ were shorted to V_{CC}, this would not prevent the Q output from going HIGH when SET goes LOW. Since Q *does not* go HIGH, this cannot be the fault. The reason that $\overline{Q}$ looks as if it is stuck HIGH is that Q is stuck LOW and that keeps $\overline{Q}$ HIGH.

Discussion Question: Suppose an ohmmeter showed a short to ground at Z1-3. How would you go about determining whether it is Z1-3, Z1-4, or Z2-2 that is internally shorted?

5-4   CLOCK SIGNALS AND CLOCKED FLIP-FLOPS

Digital systems can operate either **asynchronously** or **synchronously.** In asynchronous systems, the outputs of logic circuits can change state any time one or more of the inputs change. An asynchronous system is difficult to design and troubleshoot.

In synchronous systems, the exact times at which any output can change states are determined by a signal commonly called the **clock**. This clock signal is generally a rectangular pulse train or square wave as shown in Figure 5-14. The clock signal is distributed to all parts of the system, and most (if not all) of the system outputs can change state only when the clock makes a transition. The transitions (also called

FIGURE 5-14   Clock signals.

edges) are pointed out in Figure 5-14. When the clock changes from a 0 to a 1, this is called the **positive-going transition (PGT);** when the clock goes from 1 to 0, this is the **negative-going transition (NGT).** We will use the abbreviations PGT and NGT, since these terms appear so often throughout the text.

Most digital systems are principally synchronous (although there are always some asynchronous parts), since synchronous circuits are easier to design and troubleshoot. They are easier to troubleshoot because the circuit outputs can change only at specific instants of time. In other words, almost everything is synchronized to the clock-signal transitions.

The synchronizing action of the clock signals is accomplished through the use of *clocked flip-flops* that are designed to change states on one or the other of the clock's transitions.

Clocked Flip-Flops   There are several types of clocked FFs that are used in a wide range of applications. Before we begin our study of the different clocked FFs, we will describe the principal ideas that are common to all of them.

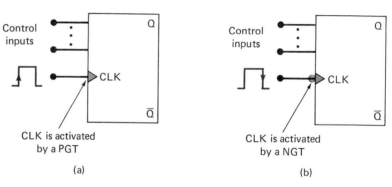

(a)   (b)

FIGURE 5-15   Clocked FFs have a clock input (*CLK*) that is active on either (a) PGT or (b) NGT. The control inputs determine the effect of the active clock transition.

1. Clocked FFs have a **clock** input that is typically labeled *CLK*, *CK*, or *CP*. We will use *CLK*, as shown in Figure 5-15. In most clocked FFs the *CLK* input is **edge-triggered,** which means that it is activated by a signal transition; this is indicated by the presence of a small triangle on the *CLK* input.

   In Figure 5-15(a) the *CLK* input is activated only when a positive-going transition (PGT) occurs; it is not affected at any other time. In Figure 5-15(b) the *CLK* input is activated only by a negative-going transition (NGT), as symbolized by the presence of the small bubble.

2. Clocked FFs also have one or more **control** inputs that can have various names, depending on their operation. The control inputs will have no effect on *Q* until the active clock transition occurs. In other words, their effect is *synchronous* with the signal applied to *CLK*. For this reason they are called **synchronous** control inputs.

   For example, the control inputs of the FF in Figure 5-15(a) will have no effect on *Q* until the PGT of the clock signal occurs. Likewise, the control inputs in Figure 5-15(b) will have no effect until the NGT of the clock signal occurs.

Teaching Tip: Here is a good analogy that clarifies the difference between the control inputs and the clock input. The control inputs are like the gearshift lever on a car. You set the gearshift to DRIVE or REVERSE to get the car ready to go forward or backward, respectively. But this alone does not make the car move in the selected direction. To get it to move, you have to step on the accelerator pedal—this is like the pulsing clock input. Once you activate the accelerator (clock pulse), the car will move according to the condition of the gearshift (control inputs).

3. In summary, we can say that the control inputs get the FF outputs ready to change, while the active transition at the *CLK* input actually *triggers* the change.

Setup and Hold Times   Two timing requirements must be met if a clocked FF is to respond reliably to its control inputs when the active *CLK* transition occurs. These requirements are illustrated in Figure 5-16 for a FF that triggers on a PGT.

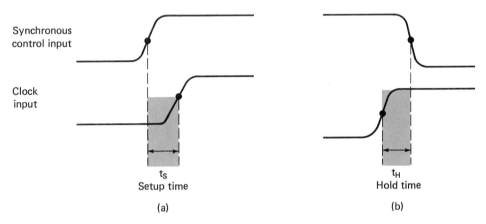

FIGURE 5-16   Control inputs have to be held stable for (a) a time, t_S, prior to active clock transition and for (b) a time, t_H, after the active clock transition.

The **setup time t_S,** is the time interval immediately preceding the active transition of the *CLK* signal during which the synchronous input has to be maintained at the proper level. IC manufacturers usually specify the minimum allowable setup time. If this time requirement is not met, the FF may not respond reliably when the clock edge occurs.

The **hold time, t_H,** is the time interval immediately following the active transition of the *CLK* signal during which the synchronous input has to be maintained at the proper level. IC manufacturers usually specify the minimum acceptable value of hold time. If this requirement is not met, the FF will not trigger reliably.

Thus, to ensure that a clocked FF will respond properly when the active clock transition occurs, the synchronous inputs must be stable (unchanging) for at least a time interval equal to t_S *prior* to the clock transition, and for at least a time interval equal to t_H *after* the clock transition.

IC flip-flops will have t_S and t_H values in the nanosecond range. Setup times are usually in the range 5 to 50 ns while hold times are generally from 0 to 10 ns. Notice that these times are measured between the 50 percent points on the transitions.

These timing requirements are very important in synchronous systems, because, as we shall see, there will be many situations where the synchronous inputs to a FF are changing at approximately the same time as the *CLK* input.

━━━ REVIEW QUESTIONS ━━━━━━━━━━━━━━━━━━━━━━━

1. What two types of inputs does a clocked FF have?
2. What is meant by the term "edge-triggered"?

5-5   CLOCKED S-C FLIP-FLOP

Figure 5-17(a) shows the logic symbol for a *clocked S-C flip-flop* that is triggered by the positive-going edge of the clock signal. This means that the FF can change states *only* when a signal applied to its clock input makes a transition from 0 to 1. The *S*

FIGURE 5-17   (a) Clocked S-C flip-flop that responds to the positive-going edge of a clock pulse; (b) truth table; (c) typical waveforms.

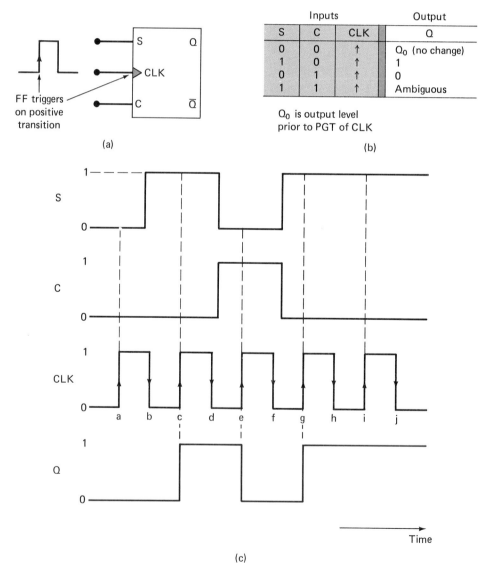

Inputs			Output
S	C	CLK	Q
0	0	↑	Q_0 (no change)
1	0	↑	1
0	1	↑	0
1	1	↑	Ambiguous

Q_0 is output level
prior to PGT of CLK

(b)

(c)

and C inputs control the state of the FF in the same manner as described earlier for the NOR-gate latch, but the FF does not respond to these inputs until the occurrence of the PGT of the clock signal.

The truth table in Figure 5-17(b) shows how the FF output will respond to the PGT at the CLK input for the various combinations of S and C inputs. This truth table uses some new nomenclature. The up arrow ($\uparrow$) indicates that a PGT is required at CLK; the label Q_0 indicates the level at Q prior to the PGT. This nomenclature is widely used by IC manufacturers on their IC data sheets.

The waveforms in Figure 5-17(c) illustrate the operation of the clocked S-C flip-flop. If we assume that the setup and hold time requirements are being met in all cases, we can analyze these waveforms as follows:

1. Initially all inputs are 0 and the Q output is assumed to be 0.

2. When the PGT of the first clock pulse occurs (point *a*), the S and C inputs are both 0, so the FF is not affected and remains in the $Q = 0$ state (i.e., $Q = Q_0$).

3. At the occurrence of the PGT of the second clock pulse (point *c*), the S input is now high, with C still low. Thus, the FF sets to the 1 state at the rising edge of this clock pulse.

4. When the third clock pulse makes its positive transition (point *e*), it finds that $S = 0$ and $C = 1$, which causes the FF to clear to the 0 state.

5. The fourth pulse sets the FF once again to the $Q = 1$ state (point *g*) because $S = 1$ and $C = 0$ when the positive edge occurs.

6. The fifth pulse also finds that $S = 1$ and $C = 0$ when it makes its positive-going transition. However, Q is already high, so it remains in that state.

7. The $S = C = 1$ condition should not be used, because it results in an ambiguous condition.

It should be noted from these waveforms that the FF is not affected by the negative-going transitions of the clock pulses. Also, note that the S and C levels have no effect on the FF, except upon the occurrence of a positive-going transition of the clock signal. The S and C inputs are synchronous *control* inputs; they control which state the FF will go to when the clock pulse occurs; the CLK input is the *trigger* input that causes the FF to change states according to what the S and C inputs are when the active clock transition occurs.

Figure 5-18 shows the symbol and truth table for a clocked S-C flip-flop that triggers on the *negative*-going transition at its CLK input. The small circle and triangle on the CLK input indicates that this FF will trigger only when the CLK input goes from 1 to 0. This FF operates in the same manner as the positive-edge FF ex-

Teaching Tip: Students will usually ask about the situation where S and/or C are changing at about the instant when the PGT occurs. This situation will be addressed in Section 5-12. For now, you can just restate that the setup- and hold-time requirements must be met in order for the FF to respond reliably.

FIGURE 5-18 Clocked S-C flip-flop that triggers on negative-going transitions.

	Inputs		Output
S	C	CLK	Q
0	0	↓	Q_0 (no change)
1	0	↓	1
0	1	↓	0
1	1	↓	Ambiguous

cept that the output can change states only on the falling edge of the clock pulses (points b, d, f, h, and j in Figure 5-17). Both positive-edge and negative-edge triggering FFs are used in digital systems.

Internal Circuitry of the Edge-Triggered S-C Flip-Flop   A detailed analysis of the internal circuitry of a clocked FF is not necessary, since all types are readily available as ICs. Although our main interest is in the FF's external operation, our understanding of this external operation can be aided by taking a look at a sim-

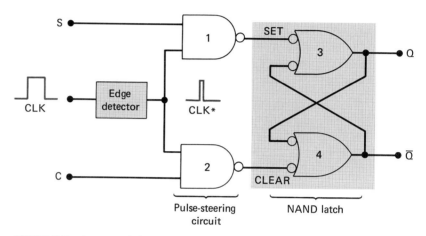

FIGURE 5-19   Simplified version of the internal circuitry for an edge-triggered S-C flip-flop.

plified version of the FF's internal circuitry. Figure 5-19 shows this for an edge-triggered S-C flip-flop.

The circuit contains three sections:

1. A basic NAND latch formed by NANDs 3 and 4.
2. A **pulse-steering** circuit formed by NANDs 1 and 2
3. An **edge-detector** circuit

The edge-detector produces a narrow positive-going spike ($CLK*$) that occurs coincident with the active transition of the CLK input pulse. The pulse-steering circuit "steers" the spike through to the SET or CLEAR input of the latch in accordance with the levels present at S and C. For example, with $S = 1$ and $C = 0$, the $CLK*$ signal is inverted and passed through NAND-1 to produce a LOW pulse at the SET input of the latch that sets $Q = 1$. With $S = 0$, $C = 1$, the $CLK*$ signal is inverted and passed through NAND-2 to produce a low pulse at the CLEAR input of the latch that resets $Q = 0$.

Figure 5-20(a) shows how the $CLK*$ signal is generated for edge-triggered FFs that trigger on a PGT. The INVERTER produces a delay of a few nanoseconds so that the transitions of $\overline{CLK}$ occur a little bit after those of CLK. The AND gate produces an output spike that is HIGH only for the few nanoseconds when CLK and $\overline{CLK}$ are both HIGH. The result is a narrow pulse at $CLK*$, which occurs on the PGT of CLK. The arrangement of Figure 5-20(b) likewise produces $CLK*$ on the

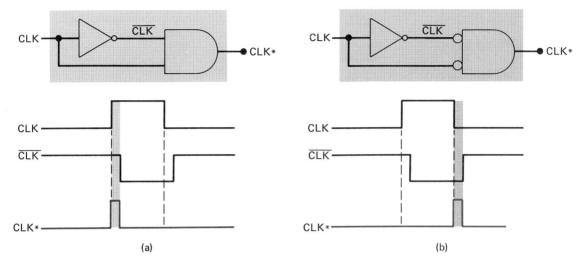

FIGURE 5-20   Implementation of edge detector circuits used in edge-triggered flip-flops: (a) PGT; (b) NGT.

NGT of *CLK* for FFs that are to trigger on a NGT.

Since the *CLK** signal is HIGH for only a few nanoseconds, *Q* is affected by the levels at *S* and *C* only for a short time during and after the occurrence of the active edge of *CLK*. This is what gives the FF its edge-triggered property.

—— REVIEW QUESTIONS ——————————————————

1. Suppose that the waveforms of Figure 5-17(c) are applied to the inputs of the FF of Figure 5-18. What will happen to *Q* at point *b*?
2. Explain why the *S* and *C* inputs affect *Q* only during the active transition of *CLK*.

5-6   CLOCKED J-K FLIP-FLOP

Figure 5-21(a) shows a *clocked J-K flip-flop* that is triggered by the positive-going edge of the clock signal. The *J* and *K* inputs control the state of the FF in the same ways as the *S* and *C* inputs do for the clocked S-C flip-flop except for one major difference: *the J = K = 1 condition does not result in an ambiguous output*. For this 1, 1 condition, the FF will always go to its *opposite* state upon the positive transition of the clock signal. This is called the **toggle** mode of operation. In this mode, if both *J* and *K* are left HIGH, the FF will change states (toggle) for each clock pulse.

The truth table in Figure 5-21(a) summarizes how the *J-K* flip-flop responds to the PGT for each combination of *J* and *K*. Notice that the truth table is the same as for the clocked S-C flip-flop (Figure 5-17) except for the *J* = *K* = 1 condition. This condition results in $Q = \overline{Q}_0$, which means that the new value of *Q* will be the inverse of the value it had prior to the PGT; this is the toggle operation.

The operation of this FF is illustrated by the waveforms in Figure 5-21(b). Once again we assume that the setup and hold time requirements are being met.

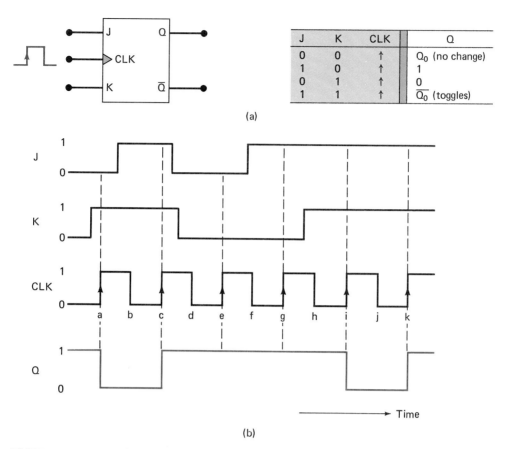

FIGURE 5-21 (a) Clocked J-K flip-flop that responds to positive edge of clock; (b) waveforms.

1. Initially all inputs are 0 and the Q output is assumed to be 1.

2. When the positive-going edge of the first clock pulse occurs (point a) the $J = 0$, $K = 1$ condition exists. Thus, the FF will be cleared to the $Q = 0$ state.

3. The second clock pulse finds $J = K = 1$ when it makes its positive transition (point c). This causes the FF to *toggle* to its opposite state, $Q = 1$.

4. At point e on the clock waveform, J and K are both 0, so that the FF does not change states on this transition.

5. At point g, $J = 1$ and $K = 0$. This is the condition that sets Q to the 1 state. However, it is already 1, and so it will remain there.

6. At point i, $J = K = 1$, and so the FF toggles to its opposite state. The same thing occurs at point k.

It should also be noted from these waveforms that the FF is not affected by the negative-going edge of the clock pulses. Also, the J and K input levels have no effect except upon the occurrence of the PGT of the clock signal. The J and K inputs by themselves cannot cause the FF to change states.

Figure 5-22 shows the symbol for a clocked J-K flip-flop that triggers on the negative-going clock-signal transitions. The small circle on the *CLK* input indicates

J	K	CLK	Q
0	0	↓	Q_0 (no change)
1	0	↓	1
0	1	↓	0
1	1	↓	$\overline{Q_0}$ (toggles)

FIGURE 5-22   J-K flip-flop that triggers on negative-going transitions.

that this FF will trigger when the *CLK* input goes from 1 to 0. This FF operates in the same manner as the positive-edge FF of Figure 5-21 except that the output can change states only on negative-going clock-signal transitions (points *b*, *d*, *f*, *h*, and *j*). Both polarities of edge-triggered J-K flip-flops are in common usage.

The J-K flip-flop is much more versatile than the S-C flip-flop because it has no ambiguous states. The $J = K = 1$ condition, which produces the toggling operation, finds extensive use in all types of binary counters. In essence, the J-K flip-flop can do anything the S-C flip-flop can do *plus* operate in the toggle mode.

Internal Circuitry of the Edge-Triggered J-K Flip-Flop

A simplified version of the internal circuitry of an edge-triggered J-K flip-flop is shown in Figure 5-23. It contains the same three sections as the edge-triggered S-C flip-flop (Figure 5-19). In fact, the only difference between the two circuits is that the Q and $\overline{Q}$ outputs are fed back to the pulse-steering NAND gates. This feedback connection is what gives the J-K flip-flop its toggle operation for the $J = K = 1$ condition.

Let's examine this toggle condition more closely by assuming that $J = K = 1$ and that Q is sitting in the LOW state when a *CLK* pulse occurs. With $Q = 0$ and $\overline{Q} = 1$, NAND gate 1 will steer *CLK** (inverted) to the SET input of the NAND latch to produce $Q = 1$. If we assume that Q is HIGH when a *CLK* pulse occurs, NAND gate 2 will steer *CLK** (inverted) to the CLEAR input of the latch to produce $Q = 0$. Thus, Q always ends up in the opposite state.

In order for the toggle operation to work as described above, the *CLK** pulse

Teaching Tip: The student may wonder what the letters *J* and *K* stand for. To the author's knowledge, they have no special meaning. It is helpful to think of a J-K flip-flop as a modified S-C flip-flop, where *J* is like *S* (set input) and *K* is like *C* (clear input), with the exception, of course, that $J = K = 1$ is allowed.

FIGURE 5-23   Internal circuit of edge-triggered J-K flip-flop.

must be very narrow. It has to return to 0 before the Q and $\overline{Q}$ outputs toggle to their new values; otherwise the new values of Q and $\overline{Q}$ will cause the $CLK*$ pulse to toggle the latch outputs again.

___ REVIEW QUESTIONS ___

1. *True or false:* A J-K flip-flop can be used as an S-C flip-flop, but an S-C flip-flop cannot be used as a J-K flip-flop.
2. Does a J-K flip-flop have any ambiguous input conditions?
3. What *J-K* input condition will always set Q upon the occurrence of the active *CLK* transition?

5-7   CLOCKED D FLIP-FLOP

Figure 5-24(a) shows the symbol and truth table for an edge-triggered D flip-flop that triggers on a PGT. Unlike the S-C and J-K flip-flops, this flip-flop has only the one synchronous control input, D, which stands for *data*. The operation of the D flip-flop is very simple; Q will go to the same state that is present on the D input when a PGT occurs at *CLK*. In other words, the level present at D will be *stored* in the flip-flop at the instant the PGT occurs. The waveforms in Figure 5-24(b) illustrate this operation.

Assume that Q is initially HIGH. When the first PGT occurs at point a, the D

FIGURE 5-24   (a) D flip-flop that triggers on positive-going transitions; (b) waveforms.

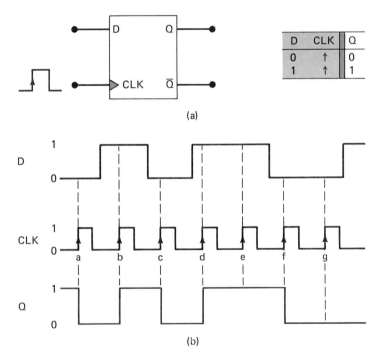

input is LOW; thus, Q will go to the 0 state. Even though the D input level changes between points a and b, it has no effect on Q; Q is storing the LOW that was on D at point a. When the PGT at b occurs, Q goes HIGH since D is HIGH at that time. Q stores this HIGH until the PGT at point c causes Q to go LOW, since D is LOW at that time. In a similar manner, the Q output takes on the levels present at D when the PGTs occur at points d, e, f, and g. Note that Q stays HIGH at point e because D is still HIGH.

Discussion Question: What will happen if the $\overline{Q}$ output is connected to the D input and pulses are applied to *CLK*? Repeat for the Q output.

Again, it is important to remember that Q can change only when a PGT occurs. The D input has no effect between PGTs.

A negative-edge-triggered D flip-flop operates in the same way just described except that Q will take on the value of D when a NGT occurs at *CLK*. The symbol for the D flip-flop that triggers on NGTs will have a bubble on the *CLK* input.

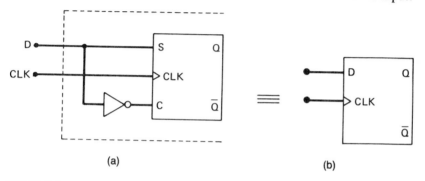

(a)   (b)

FIGURE 5-25   Edge-triggered D flip-flop implementation from S-C flip-flop.

Implementation of the D Flip-Flop   An edge-triggered D flip-flop is easily implemented by adding a single INVERTER to the edge-triggered S-C flip-flop as shown in Figure 5-25. If you try both values of D, you should see that Q takes on the level present at D when a PGT occurs.

—— EXAMPLE 5-7 ——————————————————————

How can a J-K flip-flop be modified to operate as a D flip-flop?

FIGURE 5-26   D flip-flop implementation from J-K flip-flop.

Solution   The modification, shown in Figure 5-26, is the same as was done for the S-C flip-flop in Figure 5-25.

Parallel Data Transfer   At this point you may well be wondering about the usefulness of the D flip-flop, since it appears that the Q output is the same as the D

*After occurrence of NGT

FIGURE 5-27   Parallel transfer of binary data using D flip-flops.

input. Not quite; remember, Q takes on the value of D only at certain time instances, and so it is not identical to D (e.g., see the waveforms in Figure 5-24).

In most applications of the D flip-flop, the Q output must take on the value at its D input only at precisely defined times. One example of this is illustrated in Figure 5-27. Logic-circuit outputs X, Y, Z are to be transferred to FFs Q_1, Q_2, and Q_3 for storage. Using the D flip-flops, the levels present at X, Y, and Z will be transferred to Q_1, Q_2, and Q_3 respectively, upon application of a TRANSFER pulse to the common CLK inputs. The FFs can store these values for subsequent processing. This is an example of *parallel* transfer of binary data; the three bits X, Y, and Z are all transferred *simultaneously*.

―― REVIEW QUESTIONS ――――――――――――

1. What will happen to the Q waveform in Figure 5-24(b) if the D input is held permanently LOW?
2. *True or false:* The Q output will equal the level at the D input at all times.

5-8   *D* LATCH

The edge-triggered D flip-flop uses an edge-detector circuit to ensure that the output will respond to the D input *only* when the active transition of the clock occurs. If this edge detector is not used, the resultant circuit operates somewhat differently. It is called a *D latch* and has the arrangement shown in Figure 5-28(a).

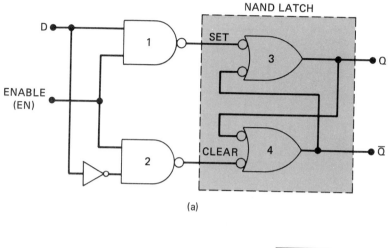

NAND LATCH

Inputs		Output
EN	D	Q
0	X	Q_0 (no change)
1	0	0
1	1	1

"X" indicates "don't care"
Q_0 is state Q just
prior to EN going LOW

(a)

(b)

(c)

FIGURE 5-28   *D* latch: (a) structure; (b) truth table; (c) logic symbol.

The circuit contains the NAND latch and the steering NAND gates 1 and 2. The common input to the steering gates is called an *enable* input (abbreviated *EN*) rather than a clock input because its effect on the Q and $\overline{Q}$ outputs is not restricted to occurring only on its transitions. The operation of the *D* latch is described as follows:

1. When *EN* is HIGH, the *D* input will produce a LOW at either the SET or CLEAR inputs of the NAND latch to cause Q to become the same level as *D*. If *D* changes while *EN* is HIGH, Q will follow the changes exactly. In other words, while *EN* = 1, the Q output will look exactly like *D*; in this mode, the *D* latch is said to be "transparent."

2. When *EN* goes LOW, the *D* input is inhibited from affecting the NAND latch since the outputs of both steering gates will be held HIGH. Thus, the Q and $\overline{Q}$ outputs will stay at whatever level they had just before *EN* went LOW. In other words, the outputs are "latched" to their current level and cannot change while *EN* is LOW even if *D* changes.

This operation is summarized in the truth table in Figure 5-28(b). The logic symbol for the *D* latch is given in Figure 5-28(c). Note that even though the *EN* input operates much like the *CLK* input of an edge-triggered FF, there is no small triangle on the *EN* input. This is because the small triangle symbol is used strictly for inputs that can cause an output change only when a transition occurs.

—— EXAMPLE 5-8 ——

Determine the Q waveform for a *D* latch with the *EN* and *D* inputs of Figure 5-29. Assume that $Q = 0$ initially.

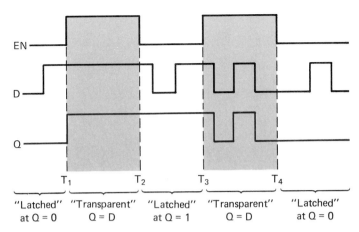

FIGURE 5-29   Waveforms for Example 5-8 showing the two modes for the D latch.

Solution   Prior to time T_1, *EN* is LOW, so that Q is "latched" at its current 0 level and cannot change even though D is changing. During the interval T_1 to T_2, *EN* is HIGH so that Q will follow the signal present at D. Thus, Q goes HIGH at T_1 and stays there since D is not changing. When *EN* returns LOW at T_2, Q will latch at the HIGH level that it has at T_2 and remain there while *EN* is LOW.

At T_3 when *EN* goes HIGH again, Q will follow the changes in the D input until T_4 when *EN* returns LOW. During the interval T_3 to T_4, the D latch is "transparent" since the variations in D go through to the output Q. At T_4 when *EN* goes LOW, Q will latch at the 0 level since that is its level at T_4. After T_4 the variations in D will have no effect on Q since it is latched (i.e., *EN* = 0).

─── REVIEW QUESTIONS ───────────

1. Describe how a D latch operates differently from an edge-triggered D flip-flop.
2. *True or false:* A D latch is in its transparent mode when *EN* = 0.
3. *True or false:* In a D latch, the D input can affect Q only when *EN* = 1.

5-9   ASYNCHRONOUS INPUTS

For the clocked flip-flops that we have been studying, the *S*, *C*, *J*, *K*, and *D* inputs have been referred to as *control* inputs. These inputs are also called **synchronous inputs,** because their effect on the FF output is synchronized with the *CLK* input. As we have seen, the synchronous control inputs must be used in conjunction with a clock signal to trigger the FF.

Most clocked FFs also have one or more **asynchronous inputs** which operate independently of the synchronous inputs and clock input. These asynchronous inputs can be used to set the FF to the 1 state or clear the FF to the 0 state at any time, regardless of the conditions at the other inputs. Stated in another way, the asyn-

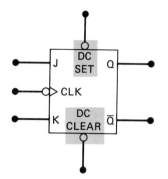

DC SET	DC CLEAR	FF response
1	1	Clocked operation*
0	1	Q = 1
1	0	Q = 0
0	0	Not used

*Q will respond to J, K, and CLK

FIGURE 5-30   Clocked J-K flip-flop with asynchronous inputs.

chronous inputs are **override** inputs, which can be used to override all the other inputs in order to place the FF in one state or the other.

Figure 5-30 shows a J-K flip-flop with two asynchronous inputs designated as DC SET and DC CLEAR. These are active-LOW inputs, as indicated by the bubbles on the FF symbol. The accompanying truth table summarizes how they affect the FF output. Let's examine the various cases.

- **DC SET = DC CLEAR = 1.** The asynchronous inputs are inactive and the FF is free to respond to the *J*, *K*, and *CLK* inputs; in other words, the clocked operation can take place.

- **DC SET = 0; DC CLEAR = 1.** The DC SET is activated and *Q* is *immediately* set to 1 no matter what conditions are present at the *J*, *K*, and *CLK* inputs. The *CLK* input cannot affect the FF while DC SET = 0.

- **DC SET = 1; DC CLEAR = 0.** The DC CLEAR is activated and *Q* is *immediately* cleared to 0 independent of the conditions on the *J*, *K*, or *CLK* inputs. The *CLK* input has no effect while DC CLEAR = 0.

- **DC SET = DC CLEAR = 0.** This condition should not be used, since it can result in an ambiguous response.

It is important to realize that these asynchronous inputs respond to dc levels. This means that if a constant 0 is held on the DC SET input, the FF will remain in the *Q* = 1 state regardless of what is occurring at the other inputs. Similarly, a constant low on the DC CLEAR input holds the FF in the *Q* = 0 state. Thus, the asynchronous inputs can be used to hold the FF in a particular state for any desired interval. Most often, however, the asynchronous inputs are used to set or clear the FF to the desired state by application of a momentary pulse.

Many clocked FFs that are available as ICs will have both of these asynchronous inputs; some will have only the DC CLEAR input. Some FFs will have asynchronous inputs that are active-HIGH rather than active-LOW. For these FFs the FF symbol would not have a bubble on the asynchronous inputs.

Designations for Asynchronous Inputs   IC manufacturers have not agreed on what nomenclature to use for these asynchronous inputs. We have called them DC SET and DC CLEAR to emphasize their response to dc levels. The most commonly used designations are given below.

Asynchronous SET Input	Asynchronous CLEAR Input
DC SET	DC CLEAR
PRESET (*PRE*)	CLEAR (*CLR*)
SET	RESET
S_D (direct set)	C_D (direct clear)

From now on, we will use the labels *PRE* and *CLR* to represent the asynchronous inputs, since these seem to be the most commonly used labels. When these asynchronous inputs are active-LOW, as they generally are, we will use the overbar to indicate their active-LOW status, that is, $\overline{PRE}$ and $\overline{CLR}$.

Although most IC flip-flops have at least one or more asynchronous inputs, there are some circuit applications where they are not used. In such cases they are held permanently at their inactive level. Often, in our use of FFs throughout the remainder of the text, we will not show a FF's unused asynchronous inputs; it will be assumed that they are permanently connected to their inactive logic level.

___ EXAMPLE 5-9 ___

Figure 5-31(a) shows an edge-triggered J-K flip-flop with active-LOW asynchronous inputs. The *J* and *K* inputs are permanently held HIGH. Determine the *Q* output response to the input waveforms at *CLK*, $\overline{PRE}$, and $\overline{CLR}$. Assume that $Q = 1$ initially.

Common Student Mistake: A common misconception is that the flip-flop will return to its prior state when the asynchronous input returns to its inactive state. For example, some students will show *Q* returning HIGH when the $\overline{CLR}$ pulse terminates in Figure 5-31. It should be re-emphasized that the flip-flop will have latched into the LOW state when $\overline{CLR}$ was LOW, and will remain there when $\overline{CLR}$ returns HIGH.

Solution   Initially, $\overline{PRE}$ and $\overline{CLR}$ are in their inactive HIGH state, so that they will have no effect on *Q*. Thus, when the first NGT of the *CLK* signal occurs at point *a*, *Q* will toggle to its opposite state; remember, $J = K = 1$ produces the toggle operation.

At point *b*, the $\overline{PRE}$ input is pulsed to its active-LOW state. This will *immediately* set $Q = 1$. Note that $\overline{PRE}$ produces $Q = 1$ without waiting for a NGT at *CLK*. The asynchronous inputs operate independent of *CLK*.

At point *c*, the NGT of *CLK* will again cause *Q* to toggle to its opposite state. Note that $\overline{PRE}$ has returned to its inactive state prior to point *c*. Likewise, the NGT of *CLK* at point *d* will toggle *Q* back HIGH.

At point *e*, the $\overline{CLR}$ input is pulsed to its active-LOW state and will *immediately* clear $Q = 0$. Again, it does this independent of *CLK*.

The NGT of *CLK* at point *f* will not toggle *Q*, because the $\overline{CLR}$ input is still active. The LOW at $\overline{CLR}$ overrides the *CLK* input and holds $Q = 0$.

When the NGT of *CLK* occurs at point *g*, it will toggle *Q* to the HIGH state since neither asynchronous input is active at that point.

These steps are summarized in Figure 5-31(b).

___ REVIEW QUESTIONS ___

1. How does the operation of an asynchronous input differ from that of a synchronous input?
2. Can a D flip-flop respond to its *D* and *CLK* inputs while $\overline{PRE} = 1$?
3. List the conditions necessary for a positive-edge-triggered J-K flip-flop with active-LOW asynchronous inputs to toggle to its opposite state.

FIGURE 5-31   Waveforms for Example 5-9 showing how a clocked flip-flop responds to asynchronous inputs.

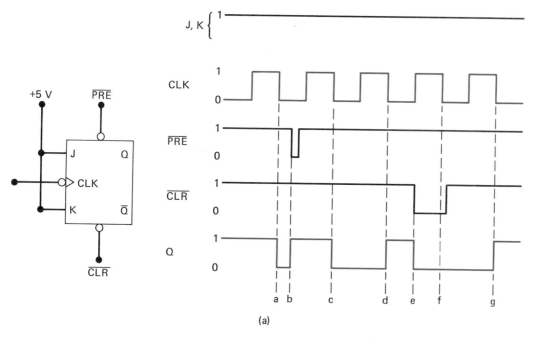

(a)

Point	Operation
a	Synchronous toggle on NGT of CLK
b	Asynchronous set on $\overline{PRE}$ = 0
c	Synchronous toggle
d	Synchronous toggle
e	Asynchronous clear on $\overline{CLR}$ = 0
f	$\overline{CLR}$ over-rides the NGT of CLK
g	Synchronous toggle

(b)

5-10   IEEE/ANSI SYMBOLS

We have been using the traditional symbols for each of the latches and FFs that have been introduced thus far, and we will continue to use these symbols in most of our circuit diagrams. In this section we will examine the new IEEE/ANSI symbols for these same devices so that you can become familiar with them.

Figure 5-32(a) shows the logic symbol for a single D latch. This is the new IEEE/ANSI symbol. It uses the letter "C" to denote the ENABLE input. As we shall see, the IEEE/ANSI symbology uses "C" for any input that *controls* when other inputs will have an effect on the output. As we know, the logic level applied to the ENABLE input controls when the D input is allowed to affect Q and $\overline{Q}$. Note that the outputs Q and $\overline{Q}$ are labeled outside the block, and note the right triangle on $\overline{Q}$ to indicate that it is the inverted output. This is standard for IEEE/ANSI. Recall that this right triangle is like the small bubble used in the older symbols.

Figure 5-32(b) shows the IEEE/ANSI symbol for a specific IC: the standard TTL 7475 quad latch. This IC contains four D latches that operate individually in the

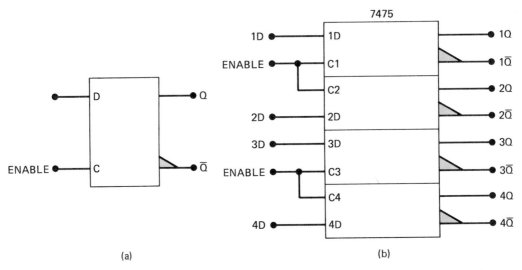

FIGURE 5-32   IEEE/ANSI symbols for (a) a single D latch and (b) an actual IC latch (7475 quad latch).

manner described earlier. This symbol also applies to the corresponding ICs in the other TTL and CMOS series: for example, the 74LS75, 74L75, 74C75, and 74HC75.

If we examine the logic symbol for the 7475 IC, several points should be noted. First, we can see that the overall symbol outline contains four smaller rectangles that represent the individual latches. Note how the inputs and outputs to each latch are labeled. For example, the D input to the top latch is labeled "$1D$," its enable input is labeled "$C1$," and its outputs are $1Q$ and $1\overline{Q}$. Finally, note that the top two latches have a common enable input; that is, $C1$ and $C2$ are connected together internally and brought out to a single pin on the IC package.

Likewise, the bottom two latches share a common enable input. Figure 5-33(a) shows the IEEE/ANSI logic symbol for a negative-edge-triggered J-K flip-flop with asynchronous inputs. The clock input is labeled "C" inside the symbol's rectangular outline. Note that there are two triangles on the clock input: the one inside the rectangle indicates that this input is edge-triggered; the one outside the rectangle indicates that it triggers on NGTs. The $\overline{PRE}$ and $\overline{CLR}$ inputs are active-LOW as symbolized by the right triangles. Interestingly, the IEEE/ANSI standard uses the labels "S" and "R" inside the rectangle to indicate asynchronous SET and RESET, which are equivalent to PRESET and CLEAR, respectively.

Figure 5-33(b) shows the IEEE/ANSI logic symbol for an IC that is part of the 74LS series of TTL devices. The 74LS112 is a dual negative-edge-triggered J-K flip-flop with preset and clear. It contains two J-K flip-flops, like the one symbolized in Figure 5-33(a). Note how the inputs and outputs are numbered. Also note that the input labels inside the rectangles are shown only for the top FF. It is understood that the inputs to the bottom FF are in the same arrangement as the top one. This same IC symbol applies to the 74S112 and 74HC112.

Figure 5-34(a) is the IEEE/ANSI symbol for a positive-edge-triggered D flip-flop with asynchronous inputs. There is no right triangle on the clock input, since this FF is clocked by PGTs.

Figure 5-34(b) is the IEEE/ANSI symbol for a 74175 IC, which contains four D flip-flops that share a common CLK input and common $\overline{CLR}$ input. The FFs do

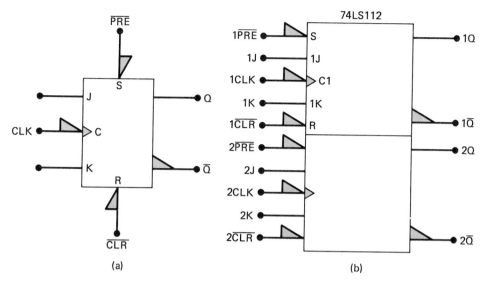

FIGURE 5-33   IEEE/ANSI symbols for (a) single edge-triggered J-K flip-flop and (b) an actual IC (74LS112 dual negative-edge-triggered J-K flip-flop).

not have a $\overline{PRE}$ input. This symbol contains a separate rectangle to represent each FF, and a special **common-control block,** which is the notched rectangle on top. The common-control block is used whenever an IC has one or more inputs that are common to more than one of the circuits on the chip. For the 74175, the *CLK* and $\overline{CLR}$ inputs are common to all four of the D flip-flops on the IC. This means that a PGT on *CLK* will cause each *Q* output to take on the level present at its *D* input; it also means that a LOW on $\overline{CLR}$ will clear all *Q* outputs to the LOW state.

FIGURE 5-34   IEEE/ANSI symbols for (a) a single edge-triggered D flip-flop and (b) an actual IC (74175 quad flip-flop with common clock and clear).

1. Explain the meaning of the two different triangles that can be part of the IEEE/ANSI symbology at a clock input.
2. Describe the meaning of the common-control block.

5-11   FLIP-FLOP TIMING CONSIDERATIONS

Manufacturers of IC flip-flops will specify several important timing parameters and characteristics that must be considered before a FF is used in any circuit application. We will describe the most important of these and then give some actual examples of specific IC flip-flops from the TTL and CMOS logic families.

Setup and Hold Times   The setup and hold times have already been discussed, and you may recall from Section 5-4 that they represent requirements that must be met for reliable FF triggering. The manufacturer's IC data sheet will always specify the *minimum* values of t_S and t_H.

Discussion Question: Why do you think maximum values of propagation delays are specified on data sheets?

Propagation Delays   Whenever a signal is to change the state of a FF's output, there is a delay from the time the signal is applied to the time when the output makes its change. Figure 5-35 illustrates the propagation delays that occur in response to a positive transition on the *CLK* input. Note that these delays are measured between the 50 percent points on the input and output waveforms. The same types of delays occur in response to signals on a FF's asynchronous inputs (PRESET and CLEAR). The manufacturer's data sheets usually specify propagation delays in response to all inputs, and they usually specify the *maximum* values for t_{PLH} and t_{PHL}.

Modern IC flip-flops have propagation delays that range from a few nanoseconds to around 1 μs: The values of t_{PLH} and t_{PHL} are generally not the same, and they increase in direct proportion to the number of loads being driven by the Q output. FF propagation delays play an important part in certain situations which we will encounter later.

FIGURE 5-35   FF propagation delays.

Delay going from
LOW to HIGH

(a)

Delay going from
HIGH to LOW

(b)

Maximum Clocking Frequency, f_{MAX}   This is the highest frequency that may be applied to the *CLK* input of a FF and still have it trigger reliably. The f_{MAX} limit will vary from FF to FF even with FFs having the same device number. For example, the manufacturer of the 7470 J-K flip-flop chip tests many of these FFs and may find that the f_{MAX} values fall in the range 20 to 35 MHz. He will then specify the *minimum* f_{MAX} as 20MHz. This may seem confusing, but a little thought should make it clear that what the manufacturer is saying is that he cannot guarantee that the 7470 FF that you put in your circuit will work above 20 MHz; many of them will, but some of them will not. If you stay below 20 MHz, however, he guarantees that they will all work.

Discussion Question: Do you think there is any relationship between a FF's propagation delay times and its maximum frequency of operation? Refer to Table 5-1.

Clock Pulse HIGH and LOW Times   The manufacturer will also specify the *minimum* time duration that the *CLK* signal must remain LOW before it goes HIGH, sometimes called $t_W(L)$, and the *minimum* time that *CLK* must be kept HIGH before it returns LOW, sometimes called $t_W(H)$. These times are defined in Figure 5-36(a). Failure to meet these minimum time requirements can result in unreliable triggering. Note that these time values are measured between the halfway points on the signal transitions.

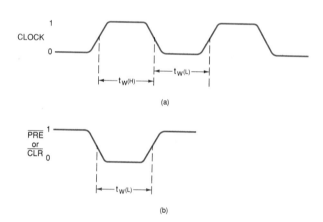

FIGURE 5-36   (a) Clock LOW and HIGH times; (b) asynchronous pulse width.

Asynchronous Active Pulse Width   The manufacturer will also specify the *minimum* time duration that a PRESET or CLEAR input has to be kept in its active state in order to reliably set or clear the FF. Figure 5-36(b) shows $t_W(L)$ for active-LOW asynchronous inputs.

Clock Transition Times   For reliable triggering, the clock waveform transition times (rise and fall times) should be kept very short. If the clock signal takes too long to make the transitions from one level to the other, the FF may trigger erratically or not at all. Manufacturers usually do not list a maximum transition-time requirement for each FF integrated circuit. Instead, it is usually given as a general

requirement for all ICs within a given logic family. For example, the transition times should generally be ≤ 50 ns for TTL devices and ≤ 200 ns for CMOS. These requirements will vary among the different manufacturers and among the various subfamilies within the broad TTL and CMOS logic families.

Actual ICs   As practical examples of these timing parameters, let's take a look at several actual integrated-circuit FFs. In particular, we will look at the following ICs:

- **7474**   Dual edge-triggered D flip-flop (TTL family)
- **74LS112** Dual edge-triggered J-K flip-flop (TTL)
- **4013B**   Dual edge-triggered D flip-flop (CMOS family)
- **74HC112** Dual edge-triggered J-K flip-flop (CMOS)

Table 5-1 lists the various timing values for each of these FFs as they appear in the manufacturers' data books. All the listed values are *minimum* values except for the propagation delays, which are *maximum* values.
Examination of Table 5-1 reveals two interesting points.

1. All the FFs have very low values of hold time; this is typical of most modern edge-triggered FFs.
2. The 74HC series of CMOS devices has timing values that are comparable to the TTL devices. The 4000 series is much slower than the 74HC series.

_____ EXAMPLE 5-10 _____

From Table 5-1 determine the following.
(a) Assume $Q = 0$. How long can it take for Q to go HIGH when a PGT occurs at the *CLK* input of a 7474?
(b) Assume $Q = 1$. How long can it take for Q to go LOW in response to the $\overline{CLR}$ input of a 74HC112?
(c) What is the narrowest pulse that should be applied to the DC CLEAR input of the 74LS112 FF to reliably clear Q?
(d) Which FF in Table 5-1 requires that the control inputs remain stable *after* the occurrence of the active clock transition?

TABLE 5-1   Flip-Flop Timing Values (in Nanoseconds)

	TTL		CMOS	
	7474	**74LS112**	**4013B**	**74HC112**
t_S	20	20	60	25
t_H	5	0	0	0
t_{PHL}—from *CLK* to Q	40	24	200	31
t_{PLH}—from *CLK* to Q	25	16	200	31
t_{PHL}—from $\overline{CLR}$ to Q	40	24	225	41
t_{PLH}—from $\overline{PRE}$ to Q	25	16	225	41
$t_W(L)$—*CLK* LOW time	37	15	100	25
$t_W(H)$—*CLK* HIGH time	30	20	100	25
$t_W(L)$—at $\overline{PRE}$ or $\overline{CLR}$	30	15	60	25
f_{MAX}—in MHz	15	30	5	20

(e) For which FFs do the control inputs have to be held stable for a minimum time prior to the active clock transition?

Solution

(a) The PGT will cause Q to go from LOW to HIGH. The delay from CLK to Q is listed as $t_{PLH} = 25$ ns for the 7474.

(b) For the 74HC112 the time required for Q to go from HIGH to LOW in response to the $\overline{CLR}$ input is listed at $t_{PHL} = 41$ ns.

(c) For the 74LS112 the narrowest pulse at the DC CLEAR input is listed as $t_W(L) = 15$ ns.

(d) The 7474 is the only FF in Table 5-1 that has a nonzero hold-time requirement.

(e) All of the FFs have a nonzero setup time requirement.

Discussion Question: A 74LS112 IC is used in a circuit that is operating at 25 MHz. Would it surprise you if the circuit operation was not adversely affected when you replaced the 74LS112 by a 74HC112?

REVIEW QUESTIONS

1. Which FF timing parameters indicate the time it takes the Q output to respond to an input?
2. *True or false:* A FF that has an f_{MAX} rating of 25 MHz can be reliably triggered by any CLK pulse waveform with a frequency below 25 MHz.

5-12 POTENTIAL TIMING PROBLEM IN FF CIRCUITS

In many digital circuits, the output of one FF is connected either directly or through logic gates to the input of another FF, and both FFs are triggered by the same clock signal. This presents a potential timing problem. A typical situation is illustrated in Figure 5-37, where the output of Q_1 is connected to the J input of Q_2 and both FFs are clocked by the same signal at their CLK inputs.

The potential timing problem is this: since Q_1 will change on the NGT of the clock pulse, the J_2 input of Q_2 will be changing as it receives the same NGT. This could lead to an unpredictable response at Q_2.

Let's assume that initially $Q_1 = 1$ and $Q_2 = 0$. Thus, the Q_1 FF has $J_1 = K_1 = 1$, and Q_2 has $J_2 = Q_1 = 1$, $K_2 = 0$ prior to the NGT of the clock pulse. When the NGT occurs, Q_1 will toggle to the LOW state, but it will not actually go LOW until after its propagation delay, t_{PHL}. The same NGT will reliably clock Q_2 to the HIGH state provided that t_{PHL} is greater than Q_2's hold-time requirement, t_H. If this condition is not met, the response of Q_2 will be unpredictable.

Fortunately, all modern edge-triggered FFs have hold-time requirements that are 5 ns or less; most have $t_H = 0$, which means that they have no hold-time requirement. For these FFs, situations like that in Figure 5-37 will not be a problem.

Unless stated otherwise, in all of the FF circuits that we encounter throughout the text, we will assume that the FF's hold-time requirement is short enough to respond reliably according to the following rule:

> ***The FF output will go to a state determined by the logic levels present at its synchronous control inputs just prior to the active clock transition.***

Teaching Tip: You might want to mention that in a well-designed system, a FF's control inputs will rarely be changing just prior to the active clock transition. If they were, then it would be necessary to ensure that the control inputs complete their change at least one setup time prior to the clock transition.

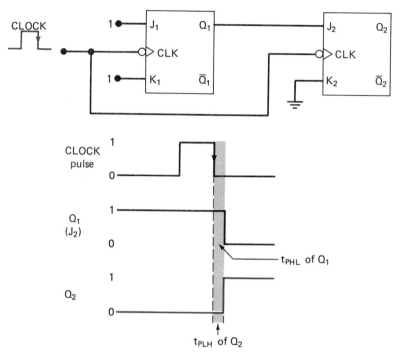

FIGURE 5-37 Q_2 will properly respond to the level present at Q_1 prior to the NGT of *CLK*, provided that Q_2's hold-time requirement, t_H, is less than Q_1's propagation delay.

If we apply this rule to Figure 5-37, it says that Q_2 will go to a state determined by the $J_2 = 1$, $K_2 = 0$ condition that is present just prior to the NGT of the clock pulse. The fact that J_2 is changing in response to the same NGT has no effect.

____ EXAMPLE 5-11 _____

Determine the Q output for a negative-edge-triggered J-K flip-flop for the input waveforms shown in Figure 5-38. Assume that $t_H = 0$ and that $Q = 0$ initially.

FIGURE 5-38 Example 5-11.

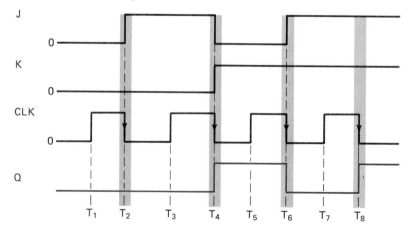

Solution The FF will respond only at times T_2, T_4, T_6, and T_8. At T_2, Q will respond to the $J = K = 0$ condition present just prior to T_2. At T_4, Q will respond to the $J = 1$, $K = 0$ condition present just prior to T_4. At T_6, Q will respond to the $J = 0$, $K = 1$ condition present just prior to T_6. At T_8, Q responds to $J = K = 1$.

5-13 MASTER/SLAVE FLIP-FLOPS

Before the development of edge-triggered FFs with little or no hold-time requirements, timing problems such as that in Figure 5-37 were often handled by using a class of FFs called *master/slave* FFs. A master/slave FF actually contains two FFs—a master and a slave. On the rising edge of the *CLK* signal, the levels on the control inputs (D, J, K) are used to determine the output of the master. When the *CLK* signal goes LOW, the state of the master is transferred to the slave, whose outputs are Q and $\overline{Q}$. Thus, Q and $\overline{Q}$ change just after the NGT of the clock. These master/slave FFs function very much like the negative-edge-triggered FFs except for one major disadvantage: the control inputs must be held stable while *CLK* is HIGH, or unpredictable operation may occur. This problem with master/slave FFs has been overcome with an improved master/slave version called a *master/slave with data lockout*.

The master/slave FF has become obsolete, although you may encounter it in older equipment. Examples of this type are the standard TTL 7473, 7476, and 74107, and the data lockout versions, 74110 and 74111. The newer IC technologies (74LS, 74AS, 74ALS, 74HC, 74HCT) do not include any master/slave FFs in their series. In fact, the 74LS76 and 74LS107 have been manufactured as edge-triggered FFs even though their standard 74 series counterparts are master/slave.

For most purposes, if you encounter a master/slave FF in a piece of equipment, you can analyze it like a negative-edge-triggered FF.

5-14 FLIP-FLOP APPLICATIONS

Earlier in this chapter we saw some examples of how the simple NAND flip-flop and NOR flip-flop were used to perform switch debouncing (Example 5-2) and event storage (Example 5-4). These simple unclocked FFs are somewhat limited in their applications. Clocked FFs offer the logic designer a group of versatile devices that have numerous applications. We will briefly introduce the more common applications in the following sections and will expand on them in subsequent chapters.

5-15 FLIP-FLOP SYNCHRONIZATION

Most digital systems are principally synchronous in their operation in that most of the signals will change states in synchronism with the clock transitions. In many cases, however, there will be an external signal that is not synchronized to the clock; in other words, it is asynchronous. Asynchronous signals often occur as a result of a human operator's actuating an input switch at some random time relative to the clock signal. This randomness can produce unpredictable and undesirable results. The following example illustrates how a FF can be used to synchronize the effect of an asynchronous input.

——— EXAMPLE 5-12 ———

Figure 5-39(a) shows a situation where input signal A is generated from a debounced switch that is actuated by an operator (a debounced switch was first introduced in Example 5-2). A goes HIGH when the operator actuates the switch and goes LOW when the operator releases the switch. This A input is used to control the passage of the clock signal through the AND gate so that clock pulses appear at output X only as long as A is HIGH.

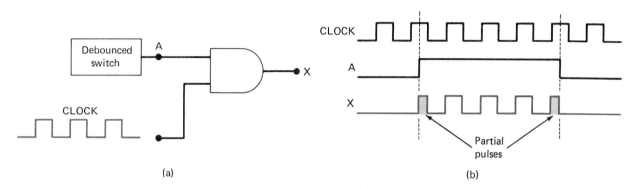

(a) (b)

FIGURE 5-39   Asynchronous signal A can produce partial pulses at X.

FIGURE 5-40   An edge-triggered D flip-flop is used to synchronize the enabling of the AND gate to the NGTs of the clock.

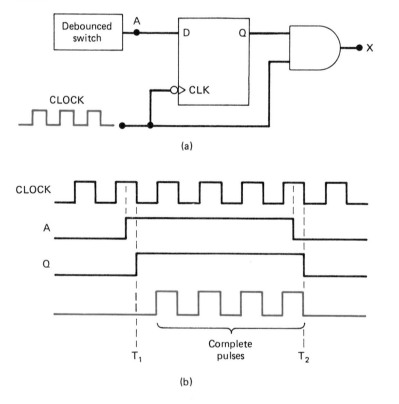

The problem with this circuit is that *A* is asynchronous; it can change states at any time relative to the clock signal because the exact times when the operator actuates or releases the switch are essentially random. This can produce *partial* clock pulses at output *X* if either transition of *A* occurs while the clock signal is HIGH, as shown in the waveforms of Figure 5-39(b).

This type of output is often not acceptable, and so a method for preventing the appearance of partial pulses at *X* must be developed. One solution is shown in Figure 5-40(a). Describe how this circuit solves the problem, and draw the *X* waveform for the same situation as in Figure 5-39(b).

Solution   The *A* signal is connected to the *D* input of FF *Q*, which is clocked by the NGT of the clock signal. Thus, when *A* goes HIGH, *Q* will not go HIGH until the next NGT of the clock at time T_1. This HIGH at *Q* will enable the AND gate to pass subsequent *complete* clock pulses to *X*, as shown in Figure 5-40(b).

When *A* returns LOW, *Q* will not go LOW until the next NGT of the clock at T_2. Thus, the AND gate will not inhibit clock pulses until the clock pulse that ends at T_2 has been passed through to *X*.

Thus, output *X* contains only complete pulses.

Discussion Question: How would the operation be affected if the FF in Figure 5-40 clocked on a PGT? Especially, be aware of a possible glitch at *X*.

5-16   DETECTING AN INPUT SEQUENCE

In many situations an output is to be activated only when the inputs are activated in a certain sequence. This cannot be accomplished using pure combinational logic but requires the storage characteristic of FFs.

For example, an AND gate can be used to determine when two inputs *A* and *B* are both HIGH, but its output will respond the same regardless of which input goes HIGH first. But suppose we want to generate a HIGH output *only* if *A* goes HIGH and then *B* goes HIGH some time later. One way to accomplish this is shown in Figure 5-41(a).

The waveforms in Figure 5-41(b) and (c) show that *X* will go HIGH only if *A* goes HIGH before *B* goes HIGH. This is because *A* must be HIGH in order for *X* to go HIGH on the PGT of *B*.

In order for this circuit to work properly, *A* has to go HIGH prior to *B* by at least an amount of time equal to the setup-time requirement of the FF.

FIGURE 5-41   Clocked J-K flip-flop used to respond to a particular sequence of inputs.

(a)

(b) A goes HIGH
before B

(c) B goes HIGH
before A

By far the most common use of flip-flops is for the storage of data or information. The data may represent numerical values (e.g., binary numbers, BCD-coded decimal numbers). These data are generally stored in groups of FFs called **registers**.

The operation most often performed on data that are stored in a FF or a register is the **transfer** operation. This involves the transfer of data from one FF or register to another. Figure 5-42 illustrates how data transfer can be accomplished between two FFs using clocked S-C, J-K, and D flip-flops. In each case, the logic value that is currently stored in FF *A* is transferred to FF *B* upon the NGT of the TRANSFER pulse. Thus, after this NGT, the *B* output will be the same as the *A* output.

The transfer operations in Figure 5-42 are examples of **synchronous transfer,** since the synchronous and *CLK* inputs are used to perform the transfer. A transfer operation can also be obtained using the asynchronous inputs of a FF. Figure 5-43 shows how an **asynchronous transfer** can be accomplished using the PRESET and CLEAR inputs of any type of FF. Here, the asynchronous inputs respond to LOW levels. When the TRANSFER ENABLE line is held LOW, the two NAND outputs are kept HIGH, with no effect on the FF outputs. When the TRANSFER ENABLE line is made HIGH, one of the NAND outputs will go LOW, depending on the state of the *A* and *Ā* outputs. This LOW will either set or clear FF *B* to the same state as FF *A*. This asynchronous transfer is done independently of the synchronous and *CLK*

Teaching Tip: The student sometimes wonders what is happening to the inputs of flip-flop *A* while its data are being transferred to flip-flop *B*. You may want to mention that in many cases the inputs to *A* will be inactive, while in other cases they are used to transfer new data into *A*. This latter situation will occur in shift registers.

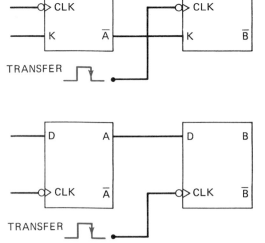

FIGURE 5-42 Synchronous data transfer operation performed by various types of FFs.

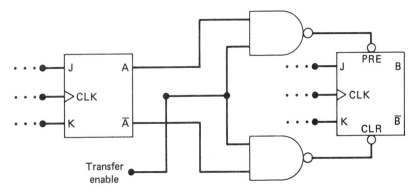

FIGURE 5-43   Asynchronous data transfer operation.

inputs of the FF. Asynchronous transfer is also called **jam transfer,** because the data can be "jammed" into FF *B* even if the synchronous inputs are active.

Parallel Data Transfer   Figure 5-44 illustrates data transfer from one register to another using *D*-type FFs. Register *X* consists of FFs X_1, X_2, and X_3; register *Y* consists of FFs Y_1, Y_2, and Y_3. Upon application of the TRANSFER pulse, the level stored in X_1 is transferred to Y_1, X_2 to Y_2, and X_3 to Y_3. The transfer of the contents of the *X* register into the *Y* register is a synchronous transfer. It is also referred to as a *parallel* transfer, since the contents of X_1, X_2, and X_3 are transferred *simultaneously* into Y_1, Y_2, and Y_3. If a *serial* transfer were performed, the contents of the *X* register

FIGURE 5-44   Parallel transfer of contents of register *X* into register *Y*.

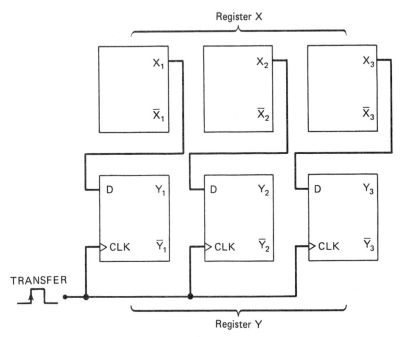

would be transferred to the Y register one bit at a time. This will be examined in the next section.

It is important to understand that parallel transfer does not change the contents of the register that is the source of data. For example, in Figure 5-44, if $X_1 X_2 X_3 = 101$ and $Y_1 Y_2 Y_3 = 011$ prior to the occurrence of the TRANSFER pulse, then both registers will be holding 101 after the TRANSFER pulse.

_____ REVIEW QUESTIONS _____

1. *True or false:* Asynchronous data transfer uses the *CLK* input.
2. Which type of FF is best suited for synchronous transfer because it requires the fewest interconnections from one FF to the other?
3. If J-K flip-flops were used in the registers of Figure 5-44, how many total interconnections would be required from register X to register Y?
4. *True or false:* Synchronous data transfer requires less circuitry than asynchronous transfer.

5-18 SERIAL DATA TRANSFER: SHIFT REGISTERS

Before we describe the serial data transfer operation, we must first examine the basic *shift-register* arrangement. A shift register is a group of FFs arranged so that the binary numbers stored in the FFs are shifted from one FF to the next for every clock pulse. You have undoubtedly seen shift registers in action in devices such as an electronic calculator, where the digits shown on the display shift over each time you key in a new digit. This is the same action taking place in a shift register.

Figure 5-45(a) shows one way to arrange J-K flip-flops to operate as a 4-bit shift register. Note that the FFs are connected so that the output of X_3 transfers into X_2, X_2 into X_1, and X_1 into X_0. What this means is that upon the occurrence of the shift pulse, each FF takes on the value stored previously in the FF on its left. Flip-flop X_3 takes on a value determined by the conditions present on its J and K inputs when the shift pulse occurs. For now, we will assume that X_3's J and K inputs are fed by the DATA IN waveform shown in Figure 5-45(b). We will also assume that all FFs are in the 0 state before shift pulses are applied.

The waveforms in Figure 5-45(b) show how the input data are shifted from left to right from FF to FF as shift pulses are applied. When the first NGT occurs at T_1, each of the FFs X_2, X_1, and X_0 will have the $J = 0$, $K = 1$ condition present at its inputs because of the state of the FF on its left. Flip-flop X_3 will have $J = 1$, $K = 0$ because of DATA IN. Thus, at T_1 only X_3 will go HIGH, while all the other FFs remain LOW. When the second NGT occurs at T_2, flip-flop X_3 will have $J = 0$, $K = 1$ because of DATA IN. Flip-flop X_2 will have $J = 1$, $K = 0$ because of the current HIGH at X_3. Flip-flops X_1 and X_0 will still have $J = 0$, $K = 1$. Thus, at T_2 only FF X_2 will go HIGH, FF X_3 will go LOW, and FFs X_1 and X_0 will remain LOW.

Similar reasoning can be used to determine how the waveforms change at T_3 and T_4. Note that on each NGT of the shift pulses, each FF output takes on the level that was present at the output of the FF on its left *prior* to the NGT. Of course, X_3 takes on the level that was present at DATA IN prior to the NGT.

Common Student Mistake: It is not uncommon for some students to think that the NGT at T_1 will cause all the FFs to go to the 1 state. You may want to explain why this cannot be.

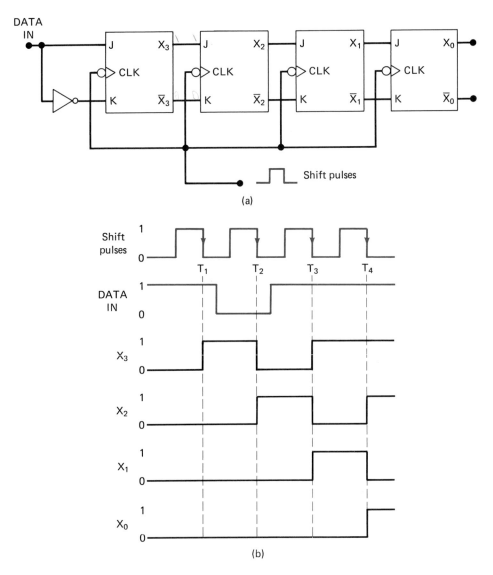

FIGURE 5-45   Four-bit shift register.

Hold-Time Requirement

Hold-Time Requirement   In this shift-register arrangement it is necessary that the FFs have a very small hold-time requirement, because there are times when the J, K inputs are changing at about the same time as the CLK transition. For example, the X_3 output switches from 1 to 0 in response to the NGT at T_2, causing the J, K inputs of X_2 to change while its CLK input is changing. Actually, because of the propagation delay of X_3, the J, K inputs of X_2 won't change for a short time after the NGT. For this reason, a shift register should be implemented using edge-triggered FFs that have a t_H value less than one FF propagation delay (CLK-to-output). This latter requirement is easily satisfied by most modern edge-triggered FFs.

Serial Transfer between Registers   Figure 5-46(a) shows two 3-bit shift registers connected so that the contents of the X register will be serially transferred (shifted) into register Y. We are using D flip-flops for each shift register, since this

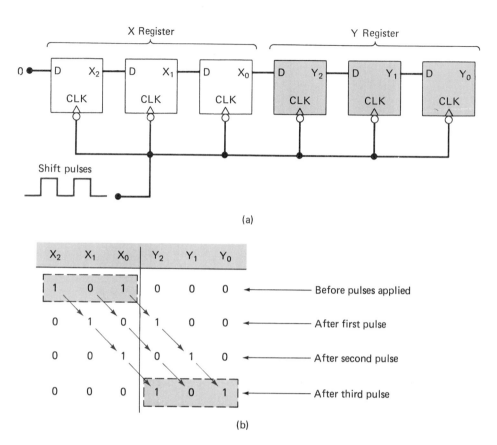

(a)

X_2	X_1	X_0	Y_2	Y_1	Y_0	
1	0	1	0	0	0	← Before pulses applied
0	1	0	1	0	0	← After first pulse
0	0	1	0	1	0	← After second pulse
0	0	0	1	0	1	← After third pulse

(b)

FIGURE 5-46   Serial transfer of information from X register into Y register.

requires fewer connections than J-K flip-flops. Notice how X_0, the last FF of register X, is connected to the input of Y_2, the first FF of register Y. Thus, as the shift pulses are applied, the information transfer takes place as follows: $X_2 \rightarrow X_1 \rightarrow X_0 \rightarrow Y_2 \rightarrow Y_1 \rightarrow Y_0$. The X_2 FF will go to a state determined by its D input. For now, D will be held LOW, so that X_2 will go LOW on the first pulse and will remain there.

To illustrate, let us assume that before any shift pulses are applied, the contents of the X register are 101 (i.e., $X_2 = 1$, $X_1 = 0$, $X_0 = 1$) and the Y register is at 000. Refer to the table in Figure 5-46(b), which shows how the states of each FF change as shift pulses are applied. The following points should be noted:

1. On the falling edge of each pulse, each FF takes on the value that was stored in the FF on its left prior to the occurrence of the pulse.

2. After *three* pulses, the 1 that was initially in X_2 is in Y_2, the 0 initially in X_1 is in Y_1, and the 1 initially in X_0 is in Y_0. In other words, the 101 stored in the X register has now been shifted into the Y register. The X register is at 000; it has lost its original data.

3. The complete transfer of the *three* bits of data requires *three* shift pulses.

_____ EXAMPLE 5-13 _____

Assume the same initial contents of the X and Y registers in Figure 5-46. What will be the contents of each FF after the occurrence of the sixth shift pulse?

Solution   If we continue the process shown in Figure 5-46(b) for three more shift pulses, we will find that all of the FFs will be in the 0 state after the sixth pulse. Another way to arrive at this result is to reason as follows: The constant 0 level at the D input of X_2 shifts in a new 0 with each pulse so that after six pulses the registers are filled up with 0s.

Discussion Question: Analyze the operation of Figure 5-46 when Y_0 is connected to the D input of X_2. Try it with $|Y| = 100$ initially.

Parallel versus Serial Transfer   In parallel transfer, all the information is transferred simultaneously upon the occurrence of a *single* transfer command pulse (Figure 5-44), no matter how many bits are being transferred. In serial transfer, as exemplified by Figure 5-46, the complete transfer of N bits of information requires N clock pulses (3 bits requires three pulses, 4 bits requires four pulses, etc.). Parallel transfer, then, is obviously much faster than serial transfer using shift registers.

In parallel transfer, the output of each FF in register X is connected to a corresponding FF input in register Y. In serial transfer, only the last FF in register X is connected to register Y. In general, then, parallel transfer requires more interconnections between the sending register (X) and the receiving register (Y) than does serial transfer. This difference becomes more critical when a greater number of bits of information are being transferred. This is an important consideration when the sending and receiving registers are remote from each other, since it determines how many lines (wires) are needed for the transmission of the information.

The choice of either parallel or serial transmission depends on the particular system application and specifications. Often, a combination of the two types is used to take advantage of the *speed* of parallel transfer and the *economy and simplicity* of serial transfer. More will be said later about information transfer.

___ REVIEW QUESTIONS ___

1. *True or false:* The fastest method for transferring data from one register to another is parallel transfer.
2. What is the major advantage of serial transfer over parallel transfer?
3. Refer to Figure 5-46. Assume that the initial contents of the registers are: $X_2 = 0$, $X_1 = 1$, $X_0 = 0$, $Y_2 = 1$, $Y_1 = 1$, $Y_0 = 0$. Also assume that the D input of X_2 is held HIGH. Determine the value of each FF output after the occurrence of the fourth shift pulse.

5-19   FREQUENCY DIVISION AND COUNTING

Refer to Figure 5-47(a). Each FF has its J and K inputs at the 1 level, so that it will change states (toggle) whenever the signal on its *CLK* input goes from HIGH to LOW. The clock pulses are applied only to the *CLK* input of FF X_0. Output X_0 is connected to the *CLK* input of FF X_1, and output X_1 is connected to the *CLK* input of FF X_2. The waveforms in Figure 5-47(b) show how the FFs change states as the pulses are applied. The following important points should be noted:

1. Flip-flop X_0 toggles on the negative-going transition of each input clock pulse. Thus, the X_0 output waveform has a frequency that is exactly one-half of the clock pulse frequency.

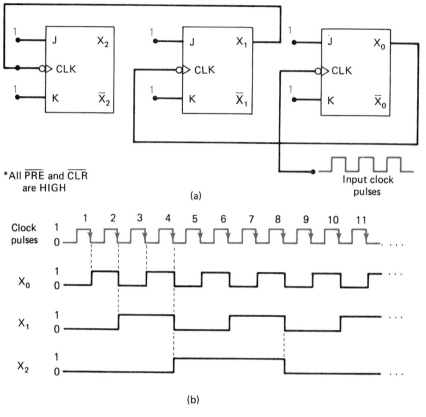

*All $\overline{PRE}$ and $\overline{CLR}$ are HIGH

(a)

(b)

FIGURE 5-47   J-K flip-flops wired as a 3-bit binary counter (MOD-8).

2. Flip-flop X_1 toggles each time the X_0 output goes from HIGH to LOW. The X_1 waveform has a frequency equal to exactly one-half the frequency of the X_0 output and therefore one-fourth of the clock frequency.

3. Flip-flop X_2 toggles each time the X_1 output goes from HIGH to LOW. Thus, the X_2 waveform has one-half the frequency of X_1 and therefore one-eighth of the clock frequency.

4. Each FF output is a square wave (50 percent duty cycle).

As described above, each FF divides the frequency of its input by 2. Thus, if we were to add a fourth FF to the chain, it would have a frequency equal to one-sixteenth of the clock frequency, and so on. Using the appropriate number of FFs, this circuit could divide a frequency by any power of 2. Specifically, using N flip-flops would produce an output frequency from the last FF which is equal to $1/2^N$ of the input frequency.

Counting Operation   In addition to functioning as a frequency divider, the circuit of Figure 5-47 also operates as a **binary counter.** This can be demonstrated by examining the sequence of states of the FFs after the occurrence of each clock pulse. Figure 5-48 presents the results in a **state table.** Let the $X_2 X_1 X_0$ values represent a binary number where X_2 is in the 2^2 position, X_1 is in the 2^1 position, and X_0 is in the 2^0 position. The first eight $X_2 X_1 X_0$ states in the table should be recognized as the binary counting sequence from 000 to 111. After the first NGT, the FFs are in the 001 state ($X_2 = 0$, $X_1 = 0$, $X_0 = 1$), which represents 001_2 (equivalent to decimal 1); after the second NGT the FFs represent 010_2, which is equivalent to 2_{10}; af-

196   CHAPTER 5 / FLIP-FLOPS AND RELATED DEVICES

$\underline{2^2}$	$\underline{2^1}$	$\underline{2^0}$	
X_2	X_1	X_0	
0	0	0	Before applying clock pulses
0	0	1	After pulse #1
0	1	0	After pulse #2
0	1	1	After pulse #3
1	0	0	After pulse #4
1	0	1	After pulse #5
1	1	0	After pulse #6
1	1	1	After pulse #7
0	0	0	After pulse #8 recycles to 000
0	0	1	After pulse #9
0	1	0	After pulse #10
0	1	1	After pulse #11
.	.	.	. . .
.	.	.	. . .
.	.	.	. . .

FIGURE 5-48 Table of flip-flop states shows binary counting sequence.

ter three pulses, $011_2 = 3_{10}$; after four pulses, $100_2 = 4_{10}$; and so on until after seven pulses, $111_2 = 7_{10}$. On the eighth NGT the FFs return to the 000 state, and the binary sequence repeats itself for succeeding pulses.

Thus, for the first seven input pulses, the circuit functions as a binary counter in which the states of the FFs represent a binary number equivalent to the number of pulses that have occurred. This counter can count as high as $111_2 = 7_{10}$, and then it returns to 000.

State Transition Diagram   Another way to show how the states of the FFs change with each applied clock pulse is to use a *state transition diagram* as illustrated in Figure 5-49. Each circle represents one possible state as indicated by the binary

FIGURE 5-49 State transition diagram shows how the states of the counter flip-flops change with each applied clock pulse.

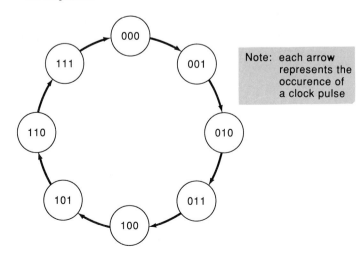

Note: each arrow represents the occurence of a clock pulse

Discussion Question: What will be the state of the counter of Figure 5-47 on power-up? How can it be placed in the desired starting state?

number inside the circle. For example, the circle containing the number 100 represents the 100 state (i.e., $X_2 = 1$, $X_1 = X_0 = 0$).

The arrows connecting one circle to another show how one state changes to another as a clock pulse is applied. By looking at a particular state circle, we can see which state precedes it and which state follows it. For example, looking at the 000 state, we see that this state is reached whenever the counter is in the 111 state and a clock pulse is applied. Likewise, we see that the 000 state is always followed by the 001 state.

We will use state transition diagrams to help describe, analyze, and design counters and other sequential FF circuits.

MOD Number   The counter of Figure 5-47 has $2^3 = 8$ different states (000 through 111). It would be referred to as a *MOD-8 counter,* where the MOD number indicates the number of states in the counting sequence. If a fourth FF were added, the sequence of states would count in binary from 0000 to 1111, a total of 16 states. This would be called a *MOD-16 counter.* In general, if N flip-flops are connected in the arrangement of Figure 5-47, the counter will have 2^N different states, and so it is a MOD-2^N counter. It would be capable of counting up to $2^N - 1$ before returning to its zero state.

The MOD number of a counter also indicates the frequency division obtained from the last FF. For instance, a 4-bit counter has four FFs, each representing one binary digit (bit), and so it is a MOD-2^4 = MOD-16 counter. It can therefore count up to 15 ($= 2^4 - 1$). It can also be used to divide the input pulse frequency by a factor of 16 (the MOD number).

We have looked only at the basic FF binary counter. We will examine counters in much more detail in Chapter 7.

___ EXAMPLE 5-14 _____

Assume that the MOD-8 counter in Figure 5-47 is in the 101 state. What will be the state (count) after 13 pulses have been applied?

Solution   Locate the 101 state on the state transition diagram. Proceed around the state diagram through eight state changes and you should be back in the 101 state. Now continue through five more state changes (for a total of 13) and you should end up in the 010 state.

Notice that since this is a MOD-8 counter with eight states, it takes eight state transitions to make one complete excursion around the diagram back to the starting state.

___ EXAMPLE 5-15 _____

Consider a counter circuit that contains six FFs wired in the arrangement of Figure 5-47, (i.e., X_5, X_4, X_3, X_2, X_1, X_0).

(a) Determine the counter's MOD number.

(b) Determine the frequency at the output of the last FF (X_5) when the input clock frequency is 1 MHz.

(c) What is the range of counting states for this counter?

(d) Assume a starting state (count) of 000000. What will be the counter's state after 129 pulses?

Solution

(a) MOD number $= 2^6 = 64$.

(b) The frequency at the last FF will equal the input clock frequency divided by the MOD number. That is,

$$f \text{ (at } X_5) = \frac{1 \text{ MHz}}{64} = 15.625 \text{ kHz}$$

(c) The counter will count from 000000_2 to 111111_2 (0 to 63_{10}) for a total of 64 states. Note that the number of states is the same as the MOD number.

(d) Since this is a MOD-64 counter, every 64 clock pulses will bring the counter back to its starting state. Therefore, after 128 pulses the count is back to 000000. The 129th pulse brings the counter to the 000001 count.

REVIEW QUESTIONS

1. A 20-kHz clock signal is applied to a J-K flip-flop with $J = K = 1$. What is the frequency of the FF output waveform?
2. How many FFs are required for a counter that will count 0 to 255_{10}?
3. What is the MOD number of this counter?
4. What is the frequency of the output of the eighth FF when the input clock frequency is 512 kHz?
5. If this counter starts at 00000000, what will be its state after 520 pulses?

Discussion Question: Examine how the counter operation changes when the FFs have clock inputs that respond to PGTs.

5-20   MICROCOMPUTER APPLICATION

Your study of digital systems is still in a relatively early stage, and you have not learned very much about microprocessors and microcomputers. However, you can get a basic idea of how FFs are employed in a typical microprocessor-controlled application without being concerned with all of the details you will need to know later.

Figure 5-50 shows a microprocessor unit (MPU) with its outputs used to transfer binary data to register X, which consists of four D flip-flops X_3, X_2, X_1, X_0. One set of MPU outputs is the *address code* made up of the eight outputs A_{15}, A_{14}, A_{13}, A_{12}, A_{11}, A_{10}, A_9, A_8. Most MPUs have at least 16 available address outputs, but they are not always all used. A second set of MPU outputs consists of the four *data lines* D_3, D_2, D_1, D_0. Most MPUs have at least eight available data lines. The other MPU output is the clock signal CP.

Recall that the MPU is the central processing unit of a microcomputer, and its main function is to execute a program of instructions stored in the computer's memory. One of the instructions it might execute could be one that tells the MPU to transfer a binary number from a storage register within the MPU to the external register X. In executing this instruction, the MPU would perform the following steps:

1. Place the binary number onto its data output lines D_3 through D_0.
2. Place the proper address code on its output lines A_{15} through A_8 to select register X as the recipient of the data.
3. Once the data and address outputs are stabilized, the MPU generates the clock pulse CP to clock the register and complete the parallel transfer of data into X.

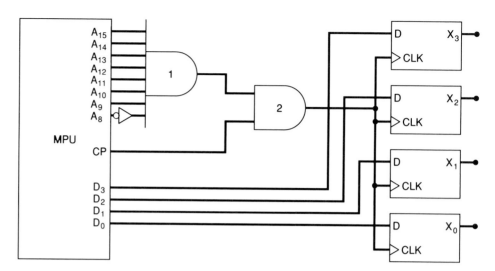

FIGURE 5-50 Example of a microprocessor transferring binary data to an external register.

Discussion Question: How would you add a second register so that the MPU can transfer data to one register or the other, but not to both at the same time?

The FFs in register X can be used for many different purposes. For instance, in situations where the MPU has performed a calculation and the result is to be displayed, each FF can be connected to a single LED so that the binary number transferred into X can be readily seen. As another example, the individual FFs can control the on/off status of electromagnetic devices such as motors and solenoids on the basis of the binary code sent from the MPU, which in turn is under the control of the program the MPU is currently executing.

_____ EXAMPLE 5-16 _____

(a) What address code must the MPU generate in order for the data to be transferred into X?

(b) Assume that X_3–X_0 = 0110, A_{15}–A_8 = 11111111, and D_3–D_0 = 1011. What will be in X after a *CP* pulse occurs?

Solution

(a) In order for the data to be transferred into X, the clock pulse must pass through AND gate 2 into the *CLK* inputs of the FFs. This will happen only if the top input of AND gate 2 is HIGH. This means that all of the inputs to AND gate 1 must be HIGH; that is, A_{15} through A_9 must be 1, and A_8 must be 0. Thus, the presence of address code 11111110 is needed to allow data to be transferred into X.

(b) With A_8 = 1, the LOW from AND gate 1 will inhibit *CP* from getting through AND gate 2, and the FFs will not be clocked. Therefore, the contents of register X will not change from 0110.

5-21 SCHMITT-TRIGGER DEVICES

A *Schmitt-trigger* circuit is not classified as a flip-flop, but it does exhibit a type of memory characteristic that makes it useful in certain special situations. One of those situations is shown in Figure 5-51(a). Here a standard INVERTER is being driven

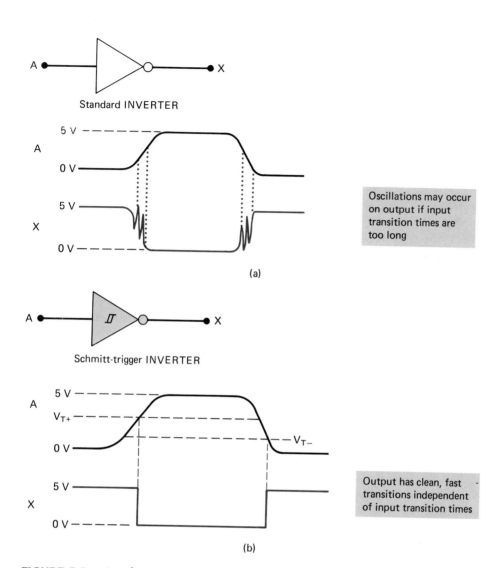

Standard INVERTER

Oscillations may occur on output if input transition times are too long

(a)

Schmitt-trigger INVERTER

Output has clean, fast transitions independent of input transition times

(b)

FIGURE 5-51 (a) If input transition times are too long, a standard logic device output might oscillate or change erratically; (b) a logic device with a Schmitt-trigger type of input will produce clean, fast output transitions.

by a logic input that has relatively slow transition times. When these transition times exceed the maximum allowed values (this depends on the particular logic family), the outputs of logic gates and INVERTERs may produce oscillations as the input signal passes through the indeterminate range. The same input conditions can also produce erratic triggering of FFs.

A device that has a Schmitt-trigger type input is designed to accept slow-changing signals and produce an output that has oscillation-free transitions. The output will generally have very rapid transition times (typically, 10 ns) that are independent of the input signal characteristics. Figure 5-51(b) shows a Schmitt-trigger INVERTER and its response to a slow-changing input.

If you examine the waveforms in Figure 5-51(b), you should note that the output does not change from HIGH to LOW until the input exceeds the **positive-going threshold** voltage, V_{T+}. Once the output goes LOW, it will remain there even when the input drops back below V_{T+} (this is its memory characteristic) until it drops all

the way down below the **negative-going threshold** voltage, V_{T-}. The values of the two threshold voltages will vary from logic family to logic family, but V_{T-} will always be less than V_{T+}.

The Schmitt-trigger INVERTER, and all devices with Schmitt-trigger inputs, use the distinctive symbol shown in Figure 5-51(b) to indicate that they can reliably respond to slow-changing input signals. Logic designers use ICs with Schmitt-trigger inputs to convert slow-changing signals to clean, fast-changing signals that can drive standard IC inputs.

There are several ICs available with Schmitt-trigger inputs. The 7414, 74LS14, and 74HC14 are hex INVERTER ICs with Schmitt-trigger inputs. The 7413, 74LS13, and 74HC13 are dual four-input NANDs with Schmitt-trigger inputs.

━━ REVIEW QUESTIONS ━━

1. What could occur when a slow-changing signal is applied to a standard logic IC?
2. How does a Schmitt-trigger logic device operate differently from a standard logic device?

5-22 ONE-SHOT (MONOSTABLE MULTIVIBRATOR)

A digital circuit that is somewhat related to the FF is the *one-shot* (abbreviated OS). Like the FF, the OS has two outputs, Q and $\overline{Q}$, which are inverses of each other. Unlike the FF, the OS has only one *stable* output state (normally, $Q = 0$, $\overline{Q} = 1$), where it remains until it is triggered by an input signal. Once triggered, the OS outputs switch to the opposite state ($Q = 1$, $\overline{Q} = 0$). It remains in this **quasi-stable** state for a fixed period of time, t_p, which is usually determined by an RC time constant which results from the values of external components connected to the OS. After a time t_p, the OS outputs return to their resting state until triggered again.

Figure 5-52(a) shows the logic symbol for a OS. The value of t_p is often indicated somewhere on the OS symbol. In practice, t_p can vary from several nanoseconds to several tens of seconds. The exact value of t_p is essentially determined by the values of external components R_T and C_T.

Two types of one-shots are available in IC form: the **nonretriggerable** and the **retriggerable**.

Nonretriggerable One-Shot   The waveforms in Figure 5-52 illustrate the operation of a nonretriggerable OS that triggers on positive-going transitions at its trigger (T) input. The important points to note are:

1. The PGTs at points a, b, c, and e will trigger the OS to its quasi-stable state for a time t_p, after which it automatically returns to the stable state.
2. The PGTs at points d and f have no effect on the OS because it has already been triggered to the quasi-stable state. The OS must return to the stable state before it can be retriggered.

FIGURE 5-52   OS symbol and typical waveforms for nonretriggerable operation.

Teaching Tip: It can be fun to refer to the OS as an "unstable relative" of the FF. When a FF is triggered to a new state, it will stay there indefinitely until an appropriate set of input conditions triggers it to its opposite state. When a OS is triggered to a new state, it will stay there only temporarily, returning to its original state on its own.

3. The OS output-pulse duration is always the same regardless of the duration of the input pulses. As stated above, t_p depends only on R_T and C_T and the internal OS circuitry. A typical OS may have a t_p given by $t_p = 0.7R_T C_T$.

Retriggerable One-Shot

The retriggerable OS operates much like the nonretriggerable OS except for one major difference: *it can be retriggered while it is in the quasi-stable state, and it will begin a new t_p interval*. Figure 5-53(a) compares the response of both types of OS using a t_p of 2 ms. Let's examine these waveforms.

Both types of OS respond to the first trigger pulse at $t = 1$ ms by going HIGH for 2 ms and then returning LOW. The second trigger pulse at $t = 5$ ms triggers both one-shots to the HIGH state. The third trigger pulse at $t = 6$ ms has no effect on the nonretriggerable OS, since it is already in its quasi-stable state. However, this trigger pulse will *retrigger* the retriggerable OS to begin a new $t_p = 2$ ms interval. Thus, it will stay HIGH for 2 ms *after* this third trigger pulse.

In effect, then, a retriggerable OS begins a new t_p interval each time a trigger pulse is applied, regardless of the current state of its Q output. In fact, trigger pulses can be applied at a rate fast enough that the OS will always be retriggered before the end of the t_p interval and Q will remain HIGH. This is shown in Figure 5-53(b), where eight pulses are applied every 1 ms. Q does not return LOW until 2 ms after the last trigger pulse.

Discussion Question: Under what conditions will a retriggerable OS operate the same as a nonretriggerable OS?

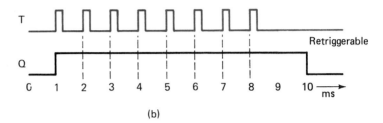

FIGURE 5-53 (a) Comparison of nonretriggerable and retriggerable OS responses for $t_p = 2$ ms. (b) Retriggerable OS begins a new t_p internal each time it receives a trigger pulse.

Actual Devices There are several one-shot ICs available in both the retriggerable and nonretriggerable versions. The 74121 and 74L121 are single nonretriggerable one-shot ICs; the 74221, 74LS221, and 74HC221 are dual nonretriggerable one-shot ICs; the 74122 and 74LS122 are single retriggerable one-shot ICs; the 74123, 74LS123, and 74HC123 are dual retriggerable one-shot ICs.

Figure 5-54(a) shows the traditional symbol for the 74121 nonretriggerable one-shot IC. Note that it contains internal logic gates to allow inputs A_1, A_2, and B to trigger the OS in a variety of ways. The B input is a Schmitt-trigger type input

FIGURE 5-54 Logic symbols for the 74121 nonretriggerable one-shot: (a) traditional; (b) IEEE/ANSI.

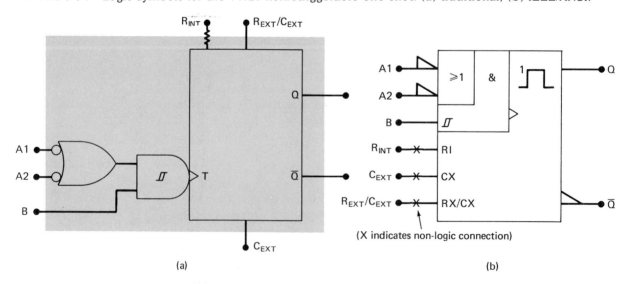

that is allowed to have slow transition times and still reliably trigger the OS. The pins labeled R_{INT}, R_{EXT}/C_{EXT}, and C_{EXT} are used to connect an external resistor and capacitor to achieve the desired output pulse duration. Figure 5-54(b) is the IEEE/ANSI symbol for the 74121 nonretriggerable OS. Note how this symbol represents the logic gates. Also notice the presence of a small pulse with 1 in front of it. This indicates that the device is a nonretriggerable OS. The IEEE/ANSI symbol for a retriggerable OS would not have the 1 in front of the pulse.

Monostable Multivibrator   One-shots are also called *monostable multivibrators* because they have only one stable state. The major applications of the OS are in timing circuits that utilize the predetermined t_p interval. Some of these applications will be introduced in the end-of-chapter problems and at various places throughout the text.

REVIEW QUESTIONS

1. In the absence of a trigger pulse, what will be the state of a OS output?
2. *True or false:* When a nonretriggerable OS is pulsed while it is in its quasi-stable state, the output is not affected.
3. What determines the t_p value for a OS?
4. Describe how a retriggerable OS operates differently from a nonretriggerable OS.

5-23 ANALYZING SEQUENTIAL CIRCUITS

Many logic circuits contain FFs, one-shots, and logic gates that are connected to perform a specific operation. Very often, a master clock signal is used to cause the logic levels in the circuit to go through a particular sequence of states. We can generally analyze these sequential circuits by following this step-by-step procedure:

1. Examine the circuit diagram and look for circuit structures such as counters or shift registers that you are familiar with. This can help to simplify the analysis.
2. Determine the logic levels that are present at the inputs of each FF *prior to* the occurrence of the first clock pulse.
3. Use these levels to determine how each FF output will change in response to 12 the first clock pulse.
4. Repeat steps 2 and 3 for each successive clock pulse.

   The following example will illustrate the procedure.

EXAMPLE 5-17

Consider the circuit of Figure 5-55. Initially, all the FFs and the OS are in the 0 state before the clock pulses are applied. These pulses have a repetition rate of 1 kHz. Determine the waveforms at X, Y, Z, W, $\overline{Q}$, A, and B for 16 cycles of the clock input.

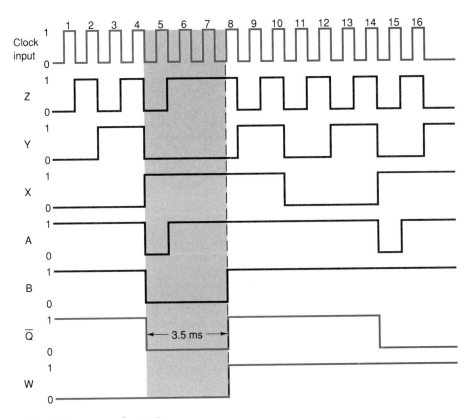

FIGURE 5-55   Example 5-17.

Solution   Initially, the FFs and OS are in the 0 state, so that $X = Y = Z = W = Q = 0$. The inputs to the NAND gate are $X = 0$, $\bar{Y} = 1$, and $\bar{Z} = 1$, and so its output $A = 1$. The inputs to the OR gate are $W = 0$ and $\bar{Q} = 1$, so that its output is $B = 1$.

As long as B remains HIGH, flip-flop Z will have its J and K inputs both HIGH, so it will operate in the toggle mode. Flip-flops X and Y are kept in the toggle mode, since their J and K inputs are held at the 1 level permanently. It should be clear, then, that FFs X, Y, and Z will operate as a 3-bit counter as long as B stays HIGH. Thus, flip-flop Z will toggle on the negative transition of each input clock pulse, flip-flop Y will toggle on the negative transition of the Z output, and flip-flop X will toggle on the negative transition of flip-flop Y. This operation continues for clock pulses 1 to 4.

When clock pulse 4 makes its negative transition, Z goes LOW, Y goes LOW, and X goes HIGH. Therefore, the inputs to the NAND gate become $X = 1$, $\bar{Y} = 1$, and $\bar{Z} = 1$, causing the NAND output A to go LOW. This negative transition at A will trigger the OS to its quasi-stable state, so that $\bar{Q}$ goes LOW for 3.5 ms. This negative transition of $\bar{Q}$ will not affect flip-flop W, since this D-type FF triggers on positive transitions at its CLK input. With $\bar{Q}$ and W both LOW, the OR gate output B goes LOW, so that flip-flop Z now has $J = 1$, $K = 0$. This means that Z will be *set* to the 1 state on the *next* clock pulse and will remain in this state for each successive clock pulse as long as $J = 1$, $K = 0$. Thus, Z goes HIGH on clock pulse 5 and stays HIGH for clock pulses 6 and 7. (*Note:* A goes back HIGH when Z goes HIGH on pulse 5 because $\bar{Z} = 0$.)

After 3.5 ms the OS returns to its stable state with $Q = 0$, $\bar{Q} = 1$. This positive transition of $\bar{Q}$ causes flip-flop W to go HIGH since $D = 1$. The 1 at W produces a 1 at B so that flip-flop Z is back in the toggle mode with $J = K = 1$. The counter will now operate properly in response to all succeeding clock pulses.

On the negative transition of clock pulse 14, the NAND output A again goes LOW thereby triggering the OS. The OS, however, will not affect the K input of Z, since W is keeping the OR output HIGH. Also, since its D input is HIGH, flip-flop W will remain in the $W = 1$ state indefinitely, allowing the counter to count normally for all succeeding clock pulses.

5-24   ASTABLE MULTIVIBRATORS

Flip-flops have two stable states; therefore, we can say that they are *bistable multivibrators*. One-shots have one stable state, and so we call them *monostable multivibrators*. A third type of multivibrator has no stable states; it is called an *astable* or *free-running multivibrator*. This type of logic circuit switches back and forth (oscillates) between two unstable output states. It is useful for providing clock signals for synchronous digital circuits.

There are several types of astable multivibrators in common use. We will present two of them without any attempt to analyze their operation. They are presented here so that you can construct a clock generator circuit if needed for a project or for testing digital circuits in the lab.

Schmitt-Trigger Oscillator   Figure 5-56 shows how a Schmitt-trigger IN-VERTER can be connected as an oscillator. The signal at V_{OUT} is an approximate

IC	Frequency	
7414	≈ 0.8/RC	(R ⩽ 500 Ω)
74LS14	≈ 0.8/RC	(R ⩽ 2 kΩ)
74HC14	≈ 1.2/RC	(R ⩽ 10 MΩ)

FIGURE 5-56   Schmitt-trigger oscillator. A 7413 Schmitt-trigger NAND may also be used.

square wave with a frequency that depends on the R and C values. The relationship between the frequency and RC values is shown in Figure 5-56 for three different Schmitt-trigger INVERTERs. Note the maximum limits on the resistance value for each device. The circuit will fail to oscillate if R is not kept below these limits.

555 Timer Used as an Astable Multivibrator   The 555 timer IC is a TTL-compatible device that can operate in several different modes. Figure 5-57 shows how external components can be connected to a 555 so that it operates as a free-

running oscillator. Its output is a repetitive rectangular waveform that switches between two logic levels with the time intervals at each logic level determined by the R and C values. The formulas for these time intervals, t_1 and t_2, and the overall period of the oscillations, T, are given in the figure. The frequency of the oscilla-

FIGURE 5-57   555 timer IC used as an astable multivibrator.

tions is, of course, the reciprocal of T. As the formulas in the diagram indicate, the t_1 and t_2 intervals cannot be equal unless R_A is made zero. This cannot be done without producing excess current through the device. This means that it is impossible to produce a perfect 50 percent duty-cycle square wave output. It is possible, however, to get very close to a 50 percent duty cycle by making $R_B \gg R_A$ (while keeping R_A greater than 1 kΩ), so that $t_1 \approx t_2$.

___ EXAMPLE 5-18 _____

Calculate the frequency and duty cycle of the 555 astable multivibrator output for: $C = 0.001$ μF, $R_A = 2.2$ kΩ, and $R_B = 100$ kΩ.

Solution

$$t_1 = 0.693(100 \text{ k}\Omega)(0.001 \ \mu\text{F}) = 69.3 \ \mu\text{s}$$
$$t_2 = 0.693(102.2 \text{ k}\Omega)(0.001 \ \mu\text{F}) = 70.7 \ \mu\text{s}$$
$$T = 69.3 + 70.7 = 140 \ \mu\text{s}$$
$$f = 1/140 \ \mu\text{s} = 7.29 \text{ kHz}$$
$$\text{duty cycle} = 70.7/140 = 50.5\%$$

Note that the duty cycle is close to 50 percent (square wave) because R_B is much greater than R_A. It can be made even closer to 50 percent by making R_B even larger compared with R_A. For instance, you should verify that if we change R_A to 1 kΩ (its minimum allowed value), the results are $f = 7.18$ kHz and duty cycle-50.3 percent.

___ REVIEW QUESTIONS _____

1. Determine the approximate frequency of a Schmitt-trigger oscillator that uses a 74HC14, $R = 10$ kΩ, and $C = 0.005$ μF.
2. Determine the approximate frequency and duty cycle of the 555 output for $R_A = R_B = 2.2$ kΩ and $C = 1000$ pF.

5-25 TROUBLESHOOTING FLIP-FLOP CIRCUITS

Flip-flop ICs are susceptible to the same kinds of internal and external faults that occur in combinatorial logic circuits. All of the troubleshooting ideas that were discussed in Chapter 4 can readily be applied to circuits that contain FFs as well as logic gates.

Because of their memory characteristic, FF circuits with one or more faults will often exhibit symptoms that would not occur in combinatorial circuits. Some of these are described below.

Open Inputs   Unconnected or floating inputs of any logic circuit are particularly susceptible to picking up spurious voltage fluctuations called *noise*. If the noise is large enough in amplitude and long enough in duration, the logic circuit's output may change states in response to the noise. In a logic gate, the output will return to its original state when the noise signal subsides. In a FF, however, the output will

remain in its new state because of its memory characteristic. Thus the effect of noise pickup at any open input is usually more critical for a FF or latch than it is for a logic gate.

The most susceptible FF inputs are those that can trigger the FF to a different state—such as the *CLK*, PRESET, and CLEAR. Whenever you see a FF output that is changing states erratically, you should consider the possibility of an open connection at one of these inputs.

━━━ EXAMPLE 5-19 ━━━━━━━━━━━━━━━━━━━━━━━━━━━━━━━━━━━━

Figure 5-58 shows a 3-bit shift register made up of TTL flip-flops. Initially, all of the FFs are in the LOW state before clock pulses are applied. As clock pulses are applied, each PGT will cause the information to shift from each FF to the one on its right. The diagram shows the "expected" sequence of FF states after each clock pulse. Since $J_2 = 1$ and $K_2 = 0$, flip-flop X_2 will go HIGH on clock pulse 1 and will stay there for all subsequent pulses. This HIGH will shift into X_1, and then X_0 on clock pulses 2 and 3, respectively. Thus, after the third pulse, all FFs will be HIGH and should remain there as pulses are continually applied.

Now let's suppose that the "actual" response of the FF states is as shown in the diagram. Here the FFs change as expected for the first three clock pulses. From then on, flip-flop X_0, instead of staying HIGH, alternates between HIGH and LOW. What possible circuit fault can produce this operation?

Solution   On the second pulse, X_1 goes HIGH. This should make $J_0 = 1$,

Discussion Question: How would the actual operation of the shift register of Figure 5-58 change if the connection to J_0 were open instead of the connection to K_0?

FIGURE 5-58   Example 5-19.

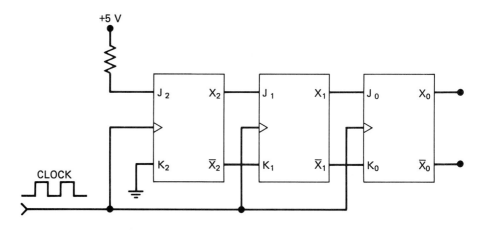

Clock pulse number	"Expected" X_2 X_1 X_0			"Actual" X_2 X_1 X_0		
0	0	0	0	0	0	0
1	1	0	0	1	0	0
2	1	1	0	1	1	0
3	1	1	1	1	1	1
4	1	1	1	1	1	0
5	1	1	1	1	1	1
6	1	1	1	1	1	0
7	1	1	1	1	1	1
8	1	1	1	1	1	0

$K_0 = 0$ so that all subsequent clock pulses should set $X_0 = 1$. Instead, we see X_0 changing states (toggling) on all pulses after the second one. This toggle operation would occur if J_0 and K_0 were both HIGH. The most probable fault is a break in the connection between $\overline{X}_1$ and K_0. Recall that a TTL device responds to an open input as if it were a logic HIGH, so that an open at K_0 is the same as a HIGH.

Shorted Outputs   The following example will illustrate how a fault in a FF circuit can cause a misleading symptom that may result in a longer time to isolate the fault.

___ EXAMPLE 5-20 ___

Consider the circuit in Figure 5-59 and examine the logic probe indications shown in the accompanying table. There is a LOW at the D input of the FF when pulses are applied to its CLK input, but the Q output fails to go to the LOW state. The technician testing this circuit considers each of the following possible circuit faults:

1. Z2-5 is internally shorted to V_{CC}.

2. Z1-4 is internally shorted to V_{CC}.

3. Z2-5 or Z1-4 is externally shorted to V_{CC}.

4. Z2-4 is internally or externally shorted to GROUND. This would keep $\overline{PRE}$ activated and would override the CLK input.

5. There is an internal failure in Z2 that prevents Q from responding properly to its inputs.

The technician, after making the necessary ohmmeter checks, rules out the first four possibilities. He also checks Z2's V_{CC} and GROUND pins and finds that they are at the proper voltages. He is reluctant to unsolder Z2 from the circuit until he is certain it is faulty, and so he decides to look at the clock signal. He uses an

FIGURE 5-59   Example 5-20.

Pin	Condition
Z1-1	HIGH
Z1-2	HIGH
Z1-3	LOW
Z2-2	LOW
Z2-3	Pulses
Z2-5	HIGH
Z1-4	HIGH

oscilloscope to check its amplitude, frequency, pulse width, and transition times. He finds that they are all within the specifications for the 74LS74. Finally, he concludes that Z2 is faulty.

He removes the 74LS74 chip and replaces it with another one. To his dismay, the circuit with the new chip behaves in exactly the same way. After scratching his head, he decides to change the NAND gate chip, although he doesn't know why. As expected, he sees no change in the circuit operation.

Becoming more puzzled, he recalls his electronics lab instructor emphasizing the value of performing a thorough visual check on the circuit board, and so he begins to examine it carefully. While he is doing that, he detects a solder bridge between pins 6 and 7 of Z2. He removes it and tests the circuit, and it functions correctly.

Explain how this fault produced the operation observed.

Solution   The solder bridge was shorting the $\overline{Q}$ output to GROUND. This means that $\overline{Q}$ is permanently stuck LOW. Recall that in all latches and FFs, the $\overline{Q}$ and Q outputs are internally cross-coupled so that the level on one will affect the other. For example, take another look at the internal circuitry for a J-K flip-flop in Figure 5-23. Note how a constant LOW at $\overline{Q}$ would keep a LOW at one input of NAND gate 3 so that Q would have to stay HIGH regardless of the conditions at J, K, and CLK.

The technician learned a valuable lesson about troubleshooting FF circuits. He learned that both outputs should be checked for faults, even those that are not connected to other devices.

Clock Skew   One of the most common timing problems in synchronous circuits is *clock skew*. One type of clock skew occurs when a clock signal, because of propagation delays, arrives at the *CLK* inputs of different FFs at different times. In many situations the skew can cause a FF to go to a wrong state. This is best illustrated with an example.

Refer to Figure 5-60(a), where the signal CLOCK1 is connected directly to FF Q_1, and indirectly to Q_2 through a NAND gate and INVERTER. Both FFs are supposed to be clocked by the occurrence of a NGT of CLOCK1 provided that X is HIGH. If we assume that initially $Q_1 = Q_2 = 0$ and $X = 1$, the NGT of CLOCK1 should set $Q_1 = 1$ and have no effect on Q_2. The waveforms in Figure 5-60(b) show how clock skew can produce incorrect triggering of Q_2.

Because of the combined propagation delays of the NAND gate and INVERTER, the transitions of the CLOCK2 signal are delayed with respect to CLOCK1 by an amount of time t_1. The NGT of CLOCK2 arrives at Q_2's *CLK* input t_1 later than the NGT of CLOCK1 appears at Q_1's *CLK* input. This t_1 is the clock skew. The NGT of CLOCK1 will cause Q_1 to go HIGH after a time t_2 which is equal to Q_1's t_{PLH} propagation delay. If t_2 is less than the skew t_1, Q_1 will be HIGH when the NGT of CLOCK2 occurs, and this may incorrectly set $Q_2 = 1$ if its setup time requirement, t_S, is met.

For example, assume that the clock skew is 40 ns and the t_{PLH} of Q_1 is 25 ns. Thus, Q_1 will go HIGH 15 ns before the NGT of CLOCK2. If Q_2's setup time requirement is smaller than 15 ns, Q_2 will respond to the HIGH at its D input when the NGT of CLOCK2 occurs and Q_2 will go HIGH. This, of course, is not the expected response of Q_2. It is supposed to remain LOW.

The effects of clock skew are not always easy to detect, because the response of the affected FF may be intermittent (sometimes it works correctly, sometimes it

(a)

t_1 = skew = combined delay of NAND gate and INVERTER
t_2 = t_{PLH} of Q_1
t_3 = t_{PLH} of Q_2

(b)

FIGURE 5-60   Clock skew occurs when two flip-flops that are supposed to be clocked simultaneously are clocked at slightly different times due to a delay in the arrival of the clock signal at the second flip-flop.

<em>Discussion Question</em>:
Would there be a clock skew problem if the FFs in Figure 5-60 responded to PGTs?

doesn't). This is because the situation is dependent on circuit propagation delays and FF timing parameters, which vary with temperature, length of connections, power supply voltage, and loading. Sometimes just connecting an oscilloscope probe to a FF or gate output will add enough load capacitance to increase the device's propagation delay so that the circuit functions correctly; then when the probe is removed, the incorrect operation reappears. This is the kind of situation that explains why some technicians are prematurely gray.

Problems caused by clock skew can be eliminated by equalizing the delays in the various paths of the clock signal so that the active transition arrives at each FF at approximately the same time. This is examined in Problem 5-49.

REVIEW QUESTION

1. What is clock skew? How can it cause a problem?

5-26   FLIP-FLOP SUMMARY

1. NOR gate LATCH (Figure 5-61)

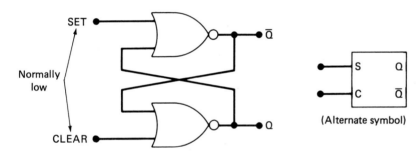

S	C	Q
0	0	No change
1	0	Q = 1
0	1	Q = 0
1	1	Invalid

FIGURE 5-61

2. NAND gate LATCH (Figure 5-62)

FIGURE 5-62

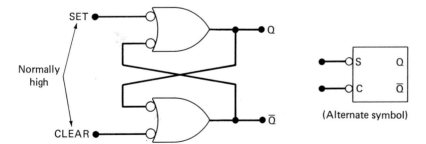

S	C	Q
0	0	Invalid
1	0	Q = 0
0	1	Q = 1
1	1	No change

3. Edge-triggered S-C flip-flop (Figure 5-63)

S	C	Q*
0	0	Q_0 (no change)
1	0	1
0	1	0
1	1	Ambiguous

*on ↑ of CLK

FIGURE 5-63

4. Edge-triggered J-K flip-flop (Figure 5-64)

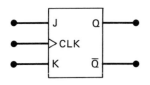

J	K	Q*
0	0	Q_0 (no change)
1	0	1
0	1	0
1	1	$\overline{Q}_0$ (toggles)

*on ↑ of CLK

FIGURE 5-64

5. Edge-triggered D flip-flop (Figure 5-65)

D	Q*
0	0
1	1

*on ↑ of CLK

FIGURE 5-65

6. D LATCH (Figure 5-66)

FIGURE 5-66

EN	D	Q*
0	X	No change
1	0	0
1	1	1

Q* follows D input
while EN is HIGH

7. Asynchronous inputs (Figure 5-67)

PRESET	CLEAR	Q*
1	1	No effect; FF can respond to J, K and CLK
1	0	Q = 0 independent of synchronous inputs
0	1	Q = 1 independent of synchronous inputs
0	0	Ambiguous (not used)

*CLK can be in any state

FIGURE 5-67

PROBLEMS

SECTIONS 5-1 to 5-3

5-1. *Q* sets when *x* goes LOW and clears when *y* goes LOW.

5-1. Assuming that $Q = 0$ initially, apply the x and y waveforms of Figure 5-68 to the SET and CLEAR inputs of a NAND latch and determine the Q and $\overline{Q}$ waveforms.

FIGURE 5-68

5-2. Invert the x and y waveforms of Figure 5-68, apply them to the SET and CLEAR inputs of a NOR latch, and determine the Q and $\overline{Q}$ waveforms. Assume that $Q = 0$ initially. Same output as P5-1.

5-3. *Q* goes HIGH when *x* first goes LOW, and *Q* returns LOW on the second LOW pulse at *y*.

5-3. The waveforms of Figure 5-68 are connected to the circuit of Figure 5-69. Assume that $Q = 0$ initially, and determine the Q waveform.

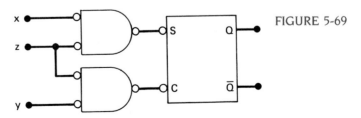

FIGURE 5-69

5-4. Modify the circuit of Figure 5-9 to use a NOR gate latch.

5-4. S and C biased normally LOW, and switched to + 5 V.

5-5. Modify the circuit of Figure 5-12 to use a NAND gate latch.

5-5. INVERTER on S input; C input biased HIGH.

5-6. Refer to the circuit of Figure 5-13. A technician tests the circuit operation by observing the outputs with a storage oscilloscope while the switch is moved from *A* to *B*. When the switch is moved from *A* to *B*, the scope display of X_B appears as shown in Figure 5-70. What circuit fault could produce this result? (*Hint:* What is the function of the NAND latch?)   Z1-4 is stuck in HIGH state.

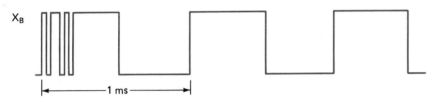

FIGURE 5-70

SECTIONS 5-4 and 5-5

5-7. A certain clocked FF is specified to have $t_S = 20$ ns and $t_H = 5$ ns. How long must the control inputs be stable prior to the active clock transition?   20 ns

5-8. Apply the *S*, *C*, and *CLK* waveforms of Figure 5-17 to the FF of Figure 5-18 and determine the *Q* waveform.

5-8. Q goes HIGH at b, LOW at f, and back HIGH at h.

5-9. A *toggle* FF is one that has a single input and operates in such a way that the FF output changes state for each pulse applied to its input. The clocked S-C flip-flop can be wired to operate in the toggle mode, as shown in Figure 5-71. The waveform applied to the *CLK* input is a 1-kHz square wave. Verify that this arrangement operates in the toggle mode; then determine the *Q* output waveform. Assume that *Q* = 0 initially.   *Q* is a 500-Hz square wave.

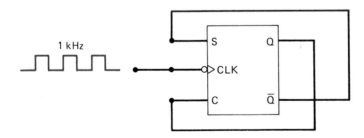

FIGURE 5-71

SECTION 5-6

5-10. Apply the *J*, *K*, and *CLK* waveforms of Figure 5-21 to the FF of Figure 5-22. Assume that *Q* = 1 initially and determine the *Q* waveform.

5-10. Q goes LOW at b, HIGH at f, LOW at h, and HIGH at j.

5-11. (a) Show how the J-K flip-flop can be operated as a toggle FF. Apply a 10-kHz square wave to its input and determine its output waveform.

5-11. (a) J = K = 1; output is 5-kHz square wave.   (b) 2.5 kHz.

(b) Connect the Q output of the FF from Problem 5-11(a) to the *CLK* input of a second J-K flip-flop that also has $J = K = 1$. Determine the frequency of the waveform at the second FF output.

5-12. See Solutions Manual.

5-12. The waveforms shown in Figure 5-72 are to be applied to two different FFs: (a) positive-edge-triggered J-K; (b) negative-edge-triggered J-K. Draw the Q waveform response for each of these FFs, assuming that $Q = 0$ initially. Assume that each FF has $t_H = 0$.

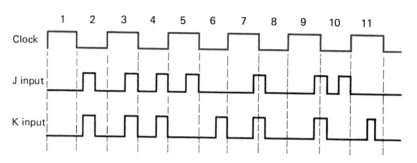

FIGURE 5-72

SECTION 5-7

5-13. A D-type FF is sometimes used to *delay* a binary waveform so that the binary information appears at the output a certain amount of time after it appears at the D input. Determine the Q waveform in Figure 5-73 and compare it with the input waveform. Note that it is delayed from the input by one clock period. How can a delay of two clock periods be obtained?   Connect a second D flip-flop to form a 2-bit shift register.

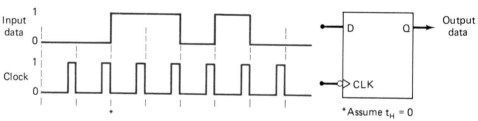

FIGURE 5-73

5-14. An edge-triggered D flip-flop can be made to operate in the toggle mode by connecting it as shown in Figure 5-74. Assume that $Q = 0$ initially and determine the Q waveform. Q is a 500-Hz square wave.

FIGURE 5-74   D flip-flop connected to toggle.

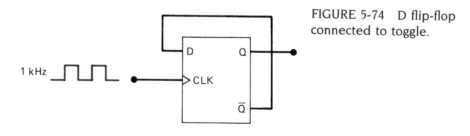

5-15. Change the circuit in Figure 5-74 so that Q is connected back to D. Then determine the Q waveform.   *Q does not change sates.*

SECTION 5-8

5-16. Compare the operation of the D latch with a negative-edge-triggered D flip-flop by applying the waveforms of Figure 5-75 to each and determining the Q waveforms.   *See Solutions Manual.*

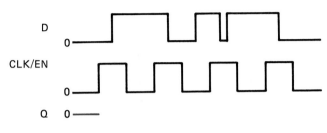

FIGURE 5-75

5-17. In Problem 5-14 we saw how an edge-triggered D flip-flop can be operated in the toggle mode. Explain why this same idea will not work for a D latch.

5-17. In a latch, if $\overline{Q}$ is connected back to D, the output will oscillate at a frequency determined by the propagation delays of the internal gates.

SECTION 5-9

5-18. Determine the Q waveform for the FF in Figure 5-76. Assume that $Q = 0$ initially and remember that the asynchronous inputs override all other inputs.

5-18. The second NGT of CLK has no effect, since DC CLR is held LOW.

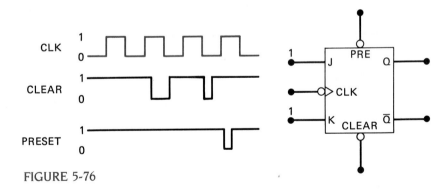

FIGURE 5-76

5-19. Apply the CLK, PRESET, and CLEAR waveforms of Figure 5-31 to a positive-edge-triggered D flip-flop with active-LOW asynchronous inputs. Assume that D is kept HIGH and Q is initially LOW. Determine the Q waveform.   *PRESET pulse has no effect, since Q is already HIGH.*

SECTION 5-11

5-20. Use Table 5-1 in Section 5-11 to determine the following:
   (a) How long can it take for the Q output of a 4013B to switch from 0 to 1 in response to an active CLK transition?   *200 ns*

(b) Which FF in Table 5-1 requires its control inputs to remain stable for the longest time after the active *CLK* transition?   7474

(c) What is the narrowest pulse that can be applied to the DC SET of a 7474 FF?   30 ns

5-21. Refer to the circuit of Figure 5-77. It shows the two J-K flip-flops on a 74LS112 chip connected so that the Q_1 output serves as the *CLK* input for the Q_2 flip-flop. Assume that $Q_1 = Q_2 = 1$ initially, and determine the *total* propagation delay between the NGT of the clock pulse and the NGT of Q_2.   48 ns

FIGURE 5-77

SECTIONS 5-15 and 5-16

5-22. Connect *A* to *J*; connect $\overline{A}$ to *K*.

5-23. (a) Correct sequence is *A*, *B*, *C*.   (b) To clear *X* and *Y* to 0.

5-22. Modify the circuit of Figure 5-40 to use a J-K flip-flop.

5-23. In the circuit of Figure 5-78 the inputs *A*, *B*, and *C* are all initially LOW. Output *Y* is supposed to go HIGH only when *A*, *B*, and *C* go HIGH in a certain sequence.
(a) Determine the sequence that will make *Y* go HIGH.
(b) Explain why the START pulse is needed.

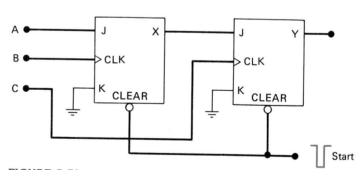

FIGURE 5-78

SECTIONS 5-17 and 5-18

5-24. Same arrangement as Figure 5-44, except connect each *X* to *J* and each $\overline{X}$ to *K* of corresponding *Y* flip-flop.

5-25. Use arrangement of Figure 5-43.

5-26. 1011, 1101, 1110, 0111, 1011, 1101, 1110, 0111, 1011

5-24. Draw a circuit diagram for the synchronous parallel transfer of data from one 3-bit register to another using J-K flip-flops.

5-25. Repeat Problem 5-24 for asynchronous parallel transfer.

5-26. A *recirculating* shift register is a shift register that keeps the binary information circulating through the register as clock pulses are applied. The shift register of Figure 5-45 can be made into a circulating regis-

ter by connecting X_0 to the DATA IN line. No external inputs are used. Assume that this circulating register starts out with 1011 stored in it (that is, $X_3 = 1$, $X_2 = 0$, $X_1 = 1$, and $X_0 = 1$). List the sequence of states that the register FFs go through as eight shift pulses are applied.

5-27. Refer to Figure 5-46, where a 3-bit number stored in register X is serially shifted into register Y. How could the circuit be modified so that at the end of the transfer operation, the original number stored in X is present in both registers? (*Hint:* See Problem 5-26.)

5-27. Connect X_0 to D input of X_3.

SECTION 5-19

5-28. Refer to the binary counter of Figure 5-47. Change it by connecting $\overline{X}_0$ to the *CLK* of flip-flop X_1, and $\overline{X}_1$ to the *CLK* of flip-flop X_2. Start out with all FFs in the 1 state and draw the various FF output waveforms (X_0, X_1, X_2) for 16 input pulses. Then list the sequence of FF states as was done in Figure 5-48. This counter is called a *down counter*. Why?   111, 110, 101, 100, 011, 010, 001, 000

5-29. Draw the state transition diagram for this down counter and compare it with the diagram of Figure 5-49. How are they different?

5-29. Arrows are in opposite direction.

5-30. Show how clocked D flip-flops can be used in a counter such as that in Figure 5-47. (*Hint:* See Problem 5-14.)

5-30. Connect each D flip-flop in the toggle mode by tying $\overline{Q}$ back to D.

5-31. (a) How many FFs are required to build a binary counter circuit that counts from 0 to 1023?   10
 (b) Determine the frequency at the output of the last FF of this counter for an input clock frequency of 2 MHz.   1953 Hz
 (c) What is the counter's MOD number?   1024
 (d) If the counter is initially at zero, what count will it hold after 2060 pulses? $12 = 1100_2$

5-32. A certain counter is being pulsed by a 256-kHz clock signal. The output frequency from the last FF is 2 kHz.
 (a) Determine the MOD number.   128
 (b) Determine the counting range.   0000000 to 1111111 (0 to 127)

5-33. A photodetector circuit is being used to generate a pulse each time a customer walks into a certain establishment. The pulses are fed to an 8-bit counter. The counter is used to count these pulses as a means for determining how many customers have entered the store. After closing the store, the proprietor checks the counter and finds that it shows a count of $00001001_2 = 9_{10}$. He knows that this is incorrect, because there were many more than nine people in his store. Assuming that the counter circuit is working properly, what could be the reason for the discrepancy?

5-33. The counter recycled back to 0 one or more times.

5-34. Draw the state transition diagram for a MOD-16 counter.

5-34. See Solutions Manual.

SECTION 5-20

5-35. Modify the circuit of Figure 5-50 so that only the presence of address code 10110110 will allow data to be transferred to register X.

5-35. Put INVERTERs on A_{14} and A_{11}.

5-36. Suppose that the circuit of Figure 5-50 is malfunctioning so that data are being transferred to X for either of the address codes 11111110 or 11111111. What are some circuit faults that could be causing this?

5-36. Bottom input of AND gate 1 is stuck HIGH or open.

5-37. Each OS output is a single pulse that begins on the NGT of the preceding output.

5-37. Determine the waveforms at Q_1, Q_2, and Q_3 in response to the single input pulse in Figure 5-79.

FIGURE 5-79

5-38. A retriggerable OS can be used as a pulse-frequency detector that detects when the frequency of a pulse input is below a predetermined value. A simple example of this application is shown in Figure 5-80. The operation begins by momentarily closing switch S_1.

(a) Describe how the circuit responds to input frequencies above 1 kHz. X and $\overline{Q}$ will stay LOW.

(b) Describe how the circuit responds to input frequencies below 1 kHz. X goes HIGH

(c) How would you modify the circuit to detect when the input frequency drops below 50 kHz?   Change t_p ro 20 μs.

FIGURE 5-80

5-39. (a) A_1 or A_2 has to be LOW when a PGT occurs at B.
(b) B and A_2 have to be HIGH when a NGT occurs at A_1.

5-39. Refer to the logic symbol for a 74121 nonretriggerable one-shot in Figure 5-54(a).

(a) What input conditions are necessary for the OS to be triggered by a signal at the B input?

(b) What input conditions are necessary for the OS to be triggered by a signal at the A_1 input?

 5-40. The output pulse width from a 74121 OS is given by the approximate formula

$$t_p \approx 0.7 R_T C_T$$

where R_T is the resistance connected between the R_{EXT}/C_{EXT} pin and

V_{CC}, and C_T is the capacitance connected between the C_{EXT} pin and the R_{EXT}/C_{EXT} pin. The value for R_T can be varied between 2 and 40 kΩ, and C_T can be as large as 1000 μF.

(a) Show how a 74121 can be connected to produce a negative-going pulse with a 5-ms duration whenever either of two logic signals (E or F) makes a NGT. Both E and F are normally in the HIGH state.

(b) Modify the circuit so that a control input signal, G, will inhibit the OS output pulse regardless of what occurs at E or F.

5-40. See Solutions Manual.

SECTION 5-23

5-41. Consider the circuit of Figure 5-81. Initially all FFs are in the 0 state. The circuit operation begins with a momentary start pulse applied to the PRESET inputs of FFs X and Y. Determine the waveforms at A, B, C, X, Y, Z, and W for 20 cycles of the clock pulses after the start pulse. State all assumptions.

5-41. Circuit outputs stop changing after 14 clock pulses.

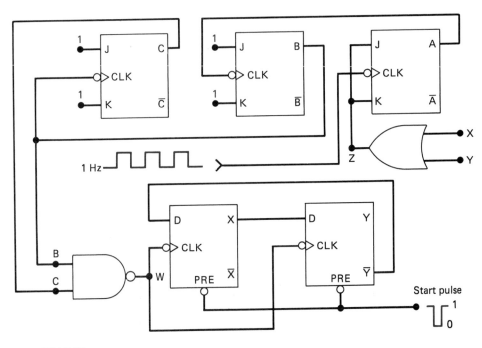

FIGURE 5-81

SECTION 5-24

5-42. Show how to use a 74LS14 Schmitt-trigger INVERTER to produce an approximate square wave with a frequency of 10 kHz.

5-42. One possibility: $R = 1$ kΩ, $C = 80$ nF.

5-43. Design a 555 free-running oscillator to produce an approximate square wave at 40 kHz. C should be kept at 100 pF or greater.

5-43. One possibility: $R_A = 1$ kΩ, $R_B = 10$ kΩ, $C = 1800$ pF.

5-44. A 555 oscillator can be combined with a J-K flip-flop to produce a perfect (50 percent duty cycle) square wave. Modify the circuit of Problem 5-43 to include a J-K flip-flop. The final output is still to be a 40-kHz square wave.

5-44. One possibility: Change C to 900 pF, and connect the 555 output to clock input of J-K flip-flop with $J = K = 1$.

5-45. CP1 and CP2 are both 500 kHz. CP2 contains every odd-numbered pulse from *CLK*; CP1 contains every even-numbered pulse.

5-45. The circuit in Figure 5-82 can be used to generate two nonoverlapping clock signals at the same frequency. These clock signals are used in some microprocessor systems that require four different clock transitions to synchronize their operations. Draw the CP1 and CP2 waveforms in response to a 1-MHz input clock frequency.

FIGURE 5-82

SECTION 5-25

 5-46. Refer to the counter circuit in Figure 5-47. Assume that all asynchronous inputs are connected to V_{CC}. When tested, the circuit waveforms appear as shown in Figure 5-83. Consider the following list of possible faults. For each one indicate "yes" or "no" as to whether it could cause the observed results. Explain each response.

(a) *CLR* input of X_2 is open.   no
(b) X_1 output's transition times are too long, possibly due to loading.   yes
(c) X_2 output shorted to ground.   no
(d) X_2's hold-time requirement is not being met.   no

FIGURE 5-83

5-47. Refer to the circuit of Figure 5-46. All FFs are TTL ICs. Assume the following initial conditions: $X_2 X_1 X_0 = 100$ and $Y_2 Y_1 Y_0 = 011$. After four shift pulses, the conditions are $X_2 X_1 X_0 = 001$ and $Y_2 Y_1 Y_0 = 111$. Subsequent shift pulses produce no change in any of the FFs. What are some of the possible causes of this faulty operation?

5-47. See *Solutions Manual*.

5-48. Consider the situation of Figure 5-60 for each of the following sets of timing values. For each indicate whether or not flip-flop Q_2 will respond correctly.

(a) *Each FF:* $t_{PLH} = 12$ ns; $t_{PHL} = 8$ ns; $t_S = 5$ ns; $t_H = 0$ ns
NAND gate: $t_{PLH} = 8$ ns; $t_{PHL} = 6$ ns
INVERTER: $t_{PLH} = 7$ ns; $t_{PHL} = 5$ ns   yes

(b) *Each FF:* $t_{PLH} = 10$ ns; $t_{PHL} = 8$ ns; $t_S = 5$ ns; $t_H = 0$ ns
NAND gate: $t_{PLH} = 12$ ns; $t_{PHL} = 10$ ns
INVERTER: $t_{PLH} = 8$ ns; $t_{PHL} = 6$ ns   no

5-49. Show and explain how the clock skew problem in Figure 5-60 can be eliminated by the appropriate insertion of two INVERTERs.

5-49. Insert two INVERTERs between CLOCK1 and *CLK* input of Q_1. Another possibility is two INVERTERs between Q_1 and D_2.

FIGURE 5-84   Problem 5-51.

(a)

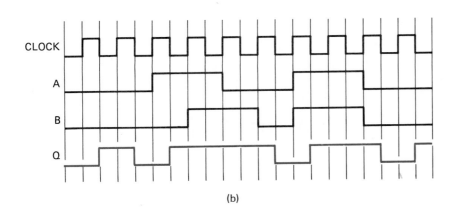

(b)

5-50. (a) Output *A* stays HIGH;
OS never triggers.   (b) LOW at
W drags $\overline{Q}$ LOW (assuming TTL)
so that *B* stays LOW and the
counter cannot count.   (c) Once
triggered, the OS output will stay
in its triggered state indefinitely
($\overline{Q} = 0$).

5-50. Refer to the circuit of Figure 5-55. Describe how the circuit operation
will change for each of the following faults:
(a) An internal short to ground at the NAND gate's top input.
(b) An internal short between the OR gate inputs.
(c) An open in the OS timing resistor.

5-51. Refer to the circuit of Figure 5-84. Assume that the ICs are of the
TTL logic family. The *Q* waveform was obtained when the circuit was
tested with the input signals shown and with the switch in the "up" po-
sition; it is not correct. Consider the following list of faults, and for
each indicate "yes" or "no" as to whether it could be the actual fault.
Explain each response.
(a) Point *X* is always LOW due to a faulty switch.   no
(b) Z1 pin 1 is internally shorted to V_{CC}.   no
(c) The connection from Z1-3 to Z2-3 is broken.   yes
(d) There is a solder bridge between pins 6 and 7 of Z1.   no

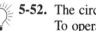

5-52. The circuit of Figure 5-85 functions as a sequential combination lock.
To operate the lock, proceed as follows:

FIGURE 5-85   Problems 5-52 and 5-53.

1. Momentarily activate the RESET switch.
2. Set the switches SWA, SWB, and SWC to the first part of the combination. Then momentarily toggle the ENTER switch back and forth.
3. Set the switches to the second part of the combination, and toggle ENTER again. This should produce a HIGH at Q_2 to open the lock.

If the incorrect combination is entered in either step, the operator must start the sequence over. Analyze the circuit and determine the correct sequence of combinations that will open the lock.

5-52. Correct sequence is 101, 000.

5-53. When the combination lock of Figure 5-85 is tested, it is found that entering the correct combination does not open the lock. A logic probe check shows that entering the correct first combination sets Q_1 HIGH, but entering the correct second combination produces only a momentary pulse at Q_2. Consider each of the following faults and indicate which one(s) could produce the observed operation. Explain each choice.
   (a) Switch bounce at SWA, SWB, or SWC.   no
   (b) *CLR* input of Q_2 is open.   no
   (c) Connection from NAND gate 4 output to NAND gate 3 input is open.   yes

DRILL QUESTIONS

5-54. For each statement indicate what type of FF is being described:
   (a) Has a SET and a CLEAR input but does not have a *CLK* input
   (b) Will toggle on each *CLK* pulse when its control inputs are both HIGH   JK
   (c) Has an ENABLE input instead of a *CLK* input   D latch
   (d) Used to easily transfer data from one FF register to another
   (e) Has only one control input   D flip-flop, D latch
   (f) Has two outputs that are complements of each other   all FFs
   (g) Can change states only on the active transition of *CLK*
   (h) Used in binary counters   JK

5-54. (a) S-C latch   (d) D flip-flop   (g) clocked SC, D, JK

5-55. Define the following terms: See Glossary.
   (a) asynchronous inputs
   (b) edge-triggered
   (c) shift register
   (d) frequency division
   (e) asynchronous (jam) transfer
   (f) state transition diagram
   (g) parallel data transfer
   (h) serial data transfer

ANSWERS TO SECTION REVIEW QUESTIONS

SECTION 5-1

1. HIGH; LOW.   **2.** $Q = 0, \overline{Q} = 1$.
3. True.   **4.** Apply a momentary LOW to SET input.

SECTION 5-2

1. LOW, HIGH.   **2.** $Q = 1$ and $\overline{Q} = 0$.
3. Make CLEAR = 1.   **4.** SET and CLEAR would both be normally in their active-LOW state.

SECTION 5-4

1. Synchronous control inputs and clock input.　　**2.** The FF output can change only when the appropriate clock transition occurs.
3. True.　　**4.** Setup time is the required interval immediately prior to the active edge of the *CLK* signal during which the control inputs must be held stable. Hold time is the required interval immediately following the active edge of *CLK* during which the control inputs must be held stable.

SECTION 5-5

1. It will go HIGH.　　**2.** Because *CLK* * is HIGH only for a few nanoseconds.

SECTION 5-6

1. True　　**2.** No　　**3.** $J = 1, K = 0$

SECTION 5-7

1. *Q* will go LOW at point *a* and remain LOW.　　**2.** False. The *D* input can change without affecting *Q* because *Q* can change only on the active *CLK* edge.

SECTION 5-8

1. In a *D* latch the *Q* output can change while *EN* is HIGH. In a D flip-flop the output can change only on the active edge of *CLK*.　　**2.** False.
3. True.

SECTION 5-9

1. Asynchronous inputs work independently of the *CLK* input.　　**2.** Yes, since $\overline{PRE}$ is active-LOW.
3. $J = K = 1$, $\overline{PRE} = \overline{CLR} = 1$, and a PGT at *CLK*.

SECTION 5-10

1. The triangle inside the rectangle indicates edge-triggered operation; the right triangle outside the rectangle indicates triggering on a NGT.
2. It is used to indicate the function of those inputs that are common to more than one circuit on the chip.

SECTION 5-11

1. t_{PLH} and t_{PHL}.　　**2.** False. The waveform must also satisfy $t_W(L)$ and $t_W(H)$ requirements.

SECTION 5-17

1. False　　**2.** D flip-flop　　**3.** Six　　**4.** True

SECTION 5-18

1. True.　　**2.** Fewer interconnections between registers.　　**3.** $X_2 X_1 X_0 = 111$; $Y_2 Y_1 Y_0 = 101$.

SECTION 5-19

1. 10 kHz　　**2.** Eight　　**3.** 256
4. 2 kHz　　**5.** $00001000_2 = 8_{10}$

SECTION 5-21

1. The output may contain oscillations.　　**2.** It will produce clean, fast output signals even for slow-changing input signals.

SECTION 5-22

1. $Q = 0, \overline{Q} = 1$.　　**2.** True.　　**3.** External *R* and *C* values.　　**4.** For a retriggerable OS, each new trigger pulse begins a new t_p interval regardless of state of *Q* output.

SECTION 5-24

1. 24 kHz　　**2.** 218.6 kHz; 66.7 percent

SECTION 5-25

1. Clock skew is the arrival of a clock signal at the *CLK* inputs of different FFs at different times. It can cause a FF to go to an incorrect state.

6

DIGITAL ARITHMETIC: OPERATIONS AND CIRCUITS

Outline

Upon completion of this chapter, you will be able to:

- Perform binary addition, subtraction, multiplication, and division on two binary numbers.
- Add and subtract hexadecimal numbers.
- Know the difference between binary addition and OR addition.
- Compare the advantages and disadvantages among three different systems of representing signed binary numbers.
- Manipulate signed binary numbers using the 2's-complement system.
- Understand the BCD adder circuit and the BCD addition process.
- Describe the basic operation of an arithmetic-logic unit.
- Employ full adders in the design of parallel binary adders.
- Cite the advantages of parallel adders with the look-ahead carry feature.
- Analyze the operation of a serial binary multiplier circuit.
- Describe some of the operations performed by an arithmetic processing unit.
- Read and understand the IEEE/ANSI symbol for a parallel adder.
- Analyze several troubleshooting case studies of adder/subtractor circuits.

Introduction

Digital computers and calculators perform the various arithmetic operations on numbers that are represented in binary form. The subject of digital arithmetic can be a very complex one if we want to understand all the various methods of computation and the theory behind them. Fortunately, this level of knowledge is not required by most technicians, at least not until they become experienced computer programmers. Our approach in this chapter will be to concentrate on those basic principles that are needed to understand how digital machines (i.e., computers) perform the basic arithmetic operations.

First we will see how the various arithmetic operations are performed on binary numbers using "pencil and paper," and then we will study the actual logic circuits that perform these operations in a digital system.

6-1 BINARY ADDITION

The addition of two binary numbers is performed in exactly the same manner as the addition of decimal numbers. In fact, binary addition is simpler, since there are fewer cases to learn. Let us first review decimal addition:

$$
\begin{array}{ccc}
3 & 7 & 6 \quad \text{LSD} \\
+4 & 6 & 1 \\
\hline
8 & 3 & 7
\end{array}
$$

The least-significant-digit (LSD) position is operated on first, producing a sum of 7. The digits in the second position are then added to produce a sum of 13, which produces a **carry** of 1 into the third position. This produces a sum of 8 in the third position.

The same general steps are followed in binary addition. However, only four cases can occur in adding the two binary digits (bits) in any position. They are:

$$0 + 0 = 0$$
$$1 + 0 = 1$$
$$1 + 1 = 10 = 0 + \text{carry of 1 into next position}$$
$$1 + 1 + 1 = 11 = 1 + \text{carry of 1 into next position}$$

The last case occurs when the two bits in a certain position are 1 and there is a carry from the previous position. Here are several examples of the addition of two binary numbers:

$$
\begin{array}{lll}
\quad 011 \ (3) & \quad 1001 \ (9) & \quad 11.011 \ (3.375) \\
+ \ \ 110 \ (6) & + \ 1111 \ (15) & + \ 10.110 \ (2.750) \\
\hline
1001 \ (9) & 11000 \ (24) & 110.001 \ (6.125)
\end{array}
$$

It is not necessary to consider the addition of more than two binary numbers at a time, because in all digital systems the circuitry that actually performs the addition can handle only two numbers at a time. When more than two numbers are to be added, the first two are added together and then their sum is added to the third number; and so on. This is not a serious drawback, since modern digital computers can typically perform an addition operation in a few microseconds or less.

Addition is the most important arithmetic operation in digital systems. As we shall see, the operations of subtraction, multiplication, and division as they are performed in most modern digital computers and calculators actually use only addition as their basic operation.

Binary Addition versus Logical OR Addition　　It is important to understand the difference between the OR-addition operation and binary addition. OR addition is the Boolean *logic* operation performed by an OR gate, which produces a 1 output whenever any one or more inputs are 1. Binary addition is an *arithmetic* operation that produces an arithmetic sum of two binary numbers. Although the + sign is used for both OR addition and binary addition, the meaning of the + sign will usually be apparent from the context in which it is being used. The major differences between OR addition and binary addition can be summarized as follows:

OR Addition	*Binary Addition*
$1 + 1 = 1$	$1 + 1 = 0 + \text{carry of 1}$
$1 + 1 + 1 = 1$	$1 + 1 + 1 = 1 + \text{carry of 1}$

1. Add the following pairs of binary numbers: (a) $10110 + 00111$, (b) $011.101 + 010.010$, (c) $10001111 + 00000001$.

6-2 REPRESENTING SIGNED NUMBERS

In digital computers, the binary numbers are represented by a set of binary storage devices (usually flip-flops). Each device represents one bit. For example, a 6-bit FF register could store binary numbers ranging from 000000 to 111111 (0 to 63 in decimal). This represents the *magnitude* of the number. Since most digital computers and calculators handle negative as well as positive numbers, some means is required for representing the *sign* of the number (+ or −). This is usually done by adding another bit to the number called the **sign bit**. In general, the common convention which has been adopted is that a 0 in the sign bit represents a positive number and a 1 in the sign bit represents a negative number. This is illustrated in Figure 6-1. Register A contains the bits 0110100. The 0 in the leftmost bit (A_6) is the sign bit that represents +. The other six bits are the magnitude of the number 110100_2, which is equal to 52 in decimal. Thus, the number stored in the A register is +52. Similarly, the number stored in the B register is −52, since the sign bit is 1, representing −.

The sign bit is used to indicate the positive or negative nature of the stored binary number. The numbers in Figure 6-1 consist of a sign bit and six magnitude bits. The magnitude bits are the true binary equivalent of the decimal value being represented. This is called the **sign-magnitude** system for representing signed binary numbers.

Although the sign-magnitude system is straightforward, calculators and computers do not normally use it, because the circuit implementation is more complex than other systems. The most commonly used system for representing signed binary numbers is the **2's-complement system.** Before we see how this is done, we must first see how to form the 1's complement and 2's complement of a binary number.

Teaching Tip: The shortcomings of the sign-magnitude system can be readily illustrated by adding the sign-magnitude representations of +5 and −5 and showing that the result does not equal zero. That is, 00101 (+5) added to 10101 (−5) gives a result of 11010 (−10).

FIGURE 6-1   Representation of signed numbers in sign-magnitude form.

1's-Complement Form

The 1's complement of a binary number is obtained by changing each 0 to a 1 and each 1 to a 0. In other words, change each bit in the number to its complement. The process is shown below.

1	0	1	1	0	1	original binary number
↓	↓	↓	↓	↓	↓	
0	1	0	0	1	0	complement each bit to form 1's complement

Thus, we say that the 1's complement of 101101 is 010010.

Teaching Tip: Another way to perform the 1's-complement operation is to subtract the binary number from the largest binary number of the same number of bits. For example, the 1's complement of 101101 can be obtained as follows:

$$
\begin{array}{r}
111111 \\
-101101 \\
\hline
010010
\end{array}
$$

2's-Complement Form

The 2's complement of a binary number is formed by taking the 1's complement of the number and adding 1 to the least significant bit position. The process is illustrated below for $101101_2 = 45_{10}$.

```
  1 0 1 1 0 1     binary equivalent of 45
  0 1 0 0 1 0     complement each bit to form 1's complement
+           1     add 1 to form 2's complement
  _____
  0 1 0 0 1 1     2's complement of original binary number
```

Thus, we say that 010011 is the 2's-complement representation of 101101.

Here's another example of converting a binary number to its 2's-complement representation:

```
  1 0 1 1 0 0     original binary number
  0 1 0 0 1 1     1's complement
+           1     add 1
  _____
  0 1 0 1 0 0     2's complement of original number
```

Representing Signed Numbers Using 2's Complement

The 2's-complement system for representing signed numbers works like this:

- If the number is positive, the magnitude is represented in its true binary form, and a sign bit of 0 is placed in front of the MSB. This is shown in Figure 6-2 for $+45_{10}$.

- If the number is negative, the magnitude is represented in its 2's-complement form, and a sign bit of 1 is placed in front of the MSB. This is shown in Figure 6-2 for -45_{10}.

Teaching Tip: You might want to add the two representations from Figure 6-2 to show that the sum is indeed zero, provided we discard the carry from the sign-bit position.

FIGURE 6-2   Representation of signed numbers in the 2's-complement system.

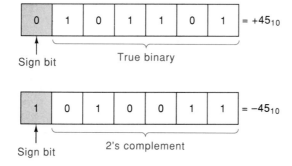

The 2's-complement system is used to represent signed numbers because, as we shall see, it allows us to perform the operation of subtraction by actually performing addition. This is significant because it means that a digital computer can use the same circuitry to both add and subtract, thereby realizing a saving in hardware.

—— EXAMPLE 6-1 ————————————————————————————

Represent each of the following signed decimal numbers as a signed binary number in the 2's-complement system. Use a total of 5 bits including the sign bit: (a) $+13$, (b) -9, (c) $+3$, (d) -2, (e) -8.

Solution

(a) Since the number is positive, the magnitude (13) will be represented in its true-magnitude form—that is, $13 = 1101_2$. Attaching the sign bit of 0, we have

$$+13 = \underline{0}1101$$

sign bit ⬏

(b) Since the number is negative, the magnitude (9) has to be represented in 2's-complement form:

$$
\begin{array}{ll}
9_{10} = 1001_2 & \\
\quad\quad 0110 & \text{(1's complement)} \\
+ \quad\quad 1 & \text{(add 1 to LSB)} \\
\hline
\quad\quad 0111 & \text{(2's complement)}
\end{array}
$$

When we attach the sign bit of 1, the complete signed number becomes

$$-9 = \mathbf{10111}$$

The procedure we have just followed required two steps. First we determined the 2's complement of the magnitude, and then we attached the sign bit. This can be accomplished in one step if we include the sign bit in the 2's-complement process. For example, to find the representation for -9, we start with the representation for $+9$, *including the sign bit,* and we take the 2's complement of it in order to obtain the representation for -9.

$$
\begin{array}{ll}
+9 = 01001 & \\
\quad\quad 10110 & \text{(1's complement of each bit including sign bit)} \\
+ \quad\quad\, 1 & \text{(add 1 to LSB)} \\
\hline
\quad\quad 10111 & \text{(2's-complement representation of } -9)
\end{array}
$$

The result is, of course, the same as before. In general, we can say that

the 2's-complement operation on a signed number will change a positive number to a negative number and vice versa.

(c) The decimal value 3 can be represented in binary using only 2 bits. However, the problem statement requires a 4-bit magnitude preceded by a sign bit. Thus, we have

$$+3_{10} = 00011$$

In many situations the number of bits is fixed by the size of the registers that will be holding the binary numbers, so that 0s may have to be added in order to fill the required number of bit positions.

(d) Start by writing $+2$ using 5 bits:

$$
\begin{array}{ll}
+2 = 00010 & \\
\quad\quad 11101 & \text{(1's complement)} \\
+ \quad\quad\, 1 & \text{(add 1)} \\
\hline
\quad\quad \mathbf{11110} & \text{(2's-complement representation of } -2)
\end{array}
$$

(e) Start with +8:

$$+8 = 01000$$

	10111	(complement each bit)
+	1	(add 1)
	11000	(2's-complement representation of -8)

Converting from Complement Form to Binary

It is a relatively simple matter to take a number that is in its 1's- or 2's-complement form and convert it back to its true binary value. To go from 1's complement to true binary simply requires complementing each bit again. Similarly, to go from 2's complement to true binary requires complementing each bit and then adding 1 to the LSB. In both cases, the conversion back to binary is the same process that was used to produce the complement to begin with. The reader should verify this for some binary numbers. Here's an example:

$$\text{Start with} \longrightarrow \mathbf{10110}$$
$$\text{2's-complement} \longrightarrow 01010$$
$$\text{2's-complement again} \longrightarrow \mathbf{10110}$$

Teaching Tip: The conversion from 2's complement to true binary can also be done by backtracking through the operations that produced the 2's complement from the binary. That is, you can subtract 1 from the 2's complement to get back to the 1's complement, and then invert each bit to bring it back to true binary.

—— EXAMPLE 6-2 ——————————————————

Each of the following numbers is a signed binary number in the 2's-complement system. Determine the decimal value in each case: (a) 01100, (b) 11010, (c) 10001.

Solution

(a) The sign bit is 0 and so the number is *positive,* and the other 4 bits represent the true magnitude of the number. That is, $1100_2 = 12_{10}$. Thus, the decimal number is **+12**.

(b) The sign bit is a 1 and so the number is *negative,* and the other 4 bits represent the 2's complement of the magnitude. To find out what this magnitude is, we must take a 2's-complement again. Remember, the 2's-complement operation will change the polarity of a signed number, so that if we take the 2's-complement of a negative binary number (including the sign bit), we will get the positive equivalent.

Teaching Tip: It is important to stress that the sign bit is included in the 2's-complement operation, so that the complementing process can convert it to a 0 representing a positive number. Of course, the positive result is not the answer. It just tells us the magnitude of the original negative number.

	11010	(original negative number)
	00101	(1's complement)
+	1	(add 1)
	00110	(+6)

Since the result of taking the 2's-complement is +6, the original number must have been -6. That is,

$$11010_2 = \mathbf{-6_{10}}$$

(c) Follow the same procedure as in (b):

	10001	(original negative number)
	01110	(1's complement)
+	1	(add 1)
	01111	(+15)

Thus, $10001 = \mathbf{-15}$.

Special Case in 2's-Complement Representation

Whenever a signed number has a 1 in the sign bit and all 0s for the magnitude bits, its decimal equiva-

lent is -2^N, where N is the number of bits in the *magnitude*. For example,

$$1000 = -2^3 = -8$$
$$10000 = -2^4 = -16$$
$$100000 = -2^5 = -32$$

and so on.

Thus, we can state that the complete range of values that can be represented in the 2's-complement system having N magnitude bits is

$$-2^N \text{ to } +(2^N - 1)$$

There are a total of 2^{N+1} different values, *including* zero.

For example, here is the list of all signed numbers that can be represented in 4 bits using the 2's-complement system:

Decimal Value	Signed Binary Using 2's Complement
$-8 = -2^3$	1000
-7	1001
-6	1010
-5	1011
-4	1100
-3	1101
-2	1110
-1	1111
$+0$	0000
$+1$	0001
$+2$	0010
$+3$	0011
$+4$	0100
$+5$	0101
$+6$	0110
$+7 = 2^3 - 1$	0111

Discussion Question: Assuming unsigned binary, which number is greater—1000 or 1111? Repeat for signed numbers in the 2's-complement system.

Note that the sequence starts with $-8 = 1000$ and proceeds to $+7 = 0111$ by adding 0001 each time as in a 4-bit counter.

—— EXAMPLE 6-3 ——————————————

What is the range of *unsigned* decimal values that can be represented by 8 bits?

Solution   With no sign bit, all 8 bits are used for the magnitude. Therefore, the values will range from

$$00000000_2 = 0_{10}$$

to

$$11111111_2 = 255_{10}$$

This is a total of 256 values.

—— EXAMPLE 6-4 ——————————————

What is the range of *signed* decimal values that can be represented by 8 bits?

Solution   The largest negative value is
$$10000000_2 = -2^7 = -128_{10}$$
The largest positive value is
$$01111111_2 = +2^7 - 1 = +127_{10}$$
Thus, the range is -128 to $+127$; this is a total of 256 different values, including zero.

—— EXAMPLE 6-5 ————————————————————————

A certain computer is storing the following two signed numbers in its memory using the 2's-complement system. While executing a program, the computer is instructed to convert each number to its opposite sign: that is, change the $+31$ to -31 and change the -12 to $+12$. How will it do this?
$$00011111_2 = +31_{10}$$
$$11110100_2 = -12_{10}$$

Solution   A signed number can have its polarity changed simply by performing the 2's-complement operation on the *complete* number, including the sign bit. The computer circuitry will take the signed number from memory, find its 2's complement, and put the result back in memory.

Teaching Tip: The 2's complement of 10000 is 10000. The same is true for all numbers with a 1 for a sign bit and all 0s for a magnitude. This is the exception to the rule that the 2's-complement operation will change the polarity of a number.

—— REVIEW QUESTIONS ————————————————————————

1. Represent each of the following values as a 5-bit signed number in the 2's-complement system: (a) $+13$, (b) -7, (c) -16.
2. Each of the following is a signed binary number in the 2's-complement system. Determine the decimal equivalent for each: (a) 100011, (b) 1000000, (c) 0111111.
3. What range of signed decimal values can be represented in 12 bits (including the sign bit)?
4. How many bits are required to represent decimal values ranging from -50 to $+50$?
5. What is the largest negative decimal value that can be represented using a total of 16 bits?
6. Perform the 2's-complement operation on each of the following: (a) 10000, (b) 10000000, (c) 1000.

6-3   ADDITION IN THE 2's-COMPLEMENT SYSTEM

The 1's-complement system and 2's-complement system are very similar. However, the 2's-complement system is generally in use because of several advantages it enjoys in circuit implementation. We will now investigate how the operations of addition and subtraction are performed in digital machines that use the 2's-complement representation for negative numbers. In the various cases to be considered, it is im-

portant to note that the sign bit of each number is operated on in the same manner as the magnitude bits.

Case I: Two Positive Numbers. The addition of two positive numbers is straightforward. Consider the addition of +9 and +4:

$$
\begin{array}{r}
+9 \longrightarrow \quad \boxed{0} \;\; 1001 \quad \text{(augend)} \\
+4 \longrightarrow \quad \;0\; \;\; 0100 \quad \text{(addend)} \\
\hline
0\; \;\; 1101 \quad \text{(sum} = +13)
\end{array}
$$

↑—sign bits

Note that the sign bits of the **augend** and **addend** are both 0 and the sign bit of the sum is 0, indicating that the sum is positive. Also note that the augend and addend are made to have the same number of bits. This must *always* be done in the 2's complement system.

Case II: Positive Number and Smaller Negative Number. Consider the addition of +9 and −4. Remember that the −4 will be in its 2's-complement form. Thus, +4 (00100) must be converted to −4 (11100)

┌—sign bits
↓

$$
\begin{array}{r}
+9 \longrightarrow \quad \boxed{0} \;\; 1001 \quad \text{(augend)} \\
-4 \longrightarrow \quad \;1\; \;\; 1100 \quad \text{(addend)} \\
\hline
1 \;\; 0\; \;\; 0101
\end{array}
$$

↑————————this carry is disregarded; the result is 00101 (sum = +5)

In this case, the sign bit of the addend is 1. Note that the sign bits also participate in the addition process. In fact, a carry is generated in the last position of addition. *This carry is always disregarded,* so that the final sum is 00101, which is equivalent to +5.

Case III: Positive Number and Larger Negative Number. Consider the addition of −9 and +4:

$$
\begin{array}{r}
-9 \longrightarrow \quad 10111 \\
+4 \longrightarrow \quad 00100 \\
\hline
11011 \quad \text{(sum} = -5)
\end{array}
$$

↑—negative sign bit

The sum here has a sign bit of 1, indicating a negative number. Since the sum is negative, it is in 2's-complement form, so that the last four bits, 1011, actually represent the 2's complement of the sum. To find the true magnitude of the sum, we must take the 2's complement of 11011; the result is 00101 = +5. Thus, 11011 represents −5.

Case IV: Two Negative Numbers

$$
\begin{array}{r}
-9 \longrightarrow \quad 10111 \\
-4 \longrightarrow \quad 11100 \\
\hline
1 \;\; 10011
\end{array}
$$

↑   ↑—sign bit
└————this carry is disregarded; the result is 10011 (sum = −13)

This final result is again negative and in 2's-complement form with a sign bit of 1. Note that the 2's-complement of 10011 is 01101 = +13.

Teaching Tip: It is important to emphasize that in case III (as well as case IV) the final result is a negative number (sign bit = 1); we converted this result back to a positive value only in order to determine what decimal magnitude it represented.

Case V: Equal and Opposite Numbers

$$-9 \longrightarrow 10111$$
$$+9 \longrightarrow \underline{01001}$$
$$\quad 0 \qquad \cancel{1}\,00000$$

⌐———disregard; the result is 00000 (sum = +0)

The result is obviously +0, as expected.

___ REVIEW QUESTIONS ___

Assume the 2's-complement system for both questions.
1. *True or false:* Whenever the sum of two signed binary numbers has a sign bit of 1, the magnitude of the sum is in 2's-complement form.
2. Add the following pairs of signed numbers. Express the sum as a signed binary number and as a decimal number: (a) 100111 + 111011, (b) 100111 + 011001.

6-4 SUBTRACTION IN THE 2's-COMPLEMENT SYSTEM

The subtraction operation using the 2's-complement system actually involves the operation of addition and is really no different from the various cases considered in Section 6-3. When subtracting one binary number (the **subtrahend**) from another binary number (the **minuend**), the procedure is as follows:

1. Take the 2's complement of the subtrahend, *including* the sign bit. If the subtrahend is a positive number, this will change it to a negative number in 2's-complement form. If the subtrahend is a negative number, this will change it to a positive number in true binary form. In other words, we are changing the sign of the subtrahend.
2. After the 2's complement of the subtrahend is taken, it is *added* to the minuend. The minuend is kept in its original form. The result of this addition represents the required *difference*. The sign bit of this difference determines whether it is + or − and whether it is in true binary form or 2's-complement form.
3. Remember, both numbers have to have the same number of bits.

Let us consider the case where +4 is to be subtracted from +9.

$$\text{minuend } (+9) \longrightarrow 01001$$
$$\text{subtrahend } (+4) \longrightarrow 00100$$

Change the subtrahend to its 2's-complement form (11100), which represents −4. Now add this to the minuend.

$$01001 \qquad (+9)$$
$$\underline{+11100} \qquad (-4)$$
$$\cancel{1}\;\;00101 \qquad (+5)$$

⌐——disregard, so the result is 00101 = +5

Common Student Mistake:
The cases where the subtrahend is negative seem to give students the most trouble, probably because the subtrahend is already in 2's-complement form. It would be helpful to go over an example of this type.

When the subtrahend is changed to its 2's complement, it actually becomes −4, so that we are *adding* +9 and −4, which is the same as subtracting +4 from +9. This is the same as case II of Section 6-3. Any subtraction operation, then, actually becomes one of addition when the 2's-complement system is used. This feature of the 2's-complement system has made it the most widely used of the methods available, since it allows addition and subtraction to be performed by the same circuitry.

The reader should verify the results of using the above procedure for the following subtractions: (a) +9 − (−4); (b) −9 − (+4); (c) −9 − (−4); (d) +4 − (−4). Remember that when the result has a sign bit of 1, it is negative and in 2's-complement form.

Arithmetic Overflow   In each of the previous addition and subtraction examples, the numbers that were added consisted of a sign bit and 4 magnitude bits. The answers also consisted of a sign bit and 4 magnitude bits. Any carry into the sixth bit position was disregarded. In all of the cases considered, the magnitude of the answer was small enough to fit into 4 bits. Let's look at the addition of +9 and +8.

$$
\begin{array}{r}
+9 \longrightarrow \quad \boxed{0} \quad 1001 \\
+8 \longrightarrow \quad 0 \quad 1000 \\
\hline
1 \quad 0001
\end{array}
$$

↑—sign bits

The answer has a negative sign bit, which is obviously incorrect. The answer should be +17, but the magnitude 17 requires more than 4 bits and therefore *overflows* into the sign-bit position. This overflow condition always produces an incorrect result, and its occurrence is detected by examining the sign bit of the result and comparing it with the sign bits of the numbers being added. In a computer, a special circuit is used to detect any overflow condition and to signal that the answer is erroneous. We will encounter such a circuit in one of the end-of-chapter problems.

────── REVIEW QUESTIONS ──────

1. Perform the subtraction on the following pairs of signed numbers using the 2's-complement system. Express the results as signed binary numbers and as decimal values: (a) 01001 − 11010, (b) 10010 − 10011.
2. How can arithmetic overflow be detected when signed numbers are being added?

6-5   MULTIPLICATION OF BINARY NUMBERS

The multiplication of binary numbers is done in the same manner as the multiplication of decimal numbers. The process is actually simpler, since the multiplier digits are either 0 or 1 and so we are always multiplying by 0 or 1 and no other digits. The following example illustrates for unsigned binary numbers.

$$1001 \longleftarrow \text{multiplicand} = 9_{10}$$
$$1011 \longleftarrow \text{multiplier} = 11_{10}$$

$$\left.\begin{array}{l} 1001 \\ 1001 \\ 0000 \\ 1001 \end{array}\right\} \quad \text{partial products}$$

$$1100011\} \qquad \text{final product} = 99_{10}$$

In this example the multiplicand and multiplier are in true binary form and no sign bits are used. The steps followed in the process are exactly the same as in decimal multiplication. First, the LSB of the multiplier is examined; in our example it is a 1. This 1 multiplies the multiplicand to produce 1001, which is written down as the first partial product. Next, the second bit of the multiplier is examined. It is a 1, and so 1001 is written for the second partial product. Note that this second partial product is *shifted* one place to the left relative to the first one. The third bit of the multiplier is 0, and 0000 is written as the third partial product; again, it is shifted one place to the left relative to the previous partial product. The fourth multiplier bit is 1, and so the last partial product is 1001 shifted again one position to the left. The four partial products are then summed to produce the final product.

Most digital machines can only add two binary numbers at a time. For this reason, the partial products formed during multiplication cannot all be added together at the same time. Instead, they are added together two at a time; that is, the first is added to the second, their sum is added to the third, and so on. This process is now illustrated for the example above:

$$\text{Add} \quad \left\{\begin{array}{l} 1001 \longleftarrow \text{first partial product} \\ 1001 \longleftarrow \text{second partial product shifted left} \end{array}\right.$$

$$\text{Add} \quad \left\{\begin{array}{l} 11011 \longleftarrow \text{sum of first two partial products} \\ 0000 \longleftarrow \text{third partial product shifted left} \end{array}\right.$$

$$\text{Add} \quad \left\{\begin{array}{l} 011011 \longleftarrow \text{sum of first three partial products} \\ 1001 \longleftarrow \text{fourth partial product shifted left} \end{array}\right.$$

$$1100011 \longleftarrow \begin{array}{l} \text{sum of four partial products which} \\ \text{equals final total product} \end{array}$$

Multiplication in the 2's-Complement System

In computers that use the 2's-complement representation, multiplication is carried on in the manner described above provided that both the multiplicand and multiplier are put in true binary form. If the two numbers to be multiplied are positive, they are already in true binary form and are multiplied as they are. The resulting product is, of course, positive and is given a sign bit of 0. When the two numbers are negative, they will be in 2's-complement form. The 2's complement of each is taken to convert it to a positive number and then they are multiplied. The product is kept as a positive number and given a sign bit of 0.

When one of the numbers is positive and the other is negative, the negative number is first converted to a positive magnitude by taking its 2's complement. The product will be in true-magnitude form. However, the product has to be negative, since the original numbers are of opposite sign. Thus, the product is then changed to 2's-complement form and given a sign bit of 1.

Discussion Question: Try multiplying the 2's-complement representations of -3 and -4 without first converting to true binary.

1. Multiply the unsigned numbers 0111 and 1110.

6-6 BINARY DIVISION

The process for dividing one binary number (the *dividend*) by another (the *divisor*) is the same as that which is followed for decimal numbers, that which we usually refer to as "long division." The actual process is simpler in binary because when we are checking to see how many times the divisor "goes into" the dividend, there are only two possibilities, 0 or 1. To illustrate, consider the following division examples:

$$
\begin{array}{r}
0011 \\
11\,\overline{)1001} \qquad (9 \;\div\; 3 = 3) \\
\underline{011} \\
0011
\end{array}
$$

$$
\begin{array}{r}
0010.1 \\
100\,\overline{)1010.0} \qquad (10 \;\div\; 4 = 2.5) \\
\underline{100} \\
100 \\
\underline{100} \\
0
\end{array}
$$

In most modern digital machines the subtractions that are part of the division operation are usually carried out using 2's-complement subtraction—that is, taking the 2's complement of the subtrahend and then adding.

The division of signed numbers is handled in the same way as multiplication. Negative numbers are made positive by complementing and the division is then carried out. If the dividend and divisor are of opposite sign, the resulting quotient is changed to a negative number by taking its 2's-complement and is given a sign bit of 1. If the dividend and divisor are of the same sign, the quotient is left as a positive number and given a sign bit of 0.

6-7 BCD ADDITION

In Chapter 2 we stated that many computers and calculators use the BCD code to represent decimal numbers. Recall that this code takes *each* decimal digit and represents it by a 4-bit code ranging from 0000 to 1001. The addition of decimal numbers that are in BCD form can be best understood by considering the two cases that can occur when two decimal digits are added.

Sum Equals 9 or Less   Consider adding 5 and 4 using BCD to represent each digit:

$$
\begin{array}{rll}
5 & 0101 & \longleftarrow \text{BCD for 5} \\
+4 & +\ 0100 & \longleftarrow \text{BCD for 4} \\
\hline
9 & 1001 & \longleftarrow \text{BCD for 9}
\end{array}
$$

The addition is carried out as in normal binary addition and the sum is 1001, which is the BCD code for 9. As another example, take 45 added to 33:

```
 45        0100  0101   ←—— BCD for 45
+33      + 0011  0011   ←—— BCD for 33
 78        0111  1000   ←—— BCD for 78
```

In this example the 4-bit codes for 5 and 3 are added in binary to produce 1000, which is BCD for 8. Similarly, adding the second-decimal-digit positions produces 0111, which is BCD for 7. The total is 0111 1000, which is the BCD code for 78.

In the examples above, none of the sums of the pairs of decimal digits exceeded 9; therefore, *no decimal carries were produced*. For these cases the BCD addition process is straightforward and is actually the same as binary addition.

Sum Greater than 9   Consider the addition of 6 and 7 in BCD:

```
  6        0110   ←—— BCD for 6
+ 7      + 0111   ←—— BCD for 7
+13        1101   ←—— invalid code group for BCD
```

The sum 1101 does not exist in the BCD code; it is one of the six forbidden or invalid 4-bit code groups. This has occurred because the sum of the two digits exceeds 9. Whenever this occurs the sum has to be corrected by the addition of six (0110) to take into account the skipping of the six invalid code groups:

```
            0110   ←—— BCD for 6
          + 0111   ←—— BCD for 7
            1101   ←—— invalid sum
            0110   ←—— add 6 for correction
    0001    0011   ←—— BCD for 13
      1       3
```

As shown above, 0110 is added to the invalid sum and produces the correct BCD result. Note that a carry is produced into the second decimal position. This addition of 0110 has to be performed whenever the sum of the two decimal digits is greater than 9.

As another example, take 47 plus 35 in BCD:

```
  47       0100     0111   ←—— BCD for 47
+ 35     + 0011     0101   ←—— BCD for 35
  82       0111     1100   ←—— invalid sum in first digit
              1←┐   0110   ←—— add 6 to correct
           1000  └─ 0010   ←—— correct BCD sum
             8         2
```

The addition of the 4-bit codes for the 7 and 5 digits results in an invalid sum and is corrected by adding 0110. Note that this generates a carry of 1, which is carried over to be added to the BCD sum of the second-position digits.

Consider the addition of 59 and 38 in BCD:

```
                 ↓┐
                 1 │
  59       0101  │ 1001   ←—— BCD for 59
+ 38     + 0011  │ 1000   ←—— BCD for 38
  97       1001  └─0001   ←—— perform addition
                   0110   ←—— add 6 to correct
           1001    0111      BCD for 97
             9       7
```

Here, the addition of the least significant digits (LSDs) produces a sum of $17 = 10001$. This generates a carry into the next digit position to be added to the codes for 5 and 3. Since $17 > 9$, a correction factor of 6 has to be added to the LSD sum. Addition of this correction does not generate a carry; the carry was already generated in the original addition.

To summarize the BCD addition procedure:

1. Add, using ordinary binary addition, the BCD code groups for each digit position.

2. For those positions where the sum is 9 or less, no correction is needed. The sum is in proper BCD form.

3. When the sum of two digits is greater than 9, a correction of 0110 should be added to that sum to get the proper BCD result. This case always produces a carry into the next digit position, either from the original addition (step 1) or from the correction addition.

Discussion Question: What happens when the carry produced from the addition of the correction factor in one digit position produces an invalid sum in the next digit position? Try 27 (BCD) added to 76 (BCD).

The procedure for BCD addition is clearly more complicated than straight binary addition. This is also true of the other BCD arithmetic operations. Readers should perform the addition of $275 + 641$. Then check the correct procedure below.

$$
\begin{array}{r}
275 \\
+\ 641 \\
\hline
916
\end{array}
\qquad
\begin{array}{r}
0010 \quad 0111 \quad 0101 \\
+\ 0110 \quad 0100 \quad 0001 \\
\hline
1000 \quad 1011 \quad 0110 \\
+\qquad\quad 0110 \\
\hline
1001 \quad 0001 \quad 0110
\end{array}
$$

⟵ BCD for 275
⟵ BCD for 641
⟵ perform addition
⟵ add 6 to correct 2nd digit
⟵ BCD for 916

—— REVIEW QUESTIONS ————————————————

1. How can you tell when a correction is needed in BCD addition?
2. Represent 135_{10} and 265_{10} in BCD and then perform BCD addition. Check your work by converting the result back to decimal.

6-8 HEXADECIMAL ARITHMETIC

Hex numbers are used extensively in machine-language computer programming and in conjunction with computer memories (i.e., addresses). When working in these areas, there will be situations where hex numbers have to be added or subtracted.

Hex Addition   Addition of hexadecimal numbers is done in much the same way as decimal addition as long as you remember that the largest hex digit is F instead of 9. The following procedure is suggested.

1. Add the two hex digits in decimal, mentally inserting the decimal equivalent for those digits larger than 9.

2. If the sum is 15 or less, it can be directly expressed as a hex digit.

3. If the sum is greater than or equal to 16, subtract 16 and carry a 1 to the next digit position.

The following examples will illustrate the procedure.

___ EXAMPLE 6-6 ___

Add the hex numbers 58 and 24.

Solution

$$\begin{array}{r} 58 \\ +\ \ 24 \\ \hline 7C \end{array}$$

Adding the LSDs (8 and 4) produces 12, which is C in hex. There is no carry into the next digit position. Adding the 5 and 2 produces 7.

___ EXAMPLE 6-7 ___

Add the hex numbers 58 and 4B.

Solution

$$\begin{array}{r} 58 \\ +\ \ 4B \\ \hline A3 \end{array}$$

Start by adding 8 and B, mentally substituting decimal 11 for B. This produces a sum of 19. Since 19 is greater than 16, subtract 16 to get 3; write down the 3 and carry a 1 into the next position. This carry is added to the 5 and 4 to produce a sum of 10_{10}, which is then converted to hexadecimal A.

___ EXAMPLE 6-8 ___

Add 3AF to 23C.

Solution

$$\begin{array}{r} 3AF \\ +\ \ 23C \\ \hline 5EB \end{array}$$

The sum of F and C is considered as $15 + 12 = 27_{10}$. Since this is greater than 16, subtract 16 to get 11_{10}, which is hexadecimal B, and carry a 1 into the second position. Add this carry to A and 3 to obtain E. There is no carry into the MSD position.

Hex Subtraction   Remember that hex numbers are just an efficient way to represent binary numbers. Thus we can subtract hex numbers using the same method we used for binary numbers. The 2's complement of the hex subtrahend will be taken and then *added* to the minuend, and any carry out of the MSD position will be disregarded.

How do we find the 2's complement of a hex number? One way is to convert it

to binary, take the 2's complement of the binary equivalent, and then convert it back to hex. This process is illustrated below.

$$
\begin{array}{ccc}
& 73A & \longleftarrow \text{ hex number} \\
0111 \quad 0011 & 1010 & \longleftarrow \text{ convert to binary} \\
1000 \quad 1100 & 0110 & \longleftarrow \text{ take 2's complement} \\
& 8C6 & \longleftarrow \text{ convert back to hex}
\end{array}
$$

Teaching Tip: Subtracting a hex digit from F is the same as subtracting the binary equivalent from 1111, which is the same as complementing each bit.

There is a quicker procedure. Subtract *each* hex digit from F, then add 1. Let's try this for the same hex number from the example above.

$$
\left.
\begin{array}{ccc}
F & F & F \\
-\ 7 & -\ 3 & -\ A \\
\hline
8 & C & 5
\end{array}
\right\} \longleftarrow \text{ subtract each digit from F}
$$

$$
\begin{array}{ccc}
& & +\ 1 \quad \longleftarrow \text{ add 1} \\
\hline
8 & C & 6 \quad \longleftarrow \text{ hex equivalent of 2's complement}
\end{array}
$$

Try either of the above procedures on the hex number E63. The correct result for the 2's complement is 19D.

___ EXAMPLE 6-9 _____

Subtract $3A5_{16}$ from 592_{16}.

Solution   First, convert the subtrahend (3A5) to its 2's-complement form by using either method presented above. The result is C5B. Then add this to the minuend (592):

$$
\begin{array}{r}
592 \\
+\ \underline{C5B} \\
\cancel{1}1ED
\end{array}
$$

$\uparrow$_____ disregard carry

Ignoring the carry out of the MSD addition, the result is 1ED. We can prove that this is correct by adding 1ED to 3A5 and checking to see that it equals 592_{16}.

___ REVIEW QUESTIONS _____

1. Add 67F + 2A4.
2. Subtract 67F − 2A4.

6-9   ARITHMETIC CIRCUITS

One essential function of most computers and calculators is the performance of arithmetic operations. These operations are all performed in the arithmetic-logic unit of a computer, where logic gates and flip-flops are combined so that they can add, subtract, multiply, and divide binary numbers. These circuits perform arithmetic operations at speeds that are not humanly possible. Typically, an addition operation will take less than 1 μs.

We will now study some of the basic arithmetic circuits that are used to perform the arithmetic operations discussed earlier. In some cases we will go through the actual design process, even though the circuits may be commmercially available in integrated-circuit form, to provide more practice in the use of the techniques learned in Chapter 4.

Arithmetic-Logic Unit   All arithmetic operations take place in the **arithmetic-logic unit** (ALU) of a computer. Figure 6-3 is a block diagram showing the major elements included in a typical ALU. The main purpose of the ALU is to accept binary data that are stored in the memory and to execute arithmetic operations on these data according to instructions from the control unit.

The arithmetic-logic unit contains at least two flip-flop registers: the **B register** and the **accumulator register.** It also contains combinatorial logic, which performs the arithmetic operations on the binary numbers that are stored in the B register and the accumulator. A typical sequence of operations may occur as follows:

1. The control unit receives an instruction (from the memory unit) specifying that a number stored in a particular memory location (address) is to be added to the number presently stored in the accumulator register.

2. The number to be added is transferred from memory to the B register.

3. The number in the B register and the number in the accumulator register are added together in the logic circuits (upon command from the control unit). The resulting sum is then sent to the accumulator to be stored.

4. The new number in the accumulator can remain there so that another number can be added to it or, if the particular arithmetic process is finished, it can be transferred to memory for storage.

These steps should make it apparent how the accumulator register derives its name. This register "accumulates" the sums that occur when performing successive additions between new numbers acquired from memory and the previously accumulated sum. In fact, for any arithmetic problem containing several steps, the accumulator always contains the results of the intermediate steps as they are completed as well as the final result when the problem is finished.

FIGURE 6-3   Functional parts of an ALU.

Arithmetic—logic unit

Computers and calculators perform the addition operation on two binary numbers at a time, where each binary number can have several binary digits. Figure 6-4 illustrates the addition of two 5-bit numbers. The *augend* is stored in the accumulator register; that is, the accumulator contains five FFs, storing the values 10101 in successive FFs. Similarly, the *addend*, the number that is to be added to the augend, is stored in the B register (in this case, 00111).

The addition process starts by adding the least significant bits (LSBs) of the augend and addend. Thus, $1 + 1 = 10$, which means that the *sum* for that position is 0 with a *carry* of 1.

This carry has to be added to the next position along with the augend and addend bits in that position. Thus, in the second position, $1 + 0 + 1 = 10$, which is again a sum of 0 and a carry of 1. This carry is added to the next position together with the augend and addend bits in that position, and so on for the remaining positions.

At each step in this addition process we are performing the addition of 3 bits: the augend bit, the addend bit, and a carry bit from the previous position. The result of the addition of these 3 bits produces 2 bits: a *sum* bit, and a *carry* bit which is to be added to the next position. It should be clear that the same process is followed for each bit position. Thus, if we can design a logic circuit that can duplicate this process, then all we have to do is to use the identical circuit for each of the bit positions. This is illustrated in Figure 6-5.

In this diagram variables A_4, A_3, A_2, A_1, and A_0 represent the bits of the augend that are stored in the accumulator (which is also called the A register). Variables B_4, B_3, B_2, B_1, and B_0 represent the bits of the addend stored in the B register. Variables C_4, C_3, C_2, C_1, and C_0 represent the carry bits into the corresponding positions. Variables S_4, S_3, S_2, S_1, S_0 are the sum output bits for each position. Corresponding bits of the augend and addend are fed to a logic circuit called a **full adder,** along with a carry bit from the previous position. For example, bits A_1 and B_1 are fed into full adder 1 along with C_1, which is the carry bit produced by the addition of the A_0 and B_0 bits. Bits A_0 and B_0 are fed into full adder 0 along with C_0. Since A_0 and B_0 are the LSBs of the augend and addend, it appears that C_0 would always have to be 0, since there can be no carry into that position. However, we shall see that there will be situations when C_0 can also be 1.

FIGURE 6-4 Typical binary addition process.

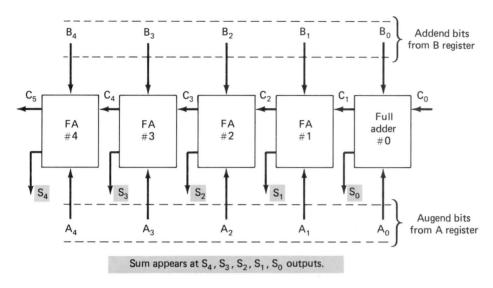

FIGURE 6-5   Block diagram of 5-bit parallel adder circuit using full adders.

The full-adder circuit used in each position has three inputs: an A bit, a B bit, and a C bit, and it produces two outputs: a sum bit and a carry bit. For example, full adder 0 has inputs A_0, B_0, and C_0, and it produces outputs S_0 and C_1. Full adder 1 has A_1, B_1, and C_1 as inputs and S_1 and C_2 as outputs; and so on. This arrangement is repeated for as many positions as there are in the augend and addend. Although this illustration is for 5-bit numbers, in modern computers the numbers usually range from 8 to 64 bits.

The arrangement in Figure 6-5 is called a **parallel adder** because all the bits of the augend and addend are present and are fed into the adder circuits *simultaneously*. This means that the additions in each position are taking place at the same time. This is different from how we add on paper, taking each position one at a time starting with the LSB. Clearly, parallel addition is extremely fast. More will be said about this later.

Teaching Tip: A *half-adder* circuit is sometimes used for the addition of the LSB position where $C_0 = 0$. This circuit accepts two inputs A and B, and produces a sum and carry output. See Problem 6-18.

___ REVIEW QUESTIONS ___

1. How many inputs does a full adder have? How many outputs?
2. Assume the following input levels in Figure 6-5: $A_4 A_3 A_2 A_1 A_0 = 01001$; $B_4 B_3 B_2 B_1 B_0 = 00111$; $C_0 = 0$.
   (a) What are the logic levels at the outputs of FA #2?
   (b) What is the logic level at the C_5 output?

6-11   DESIGN OF A FULL ADDER

Now that we know the function of the full adder, we can proceed to design a logic circuit that will perform this function. First, we must construct a truth table showing the various input and output values for all possible cases. Figure 6-6 shows the truth table having three inputs A, B, and C_{IN}, and two outputs, S and C_{OUT}. There are

Augend bit input	Addend bit input	Carry bit input	Sum bit output	Carry bit output
A	B	C_{IN}	S	C_{OUT}
0	0	0	0	0
0	0	1	1	0
0	1	0	1	0
0	1	1	0	1
1	0	0	1	0
1	0	1	0	1
1	1	0	0	1
1	1	1	1	1

FIGURE 6-6 Truth table for a full-adder circuit.

eight possible cases for the three inputs, and for each case the desired output values are listed. For example, consider the case $A = 1$, $B = 0$, and $C_{IN} = 1$. The full adder (hereafter abbreviated FA) must add these bits to produce a sum (S) of 0 and a carry (C_{OUT}) of 1. The reader should check the other cases to be sure they are understood.

Since there are two outputs, we will design the circuitry for each output individually, starting with the S output. The truth table shows that there are four cases where S is to be a 1. Using the sum-of-products method, we can write the expression for S as

$$S = \overline{A}\,\overline{B}C_{IN} + \overline{A}B\overline{C}_{IN} + A\overline{B}\,\overline{C}_{IN} + ABC_{IN} \qquad (6\text{-}1)$$

We can now try to simplify this expression by factoring. Unfortunately, none of the terms in the expression has two variables in common with any of the other terms. However, $\overline{A}$ can be factored from the first two terms and A can be factored from the last two terms:

$$S = \overline{A}(\overline{B}C_{IN} + B\overline{C}_{IN}) + A(\overline{B}\,\overline{C}_{IN} + BC_{IN})$$

The first term in parentheses should be recognized as the exclusive-OR combination of B and C_{IN}, which can be written as $B \oplus C_{IN}$. The second term in parentheses should be recognized as the exclusive-NOR of B and C_{IN}, which can be written as $\overline{B \oplus C_{IN}}$. Thus, the expression for S becomes

$$S = \overline{A}(B \oplus C_{IN}) + A(\overline{B \oplus C_{IN}})$$

If we let $X = B \oplus C_{IN}$, this can be written as

$$S = \overline{A} \cdot X + A \cdot \overline{X} = A \oplus X$$

which is simply the EX-OR of A and X. Replacing the expression for X, we have

$$S = A \oplus [B \oplus C_{IN}] \qquad (6\text{-}2)$$

Discussion Question: Simplify the expression for S by first factoring C_{IN} from the first and last terms, and $\overline{C}_{IN}$ from the middle terms. The result should be $C \oplus (A \oplus B)$.

Consider now the output C_{OUT} in the truth table of Figure 6-6. We can write the sum-of-products expression for C_{OUT} as follows:

$$C_{OUT} = \overline{A}BC_{IN} + A\overline{B}C_{IN} + AB\overline{C}_{IN} + ABC_{IN}$$

This expression can be simplified by factoring. We will employ the trick introduced in Chapter 3, whereby we will use the ABC_{IN} term *three* times since it has common factors with each of the other terms. Hence,

$$C_{OUT} = BC_{IN}(\overline{A} + A) + AC_{IN}(\overline{B} + B) + AB(\overline{C}_{IN} + C_{IN}) \qquad (6\text{-}3)$$
$$= BC_{IN} + AC_{IN} + AB$$

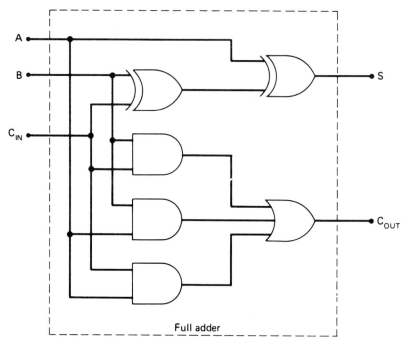

FIGURE 6-7 Complete circuitry for a full adder.

This expression cannot be simplified further.

Expressions (6-2) and (6-3) can be implemented as shown in Figure 6-7. There are several other implementations that can be used to produce the same expressions for S and C_{OUT}, none of which has any particular advantage over those shown. The complete circuit with inputs A, B, and C_{IN} and outputs S and C_{OUT} represents the full adder. Each of the FAs in Figure 6-5 contains the same circuitry (or its equivalent).

K-Map Simplification   We simplified the expressions for S and C_{OUT} using algebraic methods. The K-map method can also be used. Figure 6-8(a) shows the K

FIGURE 6-8   K mappings for the full-adder outputs.

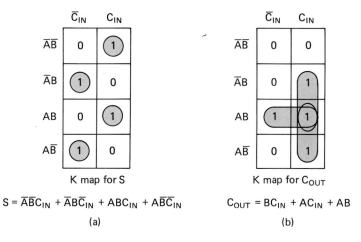

K map for S

$$S = \overline{A}\overline{B}C_{IN} + \overline{A}B\overline{C}_{IN} + ABC_{IN} + A\overline{B}\overline{C}_{IN}$$

(a)

K map for C_{OUT}

$$C_{OUT} = BC_{IN} + AC_{IN} + AB$$

(b)

map for the S output. This map has no adjacent 1s, and so there are no pairs or quads to loop. Thus, the expression for S cannot be simplified using the K map. This points out a limitation of the K-map method as compared with the algebraic method. We were able to simplify the expression for S through factoring and the use of EX-OR and EX-NOR operations.

The K map for the C_{OUT} output is shown in Figure 6-8(b). The three pairs that are looped will produce the same expression obtained from the algebraic method.

Half Adder   The FA operates on three inputs to produce a sum and carry output. In some cases a circuit is needed that will add only 2 input bits, to produce a sum and carry output. An example would be the addition of the LSB position of two binary numbers where there is no carry input to be added. A special logic circuit can be designed to take *two* input bits, A and B, and to produce sum (S) and carry (C_{OUT}) outputs. This circuit is called a **half adder** (*HA*). Its operation is similar to that of a FA except that it operates on only 2 bits. We shall leave the design of the HA as an exercise at the end of the chapter (see Problem 6-18). However, we will use the HA symbol whenever it is appropriate.

6-12   COMPLETE PARALLEL ADDER WITH REGISTERS

In a computer, the numbers that are to be added are stored in FF registers. Figure 6-9 shows the complete diagram of a 4-bit parallel adder including the storage registers. The augend bits A_3 through A_0 are stored in the accumulator (A register); the addend bits B_3 through B_0 are stored in the B register. Each of these registers is made up of D flip-flops for easy transfer of data.

The contents of the A register (i.e., the binary number stored in A_3 through A_0) is added to the contents of the B register by the four FAs, and the sum is produced at outputs S_3 through S_0. C_4 is the carry out of the fourth FA and it can be used as the carry input to a fifth FA, or as an *overflow* bit to indicate that the sum exceeds 1111.

Note that the sum outputs are connected to the D inputs of the A register. This will allow the sum to be parallel-transferred into the A register on the PGT of the TRANSFER pulse. In this way, the sum can be stored in the A register.

Also note that the D inputs of the B register are coming from the computer's memory, so that numbers from memory will be parallel-transferred into the B register on the PGT of the LOAD pulse. For now we will not be concerned with how the numbers come from memory.

Register Notation   Before we go through the complete process of how this circuit adds two binary numbers, it will be helpful to introduce some notation that makes it easy to describe the contents of a register and data transfer operations.

Whenever we want to give the levels that are present at each FF in a register or at each output of a group of outputs, we will use brackets as illustrated below:

$$[A] = 1011$$

This is the same as saying that $A_3 = 1$, $A_2 = 0$, $A_1 = 1$, $A_0 = 1$. In other words, think of [A] as representing "the contents of register A."

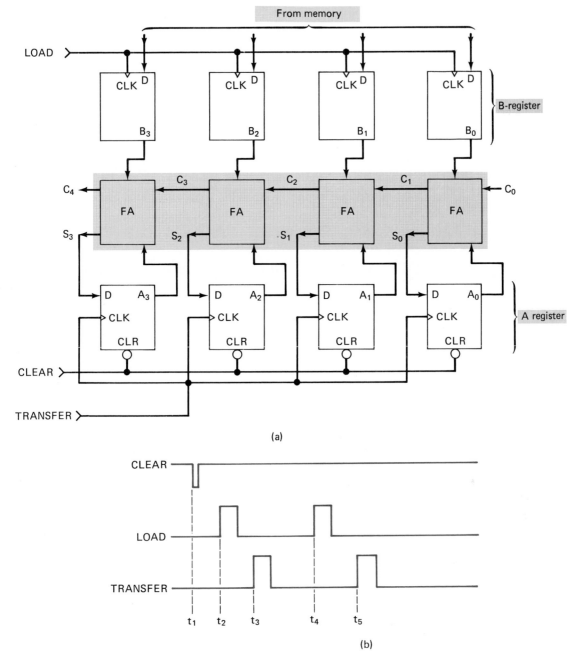

FIGURE 6-9 (a) Complete 4-bit parallel adder with registers; (b) signals used to add binary numbers from memory and store their sum in the accumulator.

Whenever we want to indicate the transfer of data to or from a register, we will use an arrow as illustrated below:

$$[B] \longrightarrow [A]$$

This means that the contents of the B register have been transferred to the A register. The old contents of the A register will be lost as a result of this operation, and the B register will be unchanged.

Sequence of Operations We will now describe the process by which the circuit of Figure 6-9 will add the binary numbers 1001 and 0101. Assume $C_0 = 0$; that is, there is no carry into the LSB position.

Teaching Tip: Students sometimes have difficulty seeing that once the sum outputs have been transferred to the *A* register, the new *A* outputs will produce a different set of sum outputs from the adders. This new sum has no meaning and it will not be used. It will be changed when the next number to be added is transferred into the *B* register.

1. **[A] = 0000.** A CLEAR pulse is applied to the asynchronous inputs (*CLR*) of each FF in register *A*. This occurs at time t_1.

2. **[M] → [B].** This first binary number is transferred from memory (*M*) to the *B* register. In this case, the binary number 1001 is loaded into register *B* on the PGT of the LOAD pulse at t_2.

3. **[S]* → [A].** With [B] = 1001 and [A] = 0000, the full adders produce a sum of 1001; that is, [S] = 1001. These sum outputs are transferred into the *A* register on the PGT of the TRANSFER pulse at t_3. This makes [A] = 1001.

4. **[M] → [B].** The second binary number, 0101, is transferred from memory into the *B* register on the PGT of the second LOAD pulse at t_4. This makes [B] = 0101.

5. **[S] → [A].** With [B] = 0101 and [A] = 1001, the FAs produce [S] = 1110. These sum outputs are transferred into the *A* register when the second TRANSFER pulse occurs at t_5. Thus, [A] = 1110.

6. At this point, the sum of the two binary numbers is present in the accumulator. In most computers the contents of the accumulator, [A], will usually be transferred to the computer's memory so that the adder circuit can be used for a new set of numbers. The circuitry that performs this [A] → [M] transfer is not shown in the diagram.

———— REVIEW QUESTIONS ————

1. Suppose that four different 4-bit numbers are to be taken from memory and added by the circuit of Figure 6-9. How many CLEAR pulses will be needed? How many TRANSFER pulses? How many LOAD pulses?
2. Determine the contents of the *A* register after the following sequence of operations: [A] = 0000, [0110] → [B], [S] → [A], [1110] → [B], [S] → [A].

6-13 CARRY PROPAGATION

The parallel adder of Figure 6-9 performs additions at a relatively high speed, since it adds the bits from each position simultaneously. However, its speed is limited by an effect called **carry propagation** or **carry ripple,** which can best be explained by considering the following addition:

$$\begin{array}{r} 0111 \\ +\ 0001 \\ \hline 1000 \end{array}$$

Addition of the LSB position produces a carry into the second position. This carry, when added to the bits of the second position, produces a carry into the third posi-

———

*Even though *S* is not a register, we will use [S] to represent the group of *S* outputs.

tion. This latter carry, when added to the bits of the third position, produces a carry into the last position. The key thing to notice in this example is that the sum bit generated in the *last* position (MSB) depended on the carry that was generated by the addition in the *first* position (LSB).

Looking at this from the viewpoint of the circuit of Figure 6-9, S_3 out of the last full adder depends on C_1 out of the first full adder. But the C_1 signal must pass through three FAs before it produces S_3. What this means is that the S_3 output will not reach its correct value until C_1 has propagated through the intermediate FAs. This represents a time delay that depends on the propagation delay produced in a FA. For example, if each FA is considered to have a propagation delay of 40 ns, then S_3 will not reach its correct level until 120 ns after C_1 is generated. This means that the add command pulse cannot be applied until 160 ns after the augend and addend numbers are present in the FF registers (the extra 40 ns is due to the delay of the LSB full adder, which generates C_1).

Obviously, the situation becomes much worse if we extend the adder circuitry to add a greater number of bits. If the adder were handling 32-bit numbers, the carry propagation delay could be 1280 ns $= 1.28$ μs. The add pulse could not be applied until at least 1.28 μs after the numbers were present in the registers.

This magnitude of delay is prohibitive for high-speed computers. Fortunately, logic designers have come up with several ingenious schemes for reducing this delay. One of the schemes, called **look-ahead carry,** utilizes logic gates to look at the lower-order bits of the augend and addend to see if a higher-order carry is to be generated. For example, it is possible to build a logic circuit with B_2, B_1, B_0, A_2, A_1, and A_0 as inputs and C_3 as an output. This logic circuit would have a shorter delay than is obtained by the carry propagation through the FAs. This scheme requires a large amount of extra circuitry but is necessary to produce high-speed adders. The extra circuitry is not a significant consideration with the present use of integrated circuits. Many high-speed adders available in integrated-circuit form utilize the look-ahead carry or a similar technique for reducing overall propagation delays.

6-14 INTEGRATED-CIRCUIT PARALLEL ADDER

Several parallel adders are available as ICs. The most common is a 4-bit parallel adder IC that contains four interconnected FAs and the look-ahead carry circuitry needed for high-speed operation. The 7483A, 74LS83A, 74283, and 74LS283 are all TTL 4-bit parallel-adder chips. The 283s are identical to the 83s except that they have V_{CC} and ground on pins 16 and 8, respectively; it has become standard on all new chips to have the power and ground pins at the corners of the chip. The 74HC283 is the high-speed CMOS version of the same 4-bit parallel adder.

Figure 6-10(a) shows the functional symbol for the 74LS83 4-bit parallel adder (and its equivalents). The inputs to this IC are two 4-bit numbers, $A_3 A_2 A_1 A_0$ and $B_3 B_2 B_1 B_0$, and the carry, C_0, into the LSB position. The outputs are the sum bits, $S_3 S_2 S_1 S_0$, and the carry, C_4, out of the MSB position. The sum bits are often labeled $\Sigma_3 \Sigma_2 \Sigma_1 \Sigma_0$, where Σ is the Greek capital letter *sigma*.

Cascading Parallel Adders   Two or more parallel-adder blocks can be connected (cascaded) to accommodate the addition of larger binary numbers. To illus-

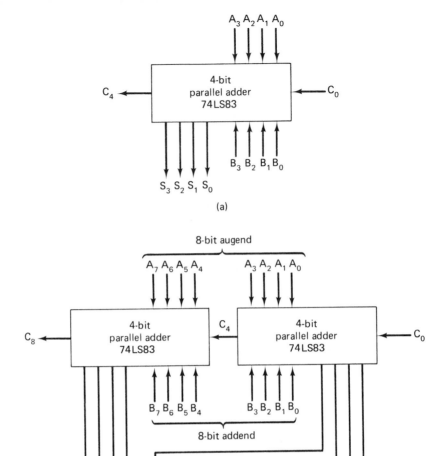

FIGURE 6-10 (a) Block symbol for the 74LS83 4-bit parallel adder; (b) cascading two 74LS83s.

trate, Figure 6-10(b) shows how two 74LS83 adders can be connected to add two 8-bit numbers. The adder on the right adds the four least significant bits of the numbers. The C_4 output of this adder is connected as the input carry to the first position of the second adder, which adds the four most significant bits of the numbers. The eight sum outputs represent the resultant sum of the two 8-bit numbers. C_8 is the carry out of the last position (MSB) of the second adder. C_8 can be used as an overflow bit or as a carry into another adder stage if larger binary numbers are to be handled.

_____ EXAMPLE 6-10 _____

Determine the logic levels at the inputs and outputs of the 8-bit adder in Figure 6-10(a) when 137_{10} is added to 72_{10}.

Solution   First convert each number to an 8-bit binary number:

$$137 = 10001001$$
$$72 = 01001000$$

These two binary values will be applied to the A and B inputs; that is, the A inputs will be 10001001 from left to right, and the B inputs will be 01001000 from left to right. The adder will produce the binary sum of the two numbers:

$$[A] = 10001001$$
$$[B] = \underline{01001000}$$
$$[S] = 11010001$$

The sum outputs will read 11010001 from left to right. There is no overflow into the C_8 bit, and so it will be a 0.

6-15   2's-COMPLEMENT SYSTEM

Most modern computers use the 2's-complement system to represent negative numbers and to perform subtraction. The operations of addition and subtraction of signed numbers can be performed using only the addition operation if we use the 2's-complement form to represent negative numbers.

Addition   Positive and negative numbers, including the sign bits, can be added together in the basic parallel-adder circuit when the negative numbers are in 2's-complement form. This is illustrated in Figure 6-11 for the addition of -3 and $+6$. The -3 is represented in its 2's-complement form as 1101, where the first 1 is the sign bit; the $+6$ is represented as 0110, with the first zero as the sign bit. These numbers are stored in their corresponding registers. The 4-bit parallel adder pro-

FIGURE 6-11   Parallel adder used to add + and − numbers in 2's-complement system.

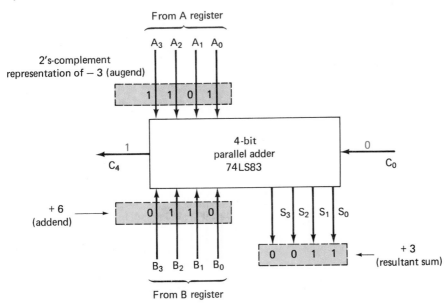

duces sum outputs of 0011, which represents +3. The C_4 output is 1 but is disregarded in the 2's-complement method.

Subtraction

When the 2's-complement system is used, the number to be subtracted (the subtrahend) is changed to its 2's complement and then *added* to the minuend (the number the subtrahend is being subtracted from). For example, we can assume that the minuend is already stored in the accumulator (*A* register). The subtrahend is then placed in the *B* register (in a computer it would be transferred here from memory) and is changed to its 2's-complement form before it is added to the number in the *A* register. The sum outputs of the adder circuit now represent the *difference* between the minuend and subtrahend.

Teaching Tip: You may want to stress that the 2's complement of the *B* number is not explicitly present in the circuit of Figure 6-12, since the addition of 1 to the LSB is accomplished internally by the adder circuit.

The parallel-adder circuit that we have been discussing can be adapted to perform the subtraction described above if we provide a means for taking the 2's complement of the *B* register number. The 2's complement of a binary number is obtained by complementing (inverting) each bit and then adding 1 to the LSB. Figure 6-12 shows how this can be accomplished. The *inverted* outputs of the *B* register are used rather than the normal outputs; that is, $\bar{B}_0$, $\bar{B}_1$, $\bar{B}_2$, and $\bar{B}_3$ are fed to the adder inputs (remember, B_3 is the sign bit). This takes care of complementing each bit of the *B* number. Also, C_0 is made a logical 1, so that it adds an extra 1 into the LSB of the adder; this accomplishes the same effect as adding 1 to the LSB of the *B* register for forming the 2's complement.

The outputs S_3 to S_0 represent the results of the subtraction operation. Of course, S_3 is the sign bit of the result and indicates whether the result is + or −. The carry output C_4 is again disregarded.

To help clarify this operation, study the following steps for subtracting +6 from +4:

1. +4 is stored in the *A* register as 0100.
2. +6 is stored in the *B* register as 0110.
3. The inverted outputs of the *B*-register FFs are fed to the adder—that is, 1001.
4. The 1001 is added to 0100 by the parallel adder along with a 1 added to the LSB position by making $C_0 = 1$. This produces sum output bits 1110 and a $C_4 = 1$, which is disregarded. This 1110 represents the required difference.

FIGURE 6-12   Parallel adder used to perform subtraction ($A - B$) using 2's-complement system. The bits of the subtrahend (*B*) are inverted, and $C_0 = 1$ to produce the 2's complement.

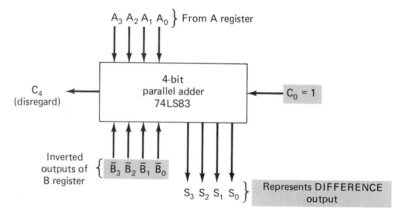

Since the sign bit $= 1$, it is a negative result and is in 2's-complement form. We can verify that 1110 represents -2_{10} by taking its 2's-complement and obtaining $+2_{10}$:

$$
\begin{array}{r}
1110 \\
0001 \\
+ \quad 1 \\
\hline
0010 = +2_{10}
\end{array}
$$

Combined Addition and Subtraction   It should now be clear that the basic parallel-adder circuit can be used to perform addition or subtraction depending on

FIGURE 6-13   Parallel adder/subtractor using 2's-complement system.

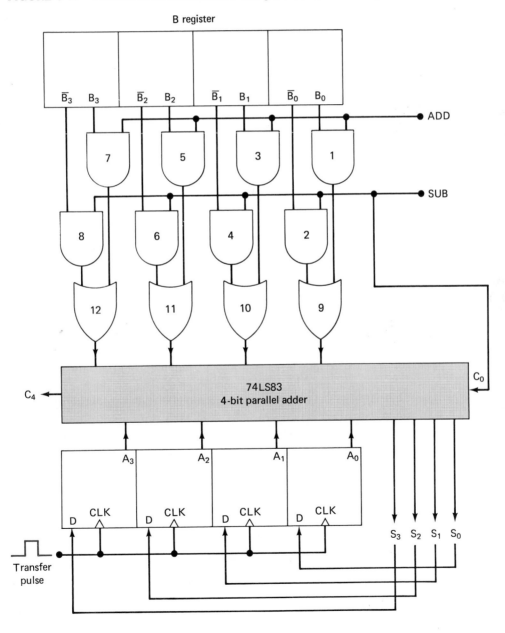

whether the B number is left unchanged or is converted to its 2's complement. A complete circuit that can perform *both* addition and subtraction in the 2's-complement system is shown in Figure 6-13 (see page 259).

This adder/subtractor circuit is controlled by the two control signals ADD and SUB. When the ADD level is HIGH, the circuit performs addition of the numbers stored in the A and B registers. When the SUB level is HIGH, the circuit subtracts the B-register number from the A-register number. The operation is described as follows:

1. Assume that ADD = 1 and SUB = 0. The SUB = 0 *disables* (inhibits) AND gates 2, 4, 6, and 8, holding their outputs at 0. The ADD = 1 *enables* AND gates 1, 3, 5, and 7, allowing their outputs to pass the B_0, B_1, B_2, and B_3 levels, respectively.

2. The levels B_0 to B_3 pass through the OR gates into the 4-bit parallel adder to be added to the bits A_0 to A_3. The *sum* appears at the outputs S_0 to S_3.

3. Note that SUB = 0 causes a carry $C_0 = 0$ into the adder.

4. Now assume that ADD = 0 and SUB = 1. The ADD = 0 inhibits AND gates 1, 3, 5, and 7. The SUB = 1 enables AND gates 2, 4, 6, and 8, so that their outputs pass the $\bar{B}_0$, $\bar{B}_1$, $\bar{B}_2$, and $\bar{B}_3$ levels, respectively.

5. The levels $\bar{B}_0$ to $\bar{B}_3$ pass through the OR gates into the adder to be added to the bits A_0 to A_3. Note also that C_0 is now 1. Thus, the B-register number has essentially been converted to its 2's complement.

6. The *difference* appears at the *outputs* S_0 to S_3.

Discussion Question: What will appear at the OR gate outputs if ADD and SUB are both HIGH?

Circuits like the adder/subtractor of Figure 6-13 are used in computers because they provide a relatively simple means for adding and subtracting signed binary numbers. In most computers the outputs present at the S output lines are usually transferred into the A register (accumulator), so that the results of the addition or subtraction always end up stored in the A register. This is accomplished by applying a TRANSFER pulse to the *CLK* inputs of register A.

―― REVIEW QUESTIONS ――――

1. Why does C_0 have to be a 1 in order to use the adder circuit in Figure 6-12 as a subtractor?
2. Assume that $[A] = 0011$ and $[B] = 0010$ in Figure 6-13. If ADD = 1 and SUB = 0, determine the logic levels at the OR-gate outputs.
3. Repeat question 2 for ADD = 0, SUB = 1.
4. *True or false:* When the adder/subtractor circuit is used for subtraction, the 2's complement of the subtrahend appears at the input of the adder.

6-16   BCD ADDER

The BCD addition process was discussed in Section 6-7 and is reviewed below:

1. Add the BCD code groups for each decimal digit position; use ordinary binary addition.

2. For those positions where the sum is 9 or less, the sum is in proper BCD form and no correction is needed.

3. When the sum of two digits is greater than 9, a correction of 0110 should be added to that sum to produce the proper BCD result. This will produce a carry to be added to the next decimal position.

A BCD adder circuit must be able to operate in accordance with the above steps. In other words, the circuit must be able to do the following:

1. Add two 4-bit BCD code groups, using straight binary addition.
2. Determine if the sum of this addition is greater than 1001 (decimal 9); if it is, add 0110 (6) to this sum and generate a carry to the next decimal position.

The first requirement is easily met by using a 4-bit binary parallel adder such as the 74LS83 IC. For example, if the two BCD code groups represented by $A_3 A_2 A_1 A_0$ and $B_3 B_2 B_1 B_0$, respectively, are applied to a 4-bit parallel adder, the adder will perform the following operation:

$$A_3 A_2 A_1 A_0 \quad \longleftarrow \text{BCD code group}$$
$$+ \; \underline{B_3 B_2 B_1 B_0} \quad \longleftarrow \text{BCD code group}$$
$$S_4 S_3 S_2 S_1 S_0 \quad \longleftarrow \text{straight binary sum}$$

S_4 is actually C_4, the carry out of the MSB.

The sum outputs $S_4 S_3 S_2 S_1 S_0$ can range anywhere from 00000 to 10010 (when both BCD code groups are 1001 = 9). The circuitry for a BCD adder must include the logic needed to detect whenever the sum is greater than 01001, so that the correction can be added in. These cases where the sum is greater than 01001 are listed below:

S_4	S_3	S_2	S_1	S_0	
0	1	0	1	0	(10)
0	1	0	1	1	(11)
0	1	1	0	0	(12)
0	1	1	0	1	(13)
0	1	1	1	0	(14)
0	1	1	1	1	(15)
1	0	0	0	0	(16)
1	0	0	0	1	(17)
1	0	0	1	0	(18)

Let's define X as a logic output that will go HIGH only when the sum is greater than 01001 (i.e., for the cases listed above). If we examine these cases, it can be reasoned that X will be HIGH for either of the following conditions:

1. Whenever $S_4 = 1$ (sums greater than 15)
2. Whenever $S_3 = 1$ and either S_2 or S_1 or both are 1 (sums 10 to 15)

This can be expressed as

$$X = S_4 + S_3(S_2 + S_1)$$

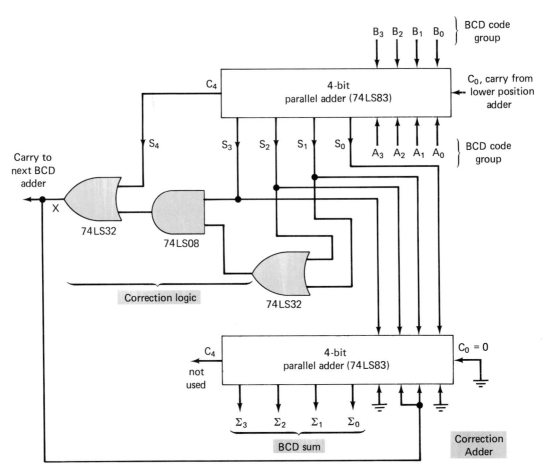

FIGURE 6-14 A BCD adder contains two 4-bit adders and a correction-detector circuit.

Whenever $X = 1$, it is necessary to add the correction 0110 to the sum bits and to generate a carry. Figure 6-14 shows the complete circuitry for a BCD adder, including the logic-circuit implementation for X.

The circuit consists of three basic parts. The two BCD code groups $A_3 A_2 A_1 A_0$ and $B_3 B_2 B_1 B_0$ are added together in the upper 4-bit adder to produce the sum $S_4 S_3 S_2 S_1 S_0$. The logic gates implement the expression for X. The lower 4-bit adder will add the correction 0110 to the sum bits *only* when $X = 1$, producing the final BCD sum output represented by $\Sigma_3 \Sigma_2 \Sigma_1 \Sigma_0$. X is also the carry output that is produced when the sum is greater than 01001. Of course, when $X = 0$, there is no carry and no addition of 0110. In such cases, $\Sigma_3 \Sigma_2 \Sigma_1 \Sigma_0 = S_3 S_2 S_1 S_0$.

To help in the understanding of the BCD adder, the reader should try several cases by following them through the circuit. The following cases would be particularly instructive:

Inputs

(a) $[A] = 0101$, $[B] = 0011$, $C_0 = 0$

(b) $[A] = 0111$, $[B] = 0110$, $C_0 = 0$

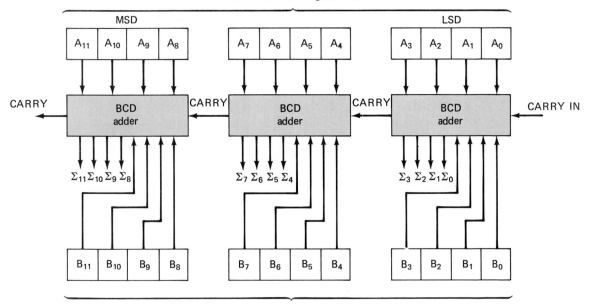

BCD codes for 3-digit number

FIGURE 6-15   Cascading BCD adders to add two 3-digit decimal numbers.

Outputs

(a) $[S] = 01000$, $X = 0$, $[\Sigma] = 1000$, CARRY $= 0$

(b) $[S] = 01101$, $X = 1$, $[\Sigma] = 0011$, CARRY $= 1$

Discussion Question: Investigate what will happen if one of the invalid BCD code groups is applied to A or B.

Cascading BCD Adders   The circuit of Figure 6-14 is used for adding two decimal digits that have been encoded in BCD code. When decimal numbers with several digits are to be added together, it is necessary to use a separate BCD adder for each digit position. Figure 6-15 is a block diagram of a circuit for the addition of two 3-digit decimal numbers. The A register contains 12 bits, which are the three BCD code groups for one of the three-digit decimal numbers; similarly, the B register contains the BCD representation of the other three-digit decimal number. The code groups A_3–A_0 and B_3–B_0 representing the least significant digits are fed to the first BCD adder. Each BCD adder block is assumed to contain the circuitry of Figure 6-14. This first BCD adder produces sum outputs $\Sigma_3 \Sigma_2 \Sigma_1 \Sigma_0$, which is the BCD code for the least significant digit of the sum. It also produces a carry output that is sent to the second BCD adder, which is adding A_7 through A_4 and B_7 through B_4, the BCD code groups for the second-decimal-digit position. The second BCD adder produces $\Sigma_7 \Sigma_6 \Sigma_5 \Sigma_4$, the BCD code for the second digit of the sum, and so on. This arrangement can, of course, be extended to decimal numbers of any size by simply adding more FFs to the registers and including a BCD adder for each digit position.

___ EXAMPLE 6-11 _____

Determine the inputs and outputs when the circuit of Figure 6-15 is used to add 247_{10} to 538_{10}.

Solution   First, the decimal numbers are represented in BCD.

$$247 = 0010 \quad 0100 \quad 0111 \quad \text{(BCD)}$$
$$538 = 0101 \quad 0011 \quad 1000 \quad \text{(BCD)}$$

These BCD numbers will be placed in the *A* and *B* registers, respectively, so that

$$[A] = 0010 \quad 0100 \quad 0111$$
$$[B] = 0101 \quad 0011 \quad 1000$$

The CARRY IN to the LSD adder will be a 0.

Once the data are in the registers, the BCD adders will begin to produce the correct BCD sums at their outputs. The LSD adder will add the 0111 (7) and 1000 (8) to produce a sum of 0101 (5) and a CARRY of 1 into the middle adder. The middle adder will add the 0100 (4) and 0011 (3) and the CARRY of 1 to produce a sum of 1000 (8) and a CARRY of 0 into the MSD adder. The MSD adder will add 0010 (2) and 0101 (5) for a sum of 0111 (7) and no CARRY out. Thus, at the sum outputs we have

$$[\Sigma] = 0111 \quad 1000 \quad 0101$$

and there is a CARRY output of 0 from the MSD adder.

___ REVIEW QUESTIONS ___

1. What are the three basic parts of a BCD adder circuit?
2. Describe how the BCD adder circuit detects the need for a correction and executes it.

6-17   BINARY MULTIPLIERS

The multiplication of two binary numbers is done with paper and pencil by performing successive additions and shifting. To illustrate:

$$
\begin{array}{r}
1011 \\
\times\ 1101 \\
\hline
1011 \\
0000 \\
1011 \\
1011 \\
\hline
10001111
\end{array}
$$

multiplicand (11)
multiplier (13)

product (143)

This process consists of examining the successive bits of the multiplier, beginning with the LSB. If the multiplier bit is a 1, the multiplicand is copied down; if it is a 0, zeros are written down. The numbers written down in successive lines are shifted one position to the left relative to the previous line. When all the multiplier bits have been examined, the various lines are *added* to produce the final product.

In the digital machines this process is modified somewhat because the binary adder is designed to add only two binary numbers at a time. Instead of adding all the lines at the end, they are added two at a time and their sum is accumulated in a register (the accumulator register). In addition, when the multiplier bit is 0, there is no need to write down and add zeros, since it does not affect the final result. The previous example is redone here showing the modified process.

FIGURE 6-16 Circuit for a 4-bit serial multiplier that uses shift registers for the multiplier and multiplicand.

multiplicand:	1011
multiplier:	<u>1101</u>

1011	LSB of multiplier = 1; write down multiplicand; shift multiplicand one position to the left (10110)
1011	2nd multiplier bit = 0; write down previous result; shift multiplicand to the left again (101100)
+ <u>101100</u>	3rd multiplier bit = 1; write down multiplicand (101100)
110111	add
	shift new multiplicand to the left (1011000)
+ <u>1011000</u>	4th multiplier bit = 1; write down new multiplicand (1011000)
10001111	add to obtain final product

This multiplication process can be performed by the **serial multiplier** circuit shown in Figure 6-16 (see page 265), which will multiply two 4-bit numbers to produce an 8-bit product. The circuit consists of the following elements:

- *X register*. A 4-bit shift register that stores the multiplier. It will shift right on the NGT of the clock. Note that 0s are shifted in from the left.
- *B register*. An 8-bit shift register that stores the multiplicand. It will shift left on the NGT of the clock. Note that 0s are shifted in from the right.
- *A register*. An 8-bit register that is the accumulator register that accumulates the partial products.
- *Adder*. An 8-bit parallel adder that produces the sum of the A and B registers. The adder outputs S_7 through S_0 are connected to the D inputs of the accumulator so that the sum can be transferred to the accumulator *only when a clock pulse gets through the AND gate*.

The circuit operation can best be described by going through each step in the multiplication of 1011 and 1101. The complete process requires four clock cycles. Refer to Figure 6-17 for the contents of each register and the adder outputs as we describe the sequence of steps.

1. *Before first clock pulse*. Prior to the occurrence of the first clock pulse, the A register is loaded with 00000000, the B register with the multiplicand 00001011, and the X register with the multiplier 1101. We can assume that each of these registers was loaded using its asynchronous inputs (i.e., PRE-SET and CLEAR). The adder outputs will be the sum of A and B, that is, 00001011.
2. *First clock pulse*. Since the LSB of the multiplier (X_0) is a 1, the first clock pulse gets through the AND gate, and its PGT transfers the sum outputs into the accumulator. The subsequent NGT causes the X and B registers to shift right and left, respectively. This, of course, produces a new sum of A and B.
3. *Second clock pulse*. The second bit of the original multiplier is now in X_0. Since this bit is a 0, the second clock pulse is inhibited from reaching the accumulator. Thus, the sum outputs are not transferred into the accumulator and the number in the accumulator does not change. The NGT of the clock pulse will again shift the X and B registers.
4. *Third clock pulse*. The third bit of the original multiplier is now in X_0. Since it is a 1, the PGT of the third clock pulse transfers the sum into the A register. The NGT shifts X and B again.

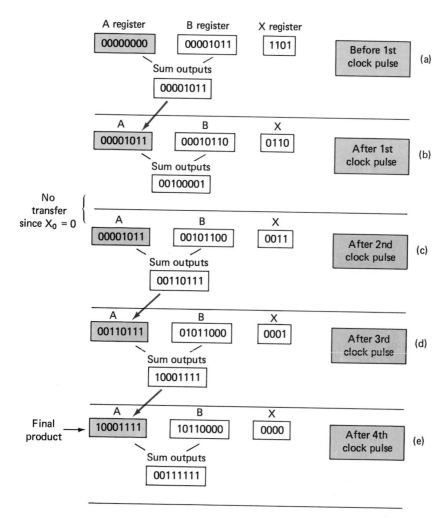

FIGURE 6-17 Contents of various registers during multiplication of 1011×1101.

5. *Fourth clock pulse.* The last bit of the original multiplier is now in X_0, and since it is a 1, the PGT of the fourth clock pulse transfers the sum into the accumulator. The accumulator now holds the final product. The NGT shifts X and B again. Note that X is now 0000, since all of the multiplier bits have been shifted out.

Discussion Question: Under what condition will this multiplier circuit work properly with A flip-flops that clock on NGTs?

Software Multiplication   Many microprocessors do not include circuitry (*hardware*) in their arithmetic-logic unit (ALU) that can perform multiplication or any other complex arithmetic operation (e.g., division, square root, trigonometric functions). For microcomputers that use these microprocessors, multiplication can be accomplished by using the program (*software*) that does the multiplication through the repeated execution of addition and shifting instructions. Similar programs can be used to perform the other complex arithmetic operations. This use of software reduces the amount of circuitry needed in the microprocessor, but, in general, it increases the time required to perform a calculation when compared with hardware.

1. Describe the sequence of events that occurs in the multiplier circuit during one clock cycle.
2. *True or false:* The outputs of the adder are transferred to the accumulator on *every* PGT of the clock.

6-18 COMPLEX ARITHMETIC INTEGRATED CIRCUITS

There are several complex ICs available in the TTL and CMOS families that can perform one or more types of arithmetic operation. We will briefly describe a few of them here to illustrate the range of available devices.

The 74LS381 and 74HC381 ICs are called *arithmetic-logic units/function generators*. They can perform eight different binary arithmetic and logic operations on two 4-bit inputs. These operations, which are selected by a set of three SELECT inputs, include addition, subtraction, ORing, EX-ORing, and ANDing.

The 74284 and 74285 are *parallel binary multipliers*. Like the serial multiplier circuit of Figure 6-16, these ICs contain the circuitry for generating the product of two 4-bit binary numbers. Unlike the serial multiplier circuit, these ICs perform parallel multiplication using only combinatorial logic to look at the two 4-bit inputs and generate the correct 8-bit output product. The total multiplication process requires only about 40 ns. The serial multiplier would take at least four times that long because of the need to shift the multiplier and multiplicands four times.

The AM9511 is an LSI device called an *arithmetic processing unit* or APU. It is a 24-pin chip that performs many complex arithmetic operations, including *addition, subtraction, multiplication, division, square roots, trigonometric functions* (e.g., *sin, cos, tan*), and *logarithms*. The APU can perform these and other operations on 16-bit or 32-bit input numbers. It uses a clock input signal (like the serial multiplier in Figure 6-16) to sequence its operations. The more complex operations require more clock cycles.

The APU is designed to receive its data and commands from the microprocessor. Once it receives the data and commands, the APU performs the requested operation in the required number of clock cycles. When it is done, it sends a signal to the microprocessor signifying the completion of the operation. The microprocessor will then typically take the output data (i.e., the answer) from the APU's output pins and put it into one of its internal registers for subsequent processing.

6-19 IEEE/ANSI SYMBOLS

Figure 6-18(a) shows the IEEE/ANSI symbol for a 1-bit adder (full adder). Note that the symbol Σ is used to indicate the addition operation. Figure 6-18(b) is the IEEE/ANSI symbol for a 4-bit parallel adder like the 7483 and 74283. Note how the letters P and Q are used to represent the two 4-bit inputs, and Σ is used for the 4-bit output sum. The P, Q, and Σ are specified by the IEEE/ANSI standard and must be used inside the symbol outline. The letters used for the inputs and outputs external to

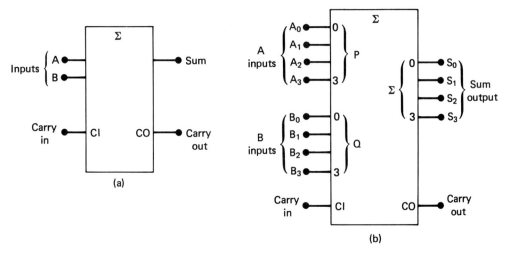

FIGURE 6-18   IEEE/ANSI symbols for (a) a full adder and (b) a 4-bit parallel adder IC (7483/74283).

the symbol outline are not specified by the standard, and so we are free to use A, B, and S, as we did in our earlier discussions.

6-20   TROUBLESHOOTING CASE STUDY

A technician is testing the adder/subtractor circuit of Figure 6-13 and records the following test results for the various operating modes:

- **Mode 1: ADD = 0, SUB = 0.** The sum outputs are always equal to the number in the A register *plus one*. For example, when $[A] = 0110$, the sum is $[S] = 0111$. This is incorrect, since the OR outputs and C_0 should all be 0 in this mode to produce $[S] = [A]$.

- **Mode 2: ADD = 1, SUB = 0.** The sum is always 1 more than it should be. For example, with $[A] = 0010$ and $[B] = 0100$, the sum output is 0111 instead of 0110.

- **Mode 3: ADD = 0, SUB = 1.** The S outputs are always equal to $[A] - [B]$, as expected.

When he examines these test results, the technician sees that the sum outputs exceed the expected results by 1 for the first two modes of operation. At first he suspects a possible fault in one of the LSB inputs to the adder, but he dismisses this because such a fault would also affect the subtraction operation, which is working correctly. Eventually, he realizes that there is another fault that could add an extra 1 to the results for the first two modes without causing an error in the subtraction mode.

Recall that C_0 is made a 1 in the subtraction mode as part of the 2's-complement operation on $[B]$. For the other modes, C_0 is to be a 0. The technician checks the connection between the SUB signal and the C_0 input to the adder and finds that it is open due to a bad solder connection. This open connection explains the observed results, since the TTL adder responds as if C_0 were a constant logic 1, causing an extra 1 to be added to the result in modes 1 and 2. The open connection would have no effect on mode 3, because C_0 is supposed to be a 1 anyway.

EXAMPLE 6-12

Consider again the adder/subtractor circuit. Suppose that there is a break in the connection path between the SUB input and the AND gates as shown in Figure 6-19. Describe the effects of this open on the circuit operation for each mode.

FIGURE 6-19 Example 6-12.

Solution   First, realize that this fault will produce a logic 1 at the affected input of AND gates 2, 4, 6, and 8, which will permanently enable each of these gates to pass its $\overline{B}$ input to the following OR gate as shown.

Mode 1: ADD = 0, SUB = 0. The fault will cause the circuit to perform subtraction—almost. The 1's complement of $[B]$ will reach the OR gate outputs and be applied to the adder along with $[A]$. With $C_0 = 0$, the 2's complement of $[B]$ will not be complete; it will be short by 1. Thus, the adder will produce $[A] - [B] - 1$. To illustrate, let's try $[A] = +6 = 0110$ and $[B] = +3 = 0011$. The adder will add as follows:

$$\begin{aligned} \text{1's complement of } [B] &= 1100 \\ [A] &= \underline{0110} \\ \text{result} &= \cancel{1}0010 \end{aligned}$$

↑——disregard carry

The result is 0010 and $+2$ instead of 0011 = $+3$, as it would be for normal subtraction.

Mode 2: ADD = 1, SUB = 0. With ADD = 1, AND gates 1, 3, 5, and 7 will pass the B inputs to the following OR gate. Thus, each OR gate will have a $\overline{B}$ and a B at its inputs, thereby producing a 1 output. For example, the inputs to OR gate 9 will be $\overline{B}_0$ coming from AND gate 2 (because of the fault), and B_0

coming from AND gate 1 (because ADD = 1). Thus, OR gate 9 will produce an output of $\bar{B}_0 + B_0 = 1$.

The adder will add the 1111 from the OR gates to the [A] to produce a sum that is 1 less than [A]. Why? Because $1111_2 = -1_{10}$.

Mode 3: Add = 0, SUB = 1. This mode will work correctly since SUB = 1 is supposed to enable AND gates 2, 4, 6, and 8 anyway.

PROBLEMS

SECTION 6-1

6-1. Add the following groups of binary numbers using binary addition.
  (a) 1010 + 1011   10101
  (b) 1111 + 0011   10010
  (c) 1011.1101 + 11.1   1111.0101
  (d) 0.1011 + 0.1111   1.1010
  (e) 10011011 + 10011101   100111000

SECTION 6-2

6-2. Represent each of the following signed decimal numbers in the 2's-complement system. Use a total of 8 bits including sign bit.
  (a) +32   00100000
  (b) −14   11110010
  (c) +63   00111111
  (d) − 104   10011000
  (e) −1   11111111
  (f) −128   10000000
  (g) +169
  (h) 0   00000000

6-2. (g) +169 is too large for 8 bits

6-3. Each of the following numbers represents a signed decimal number in the 2's-complement system. Determine the decimal value in each case.
  (a) 01101   +13
  (b) 11101   −3
  (c) 01111011   +123
  (d) 10011001   −103
  (e) 01111111   +127
  (f) 100000   −32
  (g) 11111111   −1
  (h) 10000001   −127

6-4. (a) What range of signed decimal values can be represented using 12 bits including sign bit?   −2048 to +2047
  (b) How many bits would be required to represent decimal numbers from −32,768 to + 32,767?   16

6-5. List, in order, all of the signed numbers that can be represented in 5 bits using the 2's-complement system.   10000 (−16) to 01111 (+15)

6-6. A certain computer has the following binary number stored in its memory: 10100100. Only the programmer who put this group of data in memory knows what it represents. Here are some of the possibilities.
  (a) An unsigned decimal number   164
  (b) A signed decimal number using the 2's-complement system   −92

(c) A BCD-coded decimal number   illegal BCD

(d) An ASCII code with a parity bit $

(e) A signed decimal number using sign-magnitude system   −36

For each of these possibilities, determine what the data 10100100 represent.

6-7. What is the range of unsigned decimal values that can be represented in 10 bits? What is the range of signed decimal values using the same number of bits?   (a) 0 to 1023   (b) −512 to +511

SECTIONS 6-3 and 6-4

6-8. The reason why the sign-magnitude method for representing signed numbers is not used in most computers can readily be illustrated by performing the following: +12 = 01100, −12 = 11100; sum = 101000

(a) Represent +12 in 5 bits using true magnitude and a sign bit.

(b) Represent −12 in 5 bits using true magnitude and a sign bit.

(c) Add the two binary numbers and note that the sum does not look anything like zero.

6-9. Perform the following operations in the 2's-complement system. Use 8 bits (including the sign bit) for each number. Check your results by converting the binary result back to decimal.

(a) Add + 9 to +6.   00001111

(b) Add +14 to −17.   11111101

(c) Add +19 to −24.   11111011

(d) Add −48 to −80.   10000000

(e) Subtract +16 from +17.   00000001

(f) Subtract +21 from −13.   11011110

(g) Subtract +47 from +47.   00000000

(h) Subtract −36 from −15.   00010101

(i) Add +17 to −17.   00000000

(j) Subtract −17 from −17.   00000000

6-10. Repeat Problem 6-9 for the following cases, and show that overflow occurs in each case.

(a) Add +37 to +95.   10000100

(b) Subtract +37 from −95.   01111100

SECTIONS 6-5 and 6-6

6-11. Multiply the following pairs of binary numbers:

(a) 111 × 101   100011

(b) 1011 × 1011   1111001

(c) 101.101 × 110.010   100011.00101

(d) .1101 × .1011   .10001111

6-12. Perform the following divisions:

(a) 1100 ÷ 100   11

(b) 111111 ÷ 1001   111

(c) 10111 ÷ 100   101.11

(d) 10110.1101 ÷ 1.1   1111.0011

6-13. Add the following decimal numbers after converting each to its BCD code:

(a) 74 + 23   10010111   (c) 147 + 380   010100100111

(b) 58 + 37   10010101   (d) 385 + 118   010100000011

6-14. Find the sum of each of the following pairs of hex numbers:

(a) 3E91 + 2F93   6E24

(b) 91B + 6F2   100D

(c) ABC + DEF   18AB

6-15. Perform the following subtractions on the pairs of hex numbers:

(a) 3E91 − 2F93   0EFE

(b) 91B − 6F2   229

(c) 0300 − 005A   02A6

6-16. The owner's manual for a certain personal computer states that the computer has usable memory locations at the following hex addresses: 0200 through 03FF, and 4000 through 7FD0. What is the total number of available memory locations?   16,849

SECTION 6-11

6-17. Convert the FA circuit of Figure 6-7 to all NAND gates.

6-17. Convert each gate to its NAND equivalent and cancel double inversions.

6-18. Write the truth table for a half adder (inputs A and B; outputs SUM and CARRY). From the truth table design a logic circuit that will act as a half adder.   SUM $= A \oplus B$; CARRY $= AB$

6-19. A full adder can be implemented in many different ways. Figure 6-20 shows how one may be constructed from two half adders. Construct a truth table for this arrangement and verify that it operates as a FA.

6-19. Truth table same as Figure 6-6.

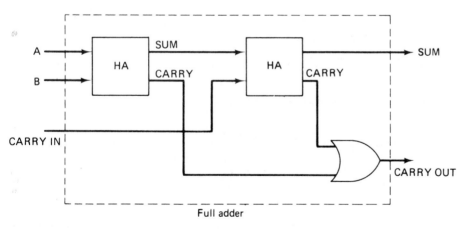

FIGURE 6-20

SECTION 6-12

6-20. Refer to Figure 6-9. Determine the contents of the A register after the following sequence of operations: $[A] = 0000$, $[0101] \longrightarrow [B]$, $[S] \longrightarrow [A]$, $[1011] \longrightarrow [B]$, $[S] \longrightarrow [A]$. $[A] = 0000$

6-21. Refer to Figure 6-9. Assume that each FF has $t_{PLH} = t_{PHL} = 30$ ns and a setup time of 10 ns, and that each FA has a propagation delay of 40 ns. What is the minimum time allowed between the PGT of the LOAD pulse and the PGT of the TRANSFER pulse for proper operation?   80 ns

6-22. In the adder and subtractor circuits discussed in this chapter we gave no consideration to the possibility of *overflow*. Overflow occurs when the two numbers being added or subtracted produce a result that contains more bits than the capacity of the accumulator. For example, using 4-bit registers, including a sign bit, numbers ranging from $+7$ to -8 (in 2's complement) can be stored. Therefore, if the result of an addition or subtraction exceeds $+7$ or -8, we would say that an overflow has occurred. When an overflow occurs, the results are useless since they cannot be stored correctly in the accumulator register. To illustrate, add $+5$ (0101) and $+4$ (0100), which results in 1001. This 1001 would be interpreted incorrectly as a negative number since there is a 1 in the sign-bit position.

In computers and calculators there are usually circuits that are used to detect an overflow condition. There are several ways to do this. One method that can be used for the adder that operates in the 2's-complement system works as follows:
1. Examine the sign bits of the two numbers being added.
2. Examine the sign bit of the result.
3. Overflow occurs whenever the numbers being added are *both positive* and the sign bit of the result is 1 *or* when the numbers are *both negative* and the sign bit of the result is 0.

This method can be verified by trying several examples. Readers should try the following cases for their own clarification: (1) $5 + 4$; (2) $-4 + (-6)$; (3) $3 + 2$. Cases 1 and 2 will produce an overflow and case 3 will not. Thus, by examining the sign bits a logic circuit can be designed that will produce a 1 output whenever the overflow condition occurs. Design this overflow circuit for the adder of Figure 6-9.   Overflow $= \bar{S}_3 A_3 B_3 + S_3 \bar{A}_3 \bar{B}_3$

SECTION 6-13

6-23. Design a look-ahead carry circuit for the adder of Figure 6-9 which generates the carry C_3 to be fed to the FA of the MSB position based on the values of A_0, B_0, C_0, A_1, B_1, A_2, and B_2. In other words, derive an expression for C_3 in terms of A_0, B_0, C_0, A_1, B_1, A_2, and B_2. (*Hint:* Begin by writing the expression for C_1 in terms of A_0, B_0, and C_0. Then write the expression for C_2 in terms of A_1, B_1, and C_1. Substitute the expression for C_1 into the expression for C_2. Then write the expression for C_3 in terms of A_2, B_2, and C_2. Substitute the expression for C_2 into the expression for C_3. Simplify the final expression for C_3 and put it in sum-of-products form. Implement the circuit.)   See Solutions Manual.

SECTION 6-14

6-24. Show the logic levels at each input and output of Figure 6-10(a) when 354_8 is added to 103_8. $[S] = 010001010111$

6-25. For the circuit of Figure 6-13, determine the sum outputs for the following cases:
  (a) *A* register = 0101 (+5), *B* register = 1110 (−2); SUBTRACT = 1, ADD = 0. 0111
  (b) *A* register = 1100 (−4), *B* register = 1110 (−2); SUBTRACT = 0, ADD = 1. 1010

6-26. Modify the circuit of Figure 6-13 so that a single control input, *X*, is used in place of ADD and SUB. The circuit is to function as an adder when *X* = 0, and as a subtractor when *X* = 1. Then simplify each set of gates. (*Hint:* Note that now each set of gates is functioning as a controlled inverter.)

> **6-26.** Use EX-ORs as controlled inverters with *X* connected as control input to each. *X* is also connected as C_0.

6-27. Assume the following inputs in Figure 6-14: [*A*] = 0101, [*B*] = 1001, C_0 = 0. Determine the logic levels at [*S*], *X*, [Σ], and CARRY.   [*S*] = 01110; *X* = 1; [Σ] = 0100; CARRY = 1

6-28. Would it make any difference in the BCD adder of Figure 6-14 if the C_0 of the upper adder was held LOW while the C_0 of the lower adder was used as the carry input? Explain.

> **6-28.** Yes, it would matter for cases where a carry input = 1 would cause the sum to be greater than 9.

6-29. Assume that the *A* register in Figure 6-15 holds the BCD code for 376, and the *B* register holds the BCD code for 469. Determine the outputs.   [Σ] = 100001000101; CARRY = 0

6-30. Refer to the circuit of Figure 6-16. Assume that the *B* register initially holds the multiplicand 0111 and the *X* register holds the multiplier 1001. Also assume [*A*] = 00000000 prior to the first clock pulse. Show the contents of registers *A*, *B*, and *X* after each clock pulse.   See Solutions Manual.

6-31. The serial multiplier circuit of Figure 6-16 is tested using various values for the multiplicand and multiplier. The results are recorded below.

Multiplicand	Multiplier	Result
4	5	60
3	4	45
15	2	225
2	15	30
1	7	15

Examination of these test data shows that the circuit is operating incorrectly. Consider each of the following possible faults, and for each indicate whether or not it could be the actual fault. Explain each answer. (*Hint:* See if there is a pattern in the recorded data.)

(a) The AND gate inputs are shorted together.   no

(b) There is a break in the X_0 connection to the AND gate.   yes

(c) The D input of flip-flop X_3 is open.   no

6-32. Assume that there is a broken connection at C_0 in the serial multiplier circuit of Figure 6-16. Show how the operation will be affected by this fault.   A 1 will always be added to the LSB position.

6-33. (a) A_4 through A_7 not being clocked—possible break in CLK line.   (b) Output A_4 might be internally or externally grounded.

6-33. The serial multiplier circuit of Figure 6-16 is tested by a technician who records the following results:

Multiplicand	Multiplier	Result
2	3	6
3	4	12
4	6	8
6	4	8
9	3	11
12	8	0
7	7	1

Examine the recorded data and try to determine one or more possible causes of the faulty operation.

6-34. The BCD adder of Figure 6-14 is tested and the results are recorded below.

	$B_3 B_2 B_1 B_0$	$A_3 A_2 A_1 A_0$	$\Sigma_3 \Sigma_2 \Sigma_1 \Sigma_0$	CARRY (X)
(1)	0 0 1 1	0 1 1 0	1 0 0 1	0
(2)	0 1 1 1	1 0 0 0	1 1 1 1	0
(3)	1 0 0 1	1 0 0 1	0 0 1 0	0

Consider each of the following possible faults, and indicate whether or not it could be the actual fault. Explain each answer.

(a) The A_1 and A_0 inputs of the correction adder are internally shorted together.   yes

(b) There is an open path from X to the correction adder.   no

(c) The upper OR gate inputs are internally shorted together.   yes

(d) The AND gate output is stuck LOW.   no

DRILL QUESTION

6-35. See Glossary.

6-35. Define each of the following terms:

(a) Full addder

(b) 2's-complement form

(c) Arithmetic-logic unit

(d) Sign bit

(e) Overflow

(f) Accumulator

(g) Parallel adder

(h) Look-ahead carry

 6-36. In a typical microprocessor ALU, the results of every arithmetic operation are usually (but not always) transferred to the accumulator register as in Figures 6-9, 6-13, and 6-16. In most microprocessor ALUs the result of each arithmetic operation is also used to control the states of several special flip-flops called *flags*. These flags are used by the microprocessor when it is making decisions during the execution of certain types of instructions. The three most common flags are:

S (sign flag). This FF is always in the same state as the sign of the last result from the ALU.

Z (zero flag). This flag is set to 1 whenever the result from an ALU operation is exactly zero. Otherwise it is cleared to 0.

C (carry flag). This FF is always in the same state as the carry from the MSB of the ALU.

Using the adder/subtractor of Figure 6-13 as the ALU, design the logic circuit that will implement these flags. The sum outputs and C_4 output are to be used to control what state each flag will go to upon the occurrence of the TRANSFER pulse. For example, if the sum is exactly zero (i.e., 0000), the Z flag should be set by the PGT of TRANSFER; otherwise it should be cleared.

6-36. Connect $S_3 + S_2 + S_1 + S_0$ to the D input of the zero FF. Connect C_4 to the D input of the carry FF. Connect S_3 to the D input of the sign FF. All FFs clocked by PGT of TRANSFER.

ANSWERS TO SECTION REVIEW QUESTIONS

SECTION 6-1

1. (a) 11101 **(b)** 101.111 **(c)** 10010000

SECTION 6-2

1. (a) 01101 **(b)** 11001 **(c)** 10000
2. (a) -29 **(b)** -64 **(c)** $+63$
3. -2048 to $+2047$ **4.** Seven
5. -32768
6. 10000, 10000000, 1000

SECTION 6-3

1. True **2. (a)** $100010_2 = -30_{10}$
(b) $000000_2 = 0_{10}$

SECTION 6-4

1. (a) $01111_2 = +15_{10};$ **(b)** $11111_2 = -1_{10}.$
2. By comparing the sign bit of the sum with the sign bits of the numbers being added.

SECTION 6-5

1. 1100010

SECTION 6-7

1. The sum of at least one decimal digit position is greater than 1001 (9). **2.** Correction factor is added to both the units and tens digit positions.

SECTION 6-8

1. 923 **2.** 3DB

SECTION 6-10

1. Three; two **2. (a)** $S_2 = 0, C_3 = 1$
(b) $C_5 = 0$

SECTION 6-12

1. One, four, four **2.** 0100

SECTION 6-15

1. To add the 1 needed to complete the 2's-complement representation of the number in the B register.
2. 0010. **3.** 1101. **4.** False; the 1's complement appears there.

SECTION 6-16

1. Two 4-bit adders and correction logic.
2. The correction logic detects a sum greater than 9 and then causes a 0110 to be added to the sum.

SECTION 6-17

1. Current sum is transferred to A; multiplicand and multiplier are shifted; adder forms new sum.
2. False; only when the multiplier bit (X_0) is HIGH.

7

COUNTERS
AND REGISTERS

Outline

Objectives

Upon completion of this chapter, you will be able to:

- Understand the operation and characteristics of synchronous and asynchronous counters.
- Construct counters with MOD numbers less than 2^N.
- Identify IEEE/ANSI symbols used in IC counters and registers.
- Construct both up and down counters.
- Analyze and evaluate various types of presettable counters.
- Design arbitrary-sequence synchronous counters.
- Understand several types of schemes used to decode different types of counters.
- Eliminate decoder spikes by employing a technique called strobing.
- Compare the major differences between the ring and Johnson counters.
- Analyze the theory of operation of a frequency counter and of a digital clock.
- Recognize and understand the operation of various types of registers.
- Apply existing troubleshooting techniques used for combinatorial logic systems to troubleshoot sequential logic systems.

Introduction

In Chapter 5 we saw how flip-flops could be connected to function as counters and registers. At that time we studied only the basic counter and register circuits. Digital systems employ many variations of these basic circuits, mostly in integrated-circuit form. In this chapter we will look at how FFs and logic gates can be combined to produce different types of counters and registers.

We will present several practical applications to illustrate the many ways in which these logic circuits can be used in digital systems. We will also look at several of the numerous counters and registers that are available as ICs.

279

Figure 7-1 shows a 4-bit binary counter circuit such as the one discussed in Chapter 5. Recall the following points concerning its operation:

D	C	B	A	Number of clock pulses
0	0	0	0	0
0	0	0	1	1
0	0	1	0	2
0	0	1	1	3
0	1	0	0	4
0	1	0	1	5
0	1	1	0	6
0	1	1	1	7
1	0	0	0	8
1	0	0	1	9
1	0	1	0	10
1	0	1	1	11
1	1	0	0	12
1	1	0	1	13
1	1	1	0	14
1	1	1	1	15
0	0	0	0	16 (recycles)
0	0	0	1	17
0	0	1	0	18
0	0	1	1	19
0	1	0	0	20
0	1	0	1	21
.	.	.	.	.
.	.	.	.	.
.	.	.	.	.

FIGURE 7-1   Four-bit asynchronous (ripple) counter.

1. The clock pulses are applied only to the *CLK* input of flip-flop *A*. Thus, flip-flop *A* will toggle (change to its opposite state) each time the clock pulses make a negative (HIGH to LOW) transition. Note that $J = K = 1$ for all FFs.

2. The normal output of flip-flop *A* acts as the *CLK* input for flip-flop *B*, and so flip-flop *B* will toggle each time the *A* output goes from 1 to 0. Similarly, flip-flop *C* will toggle when *B* goes from 1 to 0 and flip-flop *D* will toggle when *C* goes from 1 to 0.

3. The table in Figure 7-1 shows the sequence of binary states that the FFs will follow as clock pulses are continuously applied. If we let the FF outputs *D*, *C*, *B*, and *A* represent a binary number, with *D* being the MSB and *A* the LSB, then a binary counting sequence from 0000 to 1111 is produced.

4. After the fifteenth clock pulse has occurred, the counter FFs are in the 1111 condition. On the sixteenth clock pulse, flip-flop *A* goes from 1 to 0, which causes flip-flop *B* to go from 1 to 0, and so on until the counter is in the 0000 state. In other words, the counter has gone through one complete cycle (0000 through 1111) and has *recycled* back to 0000, from where it will begin a new counting cycle as subsequent clock pulses are applied.

This type of counter, where each FF output serves as the *CLK* input signal for the next FF, is referred to as an **asynchronous counter.** This is because all the FFs do *not* change states in exact synchronism with the clock pulses; only flip-flop *A* responds to the clock pulses. Flip-flop *B* has to wait for flip-flop *A* to change states before it is toggled; flip-flop *C* has to wait for flip-flop *B*; and so on. Thus, there is a delay between the responses of consecutive FFs. In modern FFs this delay may be very small (typically 10 to 40 ns), but in some cases, as we shall see, it can be troublesome. Because of the manner in which this type of counter operates, it is also commonly referred to as a **ripple counter.** In the following discussions we will use the terms "asynchronous counter" and "ripple counter" interchangeably.

Teaching Tip: The ripple counter can also be called a "chain reaction" counter or a "domino effect" counter. These terms may help the student to easily recall the operation of this type of counter.

_____ EXAMPLE 7-1 _____

The counter in Figure 7-1 starts off in the 0000 state and then clock pulses are applied. Some time later the clock pulses are removed and the counter FFs read 0011. How many clock pulses have occurred?

Solution   The apparent answer seems to be 3, since 0011 is the binary equivalent of 3. However, with the information given there is no way to tell whether the counter has recycled or not. This means that there could have been 19 clock pulses; the first 16 pulses bring the counter back to 0000 and the last 3 bring it to 0011. There could have been 35 pulses (two complete cycles and then 3 more), or 51 pulses, and so on.

_____

MOD Number   The counter in Figure 7-1 has 16 distinctly different states (0000 through 1111). Thus, it is a *MOD-16 ripple counter.* Recall that the MOD number is always equal to the number of states which the counter goes through in each complete cycle before it recycles back to its starting state. The MOD number can be increased simply by adding more FFs to the counter. That is,

$$\text{MOD number} = 2^N \qquad (7\text{-}1)$$

where *N* is the number of FFs connected in the arrangement of Figure 7-1.

_____ EXAMPLE 7-2 _____

A counter is needed that will count the number of items passing on a conveyor belt. A photocell and light source combination is used to generate a single pulse each time an item crosses its path. The counter has to be able to count as many as one thousand items. How many FFs are required?

Solution   It is a simple matter to determine what value of *N* is needed so that $2^N \geq 1000$. Since $2^9 = 512$, 9 FFs will not be enough. $2^{10} = 1024$, so 10 FFs would produce a counter that could count as high as $1111111111_2 = 1023_{10}$. Therefore, we should use 10 FFs. We could use more than 10, but it would be a waste of FFs, since any FF past the tenth one will never be toggled.

_____

SECTION 7-1 / ASYNCHRONOUS (RIPPLE) COUNTERS   281

Frequency Division   In Chapter 5 we saw that in the basic counter each FF provides an output waveform that is exactly *half* the frequency of the waveform at its *CLK* input. To illustrate, suppose that the clock signal in Figure 7-1 is 16 kHz. Figure 7-2 shows the FF output waveforms. The waveform at output *A* is an 8-kHz

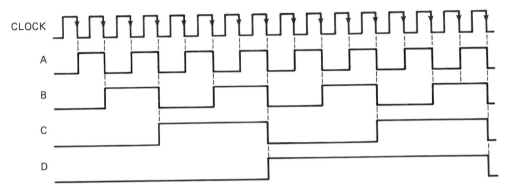

FIGURE 7-2   Counter-waveforms showing frequency division by 2 for each FF.

Discussion Question: Why are the outputs of the counter flip-flops always square waves (50 percent duty cycle)? Would this change if the CLOCK signal were not a square wave?

square wave, at output *B* it is 4 kHz, at output *C* it is 2 kHz, and at output *D* it is 1 kHz. Notice that the output of flip-flop *D* has a frequency equal to the original clock frequency divided by 16. In general, *for any counter the output from the last FF (i.e., the MSB) divides the input clock frequency by the MOD number of the counter*. For example, a MOD-16 counter could also be called a *divide-by-16 counter*.

EXAMPLE 7-3

The first step involved in building a digital clock is to take the 60-Hz power-line waveform and feed it into a shaping circuit to produce a square wave as illustrated in Figure 7-3. The 60-Hz square wave is then put into a MOD-60 counter, which is used to divide the 60-Hz frequency by exactly 60 to produce a 1-Hz waveform. This 1-Hz waveform is fed to a series of counters, which then count seconds, minutes, hours, and so on. How many FFs are required for the MOD-60 counter?

FIGURE 7-3   Example 7-3.

Solution   There is no integer power of 2 that will equal 60. The closest is $2^6 = 64$. Thus, a counter using six FFs would act as a MOD-64 counter. Obviously, this will not satisfy the requirement. It seems that there is no solution using a counter of the type shown in Figure 7-1. This is partly true; in the next section we will see how to modify this basic binary counter so that virtually *any* MOD number can be obtained and we will not be limited to values of 2^N.

1. *True or false:* In an asynchronous counter, all FFs change states at the same time.
2. Assume that the counter in Figure 7-1 is holding the count 0101. What will be the count after 27 clock pulses?

7-2 COUNTERS WITH MOD NUMBERS $< 2^N$

The basic ripple counter of Figure 7-1 is limited to MOD numbers that are equal to 2^N, where N is the number of FFs. This value is actually the maximum MOD number that can be obtained using N flip-flops. The basic counter can be modified to produce MOD numbers less than 2^N by allowing the counter to *skip states* that are normally part of the counting sequence. One of the most common methods for doing this is illustrated in Figure 7-4, where a 3-bit ripple counter is shown. Disregarding the

FIGURE 7-4 MOD-6 counter produced by clearing a MOD-8 counter when count of six ($i10$) occurs.

NAND gate for a moment, we can see that the counter is a MOD-8 binary counter which will count in sequence from 000 to 111. However, the presence of the NAND gate will alter this sequence as follows:

1. The NAND output is connected to the asynchronous CLEAR inputs of each FF. As long as the NAND output is HIGH, it will have no effect on the counter. When it goes LOW, however, it will clear all the FFs so that the counter immediately goes to the 000 state.

2. The inputs to the NAND gate are the outputs of the B and C flip-flops, and so the NAND output will go LOW whenever $B = C = 1$. This condition will occur when the counter goes from the 101 state to the 110 state (input pulse 6 on waveforms). The LOW at the NAND output will immediately (generally within a few nanoseconds) clear the counter to the 000 state. Once the FFs have been cleared, the NAND output goes back HIGH, since the $B = C = 1$ condition no longer exists.

3. The counting sequence is, therefore,

```
CBA
0 0 0 ←┐
0 0 1  │
0 1 0  │
0 1 1  │
1 0 0  │
1 0 1  │
┌─────┐│
│1 1 0│┘   (temporary state needed to clear counter)
└─────┘
```

Teaching Tip: Some students may have difficulty understanding why the 110 state is not considered as one of the counter states in determining its MOD number. You can stress the idea that the 110 state is not a stable state in which the counter resides while it waits for the next clock pulse, and is therefore not counted as part of the counting sequence.

Although the counter does go to the 110 state, it remains there for only a few nanoseconds before it recycles to 000. Thus, we can essentially say that this counter counts from 000 (zero) to 101 (five) and then recycles to 000. It essentially skips 110 and 111 so that it goes through only six different states; thus, it is a MOD-6 counter.

Notice that the waveform at the B output contains a *spike* or *glitch* caused by the momentary occurrence of the 110 state before clearing. This glitch is very narrow and so would not produce any visible indication on indicator lights or numerical displays. It could, however, cause a problem if the B output is being used to drive other circuitry outside the counter. It should also be noted that the C output has a frequency equal to one-sixth of the input frequency; in other words, this MOD-6 counter has divided the input frequency by *six*. The waveform at C is *not* a symmetrical square wave (50 percent duty cycle), because it is HIGH for only two clock cycles, whereas it is LOW for four cycles.

── EXAMPLE 7-4 ───────────────

Draw the state transition diagram for the counter of Figure 7-4.

Solution   Figure 7-5 shows how flip-flops C, B, and A change states as pulses are applied to the CLK input of flip-flop A. Recall that each circle represents one of the possible counter states, and that the arrows indicate how one state changes to another in response to an input clock pulse.

If we assume a starting count of 000, the diagram shows that the states of the

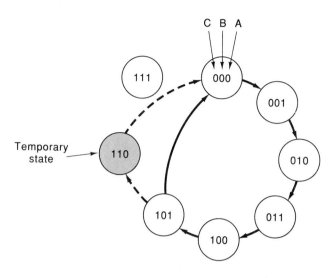

FIGURE 7-5   State transition diagram for the MOD-6 counter of Figure 7-4.

counter change normally up until the count of 101. When the next clock pulse occurs, the counter temporarily goes to the 110 count before going to the stable 000 count. The dotted lines indicate the temporary nature of the 110 state. As stated earlier, the duration of this temporary state is so short that for most purposes we can consider that the counter goes directly from 101 to 000 (solid arrow).

Note that the 111 state is never reached, not even temporarily.

Changing the MOD Number   The counter of Figure 7-4 is a MOD-6 counter because of the choice of inputs to the NAND gate. Any desired MOD number can be obtained by changing these inputs. For example, using a three-input NAND gate with inputs A, B, and C, the counter would function normally until the 111 condition was reached, at which point it would immediately reset to the 000 state. Ignoring the temporary excursion into the 111 state, the counter would go from 000 through 110 and then recycle back to 000, resulting in a MOD-7 counter (seven states).

———— EXAMPLE 7-5 ————————————————————————

Determine the MOD number of the counter in Figure 7-6(a). Also determine the frequency at the D output.

Solution   This is a 4-bit counter, which would normally count from 0000 through 1111. The NAND inputs are D, C, and B, which means that the counter will immediately recycle to 0000 when the 1110 (decimal 14) count is reached. Thus, the counter actually has 14 stable states 0000 through 1101 and is therefore a **MOD-14** counter. Since the input frequency is 30 kHz, the frequency at output D will be

$$\frac{30 \text{ kHz}}{14} = \textbf{2.14 kHz}$$

FIGURE 7-6 (a) MOD-14 ripple counter: (b) MOD-10 (decade) ripple counter.

General Procedure   To construct a counter that starts counting from all 0s and has a MOD number of X:

1. Find the smallest number of FFs such that $2^N \geq X$ and connect them as a counter. If $2^N = X$, do not do steps 2 and 3.
2. Connect a NAND gate to the asynchronous CLEAR inputs of all the FFs.
3. Determine which FFs will be in the HIGH state at a count = X; then connect the normal outputs of these FFs to the NAND gate inputs.

─── EXAMPLE 7-6 ───

Discussion Question: For the counter of Figure 7-6(b), why is it necessary that the NAND output be connected to clear all of the FFs even though C is already in the 0 state?

Construct a MOD-10 counter that will count from 0000 (zero) through 1001 (decimal 9).

Solution $2^3 = 8$ and $2^4 = 16$; thus, four FFs are required. Since the counter is to have stable operation up to the count of 1001, it must be reset to zero when the count of 1010 is reached. Therefore, FF outputs D and B must be connected as the NAND gate inputs. Figure 7-6(b) shows the arrangement.

Decade Counters/BCD Counters   The MOD-10 counter of Example 7-6 is also referred to as a *decade counter*. In fact, a decade counter is any counter that has

10 distinct states, no matter what the sequence. A decade counter such as the one in Figure 7-6(b), which counts in sequence from 0000 (zero) through 1001 (decimal 9), is also commonly called a *BCD counter,* because it uses only the 10 BCD code groups 0000, 0001, . . . , 1000, and 1001. To reiterate, any MOD-10 counter is a decade counter; and any decade counter that counts in binary from 0000 to 1001 is a BCD counter.

Decade counters, especially the BCD type, find widespread use in applications where pulses or events are to be counted and the results displayed on some type of decimal numerical readout. We shall examine this later in more detail. A decade counter is also often used for dividing a pulse frequency *exactly* by 10. The input pulses are applied to flip-flop A and the output pulses are taken from the output of flip-flop D, which has one-tenth the frequency of the input.

EXAMPLE 7-7

In Example 7-3 a MOD-60 counter was needed to divide the 60-Hz line frequency down to 1 Hz. Construct an appropriate MOD-60 counter.

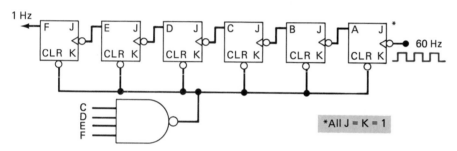

FIGURE 7-7   MOD-60 counter.

Solution $2^5 = 32$ and $2^6 = 64$, and so we need six FFs, as shown in Figure 7-7. The counter is to be cleared when it reaches the count of sixty (111100). Thus, the outputs of flip-flops C, D, E, and F must be connected to the NAND gate. The output of flip-flop F will have a frequency of 1 Hz.

REVIEW QUESTIONS
---

1. What FF outputs should be connected to the clearing NAND gate to form a MOD-13 counter?
2. *True or false:* All BCD counters are decade counters.
3. What is the output frequency of a decade counter that is clocked from a 50-kHz signal?

7-3   IC ASYNCHRONOUS COUNTERS

There are several TTL and CMOS asynchronous counter ICs. One of them is the TTL 7493 (74LS93, etc.) and its logical equivalent, the 74293 (74LS293, etc.). The 74293 is preferred in new designs because it has its power and ground at the corner

74293

*All J, K inputs are internally connected HIGH.

(a)

(b)

FIGURE 7-8 (a) Logic diagram for 7493 and 74293 asynchronous counter IC; (b) simplified symbol. (Courtesy of Fairchild, a Schlumberger company)

pins on the chip. Figure 7-8(a) shows the logic diagram for the 74293 as it would appear in the manufacturer's TTL data book. Some of the nomenclature is different from what we have been using, but it should be easy to figure out. Note the following points:

1. The 74293 has four J-K flip-flops with outputs Q_0, Q_1, Q_2, Q_3, where Q_0 is the LSB and Q_3 is the MSB. The FFs are shown arranged with the LSB on the left. This is done to satisfy the convention that the circuit input signals appear on the left. We have been drawing our counters with LSB on the right so that the order of the FFs is the same as the order of the bits in the binary count. We will continue doing it this way.

2. Each FF has a *CP* (clock pulse) input, which is just another name for the *CLK* input. The clock inputs to Q_0 and Q_1, labeled $\overline{CP}_0$ and $\overline{CP}_1$, respectively, are externally accessible. The inversion bars over these inputs indicate that they are activated by a NGT.

3. Each FF has an asynchronous CLEAR input, C_D. These are connected together to the output of a two-input NAND gate with inputs MR_1 and MR_2, where *MR* stands for *master reset*.

4. Flip-flops Q_1, Q_2, and Q_3 are already connected as a 3-bit ripple counter. Flip-flop Q_0 is not connected to anything internally. This allows the user the option of either connecting Q_0 to Q_1 to form a 4-bit counter, or using Q_0 separately if desired.

The following examples will illustrate some of the ways the 74293 can be wired to produce different counters. In these examples we will use the simplified logic symbol shown in Figure 7-8(b).

—— EXAMPLE 7-8 ——————————————————————

Show how the 74293 should be connected to operate as a MOD-16 counter with a 10-kHz clock input.

Teaching Tip: We could also achieve a MOD-16 counter by connecting Q_3 to CP_0 and using Q_0 as the MSB.

FIGURE 7-9   74293 wired as a MOD-16 counter.

Solution   A MOD-16 counter requires four FFs, and so we have to connect the Q_0 output to $\overline{CP_1}$, the clock input of flip-flop Q_1 (see Figure 7-9). The 10-kHz pulses are applied to $\overline{CP_0}$, the clock input of Q_0.

—— EXAMPLE 7-9 ——————————————————————

Show how to wire the 74293 as a MOD-10 counter.

FIGURE 7-10   74293 wired as a MOD-10 counter.

Solution   A MOD-10 counter requires four FFs, and so again we need to connect Q_0 to $\overline{CP}_1$. This time, however, we want the counter to recycle back to 0000 when it tries to go to the count of 1010 (ten). Thus, the Q_3 and Q_1 outputs have to be connected to the master reset inputs; when they both go HIGH at the count of 1010, the NAND output will immediately reset the counter to 0000.

The circuit wiring is shown in Figure 7-10. The state transition diagram is also shown. Note that the temporary 1010 state is not shown.

EXAMPLE 7-10

Show how to wire a 74293 as a MOD-14 counter.

FIGURE 7-11   An external AND gate is needed to wire the 74293 as a MOD-14 counter.

Solution   When the counter reaches the count of 1110 (14), the Q_3, Q_2, and Q_1 outputs are all HIGH. Unfortunately, the 74293's built-in reset NAND gate has only two inputs. Thus, we have to add some extra logic to ensure that the counter will reset back to 0000 when $Q_3 = Q_2 = Q_1 = 1$. In fact, all we need is a two-input AND gate as shown in Figure 7-11.

EXAMPLE 7-11

In Example 7-7 we divided the input frequency by 60 with a MOD-60 counter using six J-K flip-flops and a NAND gate. Another way to get a MOD-60 counter is shown in Figure 7-12. Explain how this circuit works.

FIGURE 7-12   Two 74293s can be combined to provide a frequency division of 60.

Solution This circuit divides the input frequency by 60 in two steps. The 74293 counter on the right is wired as a MOD-10 counter so that its output Q_3 has a frequency $= f_{in}/10$. This signal is connected to the $\overline{CP_1}$ input of the second 74293 counter, which is wired as a MOD-6 counter (note that Q_0 is not being used). Thus, the Q_3 output of the second counter will have a frequency

$$f_{out} = \frac{f_{in}/10}{6} = \frac{f_{in}}{60}$$

Discussion Question: Is the output signal of Figure 7-12 a symmetrical square wave?

This last example shows that two (or more) counters can be cascaded to produce an overall MOD number equal to the *product* of their individual MOD numbers. This can be very useful in applications where a large amount of frequency division is required.

IEEE/ANSI Symbol for 74293 Counter   Figure 7-13 shows the IEEE/ANSI symbol for the 74293. This symbol contains several new aspects of the IEEE/ANSI standard. As we describe these, you should continue to appreciate how the new IEEE/ANSI symbology is designed to tell us a lot about the IC's operation.

The symbol contains three distinct blocks. The top block (with the notches) is the common-control block. The notation "CTR" defines this IC as a counter. Recall from our discussion in Chapter 5 that the common-control block is used whenever an IC has one or more inputs that are common to more than one of the circuits on the chip. For the 74293, the MR_1 and MR_2 inputs are common to all the FFs in the counter.

MR_1 and MR_2 are shown as active-HIGH inputs that are internally combined using the AND operation as indicated by the "&" notation. This indicates that *both* MR_1 and MR_2 must be in their active states in order to clear the counter. The notation "$CT = 0$" tells us that the action of the MR inputs is to make the count equal zero.

The middle block is labeled "DIV2" to indicate that it is a MOD-2 counter, which of course is a single FF. DIV2 means that the counter will divide its clock input frequency by 2. The bottom block is labeled "DIV8" to indicate that it is a MOD-8 counter. The clock inputs to each of these blocks are shown as activated by negative-going transitions. The "+" notation on each clock input indicates that the NGT of the clock will cause the count to be *incremented by 1*. In other words, the

FIGURE 7-13   IEEE/ANSI symbol for the 74293 IC.

counter *counts up* on each NGT. A "−" notation would be used for a *down counter* (see Section 7-4).

While we will continue to use the traditional symbol for the 74293 and other counters, we will use some of the notation from the IEEE/ANSI symbology. For example, we will indicate a counter's MOD number using the DIV*n* notation, where *n* is the MOD number.

CMOS Asynchronous Counters   There are several asynchronous counters in the CMOS family. Most of them are equivalents to the TTL versions. There are, however, some CMOS asynchronous counter ICs that do not have the TTL counterpart. One of these is the 74HC4024; its logic symbol is shown in Figure 7-14. It is a

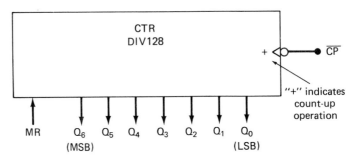

FIGURE 7-14   Logic symbol for the CMOS 74HC4024 ripple counter.

7-bit counter with one asynchronous master reset input. The seven FFs are internally connected as a MOD-128 ripple counter. The *MR* input is active-HIGH and can be used to reset all the FFs to the 0 state. Note that we have used the notation "CTR DIV128" to signify that this is a MOD-128 counter.

Another CMOS ripple counter that has no TTL counterpart is the 74HC4040, which is a 12-bit counter with a single active-HIGH master reset input. The clock input to this counter is a Schmitt-trigger type of input that permits the use of slow-changing signals without producing erratic counting.

── REVIEW QUESTIONS ──────────────────

1. A 2-kHz clock signal is applied to $\overline{CP}_1$ of a 74293. What is the frequency at Q_3?
2. What would be the final output frequency if the order of the counters were reversed in Figure 7-12?
3. What is the MOD number of a 74HC4040 counter?
4. What would the notation "DIV64" mean on a counter symbol?

7-4   ASYNCHRONOUS DOWN COUNTER

All the counters we have looked at thus far have counted *upward* from zero; that is, they were *up counters*. It is a relatively simple matter to construct asynchronous (ripple) *down counters*, which will count downward from a maximum count to zero.

All J, K inputs
are HIGH

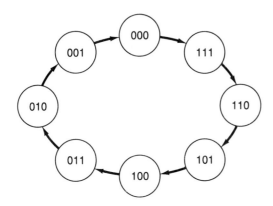

FIGURE 7-15   MOD-8 down counter.

Before looking at a ripple down counter, let us examine the count-down sequence for
a 3-bit down counter:

	CBA	CBA	CBA
(7)	1 1 1	1 1 1	1 1 1
(6)	1 1 0	1 1 0	1 1 0
(5)	1 0 1	1 0 1	. . .
(4)	1 0 0	1 0 0	. . .
(3)	0 1 1	0 1 1	. . .
(2)	0 1 0	0 1 0	etc.
(1)	0 0 1	0 0 1	
(0)	0 0 0	0 0 0	etc.

A, B, and C represent the FF output states as the counter goes through its sequence.
It can be seen that the A flip-flop (LSB) changes states (toggles) at each step in the

sequence just as it does in the up counter. The B flip-flop changes states each time A goes from LOW to HIGH; C changes states each time B goes from LOW to HIGH. Thus, in a down counter each FF, except the first, must toggle when the preceding FF goes from LOW to HIGH. If the FFs have CLK inputs that respond to negative transitions (HIGH to LOW), then an inverter can be placed in front of each CLK input; however, the same effect can be accomplished by driving each FF CLK input from the *inverted* output of the preceding FF. This is illustrated in Figure 7-15 for a MOD-8 down counter.

The input pulses are applied to the A flip-flop; The $\overline{A}$ output serves as the CLK input for the B flip-flop; the $\overline{B}$ output serves as the CLK input for the C flip-flop. The waveforms at A, B, and C show that B toggles whenever A goes LOW to HIGH (so that $\overline{A}$ goes HIGH to LOW) and C toggles whenever B goes LOW to HIGH. This results in the desired down-counting sequence at the C, B, and A outputs. The state transition diagram shows the sequence. Compare it with the diagram for the MOD-8 up counter in Figure 5-49.

Down counters are not as widely used as up counters. Their major application is in situations where it must be known when a desired number of input pulses has occurred. In these situations the down counter is *preset* to the desired number and then allowed to count down as the pulses are applied. When the counter reaches the *zero* state it is detected by a logic gate whose output then indicates that the preset number of pulses has occurred. We shall discuss presettable counters in a later section.

Teaching Tip: A down counter can also be obtained by: (1) wiring an up counter and taking the inverted FF outputs as the counter states, or (2) wiring an up counter using FFs that respond to PGTs.

. REVIEW QUESTIONS

1. What is the difference between the counting sequence of an up counter and a down counter?
2. Describe how an asynchronous down-counter circuit differs from an up-counter circuit.

7-5   PROPAGATION DELAY IN RIPPLE COUNTERS

Ripple counters are the simplest type of binary counters, since they require the fewest components to produce a given counting operation. They do, however, have one major drawback, which is caused by their basic principle of operation. Each FF is triggered by the transition at the output of the preceding FF. Because of the inherent propagation delay time (t_{pd}) of each FF, this means that the second FF will not respond until a time t_{pd} after the first FF receives an active clock transition; the third FF will not respond until a time equal to $2 \times t_{pd}$ after that clock transition; and so on. In other words, the propagation delays of the FFs accumulate so that the Nth FF cannot change states until a time equal to $N \times t_{pd}$ after the clock transition occurs. This is illustrated in Figure 7-16, where the waveforms for a 3-bit ripple counter are shown.

The first set of waveforms in Figure 7-16(a) shows a situation where an input pulse occurs every 1000 ns (the clock period $T = 1000$ ns) and it is assumed that each FF has a propagation delay of 50 ns ($t_{pd} = 50$ ns). Notice that the A flip-flop

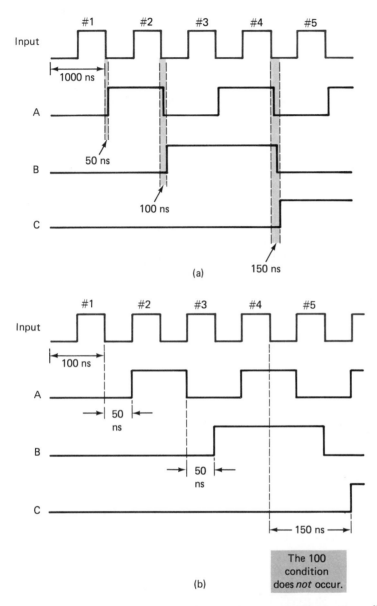

FIGURE 7-16   Waveforms of 3-bit ripple counter illustrating effects of FF propagation delays for different input pulse frequencies.

output toggles 50 ns after the NGT of each input pulse. Similarly, *B* toggles 50 ns after *A* goes from 1 to 0, and *C* toggles 50 ns after *B* goes from 1 to 0. As a result, when the fourth input pulse occurs, the *C* output goes HIGH after a delay of 150 ns. In this situation the counter does operate properly in the sense that the FFs do eventually get to their correct states, representing the binary count. However, the situation worsens if the input pulses are applied at a much higher frequency.

The waveforms in Figure 7-16(b) show what happens if the input pulses occur once every 100 ns. Again, each FF output responds 50 ns after the 1-to-0 transition at its *CLK* input (note the change in the relative time scale). Of particular interest is the situation after the falling edge of the *fourth* input pulse, where the *C* output does

not go HIGH until 150 ns later, which is the same time that the A output goes HIGH in response to the *fifth* input pulse. In other words, the condition $C = 1$, $B = A = 0$ (count of 100) never appears, because the input frequency is too high. This could cause a serious problem if this condition were supposed to be used to control some other operation in a digital system. Problems such as this can be avoided if the period between input pulses is made longer than the total propagation delay of the counter. That is,

$$T_{\text{clock}} \geq N \times t_{\text{pd}} \tag{7-2}$$

where N = number of FFs. Stated in terms of input-clock frequency, the maximum frequency that can be used is given by

$$f_{\text{max}} = \frac{1}{N \times t_{\text{pd}}} \tag{7-3}$$

For example, suppose a 4-bit ripple counter is constructed using the 74LS112 J-K flip-flop. Table 5-1 shows that the 74LS112 has $t_{\text{PLH}} = 16$ ns and $t_{\text{PHL}} = 24$ ns as the propagation delays from CLK to Q. To calculate f_{max}, we will assume the "worst case"; that is, we will use $t_{\text{pd}} = t_{\text{PHL}} = 24$ ns, so that

$$f_{\text{max}} = \frac{1}{4 \times 24 \text{ ns}} = 10.4 \text{ MHz}$$

Clearly, as the number of bits in the counter increases, the total propagation delay increases and f_{max} decreases. For example, a ripple counter that uses six 74LS112 FFs will have

$$f_{\text{max}} = \frac{1}{6 \times 24 \text{ ns}} = 6.9 \text{ MHz}$$

Thus, asynchronous counters are not useful at very high frequencies, especially for large numbers of bits. Another problem caused by propagation delays in asynchronous counters occurs when the counter outputs are *decoded*. This problem is discussed in a later section. Despite these problems, the simplicity of asynchronous counters makes them useful for applications where their frequency limitation is not critical.

REVIEW QUESTIONS

1. Explain why a ripple counter's maximum frequency limitation decreases as more FFs are added to the counter.
2. A certain J-K flip-flop has $t_{\text{pd}} = 12$ ns. What is the largest MOD counter that can be constructed from these FFs and still operate up to 10 MHz?

7-6 SYNCHRONOUS (PARALLEL) COUNTERS

The problems encountered with ripple counters are caused by the accumulated FF propagation delays; stated another way, the FFs do not all change states simultaneously in synchronism with the input pulses. These limitations can be overcome with the use of *synchronous* or *parallel* counters in which all the FFs are triggered simultaneously (in parallel) by the clock input pulses. Since the input pulses are ap-

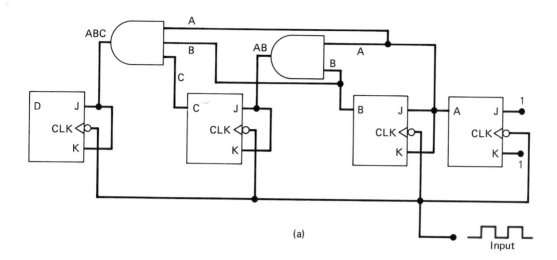

(a)

Count	D	C	B	A
0	0	0	0	0
1	0	0	0	1
2	0	0	1	0
3	0	0	1	1
4	0	1	0	0
5	0	1	0	1
6	0	1	1	0
7	0	1	1	1
8	1	0	0	0
9	1	0	0	1
10	1	0	1	0
11	1	0	1	1
12	1	1	0	0
13	1	1	0	1
14	1	1	1	0
15	1	1	1	1
0	0	0	0	0
.	.	.	.	.
.	.	.	.	.
.	.	etc.	.	.

(b)

FIGURE 7-17 Synchronous MOD-16 counter. Each FF is clocked by the NGT of the input signal so that all FFs will be toggling at the same time.

plied to all the FFs, some means must be used to control when a FF is to toggle and when it is to remain unaffected by a clock pulse. This is accomplished by using the J and K inputs and is illustrated in Figure 7-17 for a 4-bit, MOD-16 synchronous counter.

If we compare the circuit arrangement for this synchronous counter with its asynchronous counterpart in Figure 7-1, we can see the following notable differences:

- The *CLK* inputs of all the FFs are connected together so that the input clock signal is applied to each FF simultaneously.
- Only flip-flop *A*, the LSB, has its *J* and *K* inputs permanently at the HIGH level. The *J*, *K* inputs of the other FFs are driven by some combination of FF outputs.
- The synchronous counter requires more circuitry than does the asynchronous counter.

Circuit Operation   The basic principle of operation of the synchronous counter is this:

> *The J and K inputs of the FFs are connected so that only those FFs that are supposed to toggle on a given NGT will have J = K = 1 when that NGT occurs.*

Let's examine this principle for each of the FFs with the help of the counting sequence shown in Figure 7-17(b).

The counting sequence shows that the *A* flip-flop has to change states at each NGT. For this reason its *J* and *K* inputs are permanently HIGH so that it will toggle on each NGT of the clock input.

The counting sequence shows that flip-flop *B* has to change states on each NGT that occurs while $A = 1$. For example, when the count is 0001, the next NGT has to toggle *B* to the 1 state; when the count in 0011, the next NGT has to toggle *B* to the 0 state; and so on. This operation is accomplished by connecting output *A* to the *J* and *K* inputs of flip-flop *B* so that $J = K = 1$ whenever $A = 1$.

The counting sequence shows that flip-flop *C* has to change states on each NGT that occurs while $A = B = 1$. For example, when the count is 0011, the next NGT has to toggle *C* to the 1 state; when the count is 0111, the next NGT has to toggle *C* to the 0 state; and so on. This operation is ensured by connecting the signal *AB* to the *J* and *K* inputs of flip-flop *C*.

Teaching Tip: You may want to show how the counting sequence is not affected if the FFs trigger on PGTs.

In a like manner, we can see that flip-flop *D* has to toggle on each NGT that occurs while $A = B = C = 1$. When the count is 0111, the next NGT has to toggle *D* to the 1 state; when the count is 1111, the next NGT has to toggle *D* to the 0 state. This is accomplished by connecting *ABC* to the *J* and *K* inputs of flip-flop *D*.

Advantage of Synchronous Counters over Asynchronous   In a parallel counter all the FFs will change states simultaneously; that is, they are all synched to the NGTs of the input clock pulses. Thus, unlike the asynchronous counters, the propagation delays of the FFs do not add together to produce the overall delay. Instead, the total response time of a synchronous counter like the one in Figure 7-17 is the time it takes *one* FF to toggle plus the time for the new logic levels to propagate through a *single* AND gate to reach the *J*, *K* inputs. That is,

$$\text{total delay} = \text{FF } t_{\text{pd}} + \text{AND gate } t_{\text{pd}}$$

This means that a synchronous counter can operate at a much higher input frequency than an asynchronous counter with the same number of FFs. Of course, the synchronous counter requires more circuitry than the asynchronous counter.

This ability to operate at higher frequencies is the major advantage of synchronous counters. They have another advantage related to the *decoding* process that we will be discussing shortly.

Actual ICs   There are many synchronous ICs in both the TTL and CMOS logic families. Some of the most commonly used devices are:

74LS160/162, 74HC160/162: synchronous decade counters

74LS161/163, 74HC161/163: synchronous MOD-16 counters

_____ EXAMPLE 7-12 _____

(a) Determine f_{max} for the counter of Figure 7-17(a) if t_{pd} for each FF is 50 ns and t_{pd} for each AND gate is 20 ns. Compare this with f_{max} for a MOD-16 ripple counter.

(b) What has to be done to convert this counter to MOD-32?

(c) Determine f_{max} for the MOD-32 parallel counter.

Discussion Question: How is f_{max} calculated when each FF and AND gate has a different t_{pd} value?

Solution

(a) The total delay that must be allowed between input clock pulses is equal to FF t_{pd} + AND gate t_{pd}. Thus, $T_{clock} \geq 50 + 20 = 70$ ns, and so the parallel counter has

$$f_{max} = \frac{1}{70 \text{ ns}} = 14.3 \text{ MHz (parallel counter)}$$

A MOD-16 ripple counter uses four FFs with $t_{pd} = 50$ ns. Thus, f_{max} for the ripple counter is

$$f_{max} = \frac{1}{4 \times 50 \text{ ns}} = 5 \text{ MHz (ripple counter)}$$

(b) A fifth FF must be added, since $2^5 = 32$. The *CLK* input of this FF is also tied to the input pulses. Its *J* and *K* inputs are fed by the output of a four-input AND gate whose inputs are *A*, *B*, *C*, and *D*.

(c) f_{max} is still determined as in (a) regardless of the number of FFs in the parallel counter. Thus, f_{max} is still 14.3 MHz.

_____ REVIEW QUESTIONS _____

1. What is the advantage of a synchronous counter over an asynchronous counter? What is the disadvantage?

2. How many logic devices are required for a MOD-64 parallel counter?

7-7   SYNCHRONOUS DOWN AND UP/DOWN COUNTERS

In Section 7-4 we saw that a ripple counter could be made to count down by using the inverted output of each FF to drive the next FF in the counter. A parallel down counter can be constructed in a similar manner—that is, by using the inverted FF outputs to drive the following *J*, *K* inputs. For example, the parallel up counter of

Figure 7-17 can be converted to a down counter by connecting the $\overline{A}$, $\overline{B}$, and $\overline{C}$ outputs in place of A, B, and C, respectively. The counter will then proceed through the following sequence as input pulses are applied:

(15)	1	1	1	1	←
(14)	1	1	1	0	
(13)	1	1	0	1	
(12)	1	1	0	0	
	.	.	.	.	
	.	.	.	.	recycle
	.	.	.	.	
(3)	0	0	1	1	
(2)	0	0	0	1	
(1)	0	0	0	1	
(0)	0	0	0	0	←

To form a parallel up/down counter (see Figure 7-18) the control inputs (COUNT-UP and COUNT-DOWN) are used to control whether the normal FF outputs or the inverted FF outputs are fed to the J and K inputs of the following FFs. The counter in Figure 7-18 is a MOD-8 up/down counter that will count from 000 up to 111 when the COUNT-UP control input is 1 and from 111 down to 000 when the COUNT-DOWN control input is 1.

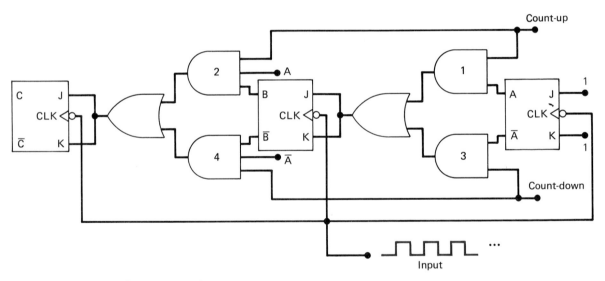

FIGURE 7-18   Synchronous up/down counter (MOD-8).

A logical 1 on the COUNT-UP line while COUNT-DOWN = 0 enables AND gates 1 and 2 and disables AND gates 3 and 4. This allows the A and B outputs through to the J and K inputs of the following FFs so that the counter will count up as pulses are applied. The opposite action takes place when COUNT-UP = 0 and COUNT-DOWN = 1.

Actual ICs   The 74LS190 and 74HC190 ICs are two of several synchronous up/down counter ICs. Each of these is an up/down decade counter.

— EXAMPLE 7-13 —

Describe how the circuit of Figure 7-18 will operate when COUNT-UP and COUNT-DOWN are both LOW, and when they are both HIGH.

Solution   When both these inputs are LOW, each AND-gate and each OR-gate output will be LOW. Thus, the J-K inputs of flip-flops B and C will be LOW, and neither of these FFs can change states. The A flip-flop will still toggle on each input pulse.

When both inputs are HIGH, each OR-gate output will be HIGH because one of the AND outputs feeding each OR gate will be HIGH, depending on the state of the corresponding FF. Thus, all three FFs will have $J = K = 1$ and will toggle on each input pulse.

7-8 PRESETTABLE COUNTERS

Many synchronous (parallel) counters that are available as ICs are designed to be *presettable;* in other words, they can be preset to any desired starting count either asynchronously (independent of the clock signal) or synchronously (on the active transition of the clock signal). This presetting operation is also referred to as *loading* the counter.

Figure 7-19 shows the logic circuit for a 3-bit presettable parallel up counter. The J, K, and CLK inputs are wired for operation as a parallel up counter. The asynchronous PRESET and CLEAR inputs are wired to perform asynchronous presetting. The counter is loaded with any desired count at any time by doing the following:

1. Apply the desired count to the parallel data inputs, P_2, P_1, and P_0.
2. Apply a LOW pulse to the PARALLEL LOAD input, $\overline{PL}$.

This procedure will perform an asynchronous transfer of the P_2, P_1, and P_0 levels into flip-flops Q_2, Q_1, and Q_0, respectively (Section 5-17). This *jam transfer* occurs independent of the J, K, and CLK inputs. The effect of the CLK input will be disabled as long as $\overline{PL}$ is in its active-LOW state, since each FF will have one of its asynchronous inputs activated while $\overline{PL} = 0$. Once $\overline{PL}$ returns HIGH, the FFs can respond to their CLK inputs and can resume the counting-up operation starting from the count that was loaded into the counter.

For example, let's say that $P_2 = 1$, $P_1 = 0$, and $P_0 = 1$. While $\overline{PL}$ is HIGH, these parallel data inputs have no effect. If clock pulses are present, the counter will perform the normal count-up operation. Now let's say that $\overline{PL}$ is pulsed LOW when the counter is at the 010 count (i.e., $Q_2 = 0$, $Q_1 = 1$, and $Q_0 = 0$). This LOW at $\overline{PL}$ will produce LOWs at the CLEAR input of Q_1 and at the PRESET inputs of Q_2 and Q_0 so that the counter will go to the 101 count *regardless of what is occurring at the CLK input.* The count will hold at 101 until $\overline{PL}$ is deactivated (returned HIGH); at that time the counter will resume counting up clock pulses from the count of 101.

Discussion Question: How can this counter be changed to be a presettable down counter?

This asynchronous presetting is used by several IC counters, such as the TTL 74190, 74191, 74192, and 74193 and the CMOS equivalents, 74HC190, 74HC191, 74HC192, and 74HC193.

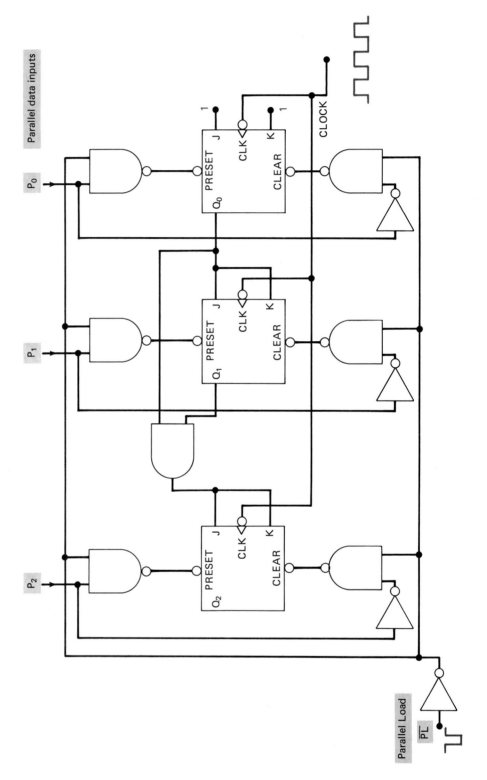

FIGURE 7-19 Presettable parallel counter with asynchronous preset.

302

Synchronous Presetting   Many IC parallel counters use *synchronous presetting* whereby the counter is preset on the active transition of the same clock signal that is used for counting. The logic level applied to the $\overline{PL}$ input determines whether the active clock transition will preset the counter or whether it will be counted as in the normal counting operation.

Examples of IC counters that use synchronous presetting include the TTL 74160, 74161, 74162, and 74163 and their CMOS equivalents, 74HC160, 74HC161, 74HC162, and 74HC163.

REVIEW QUESTIONS

1. What is meant when we say that a counter is presettable?
2. Describe the difference between asynchronous and synchronous presetting.

7-9 THE 74193 (LS193/HC193) COUNTER

Figure 7-20 shows the logic symbol and the input/output description for the 74193 counter. This counter can be described as a MOD-16, presettable up/down counter with synchronous counting, asynchronous preset, and asynchronous master reset. Let us look at the function of each input and output.

FIGURE 7-20   74193 presettable up/down counter: (a) logic symbol; (b) input/output description; (c) mode select table. (Courtesy of Fairchild, a Schlumberger company)

(a)

Mode Select

MR	$\overline{PL}$	CP_U	CP_D	Mode
H	X	X	X	Asynch. reset
L	L	X	X	Asynch. preset
L	H	H	H	No change
L	H	↑	H	Count up
L	H	H	↑	Count down

H = HIGH; L = LOW
X = Don't care; ↑ = PGT

(c)

Pin Names	Description
CP_U	Count-up clock input (active rising edge)
CP_D	Count-down clock input (active rising edge)
MR	Asynchronous master reset input (active HIGH)
$\overline{PL}$	Asynchronous parallel load input (active LOW)
P_0–P_3	Parallel data inputs
Q_0–Q_3	Flip-flop outputs
$\overline{TC_D}$	Terminal count-down (borrow) output (active LOW)
$\overline{TC_U}$	Terminal count-up (carry) output (active LOW)

(b)

Clock Inputs CP_U and CP_D   The counter will respond to the positive-going transitions at one of two clock inputs. CP_U is the *count-up clock* input. When pulses are applied to this input, the counter will increment (count up) on each PGT to a maximum count of 1111; then it recycles to 0000 and starts over. CP_D is the *count-down clock* input. When pulses are applied to this input, the counter will decrement (count down) on each PGT to a minimum count of 0000; then it recycles to 1111 and starts over. Thus, one clock input or the other will be used for counting while the other clock input is inactive (kept HIGH).

Master Reset (MR)   The master reset is an active-HIGH asynchronous input that resets the counter to the 0000 state. *MR* is a dc reset, and so it will hold the counter at 0000 as long as $MR = 1$. It also overrides *all* other inputs.

Preset Inputs   The counter FFs can be preset to the logic levels present on the parallel data inputs P_3 through P_0 by momentarily pulsing the parallel load input $\overline{PL}$ from HIGH to LOW. This is an asynchronous preset that overrides the counting operation. $\overline{PL}$ will have no effect, however, if the *MR* input is in its active HIGH state.

Count Outputs   The current count is always present at the FF outputs Q_3 through Q_0, where Q_0 is the LSB and Q_3 is the MSB.

Teaching Tip: It would be instructive to show that the MSB output (Q_3) cannot be used to directly clock a second stage because it would cause the second stage to be incremented too early in the count sequence.

Terminal Count Outputs   The terminal count outputs are used when two or more 74193s are connected as a multistage counter to produce a larger MOD number. In the count-up mode, the $\overline{TC}_U$ output of the lower-order counter is connected to the CP_U input of the next higher-order counter. In the count-down mode, the $\overline{TC}_D$ output of the lower-order counter is connected to the CP_D input of the next higher-order counter.

$\overline{TC}_U$ is the *terminal count-up* (also called the *carry*) output. It is generated on the 74193 chip using the logic shown in Figure 7-21(a). Clearly, $\overline{TC}_U$ will be LOW only when the counter is in the 1111 state and CP_U is LOW. Thus, $\overline{TC}_U$ will remain HIGH as the counter counts up from 0000 to 1110. On the next PGT of CP_U, the count goes to 1111, but $\overline{TC}_U$ does not go LOW until CP_U returns LOW. The next PGT at CP_U recycles the count to 0000 and also causes $\overline{TC}_U$ to return HIGH. This PGT at TC_U occurs when the counter recycles from 1111 to 0000, and can be used to clock a second 74193 up counter to its next higher count.

FIGURE 7-21   (a) Logic on 74193 for generating $\overline{TC}_U$; (b) logic for generating $\overline{TC}_D$.

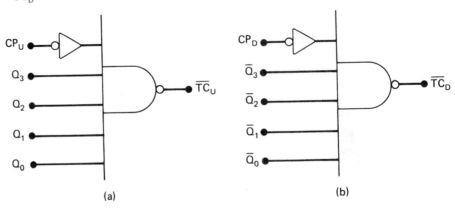

(a)   (b)

$\overline{TC}_D$ is the *terminal count-down* (also called the *borrow*) output. It is generated as shown in Figure 7-21(b). It is normally HIGH and does not go LOW until the counter has counted down to the 0000 state and CP_D is LOW. When the next PGT at CP_D recycles the counter to 1111, it causes $\overline{TC}_D$ to return HIGH. This PGT at $\overline{TC}_D$ can be used to clock a second 74193 down counter to its next lower count.

—— EXAMPLE 7-14 ————————————————————————————————

Refer to Figure 7-22(a), where a 74193 is wired as an up counter. The parallel data inputs are permanently connected as 1011, and the CP_U, $\overline{PL}$, and MR input waveforms are shown in Figure 7-22(b). Assume that the counter is initially in the 0000 state, and determine the counter output waveforms.

FIGURE 7-22   Example 7-14.

Solution   Initially (at t_0) the counter FFs are all LOW. This causes $\overline{TC}_U$ to be HIGH. Just prior to time t_1 the $\overline{PL}$ input is pulsed LOW. This immediately loads the counter with 1011 to produce $Q_3 = 1$, $Q_2 = 0$, $Q_1 = 1$, and $Q_0 = 1$. At t_1 the CP_U input makes a PGT, but the counter cannot respond to this because $\overline{PL}$ is still active at that time. At times t_2, t_3, t_4, and t_5 the counter counts up on each PGT at CP_U. After the PGT at t_5 the counter is in the 1111 state, but $\overline{TC}_U$ does not go LOW until CP_U goes LOW at t_6. When the next PGT occurs at t_7, the counter recycles to 0000, and $\overline{TC}_U$ returns HIGH.

Discussion Question: Why doesn't $\overline{TC}_D$ go LOW at t_{10}?

The counter will count up in response to the PGTs at t_8 and t_9. The PGT at t_{10} will have no effect, because the MR goes HIGH prior to t_{10} and remains active at t_{10}. This will reset all FFs to 0 and overrides the CP_U signal.

_____ EXAMPLE 7-15 _____

Figure 7-23(a) shows the 74193 wired as a down counter. The parallel data inputs are permanently wired as 0111, and the CP_D and $\overline{PL}$ waveforms are shown in Figure 7-23(b). Assume that the counter is initially in the 0000 state, and determine the output waveforms.

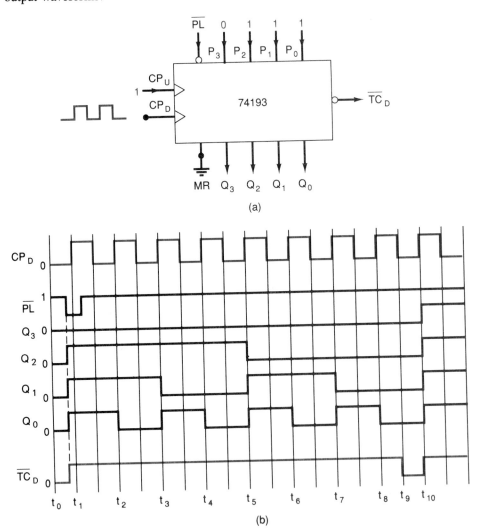

FIGURE 7-23   Example 7-15.

Solution   At t_0 all the FF outputs are LOW and CP_D is LOW. These are the conditions that produce $\overline{TC}_D = 0$. Prior to t_1 the $\overline{PL}$ input is pulsed LOW. This immediately presets the counter to 0111 and therefore causes $\overline{TC}_D$ to go HIGH. The PGT of CP_D at t_1 will have no effect, since $\overline{PL}$ is still active. The counter will respond to the PGTs at t_2 to t_8 and counts down to 0000 at t_8. $\overline{TC}_D$ does not go LOW until t_9, when

CP_D goes LOW. At t_{10} the PGT of CP_D causes the counter to recycle to 1111 and also drives $\overline{TC}_D$ back HIGH.

Variable MOD Number Using the 74193   Presettable counters can be easily wired for different MOD numbers without the need for additional logic circuitry. We will demonstrate this for the 74193 using the circuit of Figure 7-24(a). Here, the 74193 is used as a down counter with its parallel load inputs permanently connected at 0101 (5_{10}). Note that the $\overline{TC}_D$ output is connected back to the $\overline{PL}$ input.

We will begin our analysis by assuming that the counter has been counting down and is in the 0101 state at time t_0. Refer to Figure 7-24(b) for the counter waveforms.

The counter will decrement (count down) on the PGTs of CP_D at times t_1 to t_5. At t_5 the counter is in the 0000 state. When CP_D goes LOW at t_6, it drives $\overline{TC}_D$

FIGURE 7-24   (a) 74193 wired as a MOD-5 counter; (b) waveforms.

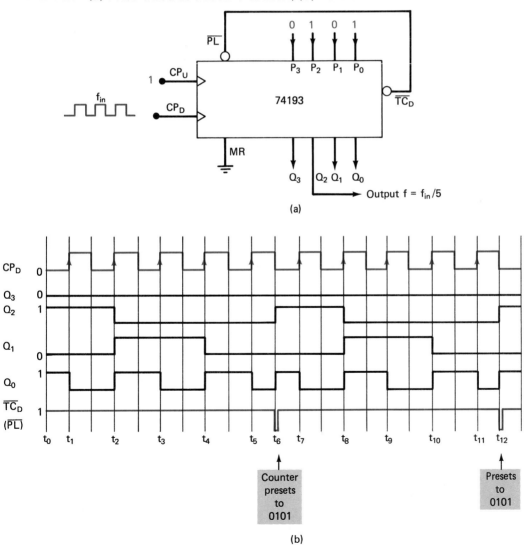

LOW. This immediately activates the $\overline{PL}$ input and presets the counter back to the 0101 state. Note that $\overline{TC}_D$ stays LOW for only a short interval because once the counter outputs go to 0101 in response to $\overline{PL} = 0$, the condition needed to keep $\overline{TC}_D = 0$ is removed. Thus, there is only a narrow glitch at $\overline{TC}_D$.

This same sequence is repeated at times t_7 to t_{12} and at equal intervals thereafter. If we examine the Q_2 waveform, we can see that it goes through one complete cycle for every *five* cycles of CP_D. For example, there are *five* clock cycles between the PGT of Q_2 at t_6 and the PGT of Q_2 at t_{11}. Thus, the frequency of the Q_2 waveform is one-fifth of the clock frequency.

It is no coincidence that the frequency-division ratio (5) is the same as the number applied to the parallel data inputs (0101 = 5). In fact, we can vary the frequency division by changing the logic levels applied to the parallel data inputs.

A *variable frequency-divider* circuit can be easily implemented by connecting switches to the parallel data inputs of the circuit in Figure 7-24. The switches can be set to a value equal to the desired frequency-division ratio.

—— REVIEW QUESTIONS ——

1. Describe the function of the inputs $\overline{PL}$ and P_0 to P_3.
2. Describe the function of the MR input.
3. *True or false:* The 74193 cannot be preset while MR is active.
4. What logic levels must be present at CP_D, $\overline{PL}$, and MR in order for the 74193 to count pulses that appear at CP_U?

7-10 MORE ON THE IEEE/ANSI DEPENDENCY NOTATION*

We can learn more about the dependency notation that is such an important part of the new IEEE/ANSI symbology by examining the IEEE/ANSI symbol for the 74193 IC shown in Figure 7-25. Each type of IC that we examine in this way will add to

FIGURE 7-25   IEEE/ANSI symbol for the 74193 IC.

*This section may be skipped without affecting the continuity of the rest of the book.

your understanding of the new symbology and will help to prepare you for the more extensive use of these symbols in the future.

Once again, it should be stated that only the labels inside the rectangular outlines are specified by the IEEE/ANSI standard. The names or labels shown outside the outlines are not standard, and in fact will vary from one IC manufacturer to another.

Some of the notation used in Figure 7-25 should be familiar. The symbol outline is divided into the common-control block that affects all of the counter FFs, and the four narrow rectangles representing the individual FFs. The bracketed number inside each FF rectangle denotes its relative weight in the counter. The label CTR DIV16 signifies that this device, when operated normally, is a counter (CTR) with 16 states (i.e., a divide-by-16 counter). The *MR* input to the common-control block has the notation $CT = 0$ to indicate that the counter will reset to zero when *MR* is HIGH.

Control Dependency (C)

The letter C in the label for an input denotes that that input *controls* the entry of data into a storage element (i.e., a FF). Usually, C is used for clock inputs that clock data into a FF on its active transition. We saw this when we looked at the IEEE/ANSI symbols for FFs in Chapter 5. In Figure 7-25, C is used for the parallel load input $\overline{PL}$, since this input controls the entry of data into the four counter FFs. Specifically, the label C3 indicates that this input will control any other input that has the digit 3 as a prefix in its label. In this case that includes inputs P_0, P_1, P_2, and P_3, since they all have the label 3D (it is shown only on the top FF block, but it is assumed to be the same for the other FF blocks). The "D" part of the label denotes "data."

What this all means is that when $\overline{PL}$ is in its active-LOW state, data from P_0 through P_3 will be entered into flip-flops Q_0 to Q_3. Since there is no edge-triggered symbol at $\overline{PL}$, it is understood that $\overline{PL}$ is in effect as long as it stays at its active-LOW level.

Counting Direction (+ or −)

The CP_U and CP_D inputs are shown as having two separate labels because they have several distinct internal effects. Let's first consider the upper label. This label for the CP_U input is 2+. The plus sign (+) indicates that a PGT at this input will increment the count by 1, in other words, cause the counter to count up. Likewise, the upper label for the CP_D input has a minus sign (−) to show that this input will decrement the count by 1, in other words, cause it to count down. The significance of the digits in front of the plus and minus signs will be explained in the following paragraphs.

AND Dependency (G)

The letter G in the label for an input denotes AND dependency. This means that an input designated by a G followed by a digit is internally ANDed with any other input or output that has the same digit as a prefix in its label. In Figure 7-25 we see that the lower label for the CP_U input is G1. This means that CP_U is internally ANDed with any input or output that has a 1 in its label. The upper label for CP_D is 1−, and so there must be an AND dependency between CP_D and CP_U. Specifically, this AND dependency tells us that CP_U must be HIGH in order for CP_D to perform its count-down function.

The lower label for CP_D is G2, which indicates that there is an AND dependency between CP_D and any input or output that has a 2 in its label. For example, the upper label for CP_U is 2+, which tells us that CP_D must be HIGH in order for CP_U to perform its count-up function.

Teaching Tip: This AND dependency notation can be confusing. It would probably be helpful to actually show the internal gates that produce the AND dependency.

Now let's look at the $\overline{TC}_D$ output label. It is $\overline{2}CT = 0$. It includes a 2 in its label, indicating that it has an AND dependency with CP_D. Actually, since it is a $\overline{2}$, the AND dependency is with $\overline{CP}_D$. Thus, the label for $\overline{TC}_D$ tells us that $\overline{TC}_D$ will go to its active-LOW state when CP_D is LOW *and* the count is zero ($CT = 0$). In a like manner, the label for $\overline{TC}_U$ tells us that $\overline{TC}_U$ will go to its active-LOW state when CP_U is LOW *and* the count is fifteen ($CT = 15$).

—— REVIEW QUESTIONS ——

1. Explain the meaning of control dependency and AND dependency.
2. Give the meaning of the following input labels: (a) +, (b) G4, (c) C5, (d) 5D.

7-11   DECODING A COUNTER

Digital counters are often used in applications where the count represented by the states of the FFs must somehow be determined or displayed. One of the simplest means for displaying the contents of a counter involves just connecting the output of each FF to a small indicator LED. In this way the states of the FFs are visibly represented by the LEDs (bright = 1, dark = 0) and the count can be mentally determined by decoding the binary states of the LEDs. For instance, suppose that this method is used for a BCD counter and the states of the LEDs are respectively dark-bright-bright-dark. This would represent 0110, which we would mentally decode as decimal 6. Other combinations of LED states would represent the other possible counts.

The indicator LED method becomes inconvenient as the size (number of bits) of the counter increases, because it is much harder to mentally decode the displayed results. For this reason it would be preferable to develop a means for *electronically* decoding the contents of a counter and displaying the results in a form that would be immediately recognizable and would require no mental operations.

An even more important reason for electronic decoding of a counter occurs because of the many applications in which counters are used to control the timing or sequencing of operations *automatically* without human intervention. For example, a certain system operation might have to be initiated when a counter reaches the 101100 state (count of 44_{10}). A logic circuit can be used to decode for or detect when this particular count is present and then initiate the operation. Many operations may have to be controlled in this manner in a digital system. Clearly, human intervention in this process would be undesirable except in extremely slow systems.

Active-HIGH Decoding   A MOD-X counter has X different states; each state is a particular pattern of 0s and 1s stored in the counter FFs. A decoding network is a logic circuit that generates X different outputs, each of which detects (decodes) the presence of one particular state of the counter. The decoder outputs can be designed to produce either a HIGH or a LOW level when the detection occurs. An active-HIGH decoder produces HIGH outputs to indicate detection. Figure 7-26 shows the complete active-HIGH decoding logic for a MOD-8 counter. The decoder consists of eight 3-input AND gates. Each AND gate produces a HIGH output for one particular state of the counter.

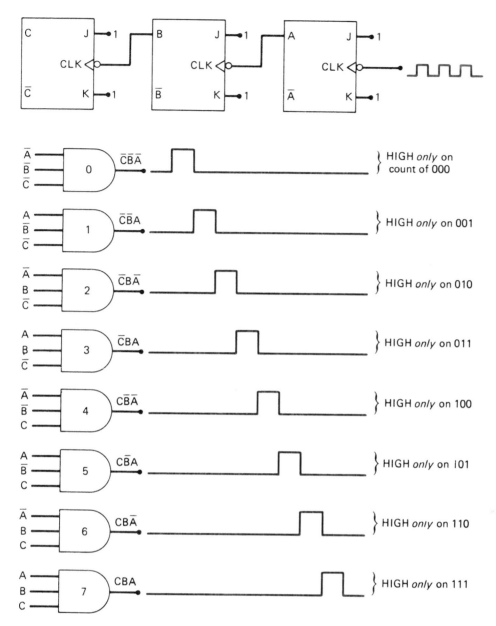

FIGURE 7-26 Using AND gates to decode a MOD-8 counter.

For example, AND gate 0 has as its inputs the FF outputs $\overline{C}$, $\overline{B}$, and $\overline{A}$. Thus its output will be LOW at all times *except* when $A = B = C = 0$—that is, on the count of 000 (zero). Similarly, AND gate 5 has as its inputs the FF outputs C, $\overline{B}$, and A, so that its output will go HIGH only when $C = 1$, $B = 0$, and $A = 1$—that is, on the count of 101 (decimal 5). The rest of the AND gates perform in the same manner for the other possible counts. At any one time only one AND-gate output is HIGH, the one which is decoding for the particular count that is present in the counter. The waveforms in Figure 7-26 show this clearly.

Discussion Question: What will be the frequency of the signals at each decoder output as compared with the clock frequency?

The eight AND outputs can be used to control eight separate indicator LEDs, which represent the decimal numbers 0 through 7. Only one LED will be on at a given time, indicating the proper count.

The AND gate decoder can be extended to counters with any number of states. The following example illustrates.

EXAMPLE 7-16

How many AND gates are required to completely decode all the states of a MOD-32 binary counter? What are the inputs to the gate that decodes for the count of 21?

Solution   A MOD-32 counter has 32 possible states. One AND gate is needed to decode for each state; therefore, the decoder requires 32 AND gates. Since $32 = 2^5$, the counter contains five FFs. Thus, each gate will have five inputs, one from each FF. To decode for the count of 21 (that is, 10101_2) requires AND-gate inputs of E, $\bar{D}$, C, $\bar{B}$, and A, where E is the MSB flip-flop.

Active-LOW Decoding   If NAND gates are used in place of AND gates, the decoder outputs will produce a normally HIGH signal, which goes LOW only when the number being decoded occurs. Both types of decoders are used, depending on the type of circuits being driven by the decoder outputs.

EXAMPLE 7-17

Figure 7-27 shows a common situation in which a counter is used to help generate a control waveform which could be applied to devices such as a motor, solenoid valve, or heater. The MOD-16 counter cycles and recycles through its counting sequence.

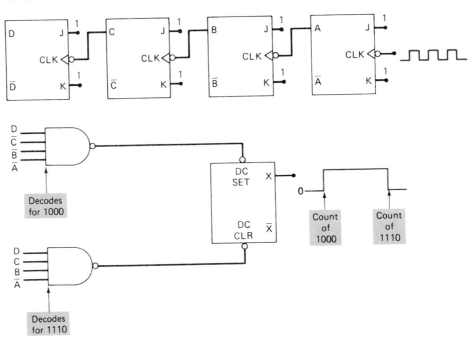

FIGURE 7-27   Example 7-17.

Each time it goes to the count of 8 (1000), the upper NAND gate will produce a LOW output, which sets flip-flop X to the 1 state. Flip-flop X stays HIGH until the counter reaches the count of 14 (1110), at which time the lower NAND gate decodes it and produces a LOW output to clear X to the 0 state. Thus, the X output is HIGH between the counts of 8 and 14 for each cycle of the counter.

BCD Counter Decoding   A BCD counter has 10 states, which can be decoded using the techniques previously described. BCD decoders provide 10 outputs corresponding to the decimal digits 0 through 9 represented by the states of the counter FFs. These 10 outputs can be used to control 10 indicator LEDs for a visual display. More often, instead of using 10 separate LEDs, a single display device is used to display the decimal numbers 0 through 9. One such device, called a *nixie tube,* contains 10 very thin numerically shaped filaments stacked on top of each other. The BCD decoder outputs control which filament is illuminated. A newer class of decimal displays contains seven small segments made of a material (usually LEDs or liquid crystal displays) which either emits light or reflects ambient light. The BCD decoder outputs control which segments are illuminated in order to produce a pattern representing one of the decimal digits.

We will go into more detail concerning these types of decoders and displays in Chapter 9. However, since BCD counters and their associated decoders and displays are very commonplace, we will use the decoder/display unit (see Figure 7-28) to represent the complete circuitry used to visually display the contents of a BCD counter as a decimal digit.

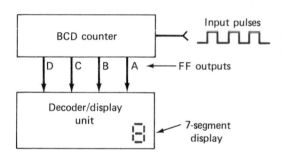

FIGURE 7-28   BCD counters usually have their count displayed on a single display device.

___ REVIEW QUESTION _____

1. Describe the decoding gate needed to produce a LOW output when a MOD-64 counter is at the count of 23.

7-12   DECODING GLITCHES

Section 7-5 discussed the effects of FF propagation delays in ripple counters. As we saw then, the accumulated propagation delays serve to essentially limit the frequency response of ripple counters. The delays between FF transitions can also cause problems when decoding a ripple counter. The problem occurs in the form of glitches or spikes at the outputs of some of the decoding gates. This is illustrated in Figure 7-29 for a MOD-4 ripple counter.

The waveforms at the outputs of each FF and decoding gate are shown in the figure. Notice the propagation delay between the clock waveform and the A output waveform and between the A waveform and the B waveform. The glitches in the X_0 and X_2 decoding waveforms are caused by the delay between the A and B wave-

MOD-4 counter

Decoder gates

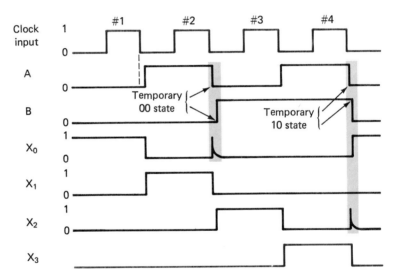

FIGURE 7-29 FF and decoding waveforms for a MOD-4 ripple counter showing glitches at X_0 and X_2 outputs.

forms. X_0 is the output of the AND gate decoding for the normal 00 count. The 00 condition also occurs momentarily as the counter goes from the 01 to the 10 count, as shown by the waveforms. This is because B cannot change states until A goes LOW. This momentary 00 state lasts only for several nanoseconds (depending on t_{pd} of flip-flop B) but can be detected by the decoding gate if the gate's response is fast enough. Hence, the spike at the X_0 output.

A similar situation produces a glitch at the X_2 output. X_2 is decoding for the 10 condition, and this condition occurs momentarily as the counter goes from 11 to 00 in response to the fourth clock pulse, as shown in the waveforms. Again, this is due to the delay of flip-flop B's response after A has gone LOW.

Although the situation is illustrated for a MOD-4 counter, the same type of situation can occur for *any* ripple counter. This is because ripple counters work on the "chain-reaction" principle, whereby each FF triggers the next one and so on. The

spikes at the decoder outputs may or may not present a problem, depending on how the counter is being used. When the counter is being used only to count pulses and display the results, the decoding spikes are of no consequence because they are very short in duration and will not even show up on the display. However, when the counter is used to control other logic circuits, such as was done in Figure 7-27, the spikes can cause improper operation. For example, in Figure 7-27, a spike at the output of either decoding NAND gate would cause flip-flop X to be set or cleared at the wrong time.

We can predict where in the asynchronous counter's sequence a temporary state will occur by stepping through a counter state transition one FF at a time. For example, let's look at the actual step-by-step process by which a ripple counter goes from 011 (3) to 100 (4):

$$
\begin{array}{ccc}
C & B & A \\
0 & 1 & 1 \quad (3) \\
\end{array}
$$

temporary states
$$
\begin{cases}
0 & 1 & 0 \quad (2) \quad \longleftarrow \text{FF } A \text{ toggles first} \\
0 & 0 & 0 \quad (0) \quad \longleftarrow \text{and causes } B \text{ to toggle}
\end{cases}
$$
$$
1 \quad 0 \quad 0 \quad (4) \quad \longleftarrow \text{which causes } C \text{ to toggle}
$$

Note the occurrence of two temporary states, 010 and 000.

In situations where the decoding spikes cannot be tolerated, there are two basic solutions to the problem. The first possibility is to use a parallel counter instead of a ripple counter. Recall that in a parallel counter the FFs are all triggered at the same time by the clock pulses so that it appears that the conditions which produced the decoder spikes cannot occur. However, even in a parallel counter the spikes may occur because the FFs will not all necessarily have the same t_{pd}, especially when some FFs may be loaded more heavily than others.

Strobing   A more reliable method for eliminating the decoder spikes is to use a technique called *strobing*. This technique uses a signal called a *strobe signal* to keep the decoding AND gates disabled (outputs at 0) until all the FFs have reached a sta-

Teaching Tip: It would be instructive to analyze a 3-bit counter for all potential decoding glitches using this process. Note how the number of possible glitches produced by a given counter state transition depends on the number of FFs that have to change states.

Teaching Tip: When one FF is driving many more loads than the other FFs, the effect of the capacitive loading is to slow down its output transition so that it changes more slowly than the others, thereby creating the conditions for possible decoding glitches.

FIGURE 7-30   Use of strobe signal to eliminate decoding spikes.

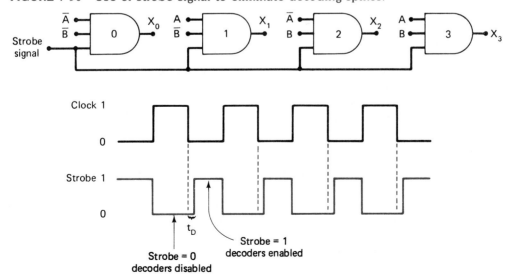

ble state in response to the negative clock transition. This is illustrated in Figure 7-30, where the strobe signal is connected as an input to each decoding gate. The accompanying waveforms show that the strobe signal goes LOW when the clock pulse goes HIGH. During the time that the strobe is LOW, the decoding gates are kept LOW. The strobe signal goes HIGH to enable the decoding gates some time t_D *after* the clock pulse goes LOW. t_D is chosen to be greater than the total time it takes the counter to reach a stable count and depends, of course, on the FF delays and the number of FFs in the counter. In this way the decoding gate outputs will not contain any spikes, because they are disabled during the time the FFs are in transition.

The strobe method is not used if a counter is used only for display purposes, since the decoding spikes are too narrow to affect the display. The strobe signal is used when the counter is used in control applications like that of Figure 7-27, where the spikes could cause erroneous operation.

REVIEW QUESTIONS

1. Explain why the decoding gates for an asynchronous counter may have glitches on their outputs.
2. How does strobing eliminate decoding glitches?

7-13 CASCADING BCD COUNTERS

BCD counters are often used whenever pulses are to be counted and the results displayed in decimal. A single BCD counter counts from 0 through 9 and then recycles to 0. To count to larger decimal values, we can cascade BCD counters as illustrated in Figure 7-31. This arrangement operates as follows:

1. Initially all counters are cleared to the zero state. Thus, the decimal display is 000.
2. As input pulses arrive, the units BCD counter advances one count per pulse. After nine pulses have occurred, the hundreds and tens BCD counters are still at zero and the units counter is at 9 (binary 1001). Thus, the decimal display reads 009.
3. On the tenth input pulse the units counter recycles to zero, causing its flip-flop D output to go from 1 to 0. This 1-to-0 transition acts as the clock input for the tens counter and causes it to advance one count. Thus, after 10 input pulses, the decimal readout is 010.
4. As additional pulses occur, the units counter advances one count per pulse, and each time the units counter recycles to zero, it advances the tens counter one count. Thus, after 99 input pulses have occurred, the tens counter is at 9, as is the units counter. The decimal readout is thus 099.
5. On the hundredth input pulse, the units counter recycles to zero, which in turn causes the tens counter to recycle to zero. The flip-flop D output of the tens counter thus makes a 1-to-0 transition which acts as the clock input for the hundreds counter and causes it to advance one count. Thus, after 100 pulses the decimal readout is 100.
6. This process continues up until 999 pulses. On the 1000th pulse, all the counters recycle back to zero.

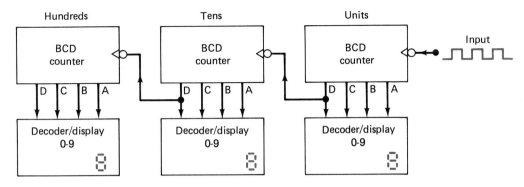

FIGURE 7-31   Cascading BCD counters to count and display numbers from 000 to 999.

It should be obvious that this arrangement can be expanded to any desired number of decimal digits simply by adding on more stages. For example, to count up to 999,999 will require six BCD counters and associated decoders and displays. In general, then, we need one BCD counter per decimal digit.

Incidentally, the BCD counters used in Figure 7-31 could be 74293s wired as MOD-10 counters, or they could be IC counters such as the 7490 or 74192 that are internally wired as BCD counters.

Discussion Question: Why won't the arrangement of Figure 7-31 work if the counters trigger on PGTs? How can it be modified to work with these counters?

7-14   SYNCHRONOUS COUNTER DESIGN*

Many different counter arrangements are available as ICs—asynchronous, synchronous, and combined asynchronous/synchronous. Most of these count in a normal binary sequence although their counting sequences can be somewhat altered using the methods we demonstrated for the 74293 and 74193 ICs. There are situations, however, where a counter is required that follows a sequence that is not counting binary; for example 000, 010, 101, 001, 110, 000, . . .

Several methods exist for designing counters that follow arbitrary sequences. We will present the details for one common method that uses J-K flip-flops in a synchronous counter configuration. The same method can be used in designs with D flip-flops. The technique is one of several design procedures that are part of an area of digital circuit design called **sequential circuit design** which is normally part of an advanced course.

Teaching Tip: An example of a nonbinary counter is one that counts through the Gray code sequence: 000, 001, 011, 010, 110, 111, 101, 100, 000, etc.

Basic Idea   In synchronous counters all of the FFs are clocked at the same time. Before each clock pulse, each J and K input of each FF in the counter must be at the correct level to ensure that each FF goes to the correct state. For example, consider the following situation:

Present State			Next State		
C	B	A	C	B	A
1	0	1	0	1	1

*This topic may be omitted without affecting the continuity of the remainder of the text.

When the next clock pulse occurs, the J and K inputs of the FFs must be at the correct levels that will cause flip-flop C to change from 1 to 0, flip-flop B from 0 to 1, and flip-flop A from 1 to 1 (i.e., no change).

The process of designing a synchronous counter, then, becomes one of designing the logic circuits that *decode* the various states of the counter to supply the logic levels to each J and K input. The inputs to these decoder circuits will come from the outputs of one or more of the FFs. To illustrate, for the synchronous counter of Figure 7-17, the AND gate that feeds the J and K inputs of flip-flop C decodes the states of flip-flops A and B. Likewise, the AND gate that feeds the J and K inputs of flip-flop D decodes the states of A, B, and C.

Teaching Tip: You may want to review the J-K flip-flop truth table here. Also stress that the excitation table is just another way to describe the J-K operation.

J-K Excitation Table

Before we begin the process of designing the decoder circuits for each J and K input, we must first review the operation of the J-K flip-flop using a different approach called an *excitation table*, shown below.

J-K Flip-flop Excitation Table

Transition at Output	PRESENT State $Q(N)$	NEXT State $Q(N + 1)$	J	K
$0 \rightarrow 0$	0	0	0	x
$0 \rightarrow 1$	0	1	1	x
$1 \rightarrow 0$	1	0	x	1
$1 \rightarrow 1$	1	1	x	0

The leftmost column of this table lists each possible FF output transition. The second and third columns list the FFs' present state, symbolized as $Q(N)$, and next state, symbolized as $Q(N + 1)$, for each transition. The last two columns list the J and K levels required to produce each transition. Let's examine each case.

$0 \rightarrow 0$ TRANSITION   The FF present state is at 0 and is to remain at 0 when a clock pulse is applied. From our understanding of how a J-K flip-flop works, this can happen when either $J = K = 0$ (no-change condition) or when $J = 0$ and $K = 1$ (clear condition). Thus, J has to be at 0, but K can be at either level. The table indicates this with a "0" under J and an "x" under K. Recall that "x" means "don't care" condition.

$0 \rightarrow 1$ TRANSITION   The present state is 0 and is to change to a 1. This can happen when either $J = 1$ and $K = 0$ (set condition) or $J = K = 1$ (toggle condition). Thus, J has to be a 1, but K can be at either level for this transition to occur.

$1 \rightarrow 0$ TRANSITION   The present state is 1 and is to change to a 0. This can happen when either $J = 0$ and $K = 1$ or $J = K = 1$. Thus, K has to be a 1 but J can be at either level.

$1 \rightarrow 1$ TRANSITION   The present state is a 1 and is to remain a 1. This can happen when either $J = K = 0$ or $J = 1$ and $K = 0$. Thus, K must be a 0 while J can be at either level.

The use of this excitation table is a principal part of the synchronous counter design procedure.

Design Procedure   We will now go through a complete synchronous counter design procedure. Although we will do it for a specific counting sequence, the same steps can be followed for any desired sequence.

> **Step 1:** Determine the desired number of bits (FFs) and the desired counting sequence.

For our example we will design a 3-bit counter that goes through the following sequence:

C	B	A
0	0	0
0	0	1
0	1	0
0	1	1
1	0	0
0	0	0
	etc.	

Notice that this sequence does not include the 101, 110, and 111 states. We will refer to these states as *undesired states*.

> **Step 2:** Draw the state transition diagram showing **all** possible states, including those that are not part of the desired counting sequence.

For our example, the state transition diagram appears as shown in Figure 7-32. The 000 through 100 states are connected in the expected sequence. The new idea used in this diagram is the inclusion of the undesired states. They have to be included in our design in case the counter accidentally gets into one of these states upon power-up or due to noise. The circuit designer can choose to have each of these undesired states go to any state upon the application of the next clock pulse. We choose to have them all go to the 000 state from where the correct sequence will be generated.

Teaching Tip: A more sophisticated design procedure would choose the next state of each undesired state so as to minimize the decoding logic. The only requirement would be that each undesired state eventually lead to one of the desired states.

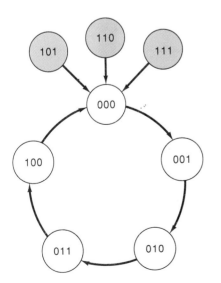

FIGURE 7-32   State transition diagram for synchronous counter design example.

Step 3: Use the state transition diagram to set up a table that lists **all** PRESENT states and their NEXT states.

For our example this table looks like this:

	PRESENT State C B A			NEXT State C B A		
line 1	0	0	0	0	0	1
2	0	0	1	0	1	0
3	0	1	0	0	1	1
4	0	1	1	1	0	0
5	1	0	0	0	0	0
6	1	0	1	0	0	0
7	1	1	0	0	0	0
8	1	1	1	0	0	0

The left-hand portion of the table lists *every* possible state, even those that are not part of the sequence. We label these as the PRESENT states. The right-hand portion lists the next state for each PRESENT state. These are obtained from the state transition diagram in Figure 7-32. For instance, line 1 shows that the PRESENT state of 000 has the NEXT state of 001, and line 5 shows the PRESENT state of 100 has the NEXT state of 000. Lines 6, 7, and 8 show that the undesired PRESENT states 101, 110, and 111 all have the NEXT state of 000.

Step 4: Add a column to this table for each J and K input. For each PRESENT state, indicate the levels required at each J and K input in order to produce the transition to the NEXT state.

Our design example uses three FFs—C, B, and A—and each one has a J and a K input. Therefore we have to add six new columns as shown below. This completed table is called the **circuit excitation table.**

Circuit Excitation Table

	PRESENT State C B A			NEXT State C B A			J_C	K_C	J_B	K_B	J_A	K_A
line 1	0	0	0	0	0	1	0	x	0	x	1	x
2	0	0	1	0	1	0	0	x	1	x	x	1
3	0	1	0	0	1	1	0	x	x	0	1	x
4	0	1	1	1	0	0	1	x	x	1	x	1
5	1	0	0	0	0	0	x	1	0	x	0	x
6	1	0	1	0	0	0	x	1	0	x	x	1
7	1	1	0	0	0	0	x	1	x	1	0	x
8	1	1	1	0	0	0	x	1	x	1	x	1

The six new columns are the J and K inputs of each FF. The entries under each J and K are obtained by using the J-K flip-flop excitation table that we developed earlier. We will demonstrate this for several of the cases, and you can verify the rest.

Let's look at line 1. The PRESENT state of 000 is to go to the NEXT state of 001 on the occurrence of a clock pulse. For this state transition, the C flip-flop goes from 0 to 0. From the J-K excitation table, we see that J_C must be at 0 and K_C at "x" for this transition to occur. The B flip-flop also goes from 0 to 0, and so $J_B = 0$ and $K_B = x$. The A flip-flop goes from 0 to 1. From the excitation table, we see that $J_A = 1$ and $K_A = x$ for this transition.

In line 4, the PRESENT state of 011 has a NEXT state of 100. For this state transition, flip-flop C goes from 0 to 1, which requires $J_C = 1$ and $K_C = x$. Flip-flops B and A are both going from 1 to 0. The J-K excitation table indicates that these two FFs need $J = x$ and $K = 1$ for this to occur.

The required J and K levels for all other lines in the table can be determined in the same manner.

Step 5: Design the logic circuits to generate the levels required at each J and K input.

The circuit excitation table lists six J, K inputs—J_C, K_C, J_B, K_B, J_A, and K_A. We have to consider each of these as an output from its own logic circuit with inputs from flip-flops C, B, and A. Then we have to design the circuit for each one. Let's design the circuit for J_A.

To do this we need to look at the PRESENT states of C, B, and A and the desired levels at J_A for each case. This information has been extracted from the circuit excitation table and presented in Figure 7-33(a). This truth table shows the levels required at J_A for each PRESENT state. Of course, for some of the cases J_A is a "don't care." To develop the logic circuit for J_A, we must first determine its expression in terms of C, B, and A. We will do this by transferring the truth table information to a three-variable Karnaugh map and performing the K-map simplification as in Figure 7-33(b).

There are only two 1s in this K map, and they can be looped to obtain the term $\overline{AC}$, but if we use the don't care conditions at $A\overline{B}\overline{C}$ and $AB\overline{C}$ as 1s we can loop a quad to obtain the simpler term $\overline{C}$. Thus, the final expression is

$$J_A = \overline{C}$$

Now let's consider K_A. We can follow the same steps as we did for J_A. How-

Teaching Tip: The choice of K-map simplification over algebraic simplification is because the don't care states are more easily handled with K mapping. You might want to show how difficult it can be to use the algebraic method to arrive at the minimum expression for K_B.

FIGURE 7-33   (a) Portion of circuit excitation table showing J_A for each PRESENT state; (b) K map used to obtain simplified expression for J_A.

PRESENT			
C	B	A	J_A
0	0	0	1
0	0	1	x
0	1	0	1
0	1	1	x
1	0	0	0
1	0	1	x
1	1	0	0
1	1	1	x

(a)

(b)

SECTION 7-14 / SYNCHRONOUS COUNTER DESIGN   **321**

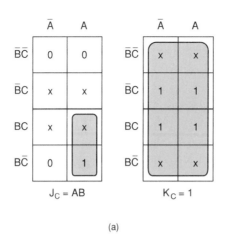

$$J_C = AB \qquad\qquad K_C = 1$$

(a)

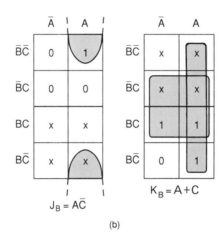

$$J_B = A\bar{C} \qquad\qquad K_B = A + C$$

(b)

FIGURE 7-34 (a) K mappings for the J_C and K_C logic circuits; (b) K mappings for the J_B and K_B logic circuits.

ever, a look at the entries under K_A in the circuit excitation table shows only 1s and don't cares. If we change all the don't cares to 1s, then K_A is always a 1. Thus, the final expression is

$$\mathbf{K_A = 1}$$

In a similar manner, we can derive the expressions for J_C, K_C, J_B, and K_B. The K maps for these expressions are given in Figure 7-34. You might want to confirm their correctness by checking them against the circuit excitation table.

Step 6: Implement the final expressions.

The decoding circuits for each J and K input are implemented from the expressions obtained from the K mapping. The complete synchronous counter design is implemented in Figure 7-35. Note that all FFs are clocked in parallel. You might want to verify that the logic for each J and K input is correct.

Final Comments   We have seen the complete design procedure for a synchronous counter using J-K flip-flops. This same procedure can be used for counters of any number of bits and any arbitrary sequence. The only restriction on the sequence is that it cannot contain the same state more than once within one complete cycle before repeating. For example, the sequence 000, 001, 011, 001, 100 will not

Teaching Tip: You may want to have the students verify the counting sequence and/or verify that an undesired starting state will always go to 000 on the next pulse.

FIGURE 7-35   Final implementation of synchronous counter design example.

work because the 001 state appears more than once in the sequence. This should make sense, since our design method cannot handle a PRESENT state of 001 that has two different NEXT states.

Teaching Tip: To obtain a sequence such as this, one would require a fourth FF that would be made to be in a different state for the two occurrences of 101.

—— REVIEW QUESTIONS ——————————————

1. List the six steps in the procedure for designing a synchronous counter.
2. What information is contained in a J-K excitation table?
3. What information is contained in the circuit excitation table?
4. *True or false:* The synchronous counter design procedure can be used for the following sequence: 0010, 0011, 0100, 0111, 1010, 1110, 1111, and repeat.

7-15   SHIFT-REGISTER COUNTERS

Shift registers can be arranged to form several types of counters. All *shift-register counters* use *feedback*, whereby the output of the last FF in the shift register is in some way connected back to the first FF. The most widely used shift-register counters are the ring counter and the Johnson counter.

Ring Counter   The simplest shift-register counter is essentially a **circulating** shift register connected so that the last FF shifts its value into the first FF. This arrangement is shown in Figure 7-36 using D-type FFs (J-K flip-flops can also be used). The FFs are connected so that information shifts from left to right and back around from Q_0 to Q_3. In most instances only a single 1 is in the register and it is made to circulate around the register as long as clock pulses are applied. For this reason it is called a *ring counter*.

The waveforms, sequence table, and state diagram in Figure 7-36 show the various states of the FFs as pulses are applied, assuming a starting state of $Q_3 = 1$

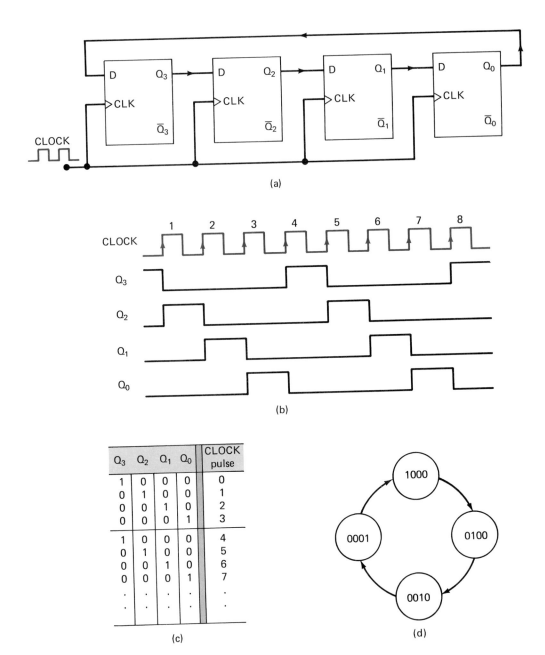

FIGURE 7-36 (a) Four-bit ring counter; (b) waveforms; (c) sequence table; (d) state diagram.

and $Q_2 = Q_1 = Q_0 = 0$. After the first pulse, the 1 has shifted from Q_3 to Q_2 so that the counter is in the 0100 state. The second pulse produces the 0010 state, and the third pulse produces the 0001 state. On the *fourth* clock pulse the 1 from Q_0 is transferred to Q_3, resulting in the 1000 state, which is, of course, the initial state. Subsequent pulses cause the sequence to repeat.

This counter functions as a MOD-4 counter, since it has *four* distinct states before the sequence repeats. Although this circuit does not progress through the normal binary counting sequence, it is still a counter because each count corresponds to a particular state of the FFs. Note that each FF output waveform has a frequency

equal to one-fourth of the clock frequency, since this is a MOD-4 ring counter.

Ring counters can be constructed for any desired MOD number; a MOD-N ring counter uses N flip-flops connected in the arrangement of Figure 7-36. In general, a ring counter will require more FFs than a binary counter for the same MOD number; for example, a MOD-8 ring counter requires eight FFs, while a MOD-8 binary counter requires only three.

Despite the fact that it is less efficient in the use of FFs, a ring counter is still useful because it can be decoded without the use of decoding gates. The decoding signal for each state is obtained at the output of its corresponding FF. Compare the FF waveforms of the ring counter with the decoding waveforms in Figure 7-26. In some cases a ring counter might be a better choice than a binary counter with its associated decoding gates. This is especially true in applications where the counter is being used to control the sequencing of operations in a system.

Starting a Ring Counter

To operate properly, a ring counter must start off with only one FF in the 1 state and all the other FFs at 0. When power is first applied to the circuit, there is only a remote possibility that the FFs will come up in such a state. Thus, it is necessary to preset the counter to the required starting state before clock pulses are applied. A simple way to accomplish this is to apply a momentary pulse to the PRESET input of one of the FFs and to the CLEAR inputs of all the others. This will place a single 1 into the ring counter.

Discussion Question: Analyze what would happen if the ring counter started in the 1010 state. What would be the MOD number?

Johnson Counter

The basic ring counter can be modified slightly to produce another type of shift-register counter, which will have somewhat different properties. The *Johnson* or *twisted-ring counter* is constructed exactly like a normal ring counter except that the *inverted* output of the last FF is connected to the input of the first FF. A 3-bit Johnson counter is shown in Figure 7-37. Note that the $\overline{Q_0}$ output is connected back to the D input of Q_2. This means that the *inverse* of the level stored in Q_0 will be transferred to Q_3 on the clock pulse.

The Johnson-counter operation is easy to analyze if we realize that on each positive clock-pulse transition the level at Q_2 shifts into Q_1, the level at Q_1 shifts into Q_0, and the *inverse* of the level at Q_0 shifts into Q_2. Using these ideas and assuming that all FFs are initially 0, the waveforms, sequence table, and state diagram of Figure 7-37 can be generated.

Examination of the waveforms and sequence table reveals the following important points:

1. This counter has six distinct states: 000, 100, 110, 111, 011, and 001 before it repeats the sequence. Thus, it is a MOD-6 Johnson counter. Note that it does not count in a normal binary sequence.
2. The waveform of each FF is a square wave (50 percent duty cycle) at one-sixth the frequency of the clock. In addition, the FF waveforms are shifted by one clock period with respect to each other.

Discussion Question: Analyze what would happen if the Johnson counter started out in the 101 state. What is its MOD number?

The MOD number of a Johnson counter will always be equal to *twice* the number of FFs. For example, if we connect five FFs in the arrangement of Figure 7-37, the result is a MOD-10 Johnson counter, where each FF output waveform is a squarewave at one-tenth the clock frequency. Thus it is possible to construct a MOD-N counter (where N is an even number) by connecting $N/2$ flip-flops in a Johnson-counter arrangement.

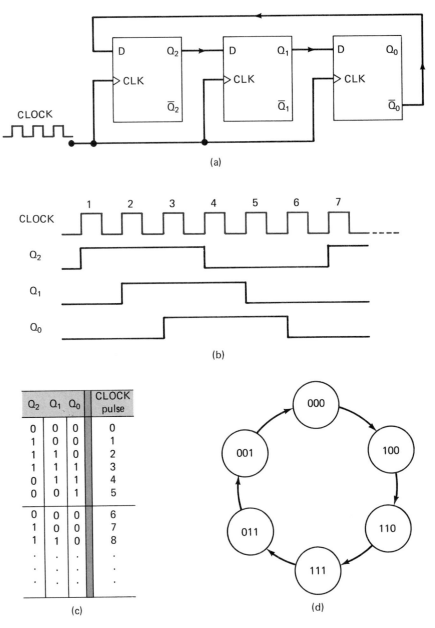

FIGURE 7-37 (a) MOD-6 Johnson counter; (b) waveform; (c) sequence table; (d) state diagram.

Decoding a Johnson Counter   For a given MOD number, a Johnson counter requires only half the number of FFs that a ring counter requires. However, a Johnson counter requires decoding gates whereas a ring counter does not. As in the binary counter, the Johnson counter uses one logic gate to decode for each count, but each gate requires only two inputs, regardless of the number of FFs in the counter. Figure 7-38 shows the decoding gates for the six states of the Johnson counter of Figure 7-37.

Notice that each decoding gate has only two inputs, even though there are three FFs in the counter. This is because for each count, two of the three FFs are in a

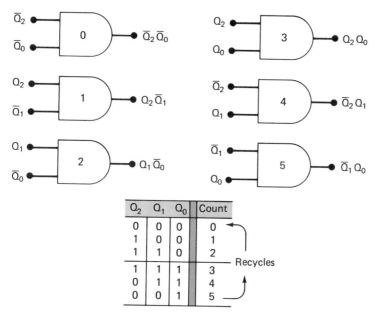

Q_2	Q_1	Q_0	Count
0	0	0	0
1	0	0	1
1	1	0	2
1	1	1	3
0	1	1	4
0	0	1	5

Recycles

FIGURE 7-38   Decoding logic for MOD-6 Johnson counter.

unique combination of states. For example, the combination $Q_2 = Q_0 = 0$ occurs only once in the counting sequence, at the count of 0. Thus, AND gate 0 with inputs $\overline{Q}_2$ and $\overline{Q}_0$ can be used to decode for this count. This same characteristic is shared by all the other states in the sequence, as the reader can verify. In fact, for *any size* Johnson counter, the decoding gates will have only two inputs

Johnson counters represent a middle ground between ring counters and binary counters. A Johnson counter requires fewer FFs than a ring counter but generally more than a binary counter; it has more decoding circuitry than a ring counter but less than a binary counter. Thus, it sometimes represents a logical choice for certain applications.

IC Shift-Register Counters   There are very few ring counters or Johnson counters available as ICs. The reason is that it is relatively simple to take a shift-register IC and to wire it as either a ring or Johnson counter. Some of the CMOS Johnson-counter ICs (74HC4017, 74HC4022) include the complete decoding circuitry on the same chip as the counter.

—— REVIEW QUESTIONS ——

1. Which shift-register counter requires the most FFs for a given MOD number?
2. Which shift-register counter requires the most decoding circuitry?
3. How can a ring counter be converted to a Johnson counter?
4. *True or false:*
   (a) The outputs of a ring counter are always square waves.
   (b) The decoding circuitry for a Johnson counter is simpler than for a binary counter.
   (c) Ring and Johnson counters are synchronous counters.

There are numerous applications for the many types of counters we have been discussing. In this section and the next we will look at two representative applications that illustrate the uses of counters in digital systems.

A *frequency counter* is a circuit that can measure and display the frequency of a signal. One of the most straightforward methods for constructing a frequency counter is shown in Figure 7-39(a) in simplified form. It contains a counter with its associated decoder/display circuitry and an AND gate. The AND gate inputs include the pulses with unknown frequency, f_x, and a SAMPLE pulse that controls how long these pulses are allowed to pass through the AND gate into the counter. The counter is usually made up of cascaded BCD counters (Figure 7-31), and the de-

FIGURE 7-39   Basic frequency-counter method.

(a)

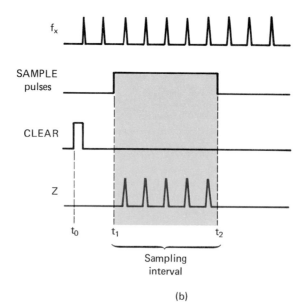

(b)

coder/display unit converts the BCD outputs into a decimal display for easy monitoring.

The waveforms in Figure 7-39(b) show that a CLEAR pulse is applied to the counter at t_0 to start the counter at zero. Prior to t_1 the SAMPLE pulse waveform is LOW, and so the AND output, Z, will be LOW and the counter will not be counting. The SAMPLE pulse goes HIGH from t_1 to t_2; this is called the **sampling interval.** During this sampling interval the unknown frequency pulses will pass through the AND gate and will be counted by the counter. After t_2 the AND output returns LOW and the counter stops counting. Thus, the counter will have counted the number of pulses that occurred during the sampling interval, and the resulting contents of the counter are a direct measure of the frequency of the pulse waveform.

—— EXAMPLE 7-18 ——

The unknown frequency is 3792 pulses per second (pps). The counter is cleared to the zero state prior to t_1. Determine the counter reading after a sampling interval of (a) 1 s, (b) 0.1 s, and (c) 10 ms.

Solution

(a) Within a sampling interval of 1 s there will be 3792 pulses entering the counter, and so after t_2 the contents of the counter will read 3792.

(b) With a 0.1-s sampling interval the number of pulses passing through the AND gate into the counter will be 3792 pulses/second $\times$ 0.1 s = 379.2. This means that either 379 or 380 pulses will be counted, depending on what part of a pulse cycle t_1 occurs in.

(c) With a 10-ms = 0.01-s sampling interval, the counter will read either 37 or 38.

The accuracy of this method depends almost entirely on the duration of the sampling interval, which must be very accurately controlled. A commonly used method for obtaining very accurate sample pulses is shown in Figure 7-40. A crystal-controlled oscillator is used to generate a very accurate 100-kHz waveform, which is shaped into square pulses and fed to a series of decade counters that are being used to successively divide this 100-kHz frequency by 10. The frequencies at the outputs of each decade counter are as accurate (percentagewise) as the crystal frequency.

The switch is used to select one of the decade-counter output frequencies to be fed to a single FF to be divided by 2. For example, in switch position 1 the 1-Hz pulses are fed to flip-flop Q, which is acting as a toggle FF so that its output will be a square wave with a period of $T = 2$ s and a pulse duration of $t_p = T/2 = 1$ s. This pulse duration is the desired 1-s sampling interval. In position 2 the sampling interval would be 0.1 s, and so on for the other positions.

—— EXAMPLE 7-19 ——

Assume that the counter in Figure 7-39 is made up of three cascaded BCD counters and their associated displays. If the unknown input frequency is between 1 and 9.99 kpps, what is the best setting for the switch position in Figure 7-40?

Solution   With three BCD counters the total capacity of the counter is 999. A 9.99 kpps frequency would produce a count of 999 if a 0.1-s sample interval were

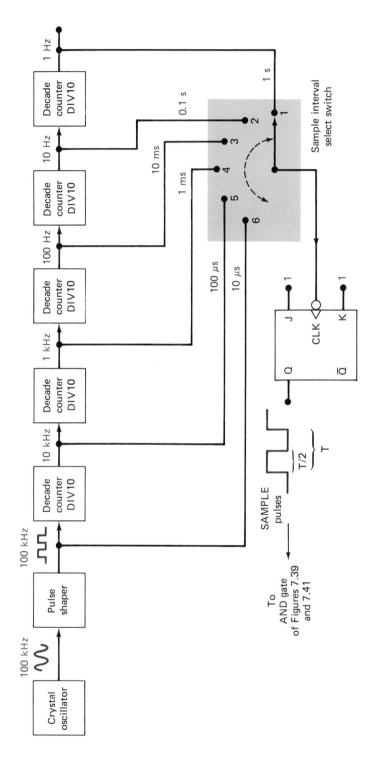

FIGURE 7-40 Method for obtaining accurate sampling intervals for frequency counter.

used. Thus, in order to use the full capacity of the counter, the switch should be set to position 2. If a 1-s sampling interval were used, the counter capacity would always be exceeded for frequencies in the specified range. If a shorter sample interval were used, the counter would count only between 0 and 99; this would give a reading to only two significant figures and would be a waste of the counter's capacity.

Complete Frequency Counter   We will now look at a more complete frequency-counter circuit in Figure 7-41(a). The circuit now contains a one-shot and a J-K flip-flop operating in the toggle mode, and the AND gate has three inputs, one of

FIGURE 7-41   Frequency counter.

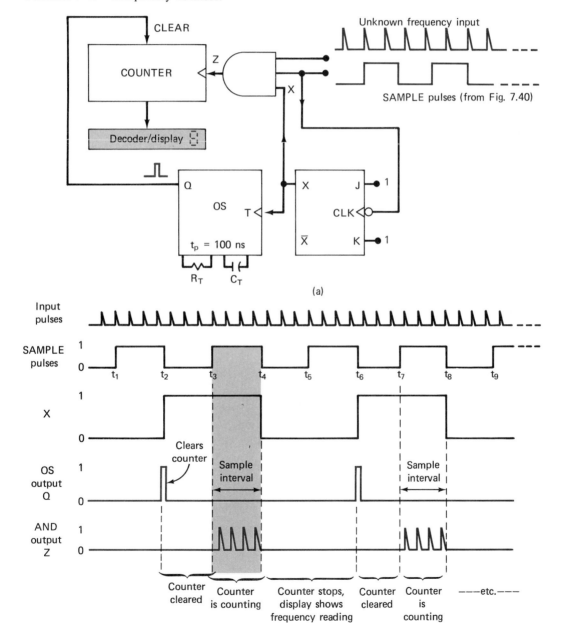

which is the FF output X. The SAMPLE pulses are connected to the AND gate and also to the *CLK* input of the FF. These SAMPLE pulses would be generated from a circuit such as that in Figure 7-40. The following step-by-step description refers to the waveforms in Figure 7-41(b).

1. Assume that flip-flop X is in the 0 state (it has toggled to 0 on the falling edge of the previous sample pulse).
2. This LOW from X is fed to the AND gate, disabling its output, so that no pulses are fed to the counter even when the first SAMPLE pulse occurs between t_1 and t_2.
3. At t_2 the NGT of the first SAMPLE pulse toggles flip-flop X to the 1 state (note that $J = K = 1$). This positive transition at X triggers the OS, which generates a 100-ns pulse to *clear* the counter. The counter now displays *zero*.
4. At t_3 the second SAMPLE pulse enables the AND gate (since X is now 1) and allows the unknown frequency into the counter to be counted until t_4.
5. At t_4 the SAMPLE pulse returns LOW and toggles X LOW, disabling the AND gate. The counter stops counting.
6. Between t_4 and t_6 the counter holds and displays the count that it had reached at t_4. Note that the third SAMPLE pulse does not enable the AND gate, because flip-flop X is LOW.
7. At t_6 the NGT of the SAMPLE pulse toggles X HIGH and the operation follows the same sequence that began at t_2.

This frequency counter, then, goes through a repetitive sequence of clearing to zero, counting, holding for display, clearing to zero, counting, and so on. For example, let's assume that the counter has three BCD stages and a three-digit display. If we use a sample interval of 1 s and the unknown frequency is 237 pps, the counter and display will go through the following sequence over and over:

- Clear to zero and display zero for 1 s [t_2 to t_3 in Figure 7-41(b)].
- Starting at zero, count unknown frequency pulses during the 1-s sample interval (t_3 to t_4); the count will stop at 237.
- Hold and display the count of 237 for 2 s (t_4 to t_6).

Discussion Question: How long will the count be visible if the switch in Figure 7-40 is set at 10 ms?

Since the display is connected directly to the counter outputs, the display will show the clearing and counting action of the counter. This makes it very difficult to read the display to determine the unknown frequency except at very slow sample intervals. This problem can be solved by inserting a *buffer register* beetween the counter and the decoder/display unit. We will consider this feature in Problem 7-36.

——— REVIEW QUESTIONS ———

1. What is the best sample interval setting to use if the pulse counter has four BCD stages and the input frequency is between 2 and 8 Mpps?
2. Describe the sequence of operations of the complete frequency counter of Figure 7-41.

7-17 COUNTER APPLICATIONS: DIGITAL CLOCK

One of the most common applications of counters is the digital clock—a time clock which displays the time of day in hours, minutes, and sometimes seconds. In order to construct an accurate digital clock, a very closely controlled basic clock frequency is required. For battery-operated digital clocks (or watches) the basic frequency is normally obtained from a quartz-crystal oscillator. Digital clocks operated from the ac power line can use the 60-Hz power frequency as the basic clock frequency. In either case, the basic frequency has to be divided down to a frequency of 1 Hz or 1 pulse per second (pps). Figure 7-42 shows the basic block diagram for a digital clock operating from 60 Hz.

The 60-Hz signal is sent through a shaping circuit to produce square pulses at the rate of 60 pps. This 60-pps waveform is fed into a MOD-60 counter which is used to divide the 60 pps down to 1 pps. The 1-pps signal is fed into the SECONDS section, which is used to count and display seconds from 0 through 59. The BCD counter advances one count per second. After 9 seconds the BCD counter recycles to 0, which triggers the MOD-6 counter and causes it to advance one count. This continues for 59 s, at which point the MOD-6 counter is at the 101 (5) count and the BCD counter is at 1001 (9), so that the display reads 59 s. The next pulse recycles the BCD counter to 0, which in turn recycles the MOD-6 counter to 0 (remember: the MOD-6 counts from 0 through 5).

The output of the MOD-6 counter in the SECONDS section has a frequency of 1 pulse per minute (the MOD-6 recycles every 60 s). This signal is fed to the MINUTES section, which counts and displays minutes from 0 through 59. The MINUTES section is identical to the SECONDS section and operates in exactly the same manner.

The output of the MOD-6 counter in the MINUTES section has a frequency of 1 pulse per hour (the MOD-6 recycles every 60 min). This signal is fed to the HOURS section, which counts and displays hours from 1 through 12. This HOURS section is different from the SECONDS and MINUTES sections in that it never goes to the zero state. The circuitry in this section is sufficiently unusual to warrant a closer investigation.

Figure 7-43 shows the detailed circuitry contained in the HOURS section. It includes a BCD counter to count units of hours, and a single FF (MOD-2) to count tens of hours. The BCD counter is a 74192, which operates exactly like the 74193 that we studied earlier except that it counts only between 0000 and 1001. In other words, the 74192 can either count up in BCD fashion (i.e., 0 to 9, recycling to 0) or count down in BCD fashion (i.e., 9 to 0, recycling to 9). Here it is used to count up in response to the 1-pulse/hour signal coming from the MINUTES section. The INVERTER on the CP_U input is needed because the 74192 responds to PGTs, and we want it to respond to the NGT that occurs when the MINUTES section recycles back to zero.

The incoming pulses advance the BCD counter once per hour. For example, at 7 o'clock this counter will be at 0111, and its decoder/display circuitry will display the numeral 7. At the same time, X will be LOW and its display will show a zero. Thus, the two displays will show "07." When the BCD counter is in the 1001 (9) state and the next input pulse occurs, it will recycle back to 0000. The NGT at Q_3 will toggle flip-flop X from 0 to 1. This produces a numeral 1 on the X display and a

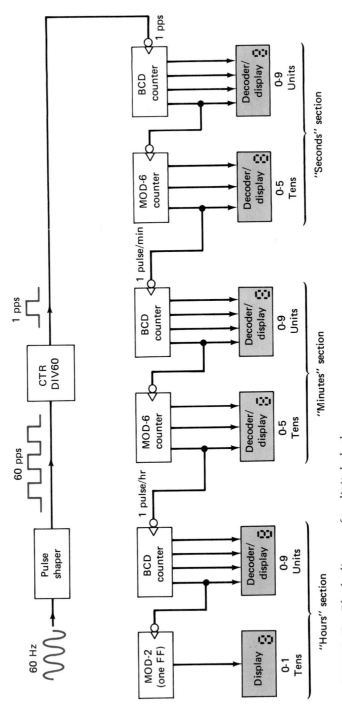

FIGURE 7-42 Block diagram for digital clock.

FIGURE 7-43 Detailed circuitry for the hours section.

numeral 0 on the BCD display so that the combined displays show "10" for 10 o'clock.

The next two pulses advance the BCD counter so that "11" and "12" are displayed at 11 o'clock and 12 o'clock, respectively. The next pulse advances the BCD counter to 0011 (3). In this state, the counter's Q_1 and Q_0 outputs are both HIGH, and X is still HIGH. Thus, the NAND gate output goes LOW and activates the DC CLEAR of flip-flop X and the $\overline{PL}$ input of the 74192. This clears X to 0 and presets the BCD counter to 0001. The result is a display of "01" for 1 o'clock. Several of the end-of-chapter problems will provide more details on the digital clock circuit.

REVIEW QUESTIONS

1. Name the basic blocks that make up a digital clock circuit.
2. Why is an INVERTER needed in Figure 7-43?

Teaching Tip: It might be interesting to show what the effect would be if the INVERTER were not used.

7-18 INTEGRATED-CIRCUIT REGISTERS

The various types of registers can be classified according to the manner in which data can be entered into the register for storage and the manner in which data are outputted from the register. The various classifications are listed below.

1. Parallel in/parallel out

2. Serial in/serial out

3. Parallel in/serial out

4. Serial in/parallel out

Each of these types is available in IC form so that a logic designer can usually find exactly what is required for a given application. In the following sections we will examine a representative IC from each of the above categories.

7-19 PARALLEL IN/PARALLEL OUT— THE 74174 AND 74178

There are actually two types of registers in this category; one is strictly parallel, the other is actually a shift register that can be loaded with parallel data and has parallel outputs available.

Discussion Question: Is it possible to wire the 74174 as a shift register?

The 74174 Figure 7-44(a) shows the logic diagram for the 74174 (also 74LS174 and 74HC174), a 6-bit register that has parallel inputs D_5 through D_0 and parallel outputs Q_5 through Q_0. Parallel data are loaded into the register on the PGT of the clock input CP. A master reset input $\overline{MR}$ can be used to asynchronously reset all the register FFs to 0.

The logic symbol for the 74174 is shown in Figure 7-44(b). This symbol is used in circuit diagrams to represent the circuitry of Figure 7-44(a).

The 74178 Figure 7-45(a) shows the logic diagram for the 74178 (also 74HC178), a 4-bit shift register that has parallel data entry (P_0 through P_3) and parallel outputs (Q_0 through Q_3). It has a serial data input, D_s, and two enable inputs: PE (parallel enable) and SE (serial enable).

Figure 7-45(b) is the mode-select table that describes the various modes of operation for this IC. The first entry gives the input conditions necessary for the shift-right operation. With $SE = 1$, the data will shift from left to right on the NGT of clock input $\overline{CP}$ regardless of the logic level at the PE input (recall that X represents the "don't care" condition). You can verify this by tracing through the logic diagram and noting that when $SE = 1$, the D_s input passes through the logic gates and appears at the D input of the Q_0 flip-flop. Likewise, Q_0 will appear at the D input of Q_1; Q_1 will appear at the D input of Q_2; and Q_2 will appear at the D input of Q_3.

The second entry in the table gives the conditions needed to produce parallel transfer from the parallel data inputs (P_0 through P_3) to the outputs (Q_0 through Q_3). When $SE = 0$ and $PE = 1$, this parallel transfer occurs on the NGT of $\overline{CP}$. Note that this is a *synchronous* transfer. Again, by tracing through the logic diagram, you

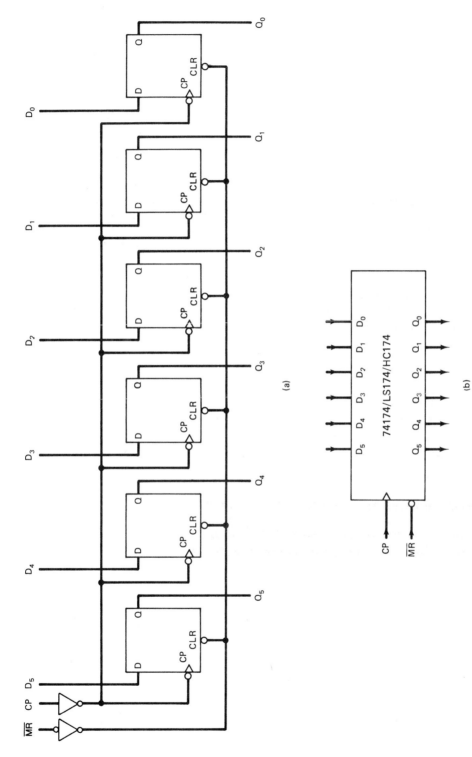

FIGURE 7-44 (a) Circuit diagram of 74174; (b) logic symbol. (Courtesy of Fairchild, a Schlumberger company)

337

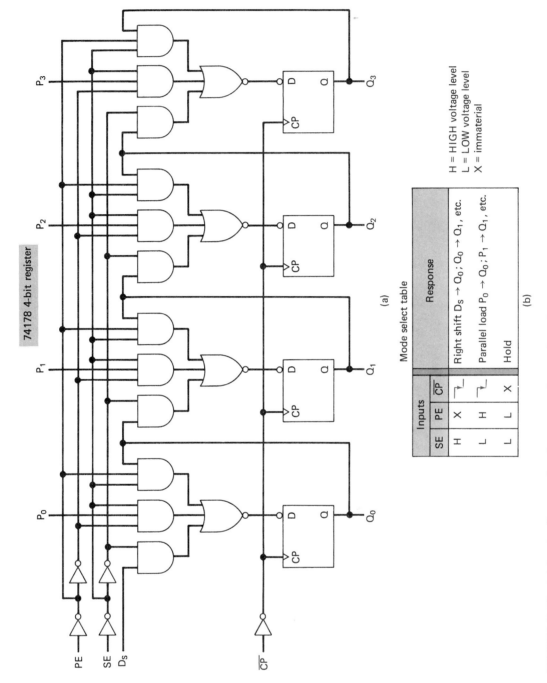

Mode select table

Inputs			Response
SE	PE	$\overline{CP}$	
H	X	⌐_	Right shift $D_S \rightarrow Q_0$; $Q_0 \rightarrow Q_1$, etc.
L	H	⌐_	Parallel load $P_0 \rightarrow Q_0$; $P_1 \rightarrow Q_1$, etc.
L	L	X	Hold

H = HIGH voltage level
L = LOW voltage level
X = immaterial

(b)

(a)

FIGURE 7-45   (a) Logic diagram for the 74178; (b) mode-select table. (Courtesy of Fairchild, a Schlumberger company)

FIGURE 7-46   Logic symbol for the 74178.

can see that when *SE* = 0 and *PE* = 1 the parallel data inputs will get through the logic gates and will appear at the *D* inputs of their respective FFs.

The final entry in the table indicates that the *SE* = 0, *PE* = 0 condition will cause the register FFs to hold their current levels regardless of what happens at the clock input. For this input condition, each FF output is allowed to pass through its respective logic gates and appear at the *D* input of the same FF. Thus, a NGT on $\overline{CP}$ will not change the state of the FF.

Figure 7-46 shows the logic symbol for the 74178/74HC178.

Teaching Tip: It would be instructive to trace D_S through the logic for *SE* = 1. Point out the bubble on the *D* inputs. This means that a LOW entering the bubble becomes a HIGH at *D* which will be transferred as a HIGH to *Q*. This is why a NOR gate is used.

━━ EXAMPLE 7-20 ━━━━━━━━━━━━━━━━━━━━━━━━━━━━━━━━━

Describe how the 74178 can be connected as a ring counter.

FIGURE 7-47   74178 wired as a ring counter.

15

Solution   The correct connections are shown in Figure 7-47. The Q_3 output is connected back around to the D_S input. The parallel data inputs are permanently set at 0001 and the *PE* input is connected permanently HIGH. The *SE* input is initially connected LOW. This allows each NGT at $\overline{CP}$ to parallel-load the shift register with 0001. Then, *SE* is switched to the HIGH state (shift mode), where it remains as the NGTs at $\overline{CP}$ cause the single 1 to shift and recirculate through the register.

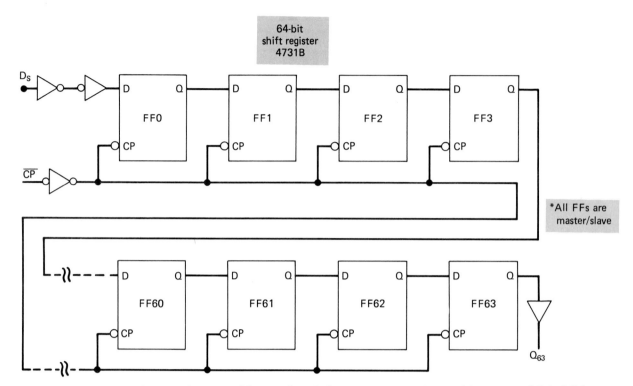

FIGURE 7-48   Logic diagram for one of four 64-bit shift registers on a 4731B. (Courtesy of Fairchild, a Schlumberger company)

7-20   SERIAL IN/SERIAL OUT—THE 4731B

The 4731B is a CMOS *quad* 64-bit shift register. It contains four identical 64-bit shift registers on one chip. Figure 7-48 shows the logic diagram for one of the 64-bit registers. It has a serial input, D_S, a clock input $\overline{CP}$ that responds to NGTs, and a serial output from the last FF, Q_{63}. This is the only output that is externally accessible. Note that this output goes through a *buffer* circuit (triangle symbol with no inversion bubble). A buffer does not change the signal's logic level; it is used to provide a greater output-current capability than normal. Also note that there is no means for parallel data entry into the register FFs.

_____ EXAMPLE 7-21 _____

How can the 4731B chip be wired as a single 256-bit shift register?

FIGURE 7-49   The four 64-bit shift registers on the 4731B chip can be connected together as a 256-bit register.

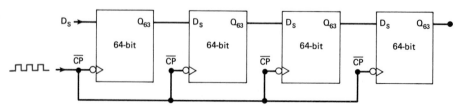

Solution   The four shift registers can be strung together as one long register by connecting the Q_{63} output of one to the D_S input of the next, and by connecting all $\overline{CP}$ inputs together as a common clock input. This is shown in Figure 7-49, where a simplified logic symbol for each shift register is used for convenience.

7-21   PARALLEL IN/SERIAL OUT—
THE 74165/74LS165/74HC165

The 74165, an 8-bit register, is shown in Figure 7-50(a). It actually has both serial data entry via D_S and parallel data entry via P_0 through P_7. The only accessible FF outputs are Q_7 and $\overline{Q}_7$.

Teaching Tip: Note the interesting way that the asynchronous inputs are wired. You may want to go through how this works.

   Note that the FFs are the clocked *SR* type (same as clocked *SC*) that respond to NGTs at their clock inputs. Each FF has asynchronous inputs, PRESET and *CL* (same as CLEAR), that are used for parallel data entry.

   Also note that there are two clock inputs, CP_1 and CP_2, either of which can be used to produce the shift operation on PGTs.

_____ EXAMPLE 7-22 _____

Determine the necessary input conditions if pulses at CP_1 are to produce the shift-right operation.

Solution   In order for the CP_1 pulses to get through the AND gate, the $\overline{PL}$ input has to be HIGH. The CP_1 pulses out of the AND gate will be inverted by the NOR gate only if the other NOR input is LOW. Thus, CP_2 must be held LOW. The inverted CP_1 pulses will trigger the FFs. The FFs trigger on the NGT of the inverted CP_1 pulses, which means that they trigger on the PGT of the pulses applied to the CP_1 input.

_____ EXAMPLE 7-23 _____

How can the 74165 be placed in the hold mode so that the register contents do not change as pulses are applied to CP_1?

Solution   If CP_2 is kept HIGH while $\overline{PL}$ is HIGH, the bottom AND gate output will be HIGH and will keep the NOR output at a constant LOW regardless of the CP_1 input.

7-22   SERIAL IN/PARALLEL OUT—
THE 74164/74LS164/74HC164

The logic diagram for the 74164 is shown in Figure 7-51(a). It is an 8-bit shift register with each FF output externally accessible. Instead of a single serial input, an AND gate combines inputs *A* and *B* to produce the serial input to flip-flop Q_0.

   The shift operation occurs on the PGTs of the clock input *CP*. The $\overline{MR}$ input provides asynchronous resetting of all FFs on a LOW level.

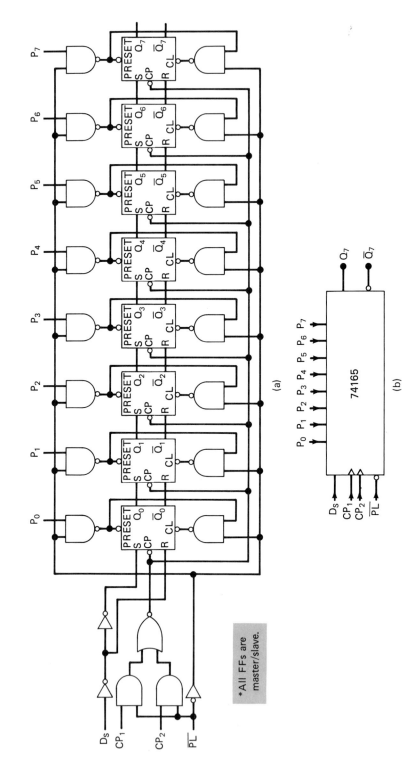

FIGURE 7-50   (a) Logic diagram of the 74165; (b) logic symbol. (Courtesy of Fairchild, a Schlumberger company)

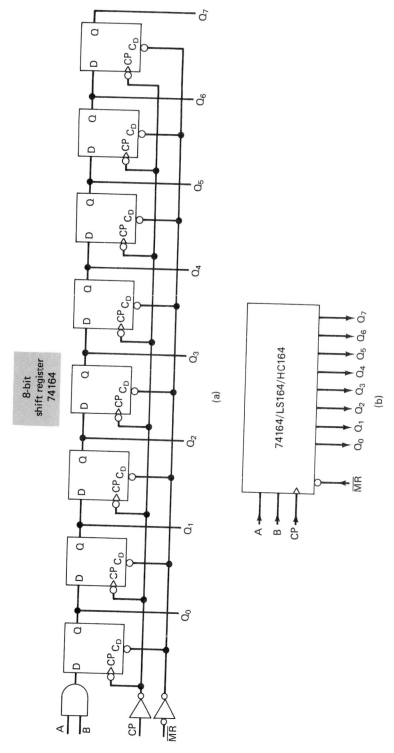

FIGURE 7-51   (a) Logic diagram for the 74164; (b) logic symbol. (Courtesy of Fairchild, a Schlumberger company)

343

Discussion Question: How can the 74164 be wired as a Johnson counter?

EXAMPLE 7-24

Assume that the initial contents of the 74164 register in Figure 7-52(a) are 00000000. Determine the sequence of states as clock pulses are applied.

(a)

Input pulse number	Q_0	Q_1	Q_2	Q_3	Q_4	Q_5	Q_6	Q_7	
0	0	0	0	0	0	0	0	0	
1	1	0	0	0	0	0	0	0	
2	1	1	0	0	0	0	0	0	
3	1	1	1	0	0	0	0	0	
4	1	1	1	1	0	0	0	0	Recycles
5	1	1	1	1	1	0	0	0	
6	1	1	1	1	1	1	0	0	
7	1	1	1	1	1	1	1	0	
8	1	1	1	1	1	1	1	1	

Temporary state

(b)

FIGURE 7-52   Example 7-24.

Solution   The correct sequence is given in Figure 7-52(b). With $A = B = 1$, the serial input is 1, so that 1s will shift into the register on each PGT of *CP*. Since Q_7 is initially at 0, the $\overline{MR}$ input is inactive.

On the eighth pulse, the register tries to go to the 11111111 state as the 1 from Q_6 shifts into Q_7. This state occurs only momentarily because $Q_7 = 1$ produces a LOW at $\overline{MR}$ that immediately resets the register back to 00000000. The sequence is then repeated on the next eight clock pulses.

The following is a list of some other register ICs that are variations on those already presented.

74194/LS194/HC194. This is a 4-bit *bidirectional universal shift register* IC which can perform shift-left, shift-right, parallel in, and parallel out operations. These operations are selected by a 2-bit mode select code applied as inputs to the device. Problem 7-47 will provide you with a chance to find out more about this versatile chip.

74373/LS373/HC373/HCT373. This is an 8-bit (octal) register containing eight *D* latches with *tristate* outputs. A tristate output is a special type of logic circuit output that allows device outputs to be safely tied together. We will introduce the characteristics of tristate outputs in the next chapter.

74374/LS374/HC374/HCT374. This is an 8-bit (octal) register containing eight edge-triggered D flip-flops with tristate outputs.

The IC registers that have been presented here are representative of the various types that are commercially available. Although there are many variations on these basic registers, most of them should now be relatively easy to understand from the manufacturers' data sheets.

We will present several register applications in the end-of-chapter problems and in the material covered in subsequent chapters.

REVIEW QUESTIONS

1. What kind of register can have a complete binary number loaded into it in one operation, and then have it shifted out one bit at a time?
2. *True or false:* A serial in/parallel out register can have all of its bits displayed at one time.
3. What type of register can have data entered into it only one bit at a time, but has all data bits available as outputs?
4. In what type of register do we have access only to the leftmost and rightmost FFs?

7-23 IEEE/ANSI REGISTER SYMBOLS*

We will present two examples of the IEEE/ANSI symbols for register ICs. First, let's consider the 74174 parallel-in/parallel-out IC whose internal logic and traditional logic symbol was shown in Figure 7-44. The IEEE/ANSI symbol for the 74174 is given in Figure 7-53(a). Its outline consists of the notched common-control block and the six narrow rectangles representing the six FFs.

The common-control block has the inputs that are common to all of the elements in the IC; in this case, the $\overline{MR}$ and *CP* inputs are common to the six FFs Q_0 through Q_5 that make up the register. The internal label for the $\overline{MR}$ input is shown as an R to indicate that its function is to *reset* each FF. The internal label for the *CP* input is C1, which tells us that this input controls the entry of data into any storage element that has a prefix of 1 in its input label. Each FF's *D* input has an internal label of 1D (shown only for Q_0, but assumed the same for each FF). The "1" in C1 and 1D establishes the dependency of the flip-flop *D* inputs on the common clock input CP.

The IEEE/ANSI symbol for the 74164 serial-in/parallel-out shift register is presented in Figure 7-53(b). Its outline consists of the common-control block and the eight FFs that make up the register. The notation SRG 8 identifies this IC as being an 8-bit shift register.

---

*This section may be omitted without affecting the continuity of the remainder of the book.

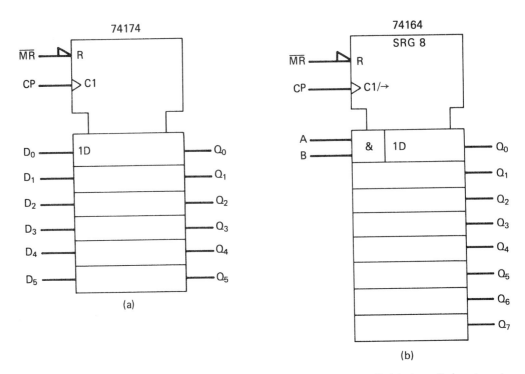

FIGURE 7-53   IEEE/ANSI symbols for the (a) 74174 parallel-in/parallel-out regis-
ter and (b) 74164 serial-in/parallel-out shift register.

The *CP* input has the internal label C1/→. The slash (/) is used to separate the
two functions C1 and → performed by this input. The C1 indicates that *CP* controls
the data entry into flip-flop Q_0 since Q_0 has the input label 1D. Note that the data bit
entered into Q_0 is indicated as the AND combination of inputs A and B. Also note
that since there are no external data inputs to Q_1 through Q_7, *CP* does not control
data entry into these FFs. The → denotes that the active transition of *CP* will pro-
duce the shift-right operation (from Q_0 toward Q_7).

——— REVIEW QUESTIONS ———

1. What does the slash (/) mean when it appears in an input label?
2. What notation would be used to describe the function performed by one of
   the registers on the 4731B IC of Figure 7-48?

7-24   TROUBLESHOOTING

Flip-flops, counters, and registers are the major components in **sequential logic sys-
tems.** A sequential logic system, because of its storage devices, has the characteris-
tic that its outputs and sequence of operations depend on both the present inputs and
the inputs that occurred earlier. Even though sequential logic systems are generally
more complex than combinatorial logic systems, the essential procedures for trou-
bleshooting apply equally well to both types of systems. Sequential systems suffer
from the same types of failures (open circuits, shorts, internal IC faults, and the
like) as do combinatorial systems.

Many of the same steps used to isolate faults in a combinatorial system can be applied to sequential systems. One of the most effective troubleshooting techniques begins with the troubleshooter observing the system operation, and, by analytical reasoning, determining the possible causes of the system malfunction. Then he or she uses available test instruments to isolate the exact fault. The following examples will show the kinds of analytical reasoning that are used in troubleshooting sequential systems. After studying these examples, you should be ready to tackle the troubleshooting problems at the end of the chapter.

_____ EXAMPLE 7-25 _____

Figure 7-54(a) shows a 74293 wired as a MOD-10 counter. A technician tests the counter operation by applying a 1-kHz clock signal and observing the Q outputs with an oscilloscope. The displayed waveforms are shown in Figure 7-54(b). Determine the possible causes for the incorrect circuit behavior.

(a)

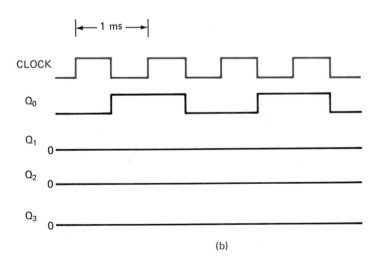

(b)

FIGURE 7-54   Example 7-25.

Solution   The waveforms show that Q_0 is toggling in response to the NGTs of the clock, but all other FFs are stuck in the LOW state. There are several possible faults that could produce this operation.

1. The Q_1 output is internally or externally shorted to ground. Referring to Figure 7-8(a), we can see that this would prevent Q_2 and Q_3 from toggling, since Q_1 is the clock signal for Q_2 and Q_2 is the clock signal for Q_3.

2. The connection from Q_0 to $\overline{CP_1}$ is open, so that Q_1 receives no clock signal.

3. There is an internal fault in the IC that prevents Q_1 from toggling.

—— EXAMPLE 7-26 ——————————————————————

After analyzing the situation described in Example 7-25, the technician proceeds to isolate the fault. He performs ohmmeter checks and verifies that Q_1 is not shorted to ground and that Q_0 is connected to $\overline{CP_1}$. This eliminates the first two possible faults. Concluding that the IC is bad, he replaces it. To his surprise, the circuit operation exhibits the same symptoms. Scratching his head, he decides to take a closer look at the FF waveforms by displaying them using a 10-ns/cm time scale. On this scale he can see a very narrow glitch occurring on the Q_1 signal at the time when Q_0 makes a NGT (see Figure 7-55). What is the probable fault?

FIGURE 7-55   Example 7-26.

Solution   When the counter is operating correctly there should be a glitch at Q_1 when the counter goes to the 1010 (ten) state, at which point the HIGHs at Q_3 and Q_1 cause the MR inputs to clear the count back to 0000. The waveforms in Figure 7-55, however, do not show Q_3 HIGH when the glitch at Q_1 occurs. The most probable fault is an open connection at MR_2, since this would be interpreted as a constant logic HIGH by the TTL integrated circuit. Thus, as soon as Q_1 goes HIGH, the MR inputs are both HIGH and the counter resets to 0000.

—— EXAMPLE 7-27 ——————————————————————

A technician tests the frequency counter of Figure 7-41 for various settings of the sample interval and for different unknown input frequencies. In all cases she finds that the displayed frequency is exactly *twice* what it should be. What is the probable cause of the malfunction?

Solution   Referring to Figure 7-41, we see that the unknown frequency is al-

lowed through the AND gate into the counter during the interval t_3 to t_4 while SAMPLE and X are both HIGH. If the middle input to the AND gate were open, it would act like a permanent HIGH (assuming TTL devices). This would allow the unknown frequency pulses through the gate while X is HIGH during the interval t_2 to t_4. This is twice the normal interval, and so the counter will count to twice the normal value.

▬ EXAMPLE 7-28 ▬

A technician wires up the digital clock of Figures 7-42 and 7-43. He observes that the seconds section is counting properly. In order to quickly test the operation of the minutes and hours sections, he bypasses the MOD-60 counter so that the counters will be pulsed at a rate that is 60 times faster than normal. He observes that the minutes section is working correctly, but the hours section counts and displays in the following manner:

Tens of Hours	Units of Hours
0	1
0	2
0	3
0	4
0	5
0	6
0	7
1	8
1	9
1	0
1	1
1	2

recycle and repeat

What is the probable cause of this incorrect sequence?

Solution   Since the problem is in the hours section, we need to refer to Figure 7-43. The sequence above is correct except that the tens digit is incremented from 0 to 1 when the units digit goes from 7 to 8 instead of when it goes from 9 to 0. This operation would occur if the CLK input of flip-flop X were mistakenly connected to Q_2 rather than Q_3 of the BCD counter. If this is the case, then when the BCD counter increments from 7 to 8, its Q_2 flip-flop makes a NGT that will toggle flip-flop X prematurely.

PROBLEMS

SECTION 7-1

7-1. An 8-MHz square wave clocks a 5-bit ripple counter. What is the frequency of the last FF? What is the duty cycle of this output waveform?   250 kHz; 50 percent duty cycle

7-2. Repeat Problem 7-1 if the input has a 20 percent duty cycle.

7-3. Assume that the 5-bit binary counter starts in the 00000 state. What will be the count after 144 input pulses? 10000_2

SECTION 7-2

7-4. Five FFs with Q_3 and Q_4 connected to a NAND with its output fed to all *CLR* inputs.

7-5. Frequency at D is 100 Hz. Output B has a glitch on occurrence of tenth clock pulse.

7-6. The NAND latch holds a LOW at the CLEAR inputs of the counter FFs for half a clock cycle.

7-4. Use J-K flip-flops and any other necessary logic to construct a MOD-24 asynchronous counter.

7-5. Draw the waveforms for all the FFs in the decade counter of Figure 7-6(b) in response to a 1-kHz clock frequency. Show any glitches that might appear on any of the FF outputs. Determine the frequency at the D output.

7-6. When a NAND gate is used to clear a counter back to zero as in the counters of Figures 7-4 to 7-7, there may be a possible problem because the propagation delays from CLEAR input to FF output may vary from FF to FF. For example, suppose that one FF clears in 10 ns and another clears in 50 ns. When the faster FF clears to 0 it will cause the NAND gate output to go back HIGH, thereby terminating the CLEAR pulse, and the slower FF may not be cleared. This problem is especially prevalent when the FF outputs are unevenly loaded, since, in general, the propagation delay of a FF increases as it drives more loads.

A possible means for eliminating this problem is shown in Figure 7-56 for a MOD-6 counter. The SET-CLEAR flip-flop is the cross-coupled NAND latch studied in Chapter 5. Analyze the operation of this circuit and explain how the clearing function is performed without the problem described above.

7-7. Refer to the counter shown in Figure 7-57. How can you tell that it is a down counter? It has been modified so that it does *not* count through the entire binary sequence 111 to 000. Determine the actual sequence it counts through. Sequence is 100, 011, 010, 001, 000, 100, 011, etc.

7-8. A counter or group of counters is often used to divide a high-frequency clock signal down to a lower-frequency output. When these

FIGURE 7-56

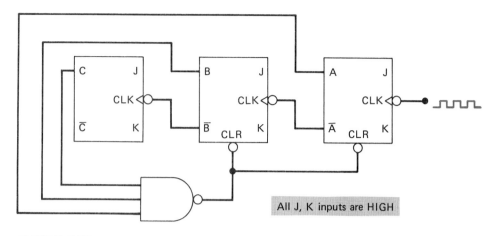

FIGURE 7-57

counters are binary counters (i.e., they count in the binary sequence), the output will not be a symmetrical square wave if the binary sequence has been shortened in order to produce the desired MOD number. For example, refer to the C waveform of the MOD-6 counter in Figure 7-4.

When a counter is being used only for frequency division, it is not necessary that it count in a binary sequence as long as it has the desired MOD number. A symmetrical square-wave output can be obtained for any *even* MOD number by breaking the MOD number into the product of two MOD numbers, one of which is a power of 2. For example, a MOD-6 counter can be formed from a MOD-3 counter and a MOD-2 counter as shown in Figure 7-58.

Here flip-flops A and B and the NAND gate make up the MOD-3 counter, whose B output has one-third the frequency of the input pulses. This B output is connected to the input of flip-flop C, which is acting as a MOD-2 counter to divide the frequency down to one-sixth the frequency of the input pulses.

FIGURE 7-58

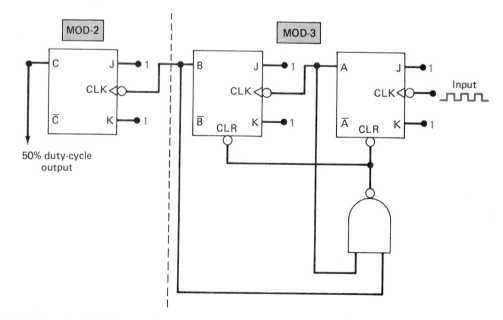

351

(a) Assume all FFs are initially LOW, and sketch the waveforms at each FF output for 12 cycles of the input.

(b) Construct the state transition diagram and show that it is not a normal binary sequence.

SECTION 7-3

7-9. Show how a 74293 counter can be used to produce a 1.2-kpps output from an 18-kpps input. *Wire it as MOD-15.*

7-10. Wire one as MOD-15 and the other as MOD-4.

7-10. Show how two 74293s can be connected to divide an input frequency by 60 while producing a symmetrical square-wave output.

7-11. Determine the frequency at output X in Figure 7-59. *60 Hz*

FIGURE 7-59

SECTION 7-4

7-12. Sequence is 1111, 1110, 1101, 1100, . . . , 0010, 0001, 0000. Starting at 0110, the counter will be at 0001 after 37 pulses.

7-12. (a) Draw the diagram for a MOD-16 *down* counter.

(b) Construct the state transition diagram.

(c) If the counter is initially in the 0110 state, what state will it be in after 37 clock pulses?

SECTION 7-5

7-13. A 4-bit ripple counter is driven by a 20-MHz clock signal. Draw the waveforms at the output of each FF if each FF has $t_{pd} = 20$ ns. Determine which counter states, if any, will not occur because of the propagation delays. *The 1000 and 0000 states will not occur.*

7-14. What is the maximum clock frequency that can be used with the counter of Problem 7-13? What could f_{max} be if the counter were expanded to 6 bits? *12.5 MHz; 8.33 MHz*

SECTION 7-6

7-15. (a) Draw the circuit diagram for a MOD-64 parallel counter.

(b) Determine f_{max} for this counter if each FF has $t_{pd} = 20$ ns and each gate has $t_{pd} = 10$ ns. *33 MHz*

(c) A more efficient gating arrangement can be used for this counter by deriving the inputs for each AND gate from the output of the preceding gate (see Figure 7-60). For example, the output of the

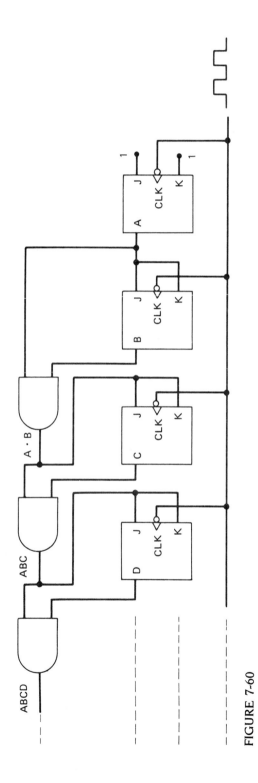

FIGURE 7-60

first AND gate (equal to $A \cdot B$) can be fed to the input of the second AND gate, together with C, to produce $A \cdot B \cdot C$. Similarly, the output of the second AND gate (equal to $A \cdot B \cdot C$) is fed to the third AND gate along with D to produce $A \cdot B \cdot C \cdot D$, and so on. This arrangement requires only two-input AND gates. However, it increases the total propagation delay, since the FF output signals must now propagate through more than one gate. Calculate the total delay for this arrangement and then calculate f_{max} and compare it with the value calculated in part (b). Assume the same delays as given in part (b). delay = 60 ns; f_{max} = 16.7 MHz

 7-16. Figure 7-61 shows a 4-bit parallel counter which is designed so that it does not sequence through the entire 16 binary states. Analyze its operation by drawing the waveforms at each FF output. Then determine the sequence which the counter goes through. Assume that all FFs are initially 0. It is a MOD-10 counter that counts 0000 through 1001.

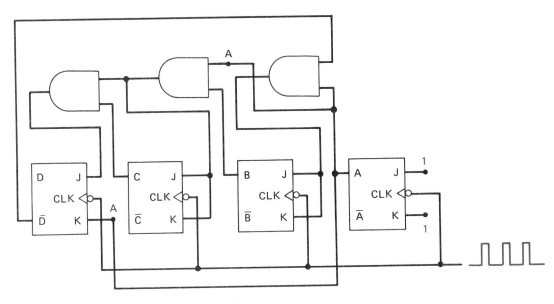

FIGURE 7-61

7-17. See Solutions Manual. **7-17.** Show how to connect two 74193s to form an 8-bit counter that divides the clock frequency by 100. Use the $\overline{TC}_D$ output of one 74193 stage as the clock for the second 74193 stage. Connect the parallel data inputs so that the combined counter presets at 100_{10}. Note that the counter is to be preset when both stages are simultaneously at their end counts.

7-18. (a) X goes HIGH on the NGT of the first clock pulse and goes LOW 6 seconds later when TC_D goes LOW. (b) It is not synched to the clock edges. (d) Flip-flop Z will not clear. Connect an OS between the START switch and the SET input of Z.

SECTIONS 7-8 and 7-9

 7-18. Figure 7-62 shows how a presettable down counter can be used in a *programmable timer* circuit. The input clock frequency is an accurate 1 Hz derived from the 60-Hz line frequency after division by 60. Switches S1 to S4 are used to preset the counter to a desired starting count when a momentary pulse is applied to $\overline{PL}$. The timer operation is initiated by depressing the START pushbutton switch. Flip-flop Z is

FIGURE 7-62

used to eliminate effects of bounce in the START switch. The OS is used to provide a very narrow pulse to the $\overline{PL}$ input. The output of flip-flop X will be a waveform that goes HIGH for a number of seconds equal to the number set on the switches.

(a) Assume that all FFs and the counter are in the 0 state and analyze and explain the circuit operation, showing waveforms when necessary, for the case where S1 and S4 are LOW and S2 and S3 are HIGH. Be sure to explain the function of flip-flop X.

(b) Why can't the timer output be taken at the $\overline{TC}_D$ output?

(c) Why can't the START switch be used to trigger the OS directly? Switch bounce.

(d) What will happen if the START switch is held down too long? Add the necessary logic needed to ensure that holding the START switch down will not affect the timer operation.

7-19. Modify the circuit of Figure 7-24 so that it functions as a MOD-10 counter. The frequency at the Q_3 output should be one-tenth the frequency of the CP_D input. Draw the waveforms at Q_3, Q_2, Q_1, Q_0, and $\overline{TC}_D$. Change the parallel data inputs to 1010.

7-20. Change the parallel data inputs in Figure 7-24 to 1001. Draw the waveforms at Q_3, Q_2, Q_1, Q_0, and $\overline{TC}_D$. What is the MOD number? It is MOD-9.

7-21. (b) the *MR* inputs clear the counter to 0000. (c) The *MS* inputs set the counter to 1001. (e) Connect Q_0 to $\overline{CP}_1$. Apply pulses to $\overline{CP}_0$. Connect all *MR* and *MS* LOW. (f) Connect Q_3 to $\overline{CP}_0$. Apply pulses to $\overline{CP}_1$. Connect all *MR* and *MS* LOW.

7-21. Figure 7-63 shows the IEEE/ANSI symbol for a 7490 or 74290 counter IC. Examine the symbol and determine the following:

(a) The overall MOD number   MOD-10

(b) The function performed by the *MR* inputs

(c) The function performed by the *MS* inputs

(d) Is this an up counter or a down counter?   Up counter.

(e) How would you connect it to function as a BCD counter? (Refer to data sheet, Appendix II.)

(f) How would you connect it to divide the clock frequency by 10 and produce a symmetrical square wave?

7490/74290   FIGURE 7-63

7-22. The 74192 counter IC operates exactly like the 74193 except for the following differences:

- The 74192 is a BCD counter that either counts up from 0 to 9, or counts down from 9 to 0.

- The $\overline{TC}_U$ output is activated when the count is 9 and the CP_U input is LOW.

Modify the IEEE/ANSI symbol of Figure 7-25 so that it represents the 74192.   Change CTR DIV16 to CTR DIV10; change $\overline{I}CT = 15$ to $\overline{I}CT = 9$.

SECTIONS 7-11 and 7-12

7-23. Draw the gates necessary to decode all the states of a MOD-16 counter using active-LOW outputs.   Sixteen 4-input NAND gates.

7-24. Draw the AND gates necessary to decode the 10 states of the BCD counter of Figure 7-6(b).   Ten 4-input AND gates.

7-25. *X* goes HIGH at count of 0100 and goes LOW at 0111. *Y* toggles on each NGT of *X*.

7-25. Figure 7-64 shows a counter being used to help generate control waveforms. Control waveforms 1 and 2 could be used for many purposes, including control of motors, solenoids, valves, and heaters. Determine the control waveforms, assuming that all FFs are initially LOW. Ignore decoding glitches. Assume that clock frequency = 1 kpps.

All J, K
inputs = 1

FIGURE 7-64

7-26. Draw the complete waveforms at the output of the decoding gates of a MOD-16 *ripple* counter, including any glitches or spikes that can occur due to the FF delays. Why are the gates which are decoding for *even* numbers the only ones that have glitches?

7-27. The circuit of Figure 7-64 might malfunction because of glitches at the outputs of the decoding NAND gates.
(a) Determine at what point(s) the glitches can cause erroneous operation.
(b) What are two ways that can be used to eliminate the possibility of erroneous operation?

SECTION 7-13

7-28. How many FFs are used in Figure 7-31? Indicate the states of each of these FFs after 795 pulses have occurred.   Twelve FFs; 0111 1001 0101.

7-29. How many cascaded BCD counters are needed to be able to count up to 8000? How many FFs does this require? Compare this with the number of FFs required for a normal binary counter to count up to 8000. Since it uses more FFs, why is the cascaded BCD method used?

SECTION 7-14

7-30. (a) Design a synchronous counter that has the following sequence: 000, 010, 101, 110, and repeat. The undesired (unused) states 001, 011, 100, and 111 must always go to 000 on the NEXT clock pulse.
(b) Redesign the counter of part (a) without any requirement on the unused states; that is, their NEXT states can be don't cares. Compare with the design from (a).

7-26. Glitches occur only when more than one FF is changing. For this to occur, the LSB flip-flop must be going from HIGH to LOW. This means any temporary states will have a 0 in the LSB, and are therefore even numbers.

7-27. (a) When the counter goes from 101 to 110, a glitch can appear at *CLK* input of flip-flop X, which might set $X = 1$ too early in the cycle.   (b) Use a parallel counter; use a strobe signal to disable the decoding gates until after the counter has changed states; redesign so that flip-flop X changes only on edges of input clock signal.

7-29. Four counters; sixteen FFs; thirteen FFs; BCD easier to decode and display.

7-30. (a) $J_A = B\overline{C}$; $K_A = 1$; $J_B = CA + \overline{C}\overline{A}$; $K_B = 1$; $J_C = A\overline{B}$; $K_C = B + \overline{A}$
(b) $J_A = B\overline{C}$; $K_A = 1$; $J_B = K_B = 1$; $J_C = K_C = B$

7-31. $J_A = K_A = 1$;
$J_B = K_B = \bar{A}$; $J_C = K_C = \overline{AB}$;
$J_D = K_D = \overline{ABC}$

7-31. Use the synchronous counter design procedure to design a 4-bit synchronous down counter that counts through all states from 1111 down to 0000. Compare your result with the synchronous down counter described in Section 7-7.

SECTION 7-15

7-32. Connect 5 J-K flip-flops as a shift register, and connect Q and $\bar{Q}$ from the last FF back to J and K, respectively, of the first.
7-33. Sequence is 100000, 010000, 001000, 000100, 000010, 100001, 010001, 001001, 000101, 000011, 100000, 010000, etc.

7-32. Draw the diagram for a 5-bit ring counter using J-K flip-flops.

7-33. Combine the ring counter of Problem 7-32 with a *single* J-K flip-flop to produce a MOD-10 counter. Determine the sequence of states for this counter. This is an example of a decade counter that is not a BCD counter.

7-34. Draw the diagram for a MOD-10 Johnson counter using J-K flip-flops and determine its counting sequence. Draw the decoding circuit needed to decode each of the 10 states. This is another example of a decade counter that is not a BCD counter.   See Solutions Manual.

7-35. Determine the frequency of the pulses at points w, x, y, and z in the circuit of Figure 7-65.   16 kHz, 1 kHz, 40 Hz, and 5 Hz.

FIGURE 7-65

SECTION 7-16

7-36. (c) The display would change from 257 to 323 at the end of the next SAMPLE interval.

7-36. As pointed out in the text, the frequency counter of Figure 7-41 has the disadvantage that the display shows all of the counter operations (resetting, counting, holding) and would therefore be confusing, if not unreadable. This can be overcome by the addition of buffer registers to store the contents of the counter at the end of each counting interval (t_3 to t_4 in Figure 7-41) and hold it for display until the end of the next counting interval (t_7 to t_8). Figure 7-66 shows this modification. A register consisting of four D flip-flops has been inserted between each BCD counter and its decoder/display unit.
(a) Analyze this circuit and describe its operation, particularly the transferring of data from the counters to the display.
(b) What would you see on a three-digit display if the unknown frequency were constant at 2570 pps and the sample interval were 0.1 s?   Display would show 257.
(c) What would you see on this display if the unknown frequency were suddenly changed to 3230 pps?

7-37. The frequency counter of Figure 7-66 uses three BCD counters and a sampling interval of 100 μs. Determine the readings on the three frequency-counter displays for each of the following input frequencies:
(a) 220 kkps   022
(b) 4.5 Mpps   450
(c) 750 pps   0 or 1

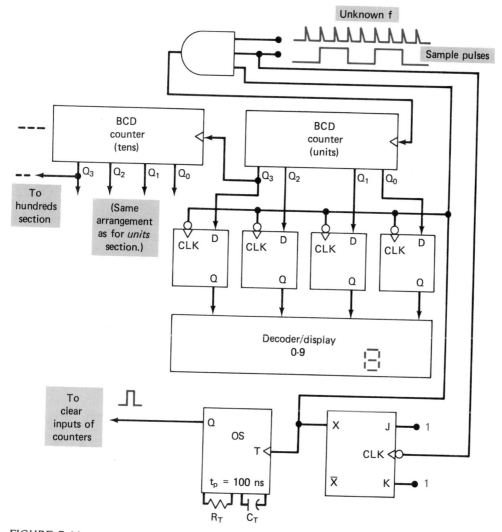

FIGURE 7-66

7-38. Design the complete circuit for the SECONDS section of the digital clock circuit of Figure 7-42. Use a 74293 for the MOD-6 and a 74290 for the BCD. Connect Q_3 of the 74290 to $\overline{CP}_1$ of the 74293.

7-39. The digital clock of Figure 7-42 has to have some means for manually setting the HOURS and MINUTES sections to the correct starting time. For example, this can be done by switching the 1-pps signal into the input of the MINUTES section when a SET MINUTES pushbutton is activated. A similar operation can be done with a SET HOURS pushbutton. Design the necessary logic to provide this capability using two pushbutton switches. See Solutions Manual.

7-40. Modify the HOURS section of the digital clock (Figure 7-43) so that it counts and displays military time (i.e., 00 to 23 hours).

7-40. Add a second J-K flip-flop for tens of hours. Clear all FFs when count is 24 = 100100.

7-42. Connect Q_3 of one to D_S of the other. Connect *PE*s HIGH. Connect 00000001 to parallel data inputs. Connect *SE*s together, and switch them from LOW to HIGH to start the shifting operation.

7-43. The 74178 is parallel-loaded on the first NGT of the sixth clock pulse. The serial signal corresponding to the parallel data (1101) appears at Q_3, LSB first.

7-41. Modify the frequency counter of Figure 7-66 so that it uses 74174 ICs. Assume that the counter contains three BCD counters and a three-digit display.   Use two 74174s for a total of 12 FFs.

7-42. Show how to connect two 74178s as an 8-bit ring counter.

7-43. Figure 7-67 shows how a 74178 can be used as a *parallel-to-serial converter*. The parallel data that are entered at P_0 through P_3 are shifted out serially so that a serial waveform appears at Q_3. The shifting out of the parallel data is controlled by the occurrence of the START pulse. This START pulse occurs asynchronously to the clock pulses, and so the D flip-flops are used to synchronize the parallel loading and shifting of the 74178.

Assume that the START pulse has been LOW and that clock pulses have been continually applied to the circuit for a long time before time t_0 (see waveforms). Draw the waveforms that appear at Q_X, $\overline{Q}_X$, Q_Y, and Q_3 in response to the START pulse shown in the figure.

FIGURE 7-67

7-44. A shift register is often used to delay a logic signal by an integral number of clock cycles. If the waveforms in Figure 7-68 are applied to the inputs of one of the 4731B shift registers, how long will it take the D_S waveform to appear at the Q_{63} output? (Assume that the t_H requirement of the register FFs is met by the D waveform.)   64 μs

FIGURE 7-68

SECTIONS 7-21 and 7-22

 7-45. Modify the circuit of Figure 7-52 so that the INVERTER output is connected to the A input instead of to $\overline{MR}$.

    (a) Draw the waveforms at each FF output in response to the input waveforms shown in Figure 7-69.

    (b) Add the necessary logic to produce a timing-signal output that goes HIGH only during time intervals t_1 to t_2 and t_8 to t_9.

    (c) Add the necessary logic to produce a timing signal that goes LOW only during the interval t_4 to t_7.

7-45. The 74164 operates as a MOD-8 Johnson counter.

FIGURE 7-69

7-46. Assume that pulses have been applied continuously to CP_1 prior to t_0 in Figure 7-70. Draw the waveform at Q_7 in response to the CP_1 and $\overline{PL}$ waveforms shown.   See Solutions Manual.

FIGURE 7-70

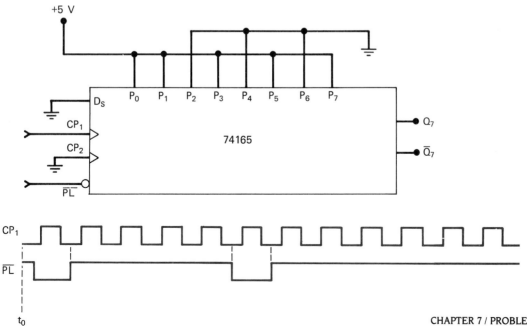

7-47. While examining the schematic for a piece of equipment, a technician or engineer will often come across an IC that is unfamiliar. In such cases, it is often necessary to consult the manufacturer's IC data book for specifications on the device. The information on the IC data sheets is always complete, but is sometimes difficult to understand especially by someone with very little experience. This problem will give you practice in obtaining information about a fairly complex IC—the 74194 bidirectional universal shift register. Consult the data sheets in Appendix II to answer the following questions. Support your answers.

(a) Is the *CLR* input asynchronous or synchronous?   asynchronous

(b) *True or false:* When *CLK* is LOW, the S_0 and S_1 levels have no effect on the register.   true

(c) Assume the following conditions:

$$Q_A\,Q_B\,Q_C\,Q_D = 1\ 0\ 1\ 1$$
$$ABCD = 0\ 1\ 1\ 0$$
$$\overline{CLR} = 1$$
$$SR\ SER = 0$$
$$SL\ SER = 1$$

7-47 (c) shifts left   (d) shifts right

If $S_0 = 0$ and $S_1 = 1$, what will be the register outputs after one *CLK* pulse? After two *CLK* pulses? After three? After four?

(d) Use the same conditions except $S_0 = 1$, $S_1 = 0$ and repeat (c).

(e) Repeat with $S_0 = S_1 = 1$.   parallel load

(f) Repeat with $S_0 = S_1 = 0$.   hold data

(g) Use the same conditions as in (c) except assume that output Q_A is connected to SL SER. What will be the register outputs after four *CLK* pulses?   recirculating left

(h) Show how to connect this IC to function as a ring counter that counts through the following $Q_A\,Q_B\,Q_C\,Q_D$ sequence: 0001, 0010, 0100, 1000, and repeat.   (h) connect as in (g) with parallel inputs = 0001

SECTION 7-24

7-48. A technician tests the counter of Figure 7-54(a) by applying a low-frequency clock signal and monitoring the FF outputs on indicator LEDs. He observes the following repetitive sequence indicated by the LEDs.

Q_3	Q_2	Q_1	Q_0
0	0	0	0
0	0	0	1
0	0	1	0
0	0	1	1
0	1	0	0
0	1	0	1
0	1	1	0
0	1	1	1

recycle and repeat

What are the possible reasons why the counter is not counting properly?   Connection from MR_1 to Q_1 is open, or Q_3 or MR_2 is shorted to ground.

7-49. DC CLR input of flip-flop X is open or stuck HIGH or not responding.

7-49. Refer to the digital clock circuit of Figures 7-42 and 7-43. A technician testing the circuit observes that the SECONDS and MINUTES

sections count properly, but the HOURS section counts as follows: 01, 02, 03, 04, 05, 06, 07, 08, 09, 10, 11, 12, 11, 12, 11, 12, . . . What is the probable cause of the malfunction?

7-50. A technician tests the digital clock circuit (Figures 7-42 and 7-43) and observes that the HOURS section does not count and the MINUTES section counts from 00 to 39, and then recycles to 00 and repeats. What are the possible causes of this incorrect behavior?

7-50. The MSB of MOD-6 is stuck LOW.

7-51. Refer to the modified frequency counter of Figure 7-66. Assume that there are three BCD counter stages with associated buffer registers. The sample interval is set at 1 s and the unknown frequency is 125 pps. Describe what will appear on the display for each of the following circuit faults.
(a) An open connection at the top input of the AND gate.
(b) A burned-out resistor R_T.   display = 000

7-51. (a) Display = 001

7-52. A technician tests the frequency counter of Figure 7-66 using a sample interval of 1 s and an unknown frequency of 125 pps. She expects to see a display of 125, but instead sees the display changing every few seconds as follows: 125, 250, 375, 500, 625, 750, 875, 000, 125, 250, etc. What can be the cause of this malfunction?

7-52. Counters are not being cleared before each SAMPLE interval.

7-53. Refer to the up/down counter of Figure 7-18. Describe how each of the following circuit faults will affect the count-up and count-down operations:
(a) AND gate 4 output is internally shorted to V_{CC}.
(b) A solder bridge is shorting AND gate 1 output to ground.

7-53. (a) Flip-flop C will toggle on every NGT of the input.
(b) Will not operate properly in the count-up mode.

7-54. A technician runs a test on the timer circuit of Figure 7-62 and records the following results:

S1	S2	S3	S4	Timer Output(s)
5 V	0 V	0 V	0 V	10
0 V	0 V	5 V	5 V	3
5 V	0 V	5 V	5 V	11
5 V	5 V	5 V	0 V	14
0 V	5 V	0 V	5 V	7
5 V	5 V	0 V	0 V	14

Examine the recorded data and determine the possible causes of the faulty operation. P_1 open, or S3 faulty.

7-55. A technician wires up the counter circuit of Figure 7-59. He applies an accurate 8.64-kpps signal to the input and measures a frequency of 54 pps at X instead of the expected 60 pps. What is the most probable wiring error that he made?

7-55. He connected Q_1 instead of Q_0 to MR_1 of the 74293 on the left so that it is a MOD-9 counter.

DRILL QUESTION

7-56. For each of the following statements, indicate the type(s) of counter being described:
(a) Each FF is clocked at the same time.   parallel
(b) Each FF divides the frequency at its *CLK* input by 2.   binary
(c) The counting sequence is 111, 110, 101, 100, 011, 010, 001, 000.   down counter

7-56. (e) asynchronous

7-56. (h) all types

(d) The counter has 10 distinct states.   <span style="font-variant:small-caps">MOD-10 or decade counter</span>

(e) The total switching delay is the sum of the individual FFs' delays.

(f) This counter requires no decoding logic.   ring

(g) The MOD number is always twice the number of FFs.   Johnson

(h) This counter divides the input frequency by its MOD number.

(i) This counter can begin its counting sequence at any desired starting state.   presettable

(j) This counter can count in either direction.   up/down

(k) Can suffer from decoding glitches.   asynchronous

(l) Counts from 0 to 9.   BCD

(m) Can be designed to count through arbitrary sequences by determining logic needed at each flip-flop's *J* and *K* input.   synchronous

MICROCOMPUTER APPLICATION

7-57. See Solutions Manual.

7-57. A microprocessor that is used in a control application must often control the timing of external events such as the turning on and off of devices like solenoids, motors, and relays. Such actions can be timed using software by repetitively executing a program loop a specific number of times. This, however, places a heavy burden on the MPU, since it can't do anything else while it is executing the loop over and over. For this reason, most timed intervals are usually generated by hardware which is *under control of the MPU*. In other words, the MPU will send data to the hardware to tell it how long a time interval to generate.

In Problem 7-18 we saw how the 74193 IC could be used in a timer circuit (Figure 7-62) to generate accurate time intervals corresponding to the binary data from four switches. This circuit can be modified so that the binary data come from an MPU instead of the switches. In Section 5-20 we saw how an MPU could transfer data to an external device using its address, data, and clock outputs (Figure 5-50).

Show how to combine these two circuits so that the timer output *X* will generate a HIGH level for a time interval equal to the binary number which the MPU presets into the 74193 counter. You can eliminate any circuitry that is not needed. Assume the MPU's *CP* signal is a 1-MHz square wave. Remember that $\overline{PL}$ is an asynchronous input.

ANSWERS TO SECTION REVIEW QUESTIONS

SECTION 7-1

1. False   **2.** 0000

SECTION 7-2

1. *D*, *C*, and *A*.   **2.** True, since a BCD counter has 10 distinct states.   **3.** 5 kHz.

SECTION 7-3

1. 250 Hz.   **2.** $f_{in}/60$.   **3.** 4096.
4. The counter is MOD-64 and divides the frequency by 64.

SECTION 7-4

1. In an up counter, the count is increased by 1 with each clock pulse; in a down counter, the count is decreased by 1 with each pulse.
2. The inverted output of each FF is connected to the *CLK* input of the following FF.

SECTION 7-5

1. Each FF adds its propagation delay to the total counter delay in response to a clock pulse.
2. MOD-256.

SECTION 7-6

1. Can operate at higher clock frequencies, and has more complex circuitry.   2. Six FFs and four AND gates.

SECTION 7-8

1. They can be preset to any desired starting count.   2. Asynchronous presetting is independent of the clock input, while synchronous presetting occurs on the active edge of the clock signal.

SECTION 7-9

1. When $\overline{PL}$ is pulsed LOW, the counter is preset to the binary number present at inputs P_0 to P_3.   2. A HIGH at MR overrides all other inputs to reset counter to 0000.   3. True.   4. 1, 0, 0, respectively.

SECTION 7-10

1. See appropriate text.   2. (a) Up-counter operation;   (b) this input is ANDed with any other input or output that has a "4" in its label; (c) this input controls the effect of any inputs that have "5" in their label;   (d) data input that is controlled by input labeled C5.

SECTION 7-11

1. A 6-input NAND gate with inputs A, B, C, $\overline{D}$, E and $\overline{F}$.

SECTION 7-12

1. The glitches would be caused by the FFs' changing states one at a time during counter state transitions.   2. The strobe signal inhibits the decoding gates until all FFs have completed their transitions.

SECTION 7-14

1. See text.   2. It shows the necessary levels at J and K to produce every possible FF state transition.   3. It shows the necessary levels at each flip-flop's J and K input to produce the counter's state transitions.   4. True.

SECTION 7-15

1. Ring counter.   2. Johnson counter. 3. The inverted output of the last FF is connected to the input of the first FF.   4. False, true, true.

SECTION 7-16

1. 1 ms.   2. Counter cleared; counter counts pulses during sample interval; counter stops and holds count for display.

SECTION 7-17

1. Pulse shaper, frequency divider, seconds counter and display, minutes counter and display, hours counter and display.   2. To change the NGT from the MINUTEs section to a PGT needed by the 74192.

SECTION 7-18 to SECTION 7-22

1. Parallel in/serial out   2. True.   3. Serial in/parallel out.   4. Serial in/serial out.

SECTION 7-23

1. It separates the two indicated functions performed by that input.   2. SRG 64

8

INTEGRATED-CIRCUIT LOGIC FAMILIES

Outline

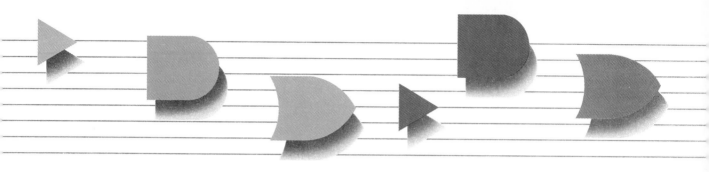

Upon completion of this chapter, you will be able to:

- Read and understand digital IC terminology as specified in manufacturers' data sheets.
- Compare the characteristics of standard TTL and the various TTL series.
- Determine the fan-out for a particular logic device.
- Use logic devices with open-collector outputs in a wired-AND arrangement.
- Analyze circuits containing tristate devices.
- Describe the major characteristics and differences among TTL, ECL, MOS, and CMOS logic families.
- Cite and implement the various considerations that are required when interfacing digital circuits from different logic families.
- Use a logic pulser and current tracer as digital circuit troubleshooting tools.

Introduction

As we described in Chapter 4, digital IC technology has advanced rapidly from small-scale integration (SSI), with fewer than 12 gates per chip, through medium-scale integration (MSI), with 12 to 99 equivalent gates per chip, and on to large-scale and very large-scale integration (LSI and VLSI), which can have tens of thousands of gates per chip and, most recently, to ULSI with over 100,000 gates per chip.

Most of the reasons that modern digital systems use integrated circuits are obvious. ICs pack a lot more circuitry in a small package, so that the overall size of almost any digital system is reduced. The cost is dramatically reduced because of the economies of mass-producing large volumes of similar devices. Some of the other advantages are not so apparent.

ICs have made digital systems more reliable by reducing the number of external interconnections from one device to another. Before we had ICs, every circuit connection was from one discrete component (transistor, diode, resistor, etc.)

to another. Now most of the connections are internal to the ICs, where they are protected from poor soldering, breaks or shorts in connecting paths on a circuit board, and other physical problems. ICs have also drastically reduced the amount of electrical power needed to perform a given function, since their miniature circuitry typically requires less power than their discrete counterparts. In addition to the savings in power-supply costs, this reduction in power has also meant that a system does not require as much cooling.

There are some things that ICs cannot do. They cannot handle very large currents or voltages because the heat generated in such small spaces would cause temperatures to rise beyond acceptable limits. In addition, ICs cannot easily implement certain electrical devices such as inductors, transformers, and large capacitors. For these reasons, ICs are principally used to perform low-power circuit operations that are commonly called *information processing*. The operations that require high power levels or devices that cannot be integrated are still handled by discrete components.

With the widespread use of ICs comes the necessity to know and understand the electrical characteristics of the most common IC logic families. Remember that the various logic families differ in the major components that they use in their circuitry. TTL and ECL use bipolar transistors as their major circuit element; PMOS, NMOS, and CMOS use unipolar MOSFET transistors as their principal component. In this chapter we will present the important characteristics of each of these IC families and their subfamilies. Once these are understood, you will be much better prepared to do analysis, troubleshooting, and some design of digital circuits that contain any combination of IC families.

8-1 DIGITAL IC TERMINOLOGY

Although there are many digital IC manufacturers, much of the nomenclature and terminology is fairly standardized. The most useful terms are defined and discussed below.

Current and Voltage Parameters (see Figure 8-1)

V_{IH}(min)—High-Level Input Voltage The voltage level required for a logical 1 at an *input*. Any voltage below this level will not be accepted as a HIGH by the logic circuit.

FIGURE 8-1   Currents and voltages in the two logic states.

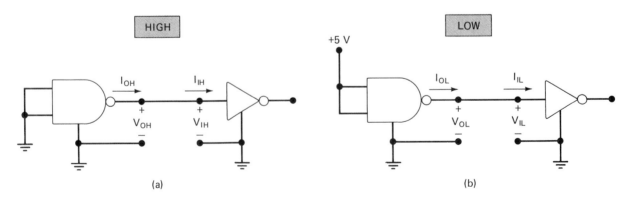

(a)

(b)

V_{IL}(max)—Low-Level Input Voltage The voltage level required for a logic 0 at an *input*. Any voltage above this level will not be accepted as a LOW by the logic circuit.

V_{OH}(min)—High-Level Output Voltage The voltage level at a logic circuit *output* in the logical 1 state. The minimum value of V_{OH} is usually specified.

V_{OL}(max)—Low-Level Output Voltage The voltage level at a logic circuit *output* in the logical 0 state. The maximum value of V_{OL} is usually specified.

I_{IH}—High-Level Input Current The current that flows into an input when a specified high-level voltage is applied to that input.

I_{IL}—Low-Level Input Current The current that flows into an input when a specified low-level voltage is applied to that input.

I_{OH}—High-Level Output Current The current that flows from an output in the logical 1 state under specified load conditions.

I_{OL}—Low-Level Output Current The current that flows from an output in the logical 0 state under specified load conditions.

Teaching Tip: Depending on the IC technology, the current directions and voltage polarities may be different from those shown in Figure 8-1. For example, I_{OL} and I_{IL} for TTL devices actually flow in the opposite direction.

Fan-Out In general, a logic-circuit output is required to drive several logic inputs. The *fan-out* (also called *loading factor*) is defined as the *maximum* number of standard logic inputs that an output can drive reliably. For example, a logic gate that is specified to have a fan-out of 10 can drive 10 standard logic inputs. If this number is exceeded, the output logic-level voltages cannot be guaranteed.

Propagation Delays A logic signal always experiences a delay in going through a circuit. The two propagation delay times are defined as

t_{PLH}: delay time in going from logical 0 to logical 1 state (LOW to HIGH)
t_{PHL}: delay time in going from logical 1 to logical 0 state (HIGH to LOW)

Figure 8-2 illustrates these propagation delays for an INVERTER. Note that t_{PHL} is the delay in the output's response as it goes from HIGH to LOW. It is measured between the 50 percent points on the input and output transitions.. The t_{PLH} value is the delay in the output's response as it goes from LOW to HIGH.

FIGURE 8-2   Propagation delays.

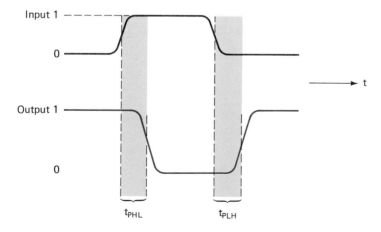

In general, t_{PHL} and t_{PLH} are not the same value, and both will vary depending on loading conditions. The values of propagation times are used as a measure of the relative speed of logic circuits. For example, a logic circuit with values of 10 ns is a faster logic circuit than one with values of 20 ns.

Power Requirements

Every IC requires a certain amount of electrical power to operate. This power is supplied by one or more power-supply voltages connected to the power pin(s) on the chip. Usually there is only one power-supply terminal on the chip, and it is labeled V_{CC} (for TTL) or V_{DD} (for MOS devices).

The amount of power that an IC requires is determined by the current, I_{CC}, that it draws from the V_{CC} supply, and the actual power is the product $I_{CC} \times V_{CC}$. For many ICs the current drain on the supply will vary depending on the logic states of the circuits on the chip. For example, Figure 8-3(a) shows a NAND chip where *all* the gate *outputs* are HIGH. The current drain on the V_{CC} supply for this case is called I_{CCH}. Likewise, Figure 8-3(b) shows the current drain when *all* the gate *outputs* are LOW. This current is called I_{CCL}.

In general, I_{CCH} and I_{CCL} will be different values. The average current is

$$I_{CC}(\text{avg}) = \frac{I_{CCH} + I_{CCL}}{2}$$

and it can be used to calculate average power drain as

$$P_D(\text{avg}) = I_{CC}(\text{avg}) \times V_{CC}$$

Speed-Power Product

Digital IC families have historically been characterized for both speed and power. It is generally more desirable to have shorter gate propagation delays (higher speed) and lower values of power dissipation. As we shall

Common Student Mistake: Students are often confused as to whether the term "input current" refers to the logic input currents, I_{IH} and I_{OH}, or the power-supply input current to the IC's power pin. The distinction should always be made clear.

FIGURE 8-3 I_{CCH} and I_{CCL}.

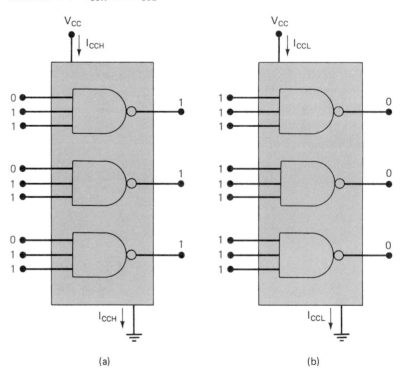

(a)   (b)

soon see, the various logic families and subfamilies provide a wide spectrum of speed and power ratings. A common means for measuring and comparing the overall performance of an IC family is the *speed-power product,* which is obtained by multiplying the gate propagation delay by the gate power dissipation. For example, suppose an IC family has an average propagation delay of 10 ns and an average power dissipation of 5 mW. The speed-power product is

$$10 \text{ ns} \times 5 \text{ mW} = 50 \times 10^{-12} \text{ watt-second}$$
$$= 50 \text{ picojoules} \quad (\text{pj})$$

Note that when propagation delay is in nanoseconds and power is in milliwatts, the speed-power product is in picojoules.

Clearly, a low value of speed-power product is desirable. IC designers are continually striving to reduce the speed-power product by increasing the speed of an IC (i.e., reducing propagation delay) or by decreasing its power dissipation. Because of the nature of transistor switching circuits, it is difficult to do both.

Noise Immunity   Stray electric and magnetic fields can induce voltages on the connecting wires between logic circuits. These unwanted, spurious signals are called *noise* and can sometimes cause the voltage at the input to a logic circuit to drop below $V_{IH}(\text{min})$ or rise above $V_{IL}(\text{max})$, which could produce unpredictable operation. The *noise immunity* of a logic circuit refers to the circuit's ability to tolerate noise voltages on its inputs. A quantitative measure of noise immunity is called **noise margin** and is illustrated in Figure 8-4.

Figure 8-4(a) is a diagram showing the range of voltages that can occur at a logic-circuit output. Any voltages greater than $V_{OH}(\text{min})$ are considered a logic 1, and any voltages lower than $V_{OL}(\text{max})$ are considered a logic 0. Voltages in the indeterminate range should not appear at a logic circuit output under normal conditions. Figure 8-4(b) shows the voltage requirements at a logic circuit input. The logic circuit will respond to any input greater than $V_{IH}(\text{min})$ as a logic 1, and will respond to voltages lower than $V_{IL}(\text{max})$ as a logic 0. Voltages in the indeterminate range will produce an unpredictable response and should not be used.

FIGURE 8-4   DC noise margins.

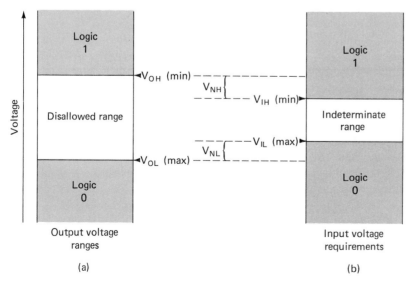

(a)

(b)

The **high-state noise margin** V_{NH} is defined as

$$V_{NH} = V_{OH}(\text{min}) - V_{IH}(\text{min}) \qquad (8\text{-}1)$$

as illustrated in Figure 8-4. V_{NH} is the difference between the lowest possible HIGH output and the minimum input voltage required for a HIGH. When a HIGH logic output is driving a logic-circuit input, any negative noise spikes greater than V_{NH} appearing on the signal line can cause the voltage to drop into the indeterminate range, where unpredictable operation can occur.

The **low-state noise margin** V_{NL} is defined as

$$V_{NL} = V_{IL}(\text{max}) - V_{OL}(\text{max}) \qquad (8\text{-}2)$$

and it is the difference between the largest possible LOW output and the maximum input voltage required for a LOW. When a LOW logic output is driving a logic input, any positive noise spikes greater than V_{NL} can cause the voltage to rise into the indeterminate range.

EXAMPLE 8-1

The input/output voltage specifications for the standard TTL family are listed below. Use these values to determine

(a) The maximum-amplitude noise spike that can be tolerated when a HIGH output is driving an input.

(b) The maximum-amplitude noise spike that can be tolerated when a LOW output is driving an input.

Discussion Question: What amplitude noise spike could cause a HIGH output to momentarily become a LOW?

Parameter	Min (V)	Typical (V)	Max (V)
V_{OH}	2.4	3.4	
V_{OL}		0.1	0.4
V_{IH}	2.0*		
V_{IL}			0.8*

*Normally only the minimum V_{IH} and maximum V_{IL} values are given.

Solution

(a) When an output is HIGH, it may be as low as $V_{OH}(\text{min}) = 2.4$ V. The minimum voltage that an input will respond to as a HIGH is $V_{IH}(\text{min}) = 2.0$ V. A negative noise spike can drive the actual voltage below 2.0 V if its amplitude is greater than

$$V_{NH} = V_{OH}(\text{min}) - V_{IH}(\text{min})$$
$$= 2.4 \text{ V} - 2.0 \text{ V} = 0.4 \text{ V}$$

(b) When an output is LOW, it may be as high as $V_{OL}(\text{max}) = 0.4$ V. The maximum voltage that an input will respond to as a LOW is $V_{IL}(\text{max}) = 0.8$ V. A positive noise spike can drive the actual voltage above the 0.8-V level if its amplitude is greater than

$$V_{NL} = V_{IL}(\text{max}) - V_{OL}(\text{max})$$
$$= 0.8 \text{ V} - 0.4 \text{ V} = 0.4 \text{ V}$$

AC Noise Margin Strictly speaking, the noise margins predicted by expressions (8-1) and (8-2) are termed dc noise margins. The term "dc noise margin" might seem somewhat inappropriate when dealing with noise, which is generally thought of as an ac signal of the transient variety. However, in today's high-speed integrated circuits, a pulse width of 1 μs is extremely long and may be treated as dc as far as the response of a logic circuit is concerned. As pulse widths decrease to the low-nanosecond region, a limit is reached where the *pulse duration* is too short for the circuit to respond. At this point, the pulse amplitude would have to be increased appreciably to produce a change in the circuit output. What this means is that a logic circuit can tolerate a large noise amplitude if the noise pulse is of a very short duration compared with the circuit's response time (i.e., propagation delays). In other words, a logic circuit's *ac noise margins* are generally substantially greater than its dc noise margins given by expressions (8-1) and (8-2).

Current-Sourcing and Current-Sinking Logic Logic families can be described according to how current flows between the output of one logic circuit and the input of another. Figure 8-5(a) illustrates *current-sourcing* action. When the output of gate 1 is in the HIGH state, it supplies a current I_{IH} to the input of gate 2, which acts essentially as a resistance to ground. Thus, the output of gate 1 is acting as a *source* of current for the gate 2 input.

Current-sinking action is illustrated in Figure 8-5(b). Here the input circuitry of gate 2 is represented as a resistance tied to $+V_{CC}$, the positive terminal of a power supply. When the gate 1 output goes to its LOW state, current will flow in the direction shown from the input circuit of gate 2 back through the output resistance of gate 1 to ground. In other words, in the LOW state the circuit driving an input of gate 2 must be able to *sink* a current, I_{IL}, coming from that input.

The distinction between current sourcing and current sinking is an important one which will become more apparent as we examine the various logic families.

Teaching Tip: The terms "source" and "sink" are always referring to the output of the driving logic circuit. It may help the student to use a plumbing analogy where you think of a source as a faucet that supplies water (current), and a sink as a means by which water (current) is led to ground.

FIGURE 8-5 Comparison of current-sourcing and current-sinking actions.

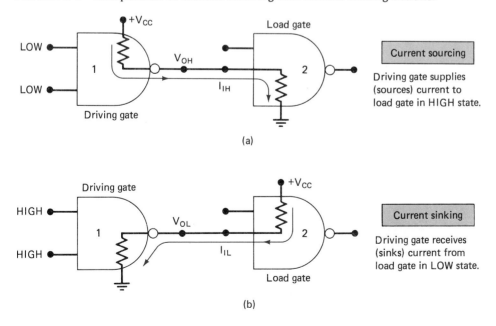

IC Complexity   The five basic levels of IC complexity are repeated below.

Complexity	Number of Gates
Small-scale integration (SSI)	Fewer than 12
Medium-scale integration (MSI)	12 to 99
Large-scale integration (LSI)	100 to 9999
Very large-scale integration (VLSI)	10,000 to 99,999
Ultra large-scale integration (ULSI)	100,000 or more

IC Packages   Integrated circuits are made in several different packages. These packages differ in the amount of circuitry contained on the silicon chip, the number of external connections to the chip, the environmental conditions under which the device is to operate, and the method to be used to mount the IC package on the circuit board. Figure 8-6 shows the most common IC packages.

The package in Figure 8-6(a) is the DIP (dual-in-line package) introduced in Chapter 4. This happens to be a 24-pin DIP. Note the presence of the notch on one end. When you orient the notch to the left, pin 1 is located below the notch. Some DIPs have a notch on both ends; a small dot on top of those packages indicates the location of pin 1.

Figure 8-6(b) is a *ceramic flat pack*. This is a hermetically sealed flat package consisting of an electrically nonconductive ceramic base, which means that it is totally immune to the effects of humidity. These packages are often used in military applications where equipment operates in extreme environmental conditions.

Figure 8-6(c) is one of the newest package types. It is a *surface-mount* package and is similar to the DIP except that its pins are bent at right angles so that they can be soldered directly to conductive pads on a circuit board. Surface-mount packages are generally smaller than DIPs because the pins can be made smaller and less rugged, since they do not have to be pushed into sockets. Surface-mount ICs also have the advantage that they can be more easily handled by automatic component-mounting equipment during the manufacture of circuit boards. Note how pin 1 is located with respect to the beveled edge of the package.

FIGURE 8-6   Common IC packages: (a) 24-pin DIP; (b) 14-pin ceramic flat pack; (c) surface-mount package.

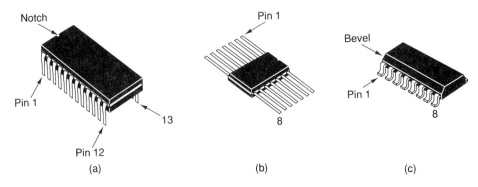

1. Define each of the following: V_{OH}, V_{IL}, I_{OL}, I_{IH}, t_{PLH}, t_{PHL}, I_{CCL}, I_{CCH}.
2. *True or false:* If a logic circuit has a fan-out of 5, the circuit has five outputs.
3. Define V_{NL} and V_{NH}.
4. *True or false:* A logic family with $t_{pd}(\text{avg}) = 12$ ns and $P_D(\text{avg}) = 15$ mW has a greater speed-power product than one with 8 ns and 30 mW.
5. Describe the difference between current sinking and current sourcing.
6. Which IC package does not have a notch to help locate pin 1?
7. Which package is designed to be immune to effects of humidity?

8-2 THE TTL LOGIC FAMILY

At this writing, the transistor-transistor logic (TTL) family still enjoys the most widespread use in applications that require SSI and MSI devices. The basic TTL logic circuit is the NAND gate. Its detailed circuit diagram, shown in Figure 8-7(a), has several distinctive characteristics. First, note that transistor Q_1 has two emitters; thus, it has two emitter-base (E-B) junctions that can be used to turn Q_1 on. This **multiple-emitter** input transistor can have up to eight emitters for an eight-input NAND gate.

Also note that on the output side of the circuit, transistors Q_3 and Q_4 are in a **totem-pole** arrangement. As we will see shortly, in normal operation either Q_3 or Q_4 will be conducting, depending on the logic state of the output.

FIGURE 8-7   (a) Basic TTL NAND gate; (b) diode equivalent for Q_1.

(a)

(b)

Circuit Operation—LOW State   Although this circuit looks extremely complex, we can simplify its analysis somewhat by using the diode equivalent of the multiemitter transistor Q_1 as shown in Figure 8-7(b). Diodes D_2 and D_3 represent the two E-B junctions of Q_1, and D_4 is the collector-base (C-B) junction. In the following analysis we will use this representation for Q_1.

First, let's consider the case where the output is LOW. Figure 8-8(a) shows this situation with inputs A and B both at $+5$ V. The $+5$ V at the cathodes of D_2 and

FIGURE 8-8   TTL NAND gate in its two output states.

Input conditions	Output conditions
A and B are both HIGH ($\geqslant 2$ V)	Q_3 OFF
Input currents are very low. $I_{IH} = 10\ \mu A$	Q_4 ON so that V_X is LOW ($\leqslant 0.4$ V)

(a)   LOW output

Input conditions	Output conditions
A or B or both are LOW ($\leqslant 0.8$ V)	Q_4 OFF
Current flows back to ground through LOW input terminal. $I_{IL} = 1.1$ mA	Q_3 acts as emitter-follower and $V_{OH} \geqslant 2.4$ V, typically 3.6 V

(b)   HIGH output

D_3 will turn these diodes off and they will conduct almost no current. The +5 V supply will push current through R_1 and D_4 into the base of Q_2, which turns on. Current from Q_2's emitter will flow into the base of Q_4 and turn Q_4 on. At the same time, the flow of Q_2 collector current produces a voltage drop across R_2 that reduces Q_2's collector voltage to a low value that is insufficient to turn Q_3 on.

The voltage at Q_2's collector is shown as approximately 0.8 V. This is because Q_2's emitter is at 0.7 V relative to ground due to Q_4's E-B forward voltage, and Q_2's collector is at 0.1 V relative to its emitter due to V_{CE}(sat). This 0.8 V at Q_3's base is not enough to forward-bias both Q_3's E-B junction and diode D_1. In fact, D_1 is needed to keep Q_3 off in this situation.

Teaching Tip: The diode equivalent circuit is valid only for purposes of dc analysis. It obviously cannot represent the transistor action that speeds up the switching times during input transitions.

With Q_4 on, the output terminal, X, will be at a very low voltage, since Q_4's on-state resistance will be low (1 to 25 Ω). Actually, the output voltage, V_{OL}, will depend on how much collector current Q_4 conducts. With Q_3 off, there is no current coming from the +5 V terminal through R_4. As we shall see, Q_4's collector current will come from the TTL inputs that terminal X is connected to.

It is important to note that the HIGH inputs at A and B will have to supply only a very small diode leakage current. Typically, this current I_{IH} is only around 10 μA at room temperature.

Circuit Operation—HIGH State

Figure 8-8(b) shows the situation where the circuit output is HIGH. This situation can be produced by connecting either or both inputs LOW. Here, input B is connected to ground. This will forward-bias D_3 so that current will flow from the +5 V source terminal, through R_1 and D_3, and through terminal B to ground. The forward voltage across D_3 will hold point Y at approximately 0.7 V. This voltage is not enough to forward-bias D_4 and the E-B junction of Q_2 sufficiently for conduction.

Teaching Tip: Students sometimes ask about the function of R_3. It provides a path for Q_2's OFF-state leakage current. Without R_3, this leakage current would flow into the base of Q_4 and turn it partially on.

With Q_2 off, there is no base current for Q_4, and it turns off. Since there is no Q_2 collector current, the voltage at Q_3's base will be large enough to forward-bias Q_3 and D_1, so that Q_3 will conduct. Actually, Q_3 acts as an emitter follower, because output terminal X is essentially at its emitter. With no load connected from point X to ground, V_{OH} will be around 3.4 to 3.8 V, because two 0.7-V diode drops (E-B of Q_3, and D_1) subtract from the 5 V applied to Q_3's base. This voltage will decrease under load because the load will draw emitter current from Q_3, which draws base current through R_2, thereby increasing the voltage drop across R_2.

It's important to note that there is a substantial current flowing back through input terminal B to ground. This current, I_{IL}, is typically around 1.1 mA. The LOW B input acts as a *sink* to ground for this current.

Current-Sinking Action

A TTL output acts as a current sink in the LOW state in that it *receives* current from the input of the gate it is driving. Figure 8-9 shows one TTL gate driving the input of another gate (the load) for both output voltage states. In the output LOW state situation depicted in Figure 8-9(a), transistor Q_4 of the driving gate is on and essentially "shorts" point X to ground. This LOW voltage at X forward-biases the emitter-base junction of Q_1 and current flows, as shown, back through Q_4. Thus, Q_4 is performing a current-sinking action that derives its current from the input current (I_{IL}) of the load gate. We will often refer to Q_4 as the **current-sinking transistor.** We will sometimes refer to it as the **pull-down transistor** because it brings the output voltage down to its LOW state.

Current-Sourcing Action

A TTL output acts as a current source in the HIGH state. This is shown in Figure 8-9(b), where transistor Q_3 is supplying the input cur-

LOW output

HIGH output

FIGURE 8-9 (a) When TTL output is in LOW state, Q_4 acts as a current sink deriving its current from the load; (b) in the output HIGH state, Q_3 acts as a current source providing current to the load gate.

rent, I_{IH}, required by the Q_1 transistor of the load gate. As stated above, this current is a small reverse-bias leakage current (typically 10 μA). We will often refer to Q_3 as the **current-sourcing transistor** or **pull-up transistor.**

Totem-Pole Output Circuit   Several points should be mentioned concerning the totem-pole arrangement of the TTL output circuit, since it is not readily apparent why it is used. The same logic could be accomplished by eliminating Q_3 and D_1 and connecting the bottom of R_4 to the collector of Q_4. But this would mean that Q_4 would conduct a fairly heavy current in its saturation state (5 V/130 $\Omega \approx$ 40 mA). With Q_3 in the circuit, there will be no current through R_4 in the output LOW state. This is important because it keeps the circuit power dissipation down.

Another advantage of this arrangement occurs in the output HIGH state. Here Q_3 is acting as an emitter follower with its associated low output impedance (typically 10 Ω). This low output impedance provides a short time constant for charging up any capacitive load on the output. This action (commonly called **active pull-up**) provides very fast rise-time waveforms at TTL outputs.

A disadvantage of the totem-pole output arrangement occurs during the transition from LOW to HIGH. Unfortunately, Q_4 turns off more slowly than Q_3 turns on, and so there is a period of a few nanoseconds during which both transistors are conducting and a relatively large current (30 to 40 mA) will be drawn from the 5-V supply. This can present a problem that will be examined later.

Summary   All TTL circuits have a structure similar to this basic NAND gate. The input of any TTL circuit will be the cathode (N-region) of a P-N junction, so that a HIGH input voltage will turn off the junction and only a small leakage current

(I_{IH}) will flow. Conversely, a LOW input voltage turns on the junction and a relatively large current (I_{IL}) will flow back through the signal source. Most, but not all, TTL circuits will have some type of totem-pole output configuration. There are some exceptions that will be discussed later.

—— REVIEW QUESTIONS ——————————————————

1. *True or false:* A TTL output acts as a current sink in the LOW state.
2. In which TTL input state does the largest amount of input current flow?
3. State the advantages and disadvantages of a totem-pole output.
4. Which TTL transistor is the pull-up transistor?

8-3 STANDARD TTL SERIES CHARACTERISTICS

In 1964 Texas Instruments Corporation introduced the first line of standard TTL ICs. The 54/74 series, as it is called, has been one of the most widely used IC logic families. We will simply refer to it as the 74 series, since the major difference between the 54 and 74 versions is that devices in the 54 series can operate over a wider range of temperatures and power-supply voltages. Many semiconductor manufacturers now produce TTL ICs. Fortunately, they all use the same numbering system, so that the basic IC number is the same from one manufacturer to another. Each manufacturer, however, usually attaches its own special prefix to the IC number. For example, Texas Instruments uses the prefix SN, National Semiconductor uses DM, and Signetics uses S. Thus, depending on the manufacturer, you may see a quad NOR-gate chip labeled as a DM7402, SN7402, S7402 or some other similar designation. The important part is the number 7402, which is the same for all manufacturers.

As we learned in Chapter 4, there are several series in the TTL family of logic devices (74, 74LS, 74S, etc.). We will first examine the electrical characteristics of the standard 74 series. Later we will introduce the other TTL series and compare their characteristics with those of the standard series.

Manufacturers' Data Sheets   To illustrate the characteristics of the standard TTL series, we will use the 7400 quad NAND-gate IC. We can find all of the information we need on any IC by consulting the manufacturer's data manual for that particular IC family. Figure 8-10 is the manufacturer's data sheet for the 5400/7400 NAND gate IC showing the recommended operating conditions, electrical characteristics, and switching characteristics. Most of the quantities discussed in the following paragraphs in this section can be found on this data sheet. As we discuss each quantity, you should refer to this data sheet to see where the information came from.

Supply Voltage and Temperature Range   Both the 74 series and 54 series use a nominal supply voltage (V_{CC}) of 5 V. The 74 series will operate reliably over the range 4.75 to 5.25 V, while the 54 series can tolerate a supply variation of 4.5 to 5.5 V. The 74 series is designed to operate properly in ambient temperatures ranging from 0 to 70°C, while the 54 series can handle −55 to +125°C. Because of its greater tolerance of voltage and temperature variations, the 54 series is more ex-

Discussion Question: Do you think the 74 series can be used in automotive applications?

recommended operating conditions

		SN5400			SN7400			UNIT
		MIN	NOM	MAX	MIN	NOM	MAX	
V_{CC}	Supply voltage	4.5	5	5.5	4.75	5	5.25	V
V_{IH}	High-level input voltage	2			2			V
V_{IL}	Low-level input voltage			0.8			0.8	V
I_{OH}	High-level output current			-0.4			-0.4	mA
I_{OL}	Low-level output current			16			16	mA
T_A	Operating free-air temperature	-55		125	0		70	$^\circ$C

electrical characteristics over recommended operating free-air temperature range (unless otherwise noted)

PARAMETER	TEST CONDITIONS†	SN5400			SN7400			UNIT
		MIN	TYP‡	MAX	MIN	TYP‡	MAX	
V_{IK}	V_{CC} = MIN, I_I = -12 mA			-1.5			-1.5	V
V_{OH}	V_{CC} = MIN, V_{IL} = 0.8 V, I_{OH} = -0.4 mA	2.4	3.4		2.4	3.4		V
V_{OL}	V_{CC} = MIN, V_{IH} = 2 V, I_{OL} = 16 mA		0.2	0.4		0.2	0.4	V
I_I	V_{CC} = MAX, V_I = 5.5 V			1			1	mA
I_{IH}	V_{CC} = MAX, V_I = 2.4 V			40			40	μA
I_{IL}	V_{CC} = MAX, V_I = 0.4 V			-1.6			-1.6	mA
I_{OS}§	V_{CC} = MAX	-20		-55	-18		-55	mA
I_{CCH}	V_{CC} = MAX, V_I = 0 V		4	8		4	8	mA
I_{CCL}	V_{CC} = MAX, V_I = 4.5 V		12	22		12	22	mA

† For conditions shown as MIN or MAX, use the appropriate value specified under recommended operating conditions.
‡ All typical values are at V_{CC} = 5 V, T_A = 25°C.
§ Not more than one output should be shorted at a time.

switching characteristics, V_{CC} = 5 V, T_A = 25°C (see note 2)

PARAMETER	FROM (INPUT)	TO (OUTPUT)	TEST CONDITIONS		MIN	TYP	MAX	UNIT
t_{PLH}	A or B	Y	R_L = 400 Ω,	C_L = 15 pF		11	22	ns
t_{PHL}						7	15	ns

NOTE 2: See General Information Section for load circuits and voltage waveforms.

FIGURE 8-10   Data sheet for 7400 NAND-gate IC. (Courtesy of Texas Instruments)

pensive. It is employed only in applications where reliable operation must be maintained over an extreme range of conditions. Examples are military and space applications.

Voltage Levels   Table 8-1 lists the required input and output voltage levels for the standard 74 series. The minimum and maximum values shown are for worst-case conditions of power supply, temperature, and loading conditions. Inspection of the table reveals a guaranteed maximum logical 0 output V_{OL} = 0.4 V, which is 400 mV less than the logical 0 voltage needed at the input V_{IL} = 0.8 V. This means that the guaranteed LOW-state dc noise margin is 400 mV. That is,

$$V_{NL} = V_{IL}(\text{max}) - V_{OL}(\text{max}) = 0.8 \text{ V} - 0.4 \text{ V} = 0.4 \text{ V} = 400 \text{ mV}$$

Similarly, the logical 1 output V_{OH} is a guaranteed minimum of 2.4 V, which is 400 mV greater than the logical 1 voltage needed at the input V_{IH} = 2.0 V. Thus, the HIGH-state dc noise margin is 400 mV.

TABLE 8-1   Standard 74 Series Voltage Levels

	Minimum	Typical	Maximum
V_{OL}	—	0.1	0.4
V_{OH}	2.4	3.4	—
V_{IL}	—	—	0.8
V_{IH}	2.0	—	—

$$V_{NH} = V_{OH}(\text{min}) - V_{IH}(\text{min}) = 2.4 \text{ V} - 2.0 \text{ V} = 0.4 \text{ V} = 400 \text{ mV}$$

Thus, the *guaranteed worst-case* dc noise margins for the 74 series are both 400 mV. In actual operation the *typical* dc noise margins are somewhat higher ($V_{NL} = 1$ V and $V_{NH} = 1.6$ V).

Maximum Voltage Ratings

The voltage values in Table 8-1 *do not include* the absolute maximum ratings beyond which the useful life of the IC may be impaired. The voltages applied to any input of a standard 74 series IC must never exceed +5.5 V. A voltage greater than +5.5 V applied to an input emitter can cause reverse breakdown of the E-B junction of Q_1.

There is also a limit on the maximum *negative* voltage that can be applied to a TTL input. This limit, -0.5 V, is caused by the fact that most TTL circuits employ protective shunt diodes on each input. These diodes were purposely left out of our earlier analysis, since they do not enter into the normal circuit operation. They are connected from each input to ground to limit the negative input voltage excursions that often occur when logic signals have excessive ringing. With these diodes, we should not apply more than -0.5 V to an input, because the protective diodes would begin to conduct and draw substantial current.

Teaching Tip: Some students may not know that "ringing" refers to the overshoot, undershoot, and damped oscillations that appear on many pulse signals.

Power Dissipation

A standard TTL NAND gate draws an average power of 10 mW. This is a result of $I_{CCH} = 4$ mA and $I_{CCL} = 12$ mA, which produces $I_{CC}(\text{avg}) = 8$ mA and $P_D(\text{avg}) = 8 \text{ mA} \times 5 \text{ V} = 40 \text{ mW}$. This 40 mW is the total power required by all four gates on the chip. Thus, one NAND gate requires an average power of 10 mW.

Propagation Delays

The standard TTL NAND gate has typical propagation delays of $t_{PLH} = 11$ ns and $t_{PHL} = 7$ ns, which is an *average* propagation delay $t_{pd}(\text{avg})$ of 9 ns.

Fan-Out

A standard TTL output can typically drive 10 standard TTL inputs. Some IC data sheets give the device's fan-out explicitly, but this is not the case for the data sheet in Figure 8-10. Later we will see how the fan-out can be determined from the input and output current values given in Figure 8-10.

Table 8-2 summarizes the characteristics of the standard 74 series NAND gate.

TABLE 8-2   Standard 74 Series Characteristics

Noise margins (worst-case)	$V_{NL} = V_{NH} = 400$ mV
Average power dissipation (basic gate)	$P_D = 10$ mW
Average propagation delay (basic gate)	9 ns
Typical fan-out	10

EXAMPLE 8-2

Common Student Mistake: Students may incorrectly use I_{IH} and I_{IL} rather than I_{CCH} and I_{CCL}.

Refer to the data sheet for the 7400 quad two-input NAND IC in Figure 8-10. Determine the *maximum* average power dissipation and *maximum* average propagation delay of a *single* gate.

Solution   Look under the electrical characteristics for the *maximum* I_{CCH} and I_{CCL} values. The values are 8 mA and 22 mA, respectively. The average I_{CC} is therefore $(8 + 22)/2 = 15$ mA. The average power is obtained by multiplying by V_{CC}. The data sheet indicates that these I_{CC} values were obtained when V_{CC} was at its maximum value (5.25 V for the 74 series). Thus we have

$$P_D(\text{avg}) = 15 \text{ mA} \times 5.25 \text{ V} = 78.75 \text{ mW}$$

as the power drawn by the *complete* IC. We can determine the power drain of one NAND gate by dividing this by 4:

$$P_D(\text{avg}) = 19.7 \text{ mW per gate}$$

Since this average power drain was calculated using the maximum current and voltage values, it is the maximum average power that a 7400 NAND gate will draw under "worst-case" conditions. Designers often use worst-case values to ensure that their circuits will work under all conditions.

The maximum propagation delays for a 7400 NAND gate are listed as

$$t_{PLH} = 22 \text{ ns}, \qquad t_{PHL} = 15 \text{ ns}$$

so that the average propagation delay is

$$t_{pd}(\text{avg}) = \frac{22 + 15}{2} = 18.5 \text{ ns}$$

Again, this is a worst-case maximum possible average propagation delay.

8-4   OTHER TTL SERIES

The standard 74 series ICs offer a combination of speed and power dissipation suited for many applications. ICs offered in this series include a wide variety of gates, flip-flops, and one-shots in the small-scale integration (SSI) line and shift registers, counters, decoders, memories, and arithmetic circuits in the medium-scale integration (MSI) line.

Several other TTL series have been developed since the introduction of the standard 74 series. These other series provide a wide choice of speed and power characteristics. We will describe these series in the following paragraphs. Notice that whenever we use the term "TTL," we are usually referring to the standard 74 series.

Teaching Tip: The 74L and 74H series are presented here for historical reasons only. Of course, these logic series may be encountered in old equipment.

Low-Power TTL, 74L Series   Low-power TTL circuits designated as the 74L series have essentially the same basic circuit as the standard 74 series except that all the resistor values are *increased*. The larger resistors reduce the power requirements but at the expense of longer propagation delays. A typical NAND gate in this series has an average power dissipation of 1 mW and an average propagation delay of 33 ns.

The 74L series is well suited for applications in which power dissipation is more critical than speed. Low-frequency, battery-operated circuits such as calculators are well suited for this TTL series. This series has the lowest power dissipation of all the TTL series.

For the most part, the 74L series has been rendered obsolete by the development of the 74LS, 74ALS, and CMOS series, each of which offers low power dissipation at a higher speed of operation than that of the 74L devices. For this reason, the 74L series is not recommended for use in any new circuit designs.

High-Speed TTL, 74H Series   The 74H series is a high-speed TTL series. The basic circuitry for this series is essentially the same as the standard 74 series except that *smaller* resistor values are used and the emitter-follower transistor Q_3 is replaced by a Darlington pair. These differences result in a much faster switching speed with an average propagation delay of 6 ns. However, the increased speed is accomplished at the expense of increased power dissipation. The basic NAND gate in this series has an average P_D of 23 mW. The 74H series has become obsolete since the development of the Schottky TTL series described in the following paragraphs, and is not recommended for new designs.

Schottky TTL, 74S Series   The 74, 74H, and 74L series all operate using saturated switching in which many of the transistors, when conducting, will be in the saturated condition. This operation causes a storage-time delay, t_S, when the transistors switch from ON to OFF, and limits the circuit's switching speed.

The 74S series reduces this storage-time delay by not allowing the transistor to go as deeply into saturation. It accomplishes this by using a Schottky barrier diode (SBD) connected between the base and collector of each transistor as shown in Figure 8-11(a). The SBD has a forward voltage of only 0.25 V. Thus, when the C-B

FIGURE 8-11   (a) Schottky-clamped transistor; (b) basic NAND gate in S-TTL series. (Courtesy of Fairchild, a Schlumberger company)

junction becomes forward-biased at the onset of saturation, the SBD will conduct and divert some of the input current away from the base. This reduces the excess base current and decreases the storage-time delay at turn-off.

As shown in Figure 8-11(a), the transistor/SBD combination is given a special symbol. This symbol is used for all the transistors in the circuit diagram for the 74S00 NAND gate shown in Figure 8-11(b). This 74S00 NAND gate has an average propagation delay of only 3 ns, which is twice as fast as the 74H00. Note the presence of shunt diodes D_1 and D_2 to limit negative input voltages.

Circuits in the 74S series also use smaller resistor values to help improve switching times. This increases the circuit average power dissipation to about 20 mW, about the same as for 74H. The 74S circuits also use a Darlington pair (Q_3 and Q_4) to provide a more rapid output rise time when switching from ON to OFF.

Thus, 74S has twice the speed of 74H at about the same power requirement. This is why 74H has become obsolete.

Low-Power Schottky TTL, 74LS Series (LS-TTL)   The 74LS series is a lower-powered, slower-speed version of the 74S series. It uses the Schottky-clamped transistor, but with larger resistor values than the 74S series (see Figure 8-12). The larger resistor values reduce the circuit power requirement, but at the expense of an increase in switching times.

A NAND gate in the 74LS series will typically have an average propagation delay of 9.5 ns and an average power dissipation of 2 mW. Since it has about the same switching speed as the standard TTL series at a much lower power require-

FIGURE 8-12   74LS NAND gate. (Courtesy of Fairchild, a Schlumberger company)

ment, the 74LS series has gradually replaced the 74 series in those applications where relatively high-speed operation is required at minimum power consumption. It has become the "mainstay" of the TTL family, and it can be found in almost all new designs that do not require maximum speed. Its leading position, however, will gradually be taken over by the new, improved 74ALS series.

Note that the 74LS NAND gate in Figure 8-12 does not use the multiple-emitter input transistor. Instead, it uses diodes D_3 and D_4, but the circuit operation remains essentially the same.

Advanced Schottky TTL, 74AS Series (AS-TTL)

Recent innovations in integrated-circuit design have led to the development of two new improved TTL series: advanced Schottky (74AS) and advanced low-power Schottky (74ALS). The 74AS series provides a considerable improvement in speed over the 74S series at a much lower power requirement. The comparison is shown below for a NAND gate in each series.

	74S	74AS
Propagation delay	3 ns	1.7 ns
Power dissipation	20 mW	8 mW
Speed-power product	60 pJ	13.6 pJ

This comparison clearly shows the advantage of the 74AS series. It is the fastest TTL series and its speed-power product is significantly lower than the 74S series. The 74AS has other improvements, including lower input current requirements (I_{IL}, I_{IH}), that result in a greater fan-out than the 74S series.

Because of these advantages, the 74AS is replacing the 74S series in all high-speed applications. As the cost of 74AS devices continues to come down, and as more functions become available in this series, the 74AS will undoubtedly make the 74S series obsolete.

Advanced Low-Power Schottky TTL, 74ALS Series

This series offers an improvement over the 74LS series in both speed and power dissipation, as the following numbers illustrate.

	74LS	74ALS
Propagation delay	9.5 ns	4 ns
Power dissipation	2 mW	1.2 mW
Speed-power product	19 pJ	4.8 pJ

The 74ALS series has the lowest speed-power product of all the TTL series and is very close to having the lowest gate power dissipation (74L has 1 mW). For this reason we will probably see the 74ALS eventually replacing the 74LS as the most widely used TTL series.

Comparison of TTL Series Characteristics

Table 8-3 gives the typical values for some of the more important characteristics of each of the TTL series. All

Teaching Tip: From Table 8-3 it is clear that the 74AS and 74ALS series are superior to the others, and will most probably be used in all new designs, except in those increasingly rare situations where some logic function is not yet available in these series. Of course, you will find 74LS and even 74 devices in older designs.

TABLE 8-3   Typical TTL Series Characteristics

	74	74L	74H	74S	74LS	74AS	74ALS
Performance ratings							
Propagation delay (ns)	9	33	6	3	9.5	1.7	4
Power dissipation (mW)	10	1	23	20	2	8	1.2
Speed-power product (pJ)	90	33	138	60	19	13.6	4.8
Max. clock rate (MHz)	35	3	50	125	45	200	70
Fan-out (same series)	10	20	10	20	20	40	20
Voltage parameters							
$V_{OH}(min)$	2.4	2.4	2.4	2.7	2.7	2.5	2.5
$V_{OL}(max)$	0.4	0.4	0.4	0.5	0.5	0.5	0.4
$V_{IH}(min)$	2.0	2.0	2.0	2.0	2.0	2.0	2.0
$V_{IL}(max)$	0.8	0.7	0.8	0.8	0.8	0.8	0.8

of the performance ratings, except for maximum clock rate, are for a NAND gate in each series. The maximum clock rate is specified as the maximum frequency that can be used to toggle a J-K flip-flop. This gives a useful measure of the frequency range over which each IC series can be operated.

—— EXAMPLE 8-3 ——

Use Table 8-3 to calculate the dc noise margins for a typical 74LS IC. How does this compare with the standard TTL noise margins obtained in Section 8-3?

Solution

$$V_{NH} = V_{OH}(min) - V_{IH}(min)$$
$$= 2.7 \text{ V} - 2.0 \text{ V}$$
$$= 0.7 \text{ V}$$

as compared with $V_{NH} = 0.4$ V for standard TTL.

$$V_{NL} = V_{OL}(max) - V_{IL}(max)$$
$$= 0.8 \text{ V} - 0.5 \text{ V}$$
$$= 0.3 \text{ V}$$

as compared with $V_{NL} = 0.4$ V for TTL.

—— EXAMPLE 8-4 ——

Which TTL series can drive the most device inputs of the same series?

Solution   The 74AS series has the highest fan-out (40). This means, for example, that a 74AS00 NAND gate can drive 40 inputs of other 74AS devices. If we want to determine the number of inputs of a *different* TTL series that an output can drive, we will need to know the input and output currents of the two series. This will be dealt with in the next section.

1. (a) Which TTL series is the best at high frequencies?
   (b) Which TTL series has the largest HIGH-state noise margin?
   (c) Which series have essentially become obsolete?
   (d) Which series use a special diode to reduce switching time?
   (e) Which series would be best for a battery-powered circuit operating at 10 MHz?
2. Assuming the same cost for each, why should you choose to use a 74ALS193 counter over a 74LS193 or 74AS193 in a circuit operating from a 40-MHz clock?

8-5 TTL LOADING AND FAN-OUT

It is important to understand what determines the fan-out or load drive capability of an IC output. Figure 8-13(a) shows a standard TTL output in the LOW state connected to drive several standard TTL inputs. Transistor Q_4 is on and is acting as a current sink for an amount of current I_{OL} that is the sum of the I_{IL} currents from each input. In its on state, Q_4's collector-emitter resistance is very small, but it is not zero, and so the current I_{OL} will produce a voltage drop V_{OL}. This voltage must not exceed the $V_{OL}(max)$ limit of the IC. This limits the maximum value of I_{OL} and thus the number of loads that can be driven.

To illustrate, suppose the ICs are in the 74 series and each I_{IL} is 1.6 mA. From Table 8-3 we see that the 74 series has $V_{OL}(max) = 0.4$ V and $V_{IL}(max) = 0.8$ V.

FIGURE 8-13 Currents when a TTL output is driving several inputs.

(a)   (b)

Let's suppose further that Q_4 can sink up to 16 mA before its output voltage reaches $V_{OL}(\text{max}) = 0.4$ V. This means that it can sink the current from up to 16 mA/1.6 mA = 10 loads. If it is connected to more than 10 loads, its I_{OL} will increase and cause V_{OL} to increase above 0.4 V. This is usually undesirable because it reduces the noise margin at the IC inputs [remember, $V_{NL} = V_{IL}(\text{max}) - V_{OL}(\text{max})$]. In fact, if V_{OL} rises above $V_{IL}(\text{max}) = 0.8$ V, it will be in the indeterminate range.

A similar situation occurs in the HIGH state depicted in Figure 8-13(b). Here Q_3 is acting as an emitter follower that is sourcing (supplying) a total current I_{OH} that is the sum of the I_{IH} currents of the different TTL inputs. If too many loads are being driven, this current I_{OH} will become large enough to cause the voltage drops across R_2, Q_3's emitter-base junction, and D_1 to bring V_{OH} below $V_{OH}(\text{min})$. This too is undesirable since it reduces the HIGH-state noise margin and could even cause V_{OH} to go into the indeterminate range.

What this all means is that a TTL output has a limit, $I_{OL}(\text{max})$, on how much current it can sink in the LOW state. It also has a limit, $I_{OH}(\text{max})$, on how much current it can source in the HIGH state. These output current limits must not be exceeded if the output voltage levels are to be maintained within their specified ranges.

Determining the Fan-Out

To determine how many different inputs an IC output can drive, you need to know the current drive capability of the output [that is, $I_{OL}(\text{max})$ and $I_{OH}(\text{max})$] and the current requirements of each input (that is, I_{IL} and I_{IH}). This information is always presented in some form on the manufacturer's IC data sheet. The following examples will illustrate one type of situation.

—— EXAMPLE 8-5 ——————————————————————

How many 7400 NAND-gate inputs can be driven by a 7400 NAND-gate output?

Teaching Tip: These current values are absolute maximums that will occur under worst-case conditions of temperature, power-supply voltage, and circuit component variations. Typical values for the 74 series, measured under normal conditions, can be much different (e.g., $I_{OL}(\text{max}) = 25$ mA, $I_{IL} = 1$ mA, $I_{OH} = 800$ μA, and $I_{IH} = 10$ μA). These would predict a much larger fan-out. Of course, good designers normally use worst-case values to ensure maximum circuit reliability.

Solution   We will consider the LOW state first as depicted in Figure 8-14. Refer to the 7400 data sheet in Figure 8-10 and find

FIGURE 8-14   Example 8-5.

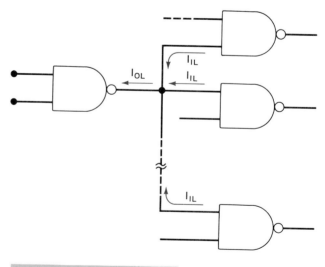

*All gates are 7400 NAND gates.

$$I_{OL}(\text{max}) = 16 \text{ mA}$$

$$I_{IL}(\text{max}) = 1.6 \text{ mA}$$

This says that a 7400 output can sink a maximum of 16 mA and that each 7400 input will source a maximum of 1.6 mA back through the driving gate's output. Thus, the number of inputs that can be driven in the LOW state is obtained as

$$\text{fan-out(LOW)} = \frac{I_{OL}(\text{max})}{I_{IL}(\text{max})}$$

$$= \frac{16 \text{ mA}}{1.6 \text{ mA}}$$

$$= 10$$

(*Note:* The entry for I_{IL} is actually -1.6 mA. The negative sign is used to indicate that this current flows *out of* the input terminal; we can ignore this sign for our purposes here.) The HIGH state is analyzed in the same manner. Refer to the data sheet to find values for I_{OH} and I_{IH}, ignoring any negative signs.

$$I_{OH}(\text{max}) = 0.4 \text{ mA} = 400 \text{ }\mu\text{A}$$

$$I_{IH}(\text{max}) = 40 \text{ }\mu\text{A}$$

Thus, the number of inputs that can be driven in the HIGH state is

$$\text{fan-out (HIGH)} = \frac{I_{OH}(\text{max})}{I_{IH}(\text{max})}$$

$$= \frac{400 \text{ }\mu\text{A}}{40\mu\text{A}}$$

$$= 10$$

These results indicate that the fan-out is 10 in both voltage states. Thus, the 7400 NAND gate can drive up to 10 other 7400 NAND gates. If fan-out (LOW) and fan-out (HIGH) are not the same, as will sometimes occur, the fan-out is chosen as the smaller of the two.

—— EXAMPLE 8-6 ——————————————————————

Refer to the data sheet in Appendix II and determine how many 74ALS20 NAND gates can be driven by the output of another 74ALS20.

Solution   The 74ALS20 data sheet gives the following values:

$$I_{OH}(\text{max}) = 0.4 \text{ mA} = 400 \text{ }\mu\text{A}$$

$$I_{OL}(\text{max}) = 8 \text{ mA}$$

$$I_{IH}(\text{max}) = 20 \text{ }\mu\text{A}$$

$$I_{IL}(\text{max}) = 0.1 \text{ mA}$$

Considering the HIGH state first, we have

$$\text{fan-out (HIGH)} = \frac{400 \text{ }\mu\text{A}}{20 \text{ }\mu\text{A}} = 20$$

For the LOW state we have

$$\text{fan-out (LOW)} = \frac{8 \text{ mA}}{0.1 \text{ mA}} = 80$$

In this case, the overall fan-out is chosen to be 20 since it is the lower of the two values. Thus, one 74ALS20 can drive 20 other 74ALS20 inputs.

TABLE 8-4   TTL Series Loading Factors

TTL Series	Input Loading		Fan-Out	
	High	Low	High	Low
74	1 UL	1 UL	10 UL	10 UL
74H	1.25 UL	1.25 UL	12.5 UL	12.5 UL
74L	0.5 UL	≈0.1 UL	10 UL	2.5 UL
74S	1.25 UL	1.24 UL	25 UL	12.5 UL
74LS	0.5 UL	≈0.25 UL	10 UL	5 UL
74AS	0.5 UL	≈0.3 UL	50 UL	12.5 UL
74ALS	0.5 UL	≈0.06 UL	10 UL	5 UL

$$1 \text{ UL} = \begin{cases} 40 \ \mu\text{A (HIGH)} \\ 1.6 \text{ mA (LOW)} \end{cases}$$

Unit Loads   Some manufacturers specify the device input and output currents in terms of a unit load (UL) where a unit load is defined as follows:

$$1 \text{ unit load (UL)} = \begin{cases} 40 \ \mu\text{A in the HIGH state} \\ 1.6 \text{ mA in the LOW state} \end{cases}$$

This simply means that when we talk about the HIGH state, UL is the same as 40 μA; when we talk about the LOW state, UL is the same as 1.6 mA. For example, if an IC is specified as having a fan-out of 10 UL in both states, this is the same as saying that

$$I_{OH}(\text{max}) = 10 \times 40 \ \mu\text{A} = 400 \ \mu\text{A}$$
$$I_{OL}(\text{max}) = 10 \times 1.6 \text{ mA} = 16 \text{ mA}$$

Similarly, if an IC input is rated as 1 UL in both states, this is the same as saying that

$$I_{IH}(\text{max}) = 1 \times 40 \ \mu\text{A} = 40 \ \mu\text{A}$$
$$I_{IL}(\text{max}) = 1 \times 1.6 \text{ mA} = 1.6 \text{ mA}$$

Table 8-4 gives the typical input and output factors for the various TTL series. These values are typical, and so there may be some variations, depending on the particular device or manufacturer. The device data sheet should be consulted to determine exact values for a specific IC.

Discussion Question: How can Table 8-4 be used to determine the number of inputs of a given series that can *typically* be driven by one output of the same series?

——— EXAMPLE 8-7 ————————————————————————————

Determine the input and output loading factors for the 7404 INVERTER CHIP (Appendix II).

Solution   This information can be found on the 7404 data sheet under the heading "INPUT LOADING/FAN-OUT." Look under "54/74(UL)" and you will find the first entry, 1.0/1.0. This means that the input requirements are 1 UL in each state. In other words, the 7404 input draws a maximum of 40 μA of current from the input signal source in the HIGH state, and sources a maximum of 1.6 mA back through a LOW input. These are the maximum worst-case input currents that flow under extreme conditions. In practice, you might measure $I_{IH} = 10 \ \mu$A and $I_{IL} = 1.1$ mA at the input. Most designers assume the worst-case values that the manufacturer provides on the data sheets.

The second entry under "54/74(UL)" is 20/10. This means that the 7404 output is rated at 20 UL in the HIGH state and 10 UL in the LOW state. In other words, the 7404 output can *supply* up to

$$20 \times 40 \ \mu\text{A} = 800 \ \mu\text{A}$$

to load devices in the HIGH state without its V_{OH} dropping below $V_{OH}(\text{min}) = 2.4$ V, and it can *sink* up to

$$10 \times 1.6 \text{ mA} = 16 \text{ mA}$$

in the LOW state without its V_{OL} rising above $V_{OL}(\text{max}) = 0.4$ V.

___ EXAMPLE 8-8 ___

Repeat for the 74LS04 IC.

Solution   Again, we look under the heading "INPUT LOADING/FAN-OUT" for the 54/74LS entry, and we see that the inputs are rated at 0.5/0.25. Thus, the 74LS04 has an input loading factor of 0.5 UL in the HIGH state, and 0.25 UL in the LOW state.

The outputs are shown rated at 10/5.0, which means that a 74LS04 output can drive 10 UL in the HIGH state, and 5 UL in the LOW state.

___ EXAMPLE 8-9 ___

The output of a 74S04 INVERTER is providing the clock signal to a parallel register made up of 74S112 J-K flip-flops. What is the maximum number of FFs that this clock signal can drive?

Discussion Question: Does this mean that, if you connected the 74S04 output to a register made up of six 74S112s, the circuit would not work?

Solution   This type of problem is solved by first determining the unit load capabilities of a 74S04 output and then determining the unit load input requirements of a 74S112 *CLK* input. The 74S04 data sheet lists the output fan-out as 25 UL (HIGH) and 12.5 UL (LOW). The 74S112 data sheet shows that its $\overline{CP}$ input requirements are 2.5 UL in both states. Thus, the number of 74S112 $\overline{CP}$ loads that a 74S04 output can drive is

$$\text{number of loads} = \frac{\text{output rating}}{\text{input rating}} = \frac{12.5 \text{ UL}}{2.5 \text{ UL}} = 5$$

Note that we used the 74S04 LOW-state output rating because it is more restrictive than its HIGH-state output rating.

___ EXAMPLE 8-10 ___

Repeat Example 8-9 using a 74LS04 and a 74LS112.

Solution   The 74LS04 has output ratings of 10 UL (HIGH) and 5 UL (LOW), while the 74LS112 clock inputs require 2 UL (HIGH) and 0.5 UL (LOW). Here it will be necessary to calculate the number of 74LS112 loads that a 74LS04 can drive in each state, and then take the smaller of the two numbers.

$$\text{number of loads (HIGH)} = \frac{10 \text{ UL}}{2 \text{ UL}} = 5$$

$$\text{number of loads (LOW)} = \frac{5 \text{ UL}}{0.5 \text{ UL}} = 10$$

Thus, a 74LS04 can drive five 74LS112 $\overline{CP}$ inputs.

___ EXAMPLE 8-11 ___

A certain IC output is rated at $I_{OH}(\text{max}) = 800$ μA and $I_{OL} = 48$ mA. Express the IC's fan-out in terms of unit loads.

Solution

> *HIGH state:* fan-out = 800 μA/40 μA = 20 UL
> *LOW state:* fan-out = 48 mA/1.6 mA = 30 UL

—— EXAMPLE 8-12 ——

A certain TTL flip-flop's *CLK* input is rated at $I_{IL} = 0.8$ mA and $I_{IH} = 10$ μA. Express its input requirements in ULs.

Solution

> *HIGH state:* input requirement = 10 μA/40 μA = 0.25 UL
> *LOW state:* input requirement = 0.8 mA/1.6 mA = 0.5 UL

These examples should make it clear that the UL rating is just a convenient way of specifying an IC's input and output current ratings. Some manufacturers prefer to list these current ratings explicitly without using ULs; others use ULs; still others provide the information in both forms. Since you will typically use ICs from several manufacturers, you should be able to convert between ULs and currents.

—— REVIEW QUESTIONS ——

1. What factors determine a device's $I_{OL}(\text{max})$ rating?
2. A certain TTL output has a fan-out of 5 UL in both states. How much current can it supply to loads in the HIGH state?
3. What can happen if a TTL output is connected to more unit loads than its output rating specification?
4. How many 74S112 $\overline{CP}$ inputs can be driven by a 74LS04 output?

8-6 OTHER TTL CHARACTERISTICS

Several other characteristics of TTL logic must be understood if one is to intelligently use TTL in a digital-system application.

Teaching Tip: In our earlier discussions of troubleshooting, we stated that the voltage measured at an unconnected TTL input would typically be in the range 1.4 to 1.8 V. You can use Figure 8-8(a) to show where this voltage comes from.

Unconnected Inputs (Floating)   Any input to a TTL circuit that is left disconnected (open) acts exactly like a logical 1 applied to that input, because in either case the emitter-base junction or diode at the input will not be forward-biased. This means that on *any* TTL IC, *all* the inputs are 1s if they are not connected to some logic signal or to ground. When an input is left unconnected, it is said to be "floating."

Unused Inputs   Frequently, all the inputs on a TTL IC are not being used in a particular application. A common example is when all the inputs to a logic gate are not needed for the required logic function. For example, suppose that we needed the logic operation $\overline{AB}$ and we were using a chip that had a three-input NAND gate. The possible ways of accomplishing this are shown in Figure 8-15.

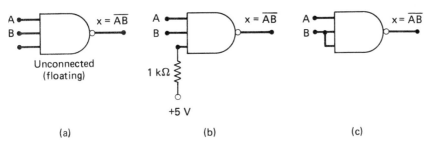

FIGURE 8-15   Three ways to handle unused logic inputs.

In Figure 8-15(a) the unused input is left disconnected, which means that it acts as a logical 1. The NAND-gate output is therefore $x = \overline{A \cdot B \cdot 1} = \overline{A \cdot B}$, which is the desired result. Although the logic is correct, it is highly undesirable to leave an input disconnected, because it will act like an antenna, which is liable to pick up stray radiated signals that could cause the gate to operate improperly. A better technique is shown in Figure 8-15(b). Here the unused input is connected to +5 V through a 1-kΩ resistor, so that the logic level is a 1. The 1-kΩ resistor is simply for current protection of the emitter-base junctions of the gate inputs in case of spikes on the power-supply line. This same technique can be used for AND gates, since a 1 on an unused input will not affect the output. As many as 30 unused inputs can share the same 1-kΩ resistor tied to V_{CC}.

A third possibility is shown in Figure 8-15(c), where the unused input is tied to a used input. This is satisfactory provided that the circuit driving input B is not going to have its fan-out exceeded. This technique can be used for *any* type of gate.

For OR gates and NOR gates, the unused inputs cannot be left disconnected or tied to +5 V since this would produce a constant-output logic level (1 for OR, 0 for NOR) regardless of the other inputs. Instead, for these gates the unused inputs must either be connected to ground (0 V) for a logic 0 or be tied to a used input as in Figure 8-15(c).

Tied-Together Inputs   When two or more TTL-gate inputs are connected together to form a common input as in Figure 8-15(c), the common input will generally have an input loading factor that is the sum of the input loading factors for the different inputs. The only exception is for NAND and AND gates. For these gates, the LOW-state input loading factor will be the *same as a single input* no matter how many inputs are tied together.

To illustrate, assume that each input of the three-input NAND gate in Figure 8-15(c) is rated at 1 UL in each state. The common B input will therefore present an input loading factor of 2 UL in the HIGH state but only 1 UL in the LOW state. The same would be true if this were an AND gate. If it were an OR or a NOR gate, the common B input would present an input loading factor of 2 UL in both states.

The reason for this characteristic can be found by looking back at the circuit diagram of the TTL NAND gate in Figure 8-8(b). The current I_{IL} is limited by the resistance R_1. Even if inputs A and B were tied together and grounded, this current would not change; it would merely divide up and flow through the parallel paths provided by diodes D_2 and D_3. The situation is different for OR and NOR gates, since they do not use multiple-emitter transistors, but rather have a separate input transistor for each input.

Teaching Tip: If you want to show the structure of TTL OR/NOR gates with their separate input transistors, refer to the data sheet for the 7402 on page 2-14 of the 1988 edition of the Texas Instruments TTL Logic Databook.

_____ EXAMPLE 8-13 _____

Determine the number of ULs that the X output is driving in Figure 8-16. Assume each gate input is rated at 1 UL in each state.

FIGURE 8-16 Example 8-13.

| Loading on gate 1 output | |
High state	Low state
2 UL — gate 2	1 UL — gate 2
1 UL — gate 3	1 UL — gate 3
3 UL — gate 4	3 UL — gate 4
6 UL — total	5 UL — total

Solution   The loading on X will be different for its two different logic states as shown in the figure. Note that the tied-together inputs of NAND gate 2 are counted as only 1 UL in the LOW state. The NOR gate inputs are treated as separate loads for either state.

Biasing TTL Inputs Low   Occasionally, the situation arises where a TTL input must be held normally LOW and then caused to go HIGH by the actuation of a mechanical switch. This is illustrated in Figure 8-17 for the input to a one-shot. This

Teaching Tip: It can be argued that choosing R_{max} to produce $V_{IL}(max)$ leaves no margin for noise. Keep in mind, however, that we are assuming a current of $I_{IL}(max)$, which is a worst-case value that never occurs in practice. Furthermore, the potential for noise pickup at this point is minimal because of the relatively low resistance to ground. If we want to be ultraconservative, we can use a value for R_{max} of half the calculated value, but this would place an unnecessary drain on the power supply when the switch is closed.

FIGURE 8-17

OS triggers on a positive transition that occurs when the switch is momentarily closed. The resistor R serves to keep the T input LOW while the switch is open. Care must be taken to keep the value of R low enough so that the voltage developed across it by the current I_{IL} that flows out of the OS input to ground will not exceed $V_{IL}(max)$. Thus, the largest value of R is given by

$$I_{IL} \times R_{max} = V_{IL}(max)$$

$$R_{max} = \frac{V_{IL}(max)}{I_{IL}} \tag{8.3}$$

R must be kept below this value to ensure that the OS input will be at an acceptable LOW level while the switch is open. The minimum value of R is determined by the current drain on the 5-V supply when the switch is closed. In practice, this current drain should be minimized by keeping R just slightly below R_{max}.

—— EXAMPLE 8-14 ————

Determine an acceptable value for R if the OS is a standard TTL IC with an input rating of 1 UL.

Solution At 1 UL, the value of I_{IL} will be a maximum of 1.6 mA. This maximum value should be used to calculate R_{max}. For standard TTL, $V_{IL}(max) = 0.8$ V. Thus, we have

$$R_{max} = \frac{0.8 \text{ V}}{1.6 \text{ mA}} = 500 \text{ } \Omega$$

A good choice here would be $R = 470$ Ω, a standard resistor value.

▬ EXAMPLE 8-15 ▬▬▬▬▬▬▬▬▬▬▬▬▬▬▬▬▬▬▬▬▬▬

Repeat Example 8-14 if the OS belongs to the 74LS series with an input rating of 0.25 UL.

Solution At 0.25 UL, we have $I_{IL}(max) = 0.4$ mA. For 74LS, $V_{IL}(max) = 0.8$ V. Thus,

$$R_{max} = \frac{0.8 \text{ V}}{0.4 \text{ ma}} = 2 \text{ k}\Omega$$

A good choice here would be 1.8 kΩ, a standard value.

Current Transients TTL logic circuits suffer from internally generated current transients or spikes because of the totem-pole output structure. When the output is switching from the LOW state to the HIGH state (see Figure 8-18), the two output transistors are changing states: Q_3 OFF to ON and Q_4 ON to OFF. Since Q_4 is changing from the saturated condition, it will take longer than Q_3 to switch states. Thus, there is a short interval of time (about 2 ns) during the switching transition when both transistors are conducting and a relatively large surge of current (30 to 50 mA) is drawn from the +5 V supply. The duration of this current transient is extended by the effects of any load capacitance on the circuit output. This capacitance consists of

FIGURE 8-18 A large current spike is drawn from V_{CC} when a totem-pole output switches from LOW to HIGH.

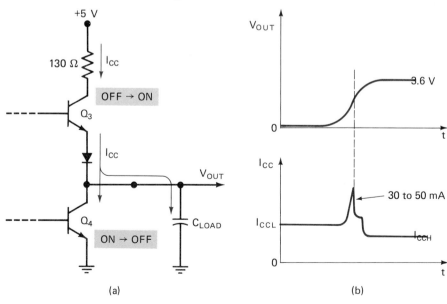

stray wiring capacitance and the input capacitance of any load circuits and must be charged up to the HIGH-state output voltage. This overall effect can be summarized as follows: *Whenever a totem-pole TTL output goes from LOW to HIGH, a high-amplitude current spike is drawn from the V_{CC} supply.*

In a complex digital circuit or system there may be many TTL outputs switching states at the same time, each one drawing a narrow spike of current from the power supply. The accumulative effect of all these current spikes will be to produce a voltage spike on the common V_{CC} line, mostly due to the distributed inductance on the supply line [remember: $V = L(di/dt)$ for inductance, and di/dt is very large for a 2-ns current spike]. This voltage spike can cause serious malfunctions during switching transitions unless some type of filtering is used. The most common technique uses small radio-frequency capacitors connected from V_{CC} to GROUND to essentially "short out" these high-frequency spikes. This is called **power-supply decoupling.**

It is standard practice to connect a $0.01\text{-}\mu\text{F}$ or $0.1\text{-}\mu\text{F}$ low-inductance, ceramic disk capacitor between V_{CC} and ground near each TTL IC on a circuit board. The capacitor leads are kept very short to minimize series inductance.

In addition, it is standard practice to connect a single large capacitor (2 to 20 μF) between V_{CC} and ground on each board to filter out relatively low-frequency variations in V_{CC} caused by the large changes in I_{CC} levels as outputs switch states.

Discussion Question: Why are such low values of capacitance used to short out these spikes? Wouldn't larger values work better since capacitive reactance will be lower and therefore present a lower-impedance path to ground?

—— REVIEW QUESTIONS ——

1. What will be the logic output of a TTL NAND gate that has all its inputs unconnected?
2. What are two acceptable ways to handle unused inputs to an AND gate?
3. Repeat for a NOR gate.
4. *True or false:* When NAND gate inputs are tied together, they are always treated as a single load on the signal source.
5. What is power-supply decoupling? Why is it used?

8-7   TTL OPEN-COLLECTOR OUTPUTS

Consider the logic circuit of Figure 8-19(a). NAND gates 4 and 5 provide the AND function, which is ANDing the outputs of NAND gates 1, 2, and 3, so that the final output x has the expression

$$x = \overline{AB} \cdot \overline{CD} \cdot \overline{EF}$$

The circuit of Figure 8-19(b) shows the same logic operation obtained by simply tying together the outputs of NAND gates 1, 2, and 3. In other words, the AND operation is performed by tying the outputs together. This can be reasoned as follows: With all outputs tied together, when any one of the gate outputs goes to the LOW state, the common output point must go LOW as a result of the "shorting-to-ground" action of the Q_4 transistor in that gate. The common output point will be HIGH only when all gate outputs are in the HIGH state. Clearly, this is the AND operation.

The arrangement in Figure 8-19(b) has an advantage over the conventional arrangement of Figure 8-19(a). It requires fewer gates to produce the desired output. This configuration is called the **wired-AND** operation because it produces the AND

Teaching Tip: The term wired-OR comes about because the expression for X can be rearranged as $X = AB + CD + EF$. The actual logic operation performed at the point where the outputs are connected is an AND operation.

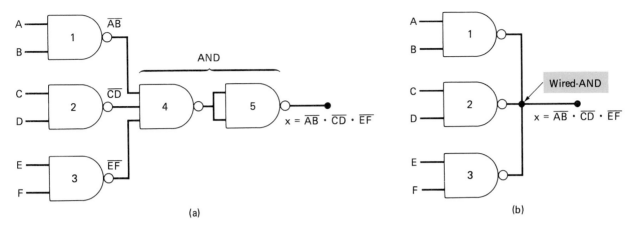

$x = \overline{AB} \cdot \overline{CD} \cdot \overline{EF}$

$x = \overline{AB} \cdot \overline{CD} \cdot \overline{EF}$

(a)   (b)

FIGURE 8-19   Wired-AND operation.

operation by connecting output wires together. It is sometimes misleadingly called the *wired-OR* operation.

Totem-Pole Outputs Cannot Be Wired-ANDed

In order to take advantage of the wired-AND configuration the outputs of two or more gates must be tied together without harmful effects. Unfortunately, the totem-pole output circuitry of conventional TTL circuits prohibits tying outputs together. This is illustrated in Figure 8-20, where the totem-pole outputs of two separate gates are connected together at point X. Suppose that the gate-A output is in the HIGH state (Q_{3A} on, Q_{4A} off) and the gate-B output is in the LOW state (Q_{3B} off, Q_{4B} on). In this situation Q_{4B} is a very low-resistance load on Q_{3A} and will draw a current which can go as high as 55 mA.

FIGURE 8-20   Totem-pole outputs tied together can produce harmful current through Q_4.

This current can easily damage Q_{3A} or Q_{4B}. The situation is even worse when more than two TTL outputs are tied together.

Sometimes totem-pole outputs are unintentionally tied together because of wiring errors or accidental shorting on a printed circuit board. When this happens, the signal at the common point will usually be the AND combination of the two output signals that have been shorted together. This is true only of TTL circuits; MOS and CMOS behave less predictably when outputs are tied together.

Open-Collector Outputs   To permit wire-ANDing, some TTL circuits are designed with *open-collector* outputs. As shown in Figure 8-21(a) the open-collector-type circuit eliminates Q_3, D_1, and R_4. The output is taken at Q_4's collector, which is unconnected. In the output LOW state, Q_4 is on (has base current), and in the HIGH state it is off (essentially an open circuit). For proper operation an external *pull-up* collector resistor R_P should be connected as shown in Figure 8-21(b), so that a high voltage level will appear at the output in the HIGH state. Without R_P, there would be no output voltage when Q_4 is off.

With open-collector outputs, the wired-AND operation can be accomplished safely. Figure 8-22 shows three 2-input *open-collector* NAND gates (7401s) that are wired-ANDed together. Notice that the open-collector NAND gates have no special symbol. Also notice the presence of the external pull-up resistor (R_P) and the sometimes-used wired-AND symbol.

Discussion Question: What logic voltage levels would appear at the output if R_P is open?

Value of R_P   The value of R_P must be chosen so that when one gate output goes LOW while the others are HIGH, the sink current through the LOW output does not exceed its $I_{OL}(max)$ limit. A value of $R_P = 1$ kΩ will produce a sink current of about 5 mA through the LOW gate's output transistor. Of course, the output is usually driving other TTL loads that will add to this sink current, whose total must not exceed $I_{OL}(max)$. It might appear that a large value of R_P would therefore be advisable. However, it must be realized that any load capacitance will be charged up through

FIGURE 8-21   (a) Open-collector TTL circuit; (b) with external pull-up resistor.

(a)

(b)

Operating states

| Q_4 ON $\rightarrow$ V_O = V_{OL} $\leqslant$ 0.4 V |
| Q_4 OFF $\rightarrow$ V_O = V_{OH} = +5 V |

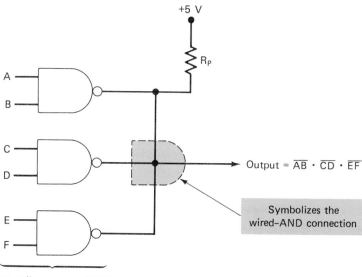

+5 V

R_P

A
B

C
D

Output = $\overline{AB} \cdot \overline{CD} \cdot \overline{EF}$

Symbolizes the
wired-AND connection

E
F

Open collector gates (7401s)

FIGURE 8-22   Wired-AND operation using open-collector gates.

R_P, so that R_P should be made as small as possible to enhance switching speed. Even with R_P minimized, this open-collector arrangement is much slower than totem-pole TTL outputs, where Q_3 acts as a low-impedance emitter follower to rapidly charge up load capacitance. For this reason, open-collector circuits should not be used in applications where switching speed is a principal consideration.

—— EXAMPLE 8-16 ——

The 7405 IC (see Appendix II) contains six inverters with open-collector outputs. These six inverters are connected in a wired-AND arrangement in Figure 8-23(a).

(a) Determine the logic expression for output X.

(b) Determine a value for R_P assuming that output X is to drive other circuits with a total loading factor of 4 UL.

Solution

(a) Each inverter output is the inverse of its input. The wired-AND connection simply ANDs each inverter output. Thus,
$$X = \overline{A} \cdot \overline{B} \cdot \overline{C} \cdot \overline{D} \cdot \overline{E} \cdot \overline{F}$$
Using DeMorgan's theorem, this is equivalent to
$$X = \overline{A + B + C + D + E + F}$$
which is the NOR operation.

(b) If we assume that only one of the INVERTERs has a HIGH input, then the output transistor of that INVERTER has to be able to sink the currents I_{RP} and I_{IL} as shown in Figure 8-23(b). Referring to the data sheet for the 7405, we find that it has a fan-out of 10 in the LOW state. Thus, it can sink a total current $I_{OL}(\text{max}) = 16$ mA. Summing the currents at X, we have
$$I_{OL}(\text{max}) = I_{RP} + I_{IL}$$
$$16 \text{ mA} = I_{RP} + 6.4 \text{ mA}$$
$$\therefore I_{RP} = 9.6 \text{ mA}$$

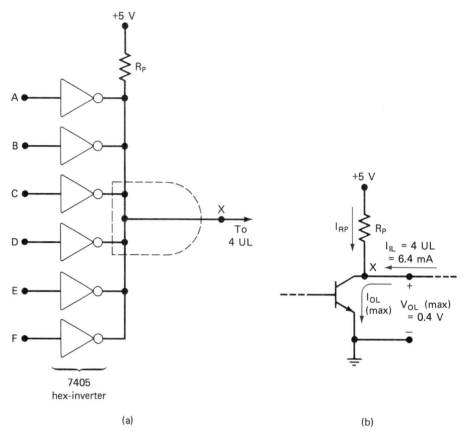

FIGURE 8-23   Example 8-16.

The voltage across R_P will be $V_{CC} - V_{OL}(\text{max})$; we use $V_{OL}(\text{max})$ since the transistor is sinking its maximum current. For standard TTL, we know that $V_{OL}(\text{max}) = 0.4$ V. Thus, we have

$$R_P(\text{min}) = \frac{V_{CC} - V_{OL}(\text{max})}{I_{RP}} = \frac{5\text{ V} - 0.4\text{ V}}{9.6\text{ mA}} = 480\ \Omega$$

In practice, we would use a value slightly larger than this to be on the safe side. Remember, we want to keep R_P as small as possible to avoid increasing the switching time. A good choice would be the standard value of 560 Ω.

—— EXAMPLE 8-17 ——————————————————

A technician breadboards the circuit of Figure 8-23 with $R_P = 10$ kΩ. He figures that since the circuit is operating at a very low frequency, there is no need to worry about the decrease in switching speed produced by increasing R_P. Besides, he wants to minimize the current drawn from V_{CC}.

He tests the circuit and finds that the voltage at point X is slightly less than 2 V in the HIGH state. He replaces the 7405 chip and observes the same result. What is causing V_{CC} to be so low?

Solution   In order for X to go HIGH, each INVERTER input has to be LOW.

This will turn off the output transistor of each open-collector output. Unfortunately, these transistors are not ideal, and they do have some leakage current in the OFF state. This leakage current, I_{OH}, is specified as a *maximum* of 250 μA for the standard TTL series, and 100 μA for LS-TTL. Under *normal* conditions it will usually be about 50 μA for TTL and 20 μA for LS-TTL. These leakage currents, however, are additive, so that the total leakage current of the six output transistors can be substantial. This current has to be supplied by V_{CC} through R_P. In addition V_{CC} has to supply the I_{IH} for each of the four ULs. As a result, there may typically be anywhere from 300 to 400 μA flowing from V_{CC} through R_P. With $R_P = 10$ kΩ, it is not surprising that around 3 V is dropped across R_P, leaving only about 2 V at point X.

Because the open-collector leakage currents and the I_{IH} of the loads have to flow through R_P, it is best to keep R_P as close as possible to the minimum value that is calculated on the basis of the open-collector output's maximum sink current, $I_{OL}(\text{max})$.

Open-Collector Buffer/Drivers   Any logic circuit that is called a *buffer*, a *driver*, or a *buffer/driver* is designed to have a greater output current and/or voltage capability than an ordinary logic circuit. Buffer/driver ICs are available with totem-pole outputs and with open-collector outputs.

The 7406 is a popular open-collector buffer/driver IC that contains six IN-VERTERs with open-collector outputs that can sink up to 40 mA in the LOW state. In addition, the 7406 can handle output voltages up to 30 V. This means that the output transistor can be connected to a voltage greater than 5 V.

This is illustrated in Figure 8-24, where a 7406 is used as a buffer between a 74LS112 flip-flop and an incandescent indicator lamp that is rated at 24 V, 25 mA. The 7406 controls the lamp's ON/OFF status to indicate the state of FF output Q. Note that the lamp is powered from +24 V, and it acts as the pull-up resistor for the open-collector output.

When $Q = 1$, the 7406 output goes LOW and its output transistor sinks the 25 mA of lamp current supplied by the 24-V source, and the lamp is on. When $Q = 0$, the 7406 output transistor turns off; there is no path for current and the lamp turns off. In this state, the full 24 V will appear across the OFF output transistor so that $V_{OH} = 24$ V, which is lower than the 7406 maximum V_{OH} rating.

Teaching Tip: Open-collector buffers are also used to interface logic outputs to relays and optocouplers that provide electrical isolation between high current/voltage loads and the low-power logic circuitry.

FIGURE 8-24   An open-collector buffer-driver drives a high-current, high-voltage load.

FIGURE 8-25 An open-collector output can be used to drive an LED indicator.

Open-collector outputs are often used to drive indicator LEDs as shown in Figure 8-25. Here either a 7405 or a 7406 can be used, depending on the LED's required current. The resistor is used to limit the current to a safe value. When the INVERTER output is LOW, its output transistor will provide a low-resistance path to ground for the LED current, so that the LED will be on. When the INVERTER output is HIGH, its output transistor will be off and there will be no path for LED current; in this state, the LED will be off.

IEEE/ANSI Symbol for Open-Collector Outputs   The traditional symbols for logic circuits with open-collector outputs are the same as for totem-pole outputs. The new IEEE/ANSI symbology, however, does use a distinctive notation to identify open-collector outputs. Figure 8-26 shows the standard IEEE/ANSI designation for an open-collector output. It is an underlined diamond. Although we will not normally use the complete IEEE/ANSI symbology in this book, we will use this underlined diamond to indicate open-collector outputs.

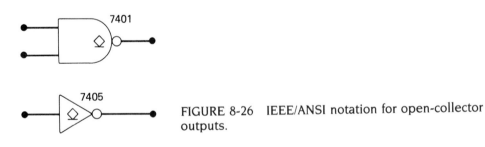

FIGURE 8-26 IEEE/ANSI notation for open-collector outputs.

REVIEW QUESTIONS

1. What is wired-ANDing?
2. Why shouldn't logic devices with totem-pole outputs be wired-ANDed?
3. How do open-collector outputs differ from totem-pole outputs?
4. Why do open-collector outputs need a pull-up resistor?
5. What factors are involved in determining a value for the pull-up resistor?
6. Why are open-collector outputs generally slower than totem-pole?
7. What is the IEEE/ANSI symbol for open-collector outputs?

8-8 TRISTATE (3-STATE) TTL

The tristate configuration is a third type of TTL output configuration. It utilizes the high-speed operation of the totem-pole arrangement while permitting outputs to be wired-ANDed (connected together). It is called tristate TTL because it allows *three* possible output states: HIGH, LOW, and high-impedance (Hi-Z). The Hi-Z state is a condition in which both transistors in the totem-pole arrangement are turned off so that the output terminal is a high impedance to ground and to V_{CC}. In other words, the output is an open or floating terminal that is neither a LOW nor a HIGH. In practice, the output terminal is not an exact open circuit, but has a resistance of several megohms or more relative to ground and V_{CC}.

The tristate operation is obtained by modifying the basic totem-pole circuit. Figure 8-27(a) shows the circuit for a tristate INVERTER where the portion en-

FIGURE 8-27   Tristate TTL INVERTER.

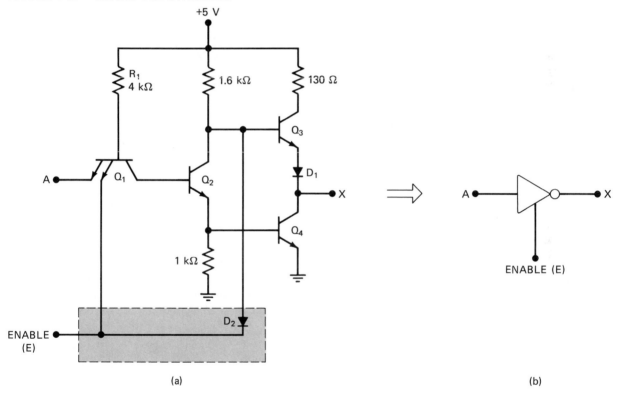

(a)

(b)

ENABLE (E)	OUTPUT
HIGH	Enabled: operates as an INVERTER
LOW	Disabled: output is Hi-Z state. Input A has no effect

Operating conditions

(c)

closed in dotted lines has been added to the basic circuit. The circuit has two inputs: A is the normal logic input, E is an ENABLE input that can produce the Hi-Z state. We will examine the operation for both states of E.

The Enabled State

With $E = 1$ the circuit operates as a normal INVERTER because the HIGH voltage at E has no effect on Q_1 or D_2. In this enabled condition, the output is simply the inverse of logic input A.

The Disabled State (Hi-Z)

When $E = 0$ the circuit goes into its Hi-Z state regardless of the state of logic input A. The LOW at E forward-biases the emitter-base junction of Q_1 and shunts the R_1 current away from Q_2 so that Q_2 turns off, which turns Q_4 off. The LOW at E also forward-biases diode D_2 to shunt current away from the base of Q_3, so that Q_3 also turns off.

With both totem-pole transistors in the nonconducting state, the output terminal is essentially an open circuit. This is shown symbolically in the table of Figure 8-27(c).

The logic symbol for the tristate INVERTER is shown in Figure 8-27(b). Note where the ENABLE input is placed on the INVERTER symbol. Also note that E is active-HIGH; that is, the INVERTER is enabled when $E = 1$.

Advantage of Tristate

The outputs of tristate ICs can be connected together (paralleled) without sacrificing switching speed. This is because a tristate output, when enabled, operates as a totem-pole output with its associated low-impedance, high-speed characteristic. It is important to realize, however, that when tristate outputs are paralleled, only one of them should be enabled at one time. Otherwise, two active totem-pole outputs would be connected together and damaging currents could flow (Figure 8-20). We will elaborate on this in our discussion of tristate buffers.

Tristate Buffers

A *tristate buffer* is a circuit that is used to control the passage of a logic signal from input to output. Some tristate buffers also invert the signal as it goes through. The circuit in Figure 8-27 can be called an *inverting tristate buffer*.

Two commonly used tristate buffer ICs are the 74LS125 and LS126. Both contain four *noninverting* tristate buffers like those shown in Figure 8-28. The 74LS125

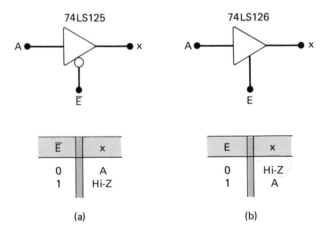

FIGURE 8-28   Tristate noninverting buffers.

$\overline{E}$	x
0	A
1	Hi-Z

(a)

E	x
0	Hi-Z
1	A

(b)

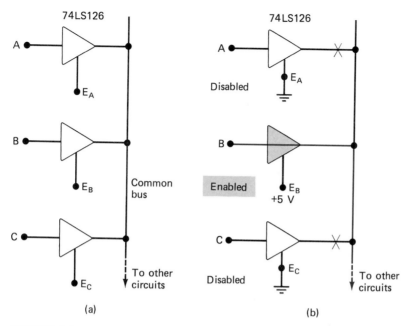

FIGURE 8-29 (a) Tristate buffers used to connect several signals to a common bus; (b) conditions for transmitting B to the bus.

and 74LS126 differ only in the active state of their ENABLE inputs. The 74LS125 allows the input signal A to reach the output when $\bar{E} = 0$, while the 74LS126 passes the input when $E = 1$.

Tristate buffers have many applications in circuits where several signals are connected to common lines (buses). We will examine some of these applications in Chapter 9, but we can get the basic idea from Figure 8-29(a). Here we have three logic signals A, B, and C connected to a common bus line through 74LS126 tristate buffers. This arrangement permits us to transmit any one of these signals over the bus line to other circuits by enabling the appropriate buffer.

For example, consider the situation in Figure 8-29(b) where $E_B = 1$ and $E_A = E_C = 0$. This disables the upper and lower buffers so that their outputs are in the Hi-Z state and are essentially disconnected from the bus. This is symbolized by the X's on the diagram. The middle buffer is enabled so that its input, B, is passed through to its output and onto the bus, from where it is routed to other circuits connected to the bus. When tristate outputs are connected together as in Figure 8-29, it is important to remember that no more than one output should be enabled at one time. Otherwise, two or more active totem-pole outputs would be connected, which could produce damaging currents. Even if damage did not occur, this situation would produce a signal on the bus that is a combination of more than one signal. This is commonly referred to as **bus contention.** In tristate bus systems the designer has to make sure that the enable signals do not allow bus contention to occur.

Discussion Question: What happens if none of the buffers is enabled?

Tristate ICs   In addition to tristate buffers, there are many ICs that are designed with tristate outputs. For example, the 74LS374 is an octal D-type FF register IC with tristate outputs. This means that it is an 8-bit register made up of D-type FFs whose outputs are connected to tristate buffers. This type of register can be connected to common bus lines along with the outputs from other, similar devices to allow efficient transfer of data over the bus. We will examine this *tristate data bus* ar-

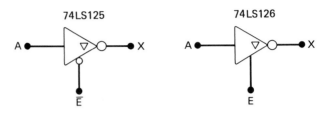

FIGURE 8-30   IEEE/ANSI notation for tristate outputs.

rangement in Chapter 9. Other types of logic devices that are available with tristate outputs include decoders, multiplexers, analog-to-digital converters, memory chips, and microprocessors.

IEEE/ANSI Symbol for Tristate Outputs   The traditional logic symbology has no special notation for tristate outputs. Figure 8-30 shows the notation used in the IEEE/ANSI symbology to indicate a tristate output. It is a triangle that points downward. Although it is not part of the traditional symbology, we will use this triangle to designate tristate outputs throughout the remainder of the book.

REVIEW QUESTIONS

1. What are the three possible output states of a tristate IC?
2. What is the state of a tristate output when it is disabled?
3. What is bus contention?
4. What conditions are necessary to transmit signal C onto the bus in Figure 8-29?
5. What is the IEEE/ANSI designation for tristate outputs?

8-9   THE ECL DIGITAL IC FAMILY

The TTL family uses transistors operating in the saturated mode. As a result, their switching speed is limited by the storage delay time associated with a transistor that is driven into saturation. Another *bipolar* logic family has been developed that prevents transistor saturation, thereby increasing overall switching speed. This logic family is called **emitter-coupled logic** (ECL), and it operates on the principle of current switching whereby a fixed bias current less than $I_{C(sat)}$ is switched from one transistor's collector to another. Because of this current-mode operation, this logic form is also referred to as **current-mode logic** (CML).

Basic ECL Circuit   The basic circuit for emitter-coupled logic is essentially the differential amplifier configuration of Figure 8-31(a). The V_{EE} supply produces an essentially fixed current I_E which remains around 3 mA during normal operation. This current is allowed to flow through either Q_1 or Q_2, depending on the voltage level at V_{IN}. In other words, this current will switch between Q_1's collector and Q_2's collector as V_{IN} switches between its two logic levels of -1.7 V (logical 0 for ECL) and -0.8 V (logical 1 for ECL). The table in Figure 8-31(a) shows the resulting output

FIGURE 8-31   (a) Basic ECL circuit; (b) with addition of emitter followers.

	Operating States	
V_{IN}	Outputs	
−1.7 V (logic 0)	V_{C1} = 0 V V_{C2} = −0.9 V }	Q_2 conducts
−0.8 V (logic 1)	V_{C1} = −0.9 V V_{C2} = 0 V }	Q_1 conducts

(a)

Input/output
logic levels
"0" = −1.7 V
"1" = −0.8 V

(b)

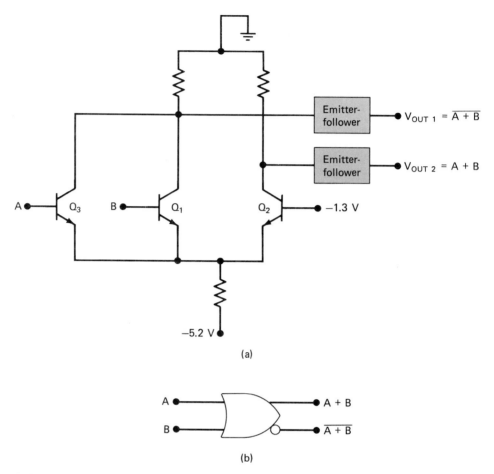

FIGURE 8-32 (a) ECL NOR/OR circuit; (b) logic symbol.

voltages for these two conditions at V_{IN}. Two important points should be noted: (1) V_{C1} and V_{C2} are the *complements* of each other, and (2) the output voltage levels are not the same as the input logic levels.

The second point noted above is easily taken care of by connecting V_{C1} and V_{C2} to emitter-follower stages (Q_3 and Q_4), as shown in Figure 8-31(b). The emitter followers perform two functions: (1) they subtract approximately 0.8 V from V_{C1} and V_{C2} to shift the output levels to the correct ECL logic levels, and (2) they provide a very low output impedance (typically 7 Ω), which provides for large fan-out and fast charging of load capacitance. This circuit produces two complementary outputs: V_{OUT1}, which equals $\overline{V_{IN}}$, and V_{OUT2}, which is equal to V_{IN}.

ECL OR/NOR Gate   The basic ECL circuit of Figure 8-31(b) can be used as an INVERTER if the output is taken at V_{OUT1}. This basic circuit can be expanded to more than one input by paralleling transistor Q_1 with other transistors for the other inputs, as in Figure 8-32(a). Here either Q_1 or Q_3 can cause the current to be switched out of Q_2, resulting in the two outputs V_{OUT1} and V_{OUT2} being the logical NOR and OR operations, respectively. This OR/NOR gate is symbolized in Figure 8-32(b) and is the fundamental ECL gate.

ECL Characteristics   The following are the most important characteristics of the ECL family of logic circuits:

TABLE 8-5

Logic Family	t_{pd} (ns)	P_D (mW)	Worst-Case Noise Margin (mV)	Maximum Clock Rate (MHz)
74	9	10	400	35
74AS	1.7	8	300	200
74ALS	4	1.2	400	70
74S	3	20	300	125
74LS	9.5	2	300	45
ECL	1	40	250	600

1. The transistors never saturate, and so switching speed is very high. Typical propagation delay time is 1 ns, which makes ECL a little faster than advanced Schottky TTL (74AS series).

2. The logic levels are nominally −0.8 V and −1.70 V for the logical 1 and 0, respectively.

3. Worst-case ECL noise margins are approximately 250 mV. These low noise margins make ECL somewhat unreliable for use in heavy industrial environments.

4. An ECL logic block usually produces an output and its complement. This eliminates the need for inverters.

5. Fan-outs are typically around 25, owing to the low-impedance emitter-follower outputs.

6. Typical power dissipation for a basic ECL gate is 40 mW, somewhat higher than the 74AS series.

7. The total current flow in an ECL circuit remains relatively constant regardless of its logic state. This helps to maintain an unvarying current drain on the circuit power supply even during switching transitions. Thus, no noise spikes will be internally generated like those produced by TTL totem-pole circuits.

Table 8-5 shows how ECL compares with the important TTL logic families.

The ECL family is not as widely used as the TTL and MOS families except in very high-frequency applications, where its speed is superior. Its relatively low noise margins and high power drain are disadvantages compared with other logic families. Another drawback is its negative supply voltage and logic levels, which are not compatible with the other logic families; this makes it difficult to use ECL in conjunction with TTL and MOS circuits.

___ REVIEW QUESTION ___

1. *True or false:*
   (a) ECL obtains high-speed operation by preventing transistor saturation.
   (b) ECL circuits usually have complementary outputs.
   (c) The noise margins for ECL circuits are larger than TTL noise margins.
   (d) ECL circuits do not generate noise spikes during state transitions.
   (e) ECL devices require less power than standard TTL.
   (f) ECL can be easily used with TTL.

MOS (metal-oxide-semiconductor) technology derives its name from the basic MOS structure of a metal electrode over an oxide insulator over a semiconductor substrate. The transistors of MOS technology are field-effect transistors called MOSFETs. Most of the MOS digital ICs are constructed entirely of MOSFETs and no other components.

The chief advantages of the MOSFET are that it is relatively simple and inexpensive to fabricate, it is small, and it consumes very little power. The fabrication of MOS ICs is approximately one-third as complex as the fabrication of bipolar ICs (TTL, ECL, etc.). In addition, MOS devices occupy much less space on a chip than bipolar transistors; typically, a MOSFET requires 1 square mil of chip area while a bipolar transistor requires about 50 square mils. More important, MOS digital ICs normally do not use the IC resistor elements that take up so much of the chip area of bipolar ICs.

All of this means that MOS ICs can accommodate a much larger number of circuit elements on a single chip than bipolar ICs. This advantage is evidenced by the fact that MOS ICs have dominated bipolar ICs in the area of large-scale integration (LSI, VLSI). The high packing density of MOS ICs results in a greater system reliability because of the reduction in the number of necessary external connections.

The principal disadvantage of MOS ICs is their relatively slow operating speed when compared with the bipolar IC families. In many applications this is not a prime consideration, and so MOS logic offers an often superior alternative to bipolar logic. We will examine the MOS logic families after a brief discussion of MOSFETs.

8-11   THE MOSFET

There are presently two general types of MOSFETs: *depletion* and *enhancement*. MOS digital ICs use enhancement MOSFETs exclusively, and so only this type will be considered in the following discussion. Furthermore, we will concern ourselves only with the operation of these MOSFETs as on/off switches.

Figure 8-33 shows the schematic symbols for the N-channel and P-channel enhancement MOSFETs, where the direction of the arrow indicates either P- or N-channel. The symbols show a broken line between the *source* and *drain* to indicate that there is *normally* no conducting channel between these electrodes. The symbol also shows a separation between the *gate* and the other terminals to indicate the very high resistance (typically greater than 10,000 MΩ) between the gate and channel.

FIGURE 8-33   Schematic symbols for enhancement MOSFETs.

N-channel   P-channel

FIGURE 8-34 N-channel MOSFET switching states.

Basic MOSFET Switch

Figure 8-34 shows the switching operation of an N-channel *MOSFET*. For the N-channel device the drain is always biased positive relative to the source. The gate-to-source voltage V_{GS} is the input voltage, which is used to control the resistance between drain and source (i.e., the channel resistance) and therefore determines whether the device is on or off.

When $V_{GS} = 0$ V, there is no conductive channel between source and drain, and the device is off. Typically the channel resistance in this OFF state is 10^{10} Ω, which for most purposes is an *open circuit*. The MOSFET will remain off as long as V_{GS} is zero or negative. As V_{GS} is made positive (gate positive relative to source), a threshold voltage (V_T) is reached, at which point a conductive channel begins to form between source and drain. Typically $V_T = +1.5$ V for N-MOSFET, and so any $V_{GS} \geq 1.5$ V will cause the MOSFET to conduct. Generally, a value of V_{GS} much larger than V_T is used to turn on the MOSFET more completely. As shown in Figure 8-34(b), when $V_{GS} = +5$ V, the channel resistance between source and drain has dropped to a value of $R_{ON} = 1000$ Ω.

Table 8-6 summarizes the P- and N-channel switching characteristics.

In essence, then, the N-MOSFET will switch from a very high resistance to a low resistance as the gate voltage switches from a LOW voltage to a HIGH voltage. It is helpful simply to think of the MOSFET as a switch that is either opened or closed between source and drain.

The P-channel MOSFET operates in exactly the same manner as the N-channel except that it uses voltages of the opposite polarity. For P-MOSFETs the drain is connected to $-V_{DD}$ so that it is biased negative relative to the source. To turn the P-MOSFET on, a negative voltage that exceeds V_T must be applied to the gate.

Teaching Tip: The value of R_{ON} is controlled by the channel geometry and can vary considerably depending on whether the MOSFET is being used as a switch or as a load resistor (always ON). It is generally not as low as the ON resistance of a bipolar transistor.

TABLE 8-6

	Drain-To-Source Bias	Gate-To-Source Voltage (V_{GS}) Needed For Conduction	R_{ON} (Ω)	R_{OFF} (Ω)
P-channel	Negative	Typically more negative than -1.5 V	1000 (typical)	10^{10}
N-channel	Positive	Typically more positive than $+1.5$ V	1000 (typical)	10^{10}

8-12 DIGITAL MOSFET CIRCUITS

Digital circuits employing MOSFETs are broken down into three categories: (1) P-MOS, which uses *only* P-channel enhancement MOSFETs; (2) N-MOS, which uses *only* N-channel enhancement MOSFETs; and (3) CMOS (complementary MOS), which uses both P- and N-channel devices.

P-MOS and N-MOS digital ICs have a greater packing density (more transistors per chip) and are therefore more economical than CMOS. N-MOS has about twice the packing density of P-MOS. In addition, N-MOS is also about twice as fast as P-MOS, owing to the fact that free electrons are the current carrier in N-MOS while holes (slower-moving positive charges) are the current carriers for P-MOS. CMOS has the greatest complexity and lowest packing density of the MOS families, but it possesses the important advantages of higher speed and much lower power dissipation.

In this section we will look at some of the basic N-MOS logic circuits, keeping in mind that the P-MOS circuits would be the same except for the voltage polarities. Since P-MOS and N-MOS find their widest applications in LSI and VLSI (microprocessors, memories, ROMs, etc.), we will defer any applications of these families until later. CMOS, which, like TTL, is widely used in MSI applications, will be covered in more detail beginning with Section 8-14.

N-MOS INVERTER   Figure 8-35 shows the basic N-MOS INVERTER circuit. It contains two N-channel MOSFETs: Q_1 is called the *load* MOSFET and Q_2 the *switching* MOSFET. Q_1 has its gate *permanently* connected to +5 V, and so it is *always* in the ON state and essentially acts as a load resistor of value R_{ON}. Q_2 will switch from ON to OFF in response to V_{IN}. The Q_1 MOSFET is designed to have a narrower channel than Q_2, and so Q_1's R_{ON} is much greater than Q_2's. Typically, R_{ON} for Q_1 is 100 kΩ and R_{ON} for Q_2 is 1 kΩ. R_{OFF} for Q_2 is usually around 10^{10} Ω.

The two states of the INVERTER are summarized in Figure 8-35(b). The best way to analyze this circuit is to consider each MOSFET channel as a resistance so

FIGURE 8-35   N-MOS INVERTER.

V_{IN}	Q_1	Q_2	$V_{OUT} = \overline{V_{IN}}$
0 V (logic 0)	R_{ON} = 100 kΩ	R_{OFF} = 10^{10} Ω	+5 V (logic 1)
+5 V (logic 1)	R_{ON} = 100 kΩ	R_{ON} = 1 kΩ	+0.05 V (logic 0)

(a)

(b)

that the output voltage is taken from a voltage divider formed by the two resistances. With $V_{IN} = 0$ V, transistor Q_2 is off, with a very large resistance of 10^{10} Ω. Since Q_1 has $R_{ON} = 100$ kΩ, the voltage-divider output will be essentially +5 V. With $V_{IN} = +5$ V, Q_2 is on, with $R_{ON} = 1$ kΩ. The voltage divider is now 100 kΩ and 1 kΩ, so that $V_{OUT} = 1/101 \times (+5$ V$) \approx 0.05$ V.

The circuit functions as an INVERTER since a LOW input produces a HIGH output and vice versa. This basic INVERTER can be modified to form NAND and NOR logic gates.

N-MOS NAND Gate   The NAND operation is performed by the circuit of Figure 8-36(a), where Q_1 is again acting as a load resistance while Q_2 and Q_3 are switches controlled by input levels A and B. If either A or B is at 0 V (logical 0), the corresponding FET is off, thereby presenting a high resistance from the output terminal to ground so that output X is HIGH (+5 V). When both A and B are +5 V (logical 1), both Q_2 and Q_3 are on, so that output X is LOW. Clearly, the output equals the NAND of the inputs ($X = \overline{AB}$).

Discussion Question: What factors do you think limit the number of inputs for an N-MOS NAND gate?

FIGURE 8-36   (a) N-MOS NAND gate; (b) NOR gate.

A	B	X
LOW (0 V)	LOW	HIGH
HIGH (+5 V)	LOW	HIGH
LOW	HIGH	HIGH
HIGH	HIGH	LOW

(a)

A	B	X
LOW	LOW	HIGH
LOW	HIGH	LOW
HIGH	LOW	LOW
HIGH	HIGH	LOW

(b)

N-MOS NOR Gate

The NOR gate of Figure 8-36(b) uses Q_2 and Q_3 as parallel switches with Q_1 again acting as a load resistance. When either input A or B is at +5 V, the corresponding MOSFET is on, forcing the output to be LOW. When both inputs are at 0 V, both Q_2 and Q_3 are off, so that the output goes HIGH. Clearly, this is the NOR operation with $X = \overline{A + B}$.

N-MOS OR gates and AND gates are easily formed by combining the NOR or NAND with inverters.

N-MOS Flip-Flops

Flip-flops are formed by using two cross-coupled NOR gates or NAND gates. The MOS flip-flop is very important in MOS memories, which will be discussed later.

8-13 CHARACTERISTICS OF MOS LOGIC

Compared with the bipolar logic families the MOS logic families are slower in operating speed; require much less power; have a better noise margin, a greater supply voltage range, and a higher fan-out; and, as was mentioned earlier, require much less "real estate" (chip area).

Operating Speed

A typical N-MOS NAND gate has a propagation delay time of 50 ns. This is due to *two* factors: the relatively high output resistance (100 kΩ) in the HIGH state and the capacitive loading presented by the inputs of the logic circuits being driven. MOS logic inputs have very high input resistance (> 10^{12} Ω), and they have a reasonably high input capacitance (MOS capacitor), typically 2 to 5 picofarads. This combination of large R_{OUT} and large C_{LOAD} serves to increase switching time.

Noise Margin

Typically, N-MOS noise margins are around 1.5 V when operated from $V_{DD} = 5$ V and will be proportionally higher for larger values of V_{DD}.

Fan-Out

Because of the extremely high input resistance at each MOSFET input, one would expect that the fan-out capabilities of MOS logic would be virtually unlimited. This is essentially true for dc or low-frequency operation. However, for frequencies greater than around 100 kHz, the gate input capacitances cause a deterioration in switching time which increases in proportion to the number of loads being driven. Even so, MOS logic can easily operate at a fan-out of 50, which is much better than most of the bipolar families.

Power Drain

MOS logic circuits draw small amounts of power because of the relatively large resistances being used. To illustrate, we can calculate the power dissipation of the INVERTER of Figure 8-35 for its two operating states.

1. $V_{\text{IN}} = 0$ V: $R_{\text{ON}(Q_1)} = 100$ kΩ; $R_{\text{OFF}(Q_2)} = 10^{10}$ Ω. Therefore, I_D, current from V_{DD} supply, ≈ 0.05 nA, and $P_D = 5$ V $\times$ 0.05 nA = 0.25 nW.
2. $V_{\text{IN}} = +5$ V: $R_{\text{ON}(Q_1)} = 100$ kΩ; $R_{\text{ON}(Q_2)} = 1$ kΩ. Therefore, $I_D = 5$ V/101 k$\Omega \approx 50$ μA and $P_D = 5$ V $\times$ 50 μA = 0.25 mW.

This gives an *average* P_D of a little over 0.1 mW for the INVERTER. The low power drain of MOS logic makes it suitable for LSI and VLSI, where many gates, FFs, and other circuits can be on one chip without causing overheating that can damage the chip.

Process Complexity

MOS logic is the simplest logic family to fabricate since it uses only one basic element, an N-MOS (or P-MOS) transistor. It requires no other elements, such as resistors or diodes. This characteristic, together with its lower P_D, makes it ideally suited for LSI (large memories, calculator chips, microprocessors) and this is where MOS logic has made its greatest impact in the digital field. The operating speed of P-MOS and N-MOS is not comparable with TTL, and so very little has been done with them in SSI and MSI applications. In fact, there are very few MOS logic circuits in the SSI or MSI categories (gates, FFs, and counters, for instance). CMOS, however, is competitive in the MSI area, which was until recently dominated by TTL.

Teaching Tip: Diodes are not required to implement the logic operations, but they are used on MOS ICs for purposes such as input overvoltage protection.

Static Sensitivity

All electronic devices, to varying degrees, are sensitive to damage by static electricity. The human body is a great storehouse of electrostatic charges. For example, when you walk across a carpet, a static charge of over 30,000 V can be built up on your body. If you then touch an electronic device, some of this sizable charge can be transferred to the device. If the transferred charge is too large, the device will be destroyed.

The MOS logic families are especially susceptible to static charge damage, while the bipolar logic families are much less so. This is a direct result of the very high impedance present at a MOSFET input. A small static charge flowing into this high impedance can develop a dangerous voltage. This same charge flowing into the relatively low impedance of a TTL or ECL input will produce a smaller, less harmful voltage.

Electrostatic discharge (ESD) is responsible for billions of dollars of damage to electronic equipment annually, and equipment manufacturers have devoted considerable attention to developing special handling procedures for all electronic devices and circuits. The following are some of the most common precautions used by most companies in their engineering labs, production facilities, and field service departments:

Common Student Mistake: Students have a hard time understanding how a human body can carry a 30,000-V static charge without feeling some unpleasant effects. The static charge, of course, can do no damage unless it has a relatively low-resistance path to ground. It would not have a chance to build up to such a high level if it had a discharge path. This same charge transferred to a MOSFET input will cause a voltage to build up until it exceeds the gate-channel breakdown voltage. This produces a low resistance to ground resulting in a damaging current flow.

1. Connect the chassis of all test instruments, soldering-iron tips, and your workbench (if metal) to earth ground (i.e., the round prong in the 120 VAC plug). This prevents the buildup of static charge on these devices that could be transferred to any circuit board or IC that they come in contact with.

2. Connect yourself to earth ground with a special wrist strap. This will allow potentially dangerous charges from your body to be discharged to ground. The wrist strap contains a 1-MΩ resistor that limits current to a nonlethal value should you accidentally touch a "live" voltage while working with the equipment.

3. Keep ICs (especially MOS) in conductive foam or aluminum foil. This will keep all IC pins shorted together so that no dangerous voltages can be developed between any two pins.

4. Store PC boards in conductive plastic or metallic envelopes.

5. Do not leave any unused IC inputs unconnected, because open inputs tend to pick up stray static charges.

Teaching Tip: Stress that these precautions are a good idea even for circuits that do not contain MOS devices.

1. Describe the relative advantages and disadvantages of N-MOS circuits as compared with TTL.
2. What determines the fan-out limitations of MOS logic?
3. What factors limit the switching speed of N-MOS circuits?
4. What makes MOS circuitry so well suited for LSI?
5. Why are MOS ICs especially sensitive to static charges?
6. List five precautions for handling ICs and circuit boards to minimize static charge damage.

8-14   COMPLEMENTARY MOS LOGIC

The *complementary MOS* (CMOS) logic family uses *both* P- and N-channel MOS-FETs in the same circuit to realize several advantages over the P-MOS and N-MOS families. Generally speaking, CMOS is faster and consumes even less power than the other MOS families. These advantages are offset somewhat by the increased complexity of the IC fabrication process and a lower packing density. Thus, CMOS cannot yet hope to compete with MOS in applications requiring the utmost in LSI.

However, CMOS logic has undergone a constant growth in the MSI area, mostly at the expense of TTL, with which it is directly competitive. The CMOS fabrication process is simpler than TTL and has a greater packing density, therefore permitting more circuitry in a given area and reducing the cost per function. CMOS uses only a fraction of the power needed for even the low-power TTL series (74L) and is thus ideally suited for applications using battery power or battery backup power. CMOS is generally slower than TTL, although the new high-speed CMOS series can compete with the 74 and 74LS series.

CMOS Inverter   The circuitry for the basic CMOS INVERTER is shown in Figure 8-37. For this diagram and those that follow, the standard symbols for the

FIGURE 8-37   Basic CMOS INVERTER.

V_{IN}	Q_1	Q_2	V_{OUT}
$+V_{DD}$ (logic 1)	OFF $R_{OFF} = 10^{10}\ \Omega$	ON $R_{ON} = 1\ k\Omega$	$\simeq 0\ V$
$0\ V$ (logic 0)	ON $R_{ON} = 1\ k\Omega$	OFF $R_{OFF} = 10^{10}\ \Omega$	$\simeq +V_{DD}$

$$V_{OUT} = \overline{V_{IN}}$$

416

MOSFETs have been replaced by blocks labeled P and N to denote a P-MOSFET and N-MOSFET, respectively. This is done simply for convenience in analyzing the circuits. The CMOS INVERTER has two MOSFETs in series in such a way that the P-channel device has its source connected to $+V_{DD}$ (a positive voltage) and the N-channel device has its source connected to ground.* The gates of the two devices are connected together as a common input. The drains of the two devices are connected together as the common output.

The logic levels for CMOS are essentially $+V_{DD}$ for logical 1 and 0 V for logical 0. Consider, first, the case where $V_{IN} = +V_{DD}$. In this situation the gate of Q_1 (P-channel) is at 0 V relative to the source of Q_1. Thus, Q_1 will be in the OFF state with $R_{OFF} \approx 10^{10}$ Ω. The gate of Q_2 (N-channel) will be at $+V_{DD}$ relative to its source. Thus, Q_2 will be on with typically $R_{ON} = 1$ kΩ. The voltage divider between Q_1's R_{OFF} and Q_2's R_{ON} will produce $V_{OUT} \approx 0$ V.

Next, consider the case where $V_{IN} = 0$ V. Q_1 now has its gate at a negative potential relative to its source while Q_2 has $V_{GS} = 0$ V. Thus, Q_1 will be on with $R_{ON} = 1$ kΩ and Q_2 off with $R_{OFF} = 10^{10}$ Ω, producing a V_{OUT} of approximately $+V_{DD}$. These two operating states are summarized in Figure 8-37(b), showing that the circuit does act as a logic INVERTER.

Discussion Question: Can you see why CMOS has such a low power dissipation?

CMOS NAND Gate Other logic functions can be constructed by modifying the basic INVERTER. Figure 8-38 shows a NAND gate formed by adding a parallel

FIGURE 8-38 CMOS NAND gate.

A	B	X
LOW	LOW	HIGH
LOW	HIGH	HIGH
HIGH	LOW	HIGH
HIGH	HIGH	LOW

*Most manufacturers label this terminal V_{SS}.

P-channel MOSFET and a series N-channel MOSFET to the basic INVERTER. To analyze this circuit it helps to realize that a 0-V input turns on its corresponding P-MOSFET and turns off its corresponding N-MOSFET, and vice versa for a $+V_{DD}$ input. Thus, it can be seen that the only time a LOW output will occur is when inputs A and B are both HIGH ($+V_{DD}$) to turn on both N-MOSFETs, thereby providing a low resistance from the output terminal to ground. For all other input conditions, at least one P-MOSFET will be on while at least one N-MOSFET will be off. This produces a HIGH output.

CMOS NOR Gate

A CMOS NOR gate is formed by adding a series P-MOSFET and a parallel N-MOSFET to the basic inverter as shown in Figure 8-39. Once again this circuit can be analyzed by realizing that a LOW at any input turns on its corresponding P-MOSFET and turns off its corresponding N-MOSFET, and vice versa for a HIGH input. It is left to the reader to verify that this circuit operates as a NOR gate.

CMOS AND and OR gates can be formed by combining NANDs and NORs with inverters.

CMOS SET-CLEAR FF

Two CMOS NOR gates or NAND gates can be cross-coupled to form a simple SET-CLEAR flip-flop. Additional gating circuitry is used to convert the basic SET-CLEAR flip-flop to clocked D and J-K flip-flops.

FIGURE 8-39 CMOS NOR gate.

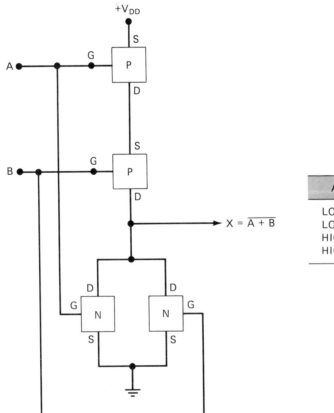

A	B	X
LOW	LOW	HIGH
LOW	HIGH	LOW
HIGH	LOW	LOW
HIGH	HIGH	LOW

1. *True or false:*
  (a) CMOS is faster but consumes more power than other MOS families.
  (b) CMOS can pack more circuits on a chip than TTL.
  (c) In a CMOS INVERTER both MOSFETs are on at the same time.
  (d) CMOS devices are ideal for battery-operated circuits.

8-15 CMOS SERIES CHARACTERISTICS

Teaching Tip: Some CMOS manufacturers such as Motorola have a 14000 series which is the same as the 4000 series.

4000/14000 Series There are several different series in the CMOS family of digital ICs. The 4000 series, which was introduced by RCA (14000 by Motorola, Inc.), was the first CMOS series. The original 4000 series is the 4000A series; an improved version is the 4000B series, which has higher output current capabilities. Both are still widely used despite the emergence of the new CMOS series. The 4000 series has been around much longer and has many functions not yet available in the newer series.

74C Series This CMOS series is compatible pin for pin and function for function with TTL devices having the same number. For example, a 74C74 is a dual edge-triggered D flip-flop that has the same pin configuration as the TTL 7474 dual edge-triggered D flip-flop IC. Many but not all functions that are available in TTL are also available in this CMOS series. This makes it possible to replace some TTL circuits by an equivalent CMOS design. The performance characteristics of the 74C series are about the same as the 4000 series.

74HC Series (High-Speed CMOS) This is an improved version of the 74C series. The main improvement is about a tenfold increase in switching speed. The speed of devices in this series is comparable to that of the 74LS TTL series. Another improvement is its higher output current capability.

74HCT Series This is also a high-speed CMOS series. The major difference between this series and the 74HC is that it is designed to be voltage-compatible with TTL devices. In other words, it can be directly driven by a TTL output. As we shall see, this is not true for the other CMOS series.

The following paragraphs will discuss some of the operating and performance characteristics that are common to all CMOS devices.

Power-Supply Voltage The 4000 series and 74C series will operate with V_{DD} values ranging from 3 to 15 V, so that power supply regulation is not critical. The 74HC and 74HCT series can operate with a supply voltage range of 2 to 6 V. When CMOS and TTL are being used together, the supply voltage is usually made 5 V so that a single 5-V supply can serve as both the V_{DD} for CMOS and the V_{CC} for TTL. In those situations where the CMOS devices are operated at a supply voltage that is greater than 5 V, special steps have to be taken to permit the CMOS and TTL devices to work together. We will examine these steps later.

TABLE 8-7 CMOS
Voltage Levels (4000B)

$V_{OL}(max) = 0$ V
$V_{OH}(min) = V_{DD}$
$V_{IL}(max) = 30\%\ V_{DD}$
$V_{IH}(min) = 70\%\ V_{DD}$

Voltage Levels   When CMOS outputs drive only CMOS inputs, the output voltage levels will be very close to 0 V for the LOW state, and $+V_{DD}$ for the HIGH state. This is because the very high CMOS input resistance draws very little current from the CMOS output that is driving it. The input voltage requirements for both logic states are expressed as a percentage of the supply voltage. $V_{IL}(max)$ is specified as 30 percent of V_{DD}, and $V_{IH}(min)$ is 70 percent of V_{DD}.* These voltage levels are summarized in Table 8-7.

To illustrate, when a CMOS IC is operated from $V_{DD} = 5$ V, it will accept any input voltage less than $V_{IL}(max) = 1.5$ V as a LOW, and any input voltage greater than $V_{IH}(min) = 3.5$ V as a HIGH.

Noise Margins   The CMOS dc noise margins can be determined from Table 8-7 as follows:

$$V_{NH} = V_{OH}(min) - V_{IH}(min)$$
$$= V_{DD} - 70\%\ V_{DD}$$
$$= 30\%\ V_{DD}$$
$$V_{NL} = V_{IL}(max) - V_{OL}(max)$$
$$= 30\%\ V_{DD} - 0$$
$$= 30\%\ V_{DD}$$

The noise margins are the same in both states and depend on V_{DD}. At $V_{DD} = 5$ V, the noise margins are both 1.5 V. This is substantially better than TTL and ECL. This makes CMOS attractive for applications that are exposed to a high-noise environment. Of course, the noise margins can be made even better by using a larger V_{DD}. This improvement in noise immunity, however, would be obtained at the expense of a higher power drain because of the larger supply voltage.

Power Dissipation   When a CMOS logic circuit is in a static state (not changing), its power dissipation is extremely low. We can see the reason by examining each of the circuits shown in Figures 8-37 to 8-39. Note that regardless of the state of the output, there is always a very high resistance between the V_{DD} terminal and ground because there is always an off MOSFET in the current path. This results in a typical CMOS dc power dissipation of only 2.5 nW per gate when $V_{DD} = 5$ V; even at $V_{DD} = 10$ V this power increases to only 10 nW. With these values for P_D, it is easy to see why CMOS is widely used in applications where power drain is a prime concern.

P_D Increases with Frequency   The power dissipation of a CMOS IC will be very low as long as it is in a dc condition. Unfortunately, P_D will increase in propor-

---

*These percentages apply to the 4000B series. The V_{IH} and V_{IL} values for the 74HC and 74HCT series will be presented in Section 8-18.

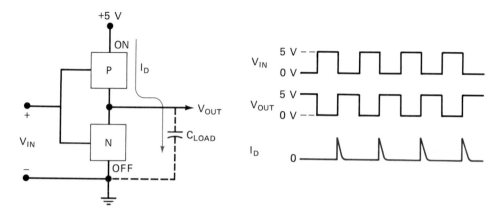

FIGURE 8-40   Current spikes are drawn from the V_{DD} supply each time the output switches from LOW to HIGH. This is due mainly to charging current of load capacitance.

tion to the frequency at which the circuits are switching states. For example, a CMOS NAND gate that has $P_D = 10$ nW under dc conditions will have $P_D = 0.1$ mW at a frequency of 100 kpps, and 1 mW at 1 MHz. The reason for this dependence on frequency is illustrated in Figure 8-40.

Each time a CMOS output switches from LOW to HIGH, a transient charging current has to be supplied to the load capacitance. This capacitance consists of the combined input capacitances of any loads being driven and the device's own output capacitance. These narrow spikes of current are supplied by V_{DD} and can have a typical amplitude of 5 mA and a duration of 20 to 30 ns. Clearly, as the switching frequency increases, there will be more of these current spikes occurring per second, and the average current drawn from V_{DD} will increase.

Thus, at higher frequencies, CMOS begins to lose some of its advantage over other logic families. As a general rule, a CMOS gate will have the same average P_D as a 74LS gate at frequencies near 2 to 3 MHz. For MSI chips, the situation is somewhat more complex than stated here, and a logic designer must do a detailed analysis to determine whether or not CMOS has a power-dissipation advantage at a particular frequency of operation.

Teaching Tip: There is another factor that contributes to this current spike. It is the same effect that occurs in TTL totem-pole outputs when the output switches from LOW to HIGH. The N-channel MOSFET will turn off more slowly than the P-channel MOSFET will turn on, so that a current path to ground exists during the switching time. This current is not as great as in TTL, because of the larger device resistances.

Fan-Out   Like N-MOS and P-MOS, CMOS inputs have an extremely large resistance (10^{12} ohms) that draws essentially no current from the signal source. Each CMOS input, however, typically presents a 5-pF load to ground. This input capacitance limits the number of CMOS inputs that one CMOS output can drive (see Figure 8-41). The CMOS output has to charge and discharge the parallel combination of all the input capacitances, so that the output switching time will be increased in proportion to the number of loads being driven. Typically, each CMOS load increases the driving circuit's propagation delay by 3 ns. For example, NAND gate 1 in Figure 8-41 might have a t_{PLH} of 25 ns if it were driving no loads; this would increase to 25 ns + 20(3 ns) = 85 ns if it were driving *twenty* loads.

Thus, CMOS fan-out depends on the permissible maximum propagation delay. Typically, CMOS outputs are limited to a fan-out of 50 for low-frequency operation (≤ 1 MHz). Of course, for higher-frequency operation the fan-out would have to be less.

Gate 1 output drives
a total C_{LOAD} of
N × 5 pF

FIGURE 8-41   Each CMOS input adds to the total
load capacitance seen by the driving gate's output.

Switching Speed   Although CMOS, like N-MOS and P-MOS, has to drive relatively large load capacitances, its switching speed is somewhat faster because of its low output resistance in each state. Recall that an N-MOS output has to charge the load capacitance through a relatively large (100 kΩ) resistance. In the CMOS circuit, the output resistance in the HIGH state is the R_{ON} of the P-MOSFET, which is typically 1 kΩ or less. This allows more rapid charging of load capacitance.

A 4000 series NAND gate will typically have an average t_{pd} of 50 ns at $V_{DD} = 5$ V, and 25 ns at $V_{DD} = 10$ V. The reason for the improvement in t_{pd} as V_{DD} is increased is that the R_{ON} on the MOSFETs decreases significantly at higher supply voltages. Thus, it appears that V_{DD} should be made as large as possible for operation at higher frequencies. Of course, the larger V_{DD} will result in increased power dissipation.

A typical NAND gate in the 74HC or 74HCT series has an average t_{pd} of around 8 ns when operated at $V_{DD} = 5$ V. This is comparable to the speed of the 74LS series.

Unused Inputs   *CMOS inputs should never be left disconnected. All CMOS inputs have to be tied either to a fixed voltage level (0 V or V_{DD}) or to another input.* This rule applies even to the inputs of extra unused logic gates on a chip. An unconnected CMOS input is susceptible to noise and static charges that could easily bias both the P- and N-channel MOSFETs in the conductive state, resulting in increased power dissipation and possible overheating.

Teaching Tip: CMOS ICs can become permanently damaged when any input is left unconnected. However, it is more common that the IC will operate in an erratic manner.

Static-Charge Susceptibility   The high input resistance of CMOS inputs makes them especially prone to static-charge buildup that can produce voltages large enough to break down the dielectric insulation between the MOSFET gate and channel. Most of the newer CMOS devices are protected against static charge damage by the inclusion of protective zener diodes on each input. These diodes are designed to turn on and limit the size of the input voltage to well below any damaging value. While these zeners usually do the job, sometimes they do not turn on quickly enough to prevent the IC from being damaged, and so it is still a good idea to observe the special handling precautions presented earlier for all ICs.

TABLE 8-8 Digital IC Series Comparison*

	74HC	4000B	74	74S	74LS	74AS	74ALS	ECL
Power dissipation per gate (mW)								
Static	2.5×10^{-3}	1.0×10^{-3}	10	20	2	8	1.2	40
At 100 kHz	0.17	0.1	10	20	2	8	1.2	40
Propagation delay (ns)	8	50	9	3	9.5	1.7	4	1
Speed-power (at 100 kHz) (pJ)	1.4	5	90	60	19	13.6	4.8	40
Maximum clock rate (MHz)	40	12	35	12.5	45	200	70	300
Worst-case noise margin (V)	0.9	1.5	0.4	0.3	0.3	0.3	0.4	0.25

*All CMOS values are for $V_{DD} = 5V$.

Comparison of CMOS and TTL Series Table 8-8 compares the typical characteristics of the principal series of digital ICs.

REVIEW QUESTIONS

1. Which CMOS series are compatible pin for pin with TTL?
2. What are the improvements of the 74HC over the 74C series?
3. How does the 74HCT series differ from the 74HC?
4. Describe what happens to each of the following CMOS characteristics as V_{DD} is increased: (a) noise margin, (b) power dissipation, (c) switching speed.
5. What factors limit CMOS fan-out?
6. What should be done with unused CMOS inputs?
7. What precautions should be taken when handling CMOS devices?
8. Which IC series is the best for battery-operated circuitry?
9. Which series is the best for operation in high-noise environments?

8-16 CMOS OPEN-DRAIN AND TRISTATE OUTPUTS

Conventional CMOS outputs should never be connected together. Figure 8-42 illustrates what can happen when two CMOS INVERTER outputs are shorted together. The output on the left is trying to go to the HIGH state and its P-channel MOSFET is conducting with $R_{ON} = 1$ kΩ. The other output is trying to go to the LOW state, and its N-channel MOSFET is conducting with $R_{ON} = 1$ kΩ. The common output terminal X will be at a voltage of approximately $V_{DD}/2$ because of the voltage-divider action between the two 1-kΩ resistances.

This voltage is in the indeterminate range (recall that $V_{IL}(max) = 30$ percent of V_{DD} and $V_{IH}(min) = 70$ percent of V_{DD}) and is therefore unacceptable for driving other devices. Furthermore, the current through the two conducting MOSFETs will be much greater than normal, especially at higher values of V_{DD}, and can damage the ICs.

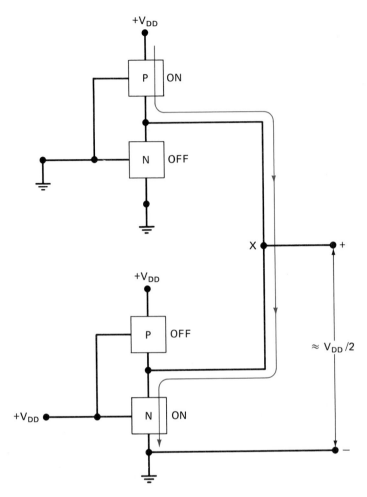

FIGURE 8-42 When CMOS outputs are shorted together, the common output terminal will be approximately $V_{DD}/2$ if the outputs are "trying" to be at different levels.

FIGURE 8-43 When two ordinary CMOS outputs are shorted together, the shorted outputs will be at $\approx V_{DD}/2$ when the outputs are trying to be different.

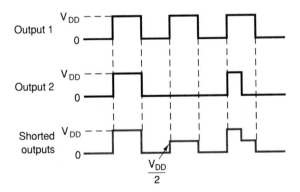

It is relatively easy to recognize when two CMOS outputs have been unintentionally wired or shorted together because the signals at the outputs will have three different levels: (1) HIGH when both outputs are trying to be HIGH; (2) LOW when both are trying to be LOW; and (3) appoximately $V_{DD}/2$ when the outputs are trying to be at different levels. This is illustrated in Figure 8-43.

Open-Drain Outputs   Some CMOS devices are available with open-drain outputs that are the counterpart to TTL open-collector outputs. In these devices, the output stage consists only of an N-channel MOSFET whose drain is unconnected since the upper P-channel MOSFET has been eliminated. An external pull-up resistor is needed to produce a HIGH-state voltage level. Like open-collector outputs, open-drain outputs can be wired-ANDed. Figure 8-44(a) shows three 74HC05 open-drain INVERTERs connected in a wired-AND arrangement.

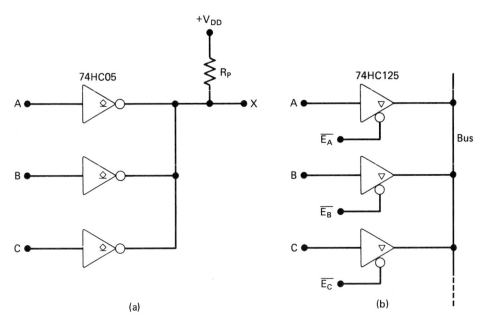

FIGURE 8-44   (a) CMOS open-drain devices in wired-AND connection; (b) CMOS tristate outputs are in bus arrangement.

Tristate Outputs   Several CMOS ICs have tristate outputs whose operation is similar to TTL tristate outputs. Everything we said about tristate TTL can be applied to CMOS tristate. CMOS tristate outputs can be connected together in a bus arrangement provided that only one output is enabled at one time. Figure 8-44(b) shows three 74HC125 tristate buffers connected in a bus arrangement.

—— REVIEW QUESTIONS ——

1. Why shouldn't conventional CMOS outputs be tied together?
2. What is an open-drain output?

8-17 CMOS TRANSMISSION GATE (BILATERAL SWITCH)

A special CMOS circuit that has no TTL or ECL counterpart is the *transmission gate* or *bilateral switch,* which acts essentially as a single-pole, single-throw switch controlled by an input logic level. This transmission gate will pass signals in both directions and is useful for digital and analog applications.

Discussion Question: What advantages and disadvantages does the CMOS bilateral switch have when compared with a relay?

Figure 8-45(a) is the basic arrangement for the bilateral switch. It consists of a P-MOSFET and an N-MOSFET in parallel so that both polarities of input voltage can be switched. The CONTROL input and its inverse are used to turn the switch on (closed) and off (open). When the CONTROL is HIGH, both MOSFETs are turned on and the switch is closed. When CONTROL is LOW, both MOSFETs are turned off and the switch is open. Ideally, this circuit operates like an electromechanical relay. In practice, however, it is not a perfect short circuit when the switch is closed; the switch resistance R_{ON} is typically 200 Ω. In the open state, the switch resistance is very large, typically 10^{12} Ω, which for most purposes is an open circuit. The symbol in Figure 8-45(b) is used to represent the bilateral switch.

This circuit is called a *bilateral* switch because the input and output terminals can be interchanged. The signals applied to the switch input can be either digital or analog signals, provided that they stay within the limits of 0 to V_{DD} volts.

Figure 8-46(a) shows the traditional logic diagram for a 4016 quad bilateral switch IC, which is also available in the 74HC series as a 74HC4016. The IC contains four bilateral switches that operate as described above. Each switch is independently controlled by its own control input. For example, the ON/OFF status of the top switch is controlled by input CONT$_A$. Since the switches are bidirectional, either switch terminal can serve as input or output, as the labeling indicates.

The IEEE/ANSI symbol for this IC is shown in Figure 8-46(b). Each small rectangle contains one bilateral switch, but, as is customary, only the top rectangle is labeled with the internal IEEE dependency notation. Looking at the top rectangle, the control input is labeled as X1. The "X" is the symbol used to denote a control input that controls the bidirectional transmission between like-numbered input/out-

FIGURE 8-45   CMOS bilateral switch (transmission gate).

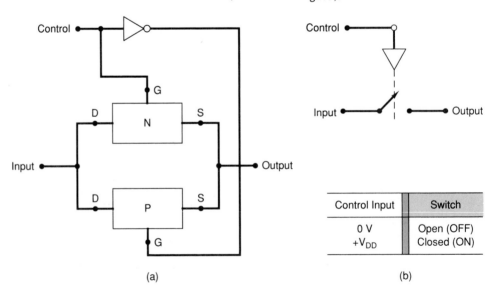

Control Input	Switch
0 V	Open (OFF)
+V_{DD}	Closed (ON)

(a)   (b)

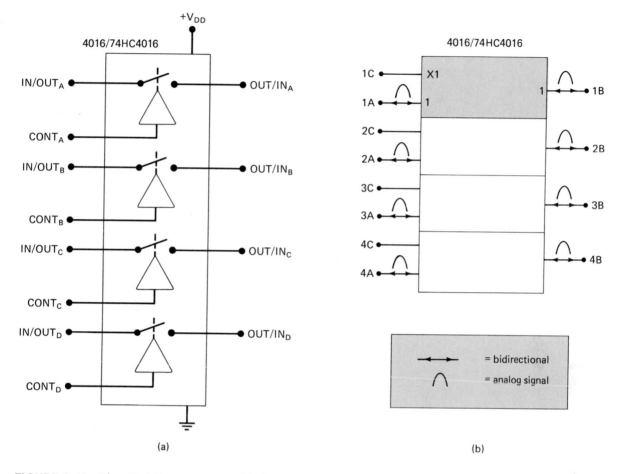

FIGURE 8-46   The 4016/74HC4016 quad bilateral switch: (a) traditional logic symbol; (b) IEEE/ANSI symbol.

put terminals. The "1" in the label X1 indicates that this control input controls the transmission between the terminals with the internal label 1. Outside the rectangle, a double-ended arrow is shown on the switch terminals to denote their bidirectional operation. The "horseshoe" is used to indicate that these terminals can handle analog signals, and are not restricted to digital levels.

_____ EXAMPLE 8-18 _____

Describe the operation of the circuit of Figure 8-47.

Solution   Here two of the bilateral switches are connected so that a common analog input signal can be switched to either output X or output Y, depending on the logic state of the OUTPUT SELECT input. When OUTPUT SELECT is LOW, the upper switch is closed and the lower one is open so that V_{IN} is connected to output X. When OUTPUT SELECT is HIGH, the upper switch is open and the lower one is closed so that V_{IN} is connected to output Y. Figure 8-47(b) shows some typical waveforms. Note that for proper operation, V_{IN} must be within the range 0 V to $+V_{DD}$.

Discussion Question: What will appear at the output if X and Y are shorted together?

The 4016/74HC4016 bilateral switch can switch only input voltages that lie between 0 V and V_{DD}, and so it could not be used for signals that were both positive and negative relative to ground. The 4316 and 74HC4316 ICs are quad bilateral

FIGURE 8-47 Example 8-18; 74HC4016 bilateral switches used to switch an analog signal to two different outputs.

switches that can switch *bipolar* analog signals. These devices have a second power-supply terminal called V_{EE}, which can be made negative with respect to ground. This permits input signals that can range from V_{EE} to V_{DD}. For example, with $V_{EE} = -5$ V and $V_{DD} = +5$ V, the analog input signal can be anywhere from -5 V to $+5$ V.

___ REVIEW QUESTIONS _____

1. Describe the operation of a CMOS bilateral switch.
2. *True or false:* There is no TTL bilateral switch.

8-18 IC INTERFACING

Interfacing means connecting the output(s) of one circuit or system to the input(s) of another circuit or system that has different electrical characteristics. Often a direct connection cannot be made because of the difference in the electrical characteristics of the *driver* circuit that is providing the output signal and the *load* circuit that is receiving the signal.

An **interface** circuit is a circuit connected between the driver and the load; its function is to take the driver output signal and condition it so that it is compatible with requirements of the load.

In the following sections we will address the problems involved in interfacing devices from one logic family to those of a different logic family. This type of interfacing occurs quite often in the more complex digital systems, where designers uti-

TABLE 8-9 Worst-Case Values for CMOS/TTL Interfacing*

Parameter	CMOS			TTL			
	4000B	**74HC**	**74HCT**	**74**	**74LS**	**74AS**	**74ALS**
V_{IH}(min)	3.5 V	3.5 V	2.0 V	2.0 V	2.0 V	2.0 V	2.0 V
V_{IL}(max)	1.5 V	1.0 V	0.8 V	0.8 V	0.8 V	0.8 V	0.8 V
V_{OH}(min)	4.95 V†	4.9 V†	4.9 V†	2.4 V	2.7 V	2.7 V	2.7 V
V_{OL}(max)	0.05 V†	0.1 V†	0.1 V†	0.4 V	0.5 V	0.5 V	0.4 V
I_{IH}(max)	1 μA	1 μA	1 μA	40 μA	20 μA	200 μA	20 μA
I_{IL}(max)	1 μA	1 μA	1 μA	1.6 mA	0.4 mA	2 mA	100 μA
I_{OH}(max)	0.4 mA	4 mA	4 mA	0.4 mA	0.4 mA	2 mA	400 μA
I_{OL}(max)	0.4 mA	4 mA	4 mA	16 mA	8 mA	20 mA	8 mA

*Supply voltage= 5 V.
†CMOS driving only CMOS inputs.

lize different logic families for different parts of the system in order to take advantage of the strong points of each family. For example, high-speed TTL (74AS, 74S) might be used in those parts of the system that are operating at the highest frequency, LS-TTL in the slower parts of the system, and N-MOS for the LSI and VLSI parts of the system.

ICs from the same logic series are designed to be connected together without any special considerations, provided that the fan-out limitation of each output is not exceeded. When you connect the output of an IC to the input of an IC from a different logic family or a different series within the same logic family, you generally have to be concerned with the voltage and current parameters of the two devices. This usually involves checking the device data sheets for values of input and output current/voltage parameters. We will use Table 8-9 to help us examine the various interfacing cases in the following sections. This table contains the worst-case values of the input and output parameters for standard devices in the CMOS and TTL series. It should be pointed out that the values in this table apply to standard devices with no special input or output circuitry. The output values may be different for devices such as buffers that use special output circuitry for increased current and voltage capabilities. The input current values may also be different for devices where an external input is internally connected to more than one gate. While the values in this table are valid for a large majority of devices, it would be advisable to check the specific IC data sheets.

8-19 TTL DRIVING CMOS

When interfacing different types of ICs, we must check that the driving device can meet the current and voltage requirements of the load device. Examination of Table 8-9 indicates that the input current values for CMOS are extremely low compared with the output current capabilities of any TTL series. Thus, TTL has no problem meeting the CMOS input current requirements.

There is a problem, however, when we compare the TTL output voltages with the CMOS input voltage requirements. We see that the V_{OH}(min) of every TTL series is too low when compared with the V_{IH}(min) requirement of the 4000B and 74HC series. For these situations, something must be done to raise the TTL output voltage to an acceptable level for CMOS.

FIGURE 8-48   External pull-up resistor is used when TTL drives CMOS.

The most common solution to this interface problem is shown in Figure 8-48, where the TTL output is connected to +5 V with a pull-up resistor. The presence of the pull-up resistor will cause the TTL output to rise to approximately 5 V in the HIGH state, thereby providing an adequate CMOS input. We determine the value of the pull-up resistor in the same way as we did for the open-collector pull-up resistor. If the TTL output is driving only CMOS, a value from 1 to 10 kΩ is normally used.

TTL Driving 74HCT

As we stated earlier, the 74HCT series is designed so that it can be driven directly by TTL outputs. In other words, its input voltage requirements are the same as standard TTL devices. This eliminates the need for the pull-up resistor of Figure 8-48. This ease of interfacing TTL to a 74HCT device comes at the cost of increased power dissipation in the CMOS device.

Teaching Tip: This is a common situation in high-noise industrial environments where high-voltage CMOS circuits are used because of their greater noise immunity, but where some TTL devices are needed to provide the particular logic function not available in CMOS.

TTL Driving High-Voltage CMOS

If the CMOS IC is operating with V_{DD} greater than 5 V, the situation becomes somewhat more difficult. For example, with $V_{DD} = 10$ V, the CMOS input requires $V_{IH}(\text{min}) = 7$ V. The outputs of many TTL devices cannot be operated at more than 5 V, and so a pull-up resistor connected to +10 V is prohibited. The LS-TTL series from certain manufacturers (e.g., Fairchild) can operate with an output pull-up to 10 V. In general, the device's data sheet should be checked before using a pull-up to more than 5 V.

When the TTL output cannot be pulled up to V_{DD}, there are some alternatives. One common solution is shown in Figure 8-49, where a 7407 open-collector buffer

FIGURE 8-49   A 7407 open-collector buffer can be used to interface TTL to high-voltage CMOS.

is used as the interface between a TTL totem-pole output and CMOS operating at $V_{DD} > 5$ V. The 7407 is the noninverting counterpart of the 7406 and has an output voltage rating of 30 V.

Another solution is to utilize a **level-translator** circuit such as the 40104. This is a CMOS chip that is designed to take a low-voltage input (e.g., from TTL) and translate it to a high-voltage output for CMOS.

—— REVIEW QUESTIONS ————————————————————————

1. What has to be done to interface a standard TTL output to a 4000B or 74HC input? Assume $V_{DD} = +5$ V.
2. What are the various ways to interface TTL to high-voltage CMOS?

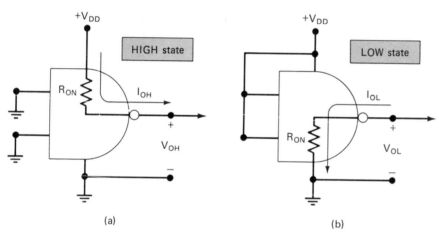

FIGURE 8-50   Equivalent CMOS output circuits for both logic states.

8-20   CMOS DRIVING TTL

Before we consider the problem of interfacing CMOS outputs to TTL inputs, it will be helpful to review the CMOS output characteristics for the two logic states.

Figure 8-50(a) shows the equivalent output circuit in the HIGH state. The R_{ON} of the P-MOSFET connects the output terminal to V_{DD} (remember, the N-MOSFET is off). Thus, the CMOS output circuit acts like a V_{DD} source with a source resistance of R_{ON}. The value of R_{ON} typically ranges from 100 to 1000 ohms.

Figure 8-50(b) shows the equivalent output circuit in the LOW state. The R_{ON} of the N-MOSFET connects the output terminal to ground (remember, the P-MOSFET is off). Thus, the CMOS output acts as a low resistance to ground; that is, it acts as a current sink.

CMOS Driving TTL in the HIGH State   Table 8-9 shows that CMOS outputs can easily supply enough voltage (V_{OH}) to satisfy the TTL input requirement in the HIGH state (V_{IH}). It also shows that CMOS outputs can supply more than enough current (I_{OH}) to meet the TTL input current requirements (I_{IH}). Thus, no special consideration is needed for the HIGH state.

CMOS Driving TTL in the LOW State   The table shows that TTL inputs have a relatively high input current in the LOW state, ranging from 100 μA to 2 mA. The 74HC and 74HCT families can sink up to 4 mA, and so they would have no trouble driving a *single* TTL load of any series. The 4000B series, however, is much more limited. Its low I_{OL} capability is not sufficient to drive even one input of the 74 or 74AS series.

———— EXAMPLE 8-19 ————————————————————

How many 74LS inputs can be driven by a 74HC output? Repeat for a 4000B output.

Solution   The 74LS series has I_{IL}(max) = 0.4 mA. The 74HC can sink up to I_{OL}(max) = 4 mA. Thus, the 74HC can drive *ten* 74LS loads (4 mA/0.4 mA = 10). The 4000B can sink only 0.4 mA, and so it can drive only *one* 74LS input.

———— EXAMPLE 8-20 ————————————————————

How many 74ALS inputs can be driven by a 74HC output? Repeat for a 4000B output.

Solution   The 74ALS series has I_{IL}(max) = 100 μA. Thus, the 74HC can drive *forty* 74ALS inputs (4mA/100 μA = 40). A 4000B can drive *four* 74ALS inputs (0.4 mA/100 μA = 4).

———— EXAMPLE 8-21 ————————————————————

What's wrong with the circuit in Figure 8-51(a)?

FIGURE 8-51A   Example 8-21.

(a)

Solution   The 74HC00 can sink 4 mA, but the three 74AS inputs require 3 × 2 mA = 6 mA.

———— EXAMPLE 8-22 ————————————————————

What's wrong with the circuit in Figure 8-51(b)?

(b)

FIGURE 8-51B   Example 8-22.

Solution   The 4001B NOR gate can sink 0.4 mA, but the three 74LS inputs require 3×0.4 mA $= 1.2$ mA.

For situations like those in Figure 8-51, some type of interface circuit is needed between the CMOS and TTL devices. The interface circuit should have a low input current requirement and a sufficiently high output current rating to drive the loads. Figure 8-52 shows two possible interfaces for the case of Figure 8-51(b). In Figure 8-52(a) the CMOS 4050B is a noninverting buffer that has an output current rating of $I_{OL}(max) = 3$ mA so that it can easily drive the three 74LS loads. In Figure 8-52(b) the 74LS125 is a permanently enabled noninverting tristate buffer which can be driven by the 4001B. Its output can easily drive the 74LS loads. In both of these circuits, the interface buffer simply passes the 4001B output signal to the 74LS loads.

74LS inputs

(a)

FIGURE 8-52   Buffers used to interface low-current CMOS to 74LS inputs.

4001B   74LS125

A

B

74LS inputs

(b)

FIGURE 8-53   A 4050B buffer can also serve as a level translator between high-voltage CMOS and TTL.

High-Voltage CMOS Driving TTL   Some IC manufacturers have produced several 74LS TTL devices that can withstand input voltages as high as 15 V. These devices can be driven directly from CMOS outputs operating at $V_{DD} = 15$ V. Most TTL inputs cannot handle more than 7 V before becoming damaged, and so an interface is necessary if they are to be driven from high-voltage CMOS. The interface functions as a **voltage-level translator** that converts the high-voltage input to a 5-V output that can be connected to TTL. Figure 8-53 shows how the 4050B performs this level translation between 15 V and 5 V.

REVIEW QUESTIONS

1. What is the function of an *interface* circuit?
2. *True or false:* All CMOS outputs can drive TTL in the HIGH state.
3. *True or false:* Any CMOS output can drive any single TTL input.
4. Which TTL series can drive CMOS inputs without a pull-up resistor?
5. How many 7400 inputs can be driven from a 74HCT00 output?
6. How is high-voltage CMOS interfaced to TTL?

8-21   TROUBLESHOOTING

A **logic pulser** is a testing and troubleshooting tool that generates a short-duration pulse when manually actuated, usually by pressing a pushbutton. The logic pulser shown in Figure 8-54 has a needle-shaped tip that is touched to the circuit node that is to be pulsed. The logic pulser is designed so that it senses the existing voltage level at the node and produces a voltage pulse in the opposite direction. In other words, if the node is LOW, the logic pulser produces a narrow positive-going pulse; if the node is HIGH, it produces a narrow negative-going pulse.

The logic pulser is used to momentarily change the logic level at a circuit node even though the output of another device may be connected to that same node. In Figure 8-54 the logic pulser is contacting node *X*, which is also connected to the output of the NAND gate. The logic pulser has a very low output impedance (typically 2 Ω or less), so that it can overcome the NAND gate's output and can change the voltage at the node. The logic pulser, however, cannot produce a voltage pulse at a node that is shorted directly to ground or V_{CC} (e.g., through a solder bridge).

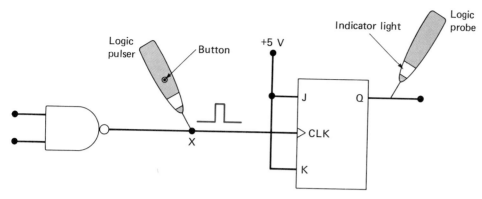

FIGURE 8-54   A logic pulser can inject a pulse at any node that is not shorted directly to ground or V$_{CC}$.

Using Logic Pulser and Probe to Test a Circuit

A logic pulser can be used to manually inject a pulse or series of pulses into a circuit in order to test the circuit's response. A logic probe is almost always used to monitor the circuit's response to the logic pulser. In Figure 8-54, the J-K flip-flop's toggle operation is being tested by applying pulses from the logic pulser and monitoring Q with the logic probe. This logic pulser/logic probe combination is very useful for checking the operation of a logic device while it is wired into a circuit. Note that the logic pulser is applied to the circuit node *without* disconnecting the output of the NAND gate that is driving that node.

Finding Shorted Nodes

The logic pulser and logic probe can be used to check for nodes that are shorted directly to ground or V_{CC}. When you touch a logic pulser and logic probe to the same node and press the logic pulser button, the logic probe should indicate the occurrence of a pulse at the node. If the probe indicates a constant LOW, the node is shorted to ground; if the probe indicates a constant HIGH, the node is shorted to V_{CC}.

The Current Tracer

A current tracer is a troubleshooting tool that can detect a *changing* current in a wire or printed-circuit-board trace without breaking the circuit. The current tracer has an insulated tip that contains a magnetic pickup coil. When the tip is placed at a point in the circuit, it senses a changing magnetic field produced by a changing current and causes a small indicator LED to flash. The current tracer does not respond to static current levels, no matter how great the current may be.

A current tracer is used with a logic pulser to trace the exact location of shorts to ground or V_{CC}. This is illustrated in Figure 8-55, where node X is shorted to ground through the internal short at gate 2's input. If the logic pulser is touched to X and its button is pressed, no voltage pulse will be detected, because of the short to ground. However, there will be a pulse of current flowing from the pulser's output to ground through the short. This current pulse can be detected by a current tracer.

In Figure 8-55(a), the current tracer is placed to the left of X and the logic pulser is pulsed. The tracer will indicate no current pulse through the path it is monitoring. In Figure 8-55(b), the tracer is moved to the other side of X. This time when

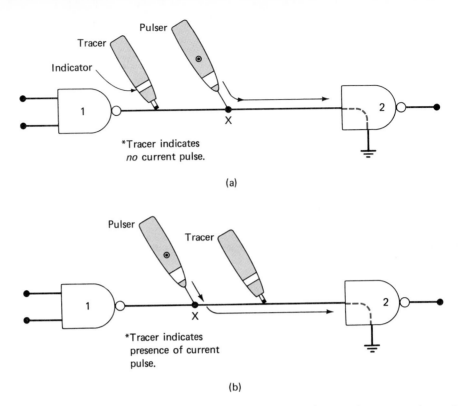

*Tracer indicates
no current pulse.

(a)

*Tracer indicates
presence of current
pulse.

(b)

FIGURE 8-55 Logic pulser and current tracer can be used to trace shorted nodes.

Discussion Question: Suppose the circuit of Figure 8-55(a) had no faults and was operating correctly. With the output of gate 1 in the LOW state, what would the current tracer indicate when the pulser is pulsed?

the logic pulser is pulsed, the current tracer will indicate the occurrence of a current pulse through the path it is monitoring. This proves that the short to ground is inside gate 2's input rather than gate 1's output.

The current tracer/logic pulser combination is very useful in troubleshooting circuits where many open-collector, open-drain, or tristate outputs are connected to a common point that is stuck LOW or HIGH. Any one of the outputs can be producing the fault, and the current tracer/logic pulser technique used above can be employed to track down the faulty output.

—— REVIEW QUESTIONS ——

1. What is the function of a logic pulser?
2. *True or false:* A logic pulser will produce a voltage pulse at any node.
3. What is the function of a current tracer?
4. *True or false:* A current tracer detects only a changing current.

PROBLEMS

(Data sheets for the ICs referred to in these problems may be found in Appendix II.)

8-1. Two different logic circuits have the following characteristics:

	Circuit A	Circuit B
V_{supply}	6 V	5 V
$V_{IH}(min)$	1.6 V	1.8 V
$V_{IL}(max)$	0.9 V	0.7 V
$V_{OH}(min)$	2.2 V	2.5 V
$V_{OL}(max)$	0.4 V	0.3 V
t_{PLH}	10 ns	18 ns
t_{PHL}	8 ns	14 ns
P_D	16 mW	10 mW

(a) Which circuit has the best LOW-state dc noise immunity? The best HIGH-state dc noise immunity?   Circuit A, circuit B
(b) Which circuit can operate at higher frequencies?   circuit A
(c) Which circuit draws the most supply current?   circuit A

8-2. Refer to Appendix II for the IC data sheets and use maximum values to determine $P_D(avg)$, $t_{pd}(avg)$, and speed-power product for one gate on each of the following TTL ICs.
(a) 7432   (b) 74S32   (c) 74LS20   (d) 74ALS20   (e) 74AS20

8-2. Speed-power products in pJ are: (a) 728.3, (b) 459.2, (c) 58.95, (d) 27.41, (e) 67.26.

8-3. A certain logic family has the following voltage parameters:
$$V_{IH}(min) = 3.5 \text{ V} \qquad V_{IL}(max) = 1.0 \text{ V}$$
$$V_{OH}(min) = 4.9 \text{ V} \qquad V_{OL}(max) = 0.1 \text{ V}$$

(a) What is the largest positive-going noise spike that can be tolerated?   0.9 V
(b) What is the largest negative-going noise spike that can be tolerated?   1.4 V

DRILL QUESTION

8-4. For each statement, indicate the term or parameter being described:
(a) Current at an input when a logic 1 is applied to that input I_{IH}
(b) Current drawn from the V_{CC} source when all outputs are LOW
(c) Time required for an output to switch from the 1 to the 0 state
(d) A common measure used to compare overall performance of different IC families   speed-power product
(e) The size of voltage spike that can be tolerated on a HIGH input without causing indeterminate operation V_{NH}
(f) An IC package that is designed for easy direct soldering to a PC board   surface-mount
(g) The action of a LOW output that receives current from the input of the circuit it is driving   current sinking
(h) Number of different inputs that an output can safely drive
(i) Arrangement of output transistors in a standard TTL circuit
(j) Term for transistor Q_4 of a TTL output circuit   pull-down transistor
(k) Range of V_{CC} values allowed for 74 series TTL   4.75 to 5.25 V
(l) $V_{OH}(min)$ and $V_{IH}(min)$ for 74 series   2.4 and 2.0 V
(m) $V_{IL}(max)$ and $V_{OL}(max)$ for 74 series   0.8 and 0.4 V

8-4. (b) I_{CCL} (c) t_{PHL}
(i) totem-pole

8-5. (a) From Table 8-3 determine noise margins when a 74LS device is driving a 74ALS input.   0.7 V, 0.3 V

(b) Repeat for a 74ALS driving a 74LS.   0.5 V, 0.4 V

(c) What will be the overall noise margin of a logic circuit that uses 74LS and 74ALS circuits in combination?   0.5 V, 0.3 V

SECTIONS 8-5 and 8-6

8-6. (d) Q_4 turns off more slowly than Q_3 turns on; voltage spikes on the V_{CC} bus; RF decoupling capacitors.

8-6. DRILL QUESTION

(a) Define unit load.   40 μA   (HIGH), 1.6 mA (LOW).

(b) In which type of gates do tied-together inputs always count as a single load in the LOW state?   AND, NAND.

(c) Define "floating" inputs.   Unconnected inputs.

(d) What causes current spikes in TTL? What undesirable effect can they produce? What can be done to reduce this effect?

8-7. (a) 0.5 UL (HIGH), 0.25 UL (LOW).   (b) Clock: 2 UL (HIGH), 0.5 UL (LOW); clear: 1.5 UL (HIGH), 0.5 UL (LOW).

8-7. Refer to the data sheet for the 74LS112 J-K flip-flop.

(a) Determine the input loading factor at the J and K inputs.

(b) Determine the input loading factor at the clock and clear inputs.

(c) How many other 74LS112s can the output of one 74LS112 drive at the clock input?   Five

8-8. Figure 8-56(a) shows a 74LS112 J-K flip-flop whose output is required to drive a total of 8 UL. Since this exceeds the fan-out of the 74LS112, a buffer of some type is needed. Figure 8-56(b) shows one possibility using one of the NAND gates from the 74LS37 quad NAND buffer, which has a much higher fan-out than the 74LS112. Note that $\overline{Q}$ is used since the NAND is acting as an INVERTER. Refer to the data sheet for the 74LS37.

(a) Determine its fan-out.   30 UL (HIGH), 15 UL (LOW);

(b) Determine its maximum sink current in the LOW state.   24 mA

(a)

(b)

FIGURE 8-56

8-9. Connect two 74LS00 gates in parallel between FF output and loads.

8-9. Buffer gates are generally more expensive than ordinary gates, and sometimes there are unused ordinary gates available that can be used to solve a loading problem such as that in Figure 8-56(a). Show how 74LS00 NAND gates can be used to solve this problem.

8-10. Refer to the logic diagram of Figure 8-57, where the 7486 exclusive-OR output is driving several 7420 inputs. Determine whether the fan-out of the 7486 is being exceeded, and explain.

8-11. Repeat Problem 8-10 using 74LS devices.

8-12. For the circuit of Figure 8-57 determine the longest time it will take for a change in the A input to be felt at output W. Use all worst-case conditions and maximum values of gate propagation delays. (*Hint:* Remember that NAND gates are inverting gates.)

FIGURE 8-57

8-13. (a) Figure 8-58 shows a 74193 counter with its active-HIGH master reset input activated by a pushbutton switch. Resistor R is used to

FIGURE 8-58

hold MR LOW while the switch is open. What is the maximum value that can be used for *R*?   500 Ω

(b) Repeat for the 74LS193.   2 kΩ

8-14. (a) The diode and voltage divider produce a positive half-cycle sine wave with reduced amplitude. The 74LS14 converts this to fast-changing pulses.   (b) The 4.7-kΩ resistor is too large.

8-14. Figure 8-59(a) shows a circuit that is used to convert a 60-Hz sine wave to a 60-pps signal that can reliably trigger FFs and counters. This type of circuit might be used in a digital clock.

(a) Explain the circuit operation.

(b) A technician is testing this circuit and observes that the 74LS14 output stays LOW. He checks the waveform at the INVERTER input and it appears as shown in Figure 8-59(b). Thinking that the INVERTER is faulty, he replaces the chip and observes the same results. What do you think is causing the problem, and how can it be fixed? (*Hint:* Examine the v_x waveform carefully.)

(a)

FIGURE 8-59

8-15. (a) Amplitude too low; (b) t_P is too small;   (c) clock LOW time is too small.

8-15. For each waveform in Figure 8-60, determine *why* it will *not* reliably trigger a 74LS112 flip-flop at its *CLK* input.

FIGURE 8-60

V_{CC} 4.97 V

1.3 V

FIGURE 8-61

8-16. A technician breadboards a logic circuit for testing. As she tests the circuit's operation, she finds that many of the FFs and counters are triggering erratically. Like any good technician, she checks the V_{CC} line with a dc meter and reads 4.97 V, which is acceptable for TTL. She then checks all circuit wiring and replaces each IC one by one, but the problem persists. Finally she decides to observe V_{CC} on the scope and sees the waveform shown in Figure 8-61. What is the probable cause of the noise on V_{CC}? What did the technician forget to include when she breadboarded the circuit?

8-16. Technician forgot decoupling capacitors.

SECTIONS 8-7 and 8-8

8-17. DRILL QUESTION
(a) Define wired-AND.   Tying logic outputs together
(b) What is a pull-up resistor? Why is it used?
(c) What types of TTL outputs can be safely tied together?
(d) What is bus contention?

8-17. (b) to produce a voltage in HIGH state for open-collector outputs; (c) open-collector and tristate; (d) more than one tri-state output enabled at the same time

8-18. The 7409 TTL IC is a quad two-input AND with open-collector outputs. Show how 7409s can be used to implement the operation $x = A \cdot B \cdot C \cdot D \cdot E \cdot F \cdot G \cdot H \cdot I \cdot J \cdot K \cdot M$.

8-18. Connect six 7409 AND gates in a wired-AND configuration.

8-19. Determine the logic expression for output X in Figure 8-62.

8-19. $X = AB + CD + FG$

FIGURE 8-62

8-20. Determine a value for R_p in Figure 8-62 if output X is driving the clear inputs of four 74LS112 flip-flops. R_P must be greater than 359 Ω.

8-21. Figure 8-63 shows a 7406 open-collector inverting buffer used to control the ON/OFF status of an LED to indicate the state of FF output Q. The LED's nominal specification is $V_F = 2.4$ V @ $I_F = 20$ mA, and $I_F(\text{max}) = 30$ mA.

(a) What voltage will appear at the 7406 output when $Q = 0$?

(b) Choose an appropriate value for the series resistor for proper operation.

FIGURE 8-63

8-22. Figure 8-64 shows how two tristate buffers can be used to construct a *bidirectional transceiver* that allows digital data to be transmitted in either direction (A to B, or B to A). Describe the circuit operation for the two states of the DIRECTION input.

FIGURE 8-64

FIGURE 8-65

8-23. The circuit of Figure 8-65 is used to provide the enable inputs for the circuit of Figure 8-29.

(a) Determine which of the data inputs (A, B, or C) will appear on the bus for each combination of inputs x and y.

(b) Explain why the circuit will not work if the NOR is changed to an EX-NOR.

8-24. A 7406 output is driving 5 ULs of TTL. What is the largest value of pull-up resistor that can be used and still maintain the TTL noise margin? (*Hint:* See Example 8-17 and 7406 data sheet.) 5.77 kΩ

8-25. The circuit of Figure 8-66 is an N-MOS logic gate. Determine what type of gate it is. Use $+16$ V = logic 1 and 0 V = logic 0. **8-25.** AND gate

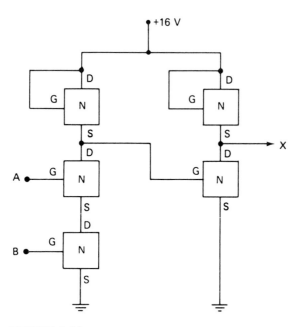

FIGURE 8-66

8-26. Draw the circuitry for an NMOS OR gate. See Solutions Manual.

8-27. Which of the following are advantages that P-MOS and N-MOS logic have over TTL? a, c, d, e, g, h
  (a) Greater packing density (f) Complementary outputs
  (b) Greater operating speed (g) Greater noise immunity
  (c) Greater fan-out (h) Simpler fabrication process
  (d) More suitable for LSI (i) More SSI and MSI functions.
  (e) Lower P_D (j) Lower supply-voltage requirement

8-28. Which of the following are advantages that CMOS generally has over TTL? a, c, e, f, g, h
  (a) Greater packing density (f) More suited for LSI
  (b) Higher speed (g) Lower P_D (below 1 MHz)
  (c) Greater fan-out (h) Transistors as only circuit element
  (d) Lower output impedance (i) Lower input capacitance
  (e) Simpler fabrication process

8-29. Which of the following operating conditions will probably result in the lowest average P_D for a CMOS logic system? Explain.
(a) $V_{DD} = 10$ V, switching frequency $f_{max} = 1$ MHz
(b) $V_{DD} = 5$ V, $f_{max} = 10$ kHz
(c) $V_{DD} = 10$ V, $f_{max} = 10$ kHz

8-30. What are the dc noise margins of the 4000B series that is operating at $V_{DD} = 12$ V?   3.6 V for both states

8-31. Refer to the data sheet for the 74HC20 NAND gate IC in Appendix II. Use maximum values to calculate P_D(avg), t_{pd}(avg), and speed-power product. Compare with the values calculated in Problem 8-2 for TTL.

SECTIONS 8-16 and 8-17

8-32. Determine the approximate values of V_{OUT} for both states of the CONTROL input in Figure 8-67.   0 V, 1.22 V

FIGURE 8-67

FIGURE 8-68

8-33. Determine the waveform at output X in Figure 8-68 for the given input waveforms. Assume $R_{ON} \approx 200\ \Omega$ for the bilateral switch.

 8-34. Determine the gain of the op-amp circuit of Figure 8-69 for the two states of the GAIN SELECT input. This circuit shows the basic principle of digitally controlled signal amplification.

8-33. Voltage at X will be a 2 $V_{p\text{-}p}$ sinewave while $C = 0$, and a 1 $V_{p\text{-}p}$ signal while $C = 1$.
8-34. Gain $= -1$ when GAIN SELECT = LOW; gain $= -2$ when GAIN SELECT = HIGH.

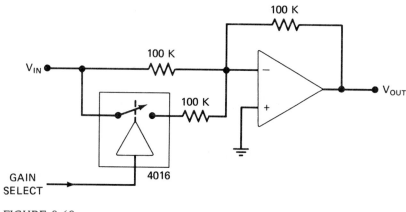

FIGURE 8-69

SECTIONS 8-18 to 8-20

8-35. DRILL QUESTION

(a) Which CMOS series can have its inputs driven directly from a TTL output?   74HCT.

(b) What is the function of a level translator? When is it used?

(c) Why is a buffer required between some CMOS outputs and TTL inputs?   CMOS cannot sink the TTL current.

(d) *True or false:* Most CMOS outputs have trouble supplying the TTL HIGH-state input current.   False.

8-35. (b) Converts between different logic voltage levels.

8-36. Refer to Figure 8-70(a), where a standard TTL output, Q, is driving a

FIGURE 8-70

(a)

(b)

445

CMOS INVERTER operating at $V_{DD} = 10$ V. The waveforms at Q and X appear as shown in Figure 8-70(b). Which of the following is a possible reason why X stays HIGH?   The choice in (c) is a possible reason.

(a) The 10-V supply is faulty.

(b) The pull-up resistor is too large.

(c) The 74LS112 output breaks down at well below 10 V and maintains a 5.5-V level in the HIGH state. This is in the indeterminate range for the CMOS input.

(d) The CMOS input is loading down the TTL output.

8-37. (a) Use Table 8-9 to determine how many 74AS inputs can typically be driven by a 4000B output.   none

(b) Repeat for a 74HC output.   two

8-38. Figure 8-71 is a logic circuit that was poorly designed. It contains at

8-38. (1) B input of OS;
(2) unused gates on 4001B;
(3) fan-out of top 7400 NAND;
(4) wired-AND connection;
(5) power-supply current;
(6) 74LS112 driving CMOS;
(7) no decoupling capacitors;
(8) 4001B driving TTL

FIGURE 8-71

least eight instances where the characteristics of the ICs have not been properly taken into account. Find as many of these as you can.

8-39. Repeat Problem 8-38 with the following changes in the circuit:
- Each TTL IC is replaced with its 74LS equivalent.
- The 4001B is replaced with a 74HCT02.

SECTION 8-21

8-40. The circuit in Figure 8-72 uses a 74HC05 IC which contains six open-drain INVERTERs. The INVERTERs are connected in a wired-AND arrangement. Using a pulser and logic probe, it is determined that node X is stuck in the HIGH state. Describe a procedure for using a current tracer to isolate the fault.

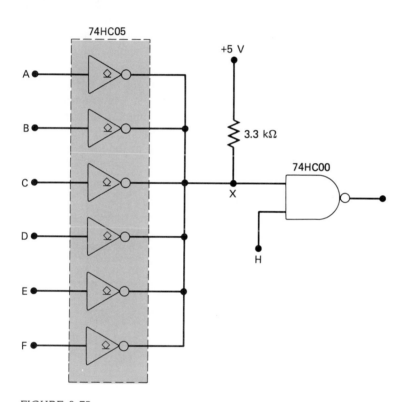

FIGURE 8-72

8-41. The circuit of Figure 8-54 has a solder bridge to ground somewhere between the output of the NAND gate and the input of the FF. Describe a procedure for tracking down the location of the solder bridge.

8-42. In Figure 8-48 a logic probe indicates that the lower end of the pull-up resistor is stuck in the LOW state. Using a logic pulser and current tracer a technician detects a current in the wire connecting the TTL output to the resistor. Which of the following is the possible fault?
(a) R_P is open.
(b) The TTL gate's current-sinking transistor has a collector-emitter short.
(c) There is a break in the connection from R_P to the CMOS gate.

8-43. The input of the NAND CMOS gate is shorted to +5 V.

8-43. In Figure 8-48 a logic probe indicates that the lower end of R_P is stuck HIGH. A logic pulser and current tracer are used to detect a current in the path connecting R_P to the CMOS gate. What is the probable fault?

MICROCOMPUTER APPLICATION

8-44. V_{OUT} starts at 0 V, steps down to −11.25 V, and comes back up to −6 V.

8-44. In Section 5-20, Figure 5-50, we saw how a microprocessor (MPU), under software control, transfers data to an external register. The diagram is repeated in Figure 8-73. Once the data are in the register, they are stored there and used for whatever purpose they are needed. Sometimes, each individual bit in the register has a unique function. For example, in automobile computers each bit could represent the status of a different physical variable being monitored by the MPU. One bit might indicate when the engine temperature is too high. Another bit might signal when oil pressure is too low. In other applications, the bits in the register are used to produce an analog output that can be used to drive devices requiring analog inputs that have many different voltage levels.

Figure 8-74 shows how we can use the MPU to generate an analog voltage by taking the register data from Figure 8-73(a) and using them to control the inputs to a summing amplifier. Assume that the MPU is executing a program that is transferring a new set of data to the register every 10 μs according to the table below. Sketch the resulting waveform at V_{OUT}.

Time (μs)	MPU Data
0	0000
10	0010
20	0100
30	0111
40	1010
50	1110
60	1111
70	1111
80	1110
90	1100
100	1000

FIGURE 8-73

FIGURE 8-74   Problem 8-44.

ANSWERS TO SECTION REVIEW QUESTIONS

SECTION 8-1

1. See text.   **2.** False.   **3.** V_{NH} is the difference between $V_{OH}(min)$ and $V_{IH}(min)$; V_{NL} is the difference between $V_{IL}(max)$ and $V_{OL}(max)$.
4. False.   **5.** Current sinking: An output actually receives (sinks) current from the input of the circuit it is driving. Current sourcing: An output supplies (sources) current to the circuit it is driving.   **6.** Surface mount.   **7.** Flat pack.

SECTION 8-2

1. True.   **2.** LOW.   **3.** Fast switching times, low power dissipation. Large current spike during switching from LOW to HIGH.   **4.** Q_3.

SECTION 8-4

1. **(a)** 74AS;   **(b)** 74S, 74LS;   **(c)** 74H, 74L;   **(d)** 74S, 74LS, 74AS, 74ALS; **(e)** 74ALS.   **2.** All three can operate at 40 MHz, but the 74ALS193 will use less power.

SECTION 8-5

1. Q_4's ON-state resistance and $V_{OL}(max)$.
2. 200 μA   **3.** Its output voltages may not remain in the allowed logic 0 and 1 ranges.
4. Two.

SECTION 8-6

1. LOW.   **2.** Connect to $+V_{CC}$ through a 1-kΩ resistor; connect to another used input.
3. Connect to ground; connect to another used input.   **4.** False; only in the LOW state.
5. Connecting small RF capacitors between V_{CC} and ground near each TTL IC to filter out voltage spikes caused by rapid current changes during output transitions from LOW to HIGH.

SECTION 8-7

1. Connecting TTL outputs together.
2. Damaging current can flow when one output is in the HIGH state and other is in the LOW state.
3. Current-sinking transistor Q_4's collector is unconnected (there is no Q_3).   **4.** To produce a V_{OH} level.   **5.** I_{OL} rating of output and the number of loads being driven.   **6.** No active pull-up transistor Q_3 **7.** See Figure 8-26.

SECTION 8-8

1. HIGH, LOW, Hi-Z.   **2.** Hi-Z.
3. When two or more tristate outputs tied to a common bus are enabled at the same time.
4. $E_A = E_B = 0$, $E_C = 1$.   **5.** See Figure 8-30.

SECTION 8-9

1. (a) True; **(b)** true; **(c)** false; **(d)** true; **(e)** false; **(f)** false

SECTION 8-13

1. Advantages—lower power; higher packing density; simple, inexpensive fabrication process. Disadvantages—slower speed, more sensitive to static electricity. **2.** Gate input capacitances that cause an increase in load capacitance for each additional load. This total load capacitance increases switching times. **3.** Large R_{OUT} and C_{LOAD}. **4.** High packing density due to its simple circuit structure. **5.** See text. **6.** See text

SECTION 8-14

1. (a) False—consumes less power; **(b)** true; **(c)** false; **(d)** true

SECTION 8-15

1. 74C, 74HC, 74HCT. **2.** Higher speed, greater output current. **3.** Can be driven directly from a TTL output. **4. (a)** Increases; **(b)** increases; **(c)** increases. **5.** Capacitances of CMOS inputs have to be charged by output of signal source. Each load adds capacitance. **6.** Connected to ground, V_{DD}, or another, used input. **7.** See Section 8-13, last paragraph. **8.** Check Table 8-8, power consumption per gate, and note that 4000B series is the lowest. **9.** In Table 8-8, the 4000B series has largest noise margins.

SECTION 8-16

1. Damaging current can flow; output voltage will be in indeterminate range. **2.** Output stage consists only of an N-channel MOSFET whose drain is unconnected.

SECTION 8-17

1. The logic level at the control input controls the open/closed status of a bidirectional switch that can pass analog signals in either direction. **2.** True.

SECTION 8-19

1. A pull-up resistor has to be connected to +5 V at the TTL output. **2.** A high-voltage buffer (7407) or a level translator (40104) should be connected between TTL output and CMOS input.

SECTION 8-20

1. It takes the output from a driver circuit and conditions it so that it is compatible with the input requirements of the load. **2.** True. **3.** False—for example, the 4000B series cannot sink I_{IL} of a 74 or 74AS device. **4.** None. **5.** Two. **6.** Through a level translator such as the 4050B.

SECTION 8-21

1. It injects a voltage pulse of selected polarity at a node that is not shorted to V_{CC} or ground. **2.** False. **3.** It indicates the occurrence of a current pulse. **4.** True.

9

MSI LOGIC CIRCUITS

Outline

Objectives

Upon completion of this chapter, you will be able to:

- Analyze and use decoders and encoders in various types of circuit applications.
- Compare the advantages and disadvantages of LEDs and LCDs.
- Utilize the observation/analysis technique for troubleshooting digital circuits.
- Understand the operation of multiplexers and demultiplexers by analyzing several circuit applications.
- Compare two binary numbers by using the magnitude comparator circuit.
- Understand the function and operation of code converters.
- Cite the precautions that must be considered when connecting digital circuits using the data-bus concept.
- Interpret the notation used on the IEEE/ANSI symbols for various MSI devices.

Introduction

Digital systems obtain binary-coded data and information that are continuously being operated on in some manner. Some of the operations include: (1) *decoding and encoding*—changing the data from one type of code to another, (2) *multiplexing*—selecting one out of several groups of data, (3) *demultiplexing*—distributing data to one of several destinations, and (4) *data busing*—transmitting data among several devices on a common bus. All these operations and others have been facilitated by the availability of numerous ICs in the MSI (medium-scale-integration) category.

In this chapter we will study many of the common types of MSI devices. For each type, we will start with a brief discussion of its basic operating principle, then introduce specific ICs. We will then show how they can be used alone or in combination with other ICs in various applications.

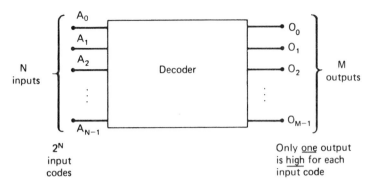

FIGURE 9-1   General decoder diagram.

9-1   DECODERS

A *decoder* is a logic circuit that converts an N-bit binary input code into $M *$ output lines such that each output line will be activated for only one of the possible combinations of inputs. Figure 9-1 shows the general decoder diagram with N inputs and M outputs. Since each of the N inputs can be 0 or 1, there are 2^N possible input combinations or codes. For each of these input combinations only one of the M outputs will be active (HIGH); all the other outputs are LOW. Many decoders are designed to produce active-LOW outputs, where only the selected output is LOW while all others are HIGH. This would be indicated by the presence of small circles on the output lines in the decoder diagram.

Some decoders do not utilize all of the 2^N possible input codes but only certain ones. For example, a BCD-to-decimal decoder has a 4-bit input code and *ten* output lines that correspond to the *ten* BCD code groups 0000 through 1001. Decoders of this type are often designed so that if any of the unused codes are applied to the input, *none* of the outputs will be activated.

In Chapter 7 we saw how decoders are used in conjunction with counters to detect the various states of the counter. In that application it was the FFs in the counter that provided the binary code inputs for the decoder. The same basic decoder circuitry is used no matter where the inputs come from. Figure 9-2 shows the circuitry for a decoder with three inputs and $2^3 = 8$ outputs. It uses all AND gates, and so the outputs are active-HIGH. For active-LOW outputs, NAND gates would be used. Note that for a given input code, the only output which is active (HIGH) is the one corresponding to the decimal equivalent of the binary input code (e.g., output O_6 goes HIGH only when $CBA = 110_2 = 6_{10}$).

This decoder can be referred to in several ways. It can be called a *3-line-to-8-line decoder*, because it has three input lines and eight output lines. It could also be called a *binary-to-octal decoder* or *converter* because it takes a 3-bit binary input code and activates the one of the eight (octal) outputs corresponding to that code. It is also referred to as a *1-of-8 decoder*, because only 1 of the 8 outputs is activated at one time.

Teaching Tip: Some IC manufacturers label the decoder inputs A, B, C, D, etc., while some use A_3, A_2, A_1, A_0, etc. The latter convention is the IEEE/ANSI standard. We will use one or the other at different times.

---

*N can be any integer and M is an integer that is less than or equal to 2^N.

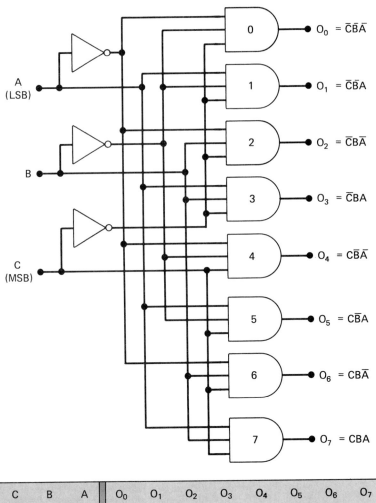

C	B	A	O_0	O_1	O_2	O_3	O_4	O_5	O_6	O_7
0	0	0	1	0	0	0	0	0	0	0
0	0	1	0	1	0	0	0	0	0	0
0	1	0	0	0	1	0	0	0	0	0
0	1	1	0	0	0	1	0	0	0	0
1	0	0	0	0	0	0	1	0	0	0
1	0	1	0	0	0	0	0	1	0	0
1	1	0	0	0	0	0	0	0	1	0
1	1	1	0	0	0	0	0	0	0	1

FIGURE 9-2   Three-line-to-8-line (or 1-of-8) decoder.

ENABLE Inputs   Some decoders have one or more ENABLE inputs that are used to control the operation of the decoder. For example, refer to the decoder in Figure 9-2 and visualize having a common ENABLE line connected to a fourth input of each gate. With this ENABLE line held HIGH the decoder will function normally and the A, B, C input code will determine which output is HIGH. With ENABLE held LOW, however, *all* the outputs will be forced to the LOW state regardless of the levels at the A, B, C inputs. Thus, the decoder is ENABLED only if ENABLE is HIGH.

Figure 9-3(a) shows the logic diagram for the 74LS138 decoder as it appears in the Fairchild *TTL Data Book*. By examining this diagram carefully, we can deter-

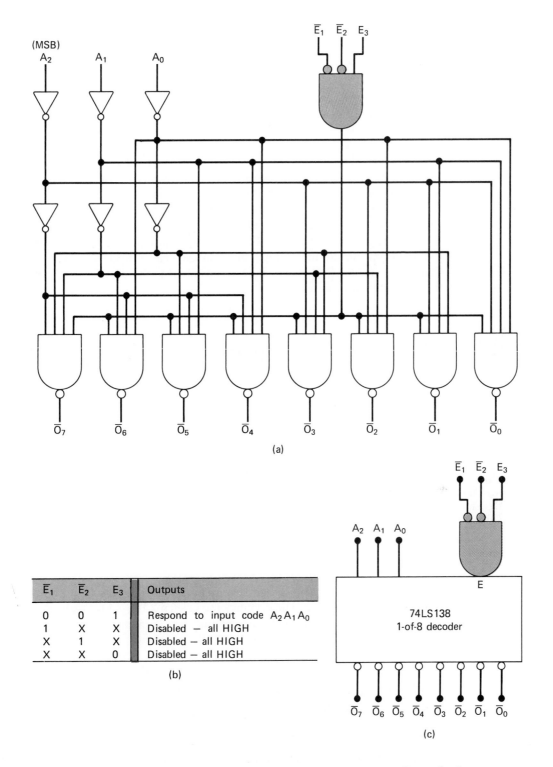

$\overline{E}_1$	$\overline{E}_2$	E_3	Outputs
0	0	1	Respond to input code $A_2 A_1 A_0$
1	X	X	Disabled — all HIGH
X	1	X	Disabled — all HIGH
X	X	0	Disabled — all HIGH

(b)

(c)

FIGURE 9-3 (a) Logic diagram for 74LS138 decoder; (b) truth table; (c) logic symbol. (Courtesy of Fairchild, a Schlumberger company)

mine exactly how this decoder functions. First, notice that it has NAND-gate outputs, so that its outputs are active-LOW. Another indication is the labeling of the outputs as $\bar{O}_7$, $\bar{O}_6$, $\bar{O}_5$, and so on; the inverting overbar indicates active-LOW outputs.

The input code is applied at A_2, A_1, and A_0, where A_2 is the MSB. With three inputs and eight outputs, this is a 3-to-8 decoder or, equivalently, a 1-of-8 decoder.

Inputs $\bar{E}_1$, $\bar{E}_2$, and E_3 are separate enable inputs that are combined in the AND gate. In order to enable the output NAND gates to respond to the input code at $A_2 A_1 A_0$, this AND-gate output has to be HIGH. This will occur only when $\bar{E}_1 = \bar{E}_2 = 0$ and $E_3 = 1$. In other words, $\bar{E}_1$ and $\bar{E}_2$ are active-LOW, E_3 is active-HIGH, and all three have to be in their active states to activate the decoder outputs. If one or more of the enable inputs is in its inactive state, the AND output will be LOW, which will force all NAND outputs to their inactive HIGH state regardless of the input code. This operation is summarized in the truth table in Figure 9-3(b). Recall that x represents the "don't care" condition.

The logic symbol for the 74LS138 is shown in Figure 9-3(c). Note how the active-LOW outputs are represented, and how the enable inputs are represented. Even though the enable AND gate is shown as external to the decoder block, it is part of the IC's internal circuitry.

____ EXAMPLE 9-1 _____

Indicate the states of the 74LS138 outputs for each of the following sets of inputs:

(a) $E_3 = \bar{E}_2 = 1$, $\bar{E}_1 = 0$, $A_2 = A_1 = 1$, $A_0 = 0$

(b) $E_3 = 1$, $\bar{E}_2 = \bar{E}_1 = 0$, $A_2 = 0$, $A_1 = A_0 = 1$

Solution

(a) With $\bar{E}_2 = 1$, the decoder is disabled and all of its outputs will be in their inactive HIGH state. This can be determined from the truth table or by following the input levels through the circuit logic.

Teaching Tip: It would be instructive to trace logic levels through the circuit for each case in the truth table.

(b) All of the enable inputs are activated, so that the decoding portion is enabled. It will decode the input code $011_2 = 3_{10}$ to activate output $\bar{O}_3$. Thus, $\bar{O}_3$ will be LOW and all other outputs will be HIGH.

____ EXAMPLE 9-2 _____

Figure 9-4 shows how four 74LS138s and an INVERTER can be arranged to function as a 1-of-32 decoder. The decoders are labeled Z1 to Z4 for easy reference, and the eight outputs from each one are combined into 32 outputs. Z1's outputs are $\bar{O}_0$ to $\bar{O}_7$, Z2's outputs $\bar{O}_0$ to $\bar{O}_7$ are renamed $\bar{O}_8$ to $\bar{O}_{15}$, respectively, Z3's outputs are renamed $\bar{O}_{16}$ to $\bar{O}_{23}$, and Z4's are renamed $\bar{O}_{24}$ to $\bar{O}_{31}$. A 5-bit input code $A_4 A_3 A_2 A_1 A_0$ will activate only one of these 32 outputs for each of the 32 possible input codes.

(a) Which output will be activated for $A_4 A_3 A_2 A_1 A_0 = 01101$?

(b) What range of input codes will activate the Z4 chip?

Solution

(a) The 5-bit input code has two distinct portions. The A_4 and A_3 bits determine which one of the decoder chips Z1 to Z4 will be enabled, while $A_2 A_1 A_0$ determine which output of the enabled chip will be activated. With $A_4 A_3 = 01$, only Z2 has all its enable inputs activated. Thus Z2 responds to the $A_2 A_1 A_0 = 101$

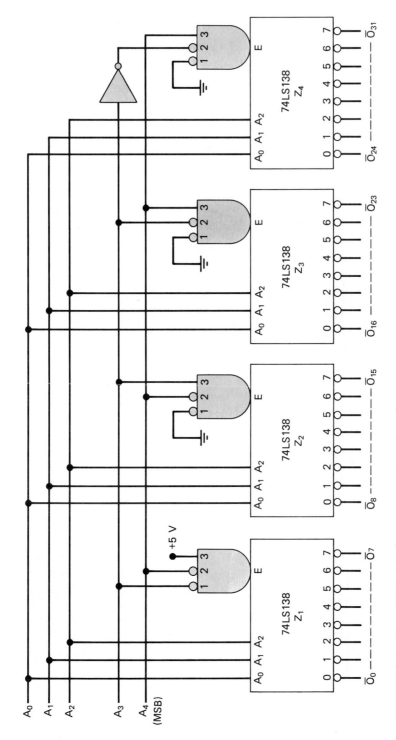

FIGURE 9-4 Four 74LS138s forming a 1-of-32 decoder.

457

code and activates its $\overline{O}_5$ output, which has been renamed $\overline{O}_{13}$. Thus the input code 01101, which is the binary equivalent of decimal 13, will cause output $\overline{O}_{13}$ to go LOW, while all others stay HIGH.

(b) To enable Z4, both A_4 and A_3 have to be HIGH. Thus all input codes ranging from 11000 (24_{10}) to 11111 (31_{10}) will activate Z4. This corresponds to outputs $\overline{O}_{24}$ to $\overline{O}_{31}$.

FIGURE 9-5 (a) Logic diagram for the 7442 BCD-to-decimal decoder; (b) logic symbol; (c) truth table. (Courtesy of Fairchild, a Schlumberger company)

(a)

(b)

Inputs				Active Output
D	C	B	A	
L	L	L	L	$\overline{O}_0$
L	L	L	H	$\overline{O}_1$
L	L	H	L	$\overline{O}_2$
L	L	H	H	$\overline{O}_3$
L	H	L	L	$\overline{O}_4$
L	H	L	H	$\overline{O}_5$
L	H	H	L	$\overline{O}_6$
L	H	H	H	$\overline{O}_7$
H	L	L	L	$\overline{O}_8$
H	L	L	H	$\overline{O}_9$
H	L	H	L	None
H	L	H	H	None
H	H	L	L	None
H	H	L	H	None
H	H	H	L	None
H	H	H	H	None

H = HIGH Voltage Level
L = LOW Voltage Level

(c)

BCD-to-Decimal Decoders

Figure 9-5(a) shows the logic diagram for a 7442 *BCD-to-decimal decoder*. It is also available as a 74LS42 and 74HC42. Each output goes LOW only when its corresponding BCD input is applied. For example, $\overline{O}_5$ will go LOW only when inputs $DCBA = 0101$; $\overline{O}_8$ will go LOW only when $DCBA = 1000$. For input combinations that are invalid for BCD, none of the outputs will be activated. This decoder can also be referred to as a *4-to-10 decoder* or a *1-of-10 decoder*. The logic symbol and truth table for the 7442 are also shown in the figure. Note that this decoder does not have an enable input. In Problem 9-7 we will see how the 7442 can be used as a 3-to-8 decoder with the D input used as an enable input.

BCD-to-Decimal Decoder/Driver

The TTL 7445 is a BCD-to-decimal decoder/*driver*. The term "driver" is added to its description because this IC has open-collector outputs that can operate at higher current and voltage limits than a normal TTL output. The 7445's outputs can sink up to 80 mA in the LOW state, and can be pulled up to 30 V in the HIGH state. This makes them suitable for directly driving loads such as indicator LEDs or lamps, relays, or dc motors.

Teaching Tip: The double inverters in Figures 9-3 and 9-5 are used to keep the loading effect on the decoder inputs at 1 UL.

Decoder Applications

Decoders are used whenever an output or group of outputs is to be activated only on the occurrence of a specific combination of input levels. These input levels are often provided by the outputs of a counter or register. When the decoder inputs come from a counter that is being continually pulsed, the decoder outputs will be activated sequentially, and they can be used as timing or sequencing signals to turn devices on or off at specific times. An example of this operation is shown in Figure 9-6 using the 74LS293 counter and the 7445 decoder/driver described above.

——— EXAMPLE 9-3 ———————————————————

Describe the operation of the circuit in Figure 9-6(a).

Solution The counter is being pulsed by a 1-pps signal so that it will sequence through the binary counts at the rate of 1 count/s. The counter FF outputs are connected as the inputs to the decoder. The 7445 open-collector outputs $\overline{O}_3$ and $\overline{O}_6$ are used to switch relays K_1 and K_2 on and off. For instance, when $\overline{O}_3$ is in its inactive

FIGURE 9-6 Example 9-3; counter/decoder combination used to provide timing and sequencing operations.

(a)

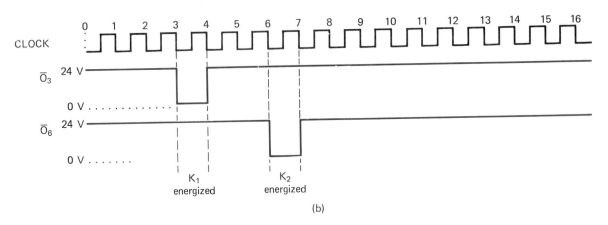

(b)

FIGURE 9-6 *(cont.)*

HIGH state, its output transistor will be off (nonconducting) so that no current can flow through relay K_1 and it will be deenergized. When $\overline{O}_3$ is in its active-LOW state, its output transistor is on and acts as a current sink for current through K_1 so that K_1 is energized. Note that the relays operate from +24 V. Also note the presence of the diodes across the relay coils; these protect the decoder's output transistors from the large "inductive kick" voltage that would be produced when coil current is abruptly stopped.

The timing diagram in Figure 9-6(b) shows the sequence of events. If we assume that the counter is in the 0000 state at time 0, then both outputs $\overline{O}_3$ and $\overline{O}_6$ are initially in the inactive HIGH state, where their output transistors are off and both relays are deenergized. As clock pulses are applied, the counter will be incremented once per second. On the NGT of the third pulse (time 3), the counter will go to the 0011 (3) state. This will activate decoder output $\overline{O}_3$ and thereby energize K_1. On the NGT of the fourth pulse, the counter goes to the 0100 (4) state. This will deactivate $\overline{O}_3$ and deenergize relay K_1.

Similarly, at time 6 the counter will go to the 0110 (6) state; this will make $\overline{O}_6 = 0$ and energize K_2. At time 7 the counter goes to 0111 (7) and deactivates $\overline{O}_6$ to deenergize K_2.

The counter will continue counting as pulses are applied. After 16 pulses, the sequence just described will start over.

Decoders are widely used in the memory system of a computer, where they respond to the address code input from the central processor to activate the memory storage location specified by the address code. We will examine this application in detail when we study memories in Chapter 11.

Another major application area for decoders is in the conversion of binary data to a form suitable for displaying on decimal readouts. We will look closely at this application in the following sections.

REVIEW QUESTIONS

1. Can more than one decoder output be activated at one time?
2. What is the function of a decoder's enable input(s)?
3. How does the 7445 differ from the 7442?
4. The 74154 is a 4-to-16 decoder with two active-LOW enable inputs. How many pins (including power and ground) does this IC have?

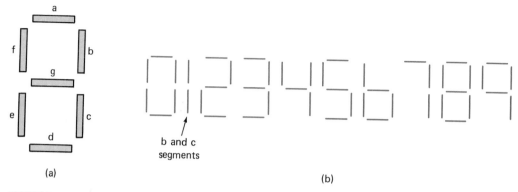

FIGURE 9-7   (a) Seven-segment arrangement; (b) active segments for each digit.

9-2 BCD-TO-7-SEGMENT DECODER/DRIVERS

Many numerical displays use a 7-segment configuration [Figure 9-7(a)] to form the decimal characters 0 through 9 and sometimes the hex characters A through F. Each segment is made up of a material that emits light when current is passed through it. Most commonly used materials include light-emitting diodes (LEDs) and incandescent filaments. Figure 9-7(b) shows the patterns of segments which are used to display the various digits. For example, to display "6," segments c, d, e, f, and g are bright while segments a and b are dark.

A *BCD-to-7-segment decoder/driver* is used to take a 4-bit BCD input and provide the outputs that will pass current through the appropriate segments to display the decimal digit. The logic for this decoder is more complicated than those we looked at previously because each output is activated for more than one combination of inputs. For example, the e segment must be activated for any of the digits 0, 2, 6, and 8, which means whenever any of the codes 0000, 0010, 0110, or 1000 occurs.

Figure 9-8(a) shows a BCD-to-7-segment decoder/driver (TTL 7446 or 7447) being used to drive a 7-segment LED readout. Each segment consists of one or two LEDs. The anodes of the LEDs are all tied to V_{CC} (+5 V). The cathodes of the LEDs are connected through current-limiting resistors to the appropriate outputs of the decoder/driver. The decoder/driver has active-LOW outputs which are open-collector driver transistors that can sink a fairly large current. This is because LED readouts may require 10 mA to 40 mA per segment, depending on their type and size.

To illustrate the operation of this circuit, let us suppose that the BCD input is $D = 0$, $C = 1$, $B = 0$, $A = 1$, which is BCD for 5. With these inputs the decoder/driver outputs $\bar{a}$, $\bar{f}$, $\bar{g}$, $\bar{c}$, and $\bar{d}$ will be driven LOW (connected to ground), allowing current to flow through the a, f, g, c, and d LED segments and thereby displaying the numeral 5. The $\bar{b}$ and $\bar{e}$ outputs will be HIGH (open), so that LED segments b and e cannot conduct.

The 7446 and 7447 decoder/drivers are designed to activate specific segments even for input codes greater than 1001 (9). Figure 9.8(b) shows which segments are activated for each of the input codes from 0000 to 1111 (15). Note that an input code of 1111 will blank out all the segments.

The LED display used in Figure 9-8 is a *common-anode* type because the anodes of all the segments are tied together to V_{CC}. Another type of 7-segment LED display uses a *common-cathode* arrangement where the cathodes of all the segments

Teaching Tip: Some students will be interested in how the logic in the BCD-to-7-segment decoder/driver is implemented. You may want to go through the design process for one of the segment outputs.

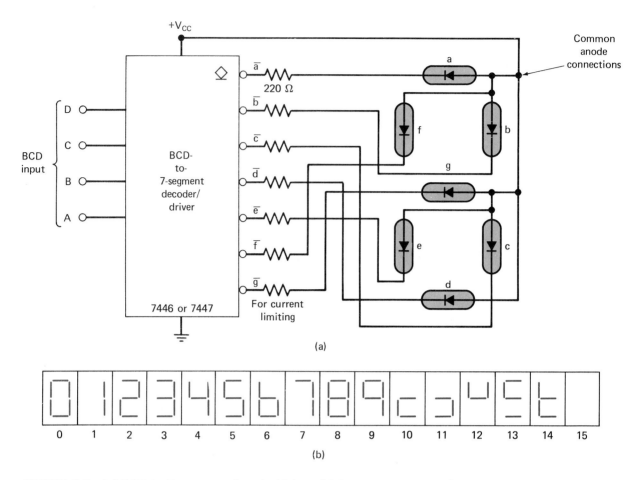

FIGURE 9-8 (a) BCD-to-7-segment decoder/driver driving a common-anode 7-segment LED display; (b) segment patterns for all possible input codes.

are tied together and connected to ground. This type of display has to be driven by a BCD-to-7-segment decoder/driver with active-HIGH outputs that apply a HIGH voltage to the anodes of those segments that are to be activated. The 7448 is a decoder/driver that can be used for this purpose.

Incandescent 7-segment Displays   Incandescent displays use what are basically light-bulb filaments as the segments. When current is made to flow through these filaments, they become hot and glow. They emit a bright white light that is easy to read. They are sometimes covered with colored filters to provide colored light. These displays are much less efficient than LED types because they require more current per segment, and therefore are not used in battery-operated devices such as calculators and multi-meters. They are, however, used in electronic cash registers and other line-operated devices where power consumption is not critical.

—— EXAMPLE 9-4 ——

Each segment of a typical 7-segment LED display is rated to operate at 10 mA at 2.7 V for normal brightness. Calculate the value of the current-limiting resistor needed to produce approximately 10 mA per segment.

Solution   Referring to Figure 9-8(a), we can see that the series resistor will have to have a voltage drop equal to the difference between $V_{CC} = 5$ V and the segment voltage of 2.7 V. This 2.3 V across the resistor must produce a current of about 10 mA. Thus we have

$$R_S = \frac{2.3 \text{ V}}{10 \text{ mA}} = 230 \ \Omega$$

A standard resistor value close to this can be used. A 220-Ω resistor would be a good choice.

REVIEW QUESTIONS

1. Which LED segments will be on for a decoder/driver input of 1001?
2. *True or false:* More than one output of a BCD-to-7-segment decoder/driver can be active at one time.

9-3   LIQUID CRYSTAL DISPLAYS

Basically, LCDs operate from a low-voltage (typically 3 to 15 V rms), low-frequency (25 to 60 Hz) ac signal and draw very little current. They are often arranged as 7-segment displays for numerical readouts as shown in Figure 9-9. The ac voltage needed to turn on a segment is applied between the segment and the *backplane,* which is common to all segments. The segment and backplane form a capacitor that draws very little current as long as the ac frequency is kept low. It is generally not lower than 25 Hz, because this would produce visible flicker.

LCDs draw much less current than LED displays and are widely used in battery-powered devices such as calculators and watches. An LCD does not emit light energy like an LED, and so it requires an external source of light.

FIGURE 9-9   Liquid crystal display.

Driving an LCD   An LCD segment will turn on when an ac voltage is applied between the segment and the backplane, and will turn off when there is no voltage between the two. Rather than generating an ac signal, it is common practice to produce the required ac voltage by applying out-of-phase square waves to the segment and backplane. This is illustrated in Figure 9-10 for one segment. A 40-Hz square wave is applied to the backplane and also to the input of a CMOS 4070 exclusive-

FIGURE 9-10   Method for driving an LCD segment. When CONTROL is LOW, segment is off. When CONTROL is HIGH, segment is on.

OR. The other input to the EX-OR is a CONTROL input that will control whether the segment is ON or OFF.

Teaching Tip: You might want to perform a more detailed analysis to show how the voltage between the segment and backplane becomes ac when CONTROL = 1.

When the CONTROL input is LOW, the EX-OR output will be exactly the same as the 40-Hz square wave, so that the signals applied to the segment and backplane are equal. Since there is no difference in voltage, the segment will be off. When the CONTROL input is HIGH, the EX-OR output will be the INVERSE of the 40-Hz square wave, so that the signal applied to the segment is out of phase with the signal applied to the backplane. As a result, the segment voltage will alternately be at +5 V and at −5 V relative to the backplane. This ac voltage will turn on the segment.

FIGURE 9-11   Method for driving 7-segment LCD.

This same idea can be extended to a complete 7-segment LCD display as shown in Figure 9-11. Here the CMOS 4511 BCD-to-7-segment decoder/driver supplies the CONTROL signals to each of seven EX-ORs for the seven segments. The 4511 has active-HIGH outputs, since a HIGH is required to turn on a segment.

In general, CMOS devices are used to drive LCDs for two reasons: (1) they require much less power than TTL and are more suited to the battery-operated applications where LCDs are used; (2) the TTL LOW-state voltage is not exactly 0 V and can be as much as 0.4 V. This will produce a dc component of voltage between the segment and backplane that considerably shortens the life of an LCD.

___ REVIEW QUESTION _____

1. Indicate which of the following statements refer to LCD displays and which refer to LED displays: (a) emit light, (b) reflect ambient light, (c) are best for low-power applications, (d) require an ac voltage, (e) use a 7-segment arrangement to produce digits, (f) require current-limiting resistors.

9-4 ENCODERS

A decoder accepts an N-bit input code and produces a HIGH (or LOW) at *one and only one* output line. In other words, we can say that a decoder identifies, recognizes, or detects a particular code. The opposite of this decoding process is called *encoding* and is performed by a logic circuit called an *encoder*. An encoder has a number of input lines, only one of which is activated at a given time, and produces an N-bit output code, depending on which input is activated. Figure 9-12 is the general diagram for an encoder with M inputs and N outputs. Here the inputs are active-HIGH, which means they are normally LOW.

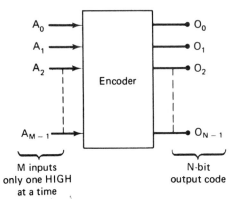

M inputs
only one HIGH
at a time

N-bit
output code

FIGURE 9-12   General encoder diagram.

We saw that a *binary-to-octal decoder (3-line-to-8-line decoder)* accepts a 3-bit input code and activates one of eight output lines corresponding to that code. An *octal-to-binary encoder (8-line-to-3-line encoder)* performs the opposite function; it accepts eight input lines and produces a 3-bit output code corresponding to the activated input. Figure 9-13 shows the logic circuit and truth table for an octal-to-binary encoder with active-LOW inputs.

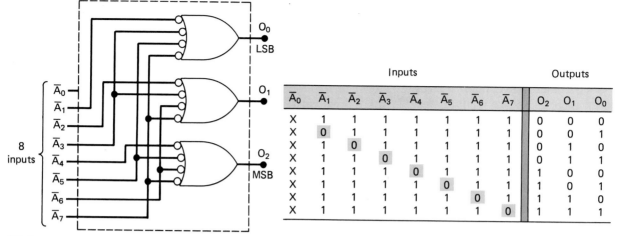

	Inputs								Outputs		
$\overline{A}_0$	$\overline{A}_1$	$\overline{A}_2$	$\overline{A}_3$	$\overline{A}_4$	$\overline{A}_5$	$\overline{A}_6$	$\overline{A}_7$	O_2	O_1	O_0
X	1	1	1	1	1	1	1	0	0	0
X	0	1	1	1	1	1	1	0	0	1
X	1	0	1	1	1	1	1	0	1	0
X	1	1	0	1	1	1	1	0	1	1
X	1	1	1	0	1	1	1	1	0	0
X	1	1	1	1	0	1	1	1	0	1
X	1	1	1	1	1	0	1	1	1	0
X	1	1	1	1	1	1	0	1	1	1

*Only one
LOW input
at a time.

FIGURE 9-13 Logic circuit for an octal-to-binary (8-line-to-3-line) encoder. For proper operation, only one input should be active at one time.

Teaching Tip: In actuality, the A_0 input is not needed to produce the 000 output condition, and is not part of the encoder per se. It may be needed, however, in applications such as Figure 9-16.

By following through the logic, you can verify that a LOW at any single input will produce the output binary code corresponding to that input. For instance, a LOW at $\overline{A}_3$ (while all other inputs are HIGH) will produce $\overline{O}_2 = 0$, $\overline{O}_1 = 1$, and $\overline{O}_0 = 1$, which is the binary code for 3. Notice that $\overline{A}_0$ is not connected to the logic gates because the encoder outputs will normally be at 000 when none of the inputs $\overline{A}_1$ to $\overline{A}_9$ is LOW.

―――― EXAMPLE 9-5 ――――――――――――――――――――――――

Determine the outputs of the encoder in Figure 9-13 when $\overline{A}_3$ and $\overline{A}_5$ are simultaneously LOW.

Solution   Following through the logic gates, we see that the LOWs at these two inputs will produce HIGHs at each output; in other words, the binary code 111. Clearly, this is not the code for either activated input.

Priority Encoders   This last example points up a drawback of the simple encoder circuit of Figure 9-13 when more than one input is activated at one time. A modified version of this circuit, called a *priority encoder*, includes the necessary logic to ensure that when two or more inputs are activated, the output code will correspond to the highest-numbered input. For example, when both $\overline{A}_3$ and $\overline{A}_5$ are LOW, the output code will be 101 (5). Similarly, when $\overline{A}_6$, $\overline{A}_2$, and $\overline{A}_0$ are all LOW, the output code is 110 (6). The 74148, 74LS148, and 74HC148 are all octal-to-binary priority encoders.

74147 Decimal-to-BCD Priority Encoder   Figure 9-14 shows the logic symbol and truth table for the 74147 (74LS147, 74HC147), which functions as a decimal-to-BCD priority encoder. It has nine active-LOW inputs representing the decimal digits 1 through 9, and produces the *inverted* BCD code corresponding to the highest-numbered activated input.

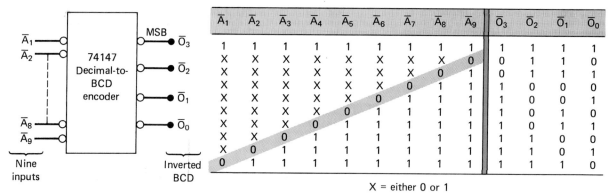

$\overline{A}_1$	$\overline{A}_2$	$\overline{A}_3$	$\overline{A}_4$	$\overline{A}_5$	$\overline{A}_6$	$\overline{A}_7$	$\overline{A}_8$	$\overline{A}_9$	$\overline{O}_3$	$\overline{O}_2$	$\overline{O}_1$	$\overline{O}_0$
1	1	1	1	1	1	1	1	1	1	1	1	1
X	X	X	X	X	X	X	X	0	0	1	1	0
X	X	X	X	X	X	X	0	1	0	1	1	1
X	X	X	X	X	X	0	1	1	1	0	0	0
X	X	X	X	X	0	1	1	1	1	0	0	1
X	X	X	X	0	1	1	1	1	1	0	1	0
X	X	X	0	1	1	1	1	1	1	0	1	1
X	X	0	1	1	1	1	1	1	1	1	0	0
X	0	1	1	1	1	1	1	1	1	1	0	1
0	1	1	1	1	1	1	1	1	1	1	1	0

X = either 0 or 1

FIGURE 9-14 74147 decimal-to-BCD priority encoder.

Let's examine the truth table to see how this IC works. The first line in the table shows all inputs in their inactive HIGH state. For this condition the outputs are 1111, which is the inverse of 0000, the BCD code for 0. The second line in the table indicates that a LOW at $\overline{A}_9$, regardless of the states of the other inputs, will produce an output code of 0110, which is the inverse of 1001, the BCD code for 9. The third line shows that a LOW at $\overline{A}_8$, provided that $\overline{A}_9$ is HIGH, will produce an output code of 0111, the inverse of 1000, the BCD code for 8. In a similar manner, the remaining lines in the table show that a LOW at any input, provided that all higher-numbered inputs are HIGH, will produce the inverse of the BCD code for that input.

The 74147 outputs will be normally HIGH when none of the inputs are activated. This corresponds to the decimal 0 input condition. There is no $\overline{A}_0$ input, since the encoder assumes the decimal 0 input state when all other inputs are HIGH. The 74147 inverted BCD outputs can be converted to normal BCD by putting each one through an INVERTER.

_____ EXAMPLE 9-6 _____

Determine the states of the outputs in Figure 9-14 when $\overline{A}_5$, $\overline{A}_7$, and $\overline{A}_3$ are LOW and all other inputs are HIGH.

Solution   The truth table shows that when $\overline{A}_7$ is LOW, the levels at $\overline{A}_5$ and $\overline{A}_3$ do not matter. Thus, the outputs will be 1000 respectively.

Switch Encoder   Figure 9-15 shows how a 74147 can be used as a *switch encoder*. The 10 switches might be the keyboard switches on a calculator representing digits 0 through 9. The switches are of the normally open type, so that the encoder inputs are all normally HIGH and the BCD output is 0000 (note the INVERTERs). When a digit key is depressed, the circuit will produce the BCD code for that digit. Since the 74147 is a priority encoder, simultaneous key depressions will produce the BCD code for the higher-numbered key.

The switch encoder of Figure 9-15 can be used whenever BCD data have to be manually entered into a digital system. A prime example would be in an electronic calculator, where the operator depresses several keyboard switches in succession to enter a decimal number. In a simple, basic calculator the BCD code for each decimal digit is entered into a 4-bit storage register. In other words, when the first key is depressed, the BCD code for that digit is sent to a 4-bit FF register; when the second

Discussion Question: What would actually occur at the outputs in Figure 9-15 when SW7 is momentarily closed and released?

FIGURE 9-15   Decimal-to-BCD switch encoder.

switch is depressed, the BCD code for that digit is sent to *another* 4-bit FF register, and so on. Thus, a calculator that can handle eight digits will have eight 4-bit registers to store the BCD codes for these digits. Each 4-bit register drives a decoder/driver and numerical display so that the eight-digit number can be displayed.

The operation described above can be accomplished with the circuit in Figure 9-16. This circuit will take three decimal digits entered from the keyboard in sequence, encode them in BCD, and store the BCD in three FF output registers. The 12 D-type flip-flops Q_0 to Q_{11} are used to receive and store the BCD codes for the digits. Q_8 to Q_{11} store the BCD code for the most significant digit (MSD), which is the first one entered on the keyboard. Q_4 to Q_7 store the second entered digit and Q_0 to Q_3 store the third entered digit. Flip-flops X, Y, and Z form a ring counter (Chapter 7) which controls the transfer of data from the encoder outputs to the appropriate output register. The OR gate produces a HIGH output any time one of the keys is depressed. This output may be affected by switch contact bounce, which would produce several pulses before settling down to the HIGH state. The OS is used to neutralize the switch bounce by triggering on the first positive transition from the OR gate and remaining HIGH for 20 ms, well past the time duration of the switch bounce. The OS output clocks the ring counter.

The circuit operation is described as follows for the case where the decimal number 309 is being entered:

FIGURE 9-16 Circuit for keyboard entry of three-digit number into storage registers.

Discussion Question: Can contact bounce of the CLEAR switch cause any problems?

1. The CLEAR key is depressed. This clears all storage flip-flops Q_0 to Q_{11} to 0. It also clears flip-flops X and Y and presets flip-flop Z to 1, so that the ring counter begins in the 001 state.

2. The CLEAR key is released and the "3" key is depressed. The encoder outputs 1100 are inverted to produce 0011, the BCD code for 3. These binary values are sent to the D inputs of the three 4-bit output registers.

3. The OR output goes HIGH (since two of its inputs are HIGH) and triggers the OS output $Q = 1$ for 20 ms. After 20 ms, Q returns LOW and clocks the ring counter to the 100 state (X goes HIGH). The positive transition at X is fed to the CLK inputs of flip-flops Q_8 to Q_{11}, so that the encoder outputs are transferred to these FFs. That is, $Q_{11} = 0$, $Q_{10} = 0$, $Q_9 = 1$, and $Q_8 = 1$. Note that flip-flops Q_0 to Q_7 are not affected, because their CLK inputs have not received a positive transition.

4. The "3" key is released and the OR gate output returns LOW. The "0" key is then depressed. This produces the BCD code of 0000, which is fed to the inputs of the storage FFs.

5. The OR output goes HIGH in response to the "0" key (note the INVERTER) and triggers the OS for 20 ms. After 20 ms the ring counter shifts to the 010 state (Y goes HIGH). The positive transition at Y is fed to the CLK inputs of Q_4 to Q_7 and transfers 0000 to these FFs. Note that flip-flops Q_0 to Q_3 and Q_8 to Q_{11} are not affected by the Y transition.

6. The "0" key is released and the OR output returns LOW. The "9" key is depressed, producing BCD outputs 1001, which are fed to the storage FFs.

Discussion Question: Discuss how the circuit might operate if the OS were not used.

7. The OR output goes HIGH again, triggering the OS, which in turn clocks the ring counter to the 001 state (Z goes HIGH). The positive transition at Z is fed to the CLK inputs of Q_0 to Q_3 and transfers the 1001 into these FFs. The other storage FFS are unaffected.

8. At this point the storage register contains 0011 0000 1001, beginning with Q_{11}. This is the BCD code of 309. These register outputs feed decoder/drivers which drive appropriate displays for indicating the decimal digits 309.

9. The storage FF outputs are also fed to other circuits in the system. In a calculator, for example, these outputs would be sent to the arithmetic section to be processed.

Several problems at the end of the chapter will deal with some other aspects of this circuit including troubleshooting exercises.

—— REVIEW QUESTIONS ——

1. How does an encoder differ from a decoder?
2. How does a priority encoder differ from an ordinary encoder?
3. What will the outputs be in Figure 9-15 when SW6, SW5, and SW2 are all closed?
4. Describe the functions of each of the following parts of the keyboard entry circuit of Figure 9-16: (a) OR gate; (b) 74147 encoder; (c) one-shot; (d) flip-flops X, Y, Z; (e) flip-flops Q_0 to Q_{11}.

9-5 IEEE/ANSI SYMBOLS

We will now look at the IEEE/ANSI symbols for several of the decoders and encoders. Figure 9-17(a) shows the IEEE/ANSI symbol for the 7442 decoder. It is fairly straightforward. Note the label BCD/DEC, which denotes that it is a BCD-to-decimal decoder. Also note the manner in which the inputs and outputs are numbered inside the symbol block.

The IEEE/ANSI symbol for the 7445 in Figure 9-17(b) is similar to the 7442 with two differences. First, each output has the underlined diamond symbol to indicate its open-collector structure. Second, the triangle in the middle of the symbol

FIGURE 9-17   IEEE/ANSI symbols for several decoders.

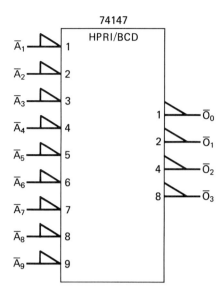

FIGURE 9-18   IEEE/ANSI symbol for the 74147 encoder.

denotes that this device is a buffer or a driver with greater-than-normal current and/or voltage capabilities.

Figure 9-17(c) is the symbol for the 74LS138 decoder. Note the BIN/OCT label to denote its binary-to-octal decoding function. Also note how the three enable inputs are combined in an AND block to produce an overall internal enable signal, EN.

Figure 9-18 shows the IEEE/ANSI symbol for the 74147 encoder IC. The label HPRI/BCD denotes that the function of this IC is to convert the active input with the highest priority to its BCD code. Again, note how the inputs and outputs are numbered inside the symbol block.

―― REVIEW QUESTIONS ――――――――――

1. What does the label BCD/DEC mean?
2. What does the label HPRI mean?
3. What does the symbol ▷ mean inside a symbol block?

9-6   TROUBLESHOOTING

As circuits and systems become more complex, the number of possible causes of failure obviously increases. While the procedure for fault isolation and correction remains essentially the same, the application of the **observation/analysis** process is more important for complex circuits because it helps the troubleshooter narrow the location of the fault to a small area of the circuit. This reduces to a reasonable amount the testing steps and resulting data that have to be analyzed. By understanding the circuit operation, observing the symptoms of the failure, and reasoning through the operation, the troubleshooter can often predict the possible faults before ever picking up a logic probe or oscilloscope. This observation/analysis process is one that inexperienced troubleshooters are hesitant to apply, probably because of the

great variety and capabilities of modern test equipment available to them. It is easy to become overly reliant on these tools while not adequately utilizing the human brain's reasoning and analytical skills.

The following examples illustrate how the observation/analysis process can be applied. Many of the end-of-chapter troubleshooting problems will provide you with the opportunity to develop your skill at applying this process.

EXAMPLE 9-7

A technician tests the circuit of Figure 9-4 by using a set of switches to apply the input code at A_4 through A_0. She runs through each possible input code and checks the corresponding decoder output to see if it is activated. She observes that all of the odd-numbered outputs respond correctly, but all of the even-numbered outputs fail to respond when their code is applied. What are the most probable faults?

Solution   In a situation where so many outputs are failing, it is unreasonable to expect that each of these outputs has a fault. It is much more likely that some faulty input condition is causing the output failures. What do all of the even-numbered outputs have in common? Let's look at the input codes for several of them.

Output	Input Code
$\overline{O}_0$	00000
$\overline{O}_4$	00100
$\overline{O}_{14}$	01110
$\overline{O}_{18}$	10010

Clearly, each even-numbered output requires an input code with an $A_0 = 0$ in order to be activated. Thus, the most probable faults would be those that prevent A_0 from going LOW. This includes:

A faulty switch connected to the A_0 output
A break in the path between the switch and the A_0 line
An external short from the A_0 line to V_{CC}
An internal short to V_{CC} at the A_0 inputs of any one of the decoder chips

Thus, the fault is narrowed to a specific area of the circuit. The exact fault can be traced with the testing and measurement techniques that we are already familiar with.

Teaching Tip: We could probably have concluded that the fault was in A_0 without writing down input codes, but it is important to impress on the student how this technique can be very helpful when troubleshooting circuits that are malfunctioning only for certain input conditions. It is often possible to find a pattern in the bits that can lead you to the area of the fault.

EXAMPLE 9-8

A technician wires the outputs from a BCD counter to the inputs of the decoder/driver of Figure 9-8. He applies pulses to the counter at a very slow rate and observes the LED display, which is shown below, as the counter counts up from 0000

COUNT	0	1	2	3	4	5	6	7	8	9
Observed display	*0*	*1*	*2*	*3*	*4*	*5*	*6*	*7*	*8*	*9*
Expected display	*0*	*1*	*2*	*3*	*4*	*5*	*6*	*7*	*8*	*9*

to 1001. Examine this observed sequence carefully and try to predict the most probable fault.

Solution Comparing the observed display with the expected display for each count, we see several important points:

For those counts where the observed display is incorrect, the observed display is not one of the segment patterns that correspond to counts greater than 1001. This rules out a faulty counter or faulty wiring from the counter to the decoder/driver.

The correct segment patterns (0, 1, 3, 6, 7, and 8) have the common property that segments e and f are either both on or both off.

The incorrect segment patterns have the common property that segments e and f are in opposite states, and if we interchange the states of these two segments, the correct pattern is obtained.

Giving some thought to those points should lead us to conclude that the technician has probably "crossed" the connections to the e and f segments.

9-7 MULTIPLEXERS (DATA SELECTORS)

A *multiplexer* or *data selector* is a logic circuit that accepts several data inputs and allows only *one* of them at a time to get through to the output. The routing of the desired data input to the output is controlled by SELECT inputs (sometimes referred to as ADDRESS inputs). Figure 9-19 shows the functional diagram of a general multiplexer (MUX). In this diagram the inputs and outputs are drawn as large arrows to indicate that they may be one or more signal lines.

The multiplexer acts like a digitally controlled multiposition switch where the digital code applied to the SELECT inputs controls which data inputs will be switched to the output. For example, output Z will equal data input I_0 for some particular SELECT input code; Z will equal I_1 for another particular SELECT input code; and so on. Stated another way, a multiplexer selects 1 out of N input data sources and transmits the selected data to a single output channel. This is called **multiplexing.**

FIGURE 9-19 Functional diagram of a digital multiplexer (MUX).

SELECT input code determines which input is transmitted to output Z.

I_0

I_1

I_{N-1}
DATA
inputs

Output
Z

MUX

SELECT
inputs

FIGURE 9-20   Two-input multiplexer.

Basic Two-Input Multiplexer   Figure 9-20 shows the logic circuitry for a two-input multiplexer with data inputs I_0 and I_1 and SELECT input S. The logic level applied to the S input determines which AND gate is enabled so that its data input passes through the OR gate to output Z. Looking at it another way, the Boolean expression for the output is

$$Z = I_0\bar{S} + I_1 S$$

With $S = 0$, this expression becomes

$$Z = I_0 \cdot 1 + I_1 \cdot 0$$
$$= I_0$$

which indicates that Z will be identical to input signal I_0, which can be a fixed logic level or a time-varying logic signal. With $S = 1$, the expression becomes

$$Z = I_0 \cdot 0 + I_1 \cdot 1 = I_1$$

showing that output Z will be identical to input signal I_1.

FIGURE 9-21   Four-input multiplexer.

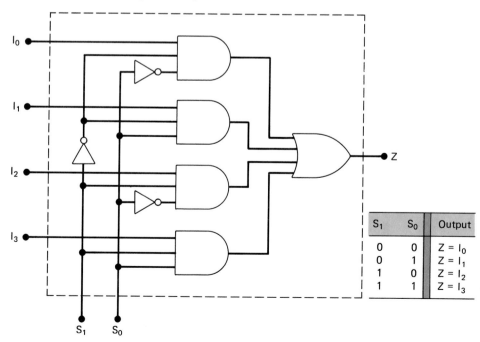

Four-Input Multiplexer   The same basic idea can be used to form the four-input multiplexer shown in Figure 9-21 (see page 475). Here there are four inputs, which are selectively transmitted to the output according to the four possible combinations of the $S_1 S_0$ select inputs. Each data input is gated with a different combination of select input levels. I_0 is gated with $\overline{S}_1 \overline{S}_0$ so that I_0 will pass through its AND gate to output Z only when $S_1 = 0$ and $S_0 = 0$. The table in the figure gives the outputs for the other three input select codes.

FIGURE 9-22   (a) Logic diagram for 74151 multiplexer; (b) truth table; (c) logic symbol. (Courtesy of Fairchild, a Schlumberger company)

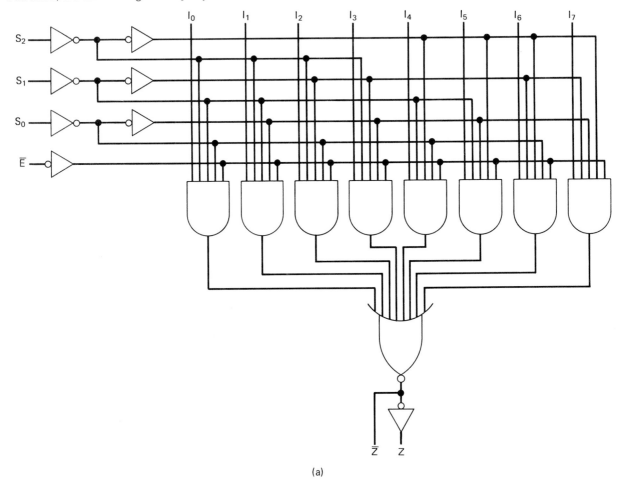

(a)

Inputs				Outputs	
$\overline{E}$	S_2	S_1	S_0	$\overline{Z}$	Z
H	X	X	X	H	L
L	L	L	L	$\overline{I}_0$	I_0
L	L	L	H	$\overline{I}_1$	I_1
L	L	H	L	$\overline{I}_2$	I_2
L	L	H	H	$\overline{I}_3$	I_3
L	H	L	L	$\overline{I}_4$	I_4
L	H	L	H	$\overline{I}_5$	I_5
L	H	H	L	$\overline{I}_6$	I_6
L	H	H	H	$\overline{I}_7$	I_7

(b)

(c)

Two-, four-, eight-, and 16-input multiplexers are readily available in the TTL and CMOS logic families. These basic ICs can be combined for multiplexing a larger number of inputs.

Eight-Input Multiplexer

Figure 9-22(a) shows the logic diagram for the 74151 (74LS151, 74HC151) eight-input multiplexer. This multiplexer has an enable input, $\bar{E}$, and provides both the normal and inverted outputs. When $\bar{E} = 0$, the select inputs $S_2 S_1 S_0$ will select one data input (from I_0 through I_7) for passage to output Z. When $\bar{E} = 1$, the multiplexer is disabled so that $Z = 0$ regardless of the select input code. This operation is summarized in Figure 9-22(b), and the 74151 logic symbol is shown in Figure 9-22(c).

Teaching Tip: Each AND gate is actually decoding for a particular SE-LECT input condition, and when it sees that condition, it passes its data input on to the OR gate. Only one AND gate will see the condition it is looking for. Thus, the OR gate will receive only one data input, which it passes on to its output.

_____ EXAMPLE 9-9 _____

The circuit in Figure 9-23 uses two 74151s, an INVERTER, and an OR gate. Describe this circuit's operation.

Solution   This circuit has a total of 16 data inputs, eight applied to each multiplexer. The two multiplexer outputs are combined in the OR gate to produce a single

FIGURE 9-23   Example 9.9: two 74151s combined to form a 16-input multiplexer.

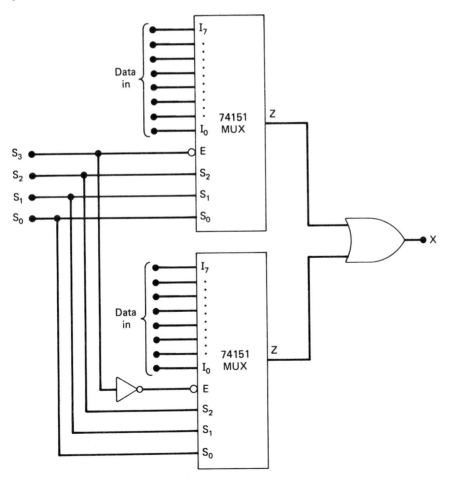

output X. The circuit functions as a 16-input multiplexer. The four select inputs $S_3 S_2 S_1 S_0$ will select one of the 16 inputs to pass through to X.

The S_3 input determines which multiplexer is enabled. When $S_3 = 0$, the top multiplexer is enabled, and the $S_2 S_1 S_0$ inputs determine which of its data inputs will appear at its output and pass through the OR gate to X. When $S_3 = 1$, the bottom multiplexer is enabled and the $S_2 S_1 S_0$ inputs select one of its data inputs for passage to output X.

FIGURE 9-24 (a) Logic diagram for 74157 multiplexer; (b) logic symbol; (c) truth table. (Courtesy of Fairchild, a Schlumberger company)

(a)

(b)

$\bar{E}$	S	Z_a	Z_b	Z_c	Z_d
H	X	L	L	L	L
L	L	I_{0a}	I_{0b}	I_{0c}	I_{0d}
L	H	I_{1a}	I_{1b}	I_{1c}	I_{1d}

(c)

Quad Two-Input MUX (74157/LS157/HC157)   The 74157 is a very useful multiplexer IC that contains four two-input multiplexers like the one in Figure 9-20. The logic diagram for the 74157 is shown in Figure 9-24(a). Note the manner in which the data inputs and outputs are labeled.

_____ EXAMPLE 9-10 _____

Determine the input conditions required for each Z output to take on the logic level of its corresponding I_0 input. Repeat for I_1.

Solution   First of all, the enable input has to be active; that is, $\bar{E} = 0$. In order for Z_a to equal I_{0a}, the select input has to be LOW. These same conditions will produce $Z_b = I_{0b}$, $Z_c = I_{0c}$, and $Z_d = I_{0d}$.

With $\bar{E} = 0$ and $S = 1$, the Z outputs will follow the set of I_1 inputs; that is, $Z_a = I_{1a}$, $Z_b = I_{1b}$, $Z_c = I_{1c}$, and $Z_d = I_{1d}$.

All the outputs will be disabled (LOW) when $\bar{E} = 1$.

It is helpful to think of this multiplexer as being a simple two-input multiplexer, but one in each input is four lines and the output is four lines. The four output lines switch back and forth between the two sets of four input lines under the control of the select input. This operation is represented by the 74157's logic symbol in Figure 9-24(b).

_____ REVIEW QUESTIONS _____

1. What is the function of a multiplexer's select inputs?
2. A certain multiplexer can switch one of 32 data inputs to its output. How many different inputs does this MUX have?

9-8   MULTIPLEXER APPLICATIONS

Multiplexer circuits find numerous and varied applications in digital systems of all types. These applications include data selection, data routing, operation sequencing, parallel-to-serial conversion, waveform generation, and logic-function generation. We shall look at some of these applications here and several more in the problems at the end of the chapter.

Data Routing   Multiplexers can route data from one of several sources to one destination. One typical application uses 74157 multiplexers to select and display the contents of either of two BCD counters using a *single* set of decoder/drivers and LED displays. The circuit arrangement is shown in Figure 9-25.

Each counter consists of two cascaded BCD stages, and each one is driven by its own clock signal. When the COUNTER SELECT line is HIGH, the outputs of counter 1 will be allowed to pass through the multiplexers to the decoder/drivers to be displayed on the LED readouts. When COUNTER SELECT = 0, the outputs of counter 2 will pass through the multiplexers to the displays. In this way the decimal contents of one counter or the other will be displayed under the control of the COUNTER SELECT input. A common situation where this might be used is in a

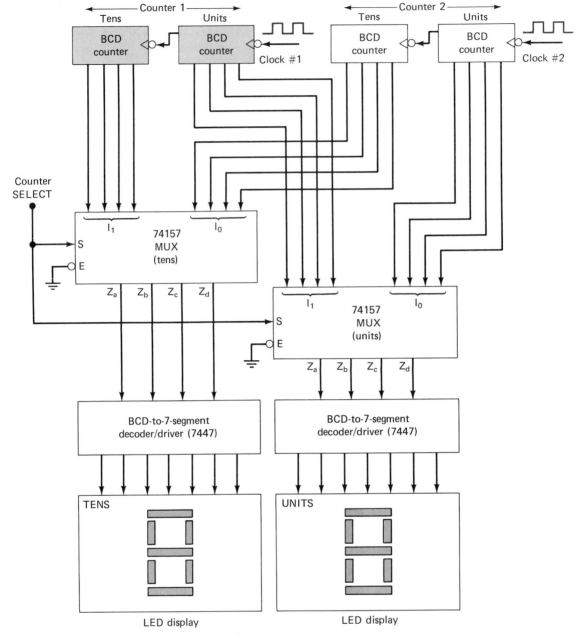

FIGURE 9-25 System for displaying two multidigit BCD counters one at a time.

digital watch. The digital watch circuitry contains many counters and registers that keep track of seconds, minutes, hours, days, months, alarm settings, and so on. A multiplexing scheme such as this one allows different data to be displayed on the limited number of decimal readouts.

The purpose of the multiplexing technique, as it is used here, is to *time-share* the decoder/drivers and display circuits between the two counters rather than have a separate set of decoder/drivers and displays for each counter. This results in a significant saving in the number of wiring connections, especially when more BCD stages are added to each counter. Even more important, it represents a significant decrease in power consumption, because decoder/drivers and LED readouts typically

draw relatively large amounts of current from the V_{CC} supply. Of course, this technique has the limitation that only one counter's contents can be displayed at a time. However, in many applications this is not a drawback. A mechanical switching arrangement could have been used to perform the function of switching first one counter and then the other to the decoder/drivers and displays, but the number of required switch contacts, the complexity of wiring, and the physical size could all be disadvantages over the completely logic method of Figure 9-25.

Parallel-to-Serial Conversion

Many digital systems process binary data in parallel form (all bits simultaneously) because it is faster. When data are to be trans-

FIGURE 9-26   (a) Parallel-to-serial converter; (b) waveforms for $X_7X_6X_5X_4X_3X_2X_1X_0 = 10110101$.

mitted over relatively long distances, however, the parallel arrangement is undesirable because it requires a large number of transmission lines. For this reason, binary data or information in parallel form is often converted to serial form before being transmitted to a remote destination. One method for performing this *parallel-to-serial conversion* uses a multiplexer, as illustrated in Figure 9-26 (see page 481).

The data are present in parallel form at the outputs of the X register and are fed to the eight-input multiplexer. A 3-bit (MOD-8) counter is used to provide the select code bits $S_2 S_1 S_0$ so that they cycle through from 000 to 111 as clock pulses are applied. In this way, the output of the multiplexer will be X_0 during the first clock period, X_1 during the second clock period, and so on. The output Z is a waveform which is a serial representation of the parallel input data. The waveforms in the figure are for the case where $X_7 X_6 X_5 X_4 X_3 X_2 X_1 X_0 = 10110101$. This conversion process takes a total of eight clock cycles. Note that X_0 (the LSB) is transmitted first and the X_7 (MSB) is transmitted last.

Discussion Question: What are two different ways to modify this parallel-to-serial converter so that the MSB is transmitted first?

Operation Sequencing

The circuit of Figure 9-27 uses an eight-input multiplexer as part of a control sequencer that steps through seven steps, each of which actuates some portion of the physical process being controlled. This process could, for example, be a large high-temperature oven which is energized by seven different heaters, which must be activated one at a time. The circuit also uses a 3-line-to-8-line decoder and a MOD-8 binary counter. The operation is described as follows:

1. Initially the counter is reset to the 000 state. The counter outputs are fed to the select inputs of the multiplexer and to the inputs of the decoder. Thus, the decoder output $\overline{O}_0 = 0$ and the others are all 1, so that all the ACTUATOR inputs of the process are LOW. The SENSOR outputs of the process all start out LOW. The multiplexer output $\overline{Z} = \overline{I}_0 = 1$, since the S inputs are 000.

2. The START pulse initiates the sequencing operation by setting flip-flop Q_0 HIGH, bringing the counter to the 001 state. This causes decoder output $\overline{O}_1$ to go LOW, thereby activating actuator 1, which is the first step in the process (it could be the turning on of a heater).

3. Some time later SENSOR output 1 goes HIGH, indicating the completion of the first step (it could be the reaching of a certain temperature level). This HIGH is now present at the I_1 input of the multiplexer. It is inverted and reaches the $\overline{Z}$ output since the select code from the counter is 001.

4. The LOW at $\overline{Z}$ is fed to the *CLK* of flip-flop Q_0. This negative transition advances the counter to the 010 state.

5. Decoder output $\overline{O}_2$ now goes LOW, activating actuator 2, which is the second step in the process. $\overline{Z}$ now equals $\overline{I}_2$ (select code is 010). Since SENSOR output 2 is still LOW, $\overline{Z}$ will go HIGH.

6. When the second process step is complete, SENSOR output 2 goes HIGH, producing a LOW at $\overline{Z}$ and advancing the counter to 011.

Teaching Tip: It may be helpful to go through a typical timing diagram using process delay times of a few seconds.

7. This same action is repeated for each of the other steps. When the seventh step is completed, SENSOR output 7 goes HIGH, causing the counter to go from 111 to 000, where it will remain until another START pulse reinitiates the sequence.

Logic Function Generation

Multiplexers can be used to implement logic functions directly from a truth table without the need for simplification. When used

FIGURE 9-27   Seven-step control sequencer.

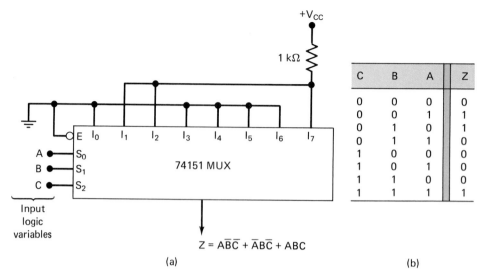

C	B	A	Z
0	0	0	0
0	0	1	1
0	1	0	1
0	1	1	0
1	0	0	0
1	0	1	0
1	1	0	0
1	1	1	1

$$Z = A\overline{B}\,\overline{C} + \overline{A}B\overline{C} + ABC$$

(a) (b)

FIGURE 9-28 Multiplexer used to implement a logic function described by truth table.

for this purpose the select inputs are used as the logic variables and each data input is connected permanently HIGH or LOW as necessary to satisfy the truth table.

Figure 9-28 illustrates how an eight-input multiplexer can be used to implement the logic circuit that satisfies the given truth table. The input variables A, B, C are connected to S_0, S_1, S_2, respectively, so that the levels on these inputs determine which data input appears at output Z. According to the truth table, Z is supposed to be LOW when $CBA = 000$. Thus, multiplexer input I_0 should be connected LOW. Likewise, Z is supposed to be LOW for $CBA = 011$, 100, 101, and 110, so that inputs I_3, I_4, I_5, and I_6 should also be connected LOW. The other sets of CBA conditions must produce $Z = 1$, and so multiplexer inputs I_1, I_2, and I_7 are connected permanently HIGH.

It is easy to see that any three-variable truth table can be implemented with this eight-input multiplexer. This method of implementation is often more efficient than using separate logic gates. For example, if we write the sum-of-products expression for the truth table in Figure 9-28, we have

$$Z = A\overline{B}\,\overline{C} + \overline{A}B\overline{C} + ABC$$

This *cannot* be simplified either algebraically or by K mapping, and so its gate implementation would require three INVERTERs and four NAND gates, for a total of two ICs.

There is an even more efficient method for using multiplexers to implement logic functions. This method will allow the logic designer to use a multiplexer with three select inputs (e.g., the 74151) to implement a *four-variable* logic function. We will present this method in Problem 9-35.

Common Student Mistake: The student can get into trouble by writing the truth table using A, B, C as the order of the variables. When this is done, care must be taken to read the binary number in the reverse order (C, B, A) to determine which inputs are to be tied HIGH.

___ REVIEW QUESTIONS ___

1. What are some of the major applications of multiplexers?
2. *True or false:* When a multiplexer is used to implement a logic function, the logic variables are applied to the multiplexer's data inputs.
3. What type of circuit provides the select inputs when a MUX is used as a parallel-to-serial converter?

FIGURE 9-29   General demultiplexer.

DATA input is transmitted only to one of the outputs as determined by select input code

9-9   DEMULTIPLEXERS (DATA DISTRIBUTORS)

A multiplexer takes several inputs and transmits *one* of them to the output. A *demultiplexer* performs the reverse operation; it takes a single input and distributes it over several outputs. Figure 9-29 shows the functional diagram for a demultiplexer (DEMUX). The large arrows for inputs and outputs can represent one or more lines. The select input code determines to which output the DATA input will be transmitted. In other words, the demultiplexer takes one input data source and selectively distributes it to 1 of N output channels just like a multiposition switch.

1-Line-to-8-Line Demultiplexer   Figure 9-30 shows the logic diagram for a demultiplexer that distributes one input line to eight output lines. The single data input line I is connected to all eight AND gates, but only one of these gates will be enabled by the SELECT input lines. For example, with $S_2 S_1 S_0 = 000$, only AND gate 0 will be enabled, and data input I will appear at output O_0. Other SELECT codes cause input I to reach the other outputs. The truth table summarizes the operation.

The demultiplexer circuit of Figure 9-30 is very similar to the 3-line-to-8-line decoder circuit in Figure 9-2 except that a fourth input (I) has been added to each gate. It was pointed out earlier that many IC decoders have an ENABLE input, which is an extra input added to the decoder gates. This type of decoder chip can therefore be used as a demultiplexer, with the binary code inputs (e.g., A, B, C in Figure 9-2) serving as the SELECT inputs and the ENABLE input serving as the data input I. For this reason, IC manufacturers often call this type of device a **decoder/demultiplexer,** and it can be used for either function.

We saw earlier how the 74LS138 is used as a 1-of-8 decoder. Figure 9-31 shows how it can be used as a demultiplexer. The enable input $\overline{E}_1$ is used as the data input I, while the other two enable inputs are held in their active states. The $A_2 A_1 A_0$ inputs are used as the select code. To illustrate the operation, let's assume that the select inputs are 000. With this input code, the only output that can be activated is $\overline{O}_0$, while all other outputs are HIGH. $\overline{O}_0$ will go LOW only if $\overline{E}_1$ goes LOW and will be HIGH if $\overline{E}_1$ goes HIGH. In other words, $\overline{O}_0$ *will follow the signal on* $\overline{E}_1$ (i.e., the data input, I) while all other outputs stay HIGH. In a similar manner, a different

Teaching Tip: It might be instructive to go through the operation of the 74LS138 as a demultiplexer by using the detailed logic diagram of Figure 9-3.

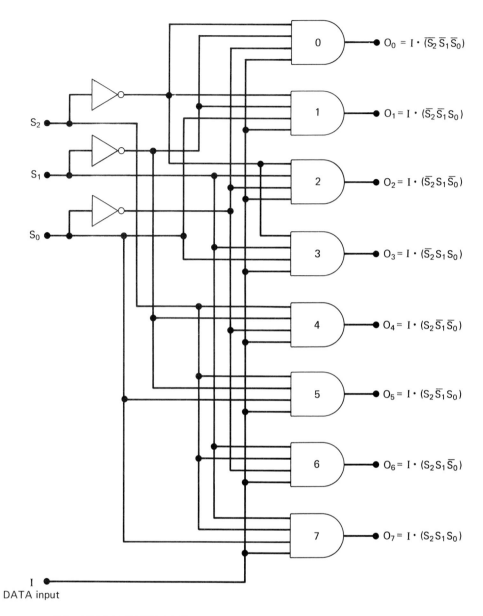

SELECT code			Outputs							
S_2	S_1	S_0	O_7	O_6	O_5	O_4	O_3	O_2	O_1	O_0
0	0	0	0	0	0	0	0	0	0	I
0	0	1	0	0	0	0	0	0	I	0
0	1	0	0	0	0	0	0	I	0	0
0	1	1	0	0	0	0	I	0	0	0
1	0	0	0	0	0	I	0	0	0	0
1	0	1	0	0	I	0	0	0	0	0
1	1	0	0	I	0	0	0	0	0	0
1	1	1	I	0	0	0	0	0	0	0

Note: I is the data input

FIGURE 9-30   One-line-to-8-line demultiplexer.

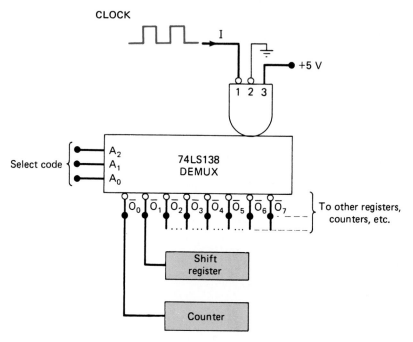

FIGURE 9-31   (a) The 74LS138 decoder can function as a demultiplexer with $\overline{E}_1$ used as the data input; (b) typical waveforms for a select code of $A_2A_1A_0 = 000$ shows $\overline{O}_0$ identical to data input.

select code applied to $A_2A_1A_0$ will cause the corresponding output to follow the data input, I.

Figure 9-31(b) shows typical waveforms for the case where $A_2A_1A_0 = 000$ selects output $\overline{O}_0$. For this case, the data signal applied to $\overline{E}_1$ will be transmitted to $\overline{O}_0$ and all other outputs will remain in their inactive HIGH state.

Clock Demultiplexer   Many applications of the demultiplexing principle are possible. Figure 9-32 shows the 74LS138 demultiplexer being used as a *clock demul-*

FIGURE 9-32   Clock demultiplexer transmits clock signal to a destination determined by select code inputs.

Discussion Question: How can the 7442 decoder (Figure 9-5) be used as a demultiplexer?

tiplexer. Under control of the SELECT lines, the clock signal is routed to the desired destination. For example, with $S_2 S_1 S_0 = 000$, the clock signal applied to I will appear at output $\overline{O}_0$. With $S_2 S_1 S_0 = 101$, the clock will appear at $\overline{O}_5$.

Security Monitoring System

Consider the case of a security monitoring system in an industrial plant where the open/closed status of many access doors is to be monitored. Each door controls the state of a switch, and it is necessary to display the state of each switch on LEDs that are mounted on a remote monitoring panel at the security guard's station. One way to do this would be to run a separate signal from each door switch to an LED on the monitoring panel. This would require running many wires over a long distance. A better approach that would reduce the amount of wiring to the monitoring panel uses a multiplexer/demultiplexer combination. Figure 9-33 shows a system that can handle eight doors, but the basic idea can be expanded to any number.

FIGURE 9-33   Security monitoring system.

EXAMPLE 9-11

Examine Figure 9-33 carefully and describe the complete operation.

Solution   The eight door switches are the data inputs to the MUX; they produce a HIGH when a door is open and a LOW when it is closed. The MOD-8 counter provides the select inputs to the MUX and also to the DEMUX on the remote monitoring panel. Each DEMUX output is connected to an indicator LED that will be on when the output is LOW. Clock pulses applied to the counter will cause the select inputs to sequence through all of the possible states 000 through 111. At each number of the counter, the switch status for the door of the same number will be inverted by the MUX and passed to output $\overline{Z}$. From there it is transmitted to the DEMUX input, which passes it through to the output corresponding to the same number.

For example, let's say that the counter is at the count of 110 (6). While the counter is in this state, let's say that door 6 is closed. The LOW at I_6 will pass through the MUX and be inverted to produce a HIGH at $\overline{Z}$. This HIGH will be passed through the DEMUX to output $\overline{O}_6$ so that LED 6 will be off, indicating that door 6 is closed. Now let's say that door 6 is open. A LOW will appear at $\overline{Z}$ and $\overline{O}_6$ so that LED 6 will be on to signal that door 6 is open. Of course, all other LEDs will be off during this time since $\overline{O}_6$ is the only active output.

As the counter is clocked through its eight states 000 through 111, the LEDs will sequentially indicate the status of the eight doors. If all the doors are closed, none of the LEDs will be on even when the corresponding DEMUX output is selected. If a door is open, its LED will turn on only during the time interval that the counter is at the appropriate count; it will be off at all other counts. Thus, the LED will be flashing on and off if its door is open. The flashing rate can be adjusted by changing the frequency of the clock.

Common Student Mistake: Students tend to associate a HIGH output with turning on an indicator LED, and so they expect the LED to be ON when the corresponding door is closed. You should remind them how the current-sinking action of the 74LS138 output in the LOW state permits current through the LED, but the current-sourcing action in the HIGH state does not.

Note that there are only four signal lines going from the "door sensing" circuitry to the remote monitoring panel: the $\overline{Z}$ output and the three select lines. This is a saving of four lines when compared with the alternative of having one line per door. The MUX/DEMUX combination is used to transmit the status of each door to its LED one at a time (serially) instead of all at once (parallel).

Synchronous Data Transmission System   Figure 9-34 shows the logic diagram for a synchronous data transmission system that is used to serially transmit four 4-bit data words from a transmitter to a remote receiver. Let's look at the transmitter circuitry first. The data words are stored in registers A, B, C, and D that are connected as recirculating shift registers with a common SHIFT (clock) input. Each register will shift right on the PGT of the SHIFT pulses from AND gate 2. The LSB of each register is connected as a data input to the 4-input multiplexer.

The two MOD-4 counters control the transmission of the data register contents to the multiplexer output Z. The *word counter* selects the register data that will appear at Z. As this counter cycles from 00 to 11, the data from each register will sequentially appear at Z. The *bit counter* makes sure that 4 data bits from each register are transmitted through the multiplexer before advancing to the next register. The bit counter advances one count for each SHIFT pulse, so that after four SHIFT pulses, it recycles to 00. The NGT at the Q_1 output of the bit counter will cause the word counter to be incremented to the next count to select the next data register for transmission. In this way the contents of each of the data registers will be transmitted to Z, one bit at a time, starting with register A (for $S_1 S_0 = 00$) and proceeding through

Teaching Tip: This same basic idea is used by the telephone company when it uses *time-division multiplexing* to place the digitized versions of hundreds of audio signals (calls) on the same cable for transmission between stations. The multiplexed signal is demultiplexed at the receiving station into the separate calls for routing to their destinations.

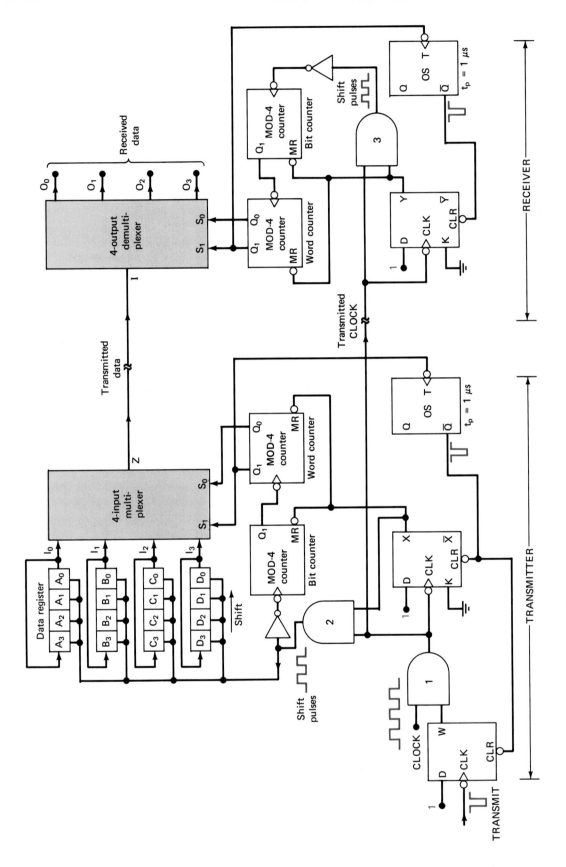

FIGURE 9-34 Synchronous data transmission system.

each register as the word counter advances one count for every four SHIFT pulses. The Z signal will thus contain 16 bits of serial data, 4 bits from each register. These data are said to be **time-division-multiplexed** because four different sets of data are appearing on the same output line at different times.

The transmission process is controlled by the two D flip-flops, AND gates 1 and 2, and the one-shot. The operation of this control logic will be described in a later paragraph.

The Receiver   The receiver circuitry contains a 1-to-4 demultiplexer that receives the Z signal from the transmitter's multiplexer and *demultiplexes* it; that is, it separates the four different sets of data and distributes them to four different outputs so that the data that came from register A will appear serially at O_0, the data from register B at O_1, and so on. The end result is almost the same as having each transmitter data register connected to its corresponding output in the receiver, except that the data are sent from one register at a time over the serial data transmission path.

The MOD-4 counters in the receiver have the same function as their counterparts in the transmitter. The word counter selects which demultiplexer output will be receiving data, and the bit counter allows 4 bits of data to reach each output before advancing the word counter to its next state. The functions of the FF, OS, and AND gate are described below.

Complete Operation   It should be clear that in order for this data transmission to work properly, there has to be some means for synchronizing the selection of the multiplexer inputs in the transmitter with the selection of the demultiplexer outputs in the receiver. We will go through a complete operation cycle to see how this synchronization is accomplished. For this illustration we will assume the following register data:

$$[A] = 0110, \quad [B] = 1001, \quad [C] = 1011, \quad [D] = 0100$$

As we go through the following steps, we will refer to the waveforms in Figure 9-35.

1. Flip-flops W and X in the transmitter and flip-flop Y in the receiver are normally LOW. The LOWs from X and Y will keep both sets of counters in the zero state. The LOW at W prevents CLOCK pulses from getting through AND gate 1.

2. With both word counters at 00, the LOW at A_0 passes through the multiplexer to Z into the demultiplexer to output O_0. All other demultiplexer outputs are LOW, since they are not selected.

3. This is the situation prior to time t_0. At t_0, the TRANSMIT pulse sets $W = 1$ to enable AND gate 1 to pass CLOCK pulses. These CLOCK pulses also become the TRANSMITTED CLOCK that is sent to the receiver along with the TRANSMITTED DATA.

4. The NGT of the first CLOCK pulse out of AND gate 1 will set X and Y both HIGH at time t_2. This removes the resets from the counters and also enables AND gates 2 and 3 to pass CLOCK pulses starting at t_3. The pulses out of AND gate 2 are the SHIFT pulses. The pulses out of AND gate 3 are exactly the same as the SHIFT pulses out of AND gate 2.

5. The PGTs of the SHIFT pulses at t_3, t_4, and t_5 will shift A_1, A_2, and A_3 into the multiplexer, out of Z, into the demultiplexer, and out of output O_0. These three PGTs are also counted by both bit counters.

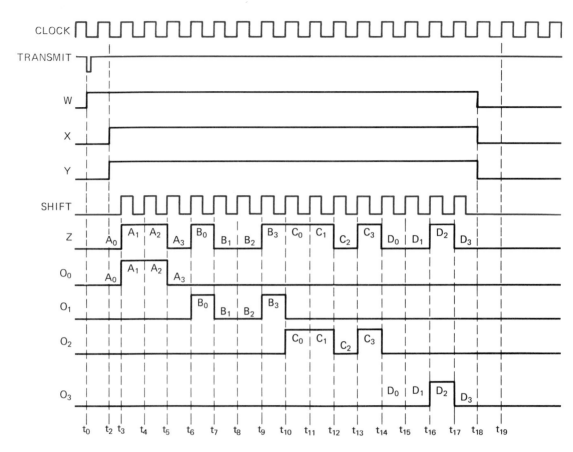

FIGURE 9-35   Waveforms during one complete transmission cycle.

6. The PGT at t_6 will shift all the registers back to their original data and will recycle the bit counters to 00. The NGT at Q_1 of the bit counters will increment the transmitter and receiver word counters to 01 to select I_1 and O_1 respectively. Thus, the HIGH at B_0 will pass through the multiplexer into the demultiplexer and out of O_1.

7. The SHIFT pulses at t_7, t_8, and t_9 will shift B_1, B_2, and B_3 into the multiplexer and out of O_1. At t_{10} the bit counters will recycle and increment the word counters to 10 to select I_2 and O_2. This places the HIGH from C_0 at Z and at O_2.

8. The SHIFT pulses at t_{11}, t_{12}, and t_{13} will shift C_1, C_2, and C_3 into the multiplexer and out of O_2. At t_{14} the bit counters recycle and increment the word counters to 11 to select I_3 and O_3. This places the LOW from D_0 at Z and at O_3.

9. The SHIFT pulses at t_{15}, t_{16}, and t_{17} will shift D_1, D_2, and D_3 into the multiplexer and out of O_3. At t_{18} the bit counters recycle and increment the word counters to 00. The NGT at Q_1 of the *word* counters will trigger their respective one-shots to produce narrow clearing pulses for FFs W, X, and Y. With these FFs all LOW, all CLOCK pulses and SHIFT pulses are inhibited, and all the counters remain in the zero state.

10. The circuit conditions are back to their original state. No more data will be transmitted until the next TRANSMIT pulse occurs.

Teaching Tip: In more sophisticated synchronous data transmission systems, the clock signal is not transmitted on a separate line, but instead is embedded into the transmitted data signal using one of several schemes. A special circuit in the receiver extracts the clock signal from the data for use in the demultiplexing process.

The waveforms at Z and O_0 to O_3 show how the register data are multiplexed onto the Z signal and then demultiplexed so that each output receives the correct data.

REVIEW QUESTIONS

1. Explain the difference between a DEMUX and a MUX.
2. *True or false:* The circuit for a DEMUX is basically the same as for a decoder.
3. For the system of Figure 9-33, what will the security guard see on the monitoring panel when all the doors are open?

9-10 MORE IEEE/ANSI SYMBOLOGY*

We will now look at the IEEE/ANSI symbols for the MUX and DEMUX integrated circuits that we have been using. First consider the symbol for the 74151 shown in Figure 9-36(a). The label $G\frac{0}{7}$ within the block denotes the AND dependency between the select inputs and each of the data inputs 0 through 7. In other words, each

FIGURE 9-36 IEEE/ANSI symbols for various multiplexers.

(a)

(b)

(c)

*This section may be skipped without affecting the continuity of the rest of the book.

data input is ANDed with one specific combination of select inputs, and when that combination occurs, the data input is routed to the output.

Figure 9-36(b) is the IEEE/ANSI symbol for the 74157 multiplexer. Recall from Figure 9-24 that this IC contains four two-input MUXes that operate identically. The common-control block shows that the enable and select inputs are common to all the MUXes. The notation G1 on the select input and the $\overline{1}$ and 1 labels on the data inputs indicate the AND dependency between the select input and the data inputs. The $\overline{1}$ label for I_{0a} means that this input will be routed to output Z_a only when the select input is a 0. Similarly, the 1 on the I_{1a} input indicates that this input will be routed to Z_a only when the select input is a 1. The other three MUXes operate the same way.

___ EXAMPLE 9-12 ___

The IEEE/ANSI symbol for the 74150 IC is shown in Figure 9-36(c). What is the function of this device?

Solution   The 74150 is a 16-input multiplexer. Note the four select inputs required to select one of the sixteen data inputs for transmission to the output. Also note that the output is inverted, so that it will be the inverse of the selected data input.

FIGURE 9-37   IEEE/ANSI symbol for the 74LS138 DEMUX.

9-11   MORE TROUBLESHOOTING

Here are three more examples to illustrate the observation/reasoning process that is such an important initial step when troubleshooting. For each case, try to determine the circuit fault before looking at the solutions.

_____ EXAMPLE 9-13 _____

Consider the circuit of Figure 9-25. A test performed on this circuit yields the following results:

		Actual Count	Displayed Count
Case 1	Counter 1	25	25
	Counter 2	37	35
Case 2	Counter 1	49	49
	Counter 2	72	79
Case 3	Counter 1	96	96
	Counter 2	14	16

What is the probable circuit fault?

Solution   In each of the test cases, the display of counter 1 matches the counter's actual count. This indicates that the I_1 inputs, all MUX outputs, and both displays are probably working correctly. On the other hand, each test case shows that counter 2's *tens* digit is displayed correctly but its *units* digit is displayed incorrectly. This could mean that there is a fault somewhere between the output of the units section of counter 2 and the I_0 inputs of the units MUX. We should compare the bit patterns of the actual and displayed values of the units for counter 2.

	Actual Units	Displayed Units
Case 1	0111 (7)	0101 (5)
Case 2	0010 (2)	1001 (9)
Case 3	0100 (4)	0110 (6)

The idea is to look for things like a bit that does not change (stuck LOW or HIGH) or two bits that are reversed (crossed connections). The data above reveal no obvious pattern.

If we take another look at the recorded test results, we see that the displayed units digit of counter 2 is always the same as the units digit of counter 1. This symptom is probably the result of a constant logic HIGH at the select input of the units MUX, since that would continually pass the units digit of counter 1 to the units MUX output. This constant HIGH at the select input is most likely caused by an open path somewhere between the select input of the tens MUX and the select input of the units MUX. It could not be caused by a short to V_{CC}, since that would also keep the select input of the tens MUX at a constant HIGH, and we know that the tens MUX is working.

Common Student Mistake: In spite of the last statement in Example 9-13, there will be some students who will not understand why an internal short to V_{CC} at the S input of the units MUX cannot be the problem. This is usually a result of a basic flaw or deficiency in their electrical background which must be rectified if they are to be effective troubleshooters.

_____ EXAMPLE 9-14 _____

The security monitoring system of Figure 9-33 is tested and the results are recorded below.

Condition	LEDs
All doors closed	All LEDs off
Door 0 open	LED 4 flashing
Door 1 open	LED 5 flashing
Door 2 open	LED 6 flashing
Door 3 open	LED 7 flashing
Door 4 open	LED 4 flashing
Door 5 open	LED 5 flashing
Door 6 open	LED 6 flashing
Door 7 open	LED 7 flashing

What are the possible faults that could produce these results?

Solution   Again, the data should be reviewed to see if there is some pattern that could help to narrow down the search for the fault to a small area of the circuit. The data above reveal that the correct LEDs flash for the open doors 4 through 7. They also show that for open doors 0 through 3 the number of the flashing LED is *four* more than the number of the door, and LEDs 0 through 3 are always off. This is most probably caused by a constant logic HIGH at A_2, the MSB of the select input of the DEMUX, since this would always make the select code 4 or greater, and it would add 4 to the select codes 0 through 3.

Thus we have two possibilities: A_2 is somehow shorted to V_{CC}, or there is an open connection at A_2. A little thought will eliminate the first choice as a possibility since this would also mean that S_2 of the MUX would also be stuck HIGH. If that were so, then the status of doors 0 through 3 would not get through the MUX and into the DEMUX. We know that this is not true because the data show that when any of these doors is open, it affects one of the DEMUX outputs.

_____ EXAMPLE 9-15 _____

A technician is troubleshooting a malfunctioning synchronous data transmission system (Figure 9-34). An oscilloscope is used to monitor the MUX and DEMUX outputs during a transmission cycle with the results shown in Figure 9-38. What are the possible causes of the malfunction?

FIGURE 9-38   Example 9-14.

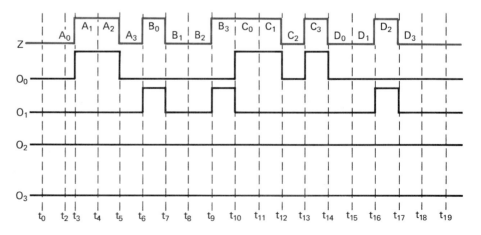

Solution Examination of the waveforms produces the following observations:

- All of the signals appear to be correct between t_0 and t_9, with the O_0 signal containing the serial data from register A, and O_1 containing the serial data from register B.
- The O_2 and O_3 outputs are never activated.
- Between t_{10} and t_{14}, the O_0 output, rather than O_2, contains the serial data from register C.
- Between t_{14} and t_{18}, the O_1 output, rather than O_3, contains the serial data from register D.

It appears that the select inputs of the receiver's DEMUX are selecting only O_0 and O_1. This would happen if S_1 were stuck in the LOW state. This stuck node could be caused by an internal short to ground at S_1, Q_1, or the T input of the one-shot, or an external short to ground. The actual location of the short can be determined using the techniques that have been presented in previous chapters.

9-12 MAGNITUDE COMPARATOR

Another useful member of the MSI category of ICs is the *magnitude comparator*. It is a combinatorial logic circuit that compares two input binary quantities and generates outputs to indicate which one has the greater magnitude. Figure 9-39 shows the logic symbol and truth table for the 74LS85 four-bit magnitude comparator, which is also available as the 7485 and 74HC85.

Data Inputs The 74LS85 compares two *unsigned* 4-bit binary numbers. One of them is $A_3A_2A_1A_0$, which is called word A; the other is $B_3B_2B_1B_0$, which is called word B. The term "word" is used in the digital computer field to designate a group of bits that represents some specific type of information. Here word A and word B represent numerical quantities.

Teaching Tip: It can be helpful to review the process we use when we look at decimal numbers (with any number of digits) to decide which one is larger.

Outputs The 74LS85 has three active-HIGH outputs. Output $O_{A>B}$ will be HIGH when the magnitude of word A is greater than the magnitude of word B. Output $O_{A<B}$ will be HIGH when the magnitude of word A is less than the magnitude of word B. Output $O_{A=B}$ will be HIGH when word A and word B are identical.

Cascading Inputs Cascading inputs provide a means for expanding the comparison operation to more than 4 bits by cascading two or more 4-bit comparators. Note that the cascading inputs are labeled the same as the outputs. When a 4-bit comparison is being made as in Figure 9-40(a), the cascading inputs should be connected as shown in order for the comparator to produce the correct outputs.

When two comparators are to be cascaded, the outputs of the lower-order comparator are connected to the corresponding inputs of the higher-order comparator. This is shown in Figure 9-40(b), where the comparator on the left is comparing the lower-order 4 bits of the two 8-bit words: $A_7A_6A_5A_4A_3A_2A_1A_0$ and $B_7B_6B_5B_4B_3B_2B_1B_0$. Its outputs are fed to the cascade inputs of the comparator on the right, which is comparing the high-order bits. The outputs of the high-order comparator are the final outputs that indicate the result of the 8-bit comparison.

Teaching Tip: The 74LS85 truth table can be somewhat imposing to the student. It would probably help to go through it carefully to show that it really is quite straightforward except for the last two cases, which have no apparent use.

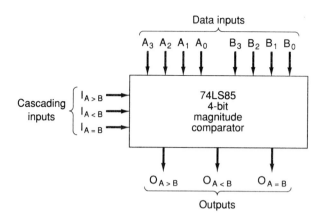

Data inputs

A_3 A_2 A_1 A_0 B_3 B_2 B_1 B_0

Cascading inputs $\begin{cases} I_{A>B} \\ I_{A<B} \\ I_{A=B} \end{cases}$

74LS85
4-bit
magnitude
comparator

$O_{A>B}$ $O_{A<B}$ $O_{A=B}$

Outputs

TRUTH TABLE

COMPARING INPUTS				CASCADING INPUTS			OUTPUTS		
A_3, B_3	A_2, B_2	A_1, B_1	A_0, B_0	$I_{A>B}$	$I_{A<B}$	$I_{A=B}$	$O_{A>B}$	$O_{A<B}$	$O_{A=B}$
$A_3 > B_3$	X	X	X	X	X	X	H	L	L
$A_3 < B_3$	X	X	X	X	X	X	L	H	L
$A_3 = B_3$	$A_2 > B_2$	X	X	X	X	X	H	L	L
$A_3 = B_3$	$A_2 < B_2$	X	X	X	X	X	L	H	L
$A_3 = B_3$	$A_2 = B_2$	$A_1 > B_1$	X	X	X	X	H	L	L
$A_3 = B_3$	$A_2 = B_2$	$A_1 < B_1$	X	X	X	X	L	H	L
$A_3 = B_3$	$A_2 = B_2$	$A_1 = B_1$	$A_0 > B_0$	X	X	X	H	L	L
$A_3 = B_3$	$A_2 = B_2$	$A_1 = B_1$	$A_0 < B_0$	X	X	X	L	H	L
$A_3 = B_3$	$A_2 = B_2$	$A_1 = B_1$	$A_0 = B_0$	H	L	L	H	L	L
$A_3 = B_3$	$A_2 = B_2$	$A_1 = B_1$	$A_0 = B_0$	L	H	L	L	H	L
$A_3 = B_3$	$A_2 = B_2$	$A_1 = B_1$	$A_0 = B_0$	X	X	H	L	L	H
$A_3 = B_3$	$A_2 = B_2$	$A_1 = B_1$	$A_0 = B_0$	L	L	L	H	H	L
$A_3 = B_3$	$A_2 = B_2$	$A_1 = B_1$	$A_0 = B_0$	H	H	L	L	L	L

H = HIGH Voltage Level
L = LOW Voltage Level
X = Immaterial

FIGURE 9-39   Logic symbol and truth table for a 74LS85 (7485, 74HC85) 4-bit magnitude comparator.

——— EXAMPLE 9-16 ———————————————————————

Describe the operation of the 8-bit comparison arrangement in Figure 9-40(b) for the following cases:

(a) $A_7 A_6 A_5 A_4 A_3 A_2 A_1 A_0$ = 10101111; $B_7 B_6 B_5 B_4 B_3 B_2 B_1 B_0$ = 10110001

(b) $A_7 A_6 A_5 A_4 A_3 A_2 A_1 A_0$ = 10101111; $B_7 B_6 B_5 B_4 B_3 B_2 B_1 B_0$ = 10101001

Solution

(a) The high-order comparator compares its inputs $A_7 A_6 A_5 A_4$ = 1010 and $B_7 B_6 B_5 B_4$ = 1011 and produces $O_{A<B}$ = 1 regardless of what levels are applied to its cascade inputs from the low-order comparator. In other words, once the high-order comparator senses a difference in the high-order bits of the two 8-bit words, it knows which 8-bit word is greater without having to look at the results of the low-order comparison.

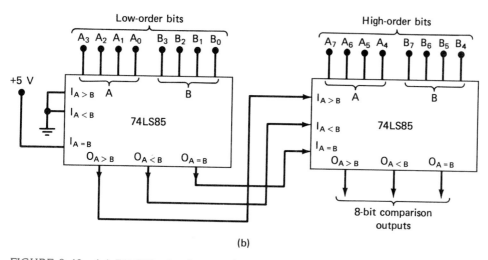

FIGURE 9-40 (a) 74LS85 wired as a 4-bit comparator; (b) two 74LS85s cascaded to perform an 8-bit comparison.

(b) The high-order comparator sees $A_7A_6A_5A_4 = B_7B_6B_5B_4 = 1010$, and so it has to look at its cascade inputs to see the result of the low-order comparison. The low-order comparator has $A_3A_2A_1A_0 = 1111$ and $B_3B_2B_1B_0 = 1001$, which produces a 1 at its $O_{A>B}$ output and the $I_{A>B}$ input of the high-order comparator. The high-order comparator senses this 1, and since its data inputs are equal, produces a HIGH at its $O_{A>B}$ to indicate the result of the 8-bit comparison.

Applications Magnitude comparators are often used as part of the address decoding circuitry used in a computer to select a specific input/output device or area of memory for the storage or retrieval of data. The comparators compare an address code generated by the central processer (CPU) with a hard-wired address code; if they are equal, the $O_{A=B}$ output of the comparator will activate the device corresponding to that address. Magnitude comparators are also useful in control applications where a binary number representing the physical variable being controlled (e.g., position, speed) is compared with a reference value. The comparator outputs are used to actuate circuitry to drive the physical variable toward the reference value. We will examine a typical comparator application in Problem 9-49.

16

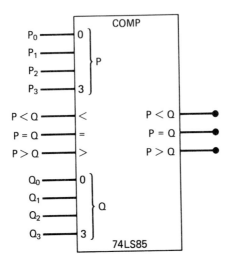

FIGURE 9-41 IEEE/ANSI symbol for comparator.

IEEE/ANSI Symbol Figure 9-41 shows the IEEE/ANSI symbol for the 74LS85 comparator. Note that the symbol uses P and Q to represent the input variables. This is the preferred designation used by the IEEE/ANSI standard.

___ REVIEW QUESTIONS ___

1. What is the purpose of the cascading inputs of the 74LS85?
2. What are the outputs of a 74LS85 with the following inputs: $A_3 A_2 A_1 A_0 = B_3 B_2 B_1 B_0 = 1001$, $I_{A>B} = I_{A<B} = 0$, and $I_{A=B} = 1$?

9-13 CODE CONVERTERS

A code converter is a logic circuit that changes data presented in one type of binary code to another type of binary code. The BCD-to-7-segment decoder/driver we presented earlier is a code converter because it changes a BCD input code to the 7-segment code needed by the LED display. Here is a partial list of some of the more common code conversions:

> BCD to 7-segment
> BCD to binary
> Binary to BCD
> Binary to Gray code
> Gray code to binary
> ASCII to EBCDIC*
> EBCDIC to ASCII
>
> ___
> * EBCDIC is an alphanumeric code developed by IBM similar to ASCII.

As an example of a code converter circuit, let's consider a BCD-to-binary converter. Before we get started on the circuit implementation, we should review the BCD representation.

Two-digit decimal values ranging from 00 to 99 can be represented in BCD by two 4-bit code groups. For example, 57_{10} is represented as

$$\overset{5}{\overline{0101}} \quad \overset{7}{\overline{0111}} \quad \text{(BCD)}$$

The straight binary representation for decimal 57 is

$$57_{10} = 111001_2$$

The largest two-digit decimal value of 99 has the following representations:

$$99_{10} = 10011001 \text{ (BCD)} = 1100011_2$$

Note that the binary representation requires only 7 bits.

Basic Idea

The diagram of Figure 9-42 shows the basic idea for a two-digit BCD-to-binary converter. The inputs to the converter are the two 4-bit code groups $D_0 C_0 B_0 A_0$, representing the 10^0 or units digit, and $D_1 C_1 B_1 A_1$, representing the 10^1 or tens digit of the decimal value. The outputs from the converter are $b_6 b_5 b_4 b_3 b_2 b_1 b_0$, the 7 bits of the binary equivalent of the same decimal value. Note the difference in the weights of the BCD bits and those of the binary bits.

A typical use of a BCD-to-binary converter would be where BCD data from an instrument like a DMM (digital multimeter) are being transferred to a computer for storage or processing. The data must be converted to binary so that they can be operated on in binary by the computer ALU, which may not have the capability of performing arithmetic operations on BCD data. The BCD-to-binary conversion can be accomplished with either hardware or software. The hardware method (which we will look at momentarily) is generally faster but requires extra circuitry. The software method uses no extra circuitry, but it takes more time, since the software does it step by step. The method chosen in a particular application depends on whether or not conversion time is an important consideration.

Teaching Tip: Remind them of these previously covered ideas: (1) BCD and straight binary are different ways of representing decimal values; (2) BCD encodes *each* decimal digit into its 4-bit binary equivalent; (3) straight binary converts the *complete decimal value* into its binary equivalent; (4) BCD requires more bits.

Teaching Tip: This may be a good time to tell them that there are many digital operations and processes that can be accomplished in either hardware (i.e., using digital ICs) or software (i.e., through CPU-executed instructions). Although the students may not yet have a feel for software, you can state the general principle that a hardware solution requires more circuitry and takes less time than a software solution. Digital system designers have to make decisions on which approach is best for the constraints they are working under.

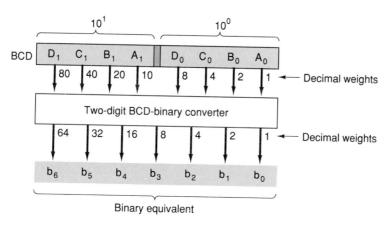

FIGURE 9-42   Basic idea of a two-digit BCD-binary converter.

Conversion Process

The bits in a BCD representation have decimal weights which are 8, 4, 2, 1 within each code group but which differ by a factor of 10 from one code group (decimal digit) to the next. Figure 9-42 shows the bit weights for the two-digit BCD representation.

The decimal weight of each bit in the BCD representation can be converted to its binary equivalent. The results are given in Table 9-1. Using these weights, we can perform the BCD-to-binary conversion by simply doing the following:

TABLE 9-1   Binary Equivalents of Decimal Weights of Each BCD Bit

BCD Bit	Decimal Weight	Binary Equivalent						
		b_6	b_5	b_4	b_3	b_2	b_1	b_0
A_0	1	0	0	0	0	0	0	1
B_0	2	0	0	0	0	0	1	0
C_0	4	0	0	0	0	1	0	0
D_0	8	0	0	0	1	0	0	0
A_1	10	0	0	0	1	0	1	0
B_1	20	0	0	1	0	1	0	0
C_1	40	0	1	0	1	0	0	0
D_1	80	1	0	1	0	0	0	0

Compute the binary sum of the binary equivalents of all those bits in the BCD representation that are 1s.

The following example will illustrate.

___ EXAMPLE 9-17 _____

Convert 01010010 (BCD for decimal 52) to binary. Repeat for 10010101 (decimal 95).

Solution   Write down the binary equivalents for all the 1s in the BCD representation. Then add them all together in binary.

Teaching Tip: You may want to go through at least one of these cases step by step.

Circuit Implementation   Clearly, one way to implement the logic circuit that performs this conversion process is to use binary adder circuits. Figure 9-43 shows how two 74LS83 four-bit parallel adders can be wired to perform the conversion. This is one of several possible adder arrangements that will work.

The two adder ICs perform the addition of the BCD bits in the proper combinations according to Table 9-1. For instance, Table 9-1 shows that A_0 is the only BCD bit that contributes to the LSB, b_0, of the binary equivalent. Since there is no carry into this bit position, A_0 is connected directly as output b_0. The table also shows that only BCD bits B_0 and A_1 contribute to bit b_1 of the binary output. These

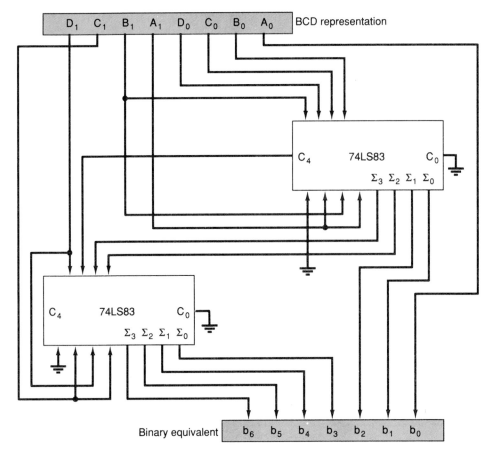

FIGURE 9-43   BCD-to-binary converter implemented with 74LS83 four-bit parallel adders.

two bits are combined in the upper adder to produce output b_1. Likewise, only BCD bits D_0, A_1, and C_1 contribute to bit b_3. The upper adder combines D_0 and A_1 to generate Σ_2, which is connected to the lower adder, where C_1 is added to it to produce b_3.

─── EXAMPLE 9-18 ───────────────────────────────

The BCD representation for decimal 56 is applied to the converter of Figure 9-43. Determine the Σ outputs from each adder and the final binary output.

Solution   Write down the bits of the BCD representation 01010110 on the circuit diagram. Since $A_0 = 0$, the b_0 bit of the output is 0.

The top inputs to the upper adder are 0011. The bottom inputs are 0101. This adder adds these to produce

$$
\begin{array}{r}
0011 \\
+0101 \\
\hline
1000
\end{array} = \Sigma_3\,\Sigma_2\,\Sigma_1\,\Sigma_0 \text{ outputs of upper adder}
$$

The Σ_1 and Σ_0 bits become binary outputs b_2 and b_1, respectively. The Σ_3 and Σ_2 bits are fed to the lower adder. The top inputs to the lower adder are therefore 0010. The

bottom inputs are 0101. This adder adds these to produce

$$0010$$
$$+\underline{0101}$$
$$0111 = \Sigma_3 \Sigma_2 \Sigma_1 \Sigma_0 \text{ outputs of lower adder}$$

These bits become $b_6 b_5 b_4 b_3$, respectively.

Thus, we have $b_6 b_5 b_4 b_3 b_2 b_1 b_0 = 0111000$ as the correct binary equivalent for decimal 56.

Other Code Converter Implementations   While all types of code converters can be made by combining logic gates, adder circuits, or other combinatorial logic, the circuitry can become quite complex, requiring many ICs. It is often more efficient to use a read-only memory (ROM) or programmable logic device (PLD) to function as a code converter. As we will see in Chapter 11, these devices contain the equivalent of hundreds of logic gates, and can be programmed to provide a wide range of logic functions.

REVIEW QUESTIONS

1. What is a code converter?
2. How many binary outputs would a three-digit BCD-to-binary converter have?

9-14   TRISTATE REGISTERS

In most modern computers the transfer of data takes place over a common set of connecting lines called a **data bus.** In these bus-organized computers, many different devices can have their outputs and inputs tied to the common data bus lines. Because of this, the devices that are tied to the data bus will often have tristate outputs, or they will be tied to the data bus through tristate buffers.

Some of the devices that are commonly connected to a data bus are (1) microprocessors, which will be discussed in Chapter 12; (2) semiconductor memory chips, which will be covered in Chapter 11; (3) digital-to-analog (DACs) converters and analog-to-digital converters (ADCs), to be covered in Chapter 10.

Almost always, the devices connected to a data bus will contain registers that hold their data. The outputs of these registers will have tristate buffers to allow data busing. There are many IC registers available that include the tristate buffers on the same chip. One of these is the TTL 74173 (also available as 74LS173 and 74HC173), whose logic diagram and truth tables are shown in Figure 9-44.

The 74173 is a 4-bit register with parallel in/parallel out capability. Note that the FF outputs are connected to tristate buffers that provide outputs O_0 through O_3. Also note that the data inputs D_0 through D_3 are connected to the D inputs of the register FFs through logic circuitry. This logic allows two modes of operation: (1) *load,* where the data at inputs D_0 to D_3 are transferred into the FFs on the PGT of the clock pulse at *CP;* (2) *hold,* where the data in the register do not change when the PGT of *CP* occurs.

	Inputs				Output
MR	CP	$\overline{IE_1}$	$\overline{IE_2}$	D_n	Q
H	X	X	X	X	L
L	L	X	X	X	Q_0
L	⌐	H	X	X	Q_0
L	⌐	X	H	X	Q_0
L	⌐	L	L	L	L
L	⌐	L	L	H	H

When either $\overline{OE_1}$ or $\overline{OE_2}$ is HIGH the output is in the OFF state (high impedance); however this does not affect the contents or sequential operating of the register

H = HIGH voltage level Q_0 = output prior to PGT
L = LOW voltage level
X = immaterial

Logic Diagram

FIGURE 9-44   Truth table and logic diagram for 74173 tristate register. (Courtesy of Fairchild, a Schlumberger company)

—— EXAMPLE 9-19 ——————————

(a) What input conditions will produce the load operation?

(b) What input conditions will produce the hold operation?

(c) What input conditions will allow the internal register outputs to appear at O_0 to O_3?

Solution

(a) The last two entries in the truth table show that each Q output takes on the value present at its D input when a PGT occurs at CP provided that MR is LOW, and both input-enable inputs, $\overline{IE}_1$ and $\overline{IE}_2$, are LOW.

(b) The third and fourth lines of the truth table state that when either $\overline{IE}$ input is HIGH, the D inputs have no effect and the Q outputs will retain their current values when the PGT occurs.

Teaching Tip: Note that in the hold condition the Q output of each FF is gated back into the D input of the FF so that the next PGT will not change Q.

(c) The output buffers are enabled when both output-enable inputs, $\overline{OE}_1$ and $\overline{OE}_2$, are LOW. This will pass the register outputs through to the external outputs O_0 to O_3. If either output-enable input is HIGH, the buffers will be disabled and the outputs will be in the Hi-Z state.

Note that the $\overline{OE}$ inputs have no effect on the data load operation. They are only used to control whether or not the register outputs are passed to the external outputs.

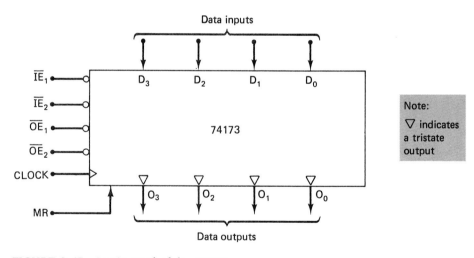

FIGURE 9-45   Logic symbol for 74173.

9-15   DATA BUSING

The data bus is very important in computer systems, and its significance will not be appreciated until our later studies of memories and microprocessors. For now, we will illustrate the data-bus concept for register-to-register data transfer. Figure 9-46

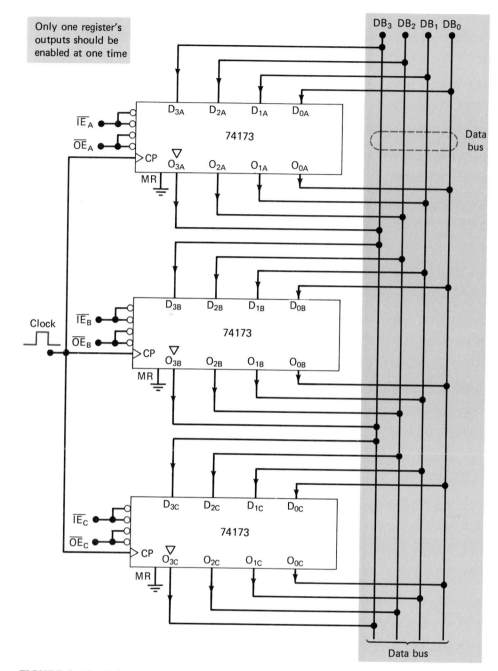

FIGURE 9-46 Tristate registers connected to data bus.

shows a bus-organized system for three 74173 tristate registers. Note that each register has its pair of $\overline{OE}$ inputs tied together as one $\overline{OE}$ input, and likewise for the $\overline{IE}$ inputs. Also note that the registers will be referred to as registers A, B, and C from top to bottom. This is indicated by the subscripts on each input and output.

In this arrangement, the data bus consists of four lines labeled DB_0 to DB_3. Corresponding outputs of each register are connected to the same data-bus line (e.g., O_{3A}, O_{3B}, and O_{3C} are connected to DB_3). Since the three registers have their outputs connected together, it is imperative that only one register have its outputs enabled

and that the other two register outputs remain in the Hi-Z state. Otherwise, there will be "bus contention" (two or more sets of outputs fighting each other) producing uncertain levels on the bus and possible damage to the register output buffers.

Corresponding register inputs are also tied to the same bus line (e.g., D_{3A}, D_{3B}, and D_{3C} are tied to DB_3). Thus, the levels on the bus will always be ready to be transferred to one or more of the registers depending on the $\overline{IE}$ inputs.

Data Transfer Operation
The contents of any one of the three registers can be parallel-transferred over the data bus to one of the other registers through the proper application of logic levels to the register enable inputs. In a typical system, the control unit of a computer (i.e., the CPU) will generate the signals that select which register will put its data on the data bus, and which one will take the data from the data bus. The following example will illustrate this.

_____ EXAMPLE 9-20 _____

Describe the input signal requirements for transferring $[A] \rightarrow [C]$.

Solution   First of all, only register A should have its outputs enabled. That is, we need

$$\overline{OE}_A = 0, \qquad \overline{OE}_B = \overline{OE}_C = 1$$

This will place the contents of register A onto the data-bus lines.

Next, only register C should have its inputs enabled. For this, we want

$$\overline{IE}_C = 0, \qquad \overline{IE}_A = \overline{IE}_B = 1$$

Discussion Question:
Would there be any
problem with having
the inputs enabled for
both registers *C* and *B*?

This will allow only register C to accept data from the data bus when the PGT of the clock signal occurs.

Finally, a clock pulse is required to actually transfer the data from the bus into the register-C flip-flops.

Bus Signals
The timing diagram in Figure 9-47 shows the various signals involved in the transfer of the data 1011 from register A to register C. The $\overline{IE}$ and $\overline{OE}$ lines that are not shown are assumed to be in their inactive HIGH state. Prior to time t_1, the $\overline{IE}_C$ and $\overline{OE}_A$ lines are also HIGH, so that all of the register outputs are disabled and none of the registers will be placing their data on the bus lines. In other words, the data bus lines are in the Hi-Z or "floating" state as represented by the hatched lines on the timing diagram. The Hi-Z state does not correspond to any particular voltage level.

Discussion Question: Can
the PGT of the clock
occur as late as t_3 and
still transfer the data
properly? Check the
74173 specs for its
hold-time requirement.

At t_1 the $\overline{IE}_C$ and $\overline{OE}_A$ inputs are activated. The outputs of register A are enabled, and they start changing the data bus lines DB_3 through DB_0 from the Hi-Z state to the logic levels 1011. After allowing time for the logic levels to stabilize on the bus, the PGT of the clock is applied at t_2. This PGT will transfer these logic levels into register C, since $\overline{IE}_C$ is active. If the PGT occurs before the data bus has valid logic levels, unpredictable data will be transferred into C.

At t_3, the $\overline{IE}_C$ and $\overline{OE}_A$ lines return to the inactive state. As a result, register A's outputs go to the Hi-Z state. This removes the register A output data from the bus lines and the bus lines return to the Hi-Z state.

Note that the data bus lines show valid logic levels only during the time interval when register A's outputs are enabled. At all other times, the data bus lines are indeterminate, and there is no way to easily predict what they would look like if displayed on an oscilloscope. A logic probe would give an "indeterminate" indication if

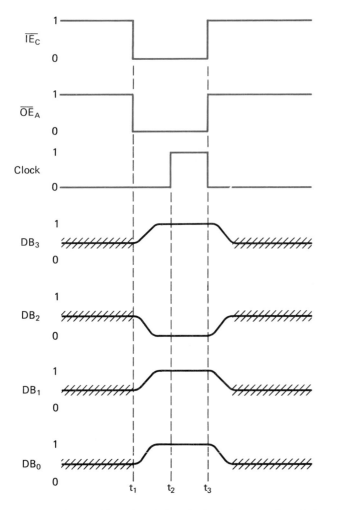

FIGURE 9-47   Signal activity during the transfer of the data 1011 from register *A* to register *C*.

it were monitoring a Hi-Z bus line. Also note the relatively slow rate at which the signals on the data bus lines are changing. Although this effect has been somewhat exaggerated in the diagram, it is a characteristic common to bus systems and is caused by the capacitive load on each line. This load consists of a combination of parasitic capacitance and the capacitances contributed by each input and output connected to the line.

Expanding the Bus   It is easy to see how we can add more registers to the data bus so that a greater number of register-to-register data transfers are possible. Of course, each register adds two more enable inputs that have to be controlled for each data transfer. Each register also adds data inputs that have to be driven by the outputs of the register that is putting data on the bus. In systems where a large number of registers (or other devices) are connected to the bus, the register outputs may have to be buffered by circuits that are called **bus drivers.** Even when the devices are all MOS or CMOS, the total capacitive load connected to each bus line will deteriorate the transition times of the level changes on each bus line. A bus-driver IC

will have very low output resistance in either logic state and will be able to charge and discharge the bus capacitances more rapidly than a normal IC.

For an illustration of bus expansion and the circuitry for selecting which devices will send and receive data, refer to the discussions in Problems 9-56 to 9-58.

Simplified Bus Representation   Usually many devices are connected to the same data bus. On a circuit schematic this can produce a confusing array of lines and connections. For this reason, a more simplified representation of data-bus connections is often used on block diagrams and in all but the most detailed circuit schematics. One type of simplified representation is shown in Figure 9-48 for an eight-line data bus.

FIGURE 9-48   Simplified representation of bus arrangement.

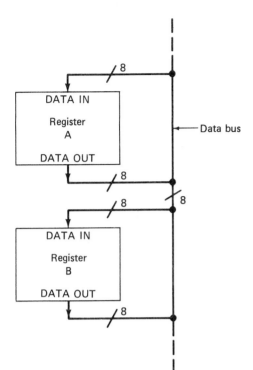

FIGURE 9-49 Simplified bus representation that uses a single line. The "/8" denotes an 8-line bus.

The connections to and from the data bus are represented by wide arrows. The numbers in brackets indicate the number of bits that each register contains, as well as the number of lines connecting the register inputs and outputs to the bus.

An even simpler bus representation that is becoming standard on many circuit diagrams is shown in Figure 9-49. It uses a single line to represent the bus and the parallel connections to the bus. The slash (/) through the line indicates that it actually represents a bus containing the designated number of lines.

Bidirectional Busing   Each register in Figure 9-46 has both its inputs and its outputs connected to the data bus, so that corresponding inputs and outputs are shorted together. For example, each register has output O_2 connected to input D_2 because of their common connection to DB_2. This, of course, would not be true if external bus drivers were connected between the register outputs and the data bus.

Because inputs and outputs are often connected together in bus systems, IC manufacturers have developed ICs that connect inputs and outputs together *internal* to the chip in order to reduce the number of IC pins and the number of connections to the bus. Figure 9-50 illustrates this for a 4-bit register. The separate data input lines (D_0 to D_3) and output lines (O_0 to O_3) have been replaced by input/output lines (I/O_0 to I/O_3).

Each I/O line will function as either an input or an output depending on the states of the enable inputs. Thus, they are called **bidirectional data lines.** The 74LS299 is an 8-bit register with common I/O lines. Many memory ICs and microprocessors have bidirectional transfer of data.

We will return to the important topic of data busing in our comprehensive coverage of memory systems in Chapter 11 and in our introduction to microprocessors and microcomputers in Chapter 12.

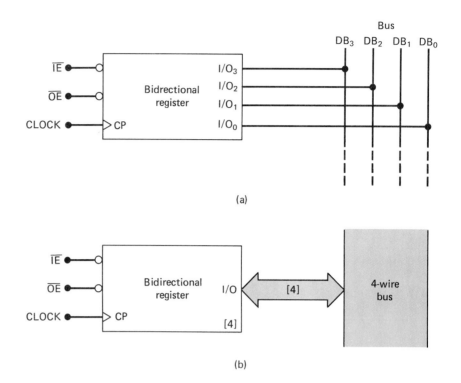

(a)

(b)

FIGURE 9-50   Bidirectional register connected to data bus.

REVIEW QUESTIONS

1. What is meant by the term "data bus"?
2. What is "bus contention," and what must be done to prevent it from occurring?
3. What is the function of a bus driver?
4. What are the reasons for having registers with common I/O lines?
5. Redraw Figure 9-50(a) using the single-line bus representation. Answer is shown in Figure 9-51.

FIGURE 9-51

PROBLEMS

SECTION 9-1

9-1.

Refer to Figure 9-3. Determine the levels at each decoder output for the following sets of input conditions:
(a) All inputs LOW   All HIGH
(b) All inputs LOW except E_3 = HIGH   output 0 LOW, all others HIGH
(c) All inputs HIGH except $\bar{E}_1 = \bar{E}_2$ = LOW
(d) All inputs HIGH   all HIGH

9-1. (c) output 7 LOW, all others HIGH;

9-2. What is the number of inputs and outputs of a decoder that accepts 64 different input combinations?   Six inputs, 64 outputs

9-3. What set of input conditions will produce a LOW at $\bar{O}_6$ of a 74LS138?

9-4. Show how to use 74LS138s to form a 1-of-16 decoder.

9-5. Figure 9-52 shows how a decoder can be used in the generation of control signals. Assume that a RESET pulse has occurred at time t_0, and determine the CONTROL waveform for 32 clock pulses.

9-3. All enable inputs active, and input code = 110

9-4. Use 4-bit input code $ABCD$ and two decoders. Connect A, B, C to A, B, C inputs of both decoders. Connect D to E_3 of one decoder and $\bar{E}_1$ of the other. Each of the other enable inputs is permanently tied to its active level.

9-5. CONTROL goes LOW at t_{28} and back HIGH at t_{30}.

FIGURE 9-52

9-6. Modify the circuit of Figure 9-52 to generate a CONTROL waveform that goes LOW from t_{20} to t_{24}. (*Hint:* The modification does not require additional logic.)

9-6. One solution is to connect $\bar{E}_2$ to ground and take CONTROL output from $\bar{O}_2$.

CHAPTER 9 / PROBLEMS **513**

FIGURE 9-53

9-7. The 7442 decoder of Figure 9-5 does not have an ENABLE input. However, we can operate it as a 1-of-8 decoder by not using outputs $\overline{O}_8$ and $\overline{O}_9$ and by using the D input as an ENABLE. This is illustrated in Figure 9-53. Describe how this arrangement works as an enabled 1-of-8 decoder, and state how the level on D either enables or disables the outputs. Decoder enabled when $D = 0$.

9-8. Consider the waveforms in Figure 9-54. Apply these signals to the 74LS138 as follows:

$$A \rightarrow A_0; \quad B \rightarrow A_1; \quad C \rightarrow A_2; \quad D \rightarrow E_3$$

Assume that $\overline{E}_1$ and $\overline{E}_2$ are tied LOW, and draw the waveforms for outputs $\overline{O}_0$, $\overline{O}_3$, $\overline{O}_6$, and $\overline{O}_7$. See Solutions Manual.

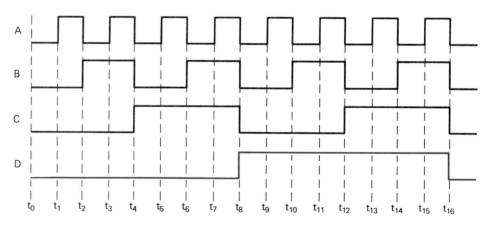

FIGURE 9-54

9-9. Connect decoder outputs 3 and 4 together, and connect outputs 6, 7, and 8 together.

9-9. Modify the circuit of Figure 9-6 so that relay K_1 stays energized from t_3 to t_5 and K_2 stays energized from t_6 to t_9. (*Hint:* This modification requires no additional circuitry.)

SECTIONS 9-2 and 9-3

9-10. Current-limiting resistors are 250 Ω. D input of MSD decoder is grounded.

9-10. Show how to connect BCD-to-7-segment decoder/drivers and LED 7-segment displays to the clock circuit of Figure 7-42 to display minutes

and hours. Assume that each segment is to operate at approximately 10 mA at 2.5 V.

9-11. (a) Refer to Figure 9-10 and draw the segment and backplane waveforms relative to ground for CONTROL = 0. Then draw the waveform of segment voltage relative to backplane voltage.
(b) Repeat for CONTROL = 1.

9-12. The BCD-to-7-segment decoder/driver of Figure 9-8 contains the logic for activating each segment for the appropriate BCD inputs. Design the logic for activating the g segment.

SECTION 9-4

9-13. DRILL QUESTION
For each statement indicate whether it is referring to a decoder or an encoder:
(a) Has more inputs than outputs.   encoder
(b) Is used to convert key actuations to a binary code.   encoder
(c) Only one output can be activated at one time.
(d) Can be used to interface a BCD input to an LED display.   decoder
(e) Can have driver-type outputs to handle large I and V.   decoder

9-14. Determine the output levels for the 74147 encoder when $\bar{A}_8 = \bar{A}_4 = 0$ and all other inputs are HIGH.   0111

9-15. Apply the signals of Figure 9-54 to the inputs of a 74147 as follows:
$$A \to \bar{A}_7, \quad B \to \bar{A}_4, \quad C \to \bar{A}_2, \quad D \to \bar{A}_1$$
Draw the waveforms for the encoder's outputs.

9-16. Figure 9-55 shows the block diagram of a logic circuit used to control

9-11. For CONTROL = 0, voltage is constant 0 V. For CONTROL = 1, voltage is a bipolar square wave with 10 $V_{p\text{-}p}$ amplitude.

9-12. Use a 1-of-10 decoder with active-HIGH outputs, and connect outputs 2, 3, 4, 5, 6, 8, and 9 to an OR gate. The OR output drives a current-sinking open-collector transistor.

9-15. When only one input is LOW, the encoder outputs reflect the inverse of that input. When more than one input is LOW, the encoder outputs reflect the inverse of the highest-numbered LOW input. For these input waveforms, output $\bar{O}_3$ stays HIGH.

9-16. EX-NORs can be used to detect a match. See Solutions Manual for one design.

FIGURE 9-55

515

the number of copies made by a copy machine. The machine operator selects the number of desired copies by closing one of the selector switches S_1 to S_9. This number is encoded in BCD by the encoder and sent to a comparator circuit. The operator then hits a momentary-contact START switch, which clears the counter and initiates a HIGH OPERATE output that is sent to the machine to signal it to make copies. As the machine makes each copy, a copy pulse is generated and fed to the BCD counter. The counter outputs are continually compared with the switch encoder outputs by the comparator. When the two BCD numbers match, indicating that the desired number of copies have been made, the comparator output $\overline{X}$ goes LOW; this causes the OPERATE level to return LOW and stop the machine so that no more copies are made. Activating the START switch will cause this process to be repeated. Design the complete logic circuitry for the comparator and control sections of this system.

9-17. The fourth key would be entered into the MSD register and displayed. One approach is to disable the OR gate once flip-flop Z makes a PGT.

9-17. The keyboard circuit of Figure 9-16 is designed to accept a three-digit decimal number. What would happen if *four* digit keys were activated (e.g., 3095)? Design the necessary logic to be added to this circuit so that after three digits have been entered, any further digits will be ignored until the CLEAR key is depressed. In other words, if 3095 is entered on the keyboard, the output registers will display 309 and ignore the 5 and any subsequent digits until the circuit is cleared.

SECTION 9-6

9-18. A technician breadboards the keyboard entry circuit of Figure 9-16 and tests its operation by trying to enter a series of three-digit numbers. He finds that sometimes the digit "0" is entered instead of the digit he pressed. He also observes that it happens with all the keys more or less randomly, although it is worse for some keys than others. He replaces all the ICs, and the malfunction persists. Which of the following circuit faults would explain his observations? Explain each choice.   Choice (b)
(a) The technician forgot to ground the unused inputs of the OR gate.
(b) He has mistakenly used $\overline{Q}$ instead of Q from the one-shot.
(c) The switch bounce from the digit keys lasts longer than 20 ms.
(d) The Y and Z outputs are shorted together.

9-19. Choice (a) or (d)

9-19. Repeat Problem 9-18 with the following symptom: the registers and displays stays at 0 no matter how many times a key is pressed.

9-20. While testing the circuit of Figure 9-16, a technician finds that every odd-numbered key results in the correct digit being entered, but every even-numbered key results in the wrong digit being entered as follows: the "0" key causes a "1" to be entered, the "2" key causes a "3" to be entered, the "4" key causes a "5" to be entered, and so on. Consider each of the following faults as possible causes of the malfunction. For each one explain why it can or cannot be the actual cause.
(a) An open connection from the output of the LSB inverter to the D inputs of the FFs.   Yes
(b) The D input of flip-flop Q_8 is internally shorted to V_{CC}.   no
(c) A solder bridge is shorting $\overline{O}_0$ to ground.   no

9-21. A technician tests the circuit of Figure 9-4 as described in Example 9-7, and obtains the following results: all of the outputs work except $\overline{O}_{16}$ to $\overline{O}_{19}$ and O_{24} to $\overline{O}_{27}$, which are permanently HIGH. What is the most probable circuit fault? A_2 bus line is open between Z2 and Z3.

9-22. A technician tests the circuit of Figure 9-4 as described in Example 9-7 and finds that the correct output is activated for each possible input code except those listed in the table below. Examine this table and determine the probable cause of the malfunction.

9-22. The node where the INVERTER output connects to E_2 of Z4 is stuck LOW.

Input Code	
$A_4A_3A_2A_1A_0$	**Activated Outputs**
1 0 0 0 0	$\overline{O}_{16}$ and $\overline{O}_{24}$
1 0 0 0 1	$\overline{O}_{17}$ and $\overline{O}_{25}$
1 0 0 1 0	$\overline{O}_{18}$ and $\overline{O}_{26}$
1 0 0 1 1	$\overline{O}_{19}$ and $\overline{O}_{27}$
1 0 1 0 0	$\overline{O}_{20}$ and $\overline{O}_{28}$
1 0 1 0 1	$\overline{O}_{21}$ and $\overline{O}_{29}$
1 0 1 1 0	$\overline{O}_{22}$ and $\overline{O}_{30}$
1 0 1 1 1	$\overline{O}_{23}$ and $\overline{O}_{31}$

9-23. Segment g would be much brighter than the rest until it burned out or until the output transistor of the decoder/driver burned out.

9-23. Suppose that a 22-Ω resistor was mistakenly used for the g segment in Figure 9-8. How would this affect the display? What possible problems could occur?

9-24. D and C inputs to decoder/driver have been wired in reverse.

9-24. Repeat Example 9-8 with the observed sequence shown below.

Count	0	1	2	3	4	5	6	7	8	9
Observed display	0	1	2	3	8	9	c	ɔ	4	5

9-25. Repeat Example 9-8 with the observed sequence shown below.

9-25. A short between decoder outputs a and b.

Count	0	1	2	3	4	5	6	7	8	9
Observed display	0	7	2	3	9	9	8	7	8	9

9-26. To test the circuit of Figure 9-11, a technician connects a BCD counter to the 4511b inputs and pulses the counter at a very slow rate. She notices that the f segment works erratically, and no particular pattern is evident. What are some of the possible causes of the malfunction? (*Hint:* Remember, the ICs are CMOS.)

9-26. One possibility is that the unused inputs of the unused EX-OR gate are unconnected. Another is an open path between output f and the EX-OR gate.

SECTIONS 9-7 and 9-8

9-27. The circuit in Figure 9-56 uses three 2-input multiplexers (Figure 9-20). Determine the function performed by this circuit.

9-27. It is a 4-to-1 multiplexer.

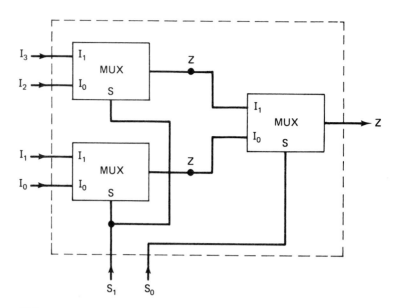

FIGURE 9-56

9-29. The outputs of the two 74157s are connected as inputs to the 74151.

9-30. The total number of connections in the circuit using the multiplexers is 63; the total using separate decoder/drivers and displays is 66.

9-28. Use the idea from the last problem to arrange several 74151 1-of-8 multiplexers to form a 1-of-64 multiplexer. Use nine 1-of-8 multiplexers.

9-29. Show how two 74157s and a 74151 can be arranged to form a 1-of-16 multiplexer with no other required logic. Label the inputs I_0 to I_{15} to show how they correspond to the select code.

9-30. (a) Expand the circuit of Figure 9-25 to display the contents of 2 three-stage BCD counters.
(b) Count the number of connections in this circuit and compare it with the number required if a separate decoder/driver and display were used for each stage of each counter.

9-31. Figure 9-57 shows how a multiplexer can be used to generate logic waveforms with any desirable pattern. The pattern is programmed using eight SPDT switches, and the waveform is repetitively produced by

FIGURE 9-57

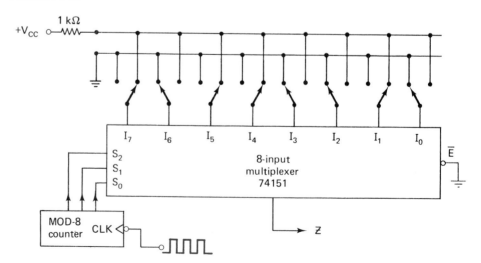

pulsing the MOD-8 counter. Draw the waveform at Z for the given switch positions.   Output Z cycles through levels I_0 through I_7 in that order.

9-32. Change the MOD-8 counter in Figure 9-57 to a MOD-16 counter and connect the MSB to the multiplexer $\overline{E}$ input. Draw the Z waveform.

9-33. Show how a 74151 can be used to generate the logic function $Z = AB + BC + AC$.

9-34. Show how a 16-input multiplexer such as the 74150 (Figure 9-36) is used to generate the function $Z = \overline{A}\,\overline{B}\,\overline{C}D + BCD + A\overline{B}\,\overline{D} + AB\overline{C}D$.

9-35. The circuit of Figure 9-58 shows how an 8-input MUX can be used to generate a four-variable logic function even though the MUX has only three SELECT inputs. Three of the logic variables A, B, and C are connected to the SELECT inputs. The fourth variable D and its inverse $\overline{D}$ are connected to selected data inputs of the MUX as required by the desired logic function. The other MUX data inputs are tied to a LOW or a HIGH as required by the function.
(a) Set up a truth table showing the output Z for the 16 possible combinations of input variables.
(b) Write the sum-of-products expression for Z and simplify it to verify that

$$Z = \overline{C}B\overline{A} + D\overline{C}\,\overline{B}\,\overline{A} + \overline{D}C\overline{B}\,\overline{A}$$

9-32. When the MSB of counter goes HIGH, it disables the MUX output so that $Z = 0$.

9-33. Connect data inputs 3, 5, 6, 7 to a HIGH, and all others LOW.

9-34. Connect data inputs 1, 5, 9, 11, 14, 15 to a HIGH, all others LOW.

9-35. Z is HIGH for $DCBA = 0010, 0100, 1000,$ and 1010.

FIGURE 9-58

9-36. The method used in Figure 9-58 can be used to generate any four-variable logic function by following these steps:
1. Set up the truth table for the desired function with Z as the output.
2. Write the sum-of-products expression for Z; do not simplify it. For example, $Z = DC\overline{B}A + \overline{D}CB\overline{A} + DCB\overline{A} + D\overline{C}BA + \overline{D}\,\overline{C}BA + \overline{D}\,\overline{C}\,\overline{B}A$.
3. Look for terms that have the same combination of C, B, and A, and factor:

$$Z = DC\overline{B}A + CB\overline{A}(\overline{D} + D) + \overline{C}BA(\overline{D} + D) + \overline{D}\,\overline{C}\,\overline{B}A$$
$$= DC\overline{B}A + CB\overline{A} + \overline{C}BA + \overline{D}\,\overline{C}\,\overline{B}A$$

4. Consider those terms that contain only C, B, and A in normal or complemented form. For each of these, connect the corresponding MUX data input to a HIGH:

$$CB\overline{A} \rightarrow \text{connect HIGH to input } I_6$$
$$\overline{C}BA \rightarrow \text{connect HIGH to input } I_3$$

5. Consider the terms that contain the D variable. Connect the D or $\overline{D}$ variable to the MUX input that corresponds to the CBA variables:

$$DC\overline{B}A \rightarrow \text{connect } D \text{ to input } I_5$$
$$\overline{D}\,\overline{C}\overline{B}A \rightarrow \text{connect } \overline{D} \text{ to input } I_1$$

9-36. (a) Design of Figure 9-58 is correct. (b) Connect data inputs 2, 3, 4, and 5 to a HIGH. Connect D to input 7. Connect $\overline{D}$ to input 0. All other data inputs are LOW.

6. Connect the remaining MUX inputs to a LOW.

    (a) Verify the design of Figure 9-58 using this method.

    (b) Use the method above to implement a function that will produce a HIGH only when the four input variables are at the same level or when the B and C variables are at different levels.

SECTION 9-9

9-37. DRILL QUESTION

For each statement indicate whether it is referring to a decoder, an encoder, a MUX, or a DEMUX:

    (a) Has more inputs than outputs.   encoder, MUX

    (b) Uses SELECT inputs.   MUX, DEMUX

    (c) Can be used in parallel-to-serial conversion.   MUX

    (d) Produces a binary code at its output.   encoder

    (e) Only one of its outputs can be active at one time.   decoder, DEMUX

    (f) Can be used to route an input signal to one of several possible outputs.   DEMUX

    (g) Can be used to generate arbitrary logic functions.   MUX

9-38. Use D as the data input; use A, B, C as the select inputs.

9-38. Show how the 7442 decoder can be used as 1-to-8 demultiplexer. (*Hint:* See Problem 9-7.)

9-39. Each DEMUX output goes LOW one at a time in sequence.

9-39. Apply the waveforms of Figure 9-54 to the inputs of the 74LS138 DEMUX of Figure 9-31(a) as follows:

$$D \rightarrow A_2, \qquad C \rightarrow A_1, \qquad B \rightarrow A_0, \qquad A \rightarrow \overline{E}_1$$

Draw the waveforms at the DEMUX outputs.

9-40. Consider the system of Figure 9-33. Assume that the clock frequency is 10 pps. Describe what the monitoring panel indications will be for each of the following cases:

    (a) All doors closed   All LEDs off

    (b) All doors open   each LED will flash

    (c) Doors 2 and 6 open   LEDs 2 and 6 will flash

9-41. Five lines going to the remote panel.

9-41. Modify the system of Figure 9-33 to handle 16 doors. Use a 74150 sixteen-input MUX and two 74LS138 DEMUXes. How many lines are going to the remote monitoring panel?

9-42. Draw the waveforms at Z, O_0, O_1, O_2, and O_3 in Figure 9-34 for the following register data: $[A] = 0011$, $[B] = 0110$, $[C] = 1001$, $[D] = 0111$.   See Solutions Manual.

9-43. Consider the control sequencer of Figure 9-27. Describe how each of the following faults will affect the operation:
(a) The I_3 input of the MUX is shorted to ground.
(b) The connections from sensors 3 and 4 to the MUX are reversed.

9-43. (a) Sequence will terminate after actuator 3 is activated.
(b) Same as (a).

9-44. Consider the circuit of Figure 9-25. A test of the circuit yields the following results:

9-44. Probable fault is a short to ground at MSB of the tens MUX.

		Actual Count	Displayed Count
Case 1	Counter 1	33	33
	Counter 2	47	47
Case 2	Counter 1	82	02
	Counter 2	64	64
Case 3	Counter 1	63	63
	Counter 2	95	15

What are the possible causes of the malfunction?

9-45. A test of the security monitoring system of Figure 9-33 produces the results recorded below.

9-45. Probable fault is that Q_0 and Q_1 of the counter are reversed before connection to MUX and DEMUX.

Condition	LEDs
All doors closed	All LEDs off
Door 0 open	LED 0 flashing
Door 1 open	LED 2 flashing
Door 2 open	LED 1 flashing
Door 3 open	LED 3 flashing
Door 4 open	LED 4 flashing
Door 5 open	LED 6 flashing
Door 6 open	LED 5 flashing
Door 7 open	LED 7 flashing

What are the possible faults that could cause this operation?

9-46. A test of the security monitoring system of Figure 9-33 produces the results recorded below.

9-46. Probable fault is a short between inputs 6 and 7 of the MUX.

Condition	LEDs
All doors closed	ALL LEDs off
Door 0 open	LED 0 flashing
Door 1 open	LED 1 flashing
Door 2 open	LED 2 flashing
Door 3 open	LED 3 flashing
Door 4 open	LED 4 flashing
Door 5 open	LED 5 flashing
Door 6 open	No LED flashing
Door 7 open	No LED flashing
Doors 6 and 7 open	LEDs 6 and 7 flashing

What are the possible faults that could cause this operation?

9-47. Suppose that the synchronous data transmission system of Figure 9-34 is malfunctioning as follows: the Z waveform is correct, but the O_0 waveform is identical to the Z waveform at all times while the other outputs are constantly LOW. What circuit faults could cause this malfunction? Assume the circuitry is TTL, and explain each choice.

9-48. The synchronous data transmission system of Figure 9-34 is malfunctioning, and the waveforms are displayed on a high-speed oscilloscope (Fig. 9-59). Note the glitches on the O_1 signal. Consider the two possible faults given below. For each one, explain whether or not it can be the cause of the malfunction.

(a) The connections to the S_1 and S_0 pins of the DEMUX are reversed.   no

(b) The connections to the Q_1 and Q_0 pins of the receiver's word counter are reversed.   yes

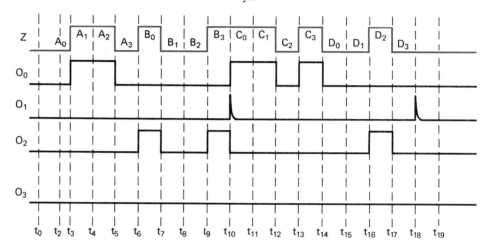

FIGURE 9-59

SECTION 9-12

9-49. Redesign the circuit of Problem 9-16 using a 74LS85 magnitude comparator. Add a "copy overflow" feature that will activate an ALARM output if the OPERATE output fails to stop the machine when the requested number of copies are done.

9-50. Show how to connect 74LS85s to compare two 10-bit numbers.

SECTION 9-13

9-51. Assume a BCD input of 69 to the code converter of Figure 9-43. Determine the levels at each Σ output and at the final binary output.

9-52. A technician tests the code converter of Figure 9-43 and observes the following results:

BCD input	Binary output
52	0110011
99	1100100
27	0011011

What is the probable circuit fault?

9-53. For the bus arrangement of Figure 9-46, describe the input signal requirements for simultaneously transferring the contents of register C to both of the other registers.

9-54. Assume that the registers in Figure 9-46 are initially $[A] = 1011$, $[B] = 1000$, and $[C] = 0111$. The signals in Figure 9-60 are applied to the register inputs.
  (a) Determine the contents of each register at times t_1, t_2, t_3, and t_4.
  (b) Describe what would happen if $\overline{IE}_A$ were LOW when the third clock pulse occurred.

9-53. Register C: $\overline{OE} = 0$, $\overline{IE} = 1$; registers A and B: $\overline{OE} = 1$, $\overline{IE} = 0$. Apply clock pulse.

9-54. (a) At t_4, $[A] = [B] = [C] = 1011$. (b) Register A would be loaded with indeterminate data.

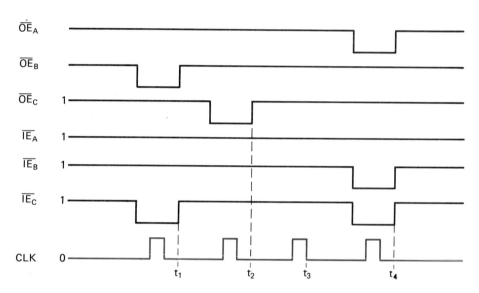

FIGURE 9-60

9-55. Assume the same initial conditions of Problem 9-54 and sketch the signal on DB_3 for the waveforms of Figure 9-60.

9-56. Figure 9-61 shows two more devices that are to be added to the data bus of Figure 9-46. One is a set of buffered switches that can be used to manually enter data into any of the bus registers. The other device is an output register that is used to latch any data that are on the bus during a data transfer operation and display them on a set of LEDs.
  (a) Assume all registers are at 0000. Outline the sequence of operations needed to load the registers with the following data from the switches: $[A] = 1011$, $[B] = 0001$, $[C] = 1110$.
  (b) What will be the state of the LEDs at the end of this sequence?

9-56. At end of sequence, LEDs will display 1110.

9-57. Now that the circuitry of Figure 9-61 has been added to Figure 9-46 a total of five devices are connected to the data bus. The circuit in Figure 9-62(a) will now be used to generate the enable signals needed to perform the different data transfers over the data bus. It uses a 74LS139 chip that contains two identical independent 1-of-4 decoders with active LOW enable. The top decoder is used to select the device that will put data on the data bus (output select), and the bottom decoder is used to select the device that is to take the data from the data

FIGURE 9-61

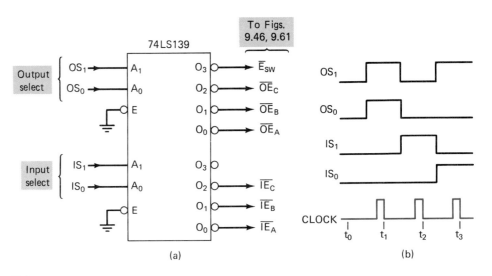

FIGURE 9-62

bus (input select). Assume that the decoder outputs are connected to the corresponding enable inputs of the devices tied to the data bus. Also assume that all registers are initially at 0000 at time t_0, and the switches are in the positions shown in Figure 9-61.

(a) Determine the contents of each register at times t_1, t_2, and t_3 in response to the waveforms in Figure 9-62(b).

(b) Can "bus contention" ever occur with this circuit?

9-57. At t_3 all registers hold 1001.

9-58. Figure 9-63 shows the basic circuitry to interface a microprocessor (MPU) to a memory module. The memory module will contain one or more memory ICs (Chapter 11) that can either receive data from the MPU (a WRITE operation) or send data to the MPU (a READ operation). The data are transferred over the 8-line data bus. The MPU's data lines and the memory's I/0 data lines are connected to this common bus. For now we will be concerned with how the MPU controls the selection of the memory module for a READ or WRITE operation.

The steps involved are:

1. The MPU places the memory address on its address output lines A_{15} to A_0.

2. The MPU generates the $R/\overline{W}$ signal to inform the memory module which operation is to be performed: $R/\overline{W} = 1$ for READ, $R/\overline{W} = 0$ for WRITE.

FIGURE 9-63   Basic microprocessor-to-memory interface circuit.

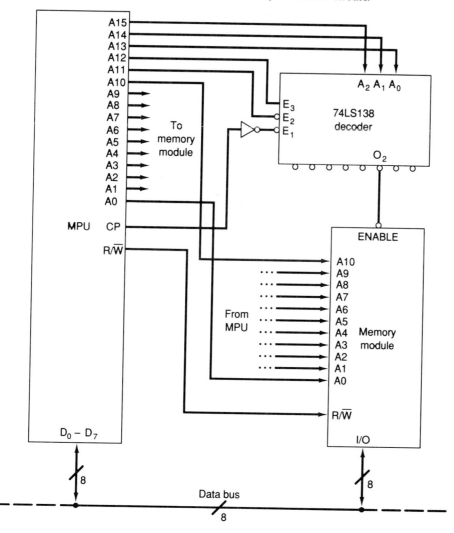

3. The upper five bits of the MPU address lines are decoded by the 74LS138, which controls the ENABLE input of the memory module. This ENABLE input must be active in order for the memory module to do a READ or WRITE operation.

4. The other 11 address bits are connected to the memory module, which uses them to select the specific *internal* memory location being accessed by the MPU, provided ENABLE is active.

In order to read from or write into the memory module, the MPU must put the correct address on the address lines to enable the memory, and then pulse *CP* to the HIGH state.

(a) Determine which, if any, of these hexadecimal addresses will activate the memory module: 607F; 57FA; 5F00.   57FA

9-58. (b) 5000 to 57FF

(b) Determine what range of hex addresses will activate the memory. (*Hint:* Inputs A_0 to A_{10} to memory can be any combination.)

(c) Assume that a second identical memory module is added to the circuit with its address, $R/\overline{W}$, and data I/0 lines connected exactly the same as the first module *except* that its ENABLE input is tied to decoder output $\overline{O}_4$. What range of hex addresses will activate this second module?   9000 to 97FF

(d) Is it possible for the MPU to read from or write to both modules at the same time? Explain.   no

ANSWERS TO SECTION REVIEW QUESTIONS

SECTION 9-1

1. No.   **2.** The enable input controls whether or not the decoder logic responds to the input binary code.   **3.** The 7445 has open-collector outputs that can handle up to 30 V and 80 mA. **4.** 24 pins: 2 enables, 16 inputs, 4 outputs, V_{CC}, and ground.

SECTION 9-2

1. *a, b, c, f, g*   **2.** True

SECTION 9-3

1. LEDs: a, e, f; LCDs: b, c, d, e

SECTION 9-4

1. An encoder produces an output code corresponding to the activated input. A decoder activates one output corresponding to an applied input code.   **2.** In a priority encoder, the output code corresponds to the *highest*-numbered input that is activated.   **3.** Normal BCD = 0110.   **4. (a)** Produces a PGT when a key is pressed.   **(b)** Converts key actuation to its BCD code.   **(c)** Generates bounce-free pulse to trigger the ring counter.   **(d)** Form a ring counter that sequentially clocks output registers. **(e)** Stores BCD codes generated by key actuations.

SECTION 9-5

1. BCD-to-decimal decoder.   **2.** High priority.   **3.** Buffer or driver.

SECTION 9-7

1. The binary number at the select inputs determines which data input will pass through to the output.   **2.** Thirty-two data inputs and five select inputs.

SECTION 9-8

1. Parallel-to-serial conversion, data routing, logic-function generation, operations sequencing.   **2.** False—they are applied to the select inputs.   **3.** Counter.

SECTION 9-9

1. A MUX selects one of many input signals to be passed to its output; a DEMUX selects one of many outputs to receive the input signal. **2.** True, provided the decoder has an ENABLE input.   **3.** The LEDs will go on and off in sequence.

SECTION 9-12

1. To provide a means for expanding the compare operations to numbers with more than 4 bits.
2. $O_{A=B} = 1$; other outputs are 0.

SECTION 9-13

1. A code converter takes input data represented in one type of binary code and converts it to another type of binary code.   **2.** Three digits can represent decimal values up to 999. To represent 999 in straight binary requires 10 bits.

SECTION 9-15

1. A set of connecting lines to which the inputs and outputs of many different devices can be connected.   **2.** Bus contention occurs when the outputs of more than one device connected to a bus are enabled at the same time. It is prevented by controlling the device enable inputs so that this cannot happen.   **3.** To provide lower output resistance to drive the many inputs connected to a bus.   **4.** Reduces the number of pins on the register IC, and the number of connections to the data bus.   **5.** See Figure 9-51.

10

INTERFACING WITH THE ANALOG WORLD

Outline

Upon completion of this chapter, you will be able to:

- Understand the theory of operation and circuit limitations of several types of digital-to-analog converters (DACs).
- Read and understand the various DAC manufacturer specifications.
- Use different test procedures to troubleshoot DAC circuits.
- Compare the advantages and disadvantages among the digital-ramp analog-to-digital converter (ADC), successive-approximation ADC, and flash ADC.
- Analyze the process by which a computer in conjunction with an ADC digitizes an analog signal and then reconstructs that analog signal from the digital data.
- Describe the basic operation of a digital voltmeter.
- Understand the need for using sample-and-hold circuits in conjunction with ADCs.
- Cite the advantages and limitations of multiplexing ADCs.

10-1 INTERFACING WITH THE ANALOG WORLD

Review of Digital versus Analog   A *digital* quantity will have a value that is specified as one of two possibilities such as 0 or 1, LOW or HIGH, true or false, and so on. In practice, a digital quantity such as a voltage may actually have a value that is anywhere within specified ranges, and we define values within a given range to have the same digital value. For example, for TTL logic we know that

$$0 \text{ V to } 0.8 \text{ V} = \text{logic } 0$$
$$2 \text{ V to } 5 \text{ V} = \text{logic } 1$$

Any voltage falling in the range 0 to 0.8 V is given the digital value 0, and any voltage in the range 2 to 5 V is assigned the digital value 1. The exact voltage values are not significant, because the digital circuits respond in the same way to all voltage values within a given range.

By contrast, an *analog* quantity can take on any value over a continuous range of values, and, most important, its exact value is significant. For example, the output

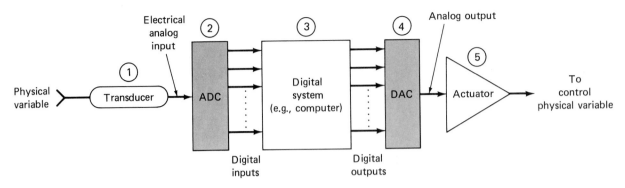

FIGURE 10-1   Analog-to-digital converter (ADC) and digital-to-analog converter (DAC) are used to interface a computer to the analog world so that the computer can monitor and control a physical variable.

of an analog temperature-to-voltage converter might be measured as 2.76 V, which may represent a specific temperature of 27.6°C. If the voltage was measured as something different, such as 2.34 V or 3.78 V, this would represent a completely different temperature. In other words, each possible value of an analog quantity has a different meaning. Another example of this is the output voltage from an audio amplifier into a speaker. This voltage is an analog quantity because each of its possible values produces a different response in the speaker.

Most physical variables are analog in nature and can take on any value within a continuous range of values. Examples include temperature, pressure, light intensity, audio signals, position, rotational speed, and flow rate. Digital systems perform all of their internal operations using digital circuitry and digital operations. Any information that has to be inputted to a digital system must first be put into digital form. Similarly, the outputs from a digital system are always in digital form. When a digital system such as a computer is to be used to monitor and/or control a physical process, we must deal with the difference between the digital nature of the computer and the analog nature of the process variables. Figure 10-1 illustrates the situation. This diagram shows the five elements that are involved when a computer is monitoring and controlling a physical variable that is assumed to be analog:

Teaching Tip: You may want to point out that many transducers require some conditioning circuitry to produce useful voltage or current values. An example is a thermistor, which is a temperature-variable resistance. It is often placed in a bridge circuit or constant-current loop. Another example is the photodiode, which produces a very small current that is proportional to light intensity. This current is usually amplified by a transistor or an op amp.

1. *Transducer*. The physical variable is normally a nonelectrical quantity. A **transducer** is a device that converts the physical variable to an electrical variable. Some common transducers include thermistors, photocells, photodiodes, flow meters, pressure transducers, and tachometers. The electrical output of the transducer is an analog current or voltage that is proportional to the physical variable it is monitoring.

    For example, the physical variable could be the temperature of water in a large tank that is being filled from cold and hot water pipes. Let's say that the water temperature varies from 80° to 150°F and that a thermistor and its associated circuitry convert this water temperature to a voltage ranging from 800 to 1500 mV. Note that the transducer's output is directly proportional to temperature; such that each 1°F produces a 10-mV output. This proportionality factor was chosen for convenience.

2. *Analog-to-digital converter (ADC)*. The transducer's electrical analog output serves as the analog input to the ADC. The ADC converts this analog input to a digital output. This digital output consists of a number of bits that represent the value of the analog input.

For example, the ADC might convert the transducer's 800- 1500-mV analog values to binary values ranging from 01010000 (80) to 10010110 (150). Note that the binary output from the ADC is proportional to the analog input voltage so that each unit of the digital output represents 10 mV.

3. *Computer*. The digital representation of the process variable is transmitted from the ADC to the digital computer, which stores the digital value and processes it according to a program of instructions that it is executing. The program might perform calculations or other operations on this digital representation of temperature to come up with a digital output that will eventually be used to control the temperature.

4. *Digital-to-analog converter (DAC)*. This digital output from the computer is connected to a DAC, which converts it to a proportional analog voltage or current. For example, the computer might produce a digital output ranging from 00000000 to 11111111, which the DAC converts to a voltage ranging from 0 to 10 V.

5. *Actuator*. The analog signal from the DAC is often connected to some device or circuit that serves as an actuator to control the physical variable. For our water temperature example, the actuator might be an electrically controlled valve that regulates the flow of hot water into the tank in accordance with the analog voltage from the DAC. The flow rate would vary in proportion to this analog voltage, with 0 V producing no flow and 10 V producing the maximum flow rate.

Thus we see that ADCs and DACs function as *interfaces* between a completely digital system, like a computer, and the analog world. This function has become increasingly more important as inexpensive microcomputers have moved into areas of process control where computer control was previously not feasible.

___ REVIEW QUESTIONS ___

1. What is the function of a transducer?
2. What is the function of an ADC?
3. What does a computer often do with the data it receives from an ADC?
4. What function does a DAC perform?
5. What is the function of an actuator?

10-2 DIGITAL-TO-ANALOG CONVERSION

We will now begin our study of digital-to-analog (D/A) and analog-to-digital (A/D) conversion. Since many A/D conversion methods utilize the D/A conversion process, we will examine D/A conversion first.

Basically, D/A *conversion* is the process of taking a value represented in *digital* code (such as straight binary or BCD) and converting it to a voltage or current which is proportional to the digital value. Figure 10-2(a) shows the symbol for a typical 4-bit D/A converter. We will not concern ourselves with the internal circuitry until later. For now, we will examine the various input/output relationships.

The digital inputs D, C, B, and A are usually derived from the output register

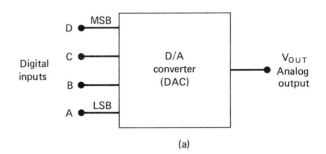

(a)

D	C	B	A	V_{OUT}	
0	0	0	0	0	volts
0	0	0	1	1	
0	0	1	0	2	
0	0	1	1	3	
0	1	0	0	4	
0	1	0	1	5	
0	1	1	0	6	
0	1	1	1	7	
1	0	0	0	8	
1	0	0	1	9	
1	0	1	0	10	
1	0	1	1	11	
1	1	0	0	12	
1	1	0	1	13	
1	1	1	0	14	
1	1	1	1	15	volts

(b)

FIGURE 10-2   Four-bit DAC with voltage output.

of a digital system. The $2^4 = 16$ different binary numbers represented by these 4 bits are listed in Figure 10-2(b). For each input number, the D/A converter output voltage is a unique value. In fact, for this case, the analog output voltage V_{OUT} is equal in volts to the binary number. It could also have been twice the binary number or some other proportionality factor. The same idea would hold true if the D/A output were a current I_{OUT}.

In general,

$$\text{analog output} = K \times \text{digital input} \qquad (10\text{-}1)$$

where K is the proportionality factor and is a constant value for a given DAC. The analog output can of course be a voltage or current. When it is a voltage, K will be in voltage units, and when the output is current, K will be in current units. For the DAC of Figure 10-2, $K = 1$ V so that

$$V_{OUT} = (1\text{ V}) \times \text{digital input}$$

We can use this to calculate V_{OUT} for any value of digital input. For example, with a digital input of $1100_2 = 12_{10}$, we obtain

$$V_{OUT} = 1\text{ V} \times 12 = 12\text{ V}$$

——— EXAMPLE 10-1A ———

A 5-bit DAC has a current output. For a digital input of 10100, an output current of 10 mA is produced. What will I_{OUT} be for a digital input of 11101?

Solution   The digital input 10100_2 is equal to decimal 20. Since $I_{OUT} = 10$ mA for this case, the proportionality factor must be 0.5 mA. Thus, we can find I_{OUT} for any digital input such as $11101_2 = 29_{10}$ as follows:

$$I_{OUT} = (0.5 \text{ mA}) \times 29$$
$$= 14.5 \text{ mA}$$

Remember, the proportionality factor, K, will vary from one DAC to another.

___

___ EXAMPLE 10-1B ___

What is the largest value of output voltage from an 8-bit DAC that produces 1.0 V for a digital input of 00110010?

Solution

$$00110010_2 = 50_{10}$$
$$1.0 \text{ V} = K \times 50$$

Therefore,

$$K = 20 \text{ mV}$$

The largest output will occur for an input of $11111111_2 = 255_{10}$.

$$V_{OUT}(\text{max}) = 20 \text{ mV} \times 255$$
$$= 5.10 \text{ V}$$

___

Analog Output   The output of a DAC is technically not an analog quantity because it can take on only specific values like the 16 possible voltage levels for V_{OUT} in Figure 10-2. Thus, in that sense, it is actually digital. However, as we will see, the number of different possible output values can be increased and the difference between successive values decreased by increasing the number of input bits. This will allow us to produce an output that is more and more like an analog quantity that varies continuously over a range of values. In other words, the DAC output is a "psuedo-analog" quantity. We will continue to refer to it as analog, keeping in mind that it is an approximation to a pure analog quantity.

Input Weights   For the DAC of Figure 10-2 it should be noted that each digital input contributes a different amount to the analog output. This is easily seen if we examine the cases where only one input is HIGH:

D	C	B	A		$V_{OUT}(V)$
0	0	0	1	$\rightarrow$	1
0	0	1	0	$\rightarrow$	2
0	1	0	0	$\rightarrow$	4
1	0	0	0	$\rightarrow$	8

The contributions of each digital input are *weighted* according to their position in the binary number. Thus, A, which is the LSB, has a *weight* of 1 V, B has a weight of 2 V, C has a weight of 4 V, and D, the MSB, has the largest weight, 8 V. The weights are successively doubled for each bit, beginning with the LSB. Thus, we can

consider V_{OUT} to be the weighted sum of the digital inputs. For instance, to find V_{OUT} for the digital input 0111 we can add the weights of the C, B, and A bits to obtain 4 V + 2 V + 1 V = 7 V.

Teaching Tip: The student may notice that the weight of the LSB is equal to the proportionality constant, K. Using $K = 0.2$ V, the value of V_{OUT} can be calculated as 0.2 V × 31 = 6.2 V.

EXAMPLE 10-2

A 5-bit D/A converter produces $V_{OUT} = 0.2$ V for a digital input of 00001. Find the value of V_{OUT} for an input of 11111.

Solution   Obviously, 0.2 V is the weight of the LSB. Thus, the weights of the other bits must be 0.4 V, 0.8 V, 1.6 V, and 3.2 V, respectively. For a digital input of 11111, then, the value of V_{OUT} will be 3.2 V + 1.6 V + 0.8 V + 0.4 V + 0.2 V = 6.2 V.

Resolution (Step Size)   *Resolution* of a D/A converter is defined as the smallest change that can occur in the analog output as a result of a change in the digital input. Referring to the table in Figure 10-2, we can see that the resolution is 1 V, since V_{OUT} can change by no less than 1 V when the digital input value is changed. The resolution is always equal to the weight of the LSB and is also referred to as the **step size,** since it is the amount that V_{OUT} will change as the digital input value is changed from one step to the next. This is illustrated better in Figure 10-3, where the outputs from a 4-bit binary counter provide the inputs to our DAC. As the counter is being continually cycled through its 16 states by the clock signal, the DAC output is a **staircase** waveform that goes up 1 V per step. When the counter is at 1111, the DAC output is at its maximum value of 15 V; this is its **full-scale output.** When the counter recycles to 0000, the DAC output returns to 0 V. The resolution or step size is the size of the jumps in the staircase waveform; in this case, each step is 1 V.

Discussion Question: What happens to the step size if full scale is kept the same while the number of input bits is increased?

Note that the staircase has 16 levels corresponding to the 16 input states, but there are only 15 steps or jumps between the 0-V level and full-scale. In general, for

FIGURE 10-3   Output waveforms of DAC as inputs are provided by a binary counter.

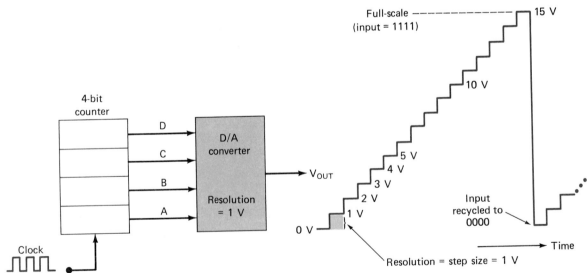

an N-bit DAC the number of different levels will be 2^N, and the number of steps will be $2^N - 1$.

You may have already figured out that resolution (step size) is the same as the proportionality factor in the DAC input/output relationship

$$\text{analog output} = K \times \text{digital input}$$

A new interpretation of this expression would be that the digital input is equal to the number of steps while K is the amount of voltage (or current) per step. The following examples will illustrate.

———— EXAMPLE 10-3A ————

What is the resolution (step size) of the DAC of Example 10-2? Describe the staircase signal out of this DAC.

Solution   The LSB for this converter has a weight of 0.2 V. This is the resolution or step size. A staircase waveform can be generated by connecting a 5-bit counter to the DAC inputs. The staircase will have 32 levels from 0 V up to a full-scale output of 6.2 V, and 31 steps of 0.2 V each.

———— EXAMPLE 10-3B ————

For the DAC of Example 10-2, determine V_{OUT} for a digital input of 10001.

Solution   The step size is 0.2 V, which is the proportionality factor K. The digital input is $10001 = 17_{10}$. Thus we have

$$V_{OUT} = (0.2 \text{ V}) \times 17$$
$$= 3.4 \text{ V}$$

Percentage Resolution   Although resolution can be expressed as the amount of voltage or current per step, it is also useful to express it as a percentage of the *full-scale output*. To illustrate, the DAC of Figure 10-3 has a maximum full-scale output of 15 V (when the digital input is 1111). The step size is 1 V, which gives a percentage resolution of

$$\% \text{ resolution} = \frac{\text{step size}}{\text{full scale (F.S.)}} \times 100\%$$

$$= \frac{1 \text{ V}}{15 \text{ V}} \times 100\% = 6.67\%$$

(10-2)

———— EXAMPLE 10-4 ————

A 10-bit DAC has a step size of 10 mV. Determine the full-scale output voltage and the percentage resolution.

Solution   With 10 bits, there will be $2^{10} - 1 = 1023$ steps of 10 mV each. The full-scale output will therefore be 10 mV $\times$ 1023 = 10.23 V and

$$\% \text{ resolution} = \frac{10 \text{ mV}}{10.23 \text{ V}} \times 100\% \approx 0.1\%$$

Example 10-4 helps to illustrate the fact that the percentage resolution becomes smaller as the number of input bits is increased. In fact, the percentage resolution can also be calculated from

$$\% \text{ resolution} = \frac{1}{\text{total number of steps}} \times 100\% \qquad (10\text{-}3)$$

For an N-bit binary input code the total number of steps is $2^N - 1$. Thus, for the previous example,

$$\% \text{ resolution} = \frac{1}{2^{10} - 1} \times 100\%$$

$$= \frac{1}{1023} \times 100\%$$

$$\approx 0.1\%$$

This means that it is *only the number of bits* which determines the *percentage* resolution. Increasing the number of bits increases the number of steps to reach full scale, so that each step is a smaller part of the full-scale voltage. Most DAC manufacturers specify resolution as the number of bits.

What Does Resolution Mean?

A DAC cannot produce a continuous range of output values, and so, strictly speaking, its output is not truly analog. A DAC produces a finite set of output values. In our example of Section 10-1 the computer generates a digital output to provide an analog voltage between 0 and 10 V to an electrically controlled valve. The DAC's resolution (number of bits) determines how many possible voltage values the computer can send to the valve. If a 6-bit DAC is used, there will be 63 possible steps of 0.159 V between 0 and 10 V. When an 8-bit DAC is used, there will be 255 possible steps of 0.039 V between 0 and 10 V. The greater the number of bits, the finer the resolution (the smaller the step size).

The system designer must decide what resolution is needed on the basis of the required system performance. The resolution limits how close the DAC output can come to a given analog value. Generally, the cost of DACs increases with the number of bits, and so the designer will use as few bits as necessary.

—— EXAMPLE 10-5 ——————————————

Figure 10-4 shows a computer controlling the speed of a motor. The 0–2 mA analog current from the DAC is amplified to produce motor speeds from 0 to 1000 rpm (revolutions per minute). How many bits should be used if the computer is to be able to produce a motor speed that is within 2 rpm of the desired speed?

FIGURE 10-4   Example 10-5.

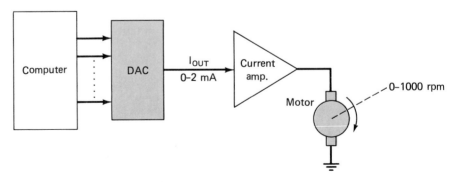

Solution   The motor speed will range from 0 to 1000 rpm as the DAC goes from zero to full scale. Each step in the DAC output will produce a step in the motor speed. We want the step size to be no greater than 2 rpm. Thus, we need at least 500 steps (1000/2). Now we must determine how many bits are required so that there are at least 500 steps from zero to full scale. We know that the number of steps is $2^N - 1$, and so we can say

$$2^N - 1 \geq 500$$

or

$$2^N \geq 501$$

Since $2^8 = 256$ and $2^9 = 512$, the smallest number of bits that will produce at least 500 steps is *nine*. We could use more than 9 bits, but this might add to the cost of the DAC.

Teaching Tip: For students who insist on solving for N using their calculator, logarithms must be used. That is, $N \log_{10} 2 = \log_{10} 501$. N must of course be rounded up.

───── EXAMPLE 10-6 ─────────────────

Using 9 bits, how close to 326 rpm can the speed be adjusted?

Solution   With 9 bits, there will be 511 steps ($2^9 - 1$). Thus, the motor speed will go up in steps of 1000 rpm/511 = 1.957 rpm. The number of steps needed to reach 326 rpm is 326/1.957 = 166.58. This is not a whole number of steps, and so we will round it to 167. The actual motor speed on the 167th step will be $167 \times 1.957 = 326.8$ rpm. Thus, the computer must output the 9-bit binary equivalent of 167_{10} to produce the desired motor speed within the resolution of the system.

In all of our examples, we have assumed that the DACs have been perfectly accurate in producing an analog output that is directly proportional to the binary input, and that the resolution is the only thing that limits how close we can come to a desired analog value. This, of course, is unrealistic, since all devices contain inaccuracies. We will examine the causes and effects of DAC inaccuracy in a later section.

BCD Input Code   The DACs we have considered thus far have used a binary input code. Many DACs use a BCD input code where 4-bit code groups are used for each decimal digit. Figure 10-5 shows the diagram of an 8-bit (two-digit) converter of this type. Each 4-bit code group can range from 0000 to 1001, and so the BCD inputs represent decimal numbers from 00 to 99. Within each code group the

FIGURE 10-5   DAC using BCD input code. This converter accepts a 2-digit input and generates 100 possible analog output values.

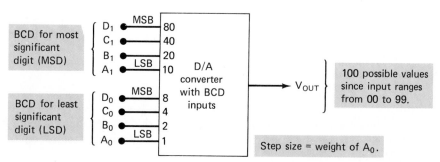

weights of the different bits are in the normal binary proportions (1, 2, 4, 8), but the relative weights of each code group are different by a factor of 10. Figure 10-5 shows the relative weights of the various bits. Note that the bits that make up the BCD code for the most significant digit (MSD) have a relative weight that is 10 times that of the corresponding bits of the LSD.

EXAMPLE 10-7A

If the weight of A_0 is 0.1 V in Figure 10-5, find the following values:
(a) Step size
(b) Full-scale output and percentage resolution
(c) V_{OUT} for $D_1 C_1 B_1 A_1 = 0101$ and $D_0 C_0 B_0 A_0 = 1000$.

Solution

(a) Step size is the weight of the LSB of the LSD, 0.1 V.
(b) There are 99 steps since there are two BCD digits. Thus, full-scale output is $99 \times 0.1 = 9.9$ V. The resolution is [using equation (10-2)]

$$\frac{\text{step size}}{\text{F.S.}} \times 100\% = \frac{0.1}{9.9} \times 100\% \approx 1\%$$

We could also have used equation (10-3) to calculate percentage resolution, since the total number of steps is 99.
(c) The exact weights in volts are

MSD				LSD			
D_1	C_1	B_1	A_1	D_0	C_0	B_0	A_0
8.0	4.0	2.0	1.0	0.8	0.4	0.2	0.1

One way to find V_{OUT} for a given input is to add the weights of all bits that are 1s. Thus, for an input of 0101 1000, we have

$$V_{OUT} = \overbrace{4 \text{ V}}^{C_1} + \overbrace{1 \text{ V}}^{A_1} + \overbrace{0.8 \text{ V}}^{D_0} = 5.8 \text{ V}$$

An easier way is to realize that the BCD input code represents 58_{10} and the step size is 0.1 V, so that

$$V_{OUT} = (0.1 \text{ V}) \times 58 = 5.8 \text{ V}$$

EXAMPLE 10-7B

A certain 12-bit BCD digital-to-analog converter has a full-scale output of 9.99 V.
(a) Determine the percent resolution.
(b) Determine the converter's step size.

Solution

(a) Twelve bits corresponds to three decimal digits; that is, decimal numbers from 000 to 999. Therefore the output of this DAC has 999 possible steps from 0 V to 9.99 V. Thus we have

$$\% \text{ resolution} = \frac{1}{\text{number of steps}} \times 100\%$$

$$= \frac{1}{999} \times 100\%$$

$$\approx \mathbf{0.1\%}$$

(b)
$$\text{step size} = \frac{\text{F.S.}}{\text{number of steps}}$$

$$= 9.99 \text{ V}/999$$

$$= \mathbf{0.01 \text{ V}}$$

Bipolar DACs   Up to this point we have assumed that the binary input to a DAC has been an unsigned number and the DAC output has been a positive voltage or current. Some DACs are designed to produce both positive and negative values, such as -10 to $+10$ V. This is generally done by using the binary input as a signed number with the MSB as the sign bit (0 for $+$ and 1 for $-$). Negative input values are often represented in 2's-complement form, although the true-magnitude form is also used by some DACs. For example, suppose that we have a 6-bit bipolar DAC that uses the 2's-complement system and has a resolution of 0.2 V. The binary input values range from 100000 (-32) to 011111 ($+31$) to produce analog outputs in the range from -6.4 to $+6.2$ V. There are 63 steps ($2^6 - 1$) of 0.2 V between these negative and positive limits.

REVIEW QUESTIONS

1. An 8-bit DAC has an output of 3.92 mA for an input of 01100010. What are the DAC's resolution and full-scale output?
2. What is the weight of the MSB of the DAC of question 1?
3. What is the percentage resolution of an 8-bit DAC?
4. How many different output voltages can a 12-bit DAC produce?
5. For the system of Figure 10-4, how many bits should be used if the computer is to control the motor speed within 0.4 rpm?
6. Consider a 12-bit DAC with BCD inputs and a resolution of 10 mV. What will its output be for an input of 100001110011?
7. *True or false:* The percentage resolution of a DAC depends *only* on the number of bits.
8. What is the advantage of a smaller (finer) resolution?

10-3 D/A-CONVERTER CIRCUITRY

There are several methods and circuits for producing the D/A operation which has been described. We shall examine several of the basic schemes, to gain an insight into the ideas used. It is not important to be familiar with all the various circuit schemes because D/A converters are available as ICs or as encapsulated packages that do not require any circuit knowledge. Instead, it is important to know the significant performance characteristics of DACs, in general, so that they can be used intelligently. These will be covered in Section 10-4.

Teaching Tip: This section can be omitted if the electronic aspects of the DAC are not part of your course objectives.

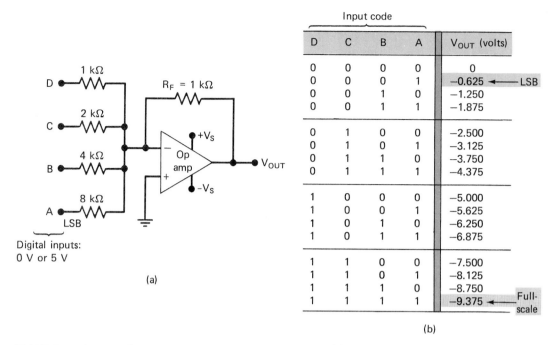

| Input code | | | | |
D	C	B	A	V_{OUT} (volts)
0	0	0	0	0
0	0	0	1	−0.625 ←——LSB
0	0	1	0	−1.250
0	0	1	1	−1.875
0	1	0	0	−2.500
0	1	0	1	−3.125
0	1	1	0	−3.750
0	1	1	1	−4.375
1	0	0	0	−5.000
1	0	0	1	−5.625
1	0	1	0	−6.250
1	0	1	1	−6.875
1	1	0	0	−7.500
1	1	0	1	−8.125
1	1	1	0	−8.750
1	1	1	1	−9.375 ←——Full-scale

(b)

FIGURE 10-6   Simple DAC using op-amp summing amplifier with binary-weighted resistors.

Figure 10-6(a) shows the basic circuit for one type of 4-bit DAC. The inputs *A*, *B*, *C*, and *D* are binary inputs which are assumed to have values of either 0 V or 5 V. The *operational amplifier* is employed as a summing amplifier, which produces the weighted sum of these input voltages. It may be recalled that the summing amplifier multiplies each input voltage by the ratio of the feedback resistor R_F to the corresponding input resistor R_{IN}. In this circuit $R_F = 1$ kΩ and the input resistors range from 1 to 8 kΩ. The *D* input has $R_{IN} = 1$ kΩ, and so the summing amplifier passes the voltage at *D* with no attenuation. The *C* input has $R_{IN} = 2$ kΩ, so that it will be attenuated by 1/2. Similarly, the *B* input will be attenuated by 1/4 and the *A* input by 1/8. The amplifier output can thus be expressed as

$$V_{OUT} = -\left(V_D + \frac{1}{2}V_C + \frac{1}{4}V_B + \frac{1}{8}V_A\right) \qquad (10\text{-}4)$$

The negative sign is present because the summing amplifier is a polarity-inverting amplifier, but it will not concern us here.

Clearly, the summing amplifier output is an analog voltage which represents a weighted sum of the digital inputs, as shown by the table in Figure 10-6(b). This table lists all the possible input conditions and the resultant amplifier output voltage. The output is evaluated for any input condition by setting the appropriate inputs to either 0 V or 5 V. For example, if the digital input is 1010, then $V_D = V_B = 5$ V and $V_C = V_A = 0$ V. Thus, using equation (10-4),

$$V_{OUT} = -\left(5\text{ V} + 0\text{ V} + \frac{1}{4} \times 5\text{ V} + 0\text{ V}\right)$$
$$= -6.25\text{ V}$$

The resolution of this D/A converter is equal to the weighting of the LSB, which is $1/8 \times 5$ V = 0.625 V. As shown in the table, the analog output increases by 0.625 V as the binary input number advances one step.

(a) Determine the weight of each input bit of Figure 10-6(a).

(b) Change R_F to 250 Ω and determine the full-scale output.

Solution

(a) The MSB passes with gain = 1, and so its weight in the output is 5 V. Thus,

$$\text{MSB} \longrightarrow 5 \text{ V}$$
$$\text{2nd MSB} \longrightarrow 2.5 \text{ V}$$
$$\text{3rd MSB} \longrightarrow 1.25 \text{ V}$$
$$\text{4th MSB = LSB} \longrightarrow 0.625 \text{ V}$$

(b) If R_F is reduced by a factor of 4, to 250 Ω, each input weight will be 4 times *smaller* than the values above. Thus, the full-scale output will be reduced by this same factor and becomes $-9.375/4 = -2.344$ V.

Discussion Question: What will happen to full-scale output as R_F is increased? What limits the maximum size of R_F?

If we look at the input resistor values in Figure 10-6, it should come as no surprise that they are **binarily weighted**. In other words, starting with the MSB resistor, the resistor values increase by a factor of 2. This of course produces the desired weighting in the voltage output.

Conversion Accuracy   The table in Figure 10-6(b) gives the *ideal* values of V_{OUT} for the various input cases. How close the circuit comes to producing these values depends primarily on two factors: (1) the precision of the input and feedback resistors, and (2) the precision of the input voltage levels. The resistors can be made

FIGURE 10-7   Complete 4-bit DAC including precision reference supply.

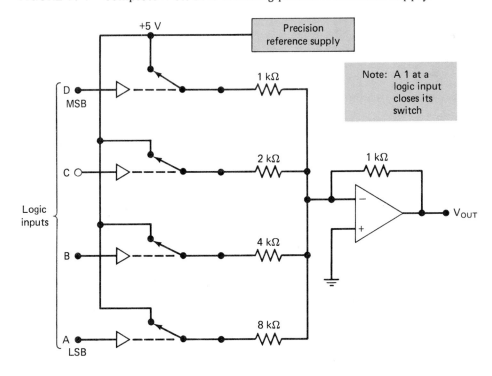

very accurate (within 0.01 percent of the desired values) by trimming, but the input voltage levels must be handled differently. It should be clear that the digital inputs cannot be taken directly from the outputs of FFs or logic gates because the output logic levels of these devices are not precise values like 0 V and 5 V but vary within given ranges. For this reason, it is necessary to add some more circuitry between each digital input and its input resistor to the summing amplifier as shown in Figure 10-7 (see page 541).

Each digital input controls a semiconductor switch like the CMOS transmission gate we studied in Chapter 8. When the input is HIGH, the switch closes and connects a *precision reference supply* to the input resistor; when the input is LOW, the switch is open. The reference supply produces a very stable, precise voltage needed to generate an accurate analog output.

Discussion Question: Discuss how the nonideal characteristics of the semiconductor switch can affect the accuracy of this DAC.

DAC with Current Output

Figure 10-8(a) shows one basic scheme for generating an analog output current proportional to a binary input. The circuit shown is a 4-bit DAC using binarily weighted resistors. The circuit uses four parallel current paths, each controlled by a semiconductor switch such as the CMOS transmission gate. The state of each switch is controlled by logic levels at the binary inputs. The current through each path is determined by an accurate reference voltage, V_{REF}, and a precision resistor in the path. The resistors are binarily weighted so that the various currents will be binarily weighted, and the total current, I_{OUT}, will be the sum of the individual currents. The MSB path has the smallest resistor, R; the next path has a

FIGURE 10-8   (a) Basic current-output DAC; (b) connected to an op-amp current-to-voltage converter.

$$I_{OUT} = B_3 \times I_0 + B_2 \times \frac{I_0}{2} + B_1 \times \frac{I_0}{4} + B_0 \times \frac{I_0}{8}$$

$$\text{where } I_0 = \frac{V_{REF}}{R}$$

$$V_{OUT} = -I_{OUT} \times R_F$$

resistor of twice the value, and so on. The output current can be made to flow through a load R_L which is much smaller than R, so that it has no effect on the value of current. Ideally, R_L should be a short to ground.

──EXAMPLE 10-9 ──────────────────────────────

Assume $V_{REF} = 10$ V and $R = 10$ kΩ. Determine the resolution and full-scale output for this DAC. Assume that $R_L \ll R$.

Solution $I_0 = V_{REF}/R = 1$ mA. This is the weight of the MSB. The other three currents will be 0.5, 0.25, and 0.125 mA. The LSB is 0.125 mA, which is also the resolution.

The full-scale output will occur when the binary inputs are all HIGH so that each current switch is closed and

$$I_{OUT} = 1 + 0.5 + 0.25 + 0.125 = 1.875 \text{ mA}$$

Note that the output current is proportional to V_{REF}. If V_{REF} is increased or decreased, I_{OUT} will change proportionally.

In order for I_{OUT} to be accurate, R_L should be a short to ground. One common way to accomplish this is to use an op amp as a current-to-voltage converter, as shown in Figure 10-8(b). Here the I_{OUT} from the DAC is connected to the op-amp's "−" input, which is virtually at ground. The op-amp negative feedback forces a current equal to I_{OUT} to flow through R_F to produce $V_{OUT} = -I_{OUT} \times R_F$. Thus, V_{OUT} will be an analog voltage that is proportional to the binary input to the DAC. This analog output can drive a wide range of loads without being loaded down.

R/2R Ladder   The DAC circuits we have looked at thus far use binary-weighted resistors to produce the proper weighting of each bit. While this method works in theory, it has some practical limitations. The biggest problem is the large difference in resistor values between the LSB and MSB, especially in high-resolution DACs (i.e., many bits). For example, if the MSB resistor is 1 kΩ in a 12-bit DAC, the

FIGURE 10-9   Basic *R/2R* ladder DAC.

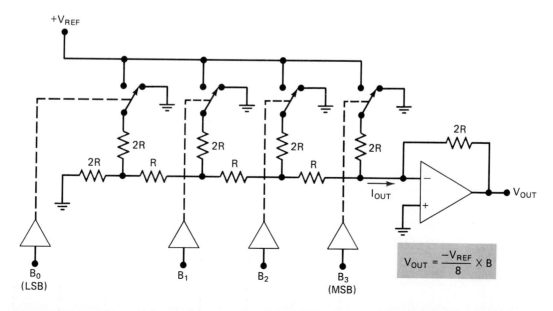

LSB resistor will be over 2 MΩ. With the current IC fabrication technology, it is very difficult to produce resistance values over a wide resistance range that maintain an accurate ratio especially with variations in temperature.

For this reason it is preferable to have a circuit that uses resistances that are fairly close in value. One of the most widely used DAC circuits that satisfies this requirement is the *R/2R ladder* network, where the resistance values span a range of only 2 to 1. One such DAC is shown in Figure 10-9 (see page 543).

Note how the resistors are arranged, and especially note that only two different values are used, R and $2R$. The current I_{OUT} depends on the positions of the four switches, and the binary inputs $B_3 B_2 B_1 B_0$ control the states of the switches. This current is allowed to flow through an op-amp current-to-voltage converter to develop V_{OUT}. We will not perform a detailed analysis of this circuit here, but it can be shown that the value of V_{OUT} is given by the expression

$$V_{OUT} = \frac{-V_{REF}}{8} \times B \tag{10-5}$$

where B is the value of the binary input, which can range from 0000 (0) to 1111 (15).

_____ EXAMPLE 10-10 _____

Assume that $V_{REF} = 5$ V for the DAC in Figure 10-9. What are the resolution and full-scale output of this converter?

Solution   The resolution is equal to the weight of the LSB, which we can determine by setting $B = 0001 = 1$ in equation (10-5):

$$resolution = \frac{-5 \text{ V} \times 1}{8}$$
$$= -0.625 \text{ V}$$

The full-scale output occurs for $B = 1111 = 15_{10}$. Again using equation (10-5),

$$full\text{-}scale = \frac{-5 \text{ V} \times 15}{8}$$
$$= -9.375 \text{ V}$$

_____ REVIEW QUESTIONS _____

1. What is the advantage of $R/2R$ ladder DACs over those that use binary-weighted resistors?
2. A certain 6-bit DAC uses binary-weighted resistors. If the MSB resistor is 20 kΩ, what is the LSB resistor?
3. What will the resolution be if the value of R_F in Figure 10-6 is changed to 800 Ω?

10-4   DAC SPECIFICATIONS

A wide variety of DACs are currently available as ICs or as self-contained encapsulated packages. One should be familiar with the more important manufacturers' specifications in order to evaluate a DAC for a particular application.

Resolution   As mentioned earlier, the percentage resolution of a DAC is dependent solely on the number of bits. For this reason, manufacturers usually specify a DAC resolution as the number of bits. A 10-bit DAC has a finer (smaller) resolution than an 8-bit DAC.

Accuracy   DAC manufacturers have several ways of specifying accuracy. The two most common are called **full-scale error** and **linearity error,** which are normally expressed as a percentage of the converter's full-scale output (%F.S.)

Full-scale error is the maximum deviation of the DAC's output from its expected (ideal) value, expressed as a percentage of full scale. For example, assume that the DAC of Figure 10-6 has an accuracy of $\pm 0.01\%$ F.S. Since this converter has a full-scale output of 9.375 V, this percentage converts to

$$\pm 0.01\% \times 9.375 \text{ V} = \pm 0.9375 \text{ mV}$$

This means that the output of this DAC can, at any time, be off by as much as 0.9375 mV from its expected value.

Linearity error is the maximum deviation in step size from the ideal step size. For example, the DAC of Figure 10-6 has an expected step size of 0.625 V. If this converter has a linearity error of $\pm 0.01\%$ F.S., this would mean that the actual *step size* could be off by as much as 0.9375 mV.

Some of the more expensive DACs have full-scale and linearity errors as low as 0.001% F.S. General-purpose DACs usually have accuracies in the 0.01 to 0.1 percentage range.

It is important to understand that accuracy and resolution of a DAC must be compatible. It is illogical to have a resolution of, say, 1 percent and an accuracy of 0.1 percent, or vice versa. To illustrate, a DAC with a resolution of 1 percent and a F.S. output of 10 V can produce an output analog voltage within 0.1 V of any desired value, assuming perfect accuracy. It makes no sense to have a costly accuracy of 0.01 percent of F.S. (or 1 mV) if the resolution already limits the closeness of the desired value to 0.1 V. The same can be said for having a resolution that is very small (many bits) while the accuracy is poor; it is a waste of input bits.

Discussion Question: Why is a DAC output more accurate as you get closer to full-scale output? A good analogy here is an analog VOM, which typically has its accuracy specified as percent of full scale.

Teaching Tip: A useful analogy to use here would be the absurdity of having a VOM that has an accuracy of 3 percent of full scale but with only three markings on its scale (i.e., 0, half scale, and full scale), or conversely a VOM that has an accuracy of 10 percent of full scale but with its scale divided into 1000 increments.

───── EXAMPLE 10-11 ─────

A certain 8-bit DAC has a full-scale output of 2 mA and a full-scale error of $\pm 0.5\%$ F.S. What is the range of possible outputs for an input of 10000000?

Solution   The step size is 2 mA/255 = 7.84 μA. Since 10000000 = 128_{10}, the ideal output should be 128 $\times$ 7.84 μA = 1004 μA. The error can be as much as

$$\pm 0.5\% \times 2 \text{ mA} = \pm 10 \text{ } \mu\text{A}$$

Thus, the actual output can deviate by this amount from the ideal 1004 μA, so the actual output can be anywhere from 994 to 1014 μA.

Setting Time   The operating speed of a DAC is usually specified by giving its *settling time,* which is the time required for the DAC output to go from zero to fullscale as the binary input is changed from all 0s to all 1s. Actually, the settling time is measured as the time for the DAC output to settle within $\pm 1/2$ step size (resolution) of its final value. For example, if a DAC has a resolution of 10 mV, settling time is measured as the time it takes the output to settle within 5 mV of its full-scale value.

Typical values for settling time range from 50 ns to 10 μs. Generally speaking, DACs with a current output will have shorter settling times than those with voltage outputs. For instance, the DAC1280 can operate as either current output or voltage output. In the current output mode, its settling time is 300 ns; in the voltage output mode, its settling time is 2.5 μs. The main reason for this difference is the response time of the op amp that is used as the current-to-voltage converter.

Offset Voltage   Ideally, the output of a DAC will be zero volts when the binary input is all 0s. In practice, however, there will be a very small output voltage for this situation; this is called **offset error**. This offset error, if not corrected, will be added to the expected DAC output for *all* input cases. For example, let's say that a 4-bit DAC has an offset error of +2 mV and a *perfect* step size of 100 mV. The table below shows the ideal and actual DAC output for several input cases. Note that the actual output is 2 mV greater than expected; this is due to the offset error. Offset error can be negative as well as positive.

Input Code	Ideal Output (mV)	Actual Output (mV)
0000	0	2
0001	100	102
1000	800	802
1111	1500	1502

Many DACs will have an external offset adjustment that allows you to zero the offset. This is usually accomplished by applying all 0s to the DAC input and monitoring the output while an *offset adjustment potentiometer* is adjusted until the output is as close to 0 V as required.

Teaching Tip: Unfortunately, offset voltages are temperature-dependent, and so they tend to drift as the device temperature changes. As a result, a small amount of offset error will always be present even when an offset adjustment has been made.

Monotonicity   A DAC is *monotonic* if its output increases as the binary input is incremented from one value to the next. Another way to describe this is that the staircase output will have no downward steps as the binary input is incremented from zero to full scale.

—— REVIEW QUESTIONS ————

1. Define full-scale error.
2. What is the meaning of settling time?
3. Describe offset error and its effect on a DAC output.
4. Why are voltage DACs generally slower than current DACs?

10-5   DAC APPLICATIONS

DACs are used whenever the output of a digital circuit has to provide an analog voltage or current to drive an analog device. Some of the most common applications are described in the following paragraphs.

Control   The digital output from a computer can be converted to an analog control signal to adjust the speed of a motor or the temperature of a furnace, or to control almost any physical variable.

Automatic Testing   Computers can be programmed to generate the analog signals (through a DAC) needed to test analog circuitry. The test circuit's analog output response will normally be converted to a digital value by an ADC and fed into the computer to be stored, displayed, and sometimes analyzed.

Signal Reconstruction   In many applications, an analog signal is **digitized,** meaning that successive points on the signal are converted to their digital equivalent and stored in memory. This conversion is performed by an analog-to-digital converter (ADC). A DAC can then be used to convert the stored digitized data back to analog—one point at a time—thereby reconstructing the original signal. This combination of digitizing and reconstruction is used in digital storage oscilloscopes, audio compact disk systems, and digital audio tape recording. We will say more on this after we learn about ADCs.

A/D Conversion   Several types of ADCs use DACs as part of their circuitry, as we shall see in Section 10-7.

10-6 TROUBLESHOOTING DACs

DACs are both digital and analog. Logic probes and pulsers can be used on the digital inputs, but a meter or an oscilloscope must be used on the analog output. There are basically two ways to test a DAC's operation: a *static accuracy test* and a *staircase test*.

The static test involves setting the binary input to a fixed value and measuring the analog output with a high-accuracy meter. This test is used to check that the output value falls within the expected range consistent with the DAC's specified accuracy. If it does not, there can be several possible causes. Here are some of them:

- Drift in the DAC's internal component values (e.g., resistor values) caused by temperature, aging, or some other factors. This can easily produce output values outside the expected accuracy range.

- Open connections or shorts in any of the binary inputs. This could either prevent an input from adding its weight to the analog output, or cause its weight to be permanently present in the output. This is especially hard to detect when the fault is in the less significant inputs.

- A faulty voltage reference. Since the analog output is directly dependent on V_{REF}, this could produce results that are way off. For DACs that use external reference sources, the reference voltage can easily be checked with a digital voltmeter. Many DACs have internal reference voltages which cannot be checked, except on some DACs that bring the reference voltage out to a pin of the IC.

- Excessive offset error caused by component aging or temperature. This would produce outputs that are off by a fixed amount. If the DAC has an external off-

set adjustment capability, this type of error can initially be zeroed out, but changes in operating temperature can cause the offset error to reappear.

The staircase test is used to check the monotonicity of the DAC; that is, it checks to see that the output increases step by step as the binary input is incremented as in Figure 10-3. The steps on the staircase must be of the same size, and there should be no missing steps or downward steps until full-scale is reached. This test can help detect internal or external faults that cause an input to have either no contribution or a permanent contribution to the analog output. The following example will illustrate.

—— EXAMPLE 10-12 ——————————————————————
How would the staircase waveform appear if the C input to the DAC of Figure 10-3 is open? Assume that the DAC inputs are TTL-compatible.

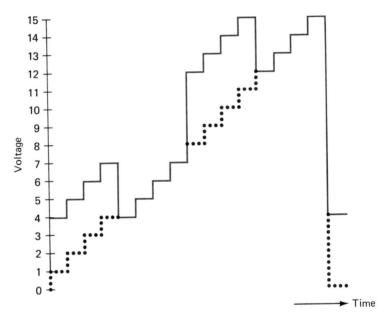

FIGURE 10-10   Example 10-12.

Discussion Question: How would the staircase appear if the D input were stuck LOW?

Solution   An open connection at C will be interpreted as a constant logic 1 by the DAC. Thus, this will contribute a constant 4 V to the DAC output so that the DAC output waveform will appear as shown in Figure 10-10. The dotted lines are the staircase as it would appear if the DAC were working correctly. Note that the faulty output waveform matches the correct one during those times when the bit C input would normally be HIGH.

10-7   ANALOG-TO-DIGITAL CONVERSION

An *analog-to-digital converter* takes an analog input voltage and after a certain amount of time produces a digital output code which represents the analog input. The A/D conversion process is generally more complex and time-consuming than

FIGURE 10-11   General diagram of one class of ADCs.

the D/A process, and many different methods have been developed and used. We shall examine several of these methods in detail, even though it may never be necessary to design or construct ADCs (they are available as completely packaged units). However, the techniques that are used provide an insight into what factors determine an ADC's performance.

Several important types of ADC utilize a DAC as part of their circuitry. Figure 10-11 is a general block diagram for this class of ADC. The timing for the operation is provided by the input clock signal. The control unit contains the logic circuitry for generating the proper sequence of operations in response to the START COMMAND, which initiates the conversion process. The op-amp comparator has two *analog* inputs and a *digital* output that switches states, depending on which analog input is greater.

The basic operation of ADCs of this type consists of the following steps:

1. The START COMMAND pulse initiates the operation.

2. At a rate determined by the clock, the control unit continually modifies the binary number that is stored in the register.

3. The binary number in the register is converted to an analog voltage, V_{AX}, by the DAC.

4. The comparator compares V_{AX} with the analog input V_A. As long as $V_{AX} < V_A$, the comparator output stays HIGH. When V_{AX} exceeds V_A by at least an amount $= V_T$ (threshold voltage), the comparator output goes LOW and stops the process of modifying the register number. At this point, V_{AX} is a close approximation to V_A. The digital number in the register, which is the digital equivalent of V_{AX}, is also the approximate digital equivalent of V_A, within the resolution and accuracy of the system.

5. The control logic activates the end-of-conversion signal, EOC, when the conversion is complete.

Teaching Tip: Stress that the digital result that ends up in the register is the digital equivalent of V_{AX}, and V_{AX} can only get as close to V_A as the resolution of the DAC will allow. As the resolution is made smaller, V_{AX} becomes a better approximation to V_A.

The several variations of this A/D conversion scheme differ mainly in the manner in which the control section continually modifies the numbers in the register. Otherwise, the basic idea is the same, with the register holding the required digital output when the conversion process is complete.

1. What is the function of the comparator in the ADC?
2. Where is the approximate digital equivalent of V_A when the conversion is complete?
3. What is the function of the EOC signal?

10-8 DIGITAL-RAMP ADC

One of the simplest versions of the general ADC of Figure 10-11 uses a binary counter as the register and allows the clock to increment the counter one step at a time until $V_{AX} \geq V_A$. It is called a *digital-ramp ADC* because the waveform at V_{AX} is a step-by-step ramp (actually a staircase) like the one shown in Figure 10-3. It is also referred to as a *counter-type* ADC.

Figure 10-12 is the diagram for a digital-ramp ADC. It contains a counter, a DAC, an analog comparator, and a control AND gate. The comparator output serves as the active-LOW end-of-conversion signal, $\overline{EOC}$. If we assume that V_A, the analog voltage to be converted, is positive, the operation proceeds as follows:

1. A START pulse is applied to reset the counter to zero. The HIGH at START also inhibits clock pulses from passing through the AND gate into the counter.

2. With all 0s at its input, the DAC's output will be $V_{AX} = 0$ V.

FIGURE 10-12 Digital-ramp ADC.

3. Since $V_A > V_{AX}$, the comparator output, $\overline{EOC}$, will be HIGH.

4. When START returns LOW, the AND gate is enabled and clock pulses get through to the counter.

5. As the counter advances, the DAC output, V_{AX}, increases one step at a time as shown in Figure 10-12(b).

6. This continues until V_{AX} reaches a step that exceeds V_A by an amount equal to or greater than V_T (typically 10 to 100 μV). At this point, $\overline{EOC}$ will go LOW and inhibit the flow of pulses into the counter and the counter will stop counting.

7. The conversion process is now complete as signaled by the HIGH-to-LOW transition at $\overline{EOC}$, and the contents of the counter are the digital representation of V_A.

8. The counter will hold the digital value until the next START pulse initiates a new conversion.

―――― EXAMPLE 10-13A ――――

Assume the following values for the ADC of Figure 10-12: clock frequency = 1 MHz; V_T = 0.1 mV; DAC has F.S. output = 10.23 V and a 10-bit input. Determine the following values:

(a) The digital equivalent obtained for V_A = 3.728 V
(b) The conversion time
(c) The resolution of this converter

Solution

(a) The DAC has a 10-bit input and a 10.23-V F.S. output. Thus, the number of total possible steps is $2^{10} - 1$ = 1023, and so the step size is

$$\frac{10.23 \text{ V}}{1023} = 10 \text{ mV}$$

This means that V_{AX} increases in steps of 10 mV as the counter counts up from zero. Since V_A = 3.728 V and V_T = 0.1 mV, V_{AX} has to reach 3.7281 V or more before the comparator switches LOW. This will require

$$\frac{3.7281 \text{ V}}{10 \text{ mV}} = 372.81 = 373 \text{ steps}$$

At the end of the conversion, then, the counter will hold the binary equivalent of 373, which is 0101110101. This is the desired digital equivalent of V_A = 3.728 V, as produced by this ADC.

(b) Three hundred seventy-three steps were required to complete the conversion. Thus, 373 clock pulses occurred at the rate of one per microsecond. This gives a total conversion time of 373 μs.

(c) The resolution of this converter is equal to step size of the DAC, which is 10 mV. In percent it is $1/1023 \times 100\% \approx 0.1\%$.

Discussion Question: What would happen if the analog input were at a negative voltage? At a voltage of 11.0 V?

―――― EXAMPLE 10-13B ――――

For the same ADC of Example 10-13A determine the approximate range of analog input voltages that will produce the same digital result of $0101110101_2 = 373_{10}$.

Solution   The table below shows the ideal DAC output voltage, V_{AX}, for several of the steps on and around the 373rd:

Step	V_{AX}
371	3.71 V
372	3.72 V
373	3.73 V
374	3.74 V
375	3.75 V

If V_A is slightly smaller than 3.72 V (by an amount $<V_T$), then $\overline{EOC}$ won't go LOW when V_{AX} reaches the 3.72-V step, but will go LOW on the 3.73 V step. If V_A is slightly smaller than 3.73 V (by an amount $< V_T$), then $\overline{EOC}$ won't go LOW until V_{AX} reaches the 3.74-V step. Thus, as long as V_A is between approximately 3.72 V and 3.73-V, $\overline{EOC}$ will go LOW when V_{AX} reaches the 3.73-V step. The exact range of V_A values is

$$3.72 \text{ V} - V_T \text{ to } 3.73 \text{ V} - V_T$$

Discussion Question: How many different analog voltages will produce the same digital equivalent?

but since V_T is so small, we can simply say that the range is approximately 3.72 V to 3.73 V—a range equal to 10 mV, the DAC's resolution. This is illustrated in Figure 10-13.

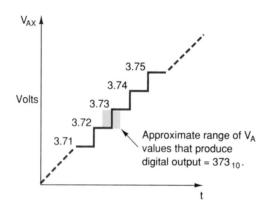

FIGURE 10-13   Example 10-13B.

A/D Resolution and Accuracy

As pointed out in Example 10-13A the resolution of the ADC is equal to the resolution of the DAC which it contains. The DAC output voltage V_{AX} is a staircase waveform that goes up in discrete steps until it exceeds V_A. Thus, V_{AX} is an approximation to the value of V_A, and the best we can expect is that V_{AX} is within 10 mV of V_A if the resolution (step size) is 10 mV. We can think of the resolution as being a built-in error that is often referred to as **quantization error.** This quantization error, which can be reduced by increasing the number of bits in the counter and DAC, is sometimes specified as an error of +1 LSB, indicating that the result could be off by as much as the weight of the LSB. In Problem 10-28 we will see how this quantization error can be modified so that it is $\pm\frac{1}{2}$LSB, which is a more common situation.

Looking at it from a different aspect, the input V_A can take on an *infinite* number of values from 0 V to F.S. The approximation V_{AX}, however, can take on only a finite number of discrete values. This means that a small range of V_A values will have

the same digital representation. In Example 10-13B we saw that values of V_A from ≈ 3.72 V to ≈ 3.73 V will require 373 steps, thereby resulting in the same digital representation. In other words, V_A must change by 10 mV (the resolution) to produce a change in digital output.

As in the DAC, *accuracy* is not related to the resolution but is dependent on the accuracy of the circuit components, such as the comparator, the DAC's precision resistors and current switches, the reference supplies, and so on. An error specification of 0.01 % F.S. indicates that the ADC result may be off by 0.01 percent of F.S., owing to nonideal components. This error is *in addition* to the quantization error due to the resolution. These two sources of error are usually of the same order of magnitude for a given ADC.

EXAMPLE 10-14

A certain 8-bit ADC has a full-scale input of 2.55 V (i.e., $V_A = 2.55$ V produces a digital output of 11111111). It has a specified error of 0.1% F.S. Determine the maximum amount by which the V_{AX} output can differ from the analog input.

Solution The step size is 2.55 V/($2^8 - 1$), which is exactly 10 mV. This means that even if the DAC has no inaccuracies, the V_{AX} output could be off by as much as 10 mV because V_{AX} can change only in 10-mV steps; this is the quantization error. The specified error of 0.1% F.S. is $0.1\% \times 2.55$ V = 2.55 mV. This means that the V_{AX} value can be off by as much as 2.55 mV because of component inaccuracies. Thus, the total possible error could be as much as 10 mV + 2.55 mV = 12.55 mV.

For example, suppose the analog input was 1.268 V. If the DAC output were perfectly accurate, the staircase would stop at the 127th step (1.27 V). But let's say V_{AX} was off by -2 mV, so it was 1.268 V at the 127th step. This would not be large enough to stop the conversion; it would stop at the 128th step. Thus, the digital output would be $10000000_2 = 128_{10}$ for an analog input of 1.268 V, an error of 12 mV.

Teaching Tip: Stress that this 12.55 mV is the maximum possible error. Most of the time, the error will be well below this.

Discussion Question: Is there any situation where the digital result can be exactly equal to the analog input? Yes there is—when the analog input voltage has a value equal to an integer multiple of the ideal step size and the DAC's full-scale error is positive.

Conversion Time, t_C The conversion time is shown in Figure 10-12(b) as the time interval between the end of the START pulse and the activation of the $\overline{\text{EOC}}$ ouput. The counter starts counting from zero and counts up until V_{AX} exceeds V_A, at which point $\overline{\text{EOC}}$ goes LOW to end the conversion process. It should be clear that the value of conversion time, t_C, depends on V_A. A larger value will require more steps before the staircase voltage exceeds V_A.

The maximum conversion time will occur when V_A is just below full scale so that V_{AX} has to go to the last step to activate $\overline{\text{EOC}}$. For an N-bit converter this will be

$$t_C(\text{max}) = 2^N - 1 \text{ clock cycles}$$

For example, the ADC in Example 10-13A would have a maximum conversion time of

$$t_C(\text{max}) = (2^{10} - 1) \times 1 \ \mu s = 1023 \ \mu s$$

Sometimes, average conversion time is specified; it is half of the maximum conversion time. For the digital-ramp converter, this would be

$$t_C(\text{avg}) = \frac{t_C(\text{max})}{2} \approx 2^{N-1} \text{ clock cyles}$$

The major disadvantage of the digital-ramp method is that conversion time essentially doubles for each bit that is added to the counter, so that resolution can be improved only at the cost of a longer t_C. This makes this type of ADC unsuitable for applications where repetitive A/D conversions of a fast-changing analog signal have

to be made. For low-speed applications, however, the relative simplicity of the digital-ramp converter is an advantage over the more complex, higher-speed ADCs.

───── EXAMPLE 10-15 ──────────────────────────────────

What will happen to the operation of a digital-ramp ADC if the analog input V_A is greater than the full-scale value?

Solution   Referring to Figure 10-12, it should be clear that the comparator output will never go LOW since the staircase voltage can never exceed V_A. Thus, pulses will be continually applied to the counter, so that the counter will repetitively count up from zero to maximum, recycle back to zero, count up, and so on. This will produce repetitive staircase waveforms at V_{AX} going from zero to full scale, and this will continue until V_A is decreased below full scale.

───── REVIEW QUESTIONS ───────────────────────────────

1. Describe the basic operation of the digital-ramp ADC.
2. Explain what is meant by "quantization error."
3. Why does conversion time increase with the value of the analog input voltage?
4. *True or false:* Everything else being equal, a 10-bit digital-ramp ADC will have a better resolution, but a longer conversion time, than an 8-bit ADC.
5. Give one advantage and one disadvantage of a digital-ramp ADC.
6. For the converter of Example 10-13, determine the digital output for $V_A = 1.345$ V. Repeat for $V_A = 1.342$ V.

10-9   DATA ACQUISITION

There are many applications in which analog data have to be *digitized* (converted to digital) and transferred into a computer's memory. The process by which the computer acquires these digitized analog data is referred to as *data acquisition*. The computer can do several different things with the data, depending on the application. In a storage application, such as digital audio recording or a digital oscilloscope, the internal microcomputer will store the data and then transfer them to a DAC at a later time to reproduce the original analog signal. In a process control application, the computer can examine the data or perform computations on them to determine what control outputs to generate.

Figure 10-14(a) shows how a microcomputer is connected to a digital-ramp ADC for the purpose of data acquisition. The computer generates the START pulses that initiate each new A/D conversion. The $\overline{\text{EOC}}$ (end-of-conversion) signal from the ADC is fed to the computer. The computer examines $\overline{\text{EOC}}$ to find out when the current A/D conversion is complete; then it transfers the digital data from the ADC output into its memory.

The waveforms in Figure 10-14(b) illustrate how the computer acquires a digital version of the analog signal, V_A. The V_{AX} staircase waveform that is generated internal to the ADC is shown superimposed on the V_A waveform for purposes of illustration. The process begins at t_0, when the computer generates a START pulse to

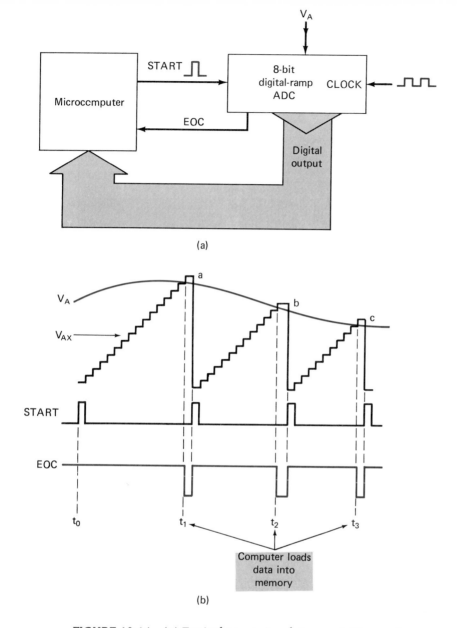

(a)

(b)

FIGURE 10-14   (a) Typical computer data acquisition system;
(b) waveforms showing how computer initiates each new conver-
sion cycle and then loads digital data into memory at end of
conversion.

start an A/D conversion cycle. The conversion is completed at t_1, when the staircase
first exceeds V_A, and $\overline{EOC}$ goes LOW. This NGT at $\overline{EOC}$ signals the computer that
the ADC has a digital output that now represents the value of V_A at point a, and the
computer will load these data into its memory.

The computer generates a new START pulse shortly after t_1 to initiate a second
conversion cycle. Note that this resets the staircase to zero and $\overline{EOC}$ back HIGH be-
cause the START pulse resets the counter in the ADC. The second conversion ends
at t_2 when the staircase again exceeds V_A. The computer then loads the digital data
corresponding to point b into its memory. These steps are repeated at t_3, t_4, and so
on.

The process whereby the computer generates a START pulse, examines $\overline{EOC}$, and loads ADC data into memory is done under the control of a program that the computer is executing. This data acquisition program will determine how many data points from the analog signal will be stored in the computer memory.

Reconstructing a Digitized Signal   In Figure 10-14(b) the ADC is operating at its maximum speed since a new START pulse is generated immediately after the computer acquires the ADC output data from the previous conversion. Note that the conversion times are not constant, because the analog input value is changing. The computer will store the digital data obtained from the various conversions so that it has a digitized version of the analog signal in its memory. For example, the digital data for the points a, b, and c might look like this:

Point	Actual Voltage (V)	Digital Equivalent
a	1.74	10101110
b	1.47	10010011
c	1.22	01111010

FIGURE 10-15   (a) Digitizing an analog signal; (b) reconstructing the signal from the digital data.

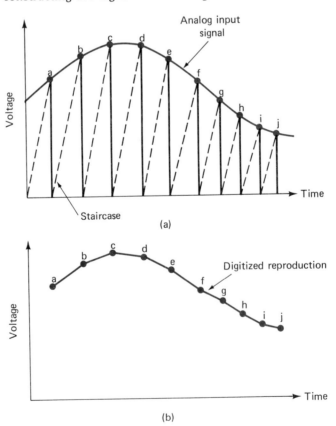

These digital data are often used to later reconstruct an approximation to the original analog signal. For example, in digital storage oscilloscopes the stored digital values are fed to a DAC to produce analog voltages that move the electron beam vertically as it is being swept horizontally by a time-base signal. The result is as if the analog signal were reconstructed by drawing straight lines (vectors) from each digitized point to the next. This is illustrated in Figure 10-15.

In Figure 10-15(a) we see how the ADC continually performs conversions to digitize the analog signal at points *a*, *b*, *c*, *d*, and so on. If these digital data are used to reconstruct the signal, the result will look like that in Figure 10-15(b). We can see that it is a fairly good reproduction of the original analog signal. This is because the analog signal does not make any rapid changes between digitized points. If the analog signal contained higher-frequency variations, the ADC would not be able to follow the variations, and the reproduced version would be much less accurate. For this reason it is important to keep the ADC conversion time short enough so that the analog signal does not change significantly between conversions. This emphasizes the need for ADCs with shorter conversion times than the simple digital-ramp converters. We will examine much faster ADC methods in the next sections.

Teaching Tip: In theory, the conversion time should meet the following criterion if the reproduced signal is to be an accurate replica of the original:

$$t_C < \frac{1}{2 \times f_H}$$

where f_H is the highest-frequency component of the analog signal.

REVIEW QUESTIONS

1. What is the meaning of the term "digitizing a signal"?
2. Describe the steps in a computer data acquisition process.

10-10 SUCCESSIVE-APPROXIMATION ADC

The successive-approximation converter is one of the most widely used types of ADC. It has more complex circuitry than the digital-ramp ADC but a much shorter conversion time. In addition, successive-approximation converters (SACs) have a fixed value of conversion time that is not dependent on the value of the analog input.

The basic arrangement, shown in Figure 10-16(a), is similar to the digital-ramp ADC. The SAC, however, does not use a counter to provide the input to the DAC block but uses a register instead. The control logic modifies the contents of the register bit by bit until the register data are the digital equivalent of the analog input V_A within the resolution of the converter. The basic sequence of operation is given by the flowchart in Figure 10-16(b). We will follow this flowchart as we go through the example illustrated in Figure 10-17.

For this example we have chosen a simple 4-bit converter with a step size of 1 V. Even though most practical SACs would have more bits and smaller resolution than our example, the operation will be exactly the same. At this point you should be able to determine that the four register bits feeding the DAC have weights of 8, 4, 2, and 1 V, respectively.

Let's assume that the analog input is $V_A = 10.4$ V. The operation begins with the control logic clearing all of the register bits to 0 so that $Q_3 = Q_2 = Q_1 = Q_0 = 0$. We will express this as $[Q] = 0000$. This makes the DAC output $V_{AX} = 0$ V, as indicated at time t_0 on the timing diagram in Figure 10-17. With $V_{AX} < V_A$, the comparator output is HIGH.

At the next step (time t_1), the control logic sets the MSB of the register to 1 so that $[Q] = 1000$. This produces $V_{AX} = 8$ V. Since $V_{AX} < V_A$, the COMP output is

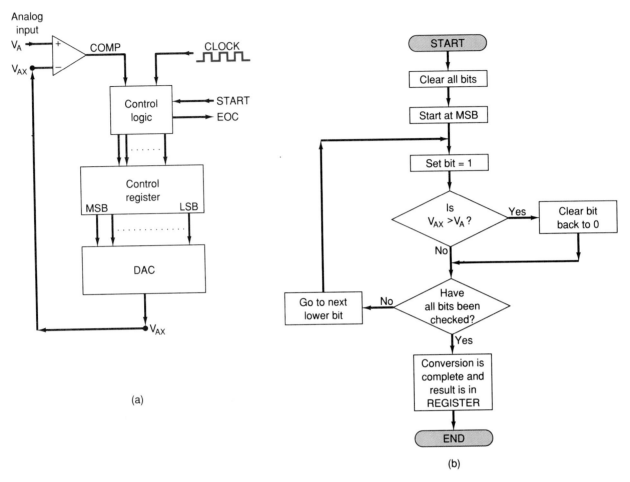

(a)

(b)

FIGURE 10-16 Successive-approximation ADC: (a) Simplified block diagram; (b) flowchart of operation.

FIGURE 10-17 Illustration of 4-bit SAC operation using DAC step size of 1 V and $V_A = 10.4$ V.

still HIGH. This HIGH tells the control logic that the setting of the MSB did not make V_{AX} exceed V_A, so that the MSB is kept at 1.

Discussion Question: Why do you think the process starts with the MSB rather than the LSB?

The control logic now proceeds to the next lower bit, Q_2. It sets Q_2 to 1 to produce $[Q] = 1100$ and $V_{AX} = 12$ V at time t_2. Since $V_{AX} > V_A$, the COMP output goes LOW. This LOW signals the control logic that the value of V_{AX} is too large, and the control logic then clears Q_2 back to 0 at t_3. Thus, at t_3, the register contents are back to 1000 and V_{AX} is back to 8 V.

The next step occurs at t_4, where the control logic sets the next lower bit Q_1 so that $[Q] = 1010$ and $V_{AX} = 10$ V. With $V_{AX} < V_A$, COMP is HIGH and tells the control logic to keep Q_1 set at 1.

The final step occurs at t_5, where the control logic sets the next lower bit Q_0 so that $[Q] = 1011$ and $V_{AX} = 11$ V. Since $V_{AX} > V_A$, COMP goes LOW to signal that V_{AX} is too large, and the control logic clears Q_0 back to 0 at t_6.

At this point, all of the register bits have been processed, the conversion is complete, and the control logic activates its $\overline{EOC}$ output to signal that the digital equivalent of V_A is now in the register. For this example, the digital output for $V_A = 10.4$ V is $[Q] = 1010$. Notice that 1010 is actually equivalent to 10 V, which is *less than* the analog input; this is a characteristic of the successive-approximation method. Recall that in the digital-ramp method the digital output was always equivalent to a voltage that was on the step above V_A.

—— EXAMPLE 10-16 ——

An 8-bit SAC has a resolution of 20 mV. What will its digital output be for an analog input of 2.17 V?

Solution

$$2.17 \text{ V}/20 \text{ mV} = 108.5$$

so that step 108 would produce $V_{AX} = 2.16$ V and step 109 would produce 2.18 V. The SAC always produces a final V_{AX} that is at the step *below* V_A. Therefore, for the case of $V_A = 2.17$ V, the digital result would be $108_{10} = 01101100_2$.

Conversion Time   In the operation just described, the control logic goes to each register bit, sets it to 1, decides whether or not to keep it at 1, and goes on to the next bit. The processing of each bit takes one clock cycle, so that the total conversion time for an N-bit SAC will be N clock cycles. That is,

$$t_C \text{ for SAC} = N \times 1 \text{ clock cycle}$$

This conversion time will be the same *regardless of the value of V_A*. This is because the control logic has to process each bit to see whether a 1 is needed or not.

—— EXAMPLE 10-17 ——

Compare the maximum conversion times of a 10-bit digital-ramp ADC and a 10-bit successive-approximation ADC if both utilize a 500-kHz clock frequency.

Solution   For the digital-ramp converter, the maximum conversion time is

$$(2^N - 1) \times (1 \text{ clock cycle}) = 1023 \times 2 \text{ } \mu s = 2046 \text{ } \mu s$$

For a 10-bit successive-approximation converter, the conversion time is always 10 clock periods or

$$10 \times 2 \text{ } \mu s = 20 \text{ } \mu s$$

Thus, it is about 100 times faster than the digital-ramp converter.

Since SACs have relatively fast conversion times, their use in data acquisition applications will permit more data values to be acquired in a given time interval. This can be very important when the analog data are changing at a relatively fast rate.

Because many SACs are available as ICs, it is rarely necessary to design the control logic circuitry, and so we will not cover it here. For those who are interested in the details of the control logic, many manufacturers' data books should provide sufficient detail.

An Actual IC: The ADC0804 Successive-Approximation ADC ADCs are available from several IC manufacturers with a wide range of operating characteristics and features. We will take a look at one of the more popular devices to get an idea of what is actually used in system applications. Figure 10-18 is the pin layout for the ADC0804, which is a 20-pin CMOS IC that performs A/D conversion using the successive-approximation method. Some of its important characteristics are as follows:

- It has two analog inputs: $V_{IN}(+)$ and $V_{IN}(-)$ to allow **differential** inputs. In other words, the actual analog input, V_{IN}, is the difference in the voltages applied to these pins [analog $V_{IN} = V_{IN}(+) - V_{IN}(-)$]. In single-ended measurements, the analog input is applied to $V_{IN}(+)$, while $V_{IN}(-)$ is connected to analog ground. During normal operation, the converter uses $V_{CC} = +5$ V as its reference voltage, and the analog input can range from 0 to 5 V full scale.
- It converts the analog input voltage to an 8-bit digital output. The digital outputs are tristate buffered so that they can be easily connected in a data-bus arrangement. With 8 bits, the resolution is 5 V/255 = 19.6 mV.
- It has an internal clock generator circuit that produces a frequency of

FIGURE 10-18   ADC0804 eight-bit successive-approximation ADC with tristate outputs. The numbers in parentheses are the IC's pin numbers.

$f = 1/(1.1RC)$, where R and C are values of externally connected components. A typical clock frequency is 606 kHz using $R = 10$ kΩ and $C = 150$ pF. An external clock signal can be used, if desired, by connecting it to the CLK IN pin.

- Using a 606-kHz clock frequency, the conversion time is approximately 100 μs.

- It has separate ground connections for digital and analog voltages. Pin 8 is the analog ground that is connected to the common reference point of the analog circuit that is generating the analog voltage. Pin 10 is the digital ground that is the one used by all the digital devices in the system. (Note the different symbols used for the different grounds.) The digital ground is inherently noisy because of the rapid current changes that occur as digital devices change states. Although it is not necessary to use a separate analog ground, doing so ensures that the noise from digital ground is prevented from causing premature switching of the analog comparator inside the ADC.

This IC is designed to be easily interfaced to a microprocessor data bus. For this reason, the names of some of the ADC0804 inputs and outputs are based on functions that are common to microprocessor-based systems. The functions of these inputs and outputs are defined as follows:

$\overline{CS}$ **(Chip Select)** This input has to be in its active-LOW state for the $\overline{RD}$ or $\overline{WR}$ inputs to have any effect. With $\overline{CS}$ HIGH, the digital outputs are in the Hi-Z state, and no conversions can take place.

$\overline{RD}$ **(Output Enable)** This input is used to enable the digital output buffers. With $\overline{CS} = \overline{RD} =$ LOW, the digital output pins will have logic levels representing the results of the *last* A/D conversion.

$\overline{WR}$ **(Start Conversion)** A LOW pulse is applied to this input to signal the start of a new conversion.

$\overline{INTR}$ **(End of Conversion)** This output signal will go HIGH at the start of a conversion, and will return LOW to signal the end of conversion.

$V_{ref}/2$ This is an optional input that can be used to reduce the internal reference voltage and thereby change the analog input range that the converter can handle. When this input is unconnected, it sits at 2.5 V ($V_{CC}/2$), since V_{CC} is being used as the reference voltage. By connecting an external voltage to this pin, the internal reference is changed to twice that voltage, and the analog input range is changed accordingly. The table below shows this.

$V_{ref}/2$	Analog Input Range (V)	Resolution (mV)
Open	0–5	19.6
2.25	0–4.5	17.6
2.0	0–4	15.7
1.5	0–3	11.8

CLK OUT A resistor is connected to this pin to use the internal clock. The clock signal appears on this pin.

CLK IN Used for external clock input, or for capacitor connection when internal clock is used.

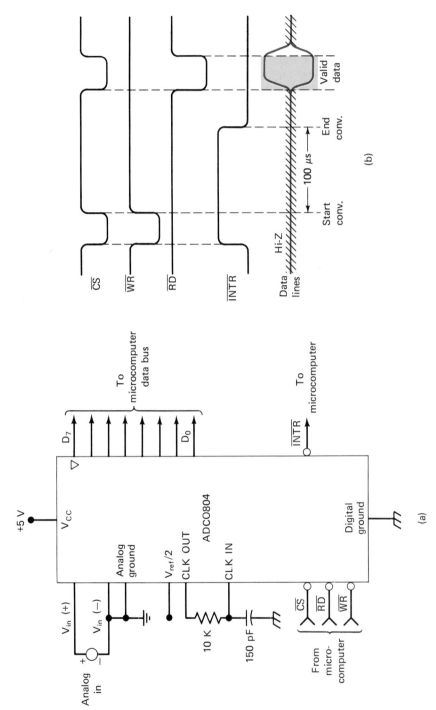

FIGURE 10-19 (a) Typical connections for interfacing ADC0804 to a microcomputer; (b) typical signals during data acquisition process.

Figure 10-19(a) shows a typical connection of the ADC0804 to a microcomputer in a data acquisition application. The microcomputer controls when a conversion is to take place by generating the $\overline{CS}$ and $\overline{WR}$ signals. It then acquires the ADC output data by generating the $\overline{CS}$ and $\overline{RD}$ signals after detecting a NGT at $\overline{INTR}$ indicating the end of conversion. The waveforms in Figure 10-19(b) show the signal activity during the data acquisition process. Note that $\overline{INTR}$ goes HIGH when $\overline{CS}$ and $\overline{WR}$ go LOW, but the conversion process does not begin until $\overline{CS}$ and $\overline{WR}$ return HIGH. Also note that the ADC output data lines are in their Hi-Z state until the microcomputer activates $\overline{CS}$ and $\overline{RD}$; at that point the ADC's data buffers are enabled so that the ADC data are sent to the microcomputer over the data bus. The data lines return to the Hi-Z state when $\overline{CS}$ and $\overline{RD}$ are returned HIGH.

Teaching Tip: At this point you may want to discuss the benefit of using differential inputs. One situation would be where the analog signal is not referenced to ground, such as the output of a bridge circuit. Another would be in high-noise environments where the common-mode noise rejection is needed.

REVIEW QUESTIONS

1. What is the main advantage of a SAC over a digital-ramp ADC?
2. What is its principal disadvantage compared with the digital-ramp converter?
3. *True or false:* The conversion time for a SAC increases as the analog voltage increases.
4. Answer the following concerning the ADC0804.
   (a) What is its resolution in bits?
   (b) What is the normal analog input voltage range?
   (c) Describe the functions of the $\overline{CS}$, $\overline{WR}$, and $\overline{RD}$ inputs.
   (d) What is the function of the $\overline{INTR}$ output?
   (e) Why does it have two separate grounds?

10-11 FLASH ADCs

The flash converter is the highest-speed ADC available, but it requires much more circuitry than the other types. For example, a 6-bit flash ADC requires 63 analog comparators, while an 8-bit unit requires 255 comparators, and a 10-bit converter requires 1023 comparators. The large number of comparators has limited the size of flash converters. IC flash converters are currently available in 2-bit to 8-bit units, and most manufacturers predict that 9- and 10-bit units will hit the market in the near future.

The principle of operation will be described for a 3-bit flash converter in order to keep the circuitry at a workable level. Once the 3-bit converter is understood, it should be easy to extend the basic idea to higher-bit flash converters.

The flash converter in Figure 10-20(a) has a 3-bit resolution and a step size of 1 V. The voltage divider sets up reference levels for each comparator so that there are seven levels corresponding to 1 V (weight of LSB), 2 V, 3 V, . . . , and 7 V (full-scale). The analog input, V_A, is connected to the other input of each comparator.

With $V_A < 1$ V, all the comparator outputs C_1 through C_7 will be HIGH. With $V_A > 1$ V, one or more of the comparator outputs will be LOW. The comparator outputs are fed into an active-LOW priority encoder that generates a binary output corresponding to the highest-numbered comparator output that is LOW. For example, when V_A is between 3 and 4 V, outputs C_1, C_2, and C_3 will be LOW and all others

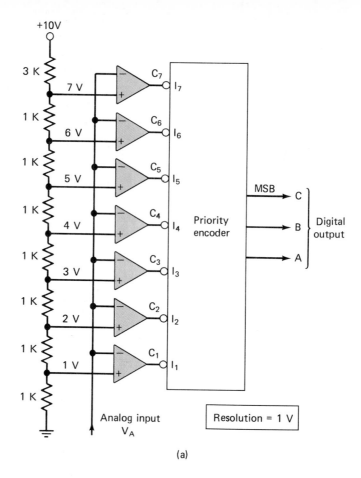

(a)

Analog in	Comparator outputs							Digital output		
V_A	C_1	C_2	C_3	C_4	C_5	C_6	C_7	C	B	A
0–1 V	1	1	1	1	1	1	1	0	0	0
1–2 V	0	1	1	1	1	1	1	0	0	1
2–3 V	0	0	1	1	1	1	1	0	1	0
3–4 V	0	0	0	1	1	1	1	0	1	1
4–5 V	0	0	0	0	1	1	1	1	0	0
5–6 V	0	0	0	0	0	1	1	1	0	1
6–7 V	0	0	0	0	0	0	1	1	1	0
> 7 V	0	0	0	0	0	0	0	1	1	1

(b)

FIGURE 10-20   (a) Three-bit flash ADC; (b) truth table.

will be HIGH. The priority encoder will respond only to the LOW at C_3 and will produce a binary output $CBA = 011$, which represents the digital equivalent of V_A, within the resolution of 1 V. When V_A is greater than 7 V, C_1 to C_7 will all be LOW, and the encoder will produce $CBA = 111$ as the digital equivalent of V_A. The table in Figure 10-20(b) shows the responses for all possible values of analog input.

The flash ADC of Figure 10-20 has a resolution of 1 V because the analog input has to change by 1 V in order to bring the digital output to its next step. To achieve finer resolutions, we would have to increase the number of input voltage levels (i.e., use more voltage-divider resistors) and the number of comparators. For example, an 8-bit flash converter would require $2^8 = 256$ voltage levels including 0 V.

This would require 256 resistors and 255 comparators (there is no comparator for the 0-V level). The 255 comparator outputs would feed a priority encoder circuit that would produce an 8-bit code corresponding to the highest-order comparator output that is LOW. In general, an N-bit flash converter would require $2^N - 1$ comparators, 2^N resistors, and the necessary encoder logic.

Discussion Question: Why must a priority encoder be used in a flash ADC?

Conversion Time   The flash converter uses no clock signal, because there is no timing or sequencing required. The conversion takes place continuously. When the value of analog input changes, the comparator outputs will change, thereby causing the encoder outputs to change. The conversion time is the time it takes for a new digital output to appear in response to a change in V_A, and depends only on the propagation delays of the comparators and encoder logic. For this reason, flash converters have extremely short conversion times. For example, the AD9002 from Analog Devices is an 8-bit flash ADC which has a rated conversion time of less than 10 ns.

REVIEW QUESTIONS

1. *True or false:* A flash ADC does not contain a DAC.
2. How many comparators would a 12-bit flash converter require? How many resistors?
3. State the major advantage and disadvantage of a flash converter.

10-12   OTHER A/D CONVERSION METHODS

There are several other methods of A/D conversion that have been in use for some time, each with its relative advantages and disadvantages. We will briefly describe some of them now.

Up/Down Digital-Ramp ADC (Tracking ADC)   As we have seen, the digital-ramp ADC is relatively slow because the counter is reset to zero at the start of each new conversion. The staircase always begins at 0 V and steps its way up to the "switching point" where V_{AX} exceeds V_A and the comparator output switches LOW. The time it takes the staircase to reset to zero and step back up to the new switching point is really wasted. The **up/down digital-ramp ADC** uses an up/down counter to reduce this wasted time.

The up/down counter replaces the up counter that feeds the DAC. It is designed to count up whenever the comparator output indicates $V_{AX} < V_A$, and to count down whenever $V_{AX} > V_A$. Thus, the DAC output is always being stepped in the direction of the V_A value. Each time the comparator output switches states, it indicates that V_{AX} has "crossed" the V_A value, the digital equivalent of V_A is in the counter, and the conversion is complete.

When a new conversion is to begin, the counter is *not reset to zero,* but begins counting either up or down from its last value, depending on the comparator output. It will count until the staircase crosses V_A again to end the conversion. The V_{AX} waveform, then, will contain both positive-going and negative-going staircase signals that "track" the V_A signal. For this reason, this ADC is often called a **tracking ADC.**

Clearly, the conversion times will generally be reduced because the counter does not start over from zero each time, but simply counts up or down from its previous value. Of course, the value of t_C will still depend on the value of V_A, and so will not be constant.

Voltage-to-Frequency ADC

The voltage-to-frequency ADC is simpler than other ADCs because it does not use a DAC. Instead it uses a *linear voltage-controlled oscillator (VCO)* that produces an output frequency that is proportional to its input voltage. The analog voltage that is to be converted is applied to the VCO to generate an output frequency. This frequency is fed to a counter to be counted for a fixed time interval. The final count is proportional to the value of the analog voltage.

To illustrate, suppose the VCO generates a frequency of 10 kHz for each volt of input (i.e., 1 V produces 10 kHz, 1.5 V produces 15 kHz, 2.73 V produces 27.3 kHz). If the analog input voltage is 4.54 V, then the VCO output will be a 45.4-kHz signal that clocks a counter for, say, 10 ms. After the 10-ms counting interval, the counter will hold the count of 454, which is the digital representation of 4.54 V.

Although this is a simple method of conversion, it is difficult to achieve a high degree of accuracy because of the difficulty in designing VCOs with accuracies of better than 0.1 percent.

Teaching Tip: This is much preferable to transmitting the parallel digital outputs from a digital-ramp or successive-approximation ADC, which would require transmitting many more signals. The trade-off is that the receiver has to convert the VCO signal back to digital.

One of the main applications of this type of converter is in noisy industrial environments where small analog signals have to be transmitted from transducer circuits to a control computer. The small analog signals can be drastically affected by noise if they are directly transmitted to the control computer. A better approach is to feed the analog signal to a VCO, which generates a digital signal whose output frequency changes according to the analog input. This digital signal is transmitted to the computer and will be much less affected by noise. Circuitry in the control computer will count the digital pulses (i.e., perform a frequency-counting function) to produce a digital value equivalent to the original analog input.

Dual-Slope ADC

The dual-slope converter has one of the slowest conversion times (typically 10 to 100 ms) but has the advantage of relatively low cost because it does not require precision components such as a DAC or VCO. The basic operation of this converter involves the *linear* charging and discharging of a capacitor using constant currents. First, the capacitor is charged up for a fixed time interval from a constant current derived from the analog input voltage, V_A. Thus, at the end of this fixed charging interval, the capacitor voltage will be proportional to V_A. At that point, the capacitor is linearly discharged from a constant current derived from a precise reference voltage, V_{ref}. When the capacitor voltage reaches zero, the linear discharging is terminated. During the discharge interval, a digital reference frequency is fed to a counter and counted. The duration of the discharge interval will be proportional to the initial capacitor voltage. Thus, at the end of the discharge interval, the counter will hold a count proportional to the initial capacitor voltage, which, as we said, is proportional to V_A.

In addition to its low cost, another advantage of the dual-slope ADC is its low sensitivity to noise and to variations in its component values caused by temperature changes. Because of its slow conversion times, the dual-slope ADC is not used in any data acquisition applications. The slow conversion times, however, are not a problem in applications such as digital voltmeters or multimeters, and this is where they find their major application.

1. How does the up/down digital-ramp ADC improve upon the digital-ramp ADC?
2. What is the main element of a voltage-to-frequency ADC?
3. Cite two advantages and one disadvantage of the dual-slope ADC.
4. Name three types of ADC that do not use a DAC.

10-13 DIGITAL VOLTMETER

A digital voltmeter (DVM) converts an analog voltage to its BCD-code representation, which is then decoded and displayed on some type of readout. Figure 10-21 shows a 3-digit DVM circuit that uses a digital-ramp ADC (shown inside the dashed

FIGURE 10-21 Continuous-conversion DVM using a digital-ramp ADC.

567

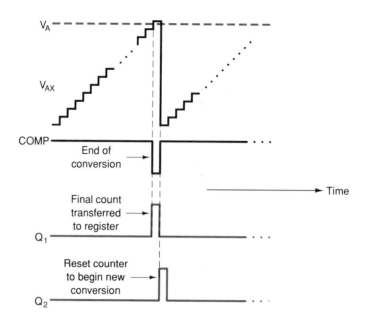

FIGURE 10-22   Waveforms for DVM.

lines). Three cascaded BCD counters provide the inputs to a 3-digit BCD-to-analog converter that has a step size of 10 mV and a full-scale output of 9.99 V. Each BCD counter stage also drives a 4-bit register that feeds a decoder/driver and display. The contents of the BCD counters are transferred to the registers at the end of each conversion cycle, so that the displays do not show the counters resetting and counting, but only display the final count that represents the unknown voltage.

As long as $V_{AX} < V_A$, the COMP output stays HIGH and clock pulses pass through the AND gate to the counter. As the counter increments, the V_{AX} signal increases at 10 mV per step until it exceeds V_A (see Figure 10-22). At that point COMP goes LOW to disable the AND gate and stop the counter, thereby ending the conversion. The NGT of COMP also triggers one-shot OS1 to produce a 1-μs pulse at Q_1. The PGT of Q_1 transfers the outputs of the BCD counter stages to their respective registers to be stored and displayed. The NGT of Q_1 triggers a second one-shot, OS2, to produce a pulse that resets all the counters to zero. This brings V_{AX} back to zero, and COMP returns HIGH, allowing pulses into the counter to begin a new conversion cycle.

Thus, this DVM will perform one conversion right after another. Of course, the storage registers will keep the displays from showing the conversion process. The display readings will change only if V_A changes, so that different counter contents are transferred to the registers at the end of the conversion cycle.

A numerical example will help illustrate this circuit's operation. Assume that V_A is 6.372 V. In order for the COMP output to switch LOW, V_{AX} must exceed 6.3721 V. Since the DAC output increases by 10 mV/step, this requires

<div style="float: left">*Discussion Question:* How can the operation of this DVM be improved by using a tracking ADC?</div>

$$\frac{6.3721 \text{ V}}{10 \text{ mV}} = 637.21 \longrightarrow 638 \text{ steps}$$

Thus, the counters will count up to 638, which will be transferred to the registers and displayed. A small LED can be used to display a decimal point so that the operator sees 6.38 V.

—— EXAMPLE 10-18 ——

What will happen if V_A is greater than 9.99 V?

Solution   With $V_A > 9.99$ V, the comparator output will stay HIGH, allowing clock pulses into the counter continuously. The counter will repetitively count up to 999 and recycle to 000. One-shot Q_1 *will never* be triggered and the register contents will not change from their previous value. A well-designed DVM would have some means of detecting this *over-range* condition and activating some type of over-range indicator. One possible method would simply add a thirteenth bit to the counter string. This bit would go HIGH only if the counter recycled from 999 to 000, and would indicate the over-range condition.

The DVM can be modified to read input voltages over several ranges by using a suitable amplifier or attenuator between V_A and the comparator. For example, if V_A were 63.72 V, it could be attenuated by a factor of 10, so that the comparator would receive 6.372 V at its + input and the counters would display 638 at the end of the conversion. The decimal-point indicator would be placed in front of the LSD, so that the display would read 63.8 V.

The DVM can be modified to operate as a DMM that can measure current and resistance as well as voltage. To measure current, the unknown current is made to flow through a fixed reference resistor to produce a voltage. To measure resistance, a fixed reference current is made to flow through the unknown resistance to develop a voltage. In each case, the resulting voltage is fed to a circuit like that of Figure 10-21 to be measured and displayed. The circuit may require an attenuation scaling factor so that the displayed value correctly represents the quantity (current or resistance) being measured.

AC voltages can be measured by this same DVM if they are first converted to dc. This is usually done with a precision peak-value-detector circuit. Again, a scaling factor may be required in order to convert the peak value to an rms value.

—— REVIEW QUESTIONS ——

1. What is the purpose of the registers in the DVM circuit?
2. What is the purpose of the two one-shots?

10-14   SAMPLE-AND-HOLD CIRCUITS

When an analog voltage is connected directly to the input of an ADC, the conversion process can be adversely affected if the analog voltage is changing during the conversion time. The stability of the conversion process can be improved by using a *sample-and-hold circuit* to hold the analog voltage constant while the A/D conversion is taking place. A simplified diagram of a sample-and-hold (S/H) circuit is shown in Figure 10-23.

The S/H circuit contains a unity-gain buffer amplifier A_1 that presents a high impedance to the analog signal and has a low output impedance that can rapidly

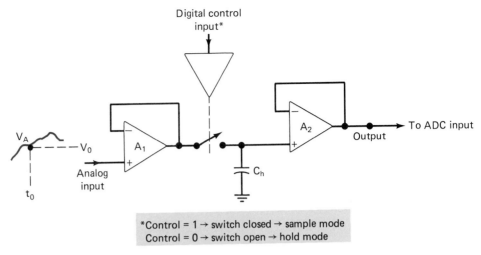

*Control = 1 → switch closed → sample mode
Control = 0 → switch open → hold mode

FIGURE 10-23 Simplified diagram of a sample-and-hold circuit.

Teaching Tip: You may
want to discuss how
the ON and OFF resis-
tances of the switch af-
fect the operation of
the S/H circuit.

charge the hold capacitor, C_h. The capacitor will be connected to the output of A_1 when the digitally controlled switch is closed. This is called the **sample** operation. The switch will be closed long enough for C_h to charge to the current value of the analog input. For example, if the switch is closed at time t_0, the A_1 output will quickly charge C_h up to a voltage V_0. When the switch opens, C_h will **hold** this voltage so that the output of A_2 will apply this voltage to the ADC. The unity-gain buffer amplifier A_2 presents a high input impedance that will not discharge the capacitor voltage appreciably during the conversion time of the ADC, and so the ADC will essentially receive a dc input voltage V_0.

In a computer-controlled data acquisition system such as the one discussed earlier, the sample-and-hold switch would be controlled by a digital signal from the computer. The computer signal would close the switch in order to charge C_h to a new sample of the analog voltage; the amount of time the switch would have to remain closed is called the **acquisition time** and it depends on the value of C_h and the characteristics of the S/H circuit. The LF198 is an integrated S/H circuit that has a typical acquisition time of 4 μs for $C_h = 1000$ pF, and 20 μs for $C_h = 0.01$ μF. The computer signal would then open the switch to allow C_h to hold its value and provide a relatively constant analog voltage at the A_2 output. For example, with the LF198, the capacitor voltage will typically discharge at the rate of only 30 mV per second for a 1000-pF capacitor.

___ REVIEW QUESTIONS ___

1. Describe the function of a sample-and-hold circuit.
2. *True or false:* The amplifiers in the S/H circuit are used to provide voltage amplification.

10-15 MULTIPLEXING

When analog inputs from several sources are to be converted, a multiplexing technique can be used so that one ADC may be time-shared. The basic scheme is illustrated in Figure 10-24 for a three-channel acquisition system. Switches S_1, S_2, and S_3

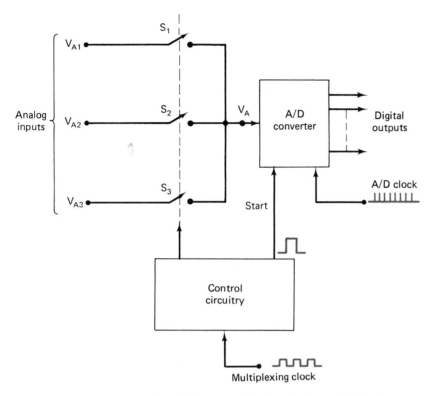

FIGURE 10-24   Conversion of three analog inputs by multiplexing through one ADC.

are used to switch each analog signal sequentially to the ADC. The control circuitry controls the actuation of these switches, which are usually semiconductor switches, so that only one switch is closed at a time. The control circuitry also generates the START pulse for the ADC. The operation proceeds in the following manner:

1. The control circuit closes S_1, which connects V_{A1} to the ADC input.
2. A START pulse is generated and the ADC converts V_{A1} to its digital equivalent. S_1 stays closed long enough to allow the conversion to be completed.
3. The ADC outputs representing V_{A1} can now be transferred to another location. In many cases this would be to the memory of a computer.
4. The control circuit opens S_1 and closes S_2 to connect V_{A2} to the ADC input.
5. Steps 2 and 3 are repeated.
6. S_2 is open and S_3 is closed to connect V_{A3} to the ADC input.
7. Steps 2 and 3 are repeated.

The multiplexing clock controls the rate at which the analog signals are sequentially switched into the ADC. The maximum rate is determined by the delay time of the switches and the conversion time of the ADC. The switch delay time can be minimized by using semiconductor switches such as the CMOS bilateral switch described in Chapter 8. It may be necessary to connect a sample-and-hold circuit at the input of the ADC if the analog inputs will change significantly during the ADC conversion time.

Discussion Question: What are the trade-offs when multiplexing is used instead of a separate ADC for each analog signal?

Many integrated ADCs contain the multiplexing circuitry on the same chip as the ADC. The ADC0808, for example, can multiplex eight different analog inputs into one ADC. It uses a 3-bit select input code to determine which analog input is connected to the ADC.

10-16 DIGITAL STORAGE OSCILLOSCOPE (DSO)

As a final example of the application of D/A and A/D converters, we will take a brief look at the digital storage oscilloscope. The DSO uses both of these devices to digitize, store, and display analog waveforms. A DSO has many advantages over the conventional storage oscilloscope (CSO), which stores the waveform image as electrical charges on a mesh screen between the CRT's electron gun and the phosphor-coated screen. Here are some of them:

- The DSO can store the waveforms for an indefinite time because the digital values are stored in digital memory (i.e., flip-flops). In a CSO, the image gradually fades as the charges on the mesh screen are slowly neutralized.

- In a DSO, the stored waveform can be positioned anywhere on the CRT screen, and can have its vertical and horizontal scales changed to suit the measurement situation. This is not possible with a CSO.

- With a DSO, it is possible to store and display portions of the waveform that occur *before* the occurrence of the trigger point. Again, this cannot be done with a CSO.

A block diagram of a DSO is shown in Figure 10-25. The overall operation is

FIGURE 10-25 Block diagram of a digital storage oscilloscope.

controlled and synchronized by the circuits in the CONTROL block, which usually contains a microprocessor executing a control program stored in ROM (read-only memory). The data acquisition portion of the system contains a sample-and-hold (S/H) and an ADC that repetitively samples and digitizes the input signal at a rate determined by the SAMPLE CLOCK, and sends the digitized data to memory for storage. The CONTROL block makes sure that successive data points are stored in successive memory locations by continually updating the memory's ADDRESS COUNTER.

When memory is full, the next data point from the ADC is stored in the first memory location, writing over the old data, and so on for successive data points. This data acquisition and storage process continues until the CONTROL block receives a trigger signal from either the input waveform (INTERNAL trigger) or an EXTERNAL trigger source. When the trigger occurs, the system stops acquiring new data and enters the display mode of operation, in which all or part of the memory data is repetitively displayed on the CRT.

The display operation uses two DACs to provide the vertical and horizontal deflection voltages for the CRT. Data from memory produce the vertical deflection of the electron beam, while the TIME-BASE COUNTER provides the horizontal deflection in the form of a staircase sweep signal. The CONTROL block synchronizes the display operation by incrementing the memory ADDRESS COUNTER and the TIME-BASE COUNTER at the same time so that each horizontal step of the electron beam is accompanied by a new data value from memory to the vertical DAC. The counters are continuously recycled so that the stored data points are repetitively replotted on the CRT screen. The screen display consists of discrete dots representing the various data points, but the number of dots is usually so large (typically 1000 or more) that they tend to blend together and appear to be a smooth continuous waveform.

The display operation is terminated when the operator presses a front-panel button that commands the DSO to begin a new data acquisition cycle.

Teaching Tip: Actually, the data acquisition may continue after the trigger event occurs if the DSO is set for the POST TRIGGER mode. In this mode, a predetermined number of samples of the input signal are taken after the trigger event. In this way, the display will show portions of the signal both before and after the trigger point. This is something that a conventional storage scope cannot do.

Teaching Tip: The rate at which data are sent from memory to the DAC to be displayed can be, and usually is, different from the rate at which the data were acquired (sample rate). In this way, slow-changing signals can be sampled at a very slow rate and then "played back" at a much faster rate to produce a nonflickering display.

REVIEW QUESTIONS

1. What are some of the advantages of a DSO over a CSO?
2. Describe the functions of the ADC and DACs that are part of the DSO.

PROBLEMS

SECTIONS 10-1 and 10-2

10-1. DRILL QUESTION

(a) What is the expression relating the output and inputs of a DAC?
(b) Define step size of a DAC.
(c) Define resolution of a DAC.
(d) Define full scale.

10-1. (a) analog output = $K \times$ digital input (b) the amount the analog output changes when the digital input is incremented by 1 (c) same as step size (d) maximum analog output

(e) Define percentage resolution.   (step size/full scale) × 100%

(f) *True or false:* A 10-bit DAC will have a smaller resolution than a 12-bit DAC for the same full-scale output.   false

(g) *True or false:* A 10-bit DAC with full-scale output of 10 V has a smaller percentage resolution than a 10-bit DAC with 12 V full scale.   false

10-2. An 8-bit DAC produces an output voltage of 2.0 V for an input code of 01100100. What will be the value of V_{OUT} for an input code of 10110011?   3.58 V

10-3. Determine the weight of each input bit for the DAC of Problem 10-2.

10-3. 20, 40, 80, 160, 320, 640, 1280, 2560 mV

10-4. What is the resolution of the DAC of Problem 10-2? Express it in volts and in percent.   20 mV, 0.04%

10-5. What is the resolution in volts of a 10-bit DAC whose F.S. output is 5 V?   Approx. 5 mV

10-6. How many bits are required for a DAC so that its F.S. output is 10 mA and its resolution is less than 40 μA?   eight

10-7. What is the percentage resolution of the DAC of Figure 10-26? What is the step size if the top step is 2 V?   14.3%; 0.286 V

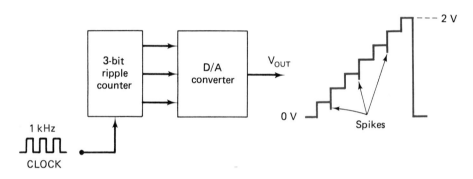

FIGURE 10-26

10-8. temporary states of ripple counter

10-8. What is the cause of the negative-going spikes on the V_{OUT} waveform of Figure 10-26? (*Hint:* Note that the counter is a ripple counter and that the spikes occur on every other step.)

10-9. Assuming a 12-bit DAC with perfect accuracy, how close to 250 rpm can the motor speed be adjusted in Figure 10-4?   250.061 RPM

10-10. 10 mV, 0.1%, 6.95 V

10-10. A 12-bit (three-digit) DAC which uses the BCD input code has a full-scale output 9.99 V. Determine the step size, the percentage resolution, and the value of V_{OUT} for an input code of 0110 1001 0101.

10-11. Binary: 3.88 mV, 0.392% BCD: 10 mV, 1.01%

10-11. Compare the step size and percentage resolution of a DAC with an 8-bit binary input with a DAC having an 8-bit BCD input. Assume 990 mV full scale for each DAC.

10-12. 10, 20, 40, 80, 100, 200, 400, 800, 1000, 2000, 4000, 8000 mV; 8.23 V

10-12. Determine the weight of each bit for the BCD DAC of Problem 10-10. What do you think the output would be if the input is 100000100011?

SECTION 10-3

10-13. 800 Ω; no

10-13. The step size of the DAC of Figure 10-6 can be changed by changing

the value of R_F. Determine the required value of R_F for a step size of 0.5 V. Will the new value of R_F change the percentage resolution?

10-14. Assume that the output of the DAC in Figure 10-8(a) is connected to the op amp of Figure 10-8(b).

    (a) With $V_{REF} = 5$ V, $R = 20$ kΩ, and $R_F = 10$ kΩ, determine the step size and full-scale voltage at V_{OUT}. -0.3125 V, -4.6875 V

    (b) Change the value of R_F so that the full-scale voltage at V_{OUT} is -2 V.   4.27 kΩ

    (c) Use this new value of R_F, and determine the proportionality factor, K, in the relationship $V_{OUT} = K(V_{REF} \times B)$. -0.0267

10-15. What is the advantage of the DAC of Figure 10-9 over that of Figure 10-8, especially for a larger number of input bits?

    10-15. Uses fewer different resistance values.

SECTIONS 10-4 and 10-5

10-16. An 8-bit DAC has a full-scale error of 0.2% F.S. If the DAC has a full-scale output of 10 mA, what is the most that it can be in error for any digital input? If the D/A output reads 50 μA for a digital input of 00000001, is this within the specified range of accuracy? (Assume no offset error.)   20 μA; yes

 10-17. The control of a positioning device may be achieved using a *servomotor*, which is a motor designed to drive a mechanical device as long as an error signal exists. Figure 10-27 shows a simple servo-controlled system which is controlled by a digital input that could be coming di-

FIGURE 10-27

rectly from a computer or from an output medium such as magnetic tape. The lever arm is moved vertically by the servomotor. The motor rotates clockwise or counterclockwise, depending on whether the voltage from the power amplifier (P.A.) is positive or negative. The motor stops when the P.A. output is zero.

The mechanical position of the lever is converted to a dc voltage by the potentiometer arrangement shown. When the lever is at its zero reference point, $V_P = 0$ V. The value of V_P increases at the rate of 1 V/inch until the lever is at its highest point (10 inches) and $V_P = 10$ V. The desired position of the lever is provided as a digital code from the computer and is then fed to a DAC, producing V_A. The *difference* between V_P and V_A (called *error*) is produced by the *differential* amplifier and amplified by the P.A. to drive the motor in the direction that causes the error signal to decrease to 0—that is, moves the lever until $V_P = V_A$.

(a) If it is desired to position the lever within a resolution of 0.1 inch, what is the number of bits needed in the digital input code? *seven*

(b) In actual operation, the lever arm might oscillate slightly around the desired position, especially if a *wire-wound* potentiometer is used. Can you explain why? *noncontinuous variation of resistance*

10-18. (a) See Figure 10-8; (b) see Figure 10-9; (c) time required for DAC output to settle within 0.5 LSB of full scale; (d) possible absolute error in analog output expressed as a percentage of full scale; (e) analog output when digital input is zeroed.

10-19. 242.5 mV not within specifications

10-18. DRILL QUESTION

(a) Define binary-weighted resistor network.
(b) Define $R/2R$ ladder network.
(c) Define DAC settling time.
(d) Define full-scale error.
(e) Define offset error.

10-19. A particular 6-bit DAC has a full-scale output of 1.260 V. Its accuracy is specified as $\pm 0.1\%$ F.S. and it has an offset error of ± 1 mV. Assume that the offset error has not been zeroed out. Consider the following measurements made on this DAC, and determine which of them are not within the device's specifications. (*Hint:* The offset error is added to the error caused by component inaccuracies.)

Input Code	Output (mV)
000010	41.5
000111	140.2
001100	242.5
111111	1.258 V

SECTION 10-6

10-20. A certain DAC has the following specifications: 8-bit resolution, full scale = 2.55 V, offset $\leq$ 2 mV; accuracy = $\pm 0.1\%$ F.S. A static test on this DAC produces the results shown below. What is the probable cause of the malfunction? *Offset out of spec*

Input Code	Output	
00000000	8	mV
00000001	18.2	mV
00000010	28.5	mV
00000100	48.3	mV
00001111	158.3	mV
10000000	1.289	V

 10-21. Repeat Problem 10-20 using the following measured data:

10-21. Bit 1 of DAC input is stuck HIGH or open.

Input Code	Output	
00000000	20.5	mV
00000001	30.5	mV
00000010	20.5	mV
00000100	60.6	mV
00001111	150.6	mV
10000000	1.300	V

10-22. A technician connects a counter to the DAC of Figure 10-3 to perform a staircase test using a 1-kHz clock. The result is shown in Figure 10-28. What is the probable cause of the incorrect staircase signal?

10-22. Bit 0 and bit 1 connections from counter to DAC have been reversed.

FIGURE 10-28

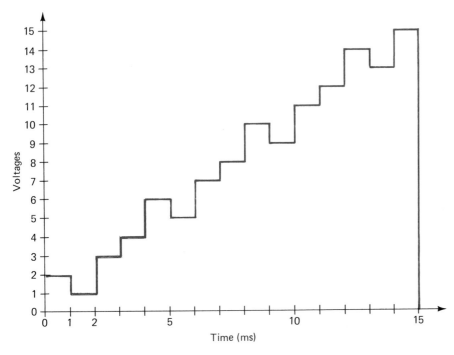

10-23. DRILL QUESTION

Fill in the blanks in the following description of the digital-ramp ADC operation. Each blank may be one or more words.

A START pulse is applied to _____reset_____ the counter and to keep _____pulses_____ from passing through the AND gate into the _____counter_____ . At this point, the DAC output, V_{AX}, is _____zero_____ and $\overline{EOC}$ is _____HIGH_____ .

When START returns _____LOW_____ , the AND gate is _____enabled_____ , and the counter is allowed to _____count_____ . The V_{AX} signal is increased one _____step_____ at a time until it _____exceeds_____ V_A. At that point, _____$\overline{EOC}$_____ goes LOW to _____prevent_____ further pulses from _____being counted_____ . This signals the end of conversion, and the digital equivalent of V_A is present at the _____counter outputs_____ .

10-24. An 8-bit digital-ramp ADC with a 40-mV resolution uses a clock frequency of 2.5 MHz and a comparator with $V_T = 1$ mV. Determine the following values:
(a) The digital output for $V_A = 6.000$ V.
(b) The digital output for 6.035 V.
(c) The maximum and average conversion times for this ADC.

10-25. Why were the digital outputs the same for parts (a) and (b) of Problem 10-24? Because the difference in the two values was less than the resolution.

10-26. What would happen in the ADC of Problem 10-24 if an analog voltage of $V_A = 10.853$ V were applied to the input? What waveform would appear at the D/A output? Incorporate the necessary logic in this ADC so that an "over-scale" indication will be generated whenever V_A is too large.

10-27. An ADC has the following characteristics: resolution, 12 bits; full-scale error, 0.03% F.S.; full scale output, +5 V.
(a) What is the quantization error in volts? 1.2 mV
(b) What is the total possible error in volts? 2.7 mV

 10-28. The quantization error of an ADC such as the one in Figure 10-12 is always positive since the V_{AX} value must exceed V_A in order for the comparator output to switch states. This means the value of V_{AX} could be as much as 1 LSB greater than V_A. This quantization error can be modified so that V_{AX} would be within $\pm 1/2$ LSB of V_A. This can be done by adding a fixed voltage equal to 1/2 LSB (1/2 step) to the value of V_{AX}. Figure 10-29 shows this symbolically for a converter that has a resolution of 10 mV/step. A fixed voltage of +5 mV is added to the D/A output in the summing amplifier and the result, V_{AY}, is fed to the comparator, which has $V_T = 1$ mV.

For this modified converter, determine the digital output for the following V_A values:
(a) $V_A = 5.022$ V   0111110110
(b) $V_A = 5.028$ V   0111110111

Determine the quantization error in each case by comparing V_{AX} and

FIGURE 10-29

V_A. Note that the error is positive in one case and negative in the other.

10-29. For the ADC of Figure 10-29, determine the range of analog input values that will produce a digital output of 0100011100.

10-29. 2.834 to 2.844 V

SECTION 10-9

10-30. Assume that the analog signal in Figure 10-30 is to be digitized by performing continuous A/D conversions using an 8-bit digital-ramp

10-30. See Solutions Manual.

FIGURE 10-30

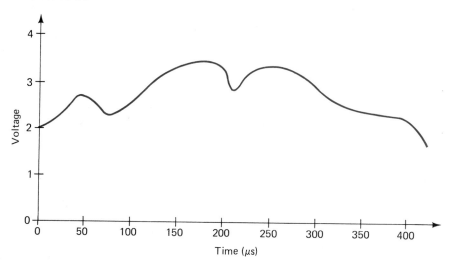

converter whose staircase rises at the rate of 1 V every 25 μs. Sketch the reconstructed signal using the data obtained during the digitizing process. Compare it with the original signal, and discuss what could be done to make it a more accurate representation.

SECTION 10-10

10-31. Digital-ramp; a, d, e, f, h;
SAC: b, c, d, e, g, h

10-31. DRILL QUESTION

Indicate whether each of the following statements refers to the digital-ramp ADC, successive-approximation ADC, or both.
(a) Produces a staircase signal at its DAC output
(b) Has a constant conversion time independent of V_A
(c) Has a shorter average conversion time
(d) Uses an analog comparator
(e) Uses a DAC
(f) Uses a counter
(g) Has complex control logic
(h) Has an $\overline{EOC}$ output

10-32. Draw the waveform for V_{AX} as the SAC of Figure 10-17 converts $V_A = 6.7$ V. V_{AX} ends up a 6 V (bits 1 and 2 set).

10-33. Repeat Problem 10-32 for $V_A = 16$ V. V_{AX} ends up at 15 V (all bits set).

10-34. A certain 8-bit successive-approximation converter has 2.55 V full scale. The conversion time for $V_A = 1$ V is 80 μs. What will be the conversion time for $V_A = 1.5$ V?   80 μs

10-35. Figure 10-31 shows the waveform at V_{AX} for a 6-bit SAC with a step size of 40 mV during a complete conversion cycle. Examine this waveform and describe what is occurring at times t_0 to t_5. Then determine the resultant digital output.   Digital output = 100101

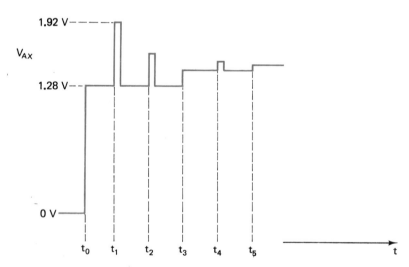

FIGURE 10-31

10-36. Refer to Figure 10-19. What is the approximate value of the analog input if the microcomputer's data bus is at 10010111 when $\overline{RD}$ is pulsed LOW?   2.96 V

10-37. Connect a 2.0-V reference source to $V_{ref}/2$ and repeat Problem 10-36.
  10-37. 2.37 V

SECTION 10-11

10-38. Discuss how a flash ADC with a conversion time of 1 μs would work for the situation of Problem 10-30.

10-39. Draw the circuit diagram for a 4-bit flash converter with BCD output and a resolution of 0.1 V. Assume that a +5 V precision supply voltage is available.

10-38. The faster conversion time would allow more closely spaced samples of the analog signal. This would produce a more accurate reconstructed signal.

10-39. Nine comparators, ten resistors, and a 74147 priority encoder.

DRILL QUESTION

10-40. For each of the following statements, indicate which type of ADC is being described—digital-ramp, SAC, or flash.
  (a) Fastest method of conversion   flash
  (b) Needs a START pulse   digital-ramp, SAC
  (c) Requires the most circuitry   flash
  (d) Does not use a DAC   flash
  (e) Generates a staircase signal   digital-ramp
  (f) Uses an analog comparator   all three
  (g) Has a relatively fixed conversion time independent of V_A SAC and flash

SECTION 10-12

DRILL QUESTION

10-41. For each statement indicate what type(s) of ADC is (are) being described:
  (a) Uses a counter that is never reset to zero   up/down digital-ramp
  (b) Uses a large number of comparators   flash
  (c) Uses a VCO   voltage-frequency ADC
  (d) Used in noisy industrial environments   voltage-frequency ADC
  (e) Uses a capacitor   dual-slope ADC
  (f) Is relatively insensitive to temperature   dual-slope ADC

SECTION 10-13

10-42. A voltage $V_A = 3.853$ V is applied to the DVM of Figure 10-21. Assume that the counters are initially reset to zero and sketch the waveforms at V_{AX}, COMP, Q_1, Q_2, and the AND gate output for *two* complete cycles. Use a 100-kHz clock and assume both one-shots have $t_p = 1$ μs. V_A steps up to 3.86 V (386 steps) before COMP goes LOW.

10-43. Refer to the DVM circuit in Figure 10-21. Assume that the circuit works correctly for V_A values up to about 7.99 V. However, when V_A is increased above 8 V, the displays do not change from their previous reading, the staircase waveform at V_{AX} steps up to only about 7.99 volts before jumping back down to zero, and COMP remains HIGH. What are some possible causes for this malfunction?

10-43. The MSB of the MSD counter is stuck LOW; the MSB input to the BCD-to-analog converter is stuck LOW; the MSB of the BCD-to-analog converter is not responding.

10-44. Add the necessary logic to the DVM of Figure 10-21 to turn on an over-scale LED whenever the analog input voltage exceeds 9.99 V. The LED should be turned off at the start of each new conversion.

10-44. Add a FF which will be clocked by NGT from MSB of MSD.

10-45. Switch is stuck closed; switch is stuck open, or capacitor is shorted.

10-45. Refer to the sample-and-hold circuit of Figure 10-23. What circuit fault would result in V_{OUT} looking exactly like V_A? What fault would cause V_{OUT} to be stuck at zero?

10-46. Use the CMOS 4016 integrated circuit (Section 8-17) for S_1, S_2, and S_3 in Figure 10-24 and design the necessary control logic so that each analog input is converted to its digital equivalent in sequence. The ADC is a 10-bit successive-approximation type using a 50-kHz clock signal and requires a 10-μs-duration start pulse to begin each conversion. The digital outputs are to remain stable for 100 μs after the conversion is complete before switching to the next analog input. Choose an appropriate multiplexing clock frequency.   See Solutions Manual.

MICROCOMPUTER APPLICATION

10-47. Figure 10-19 shows how the ADC8004 is interfaced to a microcomputer. It shows three control signals, $\overline{CS}$, $\overline{RD}$, and $\overline{WR}$, that come from the microcomputer to the ADC. These signals are used to start each new A/D conversion, and to read (transfer) the ADC data output into the microcomputer over the data bus.

Figure 10-32(a) shows how these signals may be generated from the microcomputer's address decoding and control logic circuits. The microprocessor (MPU) in this microcomputer has a single $R/\overline{W}$ output that signals whether the MPU is doing a read or a write operation. For a read operation, $R/\overline{W} = 1$; for a write operation, $R/\overline{W} = 0$.

(a) The $\overline{CS}$ signal that activates the ADC0804 is developed from the eight high-order lines of the MPU address bus. Whenever the MPU wants to communicate with the ADC0804, it places the address of the ADC0804 onto the address bus and the decoding logic drives the $\overline{CS}$ signal LOW.

Determine the address of the ADC0804.   Address is EA*xx* (hex)

(b) Figure 10-32(b) shows the signals that occur when the MPU wants to command the ADC0804 to start a new A/D conversion. During time interval t_W the MPU places the ADC0804's address on the address bus, and makes $R/\overline{W} = 0$ for a write operation. (Before and after this time interval, the MPU places different addresses on the address bus for different operations it is performing.) During t_W the decoding logic drives $\overline{CS}$ LOW as shown. The $\overline{CS}$ and $R/\overline{W}$ signals are combined in the control logic block to develop the $\overline{WR}$ signal as shown. This signal is sent to the ADC0804 to initiate a new conversion.

Design the portion of the control logic that produces $\overline{WR}$.

(c) Figure 10-32(c) shows the signals when the MPU wants to perform a read operation on the ADC0804's data outputs. The MPU places the ADC0804's address on the address bus during interval t_R and keeps $R/\overline{W}$ HIGH during this interval. Once again, the decoding logic drives $\overline{CS}$ LOW. The control logic combines $\overline{CS}$ and $R/\overline{W}$ to generate the $\overline{RD}$ signal as shown. This signal is sent to the ADC0804 to enable it to place its outputs on the data bus for transfer to the MPU.

Design the control logic that produces $\overline{RD}$.

FIGURE 10-32   (a) MPU interfaced to the ADC0804 of Figure 10-19; (b) signals for MPU write operation; (c) signals for read operation.

ANSWERS TO SECTION REVIEW QUESTIONS

SECTION 10-1

1. Converts a nonelectrical physical quantity to an electrical quantity.   2. Converts an analog voltage or current to a digital representation.
3. Stores it; performs calculation or some other operation on it.   4. Converts digital data to its analog representation.   5. Controls a physical variable according to an electrical input signal.

SECTION 10-2

1. 40 μA; 10.2 mA.   2. 5.12 mA.
3. 0.39 percent.   4. 4096.   5. Twelve.
6. 8.73 V.   7. True.   8. It produces a greater number of possible analog outputs between zero and full scale.

SECTION 10-3

1. It only uses two different sizes of resistors.
2. 640 kΩ.   3. 0.5 V.

SECTION 10-4

1. Maximum deviation of DAC output from its ideal value, expressed as percentage of full scale.   2. Time it takes DAC output to settle to within 1/2 step size of its full-scale value when the digital input changes from zero to full scale.   3. Offset error adds a small constant positive or negative value to the expected analog output for any digital input.   4. Because of the response time of the op-amp current-to-voltage converter.

SECTION 10-7

1. Tells control logic when the DAC output exceeds the analog input.   2. At outputs of register.   3. Tells us when conversion is complete and digital equivalent of V_A is at register outputs.

SECTION 10-8

1. The digital input to a DAC is incremented until the DAC staircase output exceeds the analog input.   2. The built-in error caused by the fact that V_{AX} does not continuously increase but goes up in steps equal to the DAC's resolution. The final V_{AX} can be different from V_A by as much as one step size.   3. If V_A increases it will take more steps before V_{AX} can reach the step that first exceeds V_A.   4. True.   5. Simple circuit; relatively long conversion time that changes with V_A.   6. $00100001 11_2 = 135_{10}$ for both cases.

SECTION 10-9

1. Process of converting different points on an analog signal to digital and storing the digital data for later use.   2. Computer generates START signal to begin an A/D conversion of the analog signal. When $\overline{EOC}$ goes LOW, it signals the

computer that the conversion is complete. The computer then loads the ADC output into memory. The process is repeated for the next point on the analog signal.

SECTION 10-10

1. The SAC has a shorter conversion time which doesn't change with V_A.   2. It has more complex control logic.   3. False.   4. (a) 8; (b) 0–5 V; (c) $\overline{CS}$ controls the effect of the $\overline{RD}$ and $\overline{WR}$ signals, $\overline{WR}$ is used to start a new conversion, $\overline{RD}$ enables the output buffers; (d) When LOW it signals the end of a conversion; (e) It separates the usually noisy digital ground from the analog ground so as not to contaminate the analog input signal.

SECTION 10-11

1. True.   2. 4095 comparators and 4096 resistors.   3. Major advantage is its conversion speed; disadvantage is the number of required circuit components for a practical resolution.

SECTION 10-12

1. It reduces conversion time by using an up/down counter that allows V_{AX} to track V_A without starting from zero for each conversion.   2. A VCO.   3. Advantages; low cost, temperature immunity; disadvantage: slow conversion time.   4. Flash ADC, voltage-to-frequency ADC, and dual-slope ADC.

SECTION 10-13

1. They hold for display the digital values present in the counters at the end of the last conversion.
2. The PGT of OS1 transfers the result of the conversion to the registers; the pulse at OS2 resets the counters to begin a new conversion cycle.

SECTION 10-14

1. It takes a sample of an analog voltage signal and stores it on a capacitor.   2. False; they are unity-gain buffers with high input impedance and low output impedance.

SECTION 10-16

1. Displayed waveform can be moved and manipulated; can display portions of the waveform occurring prior to the trigger; indefinite storage time.   2. The ADC digitizes the points on the input waveform for storage in memory; the vertical DAC converts the stored data points back to analog voltages to produce the vertical deflection of the electron beam; the horizontal DAC produces a staircase sweep voltage that provides the horizontal deflection of the electron beam.

11

MEMORY DEVICES

Outline

Objectives

Upon completion of this chapter, you will be able to:

- Understand and correctly use the terminology associated with memory systems.
- Describe the difference between read/write memory and read-only memory.
- Describe the difference between sequential-access and random-access memory.
- Determine the capacity of a memory device from its inputs and outputs.
- Outline the steps that occur when the CPU reads from or writes to memory.
- Distinguish among the various types of ROMs and cite some common applications.
- Understand and describe the organization and operation of static and dynamic RAMs.
- Combine memory ICs to form larger memory capacities.
- Describe the architecture of the basic types of programmable logic devices and determine the fuse pattern for a given set of logic functions.
- Describe the basic characteristics of various types of magnetic memory devices.
- Use the test results on a RAM or ROM system to determine possible faults in the memory system.

Introduction

A major advantage of digital over analog systems is the ability to easily store large quantities of digital information and data for short or long periods. This memory capability is what makes digital systems so versatile and adaptable to many situations. For example, in a digital computer the internal main memory stores instructions that tell the computer what to do under *all* possible circumstances so that the computer will do its job with a minimum amount of human intervention.

This chapter is devoted to a study of the most commonly used types of memory devices and systems. We have already become very familiar with the flip-flop, which is an electronic memory device. We have also seen how groups of FFs called registers can be used to store information and how this information can be transferred to other locations. FF registers are high-speed memory elements which are used extensively in the internal operations of a digital computer, where digital information is continually being moved from one location to another. Advances in LSI and VLSI technology have made it possible to obtain large numbers of FFs on a single chip arranged in various memory-array formats. These bipolar and MOS semiconductor memories are the fastest memory devices available, and their cost has been continuously decreasing as LSI technology improves.

Digital data can also be stored as charges on capacitors, and a very important type of semiconductor memory uses this principle to obtain high-density storage at low power-requirement levels.

Semiconductor memories are used as the *internal* main memory of a computer (Figure 11-1) where fast operation is important. The internal computer memory is in constant communication with the central processing unit (CPU) as a program of instructions is being executed. The program and any data used by the program are usually stored in the internal memory.

Although semiconductor memories are well suited for high-speed internal memory, their cost per bit of storage prohibits their use as **mass storage** devices. Mass storage refers to memory that is external to the main computer (Figure 11-1) and has the capacity to store millions of bits of data *without* the need for electrical power. Mass memory is normally much slower than internal memory and is used to store information (programs, data, etc.) not currently being used by the computer. This information is transferred to the internal memory when the computer needs it.

FIGURE 11-1   A computer system normally uses high-speed internal memory and slower external mass memory.

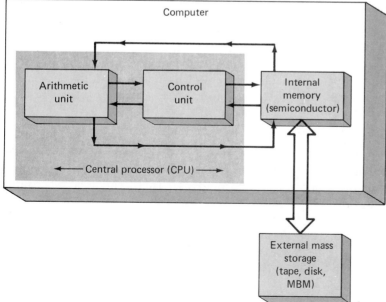

Magnetic tape and magnetic disk are popular mass storage devices that are much less expensive in cost per bit than internal memory devices. A newer entry into the mass memory category is the **magnetic bubble memory (MBM)**, a semiconductor device that uses magnetic principles to store millions of bits on one chip. The MBM is relatively slow and cannot be used as internal memory.

We will take a more detailed look at the characteristics of these memory devices after we define some of the more important terms that are common to many memory systems.

11-1 MEMORY TERMINOLOGY

The study of memory devices and systems is filled with terminology that can sometimes be overwhelming to a student. Before we get into any comprehensive discussion of memories, it would be helpful if you had the meaning of some of the more basic terms under your belt. Other new terms will be defined as they appear in the chapter.

Teaching Tip: Other examples of memory cells include: magnetic bubbles, magnetic cores, broken/intact fuses (PROMS, PLAs), holes in punched cards or paper tape.

Memory Cell A device or electrical circuit used to store a single bit (0 or 1). Examples of memory cells include a flip-flop, a charged capacitor, and a single spot on magnetic tape or disk.

Memory Word A group of bits (cells) in a memory that represents instructions or data of some type. For example, a register consisting of eight FFs can be considered to be a memory that is storing an 8-bit word. Word sizes in modern computers typically range from 4 to 64 bits, depending on the size of the computer.

Byte A special term used for an 8-bit word. A byte always consists of 8 bits, which is the most common memory word size in microcomputers.

Teaching Tip: The byte is the most common word size for microcomputer memory, and it is common to hear a microcomputer's memory size specified simply as number of words. For example, it is understood that a computer with a memory of 256K actually has a memory capacity of 256K bytes or 256K × 8.

Capacity A way of specifying how many bits can be stored in a particular memory device or complete memory system. To illustrate, suppose that we have a memory which can store 4096 twenty-bit words. This represents a total capacity of 81,920 bits. We could also express this memory's capacity as 4096 × 20. When expressed this way, the first number (4096) is the number of words and the second number (20) is the number of bits per word (word size). The number of words in a memory is often a multiple of 1024. It is common to use the designation "1K" to represent $1024 = 2^{10}$ when referring to memory capacity. Thus, a memory that has a storage capacity of 4K × 20 is actually a 4096 × 20 memory.

The development of larger memories has brought about the designation "1M" or "1 meg" to represent $2^{20} = 1,048,576$. Thus, a memory that has a capacity of 2M × 8 is actually one with a capacity of 2,097,152 × 8.

—— EXAMPLE 11-1A ——

A certain semiconductor memory chip is specified as 2K × 8. How many words can be stored on this chip? What is the word size? How many total bits can this chip store?

Solution

$$2K = 2 \times 1024 = 2048 \text{ words}$$

Each word is 8 bits (one byte). The total number of bits is therefore

$$2048 \times 8 = 16,384 \text{ bits}$$

Which memory stores the most bits: a 5M × 8 memory, or a memory that stores 1M words at a word size of 16 bits?

Solution

$$5M \times 8 = 5 \times 1{,}048{,}576 \times 8 = 41{,}943{,}040 \text{ bits}$$
$$1M \times 16 = 1{,}048{,}576 \times 16 = 16{,}777{,}216 \text{ bits}$$

The 5M × 8 memory stores more bits.

Address A number that identifies the location of a word in memory. Each word stored in a memory device or system has a unique address. Addresses are always expressed as a binary number, although octal, hexadecimal, and decimal numbers are often used for convenience. Figure 11-2 illustrates a small memory consisting of eight words. Each of these eight words has a specific address represented as a 3-bit number ranging from 000 to 111. Whenever we refer to a specific word location in memory, we use its address code to identify it.

Read Operation The operation whereby the binary word stored in a specific memory location (address) is sensed and then transferred to another device. For example, if we want to use word 4 of the memory of Figure 11-2 for some purpose, we have to perform a read operation on address 100. The read operation is often called a *fetch* operation, since a word is being fetched from memory. We will use both terms interchangeably.

Discussion Question: What is the relationship between the number of bits in a memory address and the word size?

Write Operation The operation whereby a new word is placed into a particular memory location. It is also referred to as a **store** operation. Whenever a new word is written into a memory location, it replaces the word that was previously stored there.

Access Time A measure of a memory device's operating speed. It is the amount of time required to perform a read operation. More specifically, it is the time between the memory's receiving a new address input, and the data's becoming available at the memory output. The symbol t_{ACC} is used for access time.

Volatile Memory Any type of memory that requires the application of electrical power in order to store information. If the electrical power is removed, all information stored in the memory will be lost. Many semiconductor memories are volatile, while all magnetic memories are nonvolatile.

Addresses	
0 0 0	Word 0
0 0 1	Word 1
0 1 0	Word 2
0 1 1	Word 3
1 0 0	Word 4
1 0 1	Word 5
1 1 0	Word 6
1 1 1	Word 7

FIGURE 11-2 Each word location has a specific binary address.

Random-Access Memory (RAM) Memory in which the actual physical location of a memory word has no effect on how long it takes to read from or write into that location. In other words, the access time is the same for any address in memory. Most semiconductor memories and magnetic core memories are RAMs.

Sequential-Access Memory (SAM) A type of memory in which the access time is not constant, but varies depending on the address location. A particular stored word is found by sequencing through all address locations until the desired address is reached. This produces access times which are much longer than those of random-access memories. Examples of sequential-access memory devices include magnetic tape, disk, and magnetic bubble memory. To illustrate the difference between SAM and RAM, consider the situation where you have recorded 60 minutes of songs on an audio tape cassette. When you want to get to a particular song, you have to rewind or fast-forward the tape until you find it. The process is relatively slow, and the amount of time required depends on where on the tape the desired song is recorded. This is SAM, since you have to sequence through all intervening information until you find what you are looking for. The RAM counterpart to this would be a jukebox, where you can quickly select any song by punching in the appropriate code and it takes the same time no matter what song you select.

Read/Write Memory (RWM) Any memory that can be read from or written into with equal ease.

Read-Only Memory (ROM) A broad class of semiconductor memories designed for applications where the ratio of read operations to write operations is very high. Technically, a ROM can be written into (programmed) only once, and this operation is normally performed at the factory. Thereafter information can only be read from the memory. Other types of ROM are actually read-mostly memories (RMM), which can be written into more than once, but the write operation is more complicated than the read operation and it is not performed very often. The various types of ROM will be discussed later. *All ROM is nonvolatile* and will store data when electrical power is removed.

Static Memory Devices Semiconductor memory devices in which the stored data will remain permanently stored as long as power is applied, without the need for periodically rewriting the data into memory.

Dynamic Memory Devices Semiconductor memory devices in which the stored data *will not* remain permanently stored, even with power applied, unless the data are periodically rewritten into memory. The latter operation is called a *refresh* operation.

Teaching Tip: A good analogy: static memory is like a conventional storage oscilloscope storing the waveform image on its screen without its having to be retraced; dynamic memory is like a non-storage oscilloscope in which the waveform has to be repetitively retraced for a usable display.

Internal Memory This is also referred to as the computer's *main* memory. It stores the instructions and data that the CPU is currently working on. It is the fastest memory in the computer system, and usually consists of semiconductor memory devices.

Mass Memory This type of memory is also referred to as *auxiliary* memory. It stores large amounts of information external to the computer's internal memory. It is generally slower than internal memory and is always nonvolatile. Magnetic tape and disk are the most common mass memory devices.

REVIEW QUESTIONS

1. Define the following terms: (a) memory cell, (b) memory word, (c) address, (d) byte, (e) access time.
2. A certain memory has a capacity of 8K × 16. How many bits are in each

word? How many words are being stored? How many memory cells does this memory contain?

3. Explain the difference between the read (fetch) and write (store) operations.

4. *True or false:* A volatile memory will lose its stored data when electrical power is interrupted.

5. Explain the difference between SAM and RAM.

6. Explain the difference between RWM and ROM.

7. *True or false:* A dynamic memory will hold its data as long as electrical power is applied.

11-2 GENERAL MEMORY OPERATION

Although each type of memory is different in its internal operation, certain basic operating principles are the same for all memory systems. An understanding of these basic ideas will help in our study of individual memory devices.

Every memory system requires several different types of input and output lines to perform the following functions:

1. Select the address in memory that is being accessed for a read or write operation.

2. Select either a read or a write operation to be performed.

3. Supply the input data to be stored in memory during a write operation.

4. Hold the output data coming from memory during a read operation.

5. Enable (or disable) the memory so that it will (or will not) respond to the address inputs and read/write command.

Teaching Tip: In many semiconductor memory ICs, the chip is placed in a low-power mode when disabled so that it draws much less power from the V_{CC} source than when it is enabled.

Figure 11-3(a) illustrates these basic functions in a simplified diagram of a

FIGURE 11-3   (a) Diagram of a 32 × 4 memory: (b) virtual arrangement of memory cells into thirty-two 4-bit words.

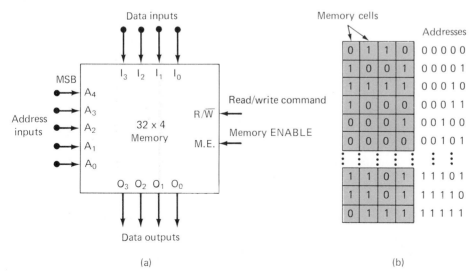

(a)

(b)

32×4 memory that stores thirty-two 4-bit words. Since the word size is 4 bits, there are four data input lines I_0 to I_3 and four data output lines O_0 to O_3. During a write operation the data to be stored into memory have to be applied to the data input lines. During a read operation the word being read from memory appears at the data output lines.

Address Inputs

Since this memory stores 32 words, it has 32 different storage locations and therefore 32 different binary addresses ranging from 00000 to 11111 (0 to 31 in decimal). Thus there are five address inputs, A_0 to A_4. To access one of the memory locations for a read or write operation, the 5-bit address code for that particular location is applied to the address inputs. In general, N address inputs are required for a memory that has a capacity of 2^N words.

We can visualize the memory of Figure 11-3(a) as an arrangement of 32 registers, with each register holding a 4-bit word as illustrated in Figure 11-3(b). Each address location is shown containing four memory cells that hold 1s and 0s that make up the data word stored at that location. For example, the data word 0110 is stored at address 00000, the data word 1001 is stored at address 00001, and so on.

The R/$\overline{W}$ Input

The read/write ($R/\overline{W}$) input line determines which memory operation is to take place. Some memory systems use two separate inputs, one for read and one for write. When a single $R/\overline{W}$ input is used, the read operation takes place for $R/\overline{W} = 1$, and the write operation takes place for $R/\overline{W} = 0$.

A simplified illustration of the read and write operations is shown in Figure 11-4. Part (a) shows the data word 0100 being written into the memory register at address location 00011. This data word would have been applied to the memory's data input lines, and replaces the data previously stored at address 00011. Part (b) shows the data word 1101 being read from address 11110. This output would appear at the memory's data output lines. After the read operation, the data word 1101 is still stored in address 11110. In other words, the read operation does not change the stored data.

FIGURE 11-4 Simplified illustration of the read and write operations on the 32×4 memory: (a) writing the data word 0100 into memory location 00011; (b) reading the data word 1101 from memory location 11110.

(a) WRITING the data word 0100 into memory location 00011.

(b) READING the data word 1101 from memory location 11110.

Memory Enable   Many memory systems have some means for completely disabling all or part of the memory so that it will not respond to the other inputs. This is represented in Figure 11-3 as the MEMORY ENABLE input, although it can have different names in the various memory systems. Here it is shown as an active-HIGH input that enables the memory to operate normally when it is kept HIGH. A LOW on this input disables the memory so that it will not respond to the address and $R/\overline{W}$ inputs. This type of input is useful when several memory modules are combined to form a larger memory. We will examine this idea later.

—— EXAMPLE 11-2 ——————————————————

Describe the conditions at each input and output when the contents of address location 00100 are to be read.

Solution

> *Address inputs:* 00100
> *Data inputs:* xxxx (not used)
> $R/\overline{W}$: HIGH
> *MEMORY ENABLE:* HIGH
> *Data outputs:* 0001

Discussion Question: What will the data outputs be if MEMORY ENABLE is LOW?

—— EXAMPLE 11-3 ——————————————————

Describe the conditions at each input and output when the data word 1110 is to be written into address location 01101.

Solution

> *Address inputs:* 01101
> *Data inputs:* 1110
> $R/\overline{W}$: LOW
> *MEMORY ENABLE:* HIGH
> *Data outputs:* xxxx (not used; usually Hi-Z)

—— EXAMPLE 11-4 ——————————————————

A certain memory has a capacity of 4K × 8.

(a) How many data input and data output lines does it have?

(b) How many address lines does it have?

(c) What is its capacity in bytes?

Solution

(a) Eight of each, since the word size is eight.

(b) The memory stores 4K = 4 × 1024 = 4096 words. Thus, there are 4096 memory addresses. Since 4096 = 2^{12}, it requires a 12-bit address code to specify one of 4096 addresses.

(c) A byte is 8 bits. This memory has a capacity of 4096 bytes.

The example memory in Figure 11-3 illustrates the important input and output functions common to most memory systems. Of course, each type of memory may

have other input and output lines that are peculiar to that memory. These will be described as we discuss the individual memory types.

━━ REVIEW QUESTIONS ━━━━━━━━━━━━━━━━━━━━━━━

1. How many address inputs, data inputs, and data outputs are required for a 16K × 12 memory?
2. What is the function of the $R/\overline{W}$ input?
3. What is the function of the MEMORY ENABLE input?

11-3   CPU-MEMORY CONNECTIONS

A major part of this chapter is devoted to semiconductor memory, which, as pointed out earlier, makes up the internal memory of most modern computers. Remember, this internal memory is in constant communication with the CPU (central processing unit). It is not necessary to be familiar with the detailed operation of a CPU at this point, and so the following simplified treatment of the CPU-memory interface will provide the perspective needed to make our study of memory devices more meaningful.

A computer's internal memory is made up of RAM and ROM ICs that are interfaced to the CPU over three groups of signal lines or buses. These are shown in Figure 11-5 as the address lines or address bus, data lines or data bus, and the control lines or control bus. Each of these buses consists of several lines (note how they are represented by a single line with a slash), and the number of lines in each bus will vary from one computer to the next. The three buses play a necessary part in allowing the CPU to write data into memory and to read data from memory.

Teaching Tip: The READ and WRITE operations always refer to actions taken by the CPU. During a READ operation, the CPU reads from memory; during a WRITE operation, the CPU writes to memory.

When a computer is executing a program of instructions, the CPU continually fetches (reads) information from those locations in memory that contain (1) the program codes representing the operations to be performed and (2) the data to be operated upon. The CPU will also store (write) data into memory locations as dictated by the program instructions. Whenever the CPU wants to write data to a particular memory location, the following steps must occur:

FIGURE 11-5   Three groups of lines (buses) connect the internal memory ICs to the CPU.

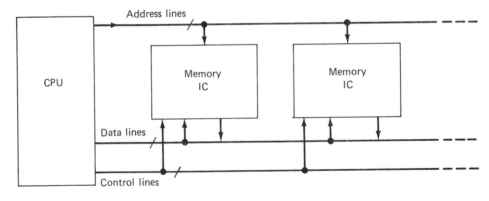

Write operation

1. The CPU supplies the binary address of the memory location where the data are to be stored. It places this address on the address bus lines.
2. The CPU places the data to be stored on the data bus lines.
3. The CPU activates the appropriate control signal lines for the memory write operation.
4. The memory ICs decode the binary address to determine which location is being selected for the store operation.
5. The data on the data bus are transferred to the selected memory location.

Whenever the CPU wants to read data from a specific memory location, the following steps must occur:

Read operation

1. The CPU supplies the binary address of the memory location from which data are to be retrieved. It places this address on the address bus lines.
2. The CPU activates the appropriate control signal lines for the memory read operation.
3. The memory ICs decode the binary address to determine which location is being selected for the read operation.
4. The memory ICs place data from the selected memory location on to the data bus, from where they are transferred to the CPU.

The steps above should make clear the function of each of the system buses:

Address Bus This is a *unidirectional* bus that carries the binary address outputs from the CPU to the memory ICs.

Data Bus This is a *bidirectional* bus that carries data between the CPU and the memory ICs.

Control Bus This bus carries control signals (such as the $R/\overline{W}$ signal) from the CPU to the memory ICs.

As we get into discussions of actual memory ICs, we will examine the signal activity that appears on these buses for the read and write operations.

_____ REVIEW QUESTIONS _____

1. Name the three groups of lines that connect the CPU and internal memory.
2. Outline the steps that take place when the CPU reads from memory.
3. Outline the steps that occur when the CPU writes to memory.

11-4 READ-ONLY MEMORIES

The read-only memory is a type of semiconductor memory that is designed to hold data that either are permanent or will not change frequently. During normal operation, no new data can be written into a ROM but data can be read from ROM. For

some ROMs the data that are stored have to be built in during the manufacturing process; for other ROMs the data can be entered electrically. The process of entering data is called **programming** or **burning** the ROM. Some ROMs cannot have their data changed once they have been programmed; others can be **erased** and reprogrammed as often as desired. We will take a detailed look later at these various types of ROMs. For now, we will assume that the ROMs have been programmed and are holding data.

ROMs are used to store data and information that are not to change during the operation of a system. A major use for ROMs is in the storage of programs in microcomputers. Since all ROMs are *nonvolatile*, these programs are not lost when the microcomputer is turned off. When the microcomputer is turned on, it can immediately begin executing the program stored in ROM. ROMs are also used for program and data storage in microprocessor-controlled equipment such as electronic cash registers, appliances, and security systems.

ROM Block Diagram   A typical block diagram for a ROM is shown in Figure 11-6(a). It has three sets of signals: address inputs, control input(s), and data outputs. From our previous discussions we can determine that this ROM is storing 16

FIGURE 11-6   (a) ROM block diagram; (b) programmed data; (c) hex data.

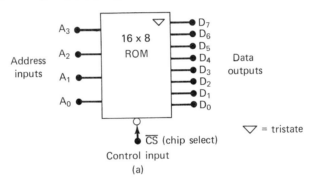

(a)

Word	A_3	A_2	A_1	A_0	D_7	D_6	D_5	D_4	D_3	D_2	D_1	D_0
0	0	0	0	0	1	1	0	1	1	1	1	0
1	0	0	0	1	0	0	1	1	1	0	1	0
2	0	0	1	0	1	0	0	0	0	1	0	1
3	0	0	1	1	1	0	1	0	1	1	1	1
4	0	1	0	0	0	0	0	1	1	0	0	1
5	0	1	0	1	0	1	1	1	1	0	1	1
6	0	1	1	0	0	0	0	0	0	0	0	0
7	0	1	1	1	1	1	1	0	1	1	0	1
8	1	0	0	0	0	0	1	1	1	1	0	0
9	1	0	0	1	1	1	1	1	1	1	1	1
10	1	0	1	0	1	0	1	1	1	0	0	0
11	1	0	1	1	1	1	0	0	0	1	1	1
12	1	1	0	0	0	0	1	0	0	1	1	1
13	1	1	0	1	0	1	1	0	1	0	1	0
14	1	1	1	0	1	1	0	1	0	0	1	0
15	1	1	1	1	0	1	0	1	1	0	1	1

(b)

Word	A_3 A_2 A_1 A_0	D_7–D_0
0	0	DE
1	1	3A
2	2	85
3	3	AF
4	4	19
5	5	7B
6	6	00
7	7	ED
8	8	3C
9	9	FF
10	A	B8
11	B	C7
12	C	27
13	D	6A
14	E	D2
15	F	5B

(c)

words, since it has $2^4 = 16$ possible addresses, and each word contains 8 bits, since there are eight data outputs. Thus, this is a 16×8 ROM. Another way to describe this ROM's capacity is to say that it stores 16 bytes of data.

The data outputs of most ROM ICs are tristate outputs to permit the connection of many ROM chips to the same data bus for memory expansion. The most common numbers of data outputs for ROMs are 4 bits and 8 bits, with 8-bit words being the most common.

The control input $\overline{CS}$ stands for **chip select.** This is essentially an enable input that enables or disables the ROM outputs. Some manufacturers use different labels for the control input, such as CE (chip enable) or OE (output enable). Many ROMs have two or more control inputs that have to be active in order to enable the data outputs so that data can be read from the selected address. In some ROM ICs, one of the control inputs (usually the CE) is used to place the ROM in a low-power standby mode when it is not being used. This reduces the current drain from the system power supply.

The $\overline{CS}$ input shown in Figure 11-6(a) is active-LOW; therefore it must be in the LOW state to enable the ROM data to appear at the data outputs. Notice that there is no R/W (read/write) input, because the ROM cannot be written into during normal operation.

The Read Operation   Let's assume that the ROM has been programmed with the data shown in the table of Figure 11-6(b). Sixteen different data words are stored at the 16 different address locations. For example, the data word stored at location 0011 is 10101111. Of course, the data are stored in binary inside the ROM, but very often we use hexadecimal notation to efficiently show the programmed data. This is done in Figure 11-6(c).

In order to read a data word from ROM, we need to do two things: apply the appropriate address inputs, and then activate the control inputs. For example, if we want to read the data stored at location 0111 of the ROM in Figure 11-6, we have to apply $A_3 A_2 A_1 A_0 = 0111$ to the address inputs, and then apply a LOW to $\overline{CS}$. The address inputs will be decoded inside the ROM to select the correct data word, 11101101, that will appear at outputs D_7 to D_0. If $\overline{CS}$ is kept HIGH, the ROM outputs will be disabled and will be in the Hi-Z state.

──── REVIEW QUESTIONS ────

1. *True or false:* All ROMs are nonvolatile.
2. Describe the procedure for reading from ROM.

11-5   ROM ARCHITECTURE

The internal architecture (structure) of a ROM IC is very complex, and we need not be familiar with all its detail. It is instructive, however, to look at a simplified diagram of the internal architecture, such as that shown in Figure 11-7 for the 16×8 ROM. There are four basic parts: *row decoder, column decoder, register array, output buffers.*

FIGURE 11-7 Architecture of 16 × 8 ROM.

Register Array The register array stores the data that have been programmed into the ROM. Each register contains a number of memory cells equal to the word size. In this case, each register stores an 8-bit word. The registers are arranged in a square matrix array that is common to many semiconductor memory chips. We can specify the position of each register as being in a specific row and specific column. For example, register 0 is in row 0, column 0, and register 9 is in row 1, column 2.

The eight data outputs of each register are connected to an internal data bus that runs through the entire circuit. Each register has two enable inputs (E); both have to be HIGH in order for the register's data to be placed on the bus.

Address Decoders The applied address code $A_3 A_2 A_1 A_0$ determines which register in the array will be enabled to place its 8-bit data word onto the bus. Address bits $A_1 A_0$ are fed to a 1-of-4 decoder which activates one row-select line, and address bits $A_3 A_2$ are fed to a second 1-of-4 decoder which activates one column-select line. Only one register will be in both the row and the column selected by the address inputs, and this one will be enabled.

_____ EXAMPLE 11-5 _____

Which register will be enabled by input address 1101?

Solution $A_3 A_2 = 11$ will cause the column decoder to activate the column 3 select line, and $A_1 A_0 = 01$ will cause the row decoder to activate the row 1 select line. This will place HIGHs at both enable inputs of register 13, thereby causing its data outputs to be placed on the bus. Note that the other registers in column 3 will have only one enable input activated; likewise for the other row 1 registers.

_____ EXAMPLE 11-6 _____

What input address will enable register 7?

Solution   The enable inputs of this register are connected to the row 3 and column 1 select lines, respectively. To select row 3, the $A_1 A_0$ inputs have to be at 11, and to select column 1, the $A_3 A_2$ inputs have to be at 01. Thus, the required address will be $A_3 A_2 A_1 A_0 = 0111$.

Output Buffers   The register that is enabled by the address inputs will place its data on the data bus. These data feed into the output buffers, which will pass the data to the external data outputs, provided that $\overline{CS}$ is LOW. If $\overline{CS}$ is HIGH, the output buffers are in the Hi-Z state, and D_7 through D_0 will be floating.

The architecture shown in Figure 11-7 is similar to that of many IC ROMs. Depending on the number of stored data words, the registers in some ROMs will not be arranged in a square array. For example, the Intel 2708 is a MOS ROM that stores 1024 eight-bit words. Its 1024 registers are arranged in a 64-by-16 array. Typically, ROM capacities range from 32×8 to $512K \times 8$.

_____ EXAMPLE 11-7 _____

Describe the internal architecture of a ROM that stores 4K bytes and uses a square register array.

Solution   4K is actually $4 \times 1024 = 4096$, and so this ROM holds 4096 eight-bit words. Each word can be thought of as being stored in an 8-bit register, and there are 4096 registers connected to a common data bus internal to the chip. Since $4096 = 64^2$, the registers are arranged in a 64-by-64 array; that is, there are 64 rows and 64 columns. This requires a 1-of-64 decoder to decode six address inputs for the row select, and a second 1-of-64 decoder to decode six other address inputs for the column select. Thus, a total of 12 address inputs are required. This makes sense, since $2^{12} = 4096$, and there are 4096 different addresses.

_____ REVIEW QUESTIONS _____

1. What input address code is required if we want to read the data from register 9 in Figure 11-7?
2. Describe the function of the row-select decoder, column-select decoder, and output buffers in the ROM architecture.

11-6 ROM TIMING

There will be a propagation delay between the application of a ROM's inputs and the appearance of the data outputs during a read operation. This time delay, called **access time,** t_{ACC}, is a measure of the ROM's operating speed. Access time is described graphically by the waveforms in Figure 11-8.

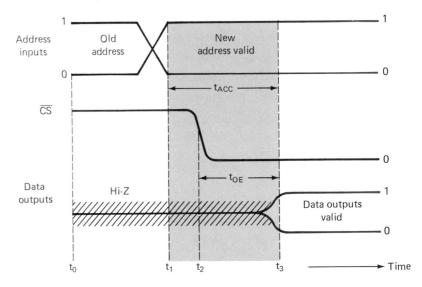

FIGURE 11-8 Typical timing for a ROM read operation.

Teaching Tip: Of course, not all of the address lines will change at t_1. In many cases, only one will change (the LSB). However, all of the data lines will change from their Hi-Z state at t_3.

Discussion Question: Can glitches at the decoder outputs (Figure 11-7) have an effect on the data outputs?

The top waveform represents the address inputs, the middle waveform is an active-LOW chip select, $\overline{CS}$, and the bottom waveform represents the data outputs. At time t_0 the address inputs are all at some specific level, some HIGH, and some LOW. $\overline{CS}$ is HIGH, so that the ROM data outputs are in their Hi-Z state (represented by hatched line).

Just prior to t_1, the address inputs are changing to a new address for a new read operation. At t_1 the new address is valid; that is, each address input is at a valid logic level. At this point the internal ROM circuitry begins to decode the new address inputs to select the register which is to send its data to the output buffers. At t_2 the $\overline{CS}$ input is activated to enable the output buffers. Finally, at t_3, the outputs change from the Hi-Z state to the valid data that represent the data stored at the specified address.

The time delay between t_1, when the new address becomes valid, and t_3, when the data outputs become valid, is the access time t_{ACC}. Typical bipolar ROMs will have access times in the range 30 to 90 ns; NMOS devices will range from 35 ns to 500 ns.

Another important timing parameter is the *output enable time,* t_{OE}, which is the delay between the $\overline{CS}$ input and the valid data output. Typical values of t_{OE} are 10 to 20 ns for bipolar ROMs and 25 to 100 ns for MOS. This timing parameter is important in situations where the address inputs are already set to their new values, but the ROM outputs have not yet been enabled. When $\overline{CS}$ goes LOW to enable the outputs, the delay will be t_{OE}.

11-7 TYPES OF ROMs

Now that we have a general understanding of the internal architecture and external operation of ROM devices, we will look at the various types of ROMs to see how they differ in the way they are programmed, erased, and reprogrammed.

Mask-Programmed ROM The mask-programmed ROM has its storage locations written into (programmed) by the manufacturer according to the customer's specifications. A photographic negative called a *mask* is used to control the electrical interconnections on the chip. A special mask is required for each different set of information to be stored in the ROM. Since these masks are expensive, this type of ROM is economical only if you need a very large quantity of the same ROM. Some ROMs of this type are available as off-the-shelf devices preprogrammed with commonly used information or data such as certain mathematical tables and character generator codes for CRT displays. A major disadvantage of this type of ROM is that it cannot be reprogrammed in the event of a design change requiring a modification of the stored data. The ROM would have to be replaced by a new one with the desired data written into it. Several types of user-programmable ROMs have been developed to overcome this disadvantage. Mask-programmed ROMs, however, still represent the most economical approach when a large quantity of identically programmed ROMs are needed.

Teaching Tip: MROMs are used only after the system has been thoroughly tested and no further design changes are anticipated. During the prototyping and testing stages, user-programmable ROMs are used so that they can be modified as needed to achieve a final acceptable design.

Mask-programmed ROMs are commonly referred to as just ROMs, but this can be confusing since the term ROM actually represents the broad category of devices that, during normal operation, are only read from. We will use the mnemonic MROM whenever we refer to mask-programmed ROMs.

Figure 11-9 shows the structure of a small bipolar MROM. It consists of 16 memory cells arranged in four rows of four cells. Each cell is an NPN bipolar transistor connected in the common-collector configuration (input at base, output at emitter). The top row of cells (ROW 0) constitutes a 4-bit register. Note how some of the transistors (Q_0 and Q_2) have their bases connected to the ROW 0 enable line, while others (Q_1 and Q_3) do not. The same is true of the cells in each of the other rows. The presence or absence of these base connections determines whether a cell is storing a 1 or 0, respectively. The condition of each base connection is controlled by the photographic mask during production based on the customer supplied data.

Note that cells that are in corresponding positions in each row (register) have their emitters connected to a common output. For instance, the emitters of Q_0, Q_4, Q_8, and Q_{12} are connected together as data output D_3. As we shall see, this will present no problem because only one row of cells will be activated at one time.

The 1-of-4 decoder is used to decode the address inputs $A_1 A_0$ to select which row (register) is to have its data read. The decoder's active-HIGH outputs provide the ROW enable lines that are the base inputs for the various rows of cells. If the decoder's enable input, $\overline{EN}$, is held HIGH, all of the decoder outputs will be in their inactive-LOW state, and all of the transistors in the array will be off because of the absence of any base voltage. For this situation, the data outputs will all be in the LOW state.

When $\overline{EN}$ is in its active-LOW state, the conditions at the address inputs determine which row (register) will be enabled so that its data can be read at the data outputs. For example, to read ROW 0, the $A_1 A_0$ inputs are set to 00. This places a

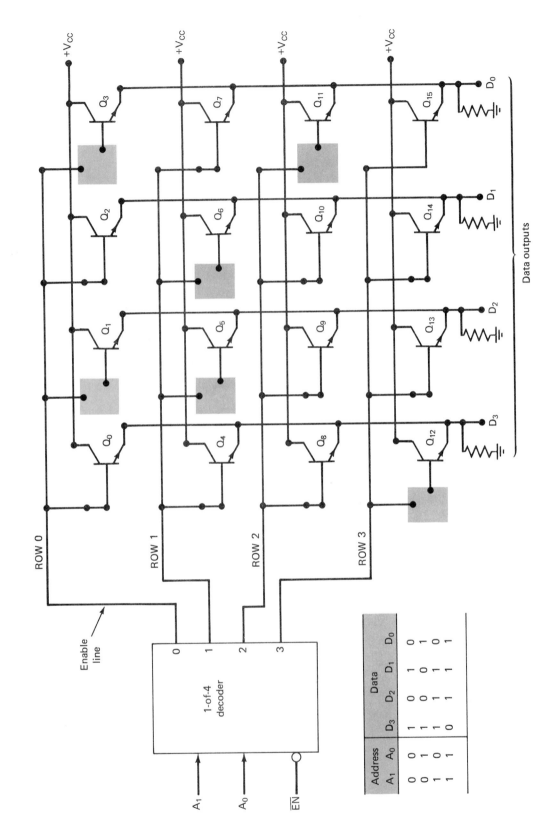

FIGURE 11-9   Structure of a bipolar MROM shows one bipolar transistor used for each memory cell. An open base connection stores a "0"; a closed base connection stores a "1".

HIGH at the ROW 0 line; all other row lines are at 0 V. This HIGH at ROW 0 turns on transistors Q_0 and Q_2, but not Q_1 and Q_3. With Q_0 and Q_2 conducting, the data outputs D_3 and D_1 will be HIGH; outputs D_2 and D_0 are still LOW. In a similar manner, application of the other address codes will produce data outputs from the corresponding register. The table in Figure 11-9 shows the data for each address. You should verify how this correlates with the base connections to the various cells.

—— EXAMPLE 11-8 ————————————————————————————

MROMs are often used to store tables of mathematical functions. Show how the MROM in Figure 11-9 can be used to store the function $y = x^2 + 3$, where the input address supplies the value for x, and the value of the output data is y.

Solution   The first step is to set up a table showing the desired output for each set of inputs. The input binary number, x, is represented by the address $A_1 A_0$. The output binary number is the desired value of y. For example, when $x = A_1 A_0 = 10_2 = 2_{10}$, the output should be $2^2 + 3 = 7_{10} = 0111_2$. The complete table is shown below.

x		$y = x^2 + 3$			
A_1	A_0	D_3	D_2	D_1	D_0
0	0	0	0	1	1
0	1	0	1	0	0
1	0	0	1	1	1
1	1	1	1	0	0

This table is supplied to the MROM manufacturer for developing the mask that will make the appropriate connections to the bases of the memory cells during the fabrication process. For instance, the first row in the table indicates that the connections to the bases of Q_0 and Q_1 will be left unconnected, while the connections to Q_2 and Q_3 will be made.

———————————————————————————————————————

Bipolar MROMs are available in several low capacities. One of the more popular ones is the 74187. It is organized as a 256 × 4 memory, and has an access time of 40 ns. Its outputs are open-collector types which require external pull-up resistors. Another bipolar ROM is the 7488A, which has a capacity of 32 × 8 and an access time of 45 ns.

MOS MROMs have an internal structure similar to Figure 11-9 except that the cells are MOSFETs rather than bipolar transistors. The TMS47256 is an NMOS version that has a capacity of 32K × 8. Its symbol is shown in Figure 11-10. Note that it has tristate outputs to permit easy interfacing to a computer data bus. In addition to the 14 address inputs, it has two enable inputs, $\overline{E}$ and $\overline{S}$. Both of these have to be LOW to enable the MROM outputs. The $\overline{E}$ input also performs a **power-down** function. When $\overline{E}$ is kept HIGH, the chip's internal circuitry is put into a low-power standby state where it draws about one-fourth of the normal supply current. The TMS47256 has an access time of 200 ns and a standby power of 82.5 mW. The CMOS version, the TMS47C256, has an access time of 150 ns and a standby power of only 2.8 mW.

FIGURE 11-10   Logic symbol for the TMS47256 MROM made using NMOS/CMOS technology.

Programmable ROMs (PROMs)

A mask-programmable ROM is very expensive and would not be used except in high-volume applications, where the cost would be spread out over many units. For lower-volume applications, manufacturers have developed **fusible-link** PROMs that are user-programmable; that is, they are not programmed during the manufacturing process but are custom-programmed by the user. Once programmed, however, a PROM is like an MROM in that it cannot be erased and reprogrammed. Thus, if the program in the PROM is faulty or has to be changed, the PROM has to be thrown away. For this reason, these devices are often referred to as "one time programmable" ROMs.

The fusible-link PROM structure is very similar to the MROM in that certain connections either are left intact or are opened in order to program a memory cell as a 1 or 0, respectively. In the MROM of Figure 11-9 these connections were from the enable lines to the transistor bases. In a PROM each of these connections is made with a thin fuse link that comes intact from the manufacturer (see Figure 11-11).

FIGURE 11-11   PROMs use fusible links that can be selectively blown open by the user to program a logic 0 into a cell.

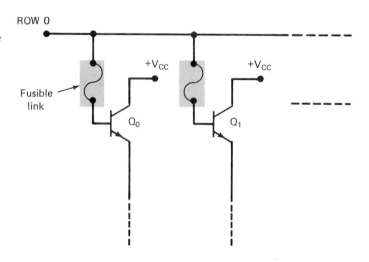

The user can selectively *blow* any of these fuse links to produce the desired stored memory data. Typically this is accomplished by applying a carefully controlled voltage to the device to produce a current flow that will cause the fuse link to open in a manner similar to blowing a fuse. Once a fuse link is broken, it cannot be reconnected.

The process of programming a PROM and then verifying the programmed data can be extremely time-consuming and tedious when done manually. A number of commercial PROM programmers are available for several hundred dollars that permit keyboard entry of the program into RWM (read/write memory) and then perform the fuse blowing and verification automatically without user intervention.

A popular bipolar PROM IC is the 74186, which is organized as sixty-four 8-bit words and has a typical access time of 50 ns. Another one is the TBP28S166, which is a 2K × 8 chip. MOS PROMs are available with much greater capacities than bipolar devices. The TMS27PC256 is a CMOS PROM with a capacity of 32K × 8 and a standby power dissipation of only 1.4 mW. It is available with maximum access times ranging from 120 to 250 ns.

Erasable Programmable ROM (EPROM)

An EPROM can be programmed by the user and it can also be *erased* and reprogrammed as often as desired. Once programmed, the EPROM is a *nonvolatile* memory that will hold its stored data indefinitely. The process for programming an EPROM involves the application of special voltage levels (typically in the 10- to 25-V range) to the appropriate chip inputs for a specified amount of time (typically 50 ms per address location). The programming process is usually performed by a special programming circuit that is separate from the circuit in which the EPROM will eventually be working. The complete programming process can take up to several minutes for one EPROM chip.

The storage cells in an EPROM are MOSFET transistors with a silicon gate that has no electrical connections (i.e., a *floating gate*). In its normal state, each transistor is off and each cell is storing a logic 1. A transistor can be turned on by the application of a high-voltage programming pulse that injects high-energy electrons into the floating-gate region. These electrons remain trapped in this region once the pulse is terminated, since there is no discharge path. This keeps the transistor on *permanently* even when power is removed from the device, and the cell is now storing a logic 0. During the programming process, the EPROM's address and data pins are used to select which memory cells will be programmed as 0s, and which ones will be left as 1s.

Once an EPROM cell has been programmed, it can be erased by exposing it to ultraviolet (UV) light applied through a window on the chip package. The UV light produces a photocurrent from the floating gate back to the silicon substrate, thereby removing the stored charges, turning the transistor off, and restoring the cell to the logic 1 state. This erasing process typically requires 15 to 30 minutes of exposure to UV rays. Unfortunately, there is no way to erase only selected cells; *the UV light erases all cells at the same time* so that an erased EPROM stores all 1s. Once erased, the EPROM can be reprogrammed.

EPROMs are available in a wide range of capacities and access times; devices with a capacity of 128K × 8 and an access time of 45 ns are commonplace. We will use the 2732 EPROM to illustrate typical EPROM operation. The 2732 is a 4K × 8 NMOS EPROM that operates from a single +5 V power source during normal operation. Figure 11-12 is the block symbol for the 2732. Note that it shows 12 address inputs, since $2^{12} = 4096$, and 8 data outputs. It has two control inputs. $\overline{CE}$ is the

Teaching Tip: An MROM is like a stone tablet that comes with information etched into it that cannot be erased. A PROM is like a blank stone tablet that you can etch yourself, but once etched the information cannot be erased.

Teaching Tip: Continuing the analogy, an EPROM is like a stone tablet onto which information can be written with chalk. The information can be erased and replaced by new information.

FIGURE 11-12   (a) Logic symbol for 2732 EPROM; (b) typical EPROM package showing ultraviolet window.

chip enable input that is used to place the device in a standby mode where its power consumption is reduced. $\overline{OE}/V_{PP}$ is a dual-purpose input whose function will depend on the device operating mode. $\overline{OE}$ is the output enable and is used to control the device's data output buffers so that the device can be connected to a microprocessor data bus without bus contention. V_{PP} is the special programming voltage required during the programming process.

The 2732 has several operating modes that are controlled by the voltages applied to the $\overline{CE}$ and $\overline{OE}/V_{PP}$ pins as presented in Figure 11-13(a). The read mode and the output disable mode are the normal operating modes for an EPROM when it is part of a computer's memory system and is frequently being read by the CPU. For these modes, the $\overline{CE}$ input is in its active-LOW state, and the level applied to $\overline{OE}$ controls whether or not the device's output pins will contain data or will be disabled.

The standby mode is achieved by holding the $\overline{CE}$ in its inactive HIGH state. This powered-down state is used when the device is not being accessed by the CPU. The EPROM will not respond to any of its other inputs when it is in the standby mode. In the standby mode, the 2732's power consumption is a maximum of 175 mW, as compared with 500 mW in the other modes.

The program mode is used to write new data into the EPROM cells. The programming process can write one 8-bit word into one address location at one time. Let's assume that the EPROM has been erased with UV light so that it is a "clean" EPROM (all cells are 1s). The steps required to program (write data into) any address location are as follows:

1. The address to be programmed is applied to the EPROM's address inputs. This is shown on the timing diagram of Figure 11-13(b) at time t_1. Prior to t_1, the device is in the standby mode with $\overline{CE}$ = HIGH.

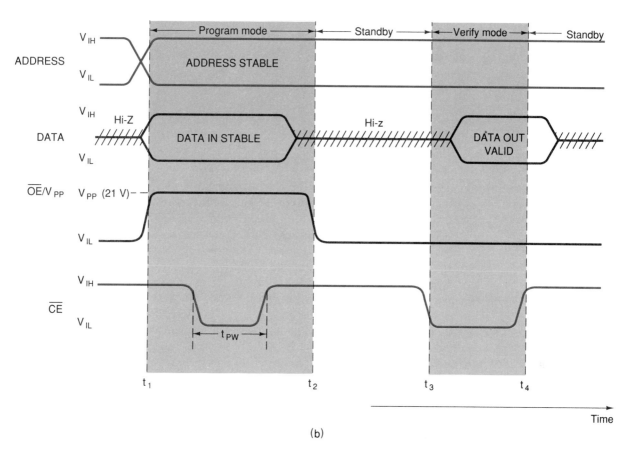

	Inputs		
Mode	$\overline{CE}$	$\overline{OE}/V_{PP}$	Outputs
READ/VERIFY	V_{IL}	V_{IL}	DATA$_{OUT}$
Output DISABLE	V_{IL}	V_{IH}	High Z
STANDBY	V_{IH}	X	High Z
PROGRAM	V_{IL}	V_{PP}	DATA$_{IN}$

Note: V_{IL} = TTL LOW
V_{IH} = TTL HIGH
X = don't care
V_{PP} = nominal 21 V

(a)

(b)

FIGURE 11-13   (a) Operating modes for 2732 EPROM; (b) timing for program and verify modes.

2. At the same time, the $\overline{OE}/V_{PP}$ pin is pulsed to the V_{PP} level (21 V).

3. At the same time, the data that are to be written into this address location are applied to the device's data pins D_7 through D_0. In the program mode, these data pins function as *inputs* because the high voltage on the $\overline{OE}/V_{PP}$ pin disables the device's data output buffers.

4. Once the address and data inputs have stabilized, the $\overline{CE}$ input is pulsed LOW for a typical duration of t_{PW} = 50 ms. During this interval each individual memory cell at the selected address location will either be left as a 1 or changed to a 0 depending on the corresponding bit of the input data.

5. When the programming process is complete, the $\overline{CE}$ input returns to its HIGH

Common Student Mistake: Students sometimes have difficulty with the idea of the EPROM's data output pins being used as inputs during the programming process. You can show how this works inside the chip using a bidirectional tristate buffer that is controlled by the $\overline{OE}/V_{PP}$ input.

state and $\overline{OE}/V_{PP}$ is returned LOW. In this standby mode, the data inputs are removed from the device's data pins and these pins become Hi-Z again.

6. The final step in the programming process is to verify that the data word has been properly written into the address location. The verify mode is really the same as the read mode. The $\overline{OE}$ is held LOW and the $\overline{CE}$ is pulsed LOW at t_3. This will cause the EPROM's internal circuitry to route the data from the selected address location to the data output buffers, where it can be read.

Teaching Tip: The 50-ms programming pulse duration is a worst-case specification. The programming process can be speeded up by using shorter-duration programming pulses. Some commercial EPROM programmers try programming each address with a shorter pulse (say, 25 ms), which is usually sufficient. If the verify step shows that the word has not been stored, the programmer automatically reprograms it using the 50-ms pulse. Since most, if not all, address locations can be burned in with a 25-ms pulse, the programming process is considerably shortened.

The programming process when done manually can take several hours. Numerous commercial EPROM programmers are available that can program and verify an entire 2732 in less than two minutes, once the user has entered the data into the EPROM programmer's RWM from a keyboard. Some EPROM programmers allow data to be transferred into the RWM from a computer. This process, called *downloading*, allows the user to first develop his program on a computer and then, when it is finished and working, transfer it from the computer's memory to the EPROM programmer, which will "burn it" into the EPROM.

Electrically Erasable PROM (EEPROM)   As noted above, EPROMs have two major disadvantages. First, they have to be removed from their sockets in order to be erased and reprogrammed. Second, erasure removes the complete memory contents; this necessitates complete reprogramming even when only one memory word has to be changed. The *electrically erasable PROM* (EEPROM) was developed around 1980 as an improvement to the EPROM.

The EEPROM takes advantage of the same floating-gate structure as the EPROM. It adds the feature of *electrical* erasability through the addition of a thin oxide region above the drain of the MOSFET memory cell. By applying a HIGH voltage (21 V) between the MOSFET's gate and drain, a charge can be induced onto the floating gate, where it will remain even when power is removed. Reversal of the same voltage causes a removal of the trapped charges from the floating gate and erases the cell. Since this charge-transport mechanism requires very low currents, programming and erasing of an EEPROM can usually be done *in circuit* (i.e., without a UV light source and PROM programmer unit).

A major advantage offered by EEPROMs over EPROMs is the ability to electrically erase and reprogram *individual* words in the memory array. Another advantage is that a complete EEPROM can be erased in about 10 ms (in circuit) versus about 30 minutes for an EPROM in external UV light. An EEPROM can also be programmed more rapidly; it requires only a 10-ms programming pulse for each data word, as compared with 50 ms for an EPROM.

Because the EEPROM can be erased and reprogrammed by applying appropriate voltages, we need not remove it from the circuit it is part of, provided that additional support components are also part of the circuitry. This support circuitry includes the 21-V programming voltage (V_{PP}), which is usually generated from the +5 V supply through a dc-to-dc converter, and circuitry to control the 10-ms timing and sequencing for the erasure and programming operations.

The Intel 2816 was the original EEPROM; it was introduced by Intel Corporation in 1981 with a 2K × 8 capacity, an access time of 250 ns, and the characteristics described above. Since then, advances in EEPROM design have resulted in many improvements. The 2864 is an 8K × 8 EEPROM that contains on-chip circuitry that generates the high voltages for the erasure and programming operations, so that the chip requires only a $+V_{CC}$ power pin. This makes the 2864 almost as easy to use as the static RWM devices we will be discussing shortly. Of course, unlike

(a)

	Inputs			
Mode	$\overline{CE}$	$\overline{OE}$	$\overline{WE}$	Outputs
READ	V_{IL}	V_{IL}	V_{IH}	$DATA_{OUT}$
WRITE	V_{IL}	V_{IH}	V_{IL}	$DATA_{IN}$
STANDBY	V_{IH}	X	X	High Z

(b)

FIGURE 11-14   (a) Symbol for 2864 EEPROM; (b) operating modes.

static RWM, the EEPROM is nonvolatile and it will hold all written data when power is turned off. On the other hand, static RWM has less complex internal circuitry and has much faster access times.

Figure 11-14 shows the symbol for the 2864 EEPROM along with its major operating modes. Note that the device's data pins are labeled as data I/O pins since they can function as either data inputs or outputs depending on the operating mode. Three control inputs determine the operating mode being used. $\overline{CE}$ is used to enable or disable the chip; when disabled, the chip will be in a low-power standby mode. $\overline{OE}$ is used to enable or disable the data output buffers. During a read operation, $\overline{OE} = 0$ enables the output buffers; during a write operation, $\overline{OE} = 1$ disables the output buffers so that data inputs can be applied to the I/O pins. $\overline{WE}$ selects either a read or a write mode. In the write mode, $\overline{WE}$ is pulsed LOW while the data to be written are applied to the I/O pins. Internal circuitry automatically erases the cells at the addressed location before writing in the new data.

— REVIEW QUESTIONS —

1. *True or false:* An MROM can be programmed by the user.
2. How does a PROM differ from an MROM? Can it be erased and reprogrammed?
3. *True or false:* PROMs are available in both bipolar and MOS versions.
4. How is an EPROM erased?
5. *True or false:* There is no way to erase only a portion of an EPROM's memory.
6. Outline the steps in programming and verifying a 2732 EPROM.
7. What are the advantages of an EEPROM over an EPROM?

11-8   ROM APPLICATIONS

ROMs can be used in any application requiring nonvolatile data storage where the data rarely or never have to change. We will briefly describe some of the most common application areas.

Microcomputer Program Storage (Firmware)

At present, microcomputer firmware is the most widespread application of ROMs. Some personal and business microcomputers use ROMs to store their operating system programs and their language interpreters (e.g., BASIC) so that the computer can be used immediately after power is turned on. Products that include a microcomputer to control their operation use ROMs to store the control programs. Examples are electronic games, electronic cash registers, electronic scales, and microcomputer-controlled automobile fuel injection.

The microcomputer programs that are stored in ROM are referred to as *firmware* because they are not subject to change; programs that are stored in RWM are referred to as *software* because they can be easily altered.

Teaching Tip: Many operating systems are very complex and require a great deal of memory, making it impractical to have them stored in ROM. The operating system typically contains many separate programs, only one of which is needed by the CPU at one time. With the operating system on disk, the CPU loads the portion of the operating system it needs into RWM for execution. This takes up much less address space than having all of the operating system programs permanently in ROM.

Bootstrap Memory

Many microcomputers and most larger computers do not have their operating system programs stored in ROM. Instead, these programs are stored in external mass memory, usually magnetic disk. How, then, do these computers know what to do when they are powered on? A relatively small program, called a *bootstrap program,* is stored in ROM. (The term "bootstrap" comes from the idea of pulling oneself up by one's own bootstraps.) When the computer is powered on, it will execute the instructions that are in this bootstrap program. These instructions typically cause the CPU to initialize the system hardware. The bootstrap program then loads the operating system programs from mass storage (disk) into its main internal memory. At that point the computer begins executing the operating system program and is ready to respond to the user commands. This start-up process is often called "booting up the system."

Data Tables

ROMs are often used to store tables of data that do not change. Some examples are the trigonometric tables (i.e., sine, cosine, etc.) and code-conversion tables.

Several standard ROM "look-up" tables are available with trig functions. One, the National Semiconductor MM4220BM, stores the sine function for angles between 0 and 90 degrees. This ROM is organized as a 128×8, with seven address inputs and eight data outputs. The address inputs represent the angle in increments of approximately 0.7°. For example, address 0000000 is 0°, address 0000001 is 0.7°, address 0000010 is 1.41°, and so on up to address 1111111, which is 89.3°. When an address is applied to the ROM, the data outputs will represent the approximate sine of the angle. For example, with input address 1000000 (representing approximately 45°) the data outputs will be 10110101. Since the sine is less than or equal to 1, these data are interpreted as a fraction; that is, .10110101, which when converted to decimal equals .707 (the sine of 45°).

Data Converter

The data converter circuit takes data expressed in one type of code and produces an output expressed in another type. Code conversion is needed, for example, when a computer is outputting data in straight binary code and we want to convert it to BCD in order to display it on 7-segment LED readouts.

One of the easiest methods of code conversion uses a ROM programmed so that the application of a particular address (the old code) produces a data output that represents the equivalent in the new code. The 74185 is a TTL ROM that stores the binary-to-BCD code conversion for a 6-bit binary input. To illustrate, a binary address input of 100110 (decimal 38) will produce a data output of 00111000, which is the BCD code for decimal 38. Problem 11.19 will deal with this ROM.

Character Generators

If you have ever looked closely at the alphanumeric characters (letters, numbers, etc.) printed on a computer's video display screen, you have noticed that each is generally made up of a group of dots. Depending on the character being displayed, some dot positions are made bright while others are dark. Each character is made to fit into a pattern of dot positions, usually arranged as a 5×7 or 7×9 matrix. The pattern of dots for each character can be represented as a binary code (i.e., bright dot = 1, dark dot = 0).

A *character-generator ROM* stores the dot pattern codes for each character at an address corresponding to the ASCII code for that character. For example, the dot pattern for the letter "A" would be stored at address 1000001, where 1000001 is the ASCII code for uppercase A. Character-generator ROMs are used extensively in any application that displays or prints out alphanumeric characters.

Function Generator

The function generator is a circuit that produces waveforms such as sine waves, sawtooth waves, triangle waves, and square waves. Figure 11-15 shows how a ROM look-up table and a DAC are used to generate a sine-wave output signal.

Discussion Question: How can a larger ROM be used to store four different waveforms? How can a particular waveform be selected for outputting at V_A? One way is to use 1K $\times$ 8 ROM with the A_9 and A_8 address inputs used to select the waveform.

FIGURE 11-15   Function generator using a ROM and a DAC.

The ROM stores 256 different 8-bit values, each one corresponding to a different waveform value (i.e., a different voltage point on the sine wave). The 8-bit counter is continuously pulsed by a clock signal to provide sequential address inputs to the ROM. As the counter cycles through the 256 different addresses, the ROM outputs the 256 data points to the DAC. The DAC output will be a waveform that steps through the 256 different analog voltage values corresponding to the data points. The low-pass filter smooths out the steps in the DAC output to produce a smooth waveform.

Circuits such as this are used in some commercial function generators. The same idea is employed in some speech synthesizers where the digitized speech waveform values are stored in the ROM.

REVIEW QUESTIONS

1. What is firmware?
2. Describe how a computer uses a bootstrap program.
3. What is a code converter?
4. What kind of information is stored in a character-generator ROM?
5. What are the main elements of a function generator?

Logic designers have a wide range of standard ICs available to them with numerous logic functions and logic circuit arrangements on a chip. In addition, these ICs are available from many manufacturers and at a reasonably low cost. For these reasons, designers have been interconnecting standard ICs to form an almost endless variety of different circuits and systems and will continue to do so in the indefinite future.

However, there are some problems with circuit and system designs that use only standard ICs. Some system designs might require hundreds or thousands of these ICs. This large number of ICs requires a considerable amount of circuit board space, and a great deal of time and cost in inserting, soldering, and testing of the ICs. The equipment manufacturer must keep a significant inventory of all the different types of ICs used in the design. Reducing the number of ICs used in a design can have several advantages: less board space and therefore fewer printed circuit boards and smaller enclosures; lower power requirements (smaller power supplies); faster and less costly assembly processes; higher reliability, since there are fewer ICs and circuit connections that can fail; and easier troubleshooting.

In order to reduce the number of ICs to be used in a design, it is necessary to put more and more logic functions on a chip. This, of course, has been done with LSI and VLSI technologies for standard functions such as memory, microprocessors, voice synthesizers, calculator chips, and so on. These devices contain hundreds and thousands of logic gates internally interconnected to operate in a predefined way. There are many design situations, however, for which there is no LSI or VLSI solution. These situations have typically required designers to interconnect several SSI and MSI chips to achieve the necessary logic functions. The recent development of *programmable logic devices (PLDs)* has presented logic designers with a way to replace a number of standard ICs with a single IC.

A PLD is an IC that contains large numbers of gates, FFs, and registers that are interconnected on the chip. Many of the connections, however, are fusible links that can be broken. The IC is said to be *programmable* because the specific function of the IC for a given application is determined by the selective breaking of some of the interconnections while leaving others intact. The "fuse blowing" process can be done by the manufacturer in accordance with the customer's instructions, or it can be done by the customer. This process is called *programming* because it produces the desired circuit pattern interconnecting the gates, FFs, registers, and so on.

Basic Idea   Figure 11-16 demonstrates the basic idea used by all programmable logic ICs. It shows an array of AND gates, and an array of OR gates that can be interconnected to generate four outputs, each of which can be any logic function of the two input variables A and B.

Each input feeds both a noninverting buffer and an inverting buffer to produce the true and inverted forms of each variable. These are the *input lines* to the AND-gate array. Each AND gate is connected to two different input lines so as to generate a unique product of the input variables. The AND outputs are called the *product lines*.

Each of the product lines is connected to one of the four inputs of each OR gate through a fusible link. With all the links initially intact, each OR output will be a constant 1. Here's the proof:

$$O_1 = \bar{A}\bar{B} + \bar{A}B + A\bar{B} + AB$$
$$= \bar{A}(\bar{B} + B) + A(\bar{B} + B)$$
$$= \bar{A} + A = 1$$

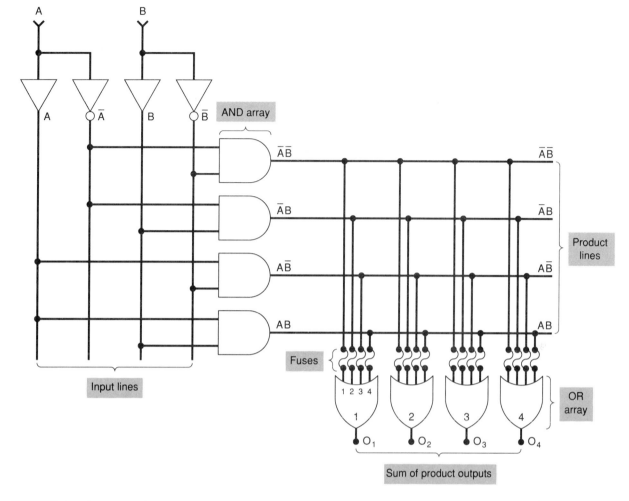

FIGURE 11-16 Example of a programmable logic device.

Each of the four outputs O_1, O_2, O_3, and O_4 can be *programmed* to be any function of A and B by selectively blowing the appropriate fuses. PLDs are designed so that a blown OR input acts as a logic 0. For example, if we blow fuses 1 and 4 at OR gate 1, the O_1 output becomes

$$O_1 = 0 + \overline{A}B + A\overline{B} + 0 = \overline{A}B + A\overline{B}$$

We can program each of the OR outputs to any desired function in a similar manner. Once all of the outputs have been programmed, the device will permanently generate the selected output functions.

PLD Symbology   The example in Figure 11-16 has only two input variables, and the circuit diagram is already quite cluttered. You can imagine how messy the diagram would be for PLDs with many more inputs. For this reason, PLD manufacturers have adopted a simplified symbology to represent the internal circuitry of these devices.

Figure 11-17 shows an example of the symbology for a 4-input AND gate. First note how the input buffers are represented as a single buffer with two outputs. Next, note that a single line is shown going into the AND gate to represent the four

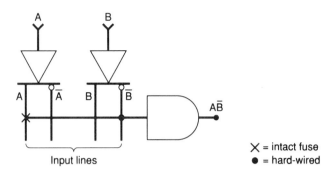

FIGURE 11-17 Simplified symbology for PLDs.

X = intact fuse
● = hard-wired

inputs. The connections from the input variable lines to the AND gate are indicated as either an *X* or a dot. An *X* represents an intact fuse. A dot represents a hard-wired connection (i.e., one that cannot be changed). The absence of either of these indicates no connection. In this example, the inputs *A* and $\bar{B}$ are connected to the AND gate to generate the product $A\bar{B}$.

Discussion Question: What would the AND output be if each connection in Figure 11-17 were represented as a dot?

PLD Architectures—the PROM   There are several common architectures used for PLDs. The first one we will examine is the PROM. Figure 11-18(a) shows a 16 × 4 PROM which can also function as a type of PLD. It has four inputs that are completely decoded by the AND array; that is, each gate generates one of the 16 possible AND products. The connections from the input lines to the AND array are hard-wired, while the connections from the AND output product lines to the OR gate inputs are all programmable.

Figure 11-18(b) shows how the PROM would be programmed to generate four specified logic functions. Let's follow the procedure for output $O_3 = AB + \overline{C}\overline{D}$. The first step is to draw a truth table showing the desired O_3 output level for all possible input combinations:

D	C	B	A	O_3	
0	0	0	0	1	$\longrightarrow$ $\overline{D}\,\overline{C}\,\overline{B}\,\overline{A}$
0	0	0	1	1	$\longrightarrow$ $\overline{D}\,\overline{C}\,\overline{B}\,A$
0	0	1	0	1	$\longrightarrow$ $\overline{D}\,\overline{C}\,B\,\overline{A}$
0	0	1	1	1	$\longrightarrow$ $\overline{D}\,\overline{C}\,B\,A$
0	1	0	0	0	
0	1	0	1	0	
0	1	1	0	0	
0	1	1	1	1	$\longrightarrow$ $\overline{D}\,C\,B\,A$
1	0	0	0	0	
1	0	0	1	0	
1	0	1	0	0	
1	0	1	1	1	$\longrightarrow$ $D\,\overline{C}\,B\,A$
1	1	0	0	0	
1	1	0	1	0	
1	1	1	0	0	
1	1	1	1	1	$\longrightarrow$ $D\,C\,B\,A$

FIGURE 11-18   (a) PROM architecture makes it suitable for PLDs; (b) fuses blown to program outputs for given functions.

Next, write down the AND products for those cases where the output is to be a 1. The O_3 output is to be the OR sum of these products. Thus, only those fuses that connect these product terms to the inputs of OR gate 3 are to be left intact. All others are to be blown, as indicated in Figure 11-18(b). This same procedure is followed to determine the status of the fuses at the other OR gate inputs.

The PROM can generate any possible logic function of the input variables because it generates every possible AND product term. In general, any application that

Teaching Tip: Verify that adding a fifth input, *E*, will increase the number of fuses from 64 to 128. In general, the number of fuses will equal $2^N \times M$, where N is the number of inputs and M is the number of outputs.

Discussion Question: How many fuses does the AM27S13 contain?

requires every input combination to be available is a good candidate for a PROM. However, PROMs become impractical when a large number of input variables have to be accommodated, because the number of fuses doubles for each added input variable.

An example of an actual PROM that is often used as a PLD is the AM27S13, which is a 512×4 PROM manufactured using high-speed Schottky TTL technology. Since $512 = 2^9$, this PROM has nine address inputs and four data outputs. Thus, the AM27S13 can be programmed to generate four outputs each of which can be any logic function of the nine different inputs.

Programmable Array Logic (PAL)

The PROM architecture is well suited for those applications where every possible input combination is required to generate the output functions. Examples are code converters and data storage tables. There are many applications, however, that do not require that all input combinations be programmable. The logic functions shown in Figure 11-18(b), for example, do not use all of the available AND product lines. This has led to the development of a class of PLDs called *programmable array logic (PAL)*. The architecture of a PAL is slightly different from that of a PROM, as shown in Figure 11-19(a).

The PAL has the same AND and OR arrays as with PROMs, but in the PAL the inputs to the AND gates are programmable while the inputs to the OR gates are hard-wired. This means that every AND gate can be programmed to generate any desired product of the four input variables and their complements. Each OR gate is hard-wired to only four AND outputs. This limits each output function to four product terms. If a function requires more than four product terms, it cannot be implemented with this PAL; one having more OR inputs would have to be used. If fewer than four product terms are required, the unneeded ones can be made 0.

Figure 11-19(b) shows how this PAL is programmed to generate four specified logic functions. Let's follow the procedure for output $O_3 = AB + \overline{C}\overline{D}$. First, we have to express this output as the OR sum of four terms because the OR gates have four inputs. We do this by putting in 0s. Thus, we have

$$O_3 = AB + \overline{C}\overline{D} + 0 + 0$$

Teaching Tip: By how much will the number of fuses increase in Figure 11-19(a) if a fifth input is added? If a fifth output is added? (Answers: 32 in each case)

Next we have to determine how to program the inputs to AND gates 1, 2, 3, and 4 so that they provide the correct product terms to OR gate 3. We do this term by term. The first term, AB, is obtained by leaving intact the fuses that connect inputs A and B to AND gate 1, and blowing all other fuses on that line. Likewise, the second term, $\overline{C}\overline{D}$, is obtained by leaving intact only the fuses that connect inputs $\overline{C}$ and $\overline{D}$ to AND gate 2. The third term is a 0. A constant 0 is produced at the output of AND gate 3 by leaving all of its input fuses intact. This would produce an output of $A\overline{A}B\overline{B}C\overline{C}D\overline{D}$, which, as we know, is 0. The fourth term is also 0, and so the input fuses to AND gate 4 are also left intact.

The inputs to the other AND gates are programmed in a like manner to generate the other output functions. Note especially that many of the AND gates have all of their input fuses intact, since they need to generate 0s.

An example of an actual PAL integrated circuit is the PAL18L8A from Texas Instruments, Inc. It is manufactured using low-power Schottky technology and has 10 logic inputs and 8 output functions. Each output OR gate is hard-wired to seven AND-gate outputs, and so it can generate functions that include up to seven terms. An added feature of this particular PAL is that six of the eight outputs are fed back into the AND array, where they can be connected as inputs to any AND gate. This makes it very useful in generating all sorts of combinational logic.

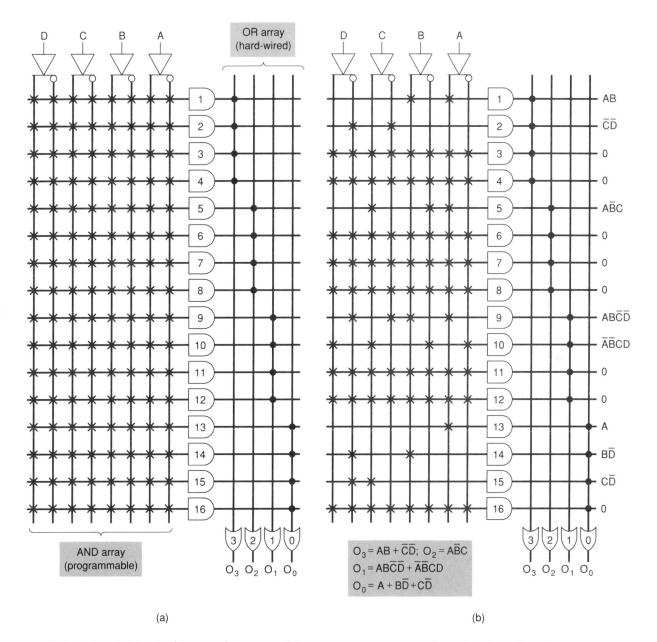

FIGURE 11-19 (a) Typical PAL architecture; (b) same PAL programmed for the given functions.

Programmable Logic Arrays (PLAs) A PLA combines the characteristics of the PROM and the PAL by providing both a programmable OR array and a programmable AND array. This feature, illustrated in Figure 11-20, makes it the most versatile of the three PLDs. However, it does have some disadvantages. Because it has two sets of fuses, it is more difficult to manufacture, program, and test than a PROM or PAL. PLAs are often referred to as *field-programmable arrays (FPLAs)*.

An example of an actual FPLA is the Texas Instruments TIFPLA840, which is specified to be a $14 \times 32 \times 6$ FPLA. It has 14 input variables, 32 AND gates to generate products of these variables, and 6 OR gates that can be any combination of

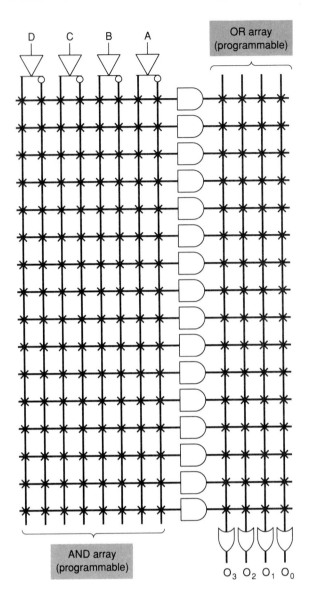

FIGURE 11-20   Architecture of a PLA shows both programmable AND array and programmable OR array for maximum versatility.

Discussion Question: Discuss whether or not an output of the TIFPLA840 can be programmed to be any function of the 14 input variables.

the 32 AND outputs. This IC also includes a *programmable output polarity* feature that permits the designer the option of inverting any of the outputs. This is illustrated in Figure 11-21, where an EX-OR gate and a polarity fuse have been added to the OR-gate output. The EX-OR gate will pass the OR gate output to O_3 with no inversion if the fuse is intact (logic 0). If the fuse is blown (i.e., a logic 1), the EX-OR gate will invert the OR gate output. This programmable output polarity is available on many PAL and FPLA devices.

Other PLD Features   Many PLDs include one or more of the following as part of their architecture: FFs, latches, input registers, and output registers. Very often the operating characteristics of these devices are programmable, as are the connections to the other devices on the chip. This gives the logic designer a great deal

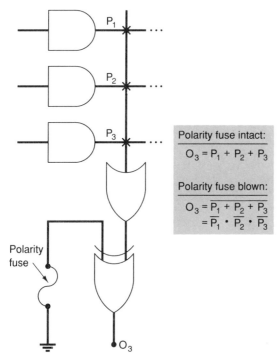

Polarity fuse intact:
$$\overline{O_3 = P_1 + P_2 + P_3}$$

Polarity fuse blown:
$$\overline{O_3 = \overline{P_1 + P_2 + P_3}}$$
$$= \overline{P_1} \cdot \overline{P_2} \cdot \overline{P_3}$$

FIGURE 11-21 Many PLDs include a programmable polarity feature that provides the option of inverting any of the output functions.

of flexibility in designing counters and other sequential logic circuits. This type of PLD is sometimes called a *programmable logic sequencer*.

Programming   When PLDs were first introduced, the logic designer would develop a *fuse map* that showed which fuses to blow and send it off to the PROM, PAL, or FPLA manufacturer. The manufacturer would then program the device according to the fuse map, test it, and ship it off to the designer. In recent years the availability of relatively inexpensive programming equipment has made it convenient for users to program their own PLDs. There are universal programmers on the market that can program most common PROMs, PALs, and FPLAs. The device to be programmed is plugged into a socket on the programmer, and the programmer will program and test the device according to data that have been supplied by the user.

The programming and test data are typically developed using commonly available software that will run on standard PCs. Using this software, the user enters data into the computer describing the logic functions he wants programmed into the PLD, as well as information on how the device is to be tested. The software then generates a fuse map and test data in a form that can be sent over a cable to the PLD programmer's memory. Once the programmer has the data, it can proceed to program and test the device. When finished, the programmer will indicate whether the device has passed or failed the testing procedure. If it passes, it can be removed from the programmer's socket and placed in the prototype circuit for further testing.

Erasable PLDs   The PLDs we have been talking about are programmed by blowing fuses. Once a fuse is blown, it cannot be reconnected. Thus, if you make a

mistake in programming, or you want to change the design, the device will no longer be useful and you might as well drop it into the wastebasket. This drawback has been addressed by several companies that have developed PLDs that can be erased and reprogrammed over and over. Called *erasable programmable logic devices (EPLDs)*, these devices are programmed and erased much like EPROMs and EEPROMs.

—— REVIEW QUESTIONS ——

1. What is a PLD?
2. What would output O_1 be in Figure 11-16 if fuses 1 and 2 were blown?
3. What does an X represent on a PLD diagram?
4. What does a dot represent on a PLD diagram?
5. Verify that the correct fuses are blown for the O_2, O_1, and O_0 functions in Figure 11-18(b).
6. How does the architecture of a PAL differ from a PROM?
7. Verify that the correct fuses are blown for the O_2, O_1, and O_0 functions in Figure 11-19(b).
8. How does the architecture of FLPAs differ from PROMs and PALs?
9. What functions are performed by a PLD programmer?

11-10 SEMICONDUCTOR RAMs

Teaching Tip: The ROMs we have studied are considered to be random-access memories since any location can be read from in the same amount of time. This can be confusing to the student who has just been told that RAM refers to semiconductor RWM.

Recall that the term RAM stands for random-access memory, meaning that any memory address location is as easily accessible as any other. Many types of memory can be classified as having random access, but when the term RAM is used with semiconductor memories it is usually taken to mean read/write memory (RWM) as opposed to ROM. Since it is common practice to use RAM to mean semiconductor RWM, we will do so throughout the following discussions.

RAMs are used in computers for the *temporary* storage of programs and data. The contents of many RAM address locations will be read from and written to as the computer executes a program. This requires fast read and write cycle times for the RAM so as not to slow down the computer operation.

A major disadvantage of RAMs is that they are volatile and will lose all stored information if power is interrupted or turned off. Some CMOS RAMs, however, use such small amounts of power in the standby mode (no read or write operations taking place) that they can be powered from batteries whenever the main power is interrupted. Of course, the main advantage of RAMs is that they can be written into and read from rapidly with equal ease.

The following discussion of RAMs will draw on some of the material covered in our treatment of ROMs, since many of the basic concepts are common to both types of memory.

11-11 RAM ARCHITECTURE

As with the ROM, it is helpful to think of the RAM as consisting of a number of registers, each storing a single data word, and each having a unique address. RAMs typically come with word capacities of 1K, 4K, 8K, 16K, 64K, 128K, 256K, and

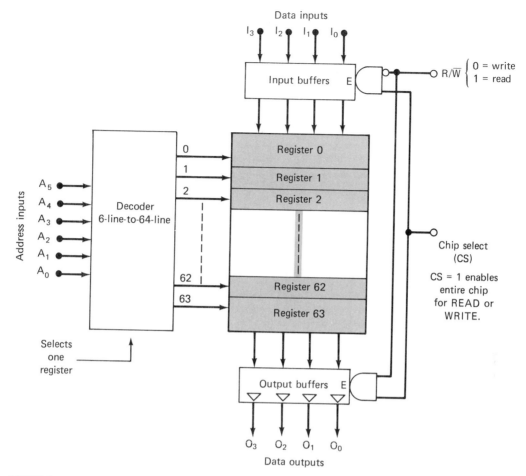

FIGURE 11-22   Internal organization of a 64 × 4 RAM.

1024K, and word sizes of 1, 4, or 8 bits. As we will see later, the word capacity and word size can be expanded by combining memory chips.

Figure 11-22 shows the simplified architecture of a RAM which stores 64 words of 4 bits each (i.e., a 64 × 4 memory). These words have addresses ranging from 0 to 63_{10}. In order to select one of the 64 address locations for reading or writing, a binary address code is applied to a decoder circuit. Since $64 = 2^6$, the decoder requires a 6-bit input code. Each address code activates one particular decoder output, which, in turn, enables its corresponding register. For example, assume an applied address code of

$$A_5 A_4 A_3 A_2 A_1 A_0 = 011010$$

Since $011010_2 = 26_{10}$, decoder output 26 will go high, selecting register 26 for either a read or write operation.

Read Operation   The address code picks out one register in the memory chip for reading or writing. In order to *read* the contents of the selected register, the READ/WRITE $(R/\overline{W})$* input must be a 1. In addition, the CHIP SELECT (CS) in-

---

*Some manufacturers use the symbol $\overline{WE}$ (write enable) or $\overline{W}$ instead of $R/\overline{W}$. In any case the operation is the same.

put must be activated (a 1 in this case). The combination of $R/\overline{W} = 1$ and $CS = 1$ enables the output buffers so that the contents of the selected register will appear at the four data outputs. $R/\overline{W} = 1$ also *disables* the input buffers so that the data inputs do not affect the memory during a read operation.

Write Operation   To write a new 4-bit word into the selected register requires $R/\overline{W} = 0$ and $CS = 1$. This combination *enables* the input buffers so that the 4-bit word applied to the data inputs will be loaded into the selected register. The $R/\overline{W} = 0$ also *disables* the output buffers, which are tristate, so that the data outputs are in their Hi-Z state during a write operation. The write operation, of course, destroys the word that was previously stored at that address.

Chip Select   Most memory chips have one or more CS inputs which are used to enable the entire chip or disable it completely. In the disabled mode all data inputs and data outputs are disabled (Hi-Z) so that neither a read nor a write operation can take place. In this mode the contents of the memory are unaffected. The reason for having CS inputs will become clear when we combine memory chips to obtain larger memories. Note that many manufacturers call these inputs CHIP ENABLE (CE). When the CS or CE inputs are in their active state, the memory chip is said to be *selected;* otherwise it is said to be *deselected.* Many memory ICs are designed to consume much lower power when they are deselected. In large memory systems, for a given memory operation, one or more memory chips will be selected while all others are deselected. More will be said on this later.

Common Input/Output Pins   In order to conserve pins on an IC package, manufacturers often combine the data input and data output functions using common input/output pins. The $R/\overline{W}$ input controls the function of these I/O pins. During a read operation, the I/O pins act as data outputs which reproduce the contents of the selected address location. During a write operation, the I/O pins act as data inputs.

We can see why this is done by considering the chip in Figure 11-22. With separate input and output pins, a total of 18 pins is required (including ground and power supply). With four common I/O pins, only 14 pins are required. The pin savings becomes even more significant for chips with larger word size.

___ EXAMPLE 11-9 ___

The 2147 is an NMOS RAM that is organized as a 4K × 1 with separate data input and output, and a single active-LOW chip-select input. Draw the logic symbol for this chip, showing all pin functions.

Solution   The logic symbol is shown in Figure 11-23(a).

___ EXAMPLE 11-10 ___

The 2168 is an NMOS RAM organized as a 4K × 4 with common input/output pins and an active-LOW chip-enable. Draw its logic symbol.

Solution   The logic symbol is shown in Figure 11-23(b).

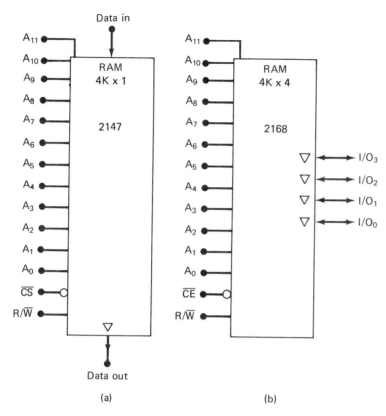

FIGURE 11-23   (a) Logic symbol for 2147 RAM; (b) for 2168 RAM.

REVIEW QUESTIONS

1. Describe the input conditions needed to read a word from a specific RAM address location.
2. Why do some RAM chips have common input/output pins?
3. How many pins are required for a 256×8 RAM with common I/O and one *CS* input?

11-12   STATIC RAM (SRAM)

The RAM operation that we have been discussing up to this point applies to a *static RAM*—one that can store data as long as power is applied to the chip. Static-RAM memory cells are essentially flip-flops that will stay in a given state (store a bit) indefinitely, provided that power to the circuit is not interrupted. Later we will describe *dynamic* RAMs, which store data as charges on capacitors. With dynamic RAMs the stored data will gradually disappear because of capacitor discharge, so that it is necessary to periodically refresh the data (i.e., recharge the capacitors).

FIGURE 11-24   Typical bipolar and NMOS static RAM cells.

Static RAMs (SRAMs) are available in bipolar and MOS technologies, although the vast majority of applications use NMOS or CMOS RAMs. As stated earlier, the bipolars have the advantage in speed (though NMOS is gradually closing the gap), and MOS devices have much greater capacities and lower power consumption. Figure 11-24 shows for comparison a typical bipolar static memory cell and a typical NMOS static memory cell. The bipolar cell contains two bipolar transistors and two resistors, while the NMOS cell contains four N-channel MOSFETs. The bipolar cell requires more chip area than the MOS cell because a bipolar transistor is more complex than a MOSFET, and because the bipolar cell requires separate resistors while the MOS cell uses MOSFETS as resistors (Q_3 and Q_4). A CMOS memory cell would be similar to the NMOS cell except that it would use P-channel MOSFETs in place of Q_3 and Q_4. This results in the lowest power consumption but increases the chip complexity.

Static RAM Timing   RAM ICs are most often used as the internal memory of a computer. The CPU (central processing unit) continually performs read and write operations on this memory at a very fast rate determined by the limitations of the CPU. The memory chips that are interfaced to the CPU have to be fast enough to respond to the CPU read and write commands, and a computer designer has to be concerned with the RAM's various timing characteristics.

Not all RAMs have the same timing characteristics, but most of them are similar, and so we will use a typical set of characteristics for illustrative purposes. The nomenclature for the different timing parameters will vary from one manufacturer to another, but the meaning of each parameter is usually easy to determine from the memory timing diagrams on the RAM data sheets. Figure 11-25 shows the timing diagrams for a complete read cycle and complete write cycle for a typical RAM chip.

Read Cycle   The waveforms in Figure 11-25(a) show how the address, $R/\overline{W}$, and chip-select inputs behave during a memory read cycle. As noted, the CPU supplies these input signals to the RAM when it wants to read data from a specific RAM

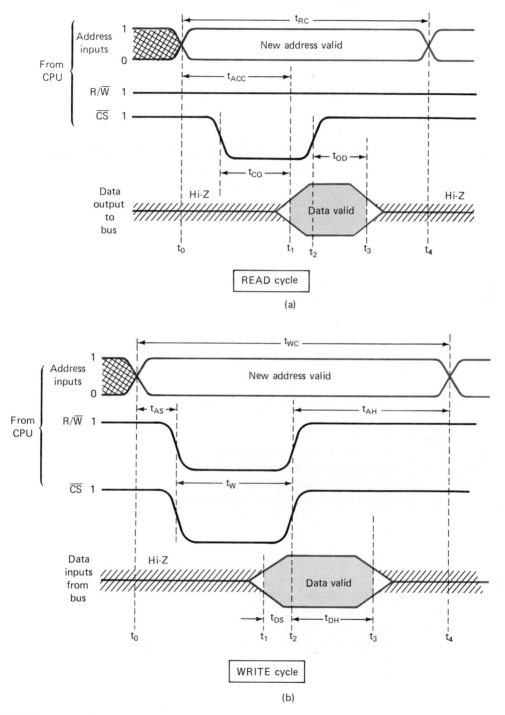

FIGURE 11-25 Typical timing for static RAM: (a) read cycle; (b) write cycle.

address location. Although a RAM may have many address inputs coming from the CPU's address bus, for clarity the diagram shows only two. The RAM's data output is also shown; we will assume that this particular RAM has one data output. Recall that the RAM's data output is connected to the CPU data bus (Figure 11-5).

The read cycle begins at time t_0. Prior to that time, the address inputs will be whatever address is on the address bus from the preceding operation. Since the RAM's chip-select is not active, it will not respond to this "old" address. Note that the $R/\overline{W}$ line is HIGH prior to t_0 and stays HIGH throughout the read cycle. In most memory systems, $R/\overline{W}$ is normally kept in the HIGH state except when it is driven LOW during a write cycle. The RAM's data output is in its Hi-Z state since $\overline{CS} = 1$.

At t_0, the CPU applies a new address to the RAM inputs; this is the address of the location to be read. After allowing time for the address signals to stabilize, the $\overline{CS}$ line is activated. The RAM responds by placing the data from the addressed location onto the data output line at t_1. The time between t_0 and t_1 is the RAM's access time, t_{ACC}, and is the time between the application of the new address and the appearance of valid output data. The timing parameter, t_{CO}, is the time it takes for the RAM output to go from Hi-Z to a valid data level once $\overline{CS}$ is activated.

At time t_2, the $\overline{CS}$ is returned HIGH, and the RAM output returns to its Hi-Z state after a time interval, t_{OD}. Thus, the RAM data will be on the data bus between t_1 and t_3. The CPU can take it from the data bus at any point during this interval. In most computers, the CPU will use the PGT of the $\overline{CS}$ signal at t_2 to latch these data into one of its internal registers.

The complete read cycle time, t_{RC}, extends from t_0 to t_4, when the CPU changes the address inputs to a different address for the next read or write cycle.

Write Cycle   Figure 11-25(b) shows the signal activity for a write cycle that begins when the CPU supplies a new address to the RAM at a time t_0. The CPU drives the $R/\overline{W}$ and $\overline{CS}$ lines LOW after waiting for a time interval t_{AS}, called the address setup time. This gives the RAM's address decoders time to respond to the new address. $R/\overline{W}$ and $\overline{CS}$ are held LOW for a time interval t_W, called the write-time interval.

During this write-time interval at time t_1 the CPU applies valid data to the data bus to be written into the RAM. These data have to be held at the RAM input for at least a time interval t_{DS} prior to, and for at least a time interval t_{DH} after, the deactivation of $R/\overline{W}$ and $\overline{CS}$ at t_2. The t_{DS} interval is called the data setup time, and t_{DH} is called the data hold time. Similarly, the address inputs have to remain stable for the address hold-time interval, t_{AH}, after t_2. If any of these setup- or hold-time requirements are not met, the write operation will not take place reliably.

The complete write-cycle time, t_{WC}, extends from t_0 to t_4, when the CPU changes the address lines to a new address for the next read or write cycle.

The read cycle time, t_{RC}, and write cycle time, t_{WC}, are what essentially determine how fast a memory chip can operate. For example, in an actual application, a CPU will often be reading successive data words from memory one right after the other. If the memory has a t_{RC} of 50 ns, the CPU can read one word every 50 ns, or 20 million words per second (20 MHz).

Actual SRAM Chip   An example of an actual SRAM IC is the CMOS 6264, which is an 8K × 8 with read and write cycle times of 100 ns and a standby power consumption of only 0.1 mW. The logic symbol for the 6264 is shown in Figure 11-26. Notice that it has 13 address inputs, since $2^{13} = 8192 = 8K$, and eight data I/O

	Inputs				
Mode	$\overline{WE}$	$\overline{CS_1}$	CS_2	$\overline{OE}$	I/O pins
READ	1	0	1	0	DATA$_{OUT}$
WRITE	0	0	1	X	DATA$_{IN}$
Output disable	1	X	X	1	High Z
Not selected	X	1	X	X	High Z
(power down)	X	X	0	X	

X = don't care

FIGURE 11-26   Symbol and mode table for CMOS 6264 SRAM.

lines. The four control inputs determine the device's operating mode according to the accompanying mode table.

The $\overline{WE}$ input is the same as the $R/\overline{W}$ input we have been using. A LOW at $\overline{WE}$ will write data into the RAM provided that the device is selected—both chip select inputs are active. Note how the & symbol is used to denote that *both* have to be active. A HIGH at $\overline{WE}$ will produce the read operation provided that the device is selected and the output buffers are enabled by $\overline{OE}$ = LOW. When deselected, the device is in its low-power mode and none of the other inputs have any effect.

Teaching Tip: Some students might ask why the $\overline{OE}$ pin does not have to be HIGH during a WRITE operation. The LOW at $\overline{WE}$ will disable the SRAM's output buffers. A HIGH at $\overline{WE}$ and a LOW at $\overline{OE}$ are required to enable the output buffers.

___ REVIEW QUESTIONS _____

1. How does a static RAM cell differ from a dynamic RAM cell?
2. Which memory technology generally uses the least power?
3. What device places data on the data bus during a read cycle?
4. What device places data on the data bus during a write cycle?
5. What RAM timing parameters determine its operating speed?
6. *True or false:* A LOW at $\overline{OE}$ will enable the output buffers of a 6264 SRAM provided both chip select inputs are active.

11-13   DYNAMIC RAM (DRAM)

Dynamic RAMs are fabricated using MOS technology and are noted for their high capacity, low power requirement, and moderate operating speed. As we stated earlier, unlike static RAMs, which store information in FFs, dynamic RAMs store 1s and 0s as charges on a small MOS capacitor (typically a few picofarads). Because of the tendency for these charges to leak off after a period of time, dynamic RAMs require periodic recharging of the memory cells; this is called **refreshing** the dynamic RAM. Typically, each cell must be refreshed at least every 2 to 10 ms or its data will be lost.

The need for refreshing is a disadvantage of dynamic RAMs as compared to static RAMs because it adds a requirement to the memory system design. Up until recently, system designers had to include added circuitry to implement the memory refresh operation during time intervals when the memory was not being accessed for a read or write operation. Now there are two alternatives available to designers to help neutralize this disadvantage. For relatively small memories (<64K words) the

integrated RAM (*iRAM*) provides a solution. An iRAM is an IC that includes the refresh circuitry on the same chip with the memory cell array. The result is a chip that externally operates just like a static RAM chip—you apply the address and collect the data—but internally uses the high-density dynamic RAM structure. The designer does not have to be concerned with the memory refresh operation; it is done internally and automatically.

For larger memory systems (>64K), a more cost-effective approach uses LSI chips called *dynamic memory controllers* which contain all of the necessary logic for refreshing the dynamic RAM chips that make up the system. This greatly reduces the amount of extra circuitry in a dynamic RAM system.

For applications where speed and reduced complexity are more critical than space and power considerations, static RAMs are still the best. They are generally faster than dynamic RAMs and require no refresh operation. They are simpler to design with, but they cannot compete with the higher capacity and lower power requirement of dynamic RAMs.

Because of their simple cell structure, DRAMs typically have 4 times the density of SRAMs. This allows 4 times as much memory capacity to be placed on a single board, or alternatively, requires one-fourth as much board space for the same amount of memory. The cost per bit of dynamic RAM storage is typically one-fifth to one-fourth that of static RAMs. A further cost saving is realized because the lower power requirements of a dynamic RAM, typically one-sixth to one-half those of a static RAM, allow the use of smaller, less expensive power supplies.

The main applications of SRAMs are in areas where only small amounts of memory are needed (up to 64K), or where high speed is required. Many microprocessor-controlled instruments and appliances have very small memory capacity requirements. Some instruments, such as digital storage oscilloscopes and logic analyzers, require very high-speed memory. For applications such as these, SRAM is normally used.

The main internal memory of most personal microcomputers (e.g., IBM PC or APPLE) uses DRAM because of its high capacity and low power consumption. These computers, however, sometimes use some small amounts of SRAM for functions requiring maximum speed such as video graphics and look-up tables.

—— REVIEW QUESTIONS ——

1. What are the main drawbacks of dynamic RAM compared with static?
2. List the advantages of dynamic RAM as compared with static RAM.
3. What are the two means available to dynamic memory system designers for handling the refresh operation without the need for a lot of extra circuitry?

11-14 DYNAMIC RAM STRUCTURE AND OPERATION

The dynamic RAM's internal architecture can be visualized as an array of single-bit cells as illustrated in Figure 11-27. Here there are 16,384 cells arranged in a 128×128 array. Each cell occupies a unique row and column position within the array. Fourteen address inputs are needed to select one of the cells ($2^{14} = 16,384$); the lower address bits, A_0 to A_6, select the row, and the higher-order bits, A_7 to A_{13},

Column address inputs

A_7 A_8 A_9 A_{10} A_{11} A_{12} A_{13}

1-of-128 decoder — Selects 1 of 128 columns

Memory cell

128 rows

Row address inputs

A_6 A_5 A_4 A_3 A_2 A_1 A_0

1-of-128 decoder

Selects 1 of 128 rows

128 columns

FIGURE 11-27   Cell arrangement in 16K × 1 dynamic RAM.

select the column. Each 14-bit address selects a unique cell to be written into or read from. The structure in Figure 11-27 is a 16K × 1 DRAM chip. DRAM chips are currently available in capacities of 1024K × 1 (1M × 1) and 256K × 4. DRAMs with a 4-bit word size have a cell arrangement similar to Figure 11-27 except that each position in the array contains four cells, and each applied address selects a group of four cells for a read or write operation. As we will see later, larger word sizes can be attained by combining several chips in the appropriate arrangement.

Figure 11-28 is a symbolic representation of a dynamic memory cell and its associated circuitry. Many of the circuit details are not shown, but this simplified diagram can be used to describe the essential ideas involved in writing to and reading from a DRAM. The switches S1 through S4 are actually MOSFETs that are controlled by various address decoder outputs and the $R/\overline{W}$ signal. The capacitor, of course, is the actual storage cell.

To write data to the cell, signals from the address decoding and read/write logic will close switches S1 and S2, while keeping S3 and S4 open. This connects the input data to C. A logic 1 at the data input charges C, and a logic 0 discharges it. Then the switches are open so that C is disconnected from the rest of the circuit.

Discussion Question: Assume that the cell in Figure 11-28 is in row 0, column 127. What conditions on the address and $R/\overline{W}$ inputs will cause S1 to close? Repeat for S2.

FIGURE 11-28   Symbolic representation of a dynamic memory cell. During a WRITE operation, semiconductor switches S1 and S2 are closed. During a read operation, all switches are closed except S1.

DATA IN

S1   S2 C S3   +   Sense amplifier   S4   DATA OUT

V_{REF} −

Ideally, C would retain its charge indefinitely, but there is always some leakage path through the off switches, so that C will gradually lose its charge.

To read data from the cell, switches S2, S3, and S4 are closed and S1 is kept open. This connects the stored capacitor voltage to the *sense amplifier*. The sense amplifier compares the voltage with some reference value to determine if it is a logic 0 or 1, and produces a solid 0V or 5V for the data output. This data output is also connected to C (S2 and S4 are closed) and refreshes the capacitor voltage by recharging or discharging. In other words, the data bit in a memory cell is refreshed each time it is read.

Discussion Question: Why do switches S1 and S2 have to be bilateral switches?

Address Multiplexing   The 16K $\times$ 1 DRAM array depicted in Figure 11-27 has 14 address inputs. A 64K $\times$ 1 DRAM array would have 16 address inputs.

FIGURE 11-29   (a) Simplified architecture of the 4116, a 16K $\times$ 1 DRAM; (b) $\overline{RAS}/\overline{CAS}$ timing.

630

High-capacity memory chips such as these would require many pins if each address input required a separate pin. In order to reduce the number of pins on their high-capacity DRAM chips, manufacturers utilize *address multiplexing* whereby each address input pin can accommodate two different address bits. The saving in pin count translates to a significant decrease in the size of the IC packages. This is very important in large-capacity memory boards, where you want to maximize the amount of memory that can fit on one board.

We will use the 4116, a 16K × 1 DRAM, to illustrate the address multiplexing idea. A simplified diagram of this chip's internal architecture is shown in Figure 11-29(a). It contains an array of cells arranged as 128 rows by 128 columns. There are a single data input line, a single data output line, and a $R/\overline{W}$ input. There are seven address inputs, and each one has a dual function (e.g., A_0/A_7 will function as both A_0 and A_7). Two address **strobe** inputs are included for clocking the row and column addresses into their respective on-chip registers. The **row address strobe** $\overline{RAS}$ clocks the 7-bit row address register, and the **column address strobe** $\overline{CAS}$ clocks the 7-bit column address register.

A 14-bit address is applied to this DRAM in two steps using $\overline{RAS}$ and $\overline{CAS}$. The timing is shown in Figure 11-29(b). Initially, $\overline{RAS}$ and $\overline{CAS}$ are both HIGH. At time t_0, the 7-bit row address (A_0 to A_6) is applied to the address inputs. After allowing time for the setup-time requirement (t_{RS}) of the row address register, the $\overline{RAS}$ input is driven LOW at t_1. This NGT loads the row address into the row address register so that A_0 to A_6 now appear at the row decoder inputs. The LOW at $\overline{RAS}$ also enables this decoder so that it can decode the row address and select one row of the array.

At time t_2, the 7-bit column address (A_7 to A_{13}) is applied to the address inputs. At t_3, the $\overline{CAS}$ input is driven LOW to clock the column address into the column address register. $\overline{CAS}$ also enables the column decoder so that it can decode the column address and select one column of the array.

At this point the two parts of the address are in their respective registers, the decoders have decoded them to select the one cell corresponding to the row and column address, and a read or a write operation can be performed on that cell just as in a static RAM.

You may have noticed that this DRAM does not have a chip-select (CS) input. The $\overline{RAS}$ and $\overline{CAS}$ signals perform the chip-select function since they must both be LOW for the decoders to select a cell for reading or writing.

Teaching Tip: Many DRAM ICs have a low-power standby mode when the $\overline{RAS}$ and $\overline{CAS}$ inputs are both HIGH.

—— EXAMPLE 11-11 ——————————————————————

How many pins are saved by using address multiplexing for the 16K × 1 DRAM?

Solution   Seven address inputs are used instead of 14; $\overline{RAS}$ and $\overline{CAS}$ are added; no CS is required. Thus, there is a net saving of *six* pins.

———————————————————————————————————

In a typical computer system, the address inputs to the memory system come from the central processing unit (CPU). When the CPU wants to access a particular memory location, it generates the complete address and places it on address lines that make up an **address bus.** Figure 11-30(a) shows this for a memory that has a capacity of 16K words and therefore requires a 14-line address bus going directly from the CPU to the memory.

This arrangement works for ROM or for static RAM, but it has to be modified for DRAM that uses multiplexed addressing. If the memory is DRAM, it will have

(a)

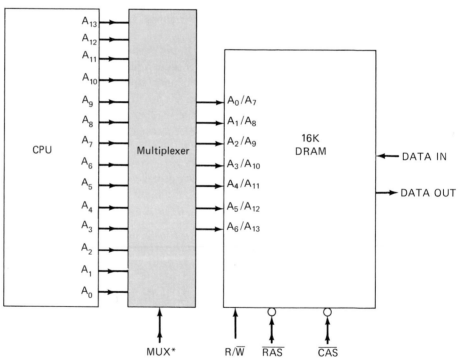

*MUX = 0 transmits CPU addresses
A_0–A_6 to DRAM. MUX = 1 transmits
A_7–A_{13} to DRAM.

(b)

FIGURE 11-30 (a) CPU address bus driving ROM or static RAM memory;
(b) CPU addresses driving a multiplexer that is used to multiplex the CPU
address lines into the DRAM.

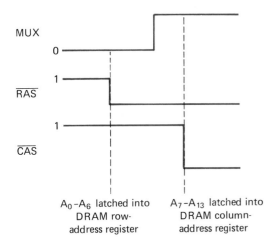

MUX

$\overline{RAS}$

$\overline{CAS}$

$A_0 - A_6$ latched into
DRAM row-
address register

$A_7 - A_{13}$ latched into
DRAM column-
address register

FIGURE 11-31   Timing required
for address multiplexing.

only 7 address inputs. This means that the 14 address lines from the CPU address bus have to be fed into a multiplexer circuit that will transmit 7 address bits at a time to the memory address inputs. This is shown symbolically in Figure 11-30(b). The multiplexer select input, labeled *MUX*, controls whether CPU address lines A_0 to A_6 or address lines A_7 to A_{13} will be present at the DRAM address inputs.

The timing of the *MUX* signal has to be synchronized with the $\overline{RAS}$ and $\overline{CAS}$ signals that clock the addresses into the DRAM. This is shown in Figure 11-31. *MUX* has to be LOW when $\overline{RAS}$ is pulsed LOW so that address lines A_0 to A_6 from the CPU will reach the DRAM address inputs to be latched on the NGT of $\overline{RAS}$. Likewise, *MUX* has to be HIGH when $\overline{CAS}$ is pulsed LOW so that A_7 to A_{13} from the CPU will be present at the DRAM inputs to be latched on the NGT of $\overline{CAS}$.

The actual multiplexing and timing circuitry will not be shown here but will be left to the end-of-chapter problems (Problems 11-25 and 11-26).

Teaching Tip: The multiplexer will generally consist of several MSI chips. The multiplexer outputs can usually drive all of the DRAMs in the memory system. The trade-off here is a few extra MSI chips to save pin count and therefore reduce package size of the DRAM ICs. Since many memory systems contain eight or more DRAM ICs, there is a net saving of board space.

REVIEW QUESTIONS

1. Describe the array structure of a 64K × 1 DRAM.
2. What is the benefit of address multiplexing?
3. How many address inputs would there be on a 1M × 1 DRAM chip?
4. What are the functions of the $\overline{RAS}$ and $\overline{CAS}$ signals?
5. What is the function of the *MUX* signal?

11-15   DRAM READ/WRITE CYCLES

The timing of the read and write operations of a DRAM is much more complex than for a static RAM, and there are many critical timing requirements that the DRAM memory designer has to consider. At this point, a detailed discussion of these requirements would probably cause more confusion than enlightenment. We will concentrate on the basic timing sequence for the read and write operations for a typical DRAM system like that of Figure 11-30(b).

DRAM Read Cycle   Figure 11-32 shows typical signal activity during the read operation. It is assumed that $R/\overline{W}$ is in its HIGH state throughout the operation. The

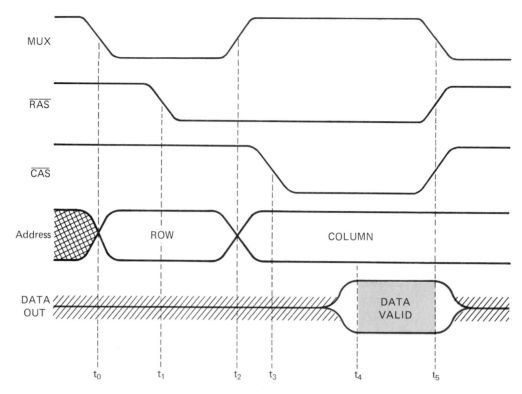

FIGURE 11-32 Signal activity for a read operation on a dynamic RAM. The R/$\overline{W}$ input (not shown) is assumed to be HIGH.

following is a step-by-step description of the events that occur at the times indicated on the diagram.

t_0: *MUX* is driven LOW to apply the row address bits (A_0 to A_6) to the DRAM address inputs.

t_1: $\overline{RAS}$ is driven LOW to latch the row address into the DRAM.

t_2: *MUX* goes HIGH to place the column address (A_7 to A_{13}) at the DRAM address inputs.

Discussion Question: Why can't the $\overline{RAS}$ signal be returned HIGH once the row address is latched into the DRAM?

t_3: $\overline{CAS}$ goes LOW to latch the column address into the DRAM.

t_4: The DRAM responds by placing valid data from the selected memory cell onto the DATA OUT line.

t_5: *MUX*, $\overline{RAS}$, $\overline{CAS}$, and DATA OUT return to their initial states.

DRAM Write Cycle Figure 11-33 shows typical signal activity during a DRAM write operation. Here is a description of the sequence of events.

t_0: The LOW at *MUX* places the row address at the DRAM inputs.

t_1: The NGT at $\overline{RAS}$ clocks the row address into the DRAM.

t_2: *MUX* goes HIGH to place the column address at the DRAM inputs.

t_3: The NGT at $\overline{CAS}$ latches the column address into the DRAM.

t_4: Data to be written are placed on the DATA IN line.

t_5: $R/\overline{W}$ is pulsed LOW to write the data into the selected cell.

t_6: Input data are removed from DATA IN.

t_7: *MUX*, $\overline{RAS}$, $\overline{CAS}$, and $R/\overline{W}$ are returned to their initial states.

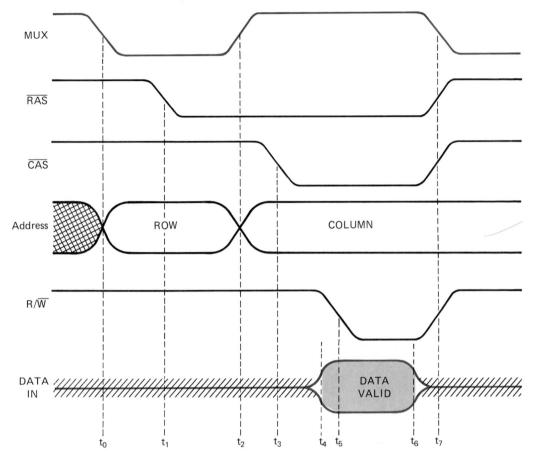

FIGURE 11-33 Signal activity for a write operation on a dynamic RAM.

11-16 DRAM REFRESHING

DRAM cells have to be refreshed periodically (typically, every 2 ms) or the stored data will be lost. As we saw in our discussion of the DRAM cell of Figure 11-28, a DRAM cell is refreshed every time a read operation is performed on that cell. It would appear then that for a 16K × 1 DRAM it would be necessary to perform read operations at the rate of one every 122 ns (2 ms/16,384 = 122 ns). This is much too fast for most currently available DRAMs, and even if the DRAMs were fast enough, it is highly unlikely that during normal operation every cell will be read from. For this reason, manufacturers have designed their DRAM chips so that

> *whenever a read operation is performed on a cell, all the cells in that same row will be refreshed.*

*R/$\overline{W}$ and $\overline{CAS}$ lines held HIGH; MUX held LOW

FIGURE 11-34   A 16K DRAM can be refreshed by sequential latching of row addresses using *RAS*.

This greatly reduces the number of read operations that must take place in order to refresh the complete memory; it is necessary to read from each of the 128 rows only once every 2 ms. It is still highly unlikely that during normal operation each of the 128 rows will be read from, and so the DRAM refresh operation has to be performed by some other means.

The most common means is provided by a 7-bit **refresh counter** which is used to cycle through the 128 different row addresses. The counter will start at 0000000 corresponding to row 0. This address is applied to the DRAM address inputs ($MUX = 0$), and $\overline{RAS}$ is pulsed LOW (see Figure 11-34) while R/$\overline{W}$ and $\overline{CAS}$ are

FIGURE 11-35   The 3242 DRAM controller performs the address multiplexing and refresh address counting for a 16K DRAM.

REFRESH ENABLE	ROW ENABLE	Controller output
HIGH	X	Refresh address (from internal counter)
LOW	HIGH	Row address (A_0–A_6 from CPU)
LOW	LOW	Column address (A_7–A_{13} from CPU)

kept HIGH. This refreshes row 0. The counter is incremented and the sequence is repeated up to row 127. This complete refresh process can be done in about 50 μs.

While the refresh counter idea seems easy enough, we must realize that the row addresses from the refresh counter cannot interfere with the addresses coming from the CPU during normal read/write operations. For this reason, the refresh counter addresses have to be multiplexed with the CPU addresses, so that the proper source of DRAM addresses is activated at the proper times.

Most manufacturers of dynamic memory ICs have developed special ICs to handle the refresh operation as well as the address multiplexing needed by DRAM systems. These ICs are called **dynamic RAM controllers.** We will look at a typical IC of this type, the Intel 3242, which is designed for use with 16K DRAMs. The block diagram in Figure 11-35 shows the 3242 used with a 16K × 1 DRAM.

The 3242 outputs a 7-bit multiplexed address to the DRAM's address inputs. There are three possible sources for this address. The 7-bit refresh counter is incremented by an external clock signal at its $\overline{COUNT}$ input. This counter supplies the row addresses to the DRAM during a refresh operation. The 3242 also takes the 14-bit address from the CPU address bus and multiplexes it into the row and column addresses that are used when the CPU performs a read or write operation on the DRAM. The logic levels applied to the REFRESH ENABLE and ROW ENABLE inputs determine which 7-bit address will appear at the controller outputs according to the table given in the diagram.

Teaching Tip: The REFRESH ENABLE signal is activated during those times when the CPU is not accessing the DRAM. In most computers, there are many times during the execution of a program when the CPU is busy doing internal operations and is not accessing memory.

_____ REVIEW QUESTIONS _____

1. What is the function of a refresh counter?
2. What functions does a DRAM controller perform?

11-17 EXPANDING WORD SIZE AND CAPACITY

In most IC memory applications the required memory capacity or word size cannot be satisfied by one memory chip. Instead, several memory chips have to be combined to provide the desired capacity and word size. We will see how this is done through several examples that illustrate all the important concepts that will be needed when we interface memory chips to a microprocessor.

Expanding Word Size   Suppose we need a memory that can store sixteen 8-bit words and all we have are RAM chips which are arranged as 16 × 4 memories with common I/O lines. We can combine two of these 16 × 4 chips to produce the desired memory. The configuration for doing so is shown in Figure 11-36. Examine this diagram carefully and see what you can find out from it before reading on.

Since each chip can store sixteen 4-bit words and we want to store sixteen 8-bit words, we are using each chip to store *half* of each word. In other words, RAM-0 stores the four *higher*-order bits of each of the 16 words, and RAM-1 stores the four *lower*-order bits of each of the 16 words. A full 8-bit word is available at the RAM outputs connected to the data bus.

Any one of the 16 words is selected by applying the appropriate address code to the four-line *address bus* (AB_3, AB_2, AB_1, AB_0). The address lines typically origi-

FIGURE 11-36 Combining two 16 × 4 RAMs for a 16 × 8 module.

nate at the CPU. Note that each address bus line is connected to the corresponding address input of each chip. This means that once an address code is placed on the address bus, this same address code is applied to both chips so that the same location in each chip is accessed at the same time.

Once the address is selected, we can read or write at this address under control of the common $R/\overline{W}$ and $\overline{CS}$ line. To read, $R/\overline{W}$ must be high and $\overline{CS}$ must be low. This causes the RAM I/O lines to act as *outputs*. RAM-0 places its selected 4-bit word on the upper four data bus lines and RAM-1 places its selected 4-bit word on the lower four data bus lines. The data bus then contains the full selected 8-bit word, which can now be transmitted to some other device (usually a register in the CPU).

To write, $R/\overline{W} = 0$ and $\overline{CS} = 0$ causes the RAM I/O lines to act as *inputs*. The 8-bit word to be written is placed on the data bus (usually by the CPU). The higher 4 bits will be written into the selected location of RAM-0 and the lower 4 bits will be written into RAM-1.

In essence, the combination of the two RAM chips acts like a single 16 × 8 memory chip. We would refer to this combination as a 16 × 8 *memory module*.

The same basic idea for expanding word size will work for many different situations. Read the following example and draw a rough diagram for what the system will look like before looking at the solution.

━━ EXAMPLE 11-12 ━━━━━━━━━━━━━━━━━━━━━━━━━━━━━━━━━━━━━━━

THE 2125A is a static RAM IC that has a capacity of 1K × 1, one active-LOW chip select input, and separate data input and output. Show how to combine several 2125A ICs to form a 1K × 8 module.

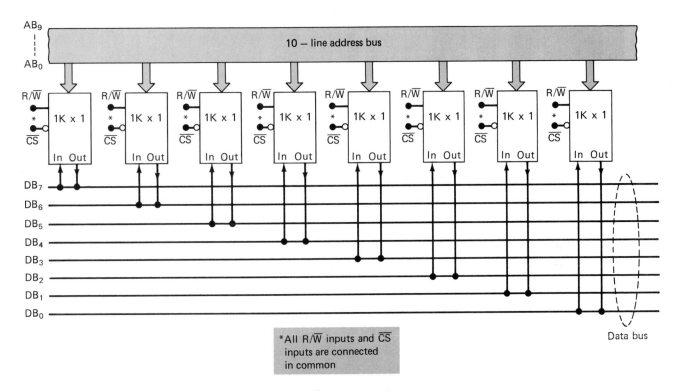

FIGURE 11-37   Eight 2125A 1K × 1 chips arranged as a 1K × 8 memory.

Solution   The arrangement is shown in Figure 11-37, where eight 2125A chips are used for a 1K × 8 module. Each chip stores one of the bits of each of the 1024 eight-bit words. Note that all of the $R/\overline{W}$ and $\overline{CS}$ inputs are wired together, and the 10-line address bus is connected to the address inputs of each chip. Also note that since the 2125A has separate data in and data out pins, both of these pins of each chip are tied to the same data bus line.

━━━

Expanding Capacity   Suppose we need a memory that can store thirty-two 4-bit words and all we have are the 16 × 4 chips. By combining two 16 × 4 chips as shown in Figure 11-38, we can produce the desired memory. Once again, examine this diagram and see what you can determine from it before reading on.

Each RAM is used to store sixteen 4-bit words. The data I/O pins of each RAM are connected to a common four-line data bus. Only one of the RAM chips can be selected (enabled) at one time so that there will be no bus-contention problems. This is ensured by driving the respective $\overline{CS}$ inputs from different logic signals.

Since the total capacity of this memory module is 32 × 4, there have to be 32 different addresses. This requires *five* address bus lines. The upper address line AB_4 is used to select one RAM or the other (via the $\overline{CS}$ inputs) as the one that will be

Address ranges: 00000 to 01111 — RAM-0
10000 to 11111 — RAM-1

Total 00000 to 11111 —(32 words)

FIGURE 11-38   Combining two 16 × 4 chips for a 32 × 4 memory.

read from or written into. The other four address lines AB_0 to AB_3 are used to select the one memory location out of 16 from the selected RAM chip.

To illustrate, when $AB_4 = 0$, the $\overline{CS}$ of RAM-0 enables this chip for read or write. Then, any address location in RAM-0 can be accessed by AB_3 through AB_0. The latter four address lines can range from 0000 to 1111 to select the desired location. Thus, the range of addresses representing locations in RAM-0 is

$$AB_4 AB_3 AB_2 AB_1 AB_0 = 00000 \text{ to } 01111$$

Note that when $AB_4 = 0$, the $\overline{CS}$ of RAM-1 is high, so that its I/O lines are disabled (High-Z) and cannot communicate (give or take data) with the data bus.

It should be clear that when $AB_4 = 1$, the roles of RAM-0 and RAM-1 are reversed. RAM-1 is now enabled and the lines AB_3 to AB_0 select one of its locations. Thus, the range of addresses located in RAM-1 is

$$AB_4 AB_3 AB_2 AB_1 AB_0 = 10000 \text{ to } 11111$$

_____ EXAMPLE 11-13 _____

It is desired to combine several 2K × 8 PROMs to produce a total capacity of 8K × 8. How many PROM chips are needed? How many address bus lines are required?

Solution   Four PROM chips are required, with each one storing 2K of the 8K words. Since 8K = 8 × 1024 = 8192 = 2^{13}, *thirteen* address lines are needed.

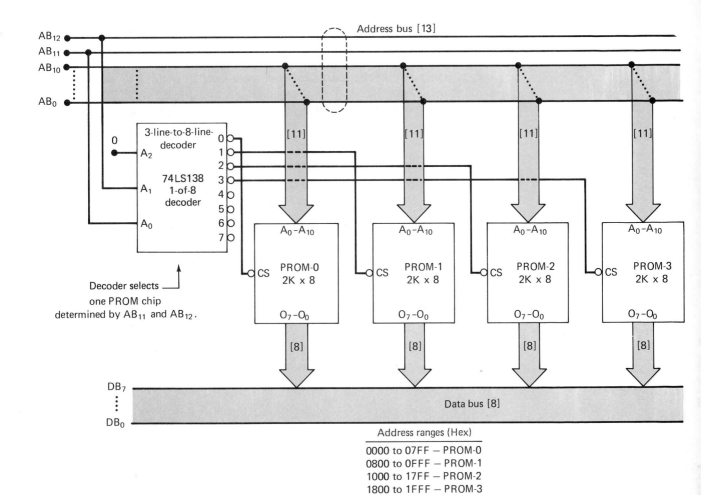

FIGURE 11-39   Four 2K × 8 PROMs arranged to form a total capacity of 8K × 8.

The configuration for the memory of Example 11-13 is similar to the 32 × 4 memory of Figure 11-38. However, it is slightly more complex, because it requires a decoder circuit for generating the $\overline{CS}$ input signals. The complete diagram for this 8192 × 8 memory is shown in Figure 11-39.

Since the total capacity is 8192 words, 13 address bus lines are required. The two highest-order lines, AB_{11} and AB_{12}, are used to select *one* of the PROM chips; the other eleven address bus lines go to each PROM to select the desired location within the selected PROM. The PROM selection is accomplished by feeding AB_{11} and AB_{12} into the 74LS138 decoder. The four possible combinations are decoded to generate active-LOW signals, which are applied to the $\overline{CS}$ inputs. For example, when $AB_{11} = AB_{12} = 0$, the 0 output of the decoder goes LOW (all others are HIGH) and enables PROM-0. This causes the PROM-0 outputs to generate the data word internally stored at the address determined by AB_0 through AB_{10}. All other PROMs are disabled, and so there is no bus contention.

While $AB_{12} = AB_{11} = 0$, the values of AB_{10} through AB_0 can range from all 0s to all 1s. Thus, PROM-0 will respond to the following range of 13-bit addresses:

$$AB_{12}\text{-}AB_0 = 0000000000000 \text{ to } 0011111111111$$

For convenience, these addresses can be more easily expressed in hexadecimal code to give a range of 0000 to 07FF.

Similarly, when $AB_{12} = 0$ and $AB_{11} = 1$, the decoder selects PROM-1, which then responds by outputting the data word it has internally stored at the address AB_{10} through AB_0. Thus, PROM-1 responds to the following range of addresses:

0100000000000 to 0111111111111 (binary)

or

0800 to 0FFF (hex)

You should verify the PROM-2 and PROM-3 address ranges given in the figure.

Discussion Question: How would this circuit operate if the A_2 input to the decoder is kept HIGH?

Clearly, address lines AB_{12} and AB_{11} are used to select one of the four PROM chips, while AB_{10} through AB_0 select the word stored in the selected PROM.

_____ EXAMPLE 11-14 _____

What size decoder would be needed to expand the memory of Figure 11-39 to 32K × 8? Describe what address lines are used.

Discussion Question: What size decoder would be needed if the PROMs were 4K × 8?

Solution   A 32K capacity will require 16 PROM chips. To select one of the 16 PROMs will require a 4-line-to-16-line decoder. Four address lines (AB_{14}, AB_{13}, AB_{12}, AB_{11}) will be connected as inputs to this decoder. The address lines AB_{10} to AB_0 are connected to the address inputs of each of the 16 PROMs. Thus, a total of 15 address lines are used. This agrees with the fact that $2^{15} = 32,768 = 32K$.

Combining DRAMs   As we stated earlier, DRAM ICs usually come with word sizes of 1 or 4 bits. In order to use these ICs in computer systems requiring word sizes of 8 or 16 bits, it is necessary to combine several of them in a similar manner to what we have done with the static RAMs and ROMs. Figure 11-40 shows how to combine eight 16K × 1 DRAMs to form a 16K × 8 memory module. The 3242 DRAM controller is used to provide the address multiplexing and refresh functions for each chip.

_____ REVIEW QUESTIONS _____

1. The 2114 is a 1K × 4 static RAM chip. How many of these are needed to form a 16K × 4 module?
2. How many 2114s are needed for a 1K × 16 module?
3. *True or false:* When memory chips are combined to form a module with a larger word size or capacity, the *CS* inputs of each chip are always connected together.
4. *True or false:* When memory chips are combined for a larger capacity, each chip is connected to the same data bus lines.

11-18   NONVOLATILE RAM

Semiconductor RAM devices have the definite advantage of high-speed operation, but they are volatile, which means that they will lose their data if power is interrupted even momentarily. ROM, of course, is nonvolatile, but it cannot be used as

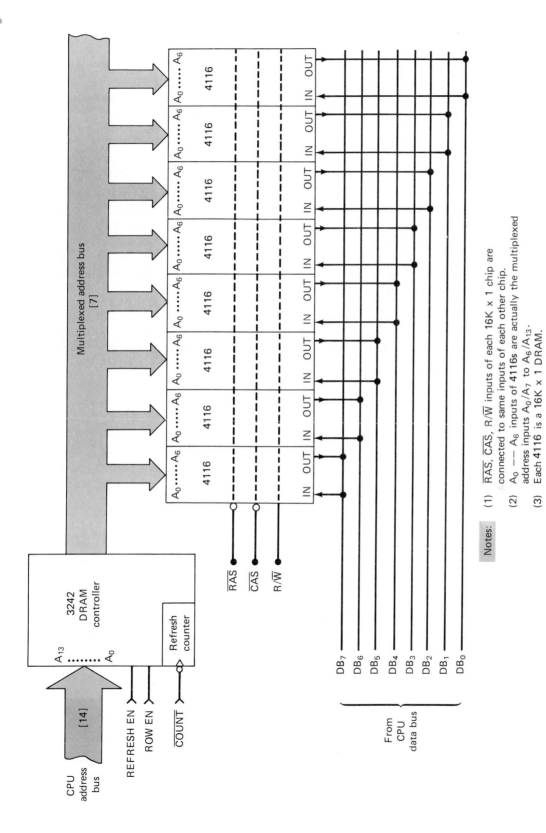

Notes:

(1) $\overline{RAS}$, $\overline{CAS}$, R/$\overline{W}$ inputs of each 16K × 1 chip are connected to same inputs of each other chip.

(2) A_0 — A_6 inputs of 4116s are actually the multiplexed address inputs A_0/A_7 to A_6/A_{13}.

(3) Each 4116 is a 16K × 1 DRAM.

FIGURE 11-40   Eight 16K × 1 DRAM chips combined to form a 16K × 8 memory module.

643

R/W memory. In some applications the volatility of RAM can mean the loss of important or crucial data in the event of a power failure. There are two solutions to this problem. The first is to use memory that can be powered from backup batteries whenever a power failure occurs. This requires memory that has a very low power consumption that will not rapidly drain the backup batteries. CMOS has the lowest power consumption of all semiconductor RAMs, and in many cases can be powered from backup batteries. Of course, when powered by batteries, the CMOS RAM chips are to be kept in their low-power standby mode for lowest power drain. Some CMOS SRAMs include a small lithium backup battery right on the chip.

Another solution is to use a device called a *nonvolatile RAM (NVRAM)*. A NVRAM contains a static RAM array and an EEPROM (electrically erasable PROM) array on the same chip. It combines the high-speed R/W operation of the static RAM with the nonvolatile storage capability of the EEPROM. Each cell in the static RAM has a corresponding cell in the EEPROM, and data can be transferred between corresponding cells in both directions. During normal operation, data are written into and read from the static RAM cells as with any conventional SRAM IC. When a power failure occurs or the power is turned off, the following sequence of events takes place:

- An external voltage-sensing circuit detects a drop in the ac source voltage, and sends a signal to the NVRAM's STORE input. Alternatively, the CPU may send the STORE signal after it has received a power-failure interrupt signal.

- This causes the NVRAM to transfer the contents of all static RAM cells to their corresponding EEPROM cells *in parallel* so that the complete transfer takes only a few milliseconds. Because of its large output capacitors, the 5-V dc supply will keep the NVRAM powered up long enough for this transfer to be completed. The EEPROM now holds a copy of the RAM data at the instant of the power failure.

- When power is restored, the NVRAM will automatically transfer the data from all EEPROM cells back into the RAM. The RAM now holds the same data it had when the power interruption occurred.

The NVRAM has the advantage of not needing a battery backup. However, an NVRAM is more complex than a normal memory chip because it contains both RAM and EEPROM cells, and the circuitry required to transfer data between the two. For this reason, NVRAMs are not available in very high capacities. When high-capacity nonvolatile memory is required, it is best to use CMOS RAM with battery backup.

REVIEW QUESTIONS

1. What are the ways of achieving nonvolatile RAM storage?
2. How many memory cells does a 512 × 8 NVRAM contain?

11-19 SEQUENTIAL MEMORIES

The semiconductor memories that we have discussed are random-access memories because the data from any memory location can be obtained quickly without sequencing through other memory locations. The high-speed operation of random-

access devices makes them suitable for use as the internal memory of a computer. Sequential-access semiconductor memories utilize shift registers to store data that can be accessed in a sequential manner. Although they are not useful as internal computer memory because of their relatively slow speed, shift-register memories find application in areas where sequential, repetitive data are required. A primary example is the storage and sequential transmission of ASCII-coded data for the characters on a video display. This data have to be supplied to the video display circuits periodically in order to refresh the displayed image on the screen. By using shift registers, the stored data can be *recirculated* to refresh the screen image periodically. Shift-register memories are also used in digital storage oscilloscopes and logic analyzers.

Recirculating Shift Register   Figure 11-41(a) shows the block diagram for a recirculating shift register. There can be any number of FFs in the shift register, and we will use four FFs for our illustrations. Data enter the shift register from the serial input D_s, which shifts into Q_3, Q_3 shifts into Q_2, Q_2 into Q_1, and Q_1 into Q_0. The Q_0 output is recirculated back to the serial input through some control logic. This logic provides two modes of operation that are controlled by the *recirculation* input *REC*. The level at *REC* determines the source of the data that will reach the serial input.

RECIRCULATE MODE *(REC = 1)*   In this mode, the upper AND gate is enabled, and the Q_0 output (DATA OUT) is applied to D_s. As clock pulses are applied, the data in the shift register will recirculate as shown below:

$$Q_3 \longrightarrow Q_2 \longrightarrow Q_1 \longrightarrow Q_0 \longrightarrow DATA\ OUT$$

While recirculating in the register, the data also appear at DATA OUT, one bit at a time. In this mode, DATA IN is inhibited and has no effect on the register data.

FIGURE 11-41   (a) Recirculating shift register; (b) simplified representation.

(a)

REC = 1:   recirculate mode

REC = 0:   data input mode

(b)

645

DATA INPUT MODE (*REC* = 0)   In this mode, the lower AND gate is enabled, and the DATA IN signal is applied to D_S. As clock pulses are applied, the data will shift as shown below:

Discussion Question: What logic function is being performed by the control logic?

DATA IN $\longrightarrow$ Q_3 $\longrightarrow$ Q_2 $\longrightarrow$ Q_1 $\longrightarrow$ Q_0 $\longrightarrow$ DATA OUT

There is no recirculation of data, since the upper AND gate is inhibited by *REC* = 0. This mode is used to enter new data at DATA IN for storage in the register.

Figure 11-41(b) is a simplified symbol that we will use for the recirculating shift register. The control logic is understood to be built into the symbol.

___ EXAMPLE 11-15 _____

Assume the shift register data $Q_3 Q_2 Q_1 Q_0 = 1011$, *REC* = 1, and DATA IN = 0. What will be the level at DATA OUT after the application of five clock pulses?

Solution   In the recirculate mode, DATA IN will have no effect. The table below shows how the register FFs change as clock pulses are applied.

Q_3	Q_2	Q_1	Q_0	
1	0	1	1	Before 1st clock pulse
1	1	0	1	After 1st clock pulse
1	1	1	0	After 2nd clock pulse
0	1	1	1	After 3rd clock pulse
1	0	1	1	After 4th clock pulse
1	1	0	1	After 5th clock pulse

After the fifth pulse Q_0 is a 1, and so DATA OUT = 1.

___ EXAMPLE 11-16 _____

Figure 11-42 shows typical input and output waveforms for the recirculating shift

FIGURE 11-42   Input and output waveforms for the two operating modes of the recirculating shift register.

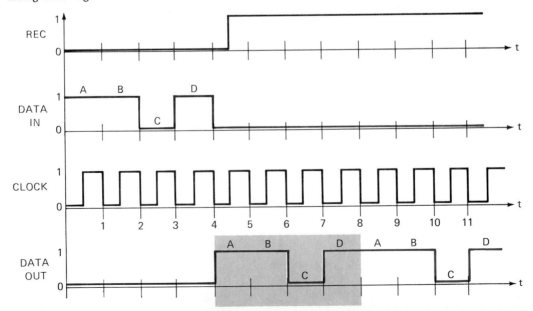

646

register. It is assumed that the register was initially reset to all 0s. Describe the sequence of events depicted by these waveforms.

Solution   During the first four clock cycles $REC = 0$ selects the data input mode. During this interval, data from the DATA IN line are shifted in one bit at a time (A, B, C, then D) on the NGTs of CLOCK. DATA OUT will stay at 0 until the fourth NGT shifts bit A into Q_0. Before the fifth NGT of CLOCK, REC goes HIGH to begin the recirculate mode. For all subsequent clock cycles the data within the register will recirculate so that the sequence of bits, A, B, C, and D will appear at DATA OUT repetitively.

Common Student Mistake: Students sometimes come up with data at the shift-register FFs that are in reverse order because they forget that the timing diagram for DATA IN shows the data bits in chronological, not spatial, order.

Figure 11-43 shows how seven recirculating shift registers are combined to store 7-bit ASCII-coded data for repetitive transmission to the circuits of a video

FIGURE 11-43   Combining several recirculating shift registers to store and repetitively transmit 7-bit data words.

647

display. The *REC* and CLOCK inputs of the registers are tied together so that all registers will be in the same mode and will shift data at the same rate. The seven DATA IN lines represent ASCII data words that are entered into the registers for storage in the data input mode (*REC* = 0). Register 0 gets the LSB of each ASCII data word, register 1 gets the second LSB of each word, and so on. As clock pulses are applied, the seven bits of a given ASCII data word shift along through the seven registers so that they are always at the same relative position in their respective registers. The seven DATA OUT lines will reflect the ASCII data words as they are shifted out of the registers and sent to refresh the video display as well as recirculated back to the input.

___ EXAMPLE 11-17 _____

A certain video display system displays 24 lines of 80 characters. How does this affect the arrangement of Figure 11-43?

Solution A total of $24 \times 80 = 1920$ characters are to be displayed on the screen. This means that 1920 ASCII codes are to be stored and recirculated by the sequential memory. Thus, each shift register has to have a length of 1920 bits.

Teaching Tip: It would be instructive to put some data in the registers of Figure 11-43 and show how the data appear at DATA OUT.

___ EXAMPLE 11-18 _____

What is the storage capacity of a shift-register memory that uses eight 512-bit registers?

Solution The number of registers is the word size. The number of bits per register is the number of words. Thus, the total capacity is 512×8, or 512 eight-bit words.

Teaching Tip: The actual operation of a FIFO is actually somewhat more complex than described here. It uses handshaking signals that signal (1) the readiness of FIFO to accept new data at DATA IN and (2) the presence of new data at DATA OUT. A good example of a relatively simple FIFO is the AMD674xx series.

First-In First-Out Memories (FIFOs) The FIFO is another type of sequential-access memory that uses shift registers. It is similar to the recirculating shift-register memory of Figure 11-43 in that the order in which the data words are entered at DATA IN is the same order in which they are read out at DATA OUT. In other words, the first word that is written in is the first word that is read out—hence the name FIFO. There are two important differences, however, between FIFOs and the recirculating shift-register memory. First, in a FIFO the output data are not recirculated; once they are shifted out, they are lost. Second, in a FIFO the operation of shifting data into the memory is completely independent from the operation of shifting data out of the memory. In fact, the rate at which data are shifted in is usually much different from the rate at which they are shifted out. This makes it particularly suitable for the transfer of data between systems that are operating at widely different speeds.

One common example is the data transfer from a computer to a printer (Figure 11-44). The computer can send data to the printer at a much more rapid rate than the printer can accept it and print it out. A FIFO can act as a **data-rate buffer** between the computer and the printer by accepting data from the computer at a fast rate (say, one word every 10 μs) and storing it; the data are then shifted out to the printer at a slow rate (say, one word every 10 ms). The FIFO can also be used as a data-rate buffer for the transmission of data from a relatively slow device like a keyboard to a much faster device like a computer. For this situation the FIFO accepts data from the

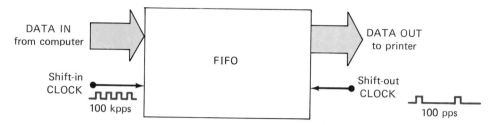

FIGURE 11-44   A FIFO is often used as a data-rate buffer between a fast device, such as a computer, and a slower device like a printer.

keyboard at a slow rate and stores them; the data are then shifted out to the computer at a fast rate. In this way, the computer can be performing other tasks while the FIFO is slowly being filled with data.

─── REVIEW QUESTIONS ───────────────

1. What are the operating modes of a recirculating shift register?
2. For the shift register of Figure 11-41 determine the level at DATA OUT after six clock pulses if $REC = 0$, DATA IN $= 1$, and the register FFs are initially all 0.
3. Repeat the last question with $REC = 1$.
4. Describe the structure of a shift-register memory that is to store 64 nine-bit words.
5. How does a FIFO differ from a recirculating shift-register memory?
6. What is a data-rate buffer?

11-20   MAGNETIC MEMORIES

We have concentrated thus far on the popular semiconductor memory devices whose cells store data in the form of electrical charge or voltage. These devices find their major application as the primary internal memory of a computer. This is mainly due to their high speed, small size, and relative simplicity. The main disadvantage of these semiconductor memories is their volatility, which prevents their use as long-term storage devices. In this section, we will present an overview of memory devices whose basic storage mechanism is magnetic rather than electronic. The characteristic common to all magnetic storage devices is their nonvolatility.

Magnetic Core   *Magnetic core* memory is a nonvolatile random-access read/write memory that was the predecessor to semiconductor memory. Almost all early computers used magnetic core for their primary internal memory before the wide availability of semiconductor memory. The basic magnetic core memory cell is a small, doughnut-shaped core made of a ferromagnetic material. These cells are called *ferrite cores* and typically have a diameter of 0.05 in.

The storage mechanism in magnetic cores is based on a small wire that is threaded through the center of the core. When a current pulse is passed through this wire, a magnetic flux is set up in the core in a direction that depends on the direction

of the current. Because of the core's magnetic retentivity, the core stays magnetized after the current pulse is terminated; in other words, it is nonvolatile. The two possible directions of magnetization are used to represent 1 and 0.

Stored data are read from a core by magnetizing it in the 0 direction (i.e., writing a 0) and using a second threaded wire as a sense wire. The size of the voltage induced in the sense wire will be greater if the core is initially in the 1 state rather than in the 0 state. The reading of a core is said to be "destructive," since it always leaves it in the 0 state. Thus, the data must be rewritten after a read operation.

Magnetic core memory systems have access times from 100 ns to 500 ns and can still be found in some old minicomputers and mainframes. But for all practical purposes, their large physical size and complex interface circuitry have made them virtually obsolete.

Magnetic Bubble Memory (MBM)

Magnetic bubble memory is a semiconductor memory that stores binary data in the form of tiny magnetic domains (bubbles) in a thin film of magnetic material. The presence or absence of a bubble in a specific position is interpreted as a 1 or a 0, respectively. Continuously changing magnetic fields are used to move the bubbles around in loops inside the magnetic material much like recirculating shift registers. The data circulates past a pickup point where it is available to the outside world at the rate of typically 50,000 bits per second. Clearly, MBMs are sequential-access devices.

The major advantage of MBMs is their nonvolatility; if electrical power is removed, the stored data are not lost, since the bubbles simply remain in their fixed positions. When power is restored, the bubbles resume circulating around their loops. Unlike other nonvolatile semiconductor memories such as ROMs, PROMs, EPROMs, and EEPROMs, the MBM can be written into and read from with equal ease. Compared with nonvolatile tape and disk memories, an MBM system has no moving parts and is therefore quieter and more reliable.

MBMs are compact and dissipate very little power (typically 1 μW per bit). A typical MBM device can store a million or more bits. With all these advantages, the MBM does not possess the speed needed for it to function as the main internal memory of a computer; typical access times are in the range 1 to 20 ms. For this reason MBMs are better suited for the external mass storage functions that have been previously dominated by magnetic tape and disk storage. At present, MBMs are about 100 times faster than floppy disks but somewhat more costly because of the required support circuitry. As their cost comes down, we may see more and more MBM systems being used in place of the slower, less reliable floppy disk systems.

Magnetic Tape and Disk

Tape and disk devices perform the recording and reading of magnetic spots on a moving surface of magnetic material. For each of these devices, a thin coating of magnetic material is applied to a smooth nonmagnetic surface. Magnetic tapes, for example, consist of a layer of magnetic material deposited on plastic tape. Disks have the magnetic material coated on both sides of a flat disk that resembles a phonograph record.

Recording and reading of binary information on tapes and disks uses the same basic principles. Figure 11-45 illustrates the fundamental concept of recording on a moving magnetic surface. The read/write head is a core of high-permeability soft iron with a coil wound around it and a small air gap (typically 0.001 in. wide). In writing, a current is driven through the coil, establishing magnetic lines of flux in

FIGURE 11-45 Fundamental parts for recording on moving magnetic surface.

the core. These flux lines remain in the core until they encounter the air gap, which has a very high reluctance to magnetic flux. This causes the flux lines to deviate and travel through the magnetic coating on the moving surface. Thus, pulses of current in the coil result in spots of magnetism on the moving surface. These spots remain magnetized after they pass the read/write head.

The read operation is just the opposite of the write operation. During the read operation the coil is used as a sense line. As the magnetic surface is moved under the read/write head, the spots that have been magnetized produce a flux up through the air gap and into the core. This change in the core's flux induces a voltage signal in the sense winding which is then amplified and interpreted as a 1 or a 0.

The coil around the core often has a center tap, so that half the coil is used as a sense winding and the other half is heavier-gauge wire used for supplying current for the write operation.

Disk and tape are used as nonvolatile mass memory systems where large quantities of data can be stored permanently, but can also be erased and written over. Since both of these use a mechanical means for recording and reading their data, their access times are much too slow to be used as a computer's primary internal memory. Their low cost per bit and high capacity make them ideal for secondary storage that can transfer data to the computer's main memory when needed, and this is where they find their principal application.

Teaching Tip: You might want to say a few words comparing the recording and playback of digital information (saturated recording) with the recording of audio (analog) information.

Hard Disk Systems   In a hard disk system, the data are recorded on concentric tracks on an aluminum disk that is covered with a magnetic coating on both sides. Hard disks come in sizes from $3\frac{1}{2}$ in. to 20 in., and are rotated by a common drive spindle at speeds from 1000 to 3600 rpm. Currently available hard disks have access times that range from under 20 ms to around 100 ms. This is the average time it takes the read/write head to be positioned over the part of the disk surface where the data are to be written or read. Hard disk data transfer rates are 1M to 10M bits/s. This is the rate at which data can be read from the disk once the head is in position.

A typical 5-in. hard disk system such as found on the IBM-PC will consist of two hard disks (four surfaces) with a total capacity of 10 to 20M bytes. Current technology has produced multidisk drives with capacities from 100M to 1000M bytes. Most hard disks are **fixed drives** in which the disks are permanently mounted on the drive mechanism and cannot be removed in normal usage. Some hard disk systems use removable disks called **disk cartridges** or **disk packs.**

Teaching Tip: The head
is also retracted when-
ever the speed of rota-
tion of the drive spin-
dle drops below some
nominal value because
of a malfunction in the
spindle motor or its
associated control cir-
cuitry. A motor failure
is bad enough without
having a disk ruined by
a head crash when the
air cushion disappears.

The high speed of rotation of hard disks produces a cushion of air a few microinches above the disk surface. The read/write head floats on this cushion so that it never touches the disk surface. This prevents a wearing down of the disk surface and the read/write head. If there are bumps or warping of the disk surface, or if there are particles of dust or dirt on the surface, a **head crash** may occur. In a head crash, the head makes contact with the disk surface, causing scratching or gouging of the surface. This results in a destruction of recorded data that usually renders the disk useless. To minimize the possibility of head crashes, most hard disk drives blow filtered air across the disks to keep them relatively clean. On power down, the read/write head is retracted before the disk is spun down. On power up, the read/write head is kept retracted until the disk has been spun up to full speed to develop a cushion of air.

Floppy Disk Systems

A floppy disk (often called a **diskette)** is a thin flexible Mylar disk that spins inside a protective paper jacket. Both surfaces of the disk are coated with a magnetic material so that data may be recorded on both sides (double-sided recording). The disks come in three sizes—8 in. (the original floppy), 5.25 in. (the minifloppy), and 3.5 in. (the microfloppy). The microfloppy actually comes in a hard plastic package. When inserted into the disk drive unit, a floppy disk is rotated at a fixed speed of 300 or 360 rpm, much slower than hard disks.

Floppy disk capacities typically range from 100K to 1M bytes, although several companies now market a 3.5-in. floppy drive with a 4M byte capacity. Floppies have access times about 10 times longer and data rates about 10 times slower than hard disks. The average access times of floppy drives is 100 to 500 ms, and the average data transfer rates range from 250K to 1M bits/s.

Although floppy disk systems are slower and have less capacity than hard disk systems, they do have the advantages of low cost and portability. They can be easily transported from one computer to another, and can be sent through the mail.

Magnetic Tape

The principal use of magnetic tape storage is as a backup for a hard disk system. Since hard disks typically store large data bases (such as inventory, employee records, or student schedules and records), a head crash could be disastrous unless the disk data had been periodically copied (backed up) to magnetic tape. The major advantage of magnetic tape is its low cost per bit and very high capacity. A small tape cassette like that used for digital audio can easily store 1000M bytes. The principal drawback of tape memory is the long access time required to find a specific block of data (just think about how long it takes to find a song on an audio tape). This drawback is not a consideration when the tape is used as a disk backup. The data will not be read from the tape very often, and when they are, the complete tape will be read from the beginning.

Optical Disk Memory

Optical disk memory is the newest mass memory technology, and one that promises to become dominant. Its operation is based on the reflection or scattering of an extremely narrow laser beam off a disk that has had microscopic pits or bubbles representing logic 1s "burned" onto its surface. One strong point of optical disk memory is its high capacity—a $5\frac{1}{2}$ inch disk with 1000M byte capacity is readily available. Other advantages are its relatively low cost and its immunity to dust. Its access times and data transfer rates are comparable to the newest hard disk systems.

Optical disk systems are available in three basic types based on their writability. Disks that can only be read from are called **optical ROM (OROM)** or **compact-disk ROM (CD-ROM).** These are used to store large fixed data bases such as dictionaries or encyclopedias. An optical disk that can be written to once is called a **write-once-read-many (WORM).** CD-ROMS and WORMs cannot be erased. The **read/write optical disk** can be written and rewritten as often as desired, and therefore operates like a magnetic hard disk. It uses a different disk surface from the other types. Its surface is coated with a magnetic material that can have its magnetic properties changed by a laser beam.

Optical disk storage has a promising future that could find it emerging as the principal mass storage device. Considerable new research is being done by many companies to develop new read/write techniques.

REVIEW QUESTIONS

1. What is the major use of disk and tape in a computer system?
2. What is a fixed hard disk drive?
3. What is a head crash?
4. What are the main disadvantages and advantages of floppy disk as opposed to hard disk storage?
5. What is the major application of tape memory?
6. What are the chief advantages of optical disk memory?
7. List the three types of optical disk storage.
8. Why aren't disks or tapes used as a computer's internal memory?

11-21 TROUBLESHOOTING RAM SYSTEMS

All computers use RAM. Many general-purpose computers and most special-purpose computers (such as microprocessor-based controllers and process-control computers) also use some form of ROM. Each RAM and ROM IC that is part of a computer's internal memory typically contains thousands of memory cells. A single faulty memory cell can cause a complete system failure (commonly referred to as a "system crash") or, at the least, unreliable system operation. The testing and troubleshooting of memory systems involves the use of techniques that are not often used on other parts of the digital system. Because memory consists of thousands of identical circuits acting as storage locations, any tests of its operation must involve checking to see exactly which locations are working and which are not. Then, by looking at the pattern of good and bad locations along with the organization of the memory circuitry, the possible causes of the memory malfunction can be determined. The problem typically can be traced to a bad memory IC; a bad decoder IC, logic gate, or signal buffer; or a problem in the circuit connections (i.e., shorts or open connections).

Because RAM has to be written to and read from, testing RAM is generally more complex than testing ROM. In this section we will look at some common procedures for testing the RAM portion of memory and interpreting the test results. We will examine ROM testing in the next section.

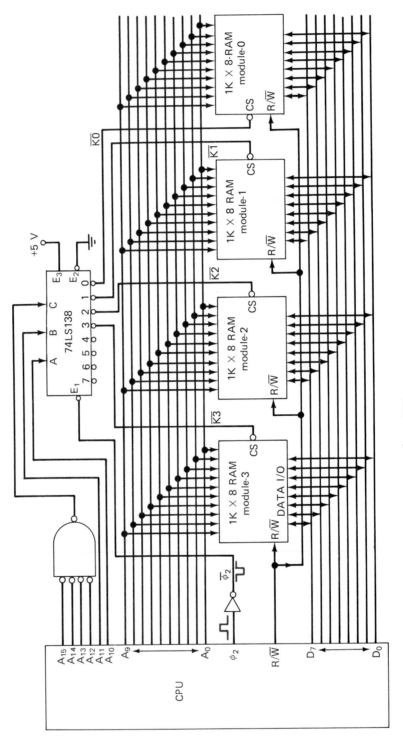

FIGURE 11-46 4K × 8 RAM memory connected to CPU.

Know the Operation   The RAM memory system shown in Figure 11-46 will be used in our examples. As we emphasized in earlier discussions, successful troubleshooting of a relatively complex circuit or system begins with a thorough knowledge of its operation. Before we can discuss the testing of this RAM system, we should first analyze it carefully so that we fully understand its operation.

Discussion Question: Can the INVERTER be eliminated by connecting ϕ_2 to E_3?

The total capacity is 4K × 8 and is made up of four 1K × 8 RAM modules. A module may be just a single IC, or it may consist of several ICs (e.g., two 1K × 4 chips). Each module is connected to the CPU through the address and data buses, and the $R/\overline{W}$ control line. The modules have common I/O data lines. During a read operation these lines become data output lines through which the selected module places its data on the bus for the CPU to read. During a write operation, these lines act as input lines to accept CPU-generated data from the data bus for writing into the selected location.

The 74LS138 decoder and the four-input OR gate combine to decode the six high-order address lines to generate the chip-select signals $\overline{K0}$, $\overline{K1}$, $\overline{K2}$, and $\overline{K3}$. These signals enable a specific RAM module for a read or write operation. The INVERTER is used to invert the CPU output clock signal ϕ_2 so that the decoder is enabled only while ϕ_2 is HIGH. The ϕ_2 pulse occurs only after allowing enough time for the address lines to stabilize following the application of a new address on the address bus. ϕ_2 will be LOW while the address and $R/\overline{W}$ lines are changing; this prevents the occurrence of decoder output glitches that could erroneously activate a memory chip and possibly cause invalid data to be stored.

Each RAM module has its address inputs connected to the CPU address bus lines A_0 through A_9. The high-order address lines A_{10} through A_{15} select one of the RAM modules. The selected module decodes address lines A_0 through A_9 to find the word location that is being addressed. The following examples will show how to determine the addresses that correspond to each module.

EXAMPLE 11-19

Assume that the CPU is performing a read operation from address 06A3 (hex). Which RAM module, if any, is being read from?

Solution   First write out the address in binary.

A_{15}	A_{14}	A_{13}	A_{12}	A_{11}	A_{10}	A_9	A_8	A_7	A_6	A_5	A_4	A_3	A_2	A_1	A_0
0	0	0	0	0	1	1	0	1	0	1	0	0	0	1	1

You should be able to verify that levels A_{15} to A_{10} will activate decoder output $\overline{K1}$ to select RAM module-1. This module internally decodes the address lines A_9 to A_0 to select the location whose data are to be placed on the data bus.

EXAMPLE 11-20

Which RAM module will have data written into it when the CPU executes a write operation to address 1C65?

Solution   Writing out the address in binary, we can see that $A_{15} = 1$. This produces a HIGH out of the OR gate and at the C input of the decoder. With $A_{11} = A_{10} = 1$, the decoder inputs are 111, which activates output 7. Outputs $\overline{K0}$ to $\overline{K3}$ will be inactive, and so none of the RAM modules will be enabled. In other words, the data placed on the data bus by the CPU will not be accepted by any of the RAMs.

——— EXAMPLE 11-21 ———

Determine the range of addresses for each module in Figure 11-46.

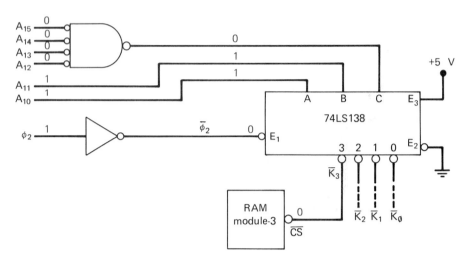

FIGURE 11-47 Example 11-21, showing address bus conditions needed to select RAM module-3.

Solution Each module stores 1024 eight-bit words. To determine the addresses of the words stored in any module, we start by determining the address bus conditions that activate that module's chip-select input. For example, module-3 will be selected when decoder output $\overline{K3}$ is LOW (Figure 11-47). $\overline{K3}$ will be LOW for $CBA = 011$. Working back to the CPU address lines A_{15} to A_{10}, we see that module-3 will be enabled when the following address is placed on the address bus:

A_{15}	A_{14}	A_{13}	A_{12}	A_{11}	A_{10}	A_9	A_8	A_7	A_6	A_5	A_4	A_3	A_2	A_1	A_0
0	0	0	0	1	1	x	x	x	x	x	x	x	x	x	x

The x's under A_9 through A_0 indicate "don't care" because these address lines are not used by the decoder to select module-3. A_9 to A_0 can be any combination ranging from 0000000000 to 1111111111, depending on which word in module-3 is being accessed. Thus, the complete range of addresses for module-3 is determined by using all 0s, and then all 1s for the x's.

Discussion Question:
Would this circuit work any differently if a 3-input OR gate were used and A_{15} were connected to E_2 instead of to the OR gate?

A_{15}	A_{14}	A_{13}	A_{12}	A_{11}	A_{10}	A_9	A_8	A_7	A_6	A_5	A_4	A_3	A_2	A_1	A_0		
0	0	0	0	1	1	0	0	0	0	0	0	0	0	0	0	$\longrightarrow$	$0C00_{16}$
0	0	0	0	1	1	1	1	1	1	1	1	1	1	1	1	$\longrightarrow$	$0FFF_{16}$

Finally, this gives us 0C00 to 0FFF as the range of hex addresses stored in module-3. When the CPU places any address in this range onto the address bus, *only* module-3 will be enabled for either a read or write, depending on the state of $R/\overline{W}$.

A similar analysis can be used to determine the address ranges for each of the other RAM modules. The results are:

Module-0: 0000–03FF

Module-1: 0400–07FF

Module-2: 0800–0BFF

Module-3: 0C00–0FFF

Note that the four modules combine for a total address range of 0000 to 0FFF.

Testing the Decoding Logic   In some situations, the decoding logic portion of the RAM circuit (Figure 11-47) can be tested using the various techniques we have applied to combinatorial circuits. It can be tested by applying signals to the six most significant address lines and ϕ_2, and monitoring the decoder outputs. To do this, it has to be possible to easily disconnect the CPU from these signal lines. If the CPU is a microprocessor chip in a socket, it can simply be removed from its socket.

Once the CPU is disconnected, you can supply the A_{10}–A_{15} and ϕ_2 signals from an external test circuit to perform a static test, using manually operated switches for each signal, or a dynamic test, using some type of counter to cycle through the various address codes. With these test signals applied, the decoder output lines can be checked for the proper response. Standard signal tracing techniques can be used to isolate any faults in the decoding logic.

EXAMPLE 11-22

A dynamic test is performed on the decoding logic of Figure 11-47 by keeping $\phi_2 = 1$ and connecting the outputs of a 6-bit counter to the address inputs A_{10} through A_{15}. The decoder outputs are monitored as the counter repetitively cycles through all 6-bit codes. A logic probe check on the decoder outputs shows pulses at $\overline{K1}$ and $\overline{K3}$, but shows $\overline{K0}$ and $\overline{K2}$ remaining HIGH. What are the most probable faults?

Teaching Tip: Using buffers between the counter outputs and the address lines will prevent any shorted address lines from affecting the counter operation.

Solution   It is possible, but highly unlikely, that $\overline{K0}$ and $\overline{K2}$ could both be stuck HIGH due to either an internal or external short to V_{CC}. A more likely fault is an open between A_{10} and the A input of the decoder, since this would act as a logic HIGH and prevent any even-numbered decoder output from being activated. It is also possible that the decoder's A input is shorted to V_{CC}, but this is also unlikely because this short would have probably affected the operation of the counter that is supplying the address inputs.

Testing the Complete RAM System   Testing and troubleshooting the decoding logic will not reveal problems with the memory chips and their connections to the CPU buses. The most common methods for testing the operation of the *complete* RAM system involve writing known patterns of 1's and 0's to each memory location and reading them back to verify that the location has stored the pattern properly. While there are many different patterns that can be used, one of the most widely used is the "checkerboard pattern." In this pattern, 1s and 0s are alternated as in 01010101. Once all locations have been tested using this pattern, the pattern is reversed (i.e., 10101010) and each location is tested again. Note that this sequence of tests will check each cell for the ability to store and read both a 1 and a 0. Because it alternates 1s and 0s, the checkerboard pattern will also detect any interactions or shorts between adjacent cells. Many other patterns can be used to detect various failure modes within RAM chips.

No memory test can catch all possible RAM faults with 100 percent accuracy, even though it may show that each cell can store and read a 0 or 1. Some faulty RAMs can be pattern-sensitive. For instance, a RAM may be able to store and read 01010101 and 10101010, but it may fail when 11100011 is stored. Even for a small RAM system, it would take a prohibitively long time to try storing and reading every possible pattern in each location. For this reason, if a RAM system passes the checkerboard test, you can conclude that it is *probably* good; if it fails the test, then it *definitely* contains a fault.

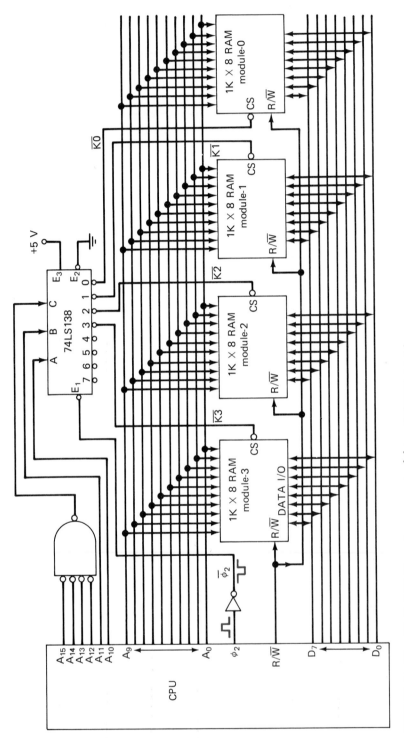

FIGURE 11-48   4K × 8 RAM system repeated from FIGURE 11-46.

Manually testing thousands of RAM locations by storing and reading checkerboard patterns would take hundreds of hours, and is obviously not feasible. RAM pattern testing is usually done automatically either by having the CPU run a memory test program or by connecting a special test instrument to the RAM system buses in place of the CPU. In fact, in many computers and microprocessor-based equipment the CPU will automatically run a memory test program every time it is powered up; this is called *a* **power-up self-test**. The self-test routine (we will call it SELF-TEST) is stored in ROM, and it is executed whenever the system is turned on, or when the operator requests it from the keyboard. As the CPU executes SELF-TEST, it will write and read test patterns to each RAM location, and will display some type of message to the user. It may be something as simple as an LED to indicate faulty memory, or it may be a descriptive message printed on the screen or printer. Typical messages might be:

Teaching Tip: It is important that the self-test routine not use any RAM locations for data storage or stack operations. Otherwise, a RAM error could cause the routine to crash.

```
RAM module-3 test OK
ALL RAM working properly
Location 027F faulty in bit positions 6 and 7
```

With messages like these and a knowledge of the RAM system operation, the troubleshooter can determine what further action is needed to isolate the fault. The following two examples will illustrate the procedure for the 4K × 8 RAM system; its diagram is repeated in Figure 11-48 for convenience. Remember from our previous examples that the addresses for each RAM module are

Module-0: 0000–03FF

Module-1: 0400–07FF

Module-2: 0800–0BFF

Module-3: 0C00–0FFF

EXAMPLE 11-23

The following messages are printed out after the SELF-TEST is run:

```
module-0 test OK
module-1 test OK
address 09A7 faulty at bit 6
module-3 test OK
```

Examine these messages and decide what to do next.

Solution   The faulty address location is in module-2. There is no circuit fault external to module-2 that could cause only that single address to be bad and no others. Thus, it appears that there is an internal fault in module-2. Before we reach this conclusion, however, the SELF-TEST program should be run several more times to see if the same messages appear. The original fault message may have been the result of noise rather than a bad memory cell; if so, further tests may produce no fault messages or different fault messages. This would point up the need to check for sources of noise in the system such as nonexistent or nonfunctioning power-supply decoupling capacitors, or crosstalk between closely spaced signal lines.

EXAMPLE 11-24

Consider the following messages from SELF-TEST:

```
module-0 test OK
address 0400 faulty at bits 0-7
address 0401 faulty at bits 0-7
address 0402 faulty at bits 0-7
         .      .      .      .      .
         .      .      .      .      .
         .      .      .      .      .
         .      .      .      .      .
address 07FE faulty at bits 0-7
address 07FF faulty at bits 0-7
module-2 test OK
module-3 test OK
```

Examine these messages, list the possible faults, and describe what to do next.

Solution   Clearly, every address in module-1 is listed as being faulty, and so module-1 is not working at all. Here are several possible faults that could cause this:

An open connection at the $R/\overline{W}$ input to module-1
An open path between $\overline{K1}$ and the module-1 chip-select
A faulty decoder output $\overline{K1}$ (stays HIGH)
The $\overline{K1}/\overline{CS}$ node externally shorted to V_{CC}
A faulty RAM module-1 (does not respond to $\overline{CS}$)
An unconnected V_{CC} or ground on module-1

Standard troubleshooting techniques can now be used to isolate the fault.

Teaching Tip: Some RAM tests simply print out a message telling how many bytes of RAM are available based on the address of the first location that failed the test. If there is a RAM fault, the number of available bytes will be fewer than expected. This information can then be used to determine the first faulty address.

As these examples have shown, the memory test program is useful in narrowing the problem to a specific area of the circuit. The troubleshooter uses the information provided from the memory test to determine possible faults and then proceeds to locate the actual fault.

It should be pointed out that there are some RAM circuit faults that will prevent the CPU from executing the SELF-TEST program on power-up. For example, any RAM circuit fault that causes a data or address line to be stuck LOW or HIGH will cause erroneous operation when the CPU tries to execute the SELF-TEST program in ROM (remember that the ROM shares the address and data buses with the RAM). When this happens, you should check for stuck bus lines and, if one is found, use the standard techniques for determining the exact cause of the stuck line.

REVIEW QUESTIONS

1. What is ϕ_2's function in the RAM circuit of Figure 11-46?
2. What is the checkerboard test? Why is it used?
3. What is a power-up self-test?

11-22 TESTING ROM

The ROM circuitry in a computer is very similar to the RAM circuitry (compare Figures 11-39 and 11-46). The ROM decoding logic can be tested in the same manner described in the preceding section for the RAM system. The ROM chips, however, must be tested differently from RAM chips, because we cannot write patterns into ROM and read them back as we can for RAM. There are several methods used to check the contents of a ROM IC.

In one approach the ROM is placed in a socket in a special test instrument which is typically microprocessor-controlled. The special test instrument can be programmed to read every location in the test ROM and print out a listing of the contents of each location. The listing can then be compared with what the ROM is supposed to contain. Except for low-capacity ROM chips, this process can be very time-consuming.

In a more efficient approach, the test instrument would have the correct data stored in its own *reference ROM* chip. The test instrument is then programmed to read the contents of each location of the test ROM and compare it with the contents of the reference ROM. This approach, of course, requires the availability of a pre-programmed reference ROM.

A third approach uses a **checksum.** This is a special code placed in the last one or two locations of the ROM chip when it is programmed. This code is derived by adding up the data words to be stored in all of the ROM locations (excluding those containing the checksum). As the test instrument reads the data from each test ROM location, it will add them up and develop its own checksum. It then compares its calculated checksum with that stored in the last ROM locations, and they should agree. If so, there is a high probability that the ROM is good (there is a very small chance that a combination of errors in the test ROM data could still produce the correct checksum value). If they do not agree, then there is a definite problem in the test ROM.

The checksum idea is illustrated in Figure 11-49(a) for a very small ROM. The data word stored in the last address is the 8-bit sum of the other seven data words

FIGURE 11-49 Checksum method for an 8 × 8 ROM: (a) ROM with correct data; (b) ROM with error in its data.

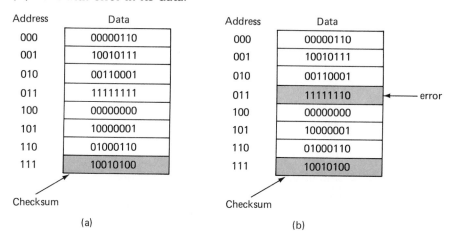

(ignoring carries from the MSB). When this ROM is programmed, the checksum is placed in the last location. Figure 11-49(b) shows the data that might actually be read from a faulty ROM that was originally programmed with the data in (a). Note the error in the word at address 011. As the test instrument reads the data from each location of the faulty ROM, it calculates its own checksum from those data. Because of the error, the calculated checksum will be 10010011. When the test instrument compares this with the checksum value stored at ROM location 111, it sees that they disagree, and a ROM error is indicated. Of course, the exact location of the error is not determined.

The checksum method can also be used by a computer or microprocessor-based equipment during an automatic power-up self-test to check out the contents of the system ROMs. Again, as in the self-test used for RAM, the CPU would execute a program on power-up that would do a checksum test on each ROM chip and would print out some type of status message. The self-test program itself will be located in a ROM, and so any error in that ROM could prevent the successful execution of the checksum tests.

___ REVIEW QUESTION ___

1. What is a checksum? What is its purpose?

PROBLEMS

SECTIONS 11-1 to 11-3

11-1. A certain memory has a capacity of 16K × 32. How many words does it store? What is the number of bits per word? How many memory cells does it contain?   16,384 words; 32 bits/word; 524,288 cells

11-2. How many different addresses are required by the memory of Problem 11-1?   16,384

11-3. What is the capacity of a memory that has 16 address inputs, four data inputs, and four data outputs?   64K × 4

11-4. 16 data inputs; 16 data outputs; 13 address lines; 16K bytes

11-4. A certain memory stores 8K sixteen-bit words. How many data input and data output lines does it have? How many address lines does it have? What is its capacity in bytes?

DRILL QUESTIONS

11-5. Define each of the following terms: See Glossary.
(a) RAM
(b) RWM
(c) ROM
(d) internal memory
(e) mass memory

11-6. (a) What are the three buses in a computer memory system?
(b) Which bus is used by the CPU to select the memory location?
(c) Which bus is used to carry data from memory to CPU during a read operation?
(d) What is the source of data on the data bus during a write operation?

11-6. (a) data, address, control (b) address (c) data bus (d) CPU

SECTIONS 11-4 AND 11-5

11-7. Refer to Figure 11-6. Determine the data outputs for each of the following input conditions:
(a) $[A] = 1011; CS = 1$ Hi-Z
(b) $[A] = 0111; CS = 0$ 11101101

11-8. Refer to Figure 11-7.
(a) What register is enabled by input address 1011?   register 11
(b) What input address code selects register 4?   0100

11-9. A certain ROM has a capacity of 16K × 4 and an internal structure like that shown in Figure 11-7.
(a) How many registers are in the array?   16,384
(b) How many bits are there per register?   4
(c) What size decoders does it require?   two 1-of-128 decoders

DRILL QUESTION

11-10. (a) *True or false:* ROMs cannot be erased.   False; some can be erased.
(b) What is meant by programming or burning a ROM?
(c) Define a ROM's access time.
(d) How many data inputs, data outputs, and address inputs are needed for a 1024 × 4 ROM?

11-10. (a) False; some can be erased. (b) Means for writing data into it. (c) Time from application of address to appearance of valid data. (d) Four data inputs, four data outputs, and ten address inputs.

SECTION 11-6

 11-11. Figure 11-50 shows how data from a ROM can be transferred to an

FIGURE 11-50

external register. The ROM has the following timing parameters: $t_{ACC} = 250$ ns and $t_{OE} = 120$ ns. Assume that the new address inputs have been applied to the ROM 500 ns before the occurrence of the TRANSFER pulse. Determine the minimum duration of the TRANSFER pulse for reliable transfer of data.   120 ns

 11-12. Repeat Problem 11-11 if the address inputs are changed 70 ns prior to the TRANSFER pulse.   180 ns

SECTION 11-7

11-13. DRILL QUESTION

For each statement below indicate the type or types of ROM being described.
(a) Can be programmed by the user, but cannot be erased.   PROM
(b) Is programmed by the manufacturer.   MROM
(c) Can be erased and reprogrammed over and over.   EPROM and EEPROM
(d) Is erased with ultraviolet light.   EPROM
(e) Selected words can be erased.   EEPROM
(f) Uses fusible links.
(g) Can be completely erased in a few milliseconds.   EEPROM

11-15. Connections to the bases of the following transistors will be open: Q_0, Q_2, Q_5, Q_6, Q_7, Q_9, and Q_{15}.

11-14. Which transistors in Figure 11-9 will be conducting when $A_1 = A_0 = 1$ and $\overline{EN} = 0$? Q_{13}, Q_{14}, and Q_{15}

11-15. Change the MROM connections in Figure 11-9 so that the MROM stores the function $y = 3x + 5$.

 11-16. Figure 11-51 shows a simple circuit for manually programming a 2732

FIGURE 11-51

EPROM. Each EPROM data pin is connected to a switch that can be set at a 1 or 0 level. The address inputs are driven by a 12-bit counter. The 50-ms programming pulse comes from a one-shot each time the PROGRAM pushbutton is actuated. Switch bounce should have no effect.

(a) Explain how this circuit can be used to sequentially program the EPROM memory locations with the desired data.

(b) Show how 74293s and a 74121 can be used to implement this circuit.

(c) Should switch bounce have any effect on the circuit operation?

SECTION 11-8

11-17. Another ROM application is the generation of timing and control signals. Figure 11-52 shows a 16×8 ROM with its address inputs driven by a MOD-16 counter so that the ROM addresses are incremented with each input pulse. Assume the ROM is programmed as in Figure 11-6, and sketch the waveforms at each ROM output as the pulses are applied. Ignore ROM delay times.

11-17. Each data output waveform will change according to the truth table as the counter sequences through the different addresses.

FIGURE 11-52

11-18. Change the program stored in the ROM of Problem 11-17 to generate the D_7 waveform of Figure 11-53.

11-18. Hex data are 5E, BA, 05, 2F, 99, FB, 00, ED, 3C, FF, B8, C7, 27, EA, 52, 5B

FIGURE 11-53

 11-19. Figure 11-54(a) shows the logic symbol for the 74185 IC, which is an "off the shelf" ROM that is programmed as a *binary-to-BCD converter*. It converts a 6-bit binary input to a two-digit BCD output.

11-19. $[A] = 110101$; $[D] = 101$; $[C] = 0011$

(a)

		Binary						BCD					
A_5	A_4	A_3	A_2	A_1	A_0	D_2	D_1	D_0	C_3	C_2	C_1	C_0
1	0	0	1	0	1	0	1	1	0	1	1	1
0	0	1	1	1	0	0	0	1	0	1	0	0
1	1	1	1	1	0	1	1	0	0	0	1	0
0	1	1	1	0	0	0	1	0	1	0	0	0
1	1	0	1	0	1							
0	0	0	1	1	0							
						0	0	1	1	0	0	1
						1	0	0	1	0	0	0

(b)

FIGURE 11-54

Note how the LSB of the 6-bit binary input is not connected to the ROM as an address input; instead, it is simply routed to the output and becomes the LSB of the LSD of the output. Figure 11-54(b) shows part of the truth table showing the contents of some of the locations in this ROM. Examine it to verify how the conversion from binary to BCD is taking place. Then fill in the missing entries in the table.

SECTION 11-9

11-20. $O_3 = A$; $O_2 = A \oplus B$; $O_1 = B \oplus C$; $O_0 = C \oplus D$

 11-20. Show how to program the PROM of Figure 11-18(a) to perform the code conversion from binary to Gray code. The inputs A, B, C, D are to represent the binary code, and the outputs are to represent the Gray code. Refer to Table 2-4 in Section 2-7.

11-21. Same expressions as P11-20.

11-21. Repeat Problem 11-20 using the PAL of Figure 11-19(a). Remember, you have to express each output as an OR sum of four terms.

SECTION 11-11

11-22. (a) Draw the logic symbol for a 51C66, which is a CMOS static RAM organized as a 16K × 1 with separate data in and data out, and an active-LOW chip enable.

(b) Draw the logic symbol for a 51C68, which is a CMOS static RAM organized as a 4K $\times$ 4 with common I/O and an active-LOW chip enable.   See Solutions Manual.

11-23. A certain static RAM has the following timing parameters (in nanoseconds):

$$t_{RC} = 100 \qquad t_{AS} = 20$$
$$t_{ACC} = 100 \qquad t_{AH} = \text{not given}$$
$$t_{CO} = 70 \qquad t_{W} = 40$$
$$t_{OD} = 30 \qquad t_{DS} = 10$$
$$t_{WC} = 100 \qquad t_{DH} = 20$$

(a) How long after the address lines stabilize will valid data appear at the outputs during a read cycle?
(b) How long will output data remain valid after $\overline{CS}$ returns HIGH?
(c) How many read operations can be performed per second?
(d) How long should $R/\overline{W}$ and $\overline{CS}$ be kept HIGH after the new address stabilizes during a write cycle?
(e) What is the minimum time that input data have to remain valid for a reliable write operation to occur?
(f) How long must the address inputs remain stable after $R/\overline{W}$ and $\overline{CS}$ return HIGH?
(g) How many write operations can be performed per second?

11-23. (a) 100 ns (b) 30 ns (c) 10 million (d) 20 ns (e) 30 ns   (f) 40 ns   (g) 10 million

11-24. Draw the logic symbol for the TMS4256, which is a 256K $\times$ 1 DRAM. How many pins are saved by using address multiplexing for this DRAM?   Net saving of 7 pins.

 11-25. Figure 11-55(a) shows a circuit that generates the $\overline{RAS}$, $\overline{CAS}$, and

FIGURE 11-55

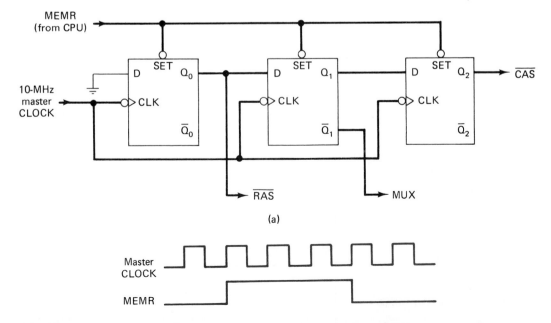

(a)

(b)

MUX signals needed for proper operation of the circuit of Figure 11-30(b). The 10-MHz master clock signal provides the basic timing for the computer. The memory request signal (*MEMR*) is generated by the CPU in synchronism with the master clock as shown in part (b) of the figure. *MEMR* is normally LOW and is driven HIGH whenever the CPU wants to access memory for a read or write operation. Determine the waveforms at Q_0, $\overline{Q}_1$, and Q_2 and compare them with the desired waveforms of Figure 11-31.   Waveforms match those in Figure 11-31.

11-26. Use one and one half 74157s.

11-26. Show how to connect two 74157 multiplexers (Section 9-7) to provide the multiplexing function required in Figure 11-30(b).

11-27. Refer to the signals in Figure 11-32. Describe what occurs at each of the labeled time points.   Refer to list in Section 11-15.

11-28. Repeat for Figure 11-33.   Refer to list in Section 11-15.

11-29. $\overline{RAS}$ and $\overline{CAS}$ signals from Figure 11-55 connect to $\overline{RAS}$ and $\overline{CAS}$ of Figure 11-35. *MUX* of Figure 11-55 is inverted and connected to ROW ENABLE of Figure 11-35.

11-29. Show how the outputs from the circuit of Figure 11-55 can be used in the circuit of Figure 11-35. Pay careful attention to the 3242's truth table.

SECTION 11-17

11-30. Common address, $R/\overline{W}$, $\overline{CE}$ inputs; separate data I/O

11-30. Show how to combine two 2168 RAM chips (Figure 11-23) to produce a 4K × 8 module.

11-31. Show how to connect two of the ROM chips symbolized in Figure 11-56 to produce a 2K × 8 ROM module. The circuit should use no other logic.   See Solutions Manual.

FIGURE 11-56

11-32. Add four PROMs; connect AB_{13} to *C* input of decoder; connect decoder outputs 4 through 7 to *CS* inputs of new PROMs.

11-32. Describe how to modify the circuit of Figure 11-39 so that it has a total capacity of 16K × 8. Use the same type of PROM chips.

11-33. Connect A_{15}, A_{14}, A_{13} to OR gate, and connect output of OR gate to decoder's A_2 input.

11-33. Modify the decoding circuit of Figure 11-39 to operate from a 16-line address bus (i.e., add A_{13}, A_{14}, and A_{15}). The four PROMs are to maintain the same hex address ranges.

11-34. Examine the memory circuit of Figure 11-57.
(a) Determine the total capacity and word size.   32 × 8

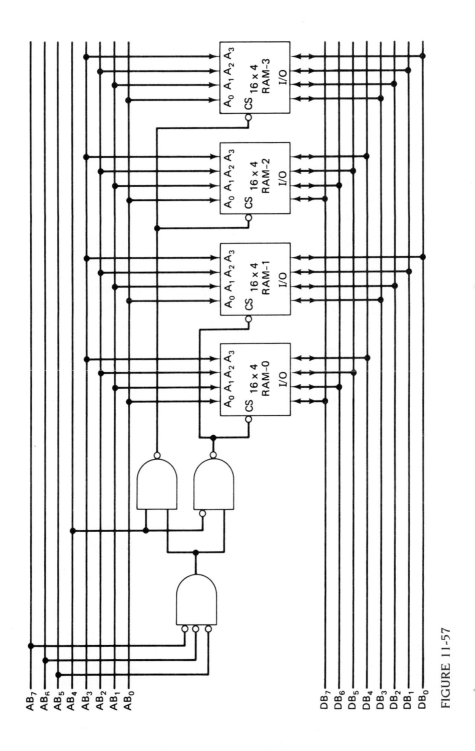

FIGURE 11-57

669

(b) Note that there are more address bus lines than are necessary to select one of the memory locations. This is not an unusual situation, especially in small computer systems, where the actual amount of memory circuitry is much less than the maximum which the computer address bus can handle. Which RAMs will put data on the data bus when $R/\overline{W} = 1$ and the address bus is at 00010110?   RAMs 2 and 3

11-34. (c) 00–0F; 10–1F

(c) Determine the range of addresses stored in the RAM-0/RAM-1 combination. Repeat for the RAM-2/RAM-3 combination.

11-35. Connect four 1K × 1 chips as a 1K × 4 module. Then connect four 1K × 4 modules as a 4K × 4 module. Use A_{10} and A_{11} to select 1 of the 4 1K × 4 modules.

11-35. Draw the complete diagram for a 4K × 4 memory that uses static RAM chips with the following specifications: 1K × 1 capacity, common input/output line, and two active-LOW chip-select inputs. [*Hint:* The circuit can be designed using only two inverters (plus memory chips).]

SECTION 11-19

11-36. Refer to the recirculating shift register of Figure 11-41. The control logic that selects the source of data that will be shifted into D_S is really just a 2-input MUX. Show how to combine four 74164 shift registers (Section 7-22) with a 74157 MUX (Section 9-7) to produce an 8 × 4 shift register memory.   See Solutions Manual.

11-37. Refer to the shift register memory of Figure 11-43. Each register is 8 bits wide. The initial contents of each register are as follows:

> *Register 0:* 01111000
> *Register 1:* 00100001
> *Register 2:* 01001000
> *Register 3:* 00000001
> *Register 4:* 00000100
> *Register 5:* 10000011
> *Register 6:* 01111100

What is the ASCII-code message that will be transmitted to the video display as clock pulses are applied?   *PEACE

SECTION 11-20

DRILL QUESTION

11-38. (a) magnetic core (b) magnetic bubble   (c) hard disk   (d) floppy disk   (e) floppy or tape   (f) tape   (g) optical disk   (h) core   (i) magnetic tape and disk   (j) Optical ROM or CD-ROM.

11-38. For each statement below, indicate what type(s) of memory is (are) being described.
(a) A magnetic, nonvolatile device used in a computer's primary memory before semiconductor memory was developed.
(b) Moves magnetic domains around a loop as in a recirculating shift register.
(c) Read/write head rides on a cushion of air.
(d) Has the advantages of low cost and portability.
(e) Often used to back up hard disks.
(f) Has very long access times.

(g) Higher capacity than hard disk with comparable access times.

(h) Has destructive readout of data.

(i) Stores data by magnetizing spots on the surface of a moving medium.

(j) A mass storage device that can only be read from.

SECTION 11-21

11-39. Modify the RAM circuit of Figure 11.46 as follows: change the OR gate to an AND gate and disconnect its output from C; connect the AND output to E_3; connect C to ground. Determine the address range for each RAM module.

11-39. F000–F3FF; F400–F7FF; F800–FBFF; FC00–FFFF

11-40. Show how to expand the system of Figure 11-46 to an 8K × 8 with addresses ranging from 0000 to 1FFF. (*Hint:* This can be done by adding the necessary memory modules and modifying the existing decoding logic.)   See Solutions Manual.

11-41. A dynamic test is performed on the decoding logic of Figure 11-46 by keeping $\phi_2 = 1$ and connecting the outputs of a 6-bit counter to address inputs A_{10} to A_{15}. The decoder outputs are monitored with an oscilloscope (or logic analyzer) as the counter is continuously pulsed by a 1-MHz clock. Figure 11-58(a) shows the displayed signals. What are the most probable faults?   The B input of the decoder is open or stuck HIGH.

FIGURE 11-58

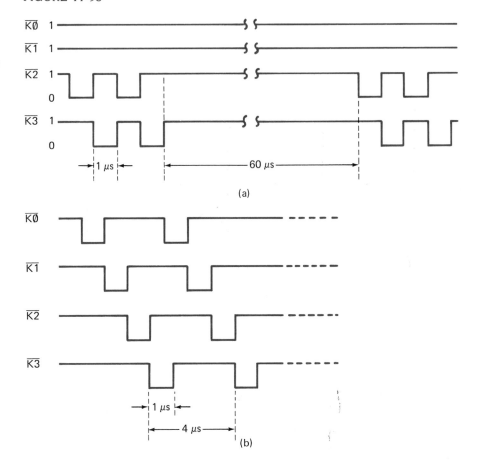

11-43. Only RAM modules 1 and 3 get tested.

11-44. The 1K × 4 RAM chip with data outputs 4 through 7 shows a fault at each bit.

11-45. Connection from output 7 of RAM module 3 to data bus is open, or output 7 of module 3 is stuck in Hi-Z state.

11-42. Repeat Problem 11-41 for the decoder outputs shown in Figure 11-58(b). The OR gate output/*C* input node is shorted to ground.

11-43. Consider the RAM system of Figure 11-46. The checkerboard pattern test will not be able to detect certain types of faults. For instance, assume that there is a break in the connection to the *A* input to the decoder. If a checkerboard pattern SELF-TEST is performed on this circuit, the displayed messages will state that the memory is OK.

(a) Explain why the circuit fault was not detected.

(b) How would you modify the SELF-TEST so that faults such as this will be detected?

11-44. Assume that the 1K × 8 modules used in Figure 11-46 are formed from two 1K × 4 RAM chips. The following messages are printed out when the power-up self-test is performed on this RAM system:

```
module-0 test OK
module-1 test OK
address 0800 faulty at bits 4-7
address 0801 faulty at bits 4-7
address 0802 faulty at bits 4-7
      .    .    .    .    .    .       .    .
      .    .    .    .    .    .       .    .
      .    .    .    .    .    .       .    .
address 0BFE faulty at bits 4-7
address 0BFF faulty at bits 4-7
module-3 test OK
```

Examine these messages and list the possible faults.

11-45. The following messages are printed out when the power-up self-test is performed on the RAM system of Figure 11-48:

```
module-0 test OK
module-1 test OK
module-2 test OK
address 0C00 faulty at bit 7
address 0C01 faulty at bit 7
address 0C02 faulty at bit 7
      .    .    .       .    .
      .    .    .       .    .
      .    .    .       .    .
address 0FFE faulty at bit 7
address 0FFF faulty at bit 7
```

Examine these messages and list the possible faults.

11-46. What messages would be printed out when a power-up self-test is performed on the RAM system of Figure 11-46 if there is a short between the decoder's $\overline{K2}$ and $\overline{K3}$ outputs? See Solutions Manual.

SECTION 11-22

11-47. Consider the 16 × 8 ROM in Figure 11-6. Replace the data word stored at address location 1111 with a checksum calculated from the other 15 data words. Checksum at address $1111_2 = 11101010_2$.

ANSWERS TO SECTION REVIEW QUESTIONS

SECTION 11-1

1. See text. **2.** Sixteen bits per word; 8192 words; 131,072 bits or cells. **3.** In a read operation a word is taken from a memory location and transferred to another device; in a write operation, a new word is placed in a memory location, replacing the one previously stored there. **4.** True. **5.** SAM: Access time is not constant, but depends on the physical location of the word being accessed. RAM: Access time is the same for any address location. **6.** RWM is memory that can be read from or written to with equal ease; ROM is memory that is mainly read from and is written into very infrequently. **7.** False; its data has to be periodically refreshed.

SECTION 11-2

1. 14, 12, 12. **2.** Commands the memory to perform either a read operation or a write operation. **3.** When in its active state, this input enables the memory to perform the read or write operation selected by the $R/\overline{W}$ input; when in its inactive state, this input disables the memory so that it cannot perform the read or write function.

SECTION 11-3

1. Address lines, data lines, control lines. **2. and 3.** See text.

SECTION 11-4

1. True. **2.** Apply desired address inputs; activate control input(s); data appear at data outputs.

SECTION 11-5

1. $A_3 A_2 A_1 A_0 = 1001$. **2.** The row-select decoder activates one of the enable inputs of all registers in the selected row; the column-select decoder activates one of the enable inputs of all registers in the selected column; the output buffers pass the data from the internal data bus to the ROM output pins when the CS input is activated.

SECTION 11-7

1. False; by the manufacturer. **2.** A PROM can be programmed once by the user. It cannot be erased and reprogrammed. **3.** True. **4.** By exposure to ultraviolet light. **5.** True. **6.** See text. **7.** An EEPROM can be programmed without removal from the circuit; individual words can be erased; erase time is only 10 ms; programming pulse is of shorter duration than for EPROM.

SECTION 11-8

1. Microcomputer programs stored in ROM. **2.** On power-up the computer executes a small bootstrap program from ROM to initialize the system hardware and to load the operating system from mass storage (disk). **3.** Circuit that takes data represented in one type of code and converts it to another type of code. **4.** Dot patterns for characters. **5.** Counter, ROM, DAC, low-pass filter.

SECTION 11-9

1. An IC that contains large numbers of gates, FFs, and registers whose interconnections can be modified by the user to perform specific functions. **2.** $O_1 = A$. **3.** An intact fuse. **4.** A hard-wired connection. **6.** The PAL has a programmable AND array; the PROM has a programmable OR array. **8.** A FLPA contains both programmable AND and OR arrays. **9.** It blows the fuses and tests the operation of a PLD according to information down-loaded to its memory.

SECTION 11-11

1. Desired address applied to address inputs; $R/\overline{W} = 1$; CS or CE activated. **2.** To reduce pin count. **3.** Twenty, including V_{CC} and ground.

SECTION 11-12

1. SRAM cells are flip-flops; DRAM cells use capacitors. **2.** CMOS. **3.** RAM. **4.** CPU. **5.** Read and write cycle times. **6.** False. When $\overline{WE}$ is LOW, the I/O pins act as data inputs regardless of the state of $\overline{OE}$ (second entry in mode table).

SECTION 11-13

1. Generally slower speed; need to be refreshed. **2.** Low power; high capacity; lower cost per bit. **3.** iRAMs; dynamic RAM controllers.

SECTION 11-14

1. 256 rows by 256 columns. **2.** It saves pins on the chip. **3.** 1M = 1024K = 1024×1024. Thus, there are 1024 rows by 1024 columns. Since $1024 = 2^{10}$, the chip needs 10 address inputs. **4.** $\overline{RAS}$ is used to latch the row address into the DRAM's row address register; $\overline{CAS}$ is used to latch the column address into the column address register. **5.** *MUX* multiplexes the full address into the row and column addresses for input to the DRAM.

SECTION 11-15

1. See text. **2.** See text.

SECTION 11-16

1. It provides row addresses to the DRAM during refresh cycles. **2.** Address multiplexing and the refresh operation.

SECTION 11-17

1. Sixteen. **2.** Four. **3.** False; when expanding memory capacity, each chip is selected by a different decoder output (see Figure 11-39). **4.** True.

SECTION 11-18

1. Battery backup for CMOS RAM, and NVRAMs. **2.** Since the SRAM and EEPROM have duplicate cells, the chip contains $2 \times 512 \times 8 = 8192$ cells.

SECTION 11-19

1. Recirculating mode; data input mode. **2.** HIGH. **3.** LOW. **4.** It contains nine 64-bit recirculating shift registers connected as in Figure 11-43. **5.** The data may be shifted in and shifted out at different rates. **6.** A FIFO used to transfer data between devices that operate at different data transfer rates.

SECTION 11-20

1. Secondary mass storage. **2.** A hard disk system in which the disk cannot be removed from its drive. **3.** Contact between the read/write head and a hard disk that damages the disk surface. **4.** Advantages: low cost, portability; disadvantages: slower speed, lower capacity. **5.** Backup for hard disk system. **6.** High capacity, low cost, immunity to dust. **7.** OROM, WORM, read/write optical disk. **8.** Their access times are too slow.

SECTION 11-21

1. Prevents decoding glitches by disabling the decoder while the address lines are changing. **2.** A way to test RAM by writing a checkerboard pattern (first 01010101, then 10101010) into each memory location and then reading it. It is used because it will detect any shorts or interactions between adjacent cells. **3.** An automatic test of RAM performed by a computer on power-up.

SECTION 11-22

1. It is a code placed in the last one or two ROM locations that represents the sum of the expected ROM data from all other locations. It is used as a means to test for errors in one or more ROM locations.

12

INTRODUCTION TO THE MICROPROCESSOR AND MICROCOMPUTER

Outline

Objectives

Upon completion of this chapter, you will be able to:

- Describe the function and operation of each one of the five basic elments of any computer organization.
- Evaluate the advantages and disadvantages among the various types of computer word formats.
- Understand the difference between a microprocessor and a microcomputer.
- Analyze the fetch and execute cycles during the execution of a machine-language program.
- Understand the operational role of the different types of buses and their signals in a microcomputer.
- Describe the sequences of events that occur during the read and write cycles during the execution of a program.

Introduction

It is no exaggeration to say that the microprocessor and microcomputer have revolutionized the electronics industry and have had a tremendous impact on many aspects of our lives. Large-scale integration (LSI) has so sharply reduced the size and cost of computers that designers routinely consider using the power and versatility of the microprocessor and microcomputer in a wide variety of applications.

In this chapter we will study the basic principles of operation of microcomputers. Since complete textbooks are devoted to the study of computers, our main objective here is only to provide a firm foundation for further study. Although we will concentrate on the microcomputer, most of the concepts and principles apply equally to computers of all sizes.

12-1 WHAT IS A DIGITAL COMPUTER?

A digital computer is a combination of digital devices and circuits that can perform a *programmed* sequence of operations with a minimum of human intervention. The sequence of operations is called a **program.** The program is a set of coded instructions that is stored in the computer's internal memory along with all the data that the program requires. When the computer is commanded to execute the program, it performs the instructions in the order that they are stored in memory until the program is completed. It does this at extremely high speeds without making any errors.

12-2 HOW DO COMPUTERS THINK?

Computers do not think! The computer *programmer* provides a *program* of instructions and data which specifies every detail of what to do, what to do it to, and when to do it. The computer is simply a high-speed machine that can manipulate data, solve problems, and make decisions, all under the control of the program. If the programmer makes a mistake in the program or puts in the wrong data, the computer will produce wrong results. A popular saying in the computer field is "garbage in/garbage out."

Perhaps a better question to ask at this point is: How does a computer go about executing a program of instructions? Typically, this question is answered by showing a diagram of a computer's architecture (arrangement of its various elements) and then going through the step-by-step process which the computer follows in executing the program. We will do this—but not yet. First, we will look at a somewhat far-fetched analogy that contains many of the concepts involved in computer operation.

12-3 SECRET AGENT 89

Secret Agent 89 is trying to find out how many days before a certain world leader is to be assassinated. His contact tells him that this information is located in a series of post office boxes. To ensure that no one else gets the information, it is spread through 10 different boxes. His contact gives him 10 keys along with the following instructions:

1. The information in each box is written in code.
2. Open box 1 first and execute the instruction located there.
3. Continue through the rest of the boxes in sequence unless instructed to do otherwise.
4. One of the boxes is wired to explode upon opening.

Agent 89 takes the 10 keys and proceeds to the post office, code book in hand.

Figure 12-1 shows the contents of the 10 post office boxes after having been decoded. Assume that you are Agent 89; begin at box 1 and go through the sequence of operations to find the number of days before the assassination attempt. Of course,

①	Add the number stored in box ⑨ to your secret agent code number.	②	Divide the previous result by the number stored in box ⑩ .
③	Subtract the number stored in box⑧	④	If the previous result is not equal to 30, go to box ⑦ . Otherwise continue to next box.
⑤	Subtract 13 from the previous result.	⑥	HALT. You now have the answer
⑦	BOMB! (too bad)	⑧	20
⑨	11	⑩	2

FIGURE 12-1   Ten post office boxes with coded message for Agent 89.

it should not be as much work for you as it was for Agent 89 because you don't have to decode the messages. The answer is given in the next paragraph.

If you have proceeded correctly, you should have ended up at box 6 with an answer of 17. If you made a mistake, you might have opened box 7, in which case you are no longer with us. As you went through the sequence of operations, you essentially duplicated the types of operations and encountered many of the concepts that are part of a computer. We will now discuss these operations and concepts in the context of the secret-agent analogy and see how they are related to actual computers.

In case you have not already guessed, the post office boxes are like the **memory** in a computer, where **instructions** and **data** are stored. Post office boxes 1 to 6 contain instructions to be executed by the secret agent and boxes 8, 9, and 10 contain the data called for by the instructions. (The contents of box 7, to our knowledge, have no counterpart in computers.) The numbers on each box are like the **addresses** of the locations in memory.

Three different classes of instructions are present in boxes 1 to 6. Boxes 1, 2, 3, and 5 are instructions that call for *arithmetic operations*. Box 4 contains a *decision-making* instruction called a *conditional jump* or *conditional branch*. This instruction calls for the agent (or computer) to decide whether to jump to address 7 or to continue to address 5, depending on the result of the previous arithmetic operation. Box 6 contains a simple control instruction that requires no data and refers to no other address (box number). This *halt* instruction tells the agent that the procedure is finished (program is completed) and to go no further.

Each of the arithmetic and conditional jump instructions consists of two parts—an **operation** and an **address.** For example, the first part of the first instruc-

tion specifies the operation of addition. The second part gives the address (box 9) of the data to be used in the addition. These data are usually called the **operand** and their address the **operand address.** The instruction in box 5 is a special case in which no operand address is specified. Instead, the operand (data) to be used in the subtraction operation is included as part of the instruction.

A computer, like the secret agent, decodes and then executes the instructions stored in memory *sequentially,* beginning with the first location. The instructions are executed in order unless some type of *branch* instruction (such as box 4) causes the operation to branch or jump to a new address location to obtain the next instruction. Once the branching occurs, instructions are executed sequentially beginning at the new address.

This is about as much information as we can extract from the secret-agent analogy. Each of the concepts we encountered will be encountered again in subsequent material. Hopefully, the analogy has furnished insights that should prove useful as you begin a more technical study of computers.

12-4 BASIC COMPUTER SYSTEM ORGANIZATION

Every computer contains five essential elements or units: the *arithmetic-logic unit* (ALU), the *memory unit,* the *control unit,* the *input unit,* and the *output unit.* The basic interconnection of these units is shown in Figure 12-2. The arrows in this diagram indicate the direction in which data, information, or control signals are flowing. Two different-size arrows are used; the larger arrows represent data or in-

Teaching Tip: Some prefer to combine the input and output units into a single input/output (I/O) unit, since many computer peripherals can perform both operations.

FIGURE 12-2 Basic computer organization.

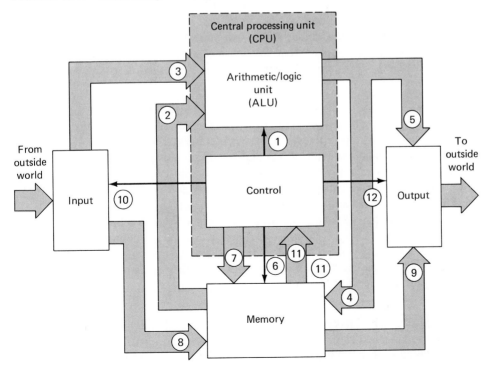

formation that actually consists of a relatively large number of parallel lines, and the smaller arrows represent control signals that are normally only one or a few lines. The various arrows are also numbered to allow easy reference to them in the following descriptions.

Arithmetic-Logic Unit   The ALU is the area of the computer in which arithmetic and logic operations are performed on data. The type of operation that is to be performed is determined by signals from the control unit (arrow 1). The data that are to be operated on by the ALU can come from either the memory unit (arrow 2) or the input unit (arrow 3). Results of operations performed in the ALU can be transferred to either the memory unit for storage (arrow 4) or the output unit (arrow 5).

Memory Unit   The memory stores groups of binary digits (words) that can represent instructions (program) that the computer is to perform and the data that are to be operated on by the program. The memory also serves as storage for intermediate and final results of arithmetic operations (arrow 4). Operation of the memory is controlled by the control unit (arrow 6), which signals for either a read or a write operation. A given location in memory is accessed by the control unit that provides the appropriate address code (arrow 7). Information can be written into the memory from the ALU or the input unit (arrow 8), again under control of the control unit. Information can be read from memory into the ALU (arrow 2) or into the output unit (arrow 9).

Input Unit   The input unit consists of all of the devices used to take information and data that are external to the computer and put them into the memory unit (arrow 8) or the ALU (arrow 3). The control unit determines where the input information is sent (arrow 10). The input unit is used to enter the program and data into the memory unit prior to starting the computer. This unit is also used to enter data into the ALU from an external device during the execution of a program. Some of the common input devices are keyboards, toggle switches, teletypewriters, punched-card readers, magnetic disk units, magnetic tape units, and analog-to-digital converters (ADCs).

Output Unit   The output unit consists of the devices used to transfer data and information from the computer to the "outside world." The output devices are directed by the control unit (arrow 12) and can receive data from memory (arrow 9) or the ALU (arrow 5); the data are then put into appropriate form for external use. Examples of common output devices are LED readouts, indicator lights, printers, disk or tape units, video monitors, and digital-to-analog converters (DACs).

Discussion Question: Would a "joy stick" or "mouse" be considered part of the input unit, or part of the output unit?

As the computer executes its program, it usually has results or control signals that it must present to the external world. For example, a computer system might have a line printer as an output device. Here, the computer sends out signals to print out the results on paper. A small microcomputer might display its results on indicator lights or on LED displays.

Interfacing   The most important aspect of the I/O units involves *interfacing*, which can be defined as the joining of dissimilar devices in such a way that they are able to function in a compatible and coordinated manner. *Computer interfacing* is

more specifically defined as the synchronization of digital information transmission between the computer and external input/output devices.

Many input/output devices are not directly compatible with the computer because of differences in such characteristics as operating speed, data format (e.g., hex, ASCII, binary), data transmission mode (e.g., serial, parallel), and logic signal level. Such I/O devices require special interface circuits which allow them to communicate with the control, memory, and ALU portions of the computer system. A common example is the video display terminal (VDT), which can operate both as an input and an output device. The VDT transmits and receives data serially (one bit at a time) while most computers handle data in parallel form. Thus, a VDT requires interface circuitry in order to send data to or receive data from a computer.

Control Unit   The function of the control unit should now be obvious. It directs the operation of all the other units by providing timing and control signals. In a sense, the control unit is like the conductor of an orchestra, who is responsible for keeping each of the orchestra members in proper synchronization. This unit contains logic and timing circuits that generate the proper signals necessary to execute each instruction in a program.

The control unit **fetches** an instruction from memory by sending an address (arrow 7) and a read command (arrow 6) to the memory unit. The instruction word stored at the memory location is then transferred to the control unit (arrow 11). This instruction word, which is in some form of binary code, is then decoded by logic circuitry in the control unit to determine which instruction is being called for. The control unit uses this information to generate the necessary signals for **executing** the instruction.

Central Processing Unit (CPU)   In Figure 12-2, the ALU and control units are shown combined into one unit called the *central processing unit (CPU)*. This is commonly done to separate the actual "brains" of the computer from the other units. We will use the CPU designation often in our work on microcomputers because, as we shall see, in microcomputers the CPU is usually implemented in an LSI chip called the **microprocessor** chip.

___ REVIEW QUESTIONS _____

1. Name the five basic units of a computer and describe the major functions of each.
2. What is the CPU?
3. What is meant by interfacing in a computer system?

12-5   BASIC μC ELEMENTS

It is important to understand the difference between the microcomputer (μC) and the microprocessor (μP). A microcomputer contains several elements, the most important of which is the microprocessor. The microprocessor is usually a single IC that contains all of the circuitry of the control and arithmetic-logic units—in other words, the CPU. It is common to refer to the microprocessor as the MPU (mi-

FIGURE 12-3   Basic elements of a microcomputer.

croprocessor unit), since it is the CPU (central processing unit) of the microcomputer. This is illustrated in Figure 12-3, where the basic elements of a microcomputer are shown.

The memory unit shows both RAM and ROM devices. The RAM section consists of one or more LSI chips arranged to provide the designed memory capacity. This section of memory is used to store programs and data, which will change often during the course of operation. It is also used as storage for intermediate and final results of operations performed during execution of a program.

The ROM section consists of one or more LSI chips to store instructions and data that do not change. For example, it might store a program that causes the μC to continually monitor a keyboard, or it might store a table of ASCII codes needed for outputting information to a VDT or printer.

The input and output sections contain the interface circuits needed to allow the I/O devices to properly communicate with the rest of the computer. In some cases these interface circuits are LSI chips designed by the MPU manufacturer to interface the MPU to a variety of I/O devices. In other cases the interface circuits may be as simple as a buffer register.

The Microprocessor (MPU)   The MPU is the heart of every microcomputer. It performs a number of functions, including:

1. Providing timing and control signals for all elements of the μC.
2. Fetching instructions and data from memory.

3. Transferring data to and from memory and I/O devices.

4. Decoding instructions.

5. Performing arithmetic and logic operations called for by instructions.

6. Responding to I/O-generated control signals such as RESET and INTER-RUPT.

The MPU contains all of the logic circuitry for performing these functions, but its internal logic is generally not externally accessible. Instead, we can control what happens inside the MPU by the *program of instructions* that we put in memory for the MPU to execute. This is what makes the MPU so versatile and flexible—when we want to change its operation, we simply change the programs stored in RAM (software) or ROM (firmware) rather than rewire the electronics (hardware).

The MPU's internal logic is extremely complex, but it can be thought of as consisting of three basic sections: the *control and timing* section, the *register* section, and the *ALU* (see Figure 12-4). Although there are definite interactions among these three sections, each has specific functions.

The main function of the timing and control section is to fetch and decode (interpret) instruction codes from program memory, and then to generate the necessary control signals required by the other MPU sections in order to carry out the execution of the instructions. This section also generates timing and control signals (e.g., R/W, clock) that are needed by external RAM, ROM, and I/O devices.

The register section contains various registers (inside the MPU) each of which performs a special function. The most important one is the **program counter (PC),** which keeps track of the addresses of the instruction codes as they are fetched from memory. We will use the PC in our subsequent description of program execution. Other MPU registers are used to perform functions such as: storing instruction codes as they are being decoded, holding data being operated on by the ALU, storing addresses of data to be fetched from memory, and many general-purpose storage and counting functions.

The ALU performs a variety of arithmetic and logic operations on data. These operations always include addition and subtraction, AND, OR, EX-OR, shifting, incrementing, and decrementing. The more advanced MPUs have ALUs that can do multiplication and division. During a microcomputer's operation, the operations that an ALU is to perform are under the control of the timing and control section, which, of course, does what it is told by the instruction codes it fetches from memory.

FIGURE 12-4 Major functional areas of a μP chip.

1. Explain the difference between a microprocessor and a microcomputer.
2. Name the basic elements of a microcomputer.
3. Name the three major sections of an MPU.

12-6 COMPUTER WORDS

The smallest unit of information in a computer is the bit. A single isolated bit, however, carries very little information. For this reason, the principal unit of information in a computer is a group of bits referred to as a **word.** The number of bits that make up a word is called the computer's **word size.** Word size is a common way of describing a computer. Computers are often described in terms of their word size, such as an 8-bit computer, a 16-bit computer, and so on. For example, a 16-bit computer is one in which the instructions and data are stored in memory as 16-bit units, and processed by the CPU in 16-bit units. The word size also indicates the size of the data bus which carries data between the CPU and memory and between the CPU and input/output devices.

Teaching Tip: Some of the newer microcomputers that have a basic word size of 16 bits can also do some 32-bit operations. The 32-bit unit is called a *long word*. That is: 8 bits = byte; 16 bits = word; 32 bits = long word.

The larger computers (mainframes, maxicomputers) have word sizes that are usually in the 32- to 64-bit range. Minicomputers have word sizes from 8 to 32 bits. Microcomputers have word sizes from 4 to 32 bits. The newer microcomputers (e.g., IBM PS/2 and APPLE MacIntosh) use word sizes of 16 and 32 bits. In general, a computer with a larger word size can execute programs of instructions at a faster rate because more data and more instruction information are stuffed into one word. The larger word sizes, however, mean more lines making up the data bus, and therefore more interconnections between the CPU and memory and input/output devices.

As we know, a group of 8 bits is called a byte. Because 8-bit microcomputers have been widely used for a long time, and because ASCII codes fit neatly into 1 byte, the byte continues to be used as a unit to describe word size and memory capacity even in computers with larger word sizes. An 8-bit computer can be said to have a word size of 1 byte. A 16-bit computer has a word size of 2 bytes, and so on. A memory that stores 128K 16-bit words can also have its capacity described as 256K bytes.

Types of Computer Words

A word stored in a computer's memory can contain either of two kinds of information—*instructions* or *data*. Data can be numerical or character information that is to be processed by a program that the CPU is executing. The data can be in many forms including unsigned or signed binary, BCD, floating-point (something like engineering notation), or ASCII codes for characters, among others. Here is an example of how the numerical value $+86_{10}$ will be stored in an 8-bit word:

$$01010110$$

Here is how the ASCII code for the character "V" would be stored in an 8-bit word.

$$01010110$$

Notice that the two words are the same. The computer doesn't know the difference between the two. It is up to the programmer to know what type of data is being stored, and to ensure that the program interprets and processes the data properly.

Here is another example where a 16-bit word is used to store two ASCII-coded characters:

$$\underbrace{0101011}_{V}\underbrace{001010111}_{W}$$

This same 16-bit word could be the representation for $+22103_{10}$. Clearly, larger word sizes can store more character data and larger numbers.

Instruction words are more complex than data words. We will discuss them in detail in the following section.

___ REVIEW QUESTIONS ___

1. What are the two types of information stored in computer words?
2. What is the advantage of a larger-word-size computer?

12-7 INSTRUCTION WORDS

The format used for *data* words varies only slightly among different computers, especially those with the same word size. This is not true, however, of the format for *instruction* words. These words contain the information necessary for a computer to execute its various operations, and the format and codes for these can vary widely from computer to computer. Depending on the computer, the information contained in an instruction word can be different. But, for most computers, the instruction words carry two basic units of information: the *operation* to be performed and the *address* of the *operand* (data) to be operated upon.

Figure 12-5 shows an example of a *single-address instruction word* for a hypothetical 20-bit computer. The 20 bits of the instruction word are divided into two parts. The first part of the word (bits 16 through 19) contains the **operation code (op code,** for short). The 4-bit op code represents the operation that the computer is being instructed to perform, such as addition or subtraction. The second part (bits 0 through 15) is the **operand address,** which represents the location in memory where the operand is stored.

With 4 bits used for the op code, there are $2^4 = 16$ different possible op codes, with each one indicating a different instruction. This means that a computer using this instruction-word format is limited to 16 different possible instructions which it can perform. A more versatile computer would have a greater number of instructions and would therefore require more bits in its op code. In any case, each instruction that a computer can perform has a specific op code which the computer (control unit) must interpret (decode).

The instruction word of Figure 12-5 has 16 bits reserved for the operand address code. With 16 bits, there are $2^{16} = 65,536$ different possible addresses. Thus,

FIGURE 12-5   Typical single-address instruction word.

this instruction word can specify 16 different instructions and 65,536 operand addresses. As an example, a 20-bit instruction word might be

| 0 | 1 | 0 | 0 | 0 | 1 | 0 | 1 | 1 | 0 | 1 | 0 | 0 | 1 | 1 | 1 | 0 | 0 | 1 | 0 |

Op   Address
code   code

The op code 0100 represents one of 16 possible operations; let's assume that it is the code for *addition* (ADD). The address code is 0101101001110010 or, more conveniently, 5A72 in hexadecimal. In fact, this complete instruction word can be expressed in hexadecimal as

$$\underbrace{4}_{\substack{\text{Op} \\ \text{code}}} \quad \underbrace{5 \quad A \quad 7 \quad 2}_{\text{Address}}$$

This complete instruction word, then, tells the computer to do the following:

> ***Fetch the data word stored in address location 5A72, send it to the ALU, and add it to the number in the accumulator register. The sum will then be stored in the accumulator. (Previous contents of accumulator are lost.)***

We will examine this and other instructions more thoroughly later.

Multiple-Address Instructions
The single-address instruction described above was once the principal type used in larger computers. The larger computers, however, have begun to use several other instruction formats which provide more information per instruction word.

Figure 12-6 shows two instruction-word formats that contain more than one address. The two-address instruction has the op code plus the addresses of *both* operands which are to take part in the specified operation. The three-address instruction has the addresses of both operands plus the address in memory where the result is to be stored.

These multiple-address instruction words have the obvious advantage that they contain more information than does a single-address instruction. This means that a computer using multiple-address instructions will require fewer instructions to exe-

FIGURE 12-6   Multiple-address instruction formats.

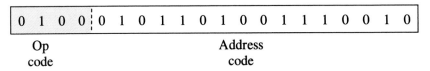

| Op code | Address of operand 1 | Address of operand 2 |

Two-address instruction

(a)

| Op code | Address of operand 1 | Address of operand 2 | Address of where to store result |

Three-address instruction

(b)

cute a particular program. Of course, the longer instruction words require a memory unit with a larger word size.

Multibyte Instructions

We have looked at instruction-word formats that contain op code and operand address information in a *single* word. In other words, a complete instruction such as those in Figure 12-5 or Figure 12-6 is stored in a *single* memory location. This is typical of computers with relatively large word sizes. For many microcomputers and minicomputers, the smaller word size makes it impossible to provide the op code and operand address in a single word.

Since a good number of microcomputers use an 8-bit (one-byte) word length, we will describe the instruction formats used in 8-bit computers. With a one-byte word size there are *three* basic instruction formats: single-byte, two-byte, and three-byte instructions. These are illustrated in Figure 12-7.

The single-byte instruction contains only an 8-bit op code, with no address portion. Clearly, this type of instruction does not specify any data from memory to be operated on. For this reason, single-byte instructions are used for operations that do not require memory data. An example would be the instruction *clear the accumulator register to zero* (CLA), which instructs the computer to clear all the FFs in the ALU's accumulator.

Teaching Tip: Other examples of single-byte instructions are INX (increment X register), HLT (halt), and ASL (shift accumulator left).

FIGURE 12-7   Instruction formats used in 8-bit microcomputers.

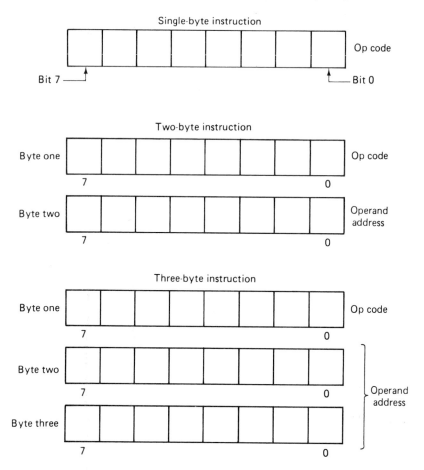

TABLE 12-1

Memory Address (Hex)	Memory Word			Description
	Binary	Hex		
0020	01001010	4A		Op code for ADD
0021	11110110	F6		Low-order address bits (LO)
0022	00110101	35		High-order address bits (HI)
.	.	.		
.	.	.		
.	.	.		
.	.	.		
.	.	.		
.	.			
35F6	01111100	7C		Operand

Teaching Tip: The second byte of some 2-byte instructions, and the second and third bytes of some 3-byte instructions, can be the actual operand (data) rather than the address of the operand.

The first byte of the two-byte instruction is an op code and the second byte is an 8-bit address code specifying the memory location of the operand. In the three-byte instruction, the second and third bytes form a 16-bit operand address. For these multibyte instructions, the two or three bytes making up the complete instruction have to be stored in successive memory locations. This is illustrated in Table 12-1 for a three-byte instruction. The left-hand column lists the address locations in memory where each byte (word) is stored. These addresses are given in hexadecimal code. The second column gives the binary word as it is actually stored in memory; the third column is the hex equivalent of this word. Examine this complete example before reading further and try to figure out what it represents.

The three bytes stored in locations 0020, 0021, and 0022 constitute the complete instruction for adding the data word stored in address location 35F6 to the accumulator. The second and third bytes hold the 8 low-order bits (LO), and 8 high-order bits (HI), respectively, of the operand address. Some microcomputers use the reverse order, with HI stored in the second byte and LO in the third byte of the instruction sequence. Memory location 35F6 is also shown; its contents are the data word (7C) which the control unit of the computer will fetch and send to the ALU for addition.

___ REVIEW QUESTIONS ___

1. What is an op code?
2. What is an operand address?
3. What information is contained in the first byte of a multibyte instruction?

12-8 EXECUTING A MACHINE-LANGUAGE PROGRAM

The instruction words we have been describing are called **machine-language** instructions because they are represented by 1s and 0s, the only language the machine (computer) understands. Many other languages can be used to program a computer,

TABLE 12-2

Mnemonic	Op Code, Binary/Hex	Description of Operation
LDA	01011011/5B	*Load accumulator:* The data stored at the operand address are loaded into the accumulator register.
ADD	01101101/6D	*Add:* The operand is added to the number stored in the accumulator and the resultant sum is stored in the accumulator.
SUB	10100010/A2	*Subtract:* The operand is subtracted from the contents of the accumulator and the result is stored in the accumulator.
STA	10001010/8A	*Store accumulator:* The contents of the accumulator are stored in memory at the location specified by the operand address.
JMP	01001100/4C	*Jump (unconditionally):* The next instruction is taken from the location specified by the operand address instead of in sequence.
JPZ	01110110/76	*Jump on zero:* The next instruction is taken from operand address *if* the accumulator contents are *zero*. Otherwise, the next instruction is taken in sequence.
HLT	00111111/3F	*Halt:* The computer operation is halted. No further instructions are executed.

and you may be familiar with some of them, such as BASIC or Pascal. These **high-level languages** are designed to make it easy to write a program. It is important to understand that these high-level programs have to be converted to machine-language instructions and placed in the computer's internal memory before the computer can execute them. In a typical microcomputer, the conversion from BASIC to machine language is accomplished through a special machine-language program stored in ROM.

In order to illustrate how a microcomputer executes a machine-language program, we will use the instructions described in Table 12-2. These instructions do not pertain to any particular microcomputer but they are typical of the kinds of instructions that most microcomputers can execute. Each instruction is accompanied by a **mnemonic** or abbreviation that is easier to remember than the op code. Read the description of each instruction carefully and keep in mind that the operand that is being referred to is the data word stored at the operand address location.

We will use some of the instructions listed in Table 12-2 to write a machine-language program that starts at address 0000 and does the following:

1. Subtracts an 8-bit data word (*X*) from another 8-bit data word (*Y*). The values of *Y* and *X* are stored in hex address locations 0300 and 0301, respectively.

2. Stores the difference (DIF) in memory location 0302.

3. If the value of DIF is zero, the program jumps to address 0400 for its next instruction. Otherwise, the program halts.

Table 12-3 shows the complete program as it would be entered into the computer's memory. Actually, the first two columns are the machine-language program; the other information is included for descriptive purposes. The first column lists the hex address of each memory location being used by the program. The second column gives the hex equivalent of the word stored in each memory location. Re-

Teaching Tip: These conversion programs are called either *interpreters* or *compilers*. An interpreter converts the high-level-language program to machine language, one line at a time, and executes it. The interpreter has to be used each time a high-level program is to be run. A compiler converts the complete high-level program to a machine-language version called the object program. The object program can then be run over and over without using the compiler. A compiler is more complex than an interpreter.

TABLE 12-3

Memory Address (Hex)	Memory Word (Hex)	Mnemonic	Description
0000	5B	LDA$0300	Load data (*Y*) from address
0001	00 ⎤		0300 into accumulator
0002	03 ⎦		
0003	A2	SUB$0301	Subtract data (*X*) at address
0004	01 ⎤		0301 from accumulator
0005	03 ⎦		
0006	8A	STA$0302	Store difference in address
0007	02 ⎤		0302
0008	03 ⎦		
0009	76	JPZ$0400	If difference (DIF) is zero,
000A	00 ⎤		jump to 0400 for next
000B	04 ⎦		instruction
000C	3F	HLT	If DIF is *not* zero, halt
			program
.	.	.	.
.	.	.	.
.	.	.	.
0300	??	*Y*	Address of data *Y*
0301	??	*X*	Address of data *X*
0302	??	DIF	Address where difference is to be stored

Teaching Tip: Students often want to know how the program gets into memory. Here are several possibilities: it may be permanently burned into a ROM; it may be entered from a hex keypad; it may be entered from a terminal; it may have been loaded from a disk or tape unit.

member, these hex values represent the actual binary addresses and instruction codes that the computer understands.

The third column gives the mnemonic and operand address (if any) associated with each instruction. The last column describes the operation performed by each instruction. For example, the first instruction in the program is a three-byte instruction. The first byte stored in memory address 0000 is the op code 5B. The second and third bytes stored in addresses 0001 and 0002 represent the operand address 0300. The mnemonic for this instruction is LDA $0300. The LDA is the abbreviation for the load accumulator operation, and the $0300 is the operand address. The $ is often used to indicate that the address is represented in hex. When the computer executes this instruction, it will read the data *Y* that are stored in location 0300 and will load them into the accumulator.

The program starts at address 0000 and ends at 000C. Address locations 0300 to 0302 are used for data storage. The contents of these latter locations are not given, because they will generally be variables. Address locations between 000C and 0300 are not being used.

Program Execution   We will now proceed through the complete execution of this program and describe what the computer does at each step. Our intent here is to outline only the main operations without getting bogged down in the details of all the activity taking place inside the computer. In particular, we will see that the

computer is always in one of two kinds of operating cycles: (1) a **fetch cycle** during which the control unit fetches the instruction codes (op code and operand address) from memory and (2) an **execute cycle** during which the control unit performs the operation called for by the op code.

The operation starts when the operator activates a START or RUN switch. This will initialize a program counter (PC) to a starting count of 0000. The PC is a counter within the control unit that keeps track of the program addresses as the computer sequences through them.

1. The control unit fetches the first byte from address 0000 as determined by the PC. This byte is 5B, which is the op code for the first instruction. Circuitry within the control unit determines that this op code calls for the LDA (load accumulator) operation and that an operand address follows the op code.

2. The control unit increments the PC to 0001 and then fetches the byte stored at this address. This byte is 00 and represents the low-order byte of the operand address. The control unit then increments PC again to 0002 and fetches the byte that is stored at this address. This byte is 03 and represents the high-order byte of the operand address.

3. This completes the first fetch cycle, and the control unit now holds the op code and operand address (0300) needed to execute the LDA instruction.

4. The control unit executes this instruction by reading the data word from address 0300 and loading it into the accumulator register in the ALU. This completes the execute cycle.

5. The PC is incremented to 0003 and the control unit begins a new fetch cycle by fetching the op code stored at this address. The control unit recognizes the op code A2 as calling for the SUB operation, which requires an operand address that is stored in the next two bytes of the program.

6. The control unit fetches the operand address 0301 in the same manner as in step 2 above. The PC is now at 0005.

7. This completes the second fetch cycle, and the control unit now holds the op code and operand address needed to perform the SUB instruction. The control unit executes this instruction by reading the data word from address 0301 and sending it to the ALU to be subtracted from the contents of the accumulator.

8. The PC is incremented to 0006 to initiate a new fetch cycle. The control unit fetches the op code (8A) stored at this address, recognizes it as calling for the STA operation, and knows that the operand address is stored in the next two bytes of the program.

9. The control unit fetches the operand address 0302. The PC is now at 0008. The contents of the accumulator are then stored in 0302.

10. The PC is incremented to 0009 to initiate a new fetch cycle, and the control unit fetches the op code (76) stored at this address. The control unit recognizes the JPZ instruction and knows that it has to make a decision based on the current contents of the accumulator.

11. If the data in the accumulator are exactly equal to zero, the control unit fetches the operand address 0400 from the next two memory addresses (000A and 000B). This operand address is loaded into the PC so that the control unit will take its next instruction from address 0400 and will continue executing instructions from that point.

Teaching Tip: The fetched operand address is stored in a CPU register typically called the data address register (DAR) or memory address register (MAR).

12. If the data in the accumulator are *not* equal to zero, the control unit increments the PC three times so that PC = 000C. The next instruction is fetched from this address, and it is recognized as the HLT instruction, which causes the control unit to stop fetching and executing instructions.

Discussion Question: How do you think a JNZ (jump if not zero) instruction works?

This simple program illustrates many of the kinds of things that take place as a computer executes a machine-language program, but it does not even begin to show the capabilities and versatility of a computer. It is important to understand that the program execution is performed in a step-by-step sequential manner starting at the first address in the program (0000 in our example). The program's address sequence can change when either a JMP or JPZ instruction is encountered (e.g., steps 10 and 11 above). Of course, even though the computer performs one step at a time, each step can be done so rapidly that the overall execution time for a program can be very short. For example, this short program would take less than 20 μs on even the slowest microcomputer.

___ REVIEW QUESTIONS ___

1. What is the difference between a machine-language program and a high-level-language program?
2. What is an instruction mnemonic?
3. What generally takes place during a fetch cycle? During an execute cycle?
4. What is the function of the program counter?

12-9   TYPICAL μC STRUCTURE

We are now prepared to take a more detailed look at μC organization. The many possible μC structures are essentially the same in principle, although they vary as to the size of the data and address buses, and the types of control signals they use. In order to provide the clearest means for learning the principles of μC operation, it is necessary to choose a single type of μC structure and study it in detail. Once a solid understanding of this typical μC is obtained, it will be relatively easy to learn about any other type. The μC structure we have chosen to present here represents a common one in use today and is shown in Figure 12-8.

This diagram shows the basic elements of an 8-bit microcomputer system and the various buses that connect them together. Although this diagram looks somewhat complex, it still does not show all the details of the μC system. For our purposes, however, it will be sufficient.

The Bus System   The μC has three buses which carry all the information and signals involved in the system operation. These buses connect the microprocessor (CPU) to each of the memory and I/O elements so that data and information can flow between the CPU and any of these other elements. In other words, the CPU is continually involved in sending or receiving information to or from a location in memory, an input device, or an output device.

In the μC, all information transfers are referenced to the CPU. When the CPU is sending data to another computer element, it is called a write operation and the

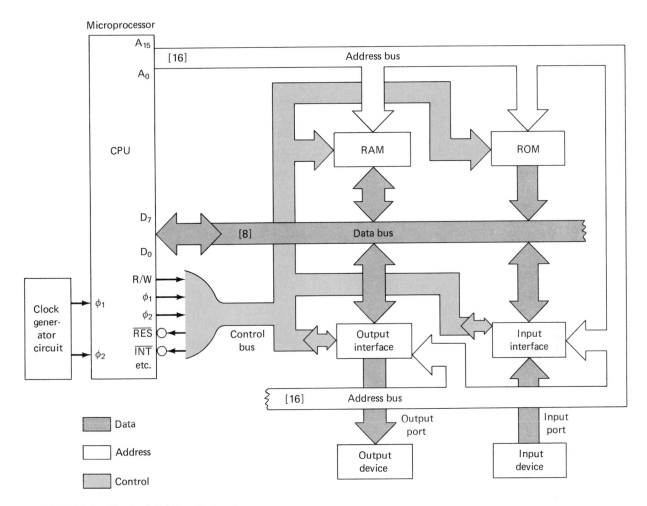

FIGURE 12-8 Typical 8-bit μC structure.

CPU is writing into the selected element. When the CPU is receiving data from another element, it is called a read operation and the CPU is reading from the selected element. It is very important to realize that the terms "read" and "write" always refer to operations performed by the CPU.

The buses involved in all the data transfers have functions that are described as follows:

Address bus. This is a *undirectional* bus, because information flows over it in only one direction, from the CPU to the memory or I/O elements. The CPU alone can place logic levels on the lines of the address bus, thereby generating $2^{16} = 65,536$ different possible addresses. Each of these addresses corresponds to one memory location or one I/O element. For example, address $20A0_{16}$ might be a location in RAM or ROM where an 8-bit word is stored, or it might be an 8-bit buffer register that is part of the interface circuitry for a keyboard input device.

When the CPU wants to communicate (read or write) with a certain memory location or I/O device, it places the appropriate 16-bit address code on its 16 address pin outputs, A_0 through A_{15}, and onto the address bus. These address bits are then *decoded* to select the desired memory location or I/O device. This decoding process usually requires decoder circuitry not shown on this diagram.

Data bus. This is a *bidirectional* bus, because data can flow to or from the CPU. The CPU's eight data pins, D_0 through D_7, can be either inputs or outputs, depending on whether the CPU is performing a read or a write operation. During a read operation they act as inputs and receive data that have been placed on the data bus by the memory or I/O element selected by the address code on the address bus. During a write operation the CPU's data pins act as outputs and place data on the data bus, which are then sent to the selected memory or I/O element. In all cases, the transmitted data words are 8 bits wide because the CPU handles 8-bit data words, making this an 8-bit μC.

Control bus. This is the set of signals that is used to synchronize the activities of the separate μC elements. Some of these control signals, such as $R/\overline{W}$, are signals the CPU sends to the other elements to tell them what type of operation is currently in progress. The I/O elements can send control signals to the CPU. An example is the reset input ($\overline{RES}$) of the CPU, which, when driven LOW, causes the CPU to reset to a particular starting state. Another example is the CPU's interrupt input ($\overline{INT}$), used by I/O devices to get the attention of the CPU when it is performing other tasks.

Clock Signals   Some microprocessors (8085, 6802, 6502) use a two-phase clock system with nonoverlapping pulses such as those shown in Figure 12-9. Other microprocessors such as the Z-80 and 8086 operate from a single clock signal. In our subsequent discussions we will use the two-phase clock system. The two clock phases, ϕ_1 and ϕ_2, are always part of the control bus. Other timing signals, derived from ϕ_1 and ϕ_2, are sometimes generated by the CPU and become part of the control bus.

FIGURE 12-9   Two-phase clock system.

I/O Ports   During the execution of a program, the CPU is constantly reading from or writing into memory. The program may also call on the CPU to read from one of the input devices or write into one of the output devices. Although the diagram of the 8-bit μC in Figure 12-7 shows only one input and one output device there can be any number of each tied to the μC bus system. Each I/O device is normally connected to the μC bus system through some type of interface circuit. The function of the interface is to make the μC and the device compatible so that data can be easily passed between them. The interface is needed whenever the I/O device uses different signal levels, signal timing, or signal format from the μC.

For example, a typical I/O device is the standard teletypewriter unit (abbreviated TTY), which sends ASCII-coded information to the computer in *serial* fashion (1 bit at a time over a single line). The μC, however, accepts data from the data bus

as 8 *parallel* bits. Thus, an interface circuit is used to convert the TTY's serial signal to an 8-bit parallel data word, and another to convert the μC's parallel output data to a serial signal for the TTY.

Although I/O devices are treated like memory locations, they are significantly different from memory in some respects. One big difference is that I/O devices can have the capability to *interrupt* the CPU while it is executing a program. What this means is that an I/O device can send a signal to the CPU's interrupt ($\overline{INT}$) input to tell the CPU that it wishes to communicate with it. The CPU will then suspend execution of the program it is currently working on and will perform the appropriate operation with the interrupting I/O device. RAM and ROM do not have interrupting capability.

___ REVIEW QUESTIONS ___

1. Describe the functions of the three buses that are part of a typical micro-computer system.
2. Which bus is unidirectional?
3. Which bus includes the clock signals?

12-10 READ AND WRITE OPERATIONS

We are now ready to take a more detailed look at how the CPU communicates with the other μC elements. Remember, the CPU contains all the control and arithmetic-logic circuitry needed to execute a program of instructions stored in RAM or ROM. The CPU is continually performing read and write operations as it executes a program. It fetches each instruction from memory with a read operation. After interpreting the instruction, it may have to perform a read operation to obtain the operand from memory, or it may have to write data into memory. In some cases, the instruction may call for the CPU to read data from an input device (such as a keyboard or an ADC) or to write data into an output device (like an LED display or a magnetic disk unit).

The Read Operation   The following steps take place during a read operation:

1. The CPU generates the proper logic level on its $R/\overline{W}$ line for initiating a read operation. Normally, $R/\overline{W} = 1$ for read. The $R/\overline{W}$ line is part of the control bus and goes to all the memory and I/O elements.
2. Simultaneously, the CPU places the 16-bit address code onto the address bus to select the particular memory location or I/O device from which the CPU wants to receive data.
3. The selected memory or I/O element places an 8-bit word on the data bus. All nonselected memory and I/O elements will not affect the data bus, because their *tristate* outputs will be in the disabled (Hi-Z) condition.
4. The CPU receives the 8-bit word from the data bus on its data pins, D_0 to D_7. These data pins act as inputs whenever $R/\overline{W} = 1$. This 8-bit word is then latched into one of the CPU's internal registers, such as the accumulator.

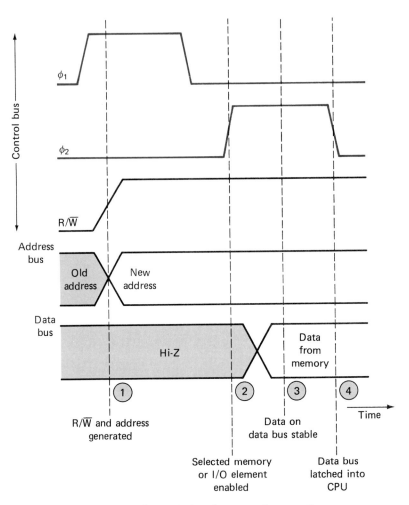

FIGURE 12-10   Typical μC timing for a read operation.

This sequence can be better understood with the help of a timing diagram showing the interrelationship between the signals on the various buses (see Figure 12-10). Everything is referenced to the ϕ_1 and ϕ_2 clock signals. The complete read operation occurs in one clock cycle. This is typically 100 ns to 1 μs for MOS microprocessors. The leading edge of ϕ_1 initiates the CPU's generation of the proper $R/\overline{W}$ and address signals. After a short delay, the $R/\overline{W}$ line goes HIGH and the address bus holds the new address code (point 1 on timing diagram). Note that the address-bus waveform shows both possible transitions (LOW-to-HIGH and HIGH-to-LOW), because some of the 16 address lines will be changing in one direction while others will be changing in the opposite direction.

Discussion Question: What determines which device is to place its data on the data bus during a READ operation?

During the ϕ_2 pulse, the selected memory or I/O device is enabled (point 2) and it proceeds to put its data word on the data bus. Prior to this, the data bus is in its Hi-Z state, since no device connected to it has been enabled. At some point during the ϕ_2 pulse, the data on the data bus become stable (point 3). Again, both possible data-line transitions are shown on the diagram. The delay between the start of the ϕ_2 pulse and the data-bus stabilizing depends on the speed of the memory and I/O elements. For memory this delay would be its *access time*. On the falling edge of ϕ_2, the data on the data bus are latched into the CPU (point 4). Clearly, then, the

memory and I/O devices must be capable of putting data on the bus prior to the falling edge of ϕ_2, or proper transfer to the CPU will not occur. Thus, it is necessary to ensure that these devices have a speed compatible with the μC clock frequency.

The Write Operation   The following steps occur during a write operation:

1. The CPU generates the proper logic level on the $R/\overline{W}$ line for initiating a write operation. Normally, $R/\overline{W} = 0$ for write.
2. Simultaneously, the CPU places the 16-bit address code onto the address bus.
3. The CPU then places an 8-bit word on the data bus via its data pins D_0 through D_7, which are now acting as outputs. This 8-bit word typically comes from an internal CPU register, such as the accumulator. All other devices connected to the data bus have their outputs disabled.
4. The selected memory or I/O element takes the data from the data bus. All nonselected memory and I/O elements will not have their inputs enabled.

This sequence has the timing diagram shown in Figure 12-11. Once again, the leading edge of ϕ_1 initiates the $R/\overline{W}$ and address-bus signals (point 1). During the ϕ_2

FIGURE 12-11   Typical μC timing for a write operation.

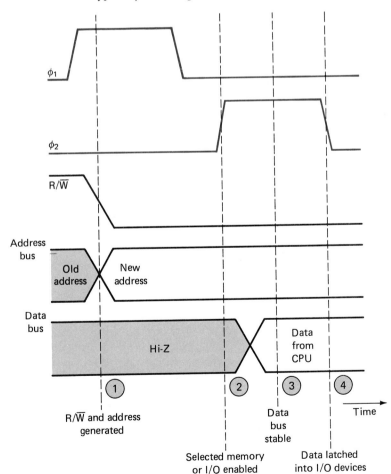

pulse, the selected memory or I/O device is enabled (point 2) and the CPU places its data on the data bus. The data-bus levels become stabilized a short time into the ϕ_2 pulse. These data are then written into the selected memory location while ϕ_2 is high. If an I/O device has been selected, it usually latches the data from the data bus on the falling edge of ϕ_2 (point 4).

The read and write operations encompass most of the μC activity that takes place outside the CPU. The following example illustrates.

——— EXAMPLE 12-1 ———

Discussion Question: What determines which device places its data on the data bus in a WRITE operation?

Below is a short program that is stored in memory locations 0020_{16} to 0029_{16} of an 8-bit μC. Note that since each word is one byte (8 bits), it takes two successive bytes to represent a 16-bit operand address code. Determine the total number of read and write operations that the CPU will perform as it executes this program.

Memory Address (Hex)	Memory Word (Hex)	Mnemonic	Description
0020	5B	LDA$5001	Load accumulator (ACC) with X.
0021	01⎫		⎡Address of⎤
0022	50⎭		⎣operand X⎦
0023	6D	ADD$5002	Add Y to contents of ACC.
0024	02⎫		⎡Address of⎤
0025	50⎭		⎣operand Y⎦
0026	8A	STA$5003	Store ACC contents.
0027	03⎫		⎡Address where ACC⎤
0028	50⎭		⎣will be stored⎦
0029	3F	HLT	Halt operation.

Solution   The CPU begins executing the program by reading the contents of memory location 0020. The word stored there (5B) is taken into the CPU and is interpreted as an instruction op code. (In other words, the CPU *always* interprets the first word of a program as an op code and the programmer must *always* adhere to this format.) The CPU decodes this op code to determine the operation to be performed and to determine if an operand address follows the op code. In this case, an operand address is required and is found in the next two successive bytes in the program. Thus, the CPU must read locations 0021 and 0022 to obtain the address of data X. Once this address (5001) has been read into the CPU, the CPU proceeds to execute the LDA instruction by reading memory location 5001 and putting its contents into the accumulator. Thus, the complete execution of this first instruction requires *four* separate read operations: one for the op code, two for the address, and one for the LDA operation.

Teaching Tip: As this example illustrates, many more read operations than write operations take place during program execution. This is why the normal state of the $R/\overline{W}$ signal is HIGH, and is pulsed LOW only during a write operation.

Similarly, *four* read operations are needed for the second instruction, which begins at 0023. This instruction also uses a two-byte operand address and requires reading the data from this address for transfer to the CPU's arithmetic unit.

The third instruction op code, at 0026, uses a two-byte operand address. The CPU uses *three* read operations to obtain these. However, the instruction STA calls for a write operation whereby the CPU transfers the contents of the accumulator to memory location 5003.

The final instruction, at 0029, is simply an op code with no operand address. The CPU reads this op code (3F), decodes it, and then halts further operations.

The total number of read operations, then, is *twelve* and the total number of write operations is *one*.

━━ REVIEW QUESTIONS ━━

1. Describe the steps that take place when the CPU performs a read operation.
2. Repeat for a write operation.
3. What device puts data on the data bus during a write operation?

12-11 FINAL COMMENTS

Our discussion has been, by necessity, only a brief introduction to the basic principles, terminology, and operations common to most microprocessers and microcomputers. We have not even begun to gain an insight into the capabilities and applications of these devices. Almost all areas of technology have started taking advantage of the inexpensive computer control which microprocessors can provide. Some typical applications include microwave ovens, traffic controllers, home computers, electronic measuring instruments, industrial process control, electronic games, automobile emission control, and a rapidly growing number of new products.

With the revolutionary impact that microprocessors have had on the electronics industry, it is not unreasonable to expect that everyone working in electronics and related areas will have to become knowledgeable in the operation of microprocessors. Hopefully, our introduction will serve as a firm foundation for further study in this important area.

ANSWERS TO SECTION REVIEW QUESTIONS

SECTION 12-4

1. Input, output, arithmetic-logic, control, memory. See text for functions. **2.** Control and arithmetic-logic units combined. **3.** The synchronization of digital information transmission between the computer and external I/O devices.

SECTION 12-5

1. The microprocessor (MPU) is the CPU portion of the microcomputer. **2.** MPU, RAM, ROM, input, output. **3.** Timing and control, register, ALU.

SECTION 12-6

1. Data and instructions. **2.** Executes programs at a faster rate.

SECTION 12-7

1. A binary code that represents the operation to be performed by the CPU. **2.** The address of the data to be operated on as the CPU executes the instruction called for by the op code. **3.** Op code.

SECTION 12-8

1. A machine-language program consists of the binary instruction codes stored in the computer's memory. A high-level-language program is written in a conversational language that has to be converted to machine language before the computer can execute it. **2.** A short abbreviation for the operation. **3.** During a fetch cycle, the CPU fetches the op code and operand address from memory. During

the execute cycle, the CPU executes the operation called for by the op code. **4.** The PC keeps track of the addresses of the instructions in memory.

SECTION 12-9

1. Data bus—carries data between CPU and memory and I/O devices; address bus—carries address code from CPU to memory and I/O devices; control bus—carries timing and synchronization signals. **2.** Address bus. **3.** Control bus.

SECTION 12-10

1. See text. **2.** See text. **3.** The CPU.

APPENDIX I
GLOSSARY

Access Time Time between the memory's receiving a new input address and the output data's becoming available in a read operation.

Accumulator Principal register of an arithmetic-logic unit (ALU).

Active-HIGH (LOW) Decoder Decoder that produces a logic HIGH (LOW) at the output when detection occurs.

Active Logic Level Logic level at which a circuit is considered active. If the symbol for the circuit includes a bubble, the circuit is active-LOW. On the other hand, if it doesn't have a bubble, then the circuit is active-HIGH.

Actuator Electrically controlled device that controls a physical variable.

Addend Number to be added to another.

Address Number that uniquely identifies the location of a word in memory.

Address Bus Unidirectional lines that carry the address code from the CPU to memory and I/O devices.

Address Multiplexing Multiplexing used in dynamic RAMs to save IC pins that involves latching the two halves of a complete address into the IC in separate steps.

Alphanumeric Codes Codes that represent numbers, letters, punctuation marks, and special characters.

Analog Representation Representation of a quantity that varies over a continuous range of values.

Analog System Combination of devices designed to manipulate physical quantities that are represented in analog form.

Analog-to-Digital Converter (ADC) Circuit that converts an analog input to a corresponding digital output.

& When used inside an IEEE/ANSI symbol, indication of an AND gate or AND function.

AND Gate Digital circuit that implements the AND operation. The output of this circuit is HIGH (logic level 1) only if all of its inputs are HIGH.

AND Operation Boolean algebra operation in which the symbol $\cdot$ is used to indicate the ANDing of two or more logic variables. The result of the AND operation will be HIGH (logic level 1) only if all variables are HIGH.

Arithmetic-Logic Unit (ALU) Digital circuit used in computers to perform various arithmetic and logic operations.

ASCII Code (American Standard Code for Information Interchange) Seven-bit alphanumeric code used by most computer manufacturers.

Asserted Term used to describe the state of a logic signal; synonymous with "active."

Astable Multivibrator Digital circuit that oscillates between two unstable output states.

Asynchronous Counter Type of counter in which each flip-flop output serves as the clock input signal for the next flip-flop in the chain.

Asynchronous Inputs Flip-flop inputs that can affect the operation of the flip-flop independent of the synchronous and clock inputs.

Asynchronous Transfer Data transfer performed without the aid of the clock.

Augend Number to which an addend is added.

Backplane Electrical connection common to all segments of an LCD.

BCD Adder Special adder containing two 4-bit parallel adders and a correction detector circuit. Whenever the addition of two BCD code groups is greater than 1001_2 (9_{10}), the correction detector circuit senses it, adds to the result the correction factor 0110_2 (6_{10}), and generates a carry to the next decimal position.

BCD Counter Binary counter that counts from 0000_2 to 1001_2 before it recycles.

BCD/DEC When used inside an IEEE/ANSI symbol, indication of a BCD-to-decimal decoding function.

701

BCD-to-Decimal Decoder Decoder that converts a BCD input into a single decimal output equivalence.

BCD-to-7-Segment Decoder/Driver Digital circuit that takes a 4-bit BCD input and activates the required outputs to display the equivalent decimal digit on a 7-segment display.

Bidirectional Data Line Term used when a data line functions as either an input or an output line depending on the states of the enable inputs.

Bilateral Switch CMOS circuit which acts like a single-pole, single-throw switch (SPST) controlled by an input logic level.

BIN/OCT When used inside an IEEE/ANSI symbol, indication of a binary-to-octal decoding function.

Binary-Coded-Decimal Code (BCD Code) Four-bit code used to represent each digit of a decimal number by its 4-bit binary equivalent.

Binary Counter Group of flip-flops connected in a special arrangement in which the states of the flip-flops represent the binary number equivalent to the number of pulses that have occurred at the input of the counter.

Binary Digit Bit.

Binary Multiplier Special digital circuit capable of performing the arithmetic operation of multiplication on two binary numbers.

Binary Point A mark which separates the integer from the fractional portion of a binary quantity.

Binary System Number system in which there are only two possible digit values, 0 and 1.

Bipolar DAC Digital-to-analog converter that accepts signed binary numbers as input and produces the corresponding positive or negative analog output value.

Bipolar ICs Integrated digital circuits in which NPN and PNP transistors are the main circuit elements.

Bistable Multivibrator Name that is sometimes used to describe a flip-flop.

Bit A digit in the binary system.

Boolean Algebra Algebraic process used as a tool in the design and analysis of digital systems. In Boolean algebra only two values are possible, "0" and "1."

Boolean Theorems Rules that can be applied to Boolean algebra to simplify logic expressions.

Bootstrap Program Program, stored in ROM, that a computer executes on power-up.

Bubbles Small circles on the input or output lines of logic-circuit symbols which represent inversion of a particular signal. If a bubble is present, the input or output is said to be active-LOW.

Buffer/Driver Circuit designed to have a greater output current and/or voltage capability than an ordinary logic circuit.

Buffer Register Register that holds digital data temporarily.

Bus Contention Situation in which the outputs of two or more active devices are placed on the same bus line at the same time.

Bus Drivers Circuits that buffer the outputs of devices connected to a common bus; used when a large number of devices share a common bus.

Byte Eight-bit word.

C When used as an input label inside an IEEE/ANSI symbol, an indication that the input controls the entry of data into a storage element.

Capacity Amount of storage space in a memory expressed as number of bits or number of words.

Carry Digit or bit that is generated when two numbers are added and the result is greater than the base for the number system being used.

Carry Propagation Intrinsic circuit delay of some parallel adders that prevents the carry bit (C_{OUT}) and the result of the addition from appearing at the output simultaneously.

Carry Ripple *See* Carry Propagation.

Central Processing Unit (CPU) The part of a computer which is composed of the arithmetic-logic unit (ALU) and the control unit.

Checksum Special data word stored in last ROM location. It is derived from the addition of all other data words in the ROM, and is used for error-checking purposes.

Circuit Excitation Table Table showing a circuit's possible PRESENT to NEXT state transitions and the required J and K levels at each flip-flop.

Circulating Shift Register Shift register in which one of the outputs of the last flip-flop is connected to the input of the first flip-flop.

CLEAR Asynchronous FF input used to make $Q = 0$ immediately.

Clear State The $Q = 0$ state of a flip-flop.

Clock Digital signal in the form of a rectangular pulse train or a square wave.

Clock Skew Arrival of a clock signal at the clock inputs of different flip-flops at different times as a result of propagation delays.

Clock Transition Times Minimum rise and fall times for the clock signal transitions used by a particular IC, specified by the IC manufacturer.

Clocked D Flip-Flop Type of flip-flop in which the D (data) input is the synchronous input.

Clocked Flip-Flops Flip-flops which have a clock input.

Clocked J-K Flip-Flop Type of flip-flop in which inputs J and K are the synchronous inputs.

Clocked S-C Flip-Flop Type of flip-flop in which the inputs SET and CLEAR are the synchronous inputs.

CMOS (Complementary Metal-Oxide-Semiconductor) Integrated-circuit technology which uses MOSFETs as the principal circuit element. This logic family belongs to the category of unipolar digital ICs.

Column Address Strobe (CAS) Signal used to latch the column address into a DRAM chip.

Combinatorial Logic Circuits Circuits made up of combinations of logic gates, with no feedback from outputs to inputs.

Common-Control Block Symbol used by the IEEE/ANSI standard to describe when one or more inputs are common to more than one of the circuits in an IC.

Complement *See* Invert.

Computer Word Group of binary bits which is the primary unit of information in a computer.

Control Bus Set of signal lines that are used to synchronize the activities of the CPU and the separate μC elements.

Control Inputs Input signals synchronized with the active clock transition that determine the output state of a flip-flop.

Control Unit That part of a computer which provides decoding of program instructions and the necessary timing and control signals for the execution of such instructions.

CT = 0 When used in the label for an input inside an IEEE/ANSI symbol, an indication that the counter will clear when that input goes active.

CTR When used inside an IEEE/ANSI symbol, an indication that the IC is a counter.

Current-Sinking Logic Logic family in which the output of a logic circuit sinks current from the input of a different logic circuit.

Current-Sinking Transistor Name given to the output transistor (Q_4) of a TTL circuit. This transistor is turned on when the output logic level is LOW.

Current-Sourcing Logic Logic family in which the output of a logic circuit sources, or supplies, current to the input of a different logic circuit.

Current-Sourcing Transistor Name given to the output transistor (Q_3) of most TTL circuits. This transistor is conducting when the output logic level is HIGH.

Current Tracer Testing tool that detects a changing current in a wire or PC-board trace.

Current Transients Current spikes generated by the totem-pole output structure of a TTL circuit, and caused when both transistors are simultaneously turned on.

D When used inside an IEEE/ANSI symbol, label indicating "data."

D Latch Circuit that contains a NAND gate latch and two steering NAND gates.

Data Acquisition Process by which a computer acquires digitized analog data.

Data Bus Bidirectional lines that carry data between the CPU and memory, or between the CPU and I/O devices.

Data Distributors *See* Demultiplexer.

Data-Rate Buffer Application of FIFOs in which sequential data are written into the FIFO at one rate and read out at a different rate.

Data Selectors *See* Multiplexer.

DC CLEAR Asynchronous flip-flop input used to clear Q immediately to 0.

DC SET Asynchronous flip-flop input used to set Q immediately to 1.

Decade Counter Any counter that is capable of going through 10 different logic states.

Decimal System Number system which uses 10 different digits or symbols to represent a quantity.

Decoder Digital circuit that converts an input binary code into a corresponding single numeric output.

DeMorgan's Theorems (1) Theorem stating that the complement of a sum (OR operation) equals the product (AND operation) of the complements, and (2) theorem stating that the complement of a product (AND operation) equals the sum (OR operation) of the complements.

Demultiplexer (DEMUX) Logic circuit that depending on the status of its select inputs will channel its data input to one of several data outputs.

Dependency Notation Method used to symbolically represent the relationship between inputs and outputs of logic circuits. This method employs the use of qualifying symbols embedded near the top center or geometric center of a symbol element.

Digital Computer System of hardware that performs arithmetic and logic operations, manipulates data, and makes decisions.

Digital Integrated Circuits Self-contained digital circuits which have been made by using one of several integrated-circuit fabrication technologies.

Digital-Ramp ADC Type of analog-to-digital converter in which an internal staircase waveform is generated and utilized for the purpose of accomplishing the conversion. The conversion time for this type of analog-to-digital converter varies depending on the value of the input analog signal.

Digital Representation Representation of a quantity that varies in discrete steps over a range of values.

Digital Storage Oscilloscope Instrument that samples, digitizes, stores, and displays analog voltage waveforms.

Digital System Combination of devices designed to manipulate physical quantities that are represented in digital form.

Digital-to-Analog Converter (DAC) Circuit that converts a digital input to a corresponding analog output.

Digitization Process by which an analog signal is converted to digital data.

DIP Dual-in-line package. The most common type of IC package.

DIVn When used inside an IEEE/ANSI symbol, indication that the circuit is a MOD-n counter.

"Don't Care" Situation when a circuit's output level for a given set of input conditions can be assigned as either a 1 or a 0.

Down Counter Counter that counts from a maximum count downward to zero.

DRAM Controller IC used to handle refresh and address multiplexing operations needed by DRAM systems.

Driver Technical term sometimes added to an IC's description to indicate that the IC's outputs can operate at higher current and/or voltage limits than a normal standard IC.

Dual-Slope Analog-to-Digital Converter Type of analog-to-digital converter that linearly charges a capacitor from a current proportional to V_A for a fixed time interval, and then

increments a counter as the capacitor is linearly discharged to zero.

Dynamic RAM (DRAM) Type of semiconductor memory that stores data as capacitor charges that need to be refreshed periodically.

ECL emitter-coupled logic; also referred to as *current-mode logic*.

Edge Detector Circuit that produces a narrow positive spike that occurs coincident with the active transition of a clock input pulse.

Edge-Triggered Manner in which a flip-flop is activated by a signal transition. A flip-flop may be either a positive or negative edge-triggered flip-flop.

8-Line-to-3-Line Encoder Digital circuit that generates a different 3-bit code depending on which one of the eight inputs is activated.

Electrically Erasable Programmable ROM (EEPROM) ROM that can be electrically programmed, erased, and reprogrammed.

Encoder Digital circuit that produces an output code depending on which of its inputs is activated.

Encoding Use of a group of symbols to represent numbers, letters, or words.

Erasable Programmable ROM (EPROM) ROM that can be electrically programmed by the user. It can be erased (usually with ultraviolet light) and reprogrammed as often as desired.

Excess-3 Code Four-bit code in which 3 is added to each decimal digit before it is encoded in binary.

Exclusive-NOR Circuit Two-input logic circuit that produces a HIGH output only when the inputs are equal.

Exclusive-OR Circuit Two-input logic circuit which produces a HIGH output only when the inputs are different.

Execute Cycle Period during which a computer's control unit performs the operation specified by the fetched op code.

Fan-Out Maximum number of standard logic inputs that the output of a digital circuit can drive reliably.

Fetch Cycle Period during which a computer's control unit obtains instruction codes from memory.

Firmware Computer programs stored in ROM.

First-In First-Out Memory (FIFO) Semiconductor sequential-access memory in which data words are read out in the same order in which they were written in.

555 Timer TTL-compatible IC that can be wired to operate in several different modes, such as a one-shot and an astable multivibrator.

Flash ADC Type of analog-to-digital converter that has the highest operating speed available.

Flip-Flop Memory device capable of storing a logic level.

Floating Input Input signal that is left disconnected in a logic circuit.

Floppy Disk Flexible magnetic disk used for mass storage.

4-to-10 Decoder *See* BCD-to-Decimal Decoder.

Frequency Counter Circuit that can measure and display the frequency of a signal.

Full Adder Logic circuit with three inputs and two outputs. The inputs are a carry bit (C_{IN}) from a previous stage, a bit from the augend, and a bit from the addend, respectively. The outputs are the sum bit and carry out bit (C_{out}) produced by the addition of the bit from the addend with the bit from the augend and C_{in}.

Full-Scale Error Term used by some digital-to-analog converter manufacturers to specify the accuracy of a digital-to-analog converter. It is defined as the maximum deviation of a digital-to-analog converter's output from its expected ideal value.

Full-Scale Output Maximum possible output value of a digital-to-analog converter.

Function Generator Circuit that produces a variety of waveforms; can be constructed using a ROM, a DAC, and a counter.

G A label used on an input inside an IEEE/ANSI symbol to indicate AND dependency.

Glitch Momentary, narrow, spurious, and sharply defined change in voltage.

Gray Code Code that never has more than one bit changing when going from one step to the next.

Half Adder Logic circuit with two inputs and two outputs. The inputs are a bit from the augend and a bit from the addend, respectively. The outputs are the sum bit produced by the addition of the bit from the addend with the bit from the augend and the resulting carry (C_{OUT}) bit, which will be added to the next stage.

Hard Disk Rigid metal magnetic disk used for mass storage.

Head Crash Contact between a R/W head and the surface of a hard disk that damages the surface of the disk.

Hexadecimal Number System Number system which has a base of 16; digits 0 through 9 plus letters A through F are used to express a hexadecimal number.

High-Level Language Computer programming language that utilizes the English language in order to facilitate the writing of a computer program.

Hold Time (t_H) Time interval immediately following the active transition of the clock signal during which the control input has to be maintained at the proper level.

HPRI/BCD When used inside an IEEE/ANSI symbol, indication that the function of an IC is to convert the active input with the highest priority to its corresponding BCD code.

Hybrid System System that employs both analog and digital techniques.

IEEE/ANSI Institute of Electrical and Electronics Engineers/American National Standards Institute.

Indeterminate Of a logic voltage level, outside the required range of voltages for either logic 0 or logic 1.

Inhibit Circuits Logic circuits that control the passage of an input signal through to the output.

Input Unit That part of a computer which facilitates the feeding of information into the computer's memory unit or ALU.

Interfacing Joining of dissimilar devices in such a way that they are able to function in a compatible and coordinated manner; connection of the output of a system to the input of a different system with different electrical characteristics.

Invert Cause a logic level to go to the opposite state.

INVERTER Also referred to as the NOT circuit; logic circuit that implements the NOT operation. An INVERTER has only one input, and its output logic level is always the opposite of this input's logic level.

Jam Transfer *See* Asynchronous Transfer.

J-K Excitation Table Table showing the required *J* and *K* input conditions for each possible state transition for a single J-K flip-flop.

Johnson Counter Shift register in which the inverted output of the last flip-flop is connected to the input of the first flip-flop.

Karnaugh Map Two-dimensional form of a truth table used to simplify a sum-of-products expression.

Latch A type of flip-flop.

LCD Liquid crystal display.

Least Significant Bit (LSB) Rightmost bit (smallest weight) of a binary expressed quantity.

Least Significant Digit (LSD) Digit that carries the least weight in a particular number.

LED Light-emitting diode.

Linearity Error Term used by some digital-to-analog converter manufacturers to specify the device's accuracy. It is defined as the maximum deviation in step size from the ideal step size.

Loading Factor *See* Fan-Out.

Logic Circuit Any circuit that behaves according to a set of logic rules.

Logic Function Generation Implementation of a logic function directly from a truth table by means of a digital IC such as a multiplexer.

Logic Level State of a voltage variable. The states 1 (HIGH) and 0 (LOW) correspond to the two usable voltage ranges of a digital device.

Logic Probe Digital troubleshooting tool which senses and indicates the logic level at a particular point in a circuit.

Logic Pulser Testing tool that generates a short-duration pulse when manually actuated.

Look-Ahead Carry Ability of some parallel adders to predict, without having to wait for the carry to propagate through the full adders, whether or not a carry bit (C_{OUT}) will be generated as a result of the addition, thus reducing the overall propagation delays.

Looping Combining of adjacent squares in a Karnaugh map containing 1s for purpose of simplification of a sum-of-products expression.

Low-Power Schottky TTL (LS-TTL) TTL subfamily that uses the identical Schottky TTL circuit but with larger resistor values.

Low-Power TTL (L-TTL) TTL subfamily which uses the basic TTL standard circuit except that all resistor values are increased.

LSI Large-scale integration (100 to 9999 gates).

Machine Language Computer programming language in which groups of 1s and 0s are used to represent instructions. Machine language is also the only language a computer actually understands.

Magnetic Bubble Memory (MBM) Solid-state, nonvolatile, sequential-access, mass storage memory device consisting of tiny magnetic domains (bubbles).

Magnetic Core Memory Nonvolatile random-access memory made up of small ferrite cores.

Magnetic Disk Memory Mass storage memory that stores data as magnetized spots on a rotating flat disk surface.

Magnetic Tape Memory Mass storage memory that stores data as magnetized spots on a magnetically coated plastic tape.

Magnitude Comparator Digital circuit that compares two input binary quantities and generates outputs to indicate whether the inputs are equal or, if not, which of the two has the greater magnitude.

Mainframe *See* Maxicomputer.

Mask-Programmed ROM (MROM) ROM that is programmed by the manufacturer according to the customer's specifications. It cannot be erased or reprogrammed.

Mass Storage Storage of large amounts of data. Not part of a computer's internal memory.

Master/Slave Flip-Flops Obsolete flip-flops which have as their internal structure two flip-flops—a master and a slave.

Maxicomputer Large computer system used to maintain and update large quantities of data and information.

Maximum Clocking Frequency (f_{MAX}) Highest frequency that may be applied to the clock input of a flip-flop and still have it trigger reliably.

Memory Ability of a circuit's output to remain at one state even after the input condition that caused that state is removed.

Memory Cell Device that stores a single bit.

Memory Unit Part of a computer which stores instructions and data received from the input unit, as well as results from the arithmetic-logic unit.

Memory Word Group of bits in memory that represents instructions or data of some type.

Microcomputer Newest member of the computer family, consisting of microprocessor chip, memory chips, and I/O interface chips. In some cases all of the aforementioned are in one single IC.

Microprocessor (MPU) LSI chip that contains the central processing unit (CPU).

Minicomputers Computers that are generally larger and faster and possess more capabilities than microcomputers.

Minuend Number from which the subtrahend is to be subtracted.

− When used inside an IEEE/ANSI symbol and on a clock input, indication that the counter will be decremented by 1 when clocked.

Mnemonic Abbreviation that represents the op code of a computer instruction.

MOD Number Number of different states that a counter can sequence through; the counter's frequency division ratio.

Monostable Multivibrator *See* One-Shot.

Monotonicity Property whereby the output of a digital-to-analog converter either increases or stays the same as the input is increased.

MOSFET Metal-oxide-semiconductor field-effect transistor.

Most Significant Bit (MSB) Leftmost binary bit (largest weight) of a binary expressed quantity.

Most Significant Digit (MSD) Digit that carries the most weight in a particular number.

MSI Medium-scale integration (12 to 99 gates).

Multibyte Instruction Computer instruction which is represented by more than one byte.

Multiple-Address Instruction Computer instruction word that contains more than one address.

Multiplexer (MUX) Logic circuit that, depending on the status of its select inputs, will channel one of several data inputs to its output.

Multiplexing Process of selecting one of several input data sources and transmitting the selected data to a single output channel.

NAND Gate Logic circuit which operates like an AND gate followed by an INVERTER. The output of a NAND gate is LOW (logic level 0) only if all inputs are HIGH (logic level 1).

NAND-Gate Latch Flip-flop constructed from two cross-coupled NAND gates.

NMOS (N-Channel Metal-Oxide-Semiconductor) Integrated-circuit technology that uses N-channel MOSFETs as the principal circuit element.

Noise Spurious voltage fluctuations that may be present in the environment and cause digital circuits to malfunction.

Noise Immunity A circuit's ability to tolerate noise voltages on its inputs.

Noise Margin Quantitative measure of noise immunity.

Nonretriggerable One-Shot Type of one-shot that will not respond to a trigger input signal while in its quasi-stable state.

Nonvolatile Memory Memory that will keep storing its information without the need for electrical power.

Nonvolatile RAM Combination of a static RAM array and an EEPROM on the same IC. The EEPROM serves as a nonvolatile backup to the SRAM.

NOR Gate Logic circuit which operates like an OR gate followed by an INVERTER. The output of a NOR gate is LOW (logic level 0) when any or all inputs are HIGH (logic level 1).

NOR-Gate Latch Flip-flop constructed from two cross-coupled NOR gates.

NOT Circuit see INVERTER.

NOT Operation Boolean algebra operation in which the overbar ( ) or the prime (′) symbol is used to indicate the inversion of one or more logic variables.

Observation/Analysis Process used to troubleshoot circuits or systems in order to predict the possible faults before ever picking up a troubleshooting instrument. When this process is used, the troubleshooter must understand the circuit operation, observe the symptoms of the failure, and then reason through the operation.

Octal Number System Number system which has a base of 8; digits from 0 to 7 are used to express an octal number.

Octal-to-Binary Encoder *See* 8-Line-to-3 Line Encoder.

Octets Groups of eight 1s that are adjacent to each other within a Karnaugh map.

Offset Error Deviation from the ideal zero volts at the output of a digital-to-analog converter when the input is all 0s. In reality, there is a very small output voltage for this situation.

1-of-10 Decoder *See* BCD-to-Decimal Decoder.

One-Shot Circuit that belongs to the flip-flop family but which has only one stable state (normally $Q = 0$).

1's-Complement Form Result obtained when each bit of a binary number is complemented.

Op Code That part of a computer instruction that defines what type of operation the computer is to execute on specified data.

Open-Collector Output Type of output structure of some TTL circuits in which only one transistor with a floating collector is used.

Operand Data that are operated on by the computer as it executes a program.

Operand Address Address in memory where operand is currently stored or is to be stored.

Optical Disk Memory Class of mass memory devices that uses a laser beam to write and read onto a specially coated disk.

OR Gate Digital circuit that implements the OR operation. The output of this circuit is HIGH (logic level 1) if any or all of its inputs are HIGH.

OR Operation Boolean algebra operation in which the symbol + is used to indicate the ORing of two or more logic variables. The result of the OR operation will be HIGH (logic level 1) if one or more variables are HIGH.

Output Unit That part of a computer that receives data from the memory unit or ALU and presents it to the outside world.

Overflow When in the process of adding signed binary numbers a carry of 1 is generated from the MSB position of the number into the sign bit position.

Override Inputs Synonymous with "asynchronous inputs."

Parallel Adder Digital circuit made from full adders and used to add all the bits from the addend and the augend together simultaneously.

Parallel Counter *See* Synchronous Counter.

Parallel Data Transfer Operation by which the entire contents of a register are transferred simultaneously to another register.

Parallel In/Parallel Out Register Type of register that can be loaded with parallel data and has parallel outputs available.

Parallel In/Serial Out Type of register that can be loaded with parallel data and has only one serial output.

Parallel-to-Serial Conversion Process by which all data bits are presented simultaneously to a circuit's input and then transmitted one bit at a time to its output.

Parallel Transmission Simultaneous transfer of all bits of a binary number from one place to another.

Parity Bit Additional bit that is attached to each code group so that total number of 1s being transmitted is always even (or always odd).

Parity Checker Circuit that takes a set of data bits (including parity bit) and checks to see if it has the correct parity.

Parity Generator Circuit that takes a set of data bits and produces the correct parity bit for the data.

Parity Method Scheme used for error detection during the transmission of data.

Percentage Resolution Ratio of the step size to the full-scale value of a digital-to-analog converter. Percentage resolution can also be defined as the reciprocal of the maximum number of steps of a digital-to-analog converter.

+ When used inside an IEEE/ANSI symbol and on a clock input, indication that the counter will be incremented by 1 when clocked.

Positional-Value System System in which the value of a digit is dependent on its relative position.

Positive-Going Transition (PGT) When a clock signal changes from a logic 0 to a logic 1.

Power-Supply Decoupling Connection of a small RF capacitor between ground and V_{CC} near each TTL integrated circuit on a circuit board.

Power-Up Self-Test Program stored in ROM and executed by the CPU on power-up to test RAM and/or ROM portions of the computer circuitry.

PRESET Asynchronous input used to immediately set $Q = 1$.

Presettable Counter Counter that can be preset to any starting count either synchronously or asynchronously.

Priority Encoder Special type of encoder that senses when two or more inputs are activated simultaneously and then generates a code corresponding to the highest-numbered input.

Program Sequence of binary-coded instructions designed to accomplish a particular task by a computer.

Program Counter (PC) CPU register that stores the address of the next instruction to be fetched.

Programmable Array Logic (PAL) Class of programmable logic devices. Its AND array is programmable, while its OR array is hard-wired.

Programmable Logic Array (PLA) Class of programmable logic devices. Both its AND and its OR arrays are programmable. Also called a *field-programmable logic array (FPLA)*.

Programmable Logic Device (PLD) IC that contains a large number of interconnected logic functions. The user can program the IC for a specific function by selectively breaking the appropriate interconnections.

Programmable ROM (PROM) ROM that can be electrically programmed by the user. It cannot be erased and reprogrammed.

Propagation Delays (t_{PLH}/t_{PHL}) Delay from the time a signal is applied to the time when the output makes its change.

Pull-Down Transistor *See* Current-Sinking Transistor.

Pull-Up Transistor *See* Current-Sourcing Transistor.

Quantization Error Error caused by the nonzero resolution of an analog-to-digital converter; this is an inherent error of the device.

Quasi-Stable State State to which a one-shot is temporarily triggered (normally, $Q = 1$) before returning to its stable state (normally, $Q = 0$).

R When used as a label for an input inside an IEEE/ANSI symbol, indication for a reset function.

R/2R Ladder DAC Type of digital-to-analog converter whose internal resistance values span a range of only 2 to 1.

Random-Access Memory (RAM) Memory in which the access time is the same for any location.

Read Term used to describe the condition when the CPU is receiving data from another element.

Read-Only Memory (ROM) Memory device designed for applications where the ratio of read operations to write operations is very high.

Read Operation Operation in which a word in a specific memory location is sensed and possibly transferred to another device.

Read/Write Head Soft "iron" core electromagnet used to read data from and write data onto a magnetic disk or tape.

Read/Write Memory (RWM) Any memory that can be read from and written into with equal ease.

Refreshing Process of recharging the cells of a dynamic memory.

Register Group of flip-flops capable of storing data.

RESET Term synonymous with "CLEAR."
The $Q = 0$ state of a flip-flop.

Resolution In a digital-to-analog converter, smallest change that can occur in the output for a change in digital input. Also called *step size*. In an analog-to-digital converter, smallest amount by which the analog input must change to produce a change in the digital output.

Retriggerable One-Shot Type of one-shot that will respond to a trigger input signal while in its quasi-stable state.

Ring Counter Shift register in which the output of the last flip-flop is connected to the input of the first flip-flop.

Ripple Counter *See* Asynchronous Counter.

Row Address Strobe (RAS) Signal used to latch the row

address into a DRAM chip.

Sample-and-Hold Circuit Type of circuit that utilizes a unity-gain buffer amplifier in conjunction with a capacitor to accomplish a more stable analog-to-digital conversion process.

Sampling Interval Time window during which a frequency counter samples and thereby determines the unknown frequency of a signal.

SBD Schottky barrier diode used in all Schottky TTL series.

Schmitt-Trigger Digital circuit that accepts a slow-changing input signal and produces a rapid, oscillation-free transition at the output.

Schottky TTL TTL subfamily that uses the basic TTL standard circuit except that it uses a Schottky barrier diode (SBD) connected between the base and collector of each transistor.

Sequential-Access Memory (SAM) Memory in which the access time will vary depending on the storage location of the data.

Sequential Logic System Logic system in which the logic output states and sequence of operations depend on both the present and past input conditions.

Serial Data Transfer Transfer of data from one place to another one bit at a time.

Serial In/Parallel Out Type of register that can be loaded with data serially and has parallel outputs available.

Serial In/Serial Out Type of register that can be loaded with data serially and has only one serial output.

Serial Transmission Transfer of binary information from one place to another a bit at a time.

SET Term synonymous with DC SET.

Set State The $Q = 1$ state of a flip-flop.

Setup Time (t_s) Time interval immediately preceding the active transition of the clock signal during which the control input has to be maintained at the proper level.

Settling Time Amount of time that it takes the output of a digital-to-analog converter to go from zero to within one-half step size of its full-scale value as the input is changed from all 0s to all 1s.

Shift Register Digital circuit that accepts binary data from some input source and then shifts these data through a chain of flip-flops one bit at a time.

Sigma Σ Greek letter that represents addition and is often used to label the sum output bits of a parallel adder.

Sign Bit Binary bit that is added to the leftmost position of a binary number to indicate whether that number represents a positive or a negative quantity.

/ When used as an output label inside an IEEE/ANSI symbol, indication of separation of two functions.

Speed-Power Product Numerical value (in joules) often used to compare different logic families. It is obtained by multiplying the propagation delay by the power dissipation of a logic circuit.

Spike *See* Glitch.

SRG$_n$ When used in the label for the common-control

block inside an IEEE/ANSI symbol, indicator that the IC is an *n*-bit shift register.

SSI Small-scale integration (fewer than 12 gates).

Staircase Test Process by which a digital-to-analog converter's digital input is incremented and its output monitored to determine whether or not it exhibits a staircase format.

Staircase Waveform Type of waveform generated at the output of a digital-to-analog converter as its digital input signal is incrementally changed.

Static Accuracy Test Test in which a fixed binary value is applied to the input of a digital-to-analog converter and the analog output is accurately measured. The measured result should fall within the expected range specified by the digital-to-analog converter's manufacturer.

Static RAM (SRAM) Semiconductor RAM that stores information in flip-flop cells that do not have to be periodically refreshed.

Step Size *See* Resolution.

Straight Binary Coding Representation of a decimal number by its equivalent binary number.

Strobing Technique often used to eliminate decoding spikes.

Substrate Piece of semiconductor material which is part of the building block of any digital IC.

Subtrahend Number that is to be subtracted from a minuend.

Successive-Approximation ADC Type of analog-to-digital converter in which an internal parallel register and complex control logic are used to perform the conversion. The conversion time for this type of analog-to-digital converter is always the same regardless of the value of the input analog signal.

Sum-of-Products Form Logic expression consisting of two or more AND terms (products) that are ORed together.

Synchronous Counter Counter in which all the flip-flops are clocked simultaneously.

Synchronous Inputs *See* Control Inputs.

Synchronous Systems Systems in which the circuit outputs can change states only on the transitions of a clock.

Synchronous Transfer Data transfer performed by using the synchronous and clock inputs of a flip-flop.

Timing Diagram Depiction of logic levels as related to time.

Toggle Mode Mode in which a flip-flop changes states for each clock pulse.

Toggling Process of changing from one binary state to the other.

Totem-Pole Output Term used to describe the way in which two bipolar transistors are arranged at the output of most TTL circuits.

Transducer Device that converts a physical variable to an electrical variable (for example, a photocell or a thermocouple).

Transmission Gate *See* Bilateral Switch.

Transparent Of a D latch, operating so that the Q output follows the D input.

Trigger Input signal to a flip-flop or one-shot which causes the output to change states depending on the conditions of the control signals.

Tristate Type of output structure which allows three types of output states: HIGH, LOW, and high-impedance (Hi-Z).

Truth Table Logic table that depicts a circuit's output response to the various combinations of the logic levels at its inputs.

TTL (Transistor-Transistor Logic) Integrated-circuit technology that uses the bipolar transistor as the principal circuit element.

2's-Complement Form Result obtained when a 1 is added to the least significant bit position of a binary number in the 1's-complement form.

ULSI Ultralarge-scale integration (100,000 or more gates).

Unasserted Term used to describe the state of a logic signal, synonymous with "inactive."

Unipolar ICs Integrated digital circuits in which unipolar field-effect transistors (MOSFETs) are the main circuit elements.

Unit Load (UL) Way that some manufacturers specify a device's input and output currents. In a standard TTL circuit, 1 UL in the high state is equal to 40 μA, and in the low state is equal to 1.6 mA.

Up Counter Counter that counts upward from zero to a maximum count.

Up/Down Digital-Ramp ADC Type of analog-to-digital converter that uses an up/down counter to step up or step down the voltage from a digital-to-analog converter until it intersects the analog input.

VLSI Very large-scale integration (10,000 to 99,999 gates).

Volatile Memory Memory requiring electrical power to keep information stored.

Voltage-Controlled Oscillator (VCO) Circuit that produces an output signal with a frequency proportional to the voltage applied to its input.

Voltage-Level-Translator Circuit that takes one set of input voltage levels and translates it to a different set of output levels.

Voltage-to-Frequency ADC Type of analog-to-digital converter that converts the analog voltage to a pulse frequency that is then counted to produce a digital output.

Wired-AND Term used to describe the logic function created when open-collector outputs are tied together.

WRITE Term used to describe the condition when the CPU is sending data to another element.

Write Operation Operation in which a new word is placed into a specific memory location.

$\rightarrow$ When used in the label for an input inside an IEEE/ANSI symbol, indication that when an active transition exists at that input, a shift-right operation will result.

APPENDIX II:
MANUFACTURERS'
IC DATA SHEETS

These data sheets are presented through the courtesy of Fairchild, a Schlumberger Company, and Texas Instruments Incorporated.

710

54/7400
54H/74H00
54S/74S00
54LS/74LS00
QUAD 2-INPUT NAND GATE

ORDERING CODE: See Section 9

PKGS	PIN OUT	COMMERCIAL GRADE $V_{CC} = +5.0$ V ±5%, $T_A = 0°$C to +70°C	MILITARY GRADE $V_{CC} = +5.0$ V ±10%, $T_A = -55°$C to +125°C	PKG TYPE
Plastic DIP (P)	A	7400PC, 74H00PC 74LS00PC, 74S00PC		9A
Ceramic DIP (D)	A	7400DC, 74H00DC 74LS00DC, 74S00DC	5400DM, 54H00DM 54LS00DM, 54S00DM	6A
Flatpak (F)	A	74LS00FC, 74S00FC	54LS00FM, 54S00FM	3I
	B	7400FC, 74H00FC	5400FM, 54H00FM	

INPUT LOADING/FAN-OUT: See Section 3 for U.L. definitions

PINS	54/74 (U.L.) HIGH/LOW	54/74H (U.L.) HIGH/LOW	54/74S (U.L.) HIGH/LOW	54/74LS (U.L.) HIGH/LOW
Inputs	1.0/1.0	1.25/1.25	1.25/1.25	0.5/0.25
Outputs	20/10	12.5/12.5	25/12.5	10/5.0 (2.5)

DC AND AC CHARACTERISTICS: See Section 3*

SYMBOL	PARAMETER	54/74 Min	54/74 Max	54/74H Min	54/74H Max	54/74S Min	54/74S Max	54/74LS Min	54/74LS Max	UNITS	CONDITIONS
I_{CCH}	Power Supply		8.0		16.8		16		1.6	mA	V_{IN} = Gnd
I_{CCL}	Current		22		40		36		4.4		V_{IN} = Open / V_{CC} = Max
t_{PLH}	Propagation Delay		22		10	2.0	4.5		10	ns	Figs. 3-1, 3-4
t_{PHL}			15		10	2.0	5.0		10		

*DC limits apply over operating temperature range; AC limits apply at T_A = +25°C and V_{CC} = +5.0 V.

54/7404
54H/74H04
54S/74S04
54S/74S04A
54LS/74LS04

HEX INVERTER

CONNECTION DIAGRAMS
PINOUT A

ORDERING CODE: See Section 9

PKGS	PIN OUT	COMMERCIAL GRADE V_{CC} = +5.0 V ±5%, T_A = 0°C to +70°C	MILITARY GRADE V_{CC} = +5.0 V ±10%, T_A = -55°C to +125°C	PKG TYPE
Plastic DIP (P)	A	7404PC, 74H04PC 74S04PC, 74S04APC 74LS04PC		9A
Ceramic DIP (D)	A	7404DC, 74H04DC 74S04DC, 74S04ADC 74LS04DC	5404DM, 54H04DM 54S04DM, 54S04ADM 54LS04DM	6A
Flatpak (F)	A	74S04FC, 74S04AFC 74LS04FC	54S04FM, 54S04AFM 54LS04FM	3I
	B	7404FC, 74H04FC	5404FM, 54H04FM	

PINOUT B

INPUT LOADING/FAN-OUT: See Section 3 for U.L. definitions

PINS	54/74 (U.L.) HIGH/LOW	54/74H (U.L.) HIGH/LOW	54/74S (U.L.) HIGH/LOW	54/74LS (U.L.) HIGH/LOW
Inputs	1.0/1.0	1.25/1.25	1.25/1.25	0.5/0.25
Outputs	20/10	12.5/12.5	25/12.5	10/5.0 (2.5)

DC AND AC CHARACTERISTICS: See Section 3*

SYMBOL	PARAMETER	54/74 Min	54/74 Max	54/74H Min	54/74H Max	54/74S Min	54/74S Max	54/74LS Min	54/74LS Max	UNITS	CONDITIONS	
I_{CCH}	Power Supply		12		26		24		2.4	mA	V_{IN} = Gnd	V_{CC} = Max
I_{CCL}	Current		33		58		54		6.6		V_{IN} = Open	
t_{PLH}	Propagation Delay		22		10	2.0	4.5		10	ns	Fig. 3-1, 3-4	
t_{PHL}			15		10	2.0	5.0		10			
t_{PLH}	Propagation Delay					1.0	3.5			ns	Fig. 3-1, 3-4	
t_{PHL}	(54/74S04A only)					1.0	4.0					

*DC limits apply over operating temperature range; AC limits apply at T_A = +25°C and V_{CC} = +5.0 V.

54/7405
54H/74H05
54S/74S05
54S/74S05A
54LS/74LS05

HEX INVERTER
(With Open-Collector Output)

CONNECTION DIAGRAMS
PINOUT A

ORDERING CODE: See Section 9

PKGS	PIN OUT	COMMERCIAL GRADE $V_{CC} = +5.0 \text{ V} \pm 5\%$, $T_A = 0°C \text{ to } +70°C$	MILITARY GRADE $V_{CC} = +5.0 \text{ V}, \pm 10\%$, $T_A = -55°C \text{ to } +125°C$	PKG TYPE
Plastic DIP (P)	A	7405PC, 74H05PC 74S05PC, 74S05APC 74LS05PC		9A
Ceramic DIP (D)	A	7405DC, 74H05DC 74S05DC, 74S05ADC 74LS05DC	5405DM, 54H05DM 54S05DM, 54S05ADM 54LS05DM	6A
Flatpak (F)	A	74S05FC, 74S05AFC 74LS05FC	54S05FM, 54S05AFM 54LS05FM	3I
	B	7405FC, 74H05FC	5405FM, 54H05FM	

PINOUT B

INPUT LOADING/FAN-OUT: See Section 3 for U.L. definitions

PINS	54/74 (U.L.) HIGH/LOW	54/74H (U.L.) HIGH/LOW	54/74S (U.L.) HIGH/LOW	54/74LS (U.L.) HIGH/LOW
Inputs	1.0/1.0	1.25/1.25	1.25/1.25	0.5/0.25
Outputs	OC**/10	OC**/12.5	OC**/12.5	OC**/5.0 (2.5)

DC AND AC CHARACTERISTICS: See Section 3*

SYMBOL	PARAMETER	54/74 Min	54/74 Max	54/74H Min	54/74H Max	54/74S Min	54/74S Max	54/74LS Min	54/74LS Max	UNITS	CONDITIONS	
I_{CCH}	Power Supply		12		26		19.8		2.4	mA	$V_{IN} = \text{Gnd}$	$V_{CC} = \text{Max}$
I_{CCL}	Current		33		58		54		6.6		$V_{IN} = \text{Open}$	
t_{PLH} t_{PHL}	Propagation Delay		55 15		18 15	2.0 2.0	7.5 7.0		22 18	ns	Fig. 3-2, 3-4	
t_{PLH} t_{PHL}	Propagation Delay (54S/74S05A only)					2.0 1.5	5.5 5.0			ns	Fig. 3-2, 3-4	

*DC limits apply over operating temperature range; AC limits apply at $T_A = +25°C$ and $V_{CC} = +5.0$ V.
**OC — Collector

54/7406

HEX INVERTER BUFFER/DRIVER
(With Open-Collector High-Voltage Output)

ORDERING CODE: See Section 9

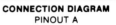

PKGS	PIN OUT	COMMERCIAL GRADE V_{CC} = +5.0 V ±5%, T_A = 0°C to +70°C	MILITARY GRADE V_{CC} = +5.0 V ±10%, T_A = -55°C to +125°C	PKG TYPE
Plastic DIP (P)	A	7406PC		9A
Ceramic DIP (D)	A	7406DC	5406DM	6A
Flatpak (F)	A	7406FC	5406FM	3I

INPUT LOADING/FAN-OUT: See Section 3 for U.L. definitions

PINS	54/74 (U.L.) HIGH/LOW
Inputs	1.0/1.0
Outputs	OC**/10

DC AND AC CHARATERISTICS: See Section 3*

SYMBOL	PARAMETER		54/74 Min	54/74 Max	UNITS	CONDITIONS	
V_{OL}	Output LOW Voltage	XC		0.7	V	I_{OL} = 40 mA	V_{CC} = Min V_{IN} = V_{IH}
		XM		0.7		I_{OL} = 30 mA	
		XC, XM		0.4		I_{OL} = 16 mA	
I_{OH}	Output HIGH Current			0.25	mA	V_{OH} = 30 V, V_{CC} = Min V_{IN} = V_{IL}	
I_{CCH}	Power Supply Current			48	mA	V_{IN} = Gnd	V_{CC} = Max
I_{CCL}				51		V_{IN} = Open	
t_{PLH}	Propagation Delay			15	ns	Fig. 3-2, 3-4	
t_{PHL}				23			

*DC limits apply over operating temperature range, AC limits apply at T_A = +25°C and V_{CC} = +5.0 V.
**OC — Open Collector

54/7414
54LS/74LS14
HEX SCHMITT TRIGGER INVERTER

ORDERING CODE: See Section 9

PKGS	PIN OUT	COMMERCIAL GRADE V_{CC} = +5.0 V ±5%, T_A = 0°C to +70°C	MILITARY GRADE V_{CC} = +5.0 V ±10%, T_A = -55°C to +125°C	PKG TYPE
Plastic DIP (P)	A	7414PC, 74LS14PC		9A
Ceramic DIP (D)	A	7414DC, 74LS14DC	5414DM, 54LS14DM	6A
Flatpak (F)	A	7414FC, 74LS14FC	5414FM, 54LS14FM	3I

INPUT LOADING/FAN-OUT: See Section 3 for U.L. definitions

PINS	54/74 (U.L.) HIGH/LOW	54/74LS (U.L.) HIGH/LOW
Inputs	1.0/1.0	0.5/0.25
Outputs	20/10	10/5.0 (2.5)

DC AND AC CHARACTERISTICS: See Section 3*

SYMBOL	PARAMETER	54/74 Min Max	54/74LS Min Max	UNITS	CONDITIONS
V_{T+}	Positive-going Threshold Voltage	1.5 2.0	1.5 2.0	V	V_{CC} = +5.0 V
V_{T-}	Negative-going Threshold Voltage	0.6 1.1	0.6 1.1	V	V_{CC} = +5.0 V
$V_{T+} - V_{T-}$	Hysteresis Voltage	0.4	0.4	V	V_{CC} = +5.0 V
I_{T+}	Input Current at Positive-going Threshold	-0.43**	-0.14**	mA	V_{CC} = +5.0 V, V_{IN} = V_{T+}
I_{T-}	Input Current at Negative-going Threshold	-0.56**	-0.18**	mA	V_{CC} = +5.0 V, V_{IN} = V_{T-}
I_{IL}	Input LOW Current	-1.2	-0.4	mA	V_{CC} = Max, V_{IN} = 0.4 V
I_{OS}	Output Short Circuit Current	-18 -55	-20 -100	mA	V_{CC} = Max, V_{OUT} = 0 V
I_{CCH}	Power Supply Current	36	16	mA	V_{IN} = Gnd V_{CC} = Max
I_{CCL}		60	21		V_{IN} = Open
t_{PLH} t_{PHL}	Propagation Delay	22 22	22 22	ns	Figs. 3-1, 3-15

*DC limits apply over operating temperature range; AC limits apply at T_A = +25°C and V_{CC} = +5.0 V. **Typical Value

54/7420
54H/74H20
54S/74S20
54LS/74LS20
DUAL 4-INPUT NAND GATE

ORDERING CODE: See Section 9

PKGS	PIN OUT	COMMERCIAL GRADE V_{CC} = +5.0 V ±5%, T_A = 0°C to +70°C	MILITARY GRADE V_{CC} = +5.0 V ±10%, T_A = -55°C to +125°C	PKG TYPE
Plastic DIP (P)	A	7420PC, 74H20PC 74S20PC, 74LS20PC		9A
Ceramic DIP (D)	A	7420DC, 74H20DC 74S20DC, 74LS20DC	5420DM, 54H20DM 54S20DM, 54LS20DM	6A
Flatpak (F)	A	74S20FC, 74LS20FC	54S20FM, 54LS20FM	3I
	B	7420FC, 74H20FC	5420FM, 54H20FM	

PINOUT B

INPUT LOADING/FAN-OUT: See Section 3 for U.L. definitions

PINS	54/74 (U.L.) HIGH/LOW	54/74H (U.L.) HIGH/LOW	54/74S (U.L.) HIGH/LOW	54/74LS (U.L.) HIGH/LOW
Inputs	1.0/1.0	1.25/1.25	1.25/1.25	0.5/0.25
Outputs	20/10	12.5/12.5	25/12.5	10/5.0 (2.5)

DC AND AC CHARACTERISTICS: See Section 3*

SYMBOL	PARAMETER	'54/74 Min	'54/74 Max	54/74H Min	54/74H Max	54/74S Min	54/74S Max	54/74LS Min	54/74LS Max	UNITS	CONDITIONS
I_{CCH}	Power Supply		4.0		8.4		8.0		0.8	mA	V_{IN} = Gnd V_{CC} = Max
I_{CCL}	Current		11		20		18		2.2		V_{IN} = Open
t_{PLH}	Propagation Delay		22		10	2.0	4.5		15	ns	Figs. 3-1, 3-4
t_{PHL}			15		10	2.0	5.0		15		

*DC limits apply over operating temperature range; AC limits apply at T_A = +25°C and V_{CC} = +5.0 V.

SN54ALS20A, SN54AS20, SN74ALS20A, SN74AS20
DUAL 4-INPUT POSITIVE-NAND GATES

D2661, APRIL 1982—REVISED MAY 1986

- Package Options Include Plastic "Small Outline" Packages, Ceramic Chip Carriers, and Standard Plastic and Ceramic 300-mil DIPs

- Dependable Texas Instruments Quality and Reliability

description

These devices contain two independent 4-input NAND gates. They perform the Boolean functions $Y = \overline{A \cdot B \cdot C \cdot D}$ or $Y = \overline{A} + \overline{B} + \overline{C} + \overline{D}$ in positive logic.

The SN54ALS20A and SN54AS20 are characterized for operation over the full military temperature range of −55°C to 125°C. The SN74ALS20A and SN74AS20 are characterized for operation from 0°C to 70°C.

FUNCTION TABLE (each gate)

INPUTS				OUTPUT
A	**B**	**C**	**D**	**Y**
H	H	H	H	L
L	X	X	X	H
X	L	X	X	H
X	X	L	X	H
X	X	X	L	H

logic symbol†

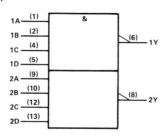

† This symbol is in accordance with ANSI/IEEE Std 91-1984 and IEC Publication 617-12.
Pin numbers shown are for D, J, and N packages.

SN54ALS20A, SN54AS20 . . . J PACKAGE
SN74ALS20A, SN74AS20 . . . D OR N PACKAGE
(TOP VIEW)

SN54ALS20A, SN54AS20 . . . FK PACKAGE
(TOP VIEW)

NC—No internal connection

logic diagram (positive logic)

Copyright © 1982, Texas Instruments Incorporated

TEXAS INSTRUMENTS
POST OFFICE BOX 655012 • DALLAS, TEXAS 75265

SN54ALS20A, SN74ALS20A
DUAL 4-INPUT POSITIVE-NAND GATES

absolute maximum ratings over operating free-air temperature range (unless otherwise noted)

Supply voltage, V_{CC} . . . . 7 V
Input voltage . . . . 7 V
Operating free-air temperature range: SN54ALS20A . . . . $-55\,^{\circ}$C to $125\,^{\circ}$C
SN74ALS20A . . . . $0\,^{\circ}$C to $70\,^{\circ}$C
Storage temperature range . . . . $-65\,^{\circ}$C to $150\,^{\circ}$C

recommended operating conditions

		SN54ALS20A			SN74ALS20A			UNIT
		MIN	NOM	MAX	MIN	NOM	MAX	
V_{CC}	Supply voltage	4.5	5	5.5	4.5	5	5.5	V
V_{IH}	High-level input voltage	2			2			V
V_{IL}	Low-level input voltage			0.7			0.8	V
I_{OH}	High-level output current			-0.4			-0.4	mA
I_{OL}	Low-level output current			4			8	mA
T_A	Operating free-air temperature	-55		125	0		70	$^{\circ}$C

electrical characteristics over recommended operating-free-air temperature range (unless otherwise noted)

PARAMETER	TEST CONDITIONS		SN54ALS20A			SN74ALS20A			UNIT
			MIN	TYP[†]	MAX	MIN	TYP[†]	MAX	
V_{IK}	$V_{CC} = 4.5$ V,	$I_I = -18$ mA			-1.5			-1.5	V
V_{OH}	$V_{CC} = 4.5$ V to 5.5 V,	$I_{OH} = -0.4$ mA	$V_{CC}-2$			$V_{CC}-2$			V
V_{OL}	$V_{CC} = 4.5$ V,	$I_{OL} = 4$ mA		0.25	0.4		0.25	0.4	V
	$V_{CC} = 4.5$ V,	$I_{OL} = 8$ mA					0.35	0.5	
I_I	$V_{CC} = 5.5$ V,	$V_I = 7$ V			0.1			0.1	mA
I_{IH}	$V_{CC} = 5.5$ V,	$V_I = 2.7$ V			20			20	μA
I_{IL}	$V_{CC} = 5.5$ V,	$V_I = 0.4$ V			-0.1			-0.1	mA
I_O[‡]	$V_{CC} = 5.5$ V,	$V_O = 2.25$ V	-30		-112	-30		-112	mA
I_{CCH}	$V_{CC} = 5.5$ V,	$V_I = 0$ V		0.22	0.4		0.22	0.4	mA
I_{CCL}	$V_{CC} = 5.5$ V,	$V_I = 4.5$ V		0.81	1.5		0.81	1.5	mA

[†] All typical values are at $V_{CC} = 5$ V, $T_A = 25\,^{\circ}$C.
[‡] The output conditions have been chosen to produce a current that closely approximates one half of the true short-circuit output current, I_{OS}.

switching characteristics (see Note 1)

PARAMETER	FROM (INPUT)	TO (OUTPUT)	$V_{CC} = 5$ V, $C_L = 50$ pF, $R_L = 500\ \Omega$, $T_A = 25\,^{\circ}$C	$V_{CC} = 4.5$ V to 5.5 V, $C_L = 50$ pF, $R_L = 500\ \Omega$, $T_A =$ MIN to MAX				UNIT
			'ALS20A	SN54ALS20A		SN74ALS20A		
			TYP	MIN	MAX	MIN	MAX	
t_{PLH}	Any	Y	7	1	18	3	11	ns
t_{PHL}	Any	Y	6	1	15	3	10	

NOTE 1: Load circuit and voltage waveforms are shown in Section 1.

Texas
Instruments

POST OFFICE BOX 655012 • DALLAS, TEXAS 75265

absolute maximum ratings over operating free-air temperature range (unless otherwise noted)

Supply voltage, V_{CC} . . . . 7 V
Input voltage . . . . 7 V
Operating free-air temperature range: SN54AS20 . . . . −55 °C to 125 °C
SN74AS20 . . . . 0 °C to 70 °C
Storage temperature range . . . . −65 °C to 150 °C

recommended operating conditions

		SN54AS20			SN74AS20			UNIT
		MIN	NOM	MAX	MIN	NOM	MAX	
V_{CC}	Supply voltage	4.5	5	5.5	4.5	5	5.5	V
V_{IH}	High-level input voltage	2			2			V
V_{IL}	Low-level input voltage			0.8			0.8	V
I_{OH}	High-level output current			−2			−2	mA
I_{OL}	Low-level output current			20			20	mA
T_A	Operating free-air temperature	−55		125	0		70	°C

electrical characteristics over recommended operating-free-air temperature range (unless otherwise noted)

PARAMETER	TEST CONDITIONS		SN54AS20			SN74AS20			UNIT
			MIN	TYP[†]	MAX	MIN	TYP[†]	MAX	
V_{IK}	$V_{CC} = 4.5$ V,	$I_I = -18$ mA			−1.2			−1.2	V
V_{OH}	$V_{CC} = 4.5$ V to 5.5 V,	$I_{OH} = -2$ mA	$V_{CC}-2$			$V_{CC}-2$			V
V_{OL}	$V_{CC} = 4.5$ V,	$I_{OL} = 20$ mA		0.35	0.5		0.35	0.5	V
I_I	$V_{CC} = 5.5$ V,	$V_I = 7$ V			0.1			0.1	mA
I_{IH}	$V_{CC} = 5.5$ V,	$V_I = 2.7$ V			20			20	µA
I_{IL}	$V_{CC} = 5.5$ V,	$V_I = 0.4$ V			−0.5			−0.5	mA
I_O[‡]	$V_{CC} = 5.5$ V,	$V_O = 2.25$ V	−30		−112	−30		−112	mA
I_{CCH}	$V_{CC} = 5.5$ V,	$V_I = 0$ V		1	1.6		1	1.6	mA
I_{CCL}	$V_{CC} = 5.5$ V,	$V_I = 4.5$ V		5.4	8.7		5.4	8.7	mA

[†] All typical values are at $V_{CC} = 5$ V, $T_A = 25$ °C.
[‡] The output conditions have been chosen to produce a current that closely approximates one half of the true short-circuit output current, I_{OS}.

switching characteristics (see Note 1)

PARAMETER	FROM (INPUT)	TO (OUTPUT)	$V_{CC} = 4.5$ V to 5.5 V, $C_L = 50$ pF, $R_L = 500$ Ω, $T_A = $ MIN to MAX				UNIT
			SN54AS20		SN74AS20		
			MIN	MAX	MIN	MAX	
t_{PLH}	Any	Y	1	5.5	1	5	ns
t_{PHL}	Any	Y	1	5	1	4.5	

NOTE 1: Load circuit and voltage waveforms are shown in Section 1.

TYPES SN54LS20, SN74LS20
DUAL 4-INPUT POSITIVE-NAND GATES

recommended operating conditions

		SN54LS20			SN74LS20			UNIT
		MIN	NOM	MAX	MIN	NOM	MAX	
V_{CC}	Supply voltage	4.5	5	5.5	4.75	5	5.25	V
V_{IH}	High-level input voltage	2			2			V
V_{IL}	Low-level input voltage			0.7			0.8	V
I_{OH}	High-level output current			-0.4			-0.4	mA
I_{OL}	Low-level output current			4			8	mA
T_A	Operating free-air temperature	-55		125	0		70	$^\circ$C

electrical characteristics over recommended operating free-air temperature range (unless otherwise noted)

PARAMETER	TEST CONDITIONS †			SN54LS20			SN74LS20			UNIT
				MIN	TYP‡	MAX	MIN	TYP‡	MAX	
V_{IK}	V_{CC} = MIN,	$I_I = -18$ mA				-1.5			-1.5	V
V_{OH}	V_{CC} = MIN,	V_{IL} = MAX,	$I_{OH} = -0.4$ mA	2.5	3.4		2.7	3.4		V
V_{OL}	V_{CC} = MIN,	V_{IH} = 2 V,	I_{OL} = 4 mA		0.25	0.4			0.4	V
	V_{CC} = MIN,	V_{IH} = 2 V,	I_{OL} = 8 mA					0.25	0.5	
I_I	V_{CC} = MAX,	V_I = 7 V				0.1			0.1	mA
I_{IH}	V_{CC} = MAX,	V_I = 2.7 V				20			20	μA
I_{IL}	V_{CC} = MAX,	V_I = 0.4 V				-0.4			-0.4	mA
I_{OS}§	V_{CC} = MAX			-20		-100	-20		-100	mA
I_{CCH}	V_{CC} = MAX,	V_I = 0 V			0.4	0.8		0.4	0.8	mA
I_{CCL}	V_{CC} = MAX,	V_I = 4.5 V			1.2	2.2		1.2	2.2	mA

† For conditions shown as MIN or MAX, use the appropriate value specified under recommended operating conditions.
‡ All typical values are at V_{CC} = 5 V, T_A = 25°C.
§ Not more than one output should be shorted at a time, and the duration of the short-circuit should not exceed one second.

switching characteristics, V_{CC} = 5 V, T_A = 25°C (see note 2)

PARAMETER	FROM (INPUT)	TO (OUTPUT)	TEST CONDITIONS		MIN	TYP	MAX	UNIT
t_{PLH}	Any	Y	R_L = 2 kΩ,	C_L = 15 pF		9	15	ns
t_{PHL}						10	15	ns

NOTE 2: See General Information Section for load circuits and voltage waveforms.

TEXAS
INSTRUMENTS

POST OFFICE BOX 225012 • DALLAS, TEXAS 75265

absolute maximum ratings over operating free-air temperature range[†]

Supply voltage range, V_{CC}	-0.5 V to 7 V
Input diode current, $I_{IK}(V_I < 0$ or $V_I > V_{CC})$	± 20 mA
Output diode current, $I_{OK}(V_O < 0$ or $V_O > V_{CC})$	± 20 mA
Continuous output current, I_O ($V_O = 0$ to V_{CC})	± 25 mA
Continuous current through V_{CC} or GND pins	± 50 mA
Lead temperature 1,6 mm (1/16 inch) from case for 60 seconds: FH, FK, or J package	300°C
Lead temperature 1,6 mm (1/16 inch) from case for 10 seconds: FN or N package	260°C
Storage temperature range	-65°C to 150°C

[†] Stresses beyond those listed under "absolute maximum ratings" may cause permanent damage to the device. These are stress ratings only and functional operation of the device at these or any other conditions beyond those indicated under "recommended operating conditions" is not implied. Exposure to absolute-maximum-rated conditions for extended periods may affect device reliability.

recommended operating conditions

			SN54HC'			SN74HC'			UNIT
			MIN	NOM	MAX	MIN	NOM	MAX	
V_{CC}	Supply voltage		2	5	6	2	5	6	V
V_{IH}	High-level input voltage	$V_{CC} = 2$ V	1.5			1.5			V
		$V_{CC} = 4.5$ V	3.15			3.15			
		$V_{CC} = 6$ V	4.2			4.2			
V_{IL}	Low-level input voltage	$V_{CC} = 2$ V	0		0.3	0		0.3	V
		$V_{CC} = 4.5$ V	0		0.9	0		0.9	
		$V_{CC} = 6$ V	0		1.2	0		1.2	
V_I	Input voltage		0		V_{CC}	0		V_{CC}	V
V_O	Output voltage		0		V_{CC}	0		V_{CC}	V
t_t	Input transition (rise and fall) times (except Schmitt-trigger inputs)	$V_{CC} = 2$ V	0		1000	0		1000	ns
		$V_{CC} = 4.5$ V	0		500	0		500	
		$V_{CC} = 6$ V	0		400	0		400	
T_A	Operating free-air temperature		-55		125	-40		85	°C

electrical characteristics over recommended operating free-air temperature range (unless otherwise noted)

PARAMETER	TEST CONDITIONS	V_{CC}	$T_A = 25$°C			SN54HC'		SN74HC'		UNIT
			MIN	TYP	MAX	MIN	MAX	MIN	MAX	
V_{OH} (Totem-pole outputs)	$V_I = V_{IH}$ or V_{IL}, $I_{OH} = -20$ μA	2 V	1.9	1.998		1.9		1.9		V
		4.5 V	4.4	4.499		4.4		4.4		
		6 V	5.9	5.999		5.9		5.9		
	$V_I = V_{IH}$ or V_{IL}, $I_{OH} = -4$ mA	4.5 V	3.98	4.30		3.7		3.84		
	$V_I = V_{IH}$ or V_{IL}, $I_{OH} = -5.2$ mA	6 V	5.48	5.80		5.2		5.34		
I_{OH} (Open-drain outputs)	$V_I = V_{IH}$ or V_{IL}, $V_O = V_{CC}$	6 V		0.01	0.5		10		5	μA
V_{OL}	$V_I = V_{IH}$ or V_{IL}, $I_{OL} = 20$ μA	2 V		0.002	0.1		0.1		0.1	V
		4.5 V		0.001	0.1		0.1		0.1	
		6 V		0.001	0.1		0.1		0.1	
	$V_I = V_{IH}$ or V_{IL}, $I_{OL} = 4$ mA	4.5 V		0.17	0.26		0.4		0.33	
	$V_I = V_{IH}$ or V_{IL}, $I_{OL} = 5.2$ mA	6 V		0.15	0.26		0.4		0.33	
V_{T+}[†]		2 V	0.8	1.2	1.5					V
		4.5 V	2	2.5	3.15					
		6 V	2.5	3.3	4.2					
V_{T-}[†]		2 V	0.3	0.6	0.8					V
		4.5 V	0.9	1.6	2					
		6 V	1.2	2	2.5					
$V_{T+} - V_{T-}$[†]		2 V	0.2	0.6	1					V
		4.5 V	0.4	0.9	1.4					
		6 V	0.5	1.3	1.7					
I_I	$V_I = 0$ to V_{CC}	6 V		± 0.1	± 100		± 1000		± 1000	nA
I_{CC}	$V_I = V_{CC}$ or 0, $I_O = 0$	6 V			2		40		20	μA
C_I		2 to 6 V		3	10		10		10	pF

[†] This parameter applies only for Schmitt-trigger inputs.

**TYPES SN54HC20, SN74HC20
DUAL 4-INPUT POSITIVE-NAND GATES**

D2684, DECEMBER 1982—REVISED MARCH 1984

- Package Options Include Both Plastic and Ceramic Chip Carriers in Addition to Plastic and Ceramic DIPs
- Dependable Texas Instruments Quality and Reliability

description

These devices contain two independent 4-input NAND gates. They perform the Boolean functions $Y = \overline{A \cdot B \cdot C \cdot D}$ or $Y = \overline{A} + \overline{B} + \overline{C} + \overline{D}$ in positive logic.

The SN54HC20 is characterized for operation over the full military temperature range of $-55\,°C$ to $125\,°C$. The SN74HC20 is characterized for operation from $-40\,°C$ to $85\,°C$.

logic symbol

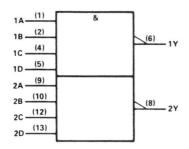

Pin numbers shown are for J and N packages.

SN54HC20 . . . J PACKAGE
SN74HC20 . . . J OR N PACKAGE
(TOP VIEW)

SN54HC20 . . . FH OR FK PACKAGE
SN74HC20 . . . FH OR FN PACKAGE
(TOP VIEW)

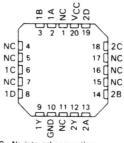

NC—No internal connection

FUNCTION TABLE (each gate)

INPUTS				OUTPUT
A	B	C	D	Y
H	H	H	H	L
L	X	X	X	H
X	L	X	X	H
X	X	L	X	H
X	X	X	L	H

maximum ratings, recommended operating conditions, and electrical characteristics

See Table I, page 2-4.

switching characteristics over recommended operating free-air temperature range (unless otherwise noted), $C_L = 50$ pF (see Note 1)

PARAMETER	FROM (INPUT)	TO (OUTPUT)	V$_{CC}$	T$_A$ = 25°C MIN	T$_A$ = 25°C TYP	T$_A$ = 25°C MAX	SN54HC20 MIN	SN54HC20 MAX	SN74HC20 MIN	SN74HC20 MAX	UNIT
t$_{pd}$	A, B, C, or D	Y	2 V		45	110		165		140	ns
			4.5 V		14	22		33		28	
			6 V		11	19		28		24	
t$_t$		Y	2 V		27	75		110		95	ns
			4.5 V		9	15		22		19	
			6 V		7	13		19		16	

C$_{pd}$	Power dissipation capacitance per gate	No load, T$_A$ = 25°C	25 pF typ

NOTE 1: For load circuit and voltage waveforms, see page 1-14.

**TEXAS
INSTRUMENTS**
POST OFFICE BOX 225012 ● DALLAS, TEXAS 75265

54/7432
54S/74S32
54LS/74LS32

QUAD 2-INPUT OR GATE

ORDERING CODE: See Section 9

PKGS	PIN OUT	COMMERCIAL GRADE Vcc = +5.0 V ±5%, TA = 0°C to +70°C	MILITARY GRADE Vcc = +5.0 V ±10%, TA = -55°C to +125°C	PKG TYPE
Plastic DIP (P)	A	7432PC, 74S32PC 74LS32PC		9A
Ceramic DIP (D)	A	7432DC, 74S32DC 74LS32DC	5432DM, 54S32DM 54LS32DM	6A
Flatpak (F)	A	7432FC, 74S32FC 74LS32FC	5432FM, 54S32FM 54LS32FM	3I

INPUT LOADING/FAN-OUT: See Section 3 for U.L. definitions

PINS	54/74 (U.L.) HIGH/LOW	54/74S (U.L.) HIGH/LOW	54/74LS (U.L.) HIGH/LOW
Inputs	1.0/1.0	1.25/1.25	0.5/0.25
Outputs	20/10	25/12.5	10/5.0 (2.5)

DC AND AC CHARACTERISTICS: See Section 3 for U.L. definitions

SYMBOL	PARAMETER	54/74 Min	54/74 Max	54/74S Min	54/74S Max	54/74LS Min	54/74LS Max	UNITS	CONDITIONS
Iссн	Power Supply Current		22		32		6.2	mA	VIN = Open, Vcc = Max
Iссl			38		68		9.8		VIN = Gnd
tPLH	Propagation Delay		15	2.0	7.0		15	ns	Figs. 3-1, 3-5
tPHL			22	2.0	7.0		15		

*DC limits apply over operating temperature range; AC limits apply at TA = +25°C and Vcc = +5.0 V.

54/7437
54LS/74LS37
QUAD 2-INPUT NAND BUFFER

ORDERING CODE: See Section 9

PKGS	PIN OUT	COMMERCIAL GRADE Vcc = +5.0 V ±5%, TA = 0°C to +70°C	MILITARY GRADE Vcc = +5.0 V ±10%, TA = -55°c to +125°C	PKG TYPE
Plastic DIP (P)	A	7437PC, 74LS37PC		9A
Ceramic DIP (D)	A	7437DC, 74LS37DC	5437DM, 54LS37DM	6A
Flatpak (F)	A	7437FC, 74LS37FC	5437FM, 54LS37FM	3I

INPUT LOADING/FAN-OUT: See Section 3 for U.L. definitions

PINS	54/74 (U.L.) HIGH/LOW	54/74LS (U.L.) HIGH/LOW
Inputs	1.0/1.0	0.5/0.25
Outputs	30/30	30/15 (7.5)

DC AND AC CHARACTERISTICS: See Section 3*

SYMBOL	PARAMETER		54/74 Min Max	54/74LS Min Max	UNITS	CONDITIONS
VOH	Output HIGH Voltage	XM	2.4	2.5	V	Vcc = Max, IOH = -1.2 mA VIN = VIL
		XC	2.4	2.7		
VOL	Output LOW Voltage	XM, XC	0.4		V	IOL = 48 mA, Vcc = Min
		XM		0.4		IOL = 12 mA, VIN = 2.0 V
		XC		0.5		IOL = 24 mA
IOS	Output Short Circuit Current	XM	-20 -70	-30 -130	mA	Vcc = Min, VOUT = 0 V
		XC	-18 -70	-30 -130		
ICCH	Power Supply Current		15.5	2.0	mA	VIN = Gnd, Vcc = Max
ICCL			54	12		VIN = Open
tPLH	Propagation Delay		22	20	ns	Figs. 3-1, 3-4
tPHL			15	20		

*DC limits apply over operating temperature range; AC limits apply at TA = +25°C and Vcc = +5.0 V.

54/7486
54S/74S86
54LS/74LS86

QUAD 2-INPUT EXCLUSIVE-OR GATE

ORDERING CODE: See Section 9

PKGS	PIN OUT	COMMERCIAL GRADE V_{CC} = +5.0 V ±5%, T_A = 0°C to +70°C	MILITARY GRADE V_{CC} = +5.0 V ±10%, T_A = -55°C to +125°C	PKG TYPE
Plastic DIP (P)	A	7486PC, 74S86PC 74LS86PC		9A
Ceramic DIP (D)	A	7486DC, 74S86DC 74LS86DC	5486DM, 54S86DM 54LS86DM	6A
Flatpak (F)	A	7486FC, 74S86FC 74LS86FC	5486FM, 54S86FM 54LS86FM	3I

INPUT LOADING/FAN-OUT: See Section 3 for U.L. definitions

PINS	54/74 (U.L.) HIGH/LOW	54/74S (U.L.) HIGH/LOW	54/74LS (U.L.) HIGH/LOW
Inputs	1.0/1.0	1.25/1.25	1.0/0.375
Outputs	20/10	25/12.5	10/5.0 (2.5)

DC AND AC CHARACTERISTICS: See Section 3*

SYMBOL	PARAMETER		54/74 Min	54/74 Max	54/74S Min	54/74S Max	54/74LS Min	54/74LS Max	UNITS	CONDITIONS
I_{CC}	Power Supply Current	XM		43		75		10	mA	V_{CC} = Max, V_{IN} = Gnd
		XC		50		75		10		
t_{PLH}	Propagation Delay			23	3.5	10.5		12	ns	Other Input LOW Figs. 3-1, 3-5
t_{PHL}				17	3.0	10		17		
t_{PLH}	Propagation Delay			30	3.5	10.5		13	ns	Other Input HIGH Figs. 3-1, 3-4
t_{PHL}				22	3.0	10		12		

*DC limits apply over operating temperature range; AC limits apply at T_A = +25°C and V_{CC} = +5.0 V.

'290, 'LS290 . . . DECADE COUNTERS
'293, 'LS293 . . . 4-BIT BINARY COUNTERS

- GND and V$_{CC}$ on Corner Pins
  (Pins 7 and 14 Respectively)

description

The SN54290/SN74290, SN54LS290/SN74LS290, SN54293/SN74293, and SN54LS293/SN74LS293 counters are electrically and functionally identical to the SN5490A/SN7490A, SN54LS90/SN74LS90, SN5493A/SN7493A, and SN54LS93/SN74LS93, respectively. Only the arrangement of the terminals has been changed for the '290, 'LS290, '293, and 'LS293.

Each of these monolithic counters contains four master-slave flip-flops and additional gating to provide a divide-by-two counter and a three-stage binary counter for which the count cycle length is divide-by-five for the '290 and 'LS290 and divide-by-eight for the '293 and 'LS293.

All of these counters have a gated zero reset and the '290 and 'LS290 also have gated set-to-nine inputs for use in BCD nine's complement applications.

To use the maximum count length (decade or four-bit binary) of these counters, the B input is connected to the Q$_A$ output. The input count pulses are applied to input A and the outputs are as described in the appropriate function table. A symmetrical divide-by-ten count can be obtained from the '290 and 'LS290 counters by connecting the Q$_D$ output to the A input and applying the input count to the B input which gives a divide-by-ten square wave at output Q$_A$.

SN54290, SN54LS290, SN54293,
SN54LS293 . . . J OR W PACKAGE
SN74290, SN74293 . . . N PACKAGE
SN74LS290, SN74LS293 . . . D OR N PACKAGE
(TOP VIEW)

'290

R9(1)	1	14 V$_{CC}$
NC	2	13 R0(2)
R9(2)	3	12 R0(1)
Q$_C$	4	11 CKB
Q$_B$	5	10 CKA
NC	6	9 Q$_A$
GND	7	8 Q$_D$

'293

NC	1	14 V$_{CC}$
NC	2	13 R0(2)
NC	3	12 R0(1)
Q$_C$	4	11 CKB
Q$_B$	5	10 CKA
NC	6	9 Q$_A$
GND	7	8 Q$_D$

SN54LS290, SN54LS293 . . . FK PACKAGE
(TOP VIEW)

'LS290

'LS293

NC - No internal connection

TEXAS
INSTRUMENTS
POST OFFICE BOX 655012 • DALLAS, TEXAS 75265

'290, 'LS290 BCD COUNT SEQUENCE (See Note A)

COUNT	OUTPUT			
	Q_D	Q_C	Q_B	Q_A
0	L	L	L	L
1	L	L	L	H
2	L	L	H	L
3	L	L	H	H
4	L	H	L	L
5	L	H	L	H
6	L	H	H	L
7	L	H	H	H
8	H	L	L	L
9	H	L	L	H

'290, 'LS290 BI-QUINARY (5-2) (See Note B)

COUNT	OUTPUT			
	Q_A	Q_D	Q_C	Q_B
0	L	L	L	L
1	L	L	L	H
2	L	L	H	L
3	L	L	H	H
4	L	H	L	L
5	H	L	L	L
6	H	L	L	H
7	H	L	H	L
8	H	L	H	H
9	H	H	L	L

'290, 'LS290 RESET/COUNT FUNCTION TABLE

RESET INPUTS				OUTPUT			
$R_{0(1)}$	$R_{0(2)}$	$R_{9(1)}$	$R_{9(2)}$	Q_D	Q_C	Q_B	Q_A
H	H	L	X	L	L	L	L
H	H	X	L	L	L	L	L
X	X	H	H	H	L	L	H
X	L	X	L	COUNT			
L	X	L	X	COUNT			
L	X	X	L	COUNT			
X	L	L	X	COUNT			

'293, 'LS293 COUNT SEQUENCE (See Note C)

COUNT	OUTPUT			
	Q_D	Q_C	Q_B	Q_A
0	L	L	L	L
1	L	L	L	H
2	L	L	H	L
3	L	L	H	H
4	L	H	L	L
5	L	H	L	H
6	L	H	H	L
7	L	H	H	H
8	H	L	L	L
9	H	L	L	H
10	H	L	H	L
11	H	L	H	H
12	H	H	L	L
13	H	H	L	H
14	H	H	H	L
15	H	H	H	H

'293, 'LS293 RESET/COUNT FUNCTION TABLE

RESET INPUTS		OUTPUT			
$R_{0(1)}$	$R_{0(2)}$	Q_D	Q_C	Q_B	Q_A
H	H	L	L	L	L
L	X	COUNT			
X	L	COUNT			

NOTES: A. Output Q_A is connected to input B for BCD count.
B. Output Q_D is connected to input A for bi-quinary count.
C. Output Q_A is connected to input B.
D. H = high level, L = low level, X = irrelevant

logic diagrams (positive logic)

'290, 'LS290

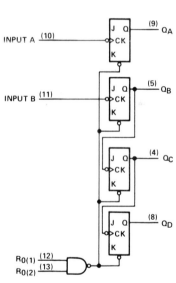

'293, 'LS293

Pin numbers shown are for D, J, N, and W packages.
The J and K inputs shown without connection are for reference only and are functionally at a high level.

TEXAS
INSTRUMENTS
POST OFFICE BOX 655012 • DALLAS, TEXAS 75265

54S/74S112
54LS/74LS112

DUAL JK NEGATIVE
EDGE-TRIGGERED FLIP-FLOP

DESCRIPTION — The '112 features individual J, K, Clock and asynchronous Set and Clear inputs to each flip-flop. When the clock goes HIGH, the inputs are enabled and data will be accepted. The logic level of the J and K inputs may change when the clock is HIGH and the bistable will perform according to the Truth Table as long as minimum setup and hold times are observed. Input data is transferred to the outputs on the falling edge of the clock pulse.

TRUTH TABLE

INPUTS		OUTPUT
@ t_n		@ t_{n+1}
J	K	Q
L	L	Q_n
L	H	L
H	L	H
H	H	$\bar{Q}_n$

Asynchronous Inputs:
    LOW input to $\bar{S}_D$ sets Q to HIGH level
    LOW input to $\bar{C}_D$ sets Q to LOW level
    Clear and Set are independent of clock
    Simultaneous LOW on $\bar{C}_D$ and $\bar{S}_D$
    makes both Q and $\bar{Q}$ HIGH

t_n = Bit time before clock pulse.
t_{n+1} = Bit time after clock pulse.
H = HIGH Voltage Level
L = LOW Voltage Level

LOGIC SYMBOL

V_{CC} = Pin 16
GND = Pin 8

ORDERING CODE: See Section 9

PKGS	PIN OUT	COMMERCIAL GRADE V_{CC} = +5.0 V ±5%, T_A = 0°C to +70°C	MILITARY GRADE V_{CC} = +5.0 V ±10%, T_A = -55°C to +125°C	PKG TYPE
Plastic DIP (P)	A	74S112PC, 74LS112PC		9B
Ceramic DIP (D)	A	74S112DC, 74LS112DC	54S112DM, 54LS112DM	6B
Flatpak (F)	A	74S112FC, 74LS112FC	54S112FM, 54LS112FM	4L

INPUT LOADING/FAN-OUT: See Section 3 for U.L. definitions

PIN NAMES	DESCRIPTION	54/74S (U.L.) HIGH/LOW	54/74LS (U.L.) HIGH/LOW
J_1, J_2, K_1, K_2	Data Inputs	1.25/1.0	0.5/0.25
$\bar{CP}_1$, $\bar{CP}_2$	Clock Pulse Inputs (Active Falling Edge)	2.5/2.5	2.0/0.5
$\bar{C}_{D1}$, $\bar{C}_{D2}$	Direct Clear Inputs (Active LOW)	2.5/4.375	1.5/0.5
$\bar{S}_{D1}$, $\bar{S}_{D2}$	Direct Set Inputs (Active LOW)	2.5/4.375	1.5/0.5
Q_1, Q_2, $\bar{Q}_1$, $\bar{Q}_2$	Outputs	25/12.5	10/5.0 (2.5)

112

LOGIC DIAGRAM (one half shown)

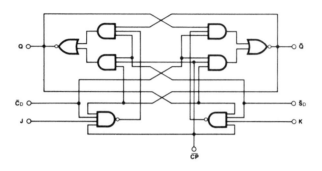

DC CHARACTERISTICS OVER OPERATING TEMPERATURE RANGE (unless otherwise specified)

SYMBOL	PARAMETER	54/74S		54/74LS		UNITS	CONDITIONS
		Min	Max	Min	Max		
I_{CC}	Power Supply Current		50		8.0	mA	V_{CC} = Max, V_{CP} = 0 V

AC CHARACTERISTICS: V_{CC} = +5.0 V, T_A = +25°C (See Section 3 for waveforms and load configurations)

SYMBOL	PARAMETER	54/74S C_L = 15 pF R_L = 280 Ω		54/74LS C_L = 15 pF		UNITS	CONDITIONS
		Min	Max	Min	Max		
f_{max}	Maximum Clock Frequency	80		30		MHz	Figs. 3-1, 3-9
t_{PLH} t_{PHL}	Propagation Delay $\overline{CP}_n$ to Q_n or $\overline{Q}_n$		7.0 7.0		16 24	ns	Figs. 3-1, 3-9
t_{PLH} t_{PHL}	Propagation Delay $\overline{C}_{Dn}$ or $\overline{S}_{Dn}$ to Q_n or $\overline{Q}_n$		7.0 7.0		16 24	ns	Figs. 3-1, 3-10

AC OPERATING REQUIREMENTS: V_{CC} = +5.0 V, T_A =+25°C

SYMBOL	PARAMETER	54/74S		54/74LS		UNITS	CONDITIONS
		Min	Max	Min	Max		
t_s (H) t_s (L)	Setup Time J_n or K_n to $\overline{CP}_n$	7.0 7.0		20 15		ns	Fig. 3-7
t_h (H) t_h (L)	Hold Time J_n or K_n to $\overline{CP}_n$	0 0		0 0		ns	
t_w (H) t_w (L)	$\overline{CP}_n$ Pulse Width	6.0 6.5		20 15		ns	Fig. 3-9
t_w (L)	$\overline{C}_{Dn}$ or $\overline{S}_{Dn}$ Pulse Width LOW	8.0		15		ns	Fig. 3-10

54/74121
MONOSTABLE MULTIVIBRATOR

DESCRIPTION — The '121 features positive and negative dc level triggering inputs and complementary outputs. Input pin 5 directly activates a Schmitt circuit which provides temperature compensated level detection, increases immunity to positive-going noise and assures jitter-free response to slowly rising triggers.

When triggering occurs, internal feedback latches the circuit, prevents re-triggering while the output pulse is in progress and increases immunity to negative-going noise. Noise immunity is typically 1.2 V at the inputs and 1.5 V on V_{CC}.

Output pulse width stability is primarily a function of the external R_x and C_x chosen for the application. A 2 kΩ internal resistor is provided for optional use where output pulse width stability requirements are less stringent. Maximum duty cycle capability ranges from 67% with a 2 kΩ resistor to 90% with a 40 kΩ resistor. Duty cycles beyond this range tend to reduce the output pulse width. Otherwise, output pulse width follows the relationship:

$$t_w = 0.69\ R_x C_x$$

LOGIC SYMBOL

V_{CC} = Pin 14
GND = Pin 7
NC = Pins 2,8,12,13

ORDERING CODE: See Section 9

PKGS	PIN OUT	COMMERCIAL GRADE V_{CC} = +5.0 V ±5%, T_A = 0°C to +70°C	MILITARY GRADE V_{CC} = +5.0 V ±10%, T_A = -55°C to +125°C	PKG TYPE
Plastic DIP (P)	A	74121PC		9A
Ceramic DIP (D)	A	74121DC	54121DM	6A
Flatpak (F)	A	74121FC	54121FM	3I

INPUT LOADING/FAN-OUT: See Section 3 for U.L. definitions

PIN NAMES	DESCRIPTION	54/74 (U.L.) HIGH/LOW
$\overline{A}_1$, $\overline{A}_2$	Trigger Inputs (Active Falling Edge)	1.0/1.0
B	Schmitt Trigger Input (Active Rising Edge)	2.0/2.0
Q, $\overline{Q}$	Outputs	20/10

54/74193
54LS/74LS193

UP/DOWN BINARY COUNTER
(With Separate Up/down Clocks)

DESCRIPTION — The '193 is an up/down modulo-16 binary counter. Separate Count Up and Count Down Clocks are used and in either counting mode the circuits operate synchronously. The outputs change state synchronous with the LOW-to-HIGH transitions on the clock inputs. Separate Terminal Count Up and Terminal Count Down outputs are provided which are used as the clocks for subsequent stages without extra logic, thus simplifying multistage counter designs. Individual preset inputs allow the circuits to be used as programmable counters. Both the Parallel Load ($\overline{PL}$) and the Master Reset (MR) inputs asynchronously override the clocks. For functional description and detail specifications please refer to the '192 data sheet.

CONNECTION DIAGRAM
PINOUT A

P1	1	16	Vcc
Q1	2	15	P0
Q0	3	14	MR
CPD	4	13	$\overline{TC}_D$
CPU	5	12	$\overline{TC}_U$
Q2	6	11	$\overline{PL}$
Q3	7	10	P2
GND	8	9	P3

LOGIC SYMBOL

Vcc = Pin 16
GND = Pin 8

ORDERING CODE: See Section 9

PKGS	PIN OUT	COMMERCIAL GRADE Vcc = +5.0 V ±5%, TA = 0°C to +70°C	MILITARY GRADE Vcc = +5.0 V ±10%, TA = -55°C to +125°C	PKG TYPE
Plastic DIP (P)	A	74193PC, 74LS193PC		9B
Ceramic DIP (D)	A	74193DC, 74LS193DC	54193DM, 54LS193DM	6B
Flatpak (F)	A	74193FC, 74LS193FC	54193FM, 54LS193FM	4L

INPUT LOADING/FAN-OUT: See Section 3 for U.L. definitions

PIN NAMES	DESCRIPTION	54/74 (U.L.) HIGH/LOW	54/74LS (U.L.) HIGH/LOW
CPU	Count Up Clock Input (Active Rising Edge)	1.0/1.0	0.5/0.25
CPD	Count Down Clock Input (Active Rising Edge)	1.0/1.0	0.5/0.25
MR	Asynchronous Master Reset Input (Active HIGH)	1.0/1.0	0.5/0.25
$\overline{PL}$	Asynchronous Parallel Load Input (Active LOW)	1.0/1.0	0.5/0.25
P0 — P3	Parallel Data Inputs	1.0/1.0	0.5/0.25
Q0 — Q3	Flip-flop Outputs	20/10	10/5.0 (2.5)
$\overline{TC}_D$	Terminal Count Down (Borrow) Output (Active LOW)	20/10	10/5.0 (2.5)
$\overline{TC}_U$	Terminal Count Up (Carry) Output (Active LOW)	20/10	10/5.0 (2.5)

193

MODE SELECT TABLE

MR	$\overline{PL}$	CP_U	CP_D	MODE
H	X	X	X	Reset (Asyn.)
L	L	X	X	Preset (Asyn.)
L	H	H	H	No Change
L	H	⌐	H	Count Up
L	H	H	⌐	Count Down

H = HIGH Voltage Level
L = LOW Voltage Level
X = Immaterial
Z = High Impedance

STATE DIAGRAM

COUNT UP
COUNT DOWN

**LOGIC EQUATIONS
FOR TERMINAL COUNT**

$$\overline{TC}_U = Q_0 \bullet Q_1 \bullet Q_2 \bullet Q_3 \bullet \overline{CP}_U$$
$$\overline{TC}_D = \overline{Q}_0 \bullet \overline{Q}_1 \bullet \overline{Q}_2 \bullet \overline{Q}_3 \bullet \overline{CP}_D$$

LOGIC DIAGRAM

- **Parallel Inputs and Outputs**
- **Four Operating Modes:**
  Synchronous Parallel Load
  Right Shift
  Left Shift
  Do Nothing
- **Positive Edge-Triggered Clocking**
- **Direct Overriding Clear**

TYPE	TYPICAL MAXIMUM CLOCK FREQUENCY	TYPICAL POWER DISSIPATION
'194	36 MHz	195 mW
'LS194A	36 MHz	75 mW
'S194	105 MHz	425 mW

SN54194, SN54LS194A, SN54S194 . . . J OR W PACKAGE
SN74194 . . . N PACKAGE
SN74LS194A, SN74S194 . . . D OR N PACKAGE
(TOP VIEW)

SN54LS194A, SN54S194 . . . FK PACKAGE
(TOP VIEW)

NC - No internal connection

description

These bidirectional shift registers are designed to incorporate virtually all of the features a system designer may want in a shift register. The circuit contains 46 equivalent gates and features parallel inputs, parallel outputs, right-shift and left-shift serial inputs, operating-mode-control inputs, and a direct overriding clear line. The register has four distinct modes of operation, namely:

Inhibit clock (do nothing)
Shift right (in the direction Q_A toward Q_D)
Shift left (in the direction Q_D toward Q_A)
Parallel (broadside) load

Synchronous parallel loading is accomplished by applying the four bits of data and taking both mode control inputs, S0 and S1, high. The data are loaded into the associated flip-flops and appear at the outputs after the positive transition of the clock input. During loading, serial data flow is inhibited.

Shift right is accomplished synchronously with the rising edge of the clock pulse when S0 is high and S1 is low. Serial data for this mode is entered at the shift-right data input. When S0 is low and S1 is high, data shifts left synchronously and new data is entered at the shift-left serial input.

Clocking of the shift register is inhibited when both mode control inputs are low. The mode controls of the SN54194/SN74194 should be changed only while the clock input is high.

logic symbol†

†This symbol is in accordance with ANSI/IEEE Std. 91-1984 and IEC Publication 617-12.
Pin numbers shown are for D, J, N, and W packages.

TEXAS INSTRUMENTS

logic diagrams (positive logic)

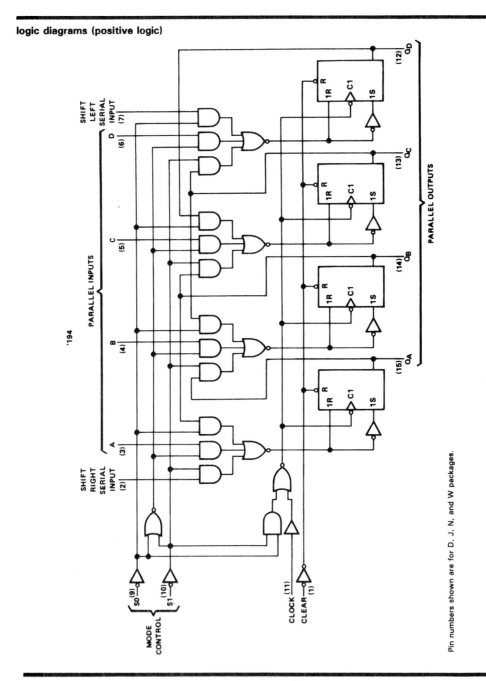

Pin numbers shown are for D, J, N, and W packages.

typical clear, load, right-shift, left-shift, inhibit, and clear sequences

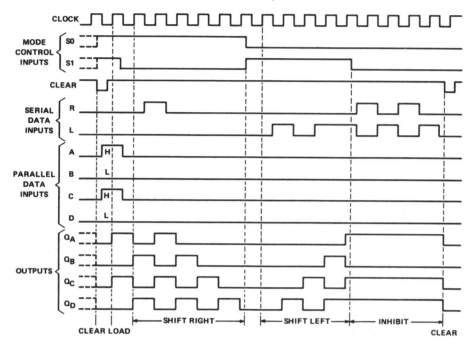

FUNCTION TABLE

INPUTS										OUTPUTS			
CLEAR	MODE		CLOCK	SERIAL		PARALLEL				Q_A	Q_B	Q_C	Q_D
	S1	S0		LEFT	RIGHT	A	B	C	D				
L	X	X	X	X	X	X	X	X	X	L	L	L	L
H	X	X	L	X	X	X	X	X	X	Q_{A0}	Q_{B0}	Q_{C0}	Q_{D0}
H	H	H	↑	X	X	a	b	c	d	a	b	c	d
H	L	H	↑	X	H	X	X	X	X	H	Q_{An}	Q_{Bn}	Q_{Cn}
H	L	H	↑	X	L	X	X	X	X	L	Q_{An}	Q_{Bn}	Q_{Cn}
H	H	L	↑	H	X	X	X	X	X	Q_{Bn}	Q_{Cn}	Q_{Dn}	H
H	H	L	↑	L	X	X	X	X	X	Q_{Bn}	Q_{Cn}	Q_{Dn}	L
H	L	L	X	X	X	X	X	X	X	Q_{An}	Q_{Bn}	Q_{Cn}	Q_{D0}

TEXAS
INSTRUMENTS

4001B   4002B
QUAD 2-INPUT NOR GATE ● DUAL 4-INPUT NOR GATE

DESCRIPTION — These CMOS logic elements provide the positive input NOR function. The outputs are fully buffered for highest noise immunity and pattern insensitivity of output impedance.

4001B
LOGIC AND CONNECTION DIAGRAM
DIP (TOP VIEW)

NOTE:
The Flatpak versions have the same pinouts (Connection Diagram) as the Dual In-line Package.

4002B
LOGIC AND CONNECTION DIAGRAM
DIP (TOP VIEW)

DC CHARACTERISTICS: V_{DD} as shown, V_{SS} = 0 V

SYMBOL	PARAMETER		LIMITS									UNITS	TEMP	TEST CONDITIONS See Note 1
			V_{DD} = 5 V			V_{DD} = 10 V			V_{DD} = 15 V					
			MIN	TYP	MAX	MIN	TYP	MAX	MIN	TYP	MAX			
I_{DD}	Quiescent Power Supply Current	XC			1			2			4	μA	MIN, 25°C	All inputs at 0 V or V_{DD}
					7.5			15			30		MAX	
		XM			0.25			0.5			1	μA	MIN, 25°C	
					7.5			15			30		MAX	

AC CHARACTERISTICS: V_{DD} as shown, V_{SS} = 0 V, T_A = 25°C, 4001B only (See Note 2)

SYMBOL	PARAMETER		LIMITS									UNITS	TEST CONDITIONS See Note 2
			V_{DD} = 5 V			V_{DD} = 10 V			V_{DD} = 15V				
			MIN	TYP	MAX	MIN	TYP	MAX	MIN	TYP	MAX		
t_{PLH}	Propagation Delay			60	110		25	60		20	48	ns	C_L = 50 pF, R_L = 200 kΩ
t_{PHL}				60	110		25	60		20	48	ns	
t_{TLH}	Output Transition Time			60	135		30	70		20	45	ns	Input Transition Times ≤ 20 ns
t_{THL}				60	135		30	70		20	45	ns	

AC CHARACTERISTICS: V_{DD} as shown, V_{SS} = 0 V, T_A = 25°C, 4002B only

SYMBOL	PARAMETER		LIMITS									UNITS	TEST CONDITIONS See Note 2
			V_{DD} = 5 V			V_{DD} = 10 V			V_{DD} = 15V				
			MIN	TYP	MAX	MIN	TYP	MAX	MIN	TYP	MAX		
t_{PLH}	Propagation Delay			65	110		30	60		20	48	ns	C_L = 50 pF, R_L = 200 kΩ
t_{PHL}				70	110		30	60		23	48	ns	
t_{TLH}	Output Transition Time			75	135		40	70		30	45	ns	Input Transition Times ≤ 20 ns
t_{THL}				60	135		23	70		15	45	ns	

NOTES:
1. Additional DC Characteristics are listed in this section under 4000B Series CMOS Family Characteristics.
2. Propagation Delays and Output Transition Times are graphically described in this section under 4000B Series CMOS Family Characteristics.

4001B   FAIRCHILD 4000B SERIES CMOS FAMILY CHARACTERISTICS

DC CHARACTERISTICS FOR THE 4000B SERIES CMOS FAMILY — Parametric Limits listed below are guaranteed for the entire Fairchild CMOS Family unless otherwise specified on the individual data sheets.

DC CHARACTERISTICS: V_{DD} = 5 V, V_{SS} = 0 V

SYMBOL	PARAMETER			LIMITS			UNITS	TEMP	TEST CONDITIONS	
				MIN	TYP	MAX				
V_{IH}	Input HIGH Voltage			3.5			V	All	Guaranteed Input HIGH Voltage	
V_{IL}	Input LOW Voltage					1.5	V	All	Guaranteed Input Low Voltage	
V_{OH}	Output HIGH Voltage			4.95			V	Min, 25°C	I_{OH} < 1 μA, Inputs at 0 or 5 V per	
				4.95				MAX	the Logic Function or Truth Table	
				4.5			V	All	I_{OH} < 1 μA, Inputs at 1.5 or 3.5 V	
V_{OL}	Output LOW Voltage					0.05	V	MIN, 25°C	I_{OL} < 1 μA, Inputs at 0 or 5 V per	
						0.05		MAX	the Logic Function or Truth Table	
						0.5	V	All	I_{OL} < 1 μA, Inputs at 1.5 or 3.5 V	
I_{OH}	Output HIGH Current			−0.63			mA	MIN, 25°C	V_{OUT} = 4.6 V	Inputs at 0 or 5 V per the Logic Function or Truth Table
				−0.36				MAX		
I_{OL}	Output LOW Current			1			mA	MIN, 25°C	V_{OUT} = 0.4 V	
				0.8						
				0.4				MAX		
C_{IN}	Input Capacitance Per Unit Load					7.5	pF	25°C	Any Input	
I_{DD}	Quiescent Power Supply Current	Gates	XC			1	μA	MIN, 25°C		
						7.5		MAX		
			XM			0.25	μA	MIN, 25°C		
						7.5		MAX		
		Buffers and Flip-Flops	XC			4	μA	MIN, 25°C	All Inputs at 0 V or V_{DD} for all Valid Input Combinations	
						30		MAX		
			XM			1	μA	MIN, 25°C		
						30		MAX		
		MSI	XC			20	μA	MIN, 25°C		
						150		MAX		
			XM			5	μA	MIN, 25°C		
						150		MAX		

4049B • 4050B
4049B HEX INVERTING BUFFER • 4050B HEX NON-INVERTING BUFFER

DESCRIPTION — These CMOS buffers provide high current output capability suitable for driving TTL or high capacitance loads. Since input voltages in excess of the buffers' supply voltage are permitted, these buffers may also be used to convert logic levels of up to 15 V to standard TTL levels. The 4049B provides six inverting buffers, the 4050B six non-inverting buffers. Their guaranteed fan out into common bipolar logic elements is shown in Table 1.

4049B LOGIC AND CONNECTION DIAGRAM DIP (TOP VIEW)

4050B LOGIC AND CONNECTION DIAGRAM DIP (TOP VIEW)

NOTE:
The Flatpak version has the same pinouts (Connection Diagram) as the Dual In-line Package.

INPUT PROTECTION

INPUT 200 Ω NOMINAL
INPUT ○——▷ TO LOGIC TRANSISTORS
D1
TO V_{SS}
NOTE: Typical Breakdown Voltage of Diode D1 is 20 V.

TABLE 1
Guaranteed fan out of 4049B, 4050B into common logic families

DRIVEN ELEMENT	GUARANTEED FAN OUT
Standard TTL, DTL	2
9LS, 93L, 74LS	9
74L	16

Conditions: $V_{DD} = V_{CC} = 5.0 \pm 0.25$ V
$V_{OL} \leq 0.5$ V, $T_A = 0$ to 75 C

DC CHARACTERISTICS: V_{DD} as shown, $V_{SS} = 0$ V, 4049BXM and 4050BXM (See Note 1)

SYMBOL	PARAMETER	V_{DD} = 5 V			V_{DD} = 10 V			V_{DD} = 15 V			UNITS	TEMP	TEST CONDITIONS
		MIN	TYP	MAX	MIN	TYP	MAX	MIN	TYP	MAX			
I_{OH}	Output HIGH Current	-1.85 -1.25 -0.9	-2.5								mA	MIN, 25°C MAX	$V_{OUT} = 2.5$ V for $V_{DD} = 5$ V Inputs at 0 or V_{DD} per Function
		-0.62 -0.5 -0.35	-1		-1.85 -1.25 -0.9	-2.5		-5.5 -3.75 -2.7	-7.5		mA	MIN, 25°C MAX	$V_{OUT} = 4.5$ V for $V_{DD} = 5$ V $V_{OUT} = 9.5$ V for $V_{DD} = 10$ V $V_{OUT} = 13.5$ V for $V_{DD} = 15$ V Inputs at 0 or V_{DD} per Function
I_{OL}	Output LOW Current	3.75 3 2.1	6		10 8 5.6	16		30 24 16.8	48		mA	MIN, 25°C MAX	$V_{OUT} = 0.4$ V for $V_{DD} = 5$ V $V_{OUT} = 0.5$ V for $V_{DD} = 10$ V $V_{OUT} = 1.5$ V for $V_{DD} = 15$ V Inputs at 0 or V_{DD} per Function
		3.3 2.6 1.8	5.2								mA	MIN, 25°C MAX	$V_{OUT} = 0.4$ V for $V_{DD} = 4.5$ V Inputs at 0 V or V_{DD} per Function
I_{DD}	Quiescent Power Supply Current		1 30			2 60			4 120		μA	MIN, 25°C MAX	All Inputs at 0 V or V_{DD}

Notes on the following page.

DC CHARACTERISTICS: V_{DD} as shown, V_{SS} = 0 V, 4049BXC and 4050BXC (Cont'd) (See Note 1)

SYMBOL	PARAM- ETER	LIMITS									UNITS	TEMP	TEST CONDITIONS
		V_{DD} = 5 V			V_{DD} = 10 V			V_{DD} = 15 V					
		MIN	TYP	MAX	MIN	TYP	MAX	MIN	TYP	MAX			
I_{OH}	Output HIGH Current	-1.5 -1.25 -1.0	-2.5								mA mA mA	MIN 25°C MAX	V_{OUT} = 2.5 V for V_{DD} = 5 V Inputs at 0 or V_{DD} per Function
		-0.6 -0.5 -0.4	-1		-1.5 -1.25 -1.0	-2.5		-4.5 -3.75 -3	-7.5		mA mA mA	MIN 25°C MAX	V_{OUT} = 4.5 V for V_{DD} = 5 V V_{OUT} = 9.5 V for V_{DD} = 10 V V_{OUT} = 13.5 V for V_{DD} = 15 V Inputs at 0 or V_{DD} per Function
I_{OL}	Output LOW Current	3.6 3.0 2.5	6		9.6 8 6.6	16		28 24 19	48		mA mA mA	MIN 25°C MAX	V_{OUT} = 0.4 V for V_{DD} = 5 V V_{OUT} = 0.5 V for V_{DD} = 10 V V_{OUT} = 1.5 V for V_{DD} = 15 V Inputs at 0 or V_{DD} per Function
		3.1 2.6 2.1	5.2								mA mA mA	MIN 25°C MAX	V_{OUT} = 0.4 V for V_{DD} = 4.5 V Inputs at 0 V or V_{DD} per Function
I_{DD}	Quiescent Power Supply Current			4 30			8 60			16 120	μA	MIN,25°C MAX	All inputs at 0 V or V_{DD}

AC CHARACTERISTICS AND SET-UP REQUIREMENTS: V_{DD} as shown, V_{SS} = 0 V, T_A = 25°C, 4049B only (See Note 2)

SYMBOL	PARAMETER	LIMITS									UNITS	TEST CONDITIONS
		V_{DD} = 5 V			V_{DD} = 10 V			V_{DD} = 15 V				
		MIN	TYP	MAX	MIN	TYP	MAX	MIN	TYP	MAX		
t_{PLH} t_{PHL}	Propagation Delay		65 50	130 105		30 25	65 50		29 17	52 40	ns	C_L = 50 pF, R_L = 200 kΩ
t_{TLH} t_{THL}	Output Transition Time		73 33	145 65		40 13	80 25		30 9	60 20	ns	Input Transition Times ≤ 20 ns

NOTES:
1. Additional DC Characteristics are listed in this section under 4000B Series CMOS Family Characteristics.
2. Propagation Delays and Output Transition Times are graphically described in this section under 4000B Series CMOS Family Characteristics.

ANSWERS TO SELECTED PROBLEMS

CHAPTER ONE

1-2. (a) 25 (b) 9.5625 (c) 1241.6875
1-4. 1023
1-5. Nine bits
1-6. (a) four circuits for parallel (b) one circuit for serial

CHAPTER TWO

2-1. (a) 22 (b) 141 (c) 2313 (d) 983 (e) 191
2-2. (a) 100101 (b) 1110 (c) 10111101 (d) 11001101
(e) 100100001001 (f) 111111111
2-3. 255; 65,535
2-4. (a) 483 (b) 30 (c) 2047 (d) 175 (e) 644
2-5. (a) 73_8 (b) 564_8 (c) 1627_8 (d) 200000_8 (e) 377_8
2-6. (a) 111100011 (b) 011110 (c) 011111111111
(d) 010101111 (e) 001010000100
2-7. (a) 26_8 (b) 215_8 (c) 4411_8 (d) 1727_8 (e) 277_8
2-8. 165, 166, 167, 170, 171, 172, 173, 174, 175, 176,
177, 200
2-9. $100100001001_2 = 4411_8$
2-10. (a) 146 (b) 422 (c) 14,333 (d) 704 (e) 2047
2-11. (a) 4B (b) 13A (c) 800 (d) 6413 (e) FFF
2-12. (a) 16 (b) 8D (c) 909 (d) 3D7 (e) BF
2-13. (a) 10010010 (b) 000110100110 (c) 0011011111111101
(d) 001011000000 (e) 011111111111
2-14. (a) 65,536 addresses (b) 000 to FFF (hex)
2-15. 280, 281, 282, 283, 284, 285, 286, 287, 288, 289,
28A, 28B, 28C, 28D, 28E, 28F, 290, 291, 292, 293, 294. 295,
296, 297, 298, 299, 29A, 29B, 29C, 29D, 29E, 29F, 2A0
2-16. (a) 01000111 (b) 100101100010 (c) 000110000111
(d) 01000010011010001001011000100111
2-17. 10 bits for binary, 12 bits for BCD
2-18. (a) 9752 (b) 184 (c) 7775 (d) 492
2-19. 11011000(X); 10111101 (=); 10110010 (2); 00110101 (5);
10101111 (/); 01011001 (Y)
2-20. D8, BD, B2, 35, AF, 59
2-21. (a) 110110110 (parity bit is leftmost bit) (b) 000101000
(c) 111110111
2-22. (a) 101110100 (parity bit on the left) (b) 000111000
(c) 0000101100101 (d) 11001001000000001

2-23. (a) no single-bit error (b) single-bit error
(c) double error (d) no single-bit error
2-25. (a) 10110001001 (b) 11111111 (c) 209 (d) 59,943
(e) 4701 (f) 777 (g) 157 (h) 2254 (i) 1961 (j)15,900
(k) 640 (l) 952B (m) 100001100101 (n) 947 (o) 135_{16}
(p) 5464_8 (q) 1001010 (r) 01011000 (BCD)
2-26. (a) 100101 (b) 00110111 (c) 01101010
(d) 0110011 0110111 (e) 45 (f) 25
2-27. (a) octal (b) 16 (c) digit (d) Gray
(e) Parity; single-bit errors (f) ASCII (g) octal; hex

CHAPTER THREE

3-1.

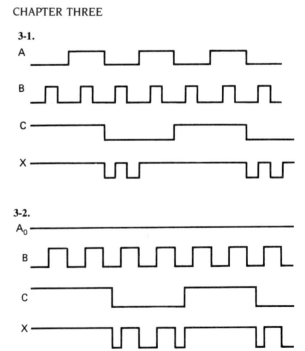

3-2.

3-3. x will be a constant HIGH.
3-5. 31
3-6. (a) x is HIGH only when A, B, and C are all HIGH.
(b) x is a constant LOW.
(c) x is HIGH only when B and C are simultaneously HIGH.
3-7. Change the OR gate to an AND gate.
3-8. OUT is always LOW.
3-10. $x = (\overline{A} + \overline{B})BC$
3-11. $x = \overline{A}\,\overline{B}\,\overline{C} + A\overline{B}\,\overline{C} + \overline{A}\,\overline{B}D$
3-12. $x = D \cdot (AB + C) + E$
3-13. $x = 1$ only when $B = C = 1$ and $A = D = 0$.
3-14. (a)

3-15.–3-18.

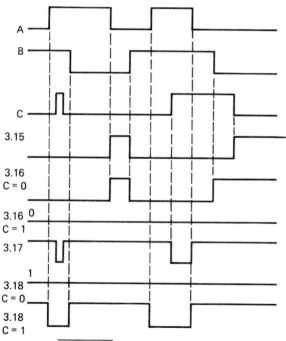

3-19. $x = \overline{(\overline{A} + B)} \cdot \overline{(BC)}$
3-20. $x = 0$ only when $A = B = 0$, $C = 1$.
3-21. (c)

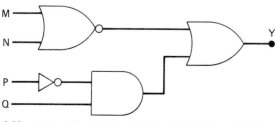

3-22. (a) 1 (b) A (c) 0 (d) C (e) 0 (f) D (g) D
(h) 1 (i) G (j) y

3-24. (a) $MP\overline{N} + \overline{M}\,P N$ (b) $B\overline{C}$
3-26. (a) $A + \overline{B} + C$ (b) $A \cdot (B + \overline{C})$ (c) $\overline{A} + \overline{B} + CD$
(d) $\overline{A} + B + \overline{C} + \overline{D}$ (e) $\overline{M}N + M\overline{N}$ (f) $(\overline{A} + \overline{B})C + \overline{D}$
3-27. $A + B + \overline{C}$
3-33. (a) NOR (b) AND (c) NAND
3-34.

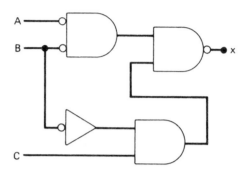

3-35. (a) Z goes HIGH only when $A = B = 0$ and $C = D = 1$.
(b) Z goes LOW when A or B is HIGH or when C or D is LOW.
3-37. X will go HIGH when $E = 1$, or $D = 0$, or $B = C = 0$, or when $B = 1$ and $A = 0$.
3-38. X is LOW for $EDCBA = 01011, 01100, 01101,$ and 01111.
3-39. (a) HIGH (b) LOW
3-40. $LIGHT = 0$ when $A = B = 0$ or $A = B = 1$.
3-41.

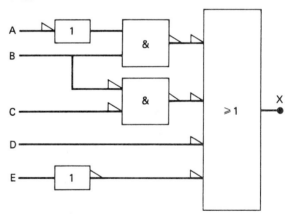

3-42. $Z = \overline{\overline{(AB + \overline{C})}F} + \overline{C} + \overline{(D + E)}$
3-45. Put INVERTERs on the A_7, A_5, A_4, A_2 inputs to the 74C30.

CHAPTER FOUR

4-1. (a) $C\overline{A} + CB$ (b) $\overline{Q}R + Q\overline{R}$ (c) $C + \overline{A}$ (d) $\overline{R}\,\overline{S}\,\overline{T}$
(e) $BC + \overline{B}(\overline{C} + A)$ (f) $BC + \overline{B}(\overline{C} + A)$ or $BC + \overline{B}\,\overline{C} + \overline{A}\,\overline{C}$
(g) $\overline{D} + AB\overline{C} + \overline{A}\,\overline{B}C$
4-2. $Q(M + N)$
4-3. $MN + Q$
4-4. One solution: $\overline{x} = \overline{B}C + AB\overline{C}$ another:
$x = \overline{A}B + \overline{B}\,\overline{C} + BC$; another $BC + \overline{B}\,\overline{C} + \overline{A}\,\overline{C}$
4-5. $x = \overline{A}\,\overline{B} + \overline{A}\,\overline{C} + \overline{B}\,\overline{C}$
4-6. $x = \overline{A}\,\overline{B}\,\overline{C} + ABC$
4-7. $x = \overline{A}_3(A_2 + A_1 A_0)$
4-8. $Alarm = ID + \overline{I}L$

4-9.

4-11.

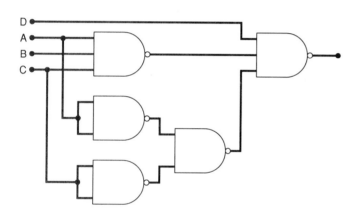

4-12. $x = BC + \bar{B}\bar{C} + AC$; or $BC + \bar{B}\bar{C} + A\bar{B}$
4-13. $y = \bar{D} + \bar{A}\bar{B}C + A\bar{B}\bar{C}$

	$\bar{C}\bar{D}$	$\bar{C}D$	CD	$C\bar{D}$
$\bar{A}\bar{B}$	1		1	1
$\bar{A}B$	1			1
AB	1			
$A\bar{B}$	1	1		1

4-14. $x = \bar{A}_3 A_2 + \bar{A}_3 A_1 A_0$

4-15. (a) $x = \bar{A}\bar{B} + \bar{A}\bar{C} + \bar{B}D + ABC\bar{D}$

	$\bar{C}\bar{D}$	$\bar{C}D$	CD	$C\bar{D}$
$\bar{A}\bar{B}$	1	1	1	1
$\bar{A}B$	1	1		
AB			1	
$A\bar{B}$		1	1	

(b) $x = \bar{A}\bar{D} + \bar{B}C + \bar{B}\bar{D}$ (c) $y = \bar{B} + AC$

	C	$\bar{C}$
$\bar{A}\bar{B}$	1	1
$\bar{A}B$	0	0
AB	1	0
$A\bar{B}$	1	1

4-16. Best solution: $x = B\bar{C} + AD$
4-17. $x = \bar{S}_1\bar{S}_2 + \bar{S}_1\bar{S}_3 + \bar{S}_3\bar{S}_4 + \bar{S}_2\bar{S}_3 + \bar{S}_2\bar{S}_4$
4-18. (a)

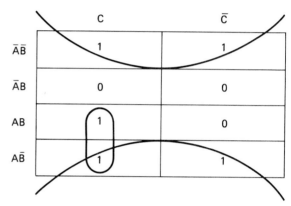

(b) $x = A$ (c) $x = \bar{A}$

4-19. $A = 0, B = C = 1$
4-20. One possibility:

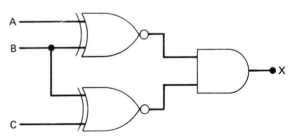

4-22. Four outputs where z_3 is the MSB

$z_3 = y_1 y_0 x_1 x_0$

$z_2 = y_1 x_1 (\overline{y_0} + \overline{x_0})$

$z_1 = y_0 x_1 (\overline{y_1} + \overline{x_0}) + y_1 x_0 (\overline{y_0} + \overline{x_1})$

$z_0 = y_0 x_0$

4-23. $x = A_3 A_2 + A_3 A_1$

4-24. $x = AB \overline{(C \oplus D)}$

4-25. $(A + B)(C + D)$

4-26. N-S $= \overline{C}\,\overline{D}(A + B) + AB(\overline{C} + \overline{D})$; E-W $= \overline{\text{N-S}}$

4-27. Change last EX-OR to EX-NOR.

4-29. $x = A\overline{B}C$

4-30. $x = A + BCD$

4-31. $x = A + (B \oplus C)$

4-32. $z = A_0 \overline{S} + A_1 S$

4-33. $z = x_1 x_0 y_1 y_0 + x_1 \overline{x_0} y_1 \overline{y_0} + \overline{x_1} x_0 \overline{y_1} y_0 + \overline{x_1}\overline{x_0}\overline{y_1}\overline{y_0}$

No pairs, quads, or octets.

4-34. (a) indeterminate   (b) 1.4–1.8 V   (c)

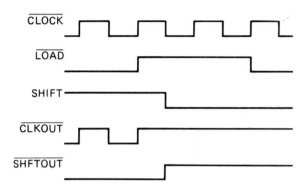

4-38. Possible faults:

faulty V_{CC} or ground on Z2;

Z2-1 or Z2-2 open internally or externally;

Z2-3 internally open

4-39. yes: c, e, f   no: a, b, d, g

4-41. Z2-6 and Z2-11 shorted together

4-43. Most likely faults:

faulty ground or V_{CC} on Z1

Z1 plugged in backwards

Z1 internally damaged

4-44. Possible faults:

Z2-13 shorted to V_{CC}

Z2-8 shorted to V_{CC}

broken connection to Z2-13

Z2-3, Z2-6, Z2-9, or Z2-10 shorted to ground

4-46. (a) 00_{16} to EF_{16} (b) F0   (c) F8 to FF

CHAPTER FIVE

5-1.

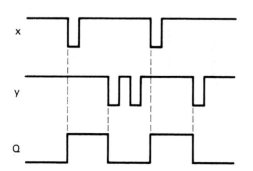

5-2. Same as 5-1

5-3.

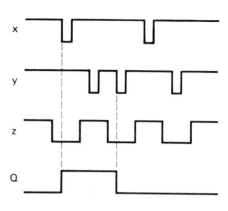

5-6. Z1-4 stuck HIGH

5-7. 20 ns

5-8.

5-9. Q is a 500-Hz square wave.

5-10.

5-11. (a) 5-kHz square wave   (b) 2.5 kHz

5-13.

5-14. 500-Hz square wave

5-15. Q stays LOW

5-18.

5-19.

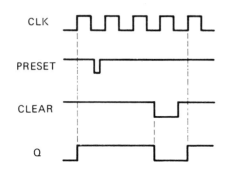

5-20. (a) 200 ns   (b) 7474   (c) 30 ns
5-21. 48 ns
5-22. Connect A to J, $\bar{A}$ to K.
5-23. (a) A, B, C
5-24. Connect X to J, $\bar{X}$ to K.
5-25. Use arrangement of Figure 5-43.
5-26. 1011, 1101, 1110, 0111, 1011, 1101, 1110, 0111, 1011
5-27. Connect X_0 to D input of X_3.
5-31. (a) 10   (b) 1953 Hz   (c) 1024   (d) 12
5-32. (a) 128   (b) 0 to 127
5-35. Put INVERTERs on A_{11} and A_{14}.
5-38. (a) $\bar{Q}$ will stay LOW.   (b) PGT at $\bar{Q}$ will set $X = 1$.
(c) make $t_p = 20$ μs.
5-39. (a) A_1 or A_2 has to be LOW when a PGT occurs at B.
(b) B and A_2 have to be HIGH when a NGT occurs at A_1.
5-41. After 14 pulses, all circuit outputs stop changing with
$A = W = X = Y = Z = 0$ and $B = C = 1$.
5-42. One possibility is $R = 1$ kΩ and $C = 80$ nF.
5-43. One possibility is $R_A = 1$ kΩ, $R_B = 10$ kΩ, $C = 1800$ pF.
5-45.

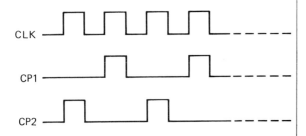

5-46. (a) no   (b) yes   (c) no   (d) no
5-47. Flip-flop X_0 has an open D input.
5-48. (a) yes   (b) no
5-49. Two cascaded INVERTERs between Q_1 and Q_2 add enough
delay to increase the effective t_{PLH} of Q_1 so that by the time Q_2
gets clocked, the Q_1 signal hasn't reached D_2.
5-51. (a) no   (b) no   (c) yes   (d) no
5-52. first combination: 101   second combination: 010
5-53. (a) no   (b) no   (c) yes
5-54. (a) S-C latch   (b) JK   (c) D latch   (d) D flip-flop
(e) D flip-flop and latch   (f) all FFs   (g) clocked SC, JK, D
(h) JK

CHAPTER SIX

6-1. (a) 10101   (b) 10010   (c) 1111.0101   (d) 1.1010
(e) 100111000
6-2. (a) 00100000 (incl. sign bit)   (b) 11110010
(c) 00111111   (d) 10011000   (e) 11111111   (f) 10000000
(g) can't be done   (h) 00000000

6-3. (a) +13   (b) −3   (c) +123   (d) −103   (e) +127
(f) −32   (g) −1   (h) −127
6-4. (a) −2048 to +2047   (b) sixteen bits including sign bit
6-5. -16_{10} to $+15_{10}$
6-6. (a) 164   (b) −92   (c) illegal BCD   (d) \$ (e) −36
6-7. 0 to 1023; −512 to +511
6-9. (a) 00001111   (b) 11111101   (c) 11111011
(d) 10000000   (e) 00000001   (f) 11011110   (g) 00000000
(h) 00010101
6-11. (a) 100011   (b) 1111001   (c) 100011.00101
(d) .10001111
6-12. (a) 11   (b) 111   (c) 101.11   (d) 1111.0011
6-13. (a) 10010111 (BCD)   (b) 10010101 (BCD)
6-14. (a) 6E24   (b) 100D
6-15. (a) 0EFE   (b) 229   (c) 02A6
6-16. 16,849 locations
6-18. SUM $= A \oplus B$ CARRY $= AB$
6-20. [A] $= 0000$
6-21. 200 ns
6-22. Overflow $= \bar{A}_3 \bar{B}_3 S_3 + A_3 B_3 \bar{S}_3$
6-23. $C_3 = A_2 B_2 + (A_2 + B_2)\{A_1 B_1 + (A_1 + B_1)[A_0 B_0 + A_0 C_0 + B_0 C_0]\}$
6-25. (a) SUM $= 0111$ (b) SUM $= 1010$

6-26.

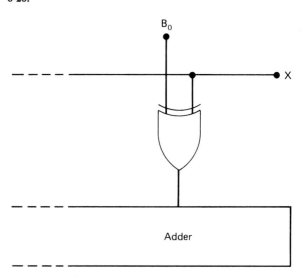

6-27. [S] $= 01110$; $x = 1$; [Σ] $= 0100$
6-28. Yes
6-29. [Σ] $= 100001000101$ (BCD)
6-30. After fourth clock pulse: [A] $= 00111111$;
[B] $= 01110000$ [X] $= 0000$; [S] $= 10101111$
6-31. (a) no   (b) yes   (c) no
6-34. (a) yes   (b) no   (c) yes   (d) no
6-36. Use D flip-flops. Connect $S_3 + S_2 + S_1 + S_0$ to the D
input of the zero FF; C_4 to the D input of the carry FF; and S_3 to
the D input of the sign FF.

CHAPTER SEVEN

7-1. 250 kHz, 50 percent
7-2. Same as 7-1
7-3. 10000_2
7-4. Five FFs are required: Q_0–Q_4 with Q_4 as the MSB.
Connect Q_3 and Q_4 outputs to a NAND gate whose output is
connected to all CLRs.

7-5.

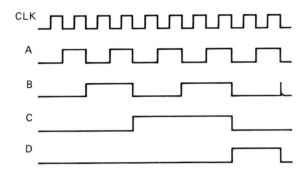

7-7. 100, 011, 010, 001, 000, and repeats
7-8. (b) 000, 001, 010, 100, 101, 110, and repeats
7-9.

7-10. A MOD-15 (from Problem 7-9) driving a MOD-4.
7-11. 60 Hz
7-12. (c) 0001
7-13. 1000 and 0000 states never occur.
7-14. (a) 12.5 MHz   (b) 8.33 MHz
7-15. (a) Add two FFs (E and F) to Figure 7-60. Connect AND

gates below to appropriate FF inputs.   (b) 33 MHz
(c) 16.7 MHz

7-16. 0000, 0001, 0010, 0011, 0100, 0101, 0110, 0111, 1000, 1001, and repeats
7-18. (d) Z will not be cleared and timer cannot be restarted.
7-19. Change the parallel data input to 1010.
7-20. Nine
7-21. (a) ten (b) clears counter to 0000
(c) sets counter to 1001   (d) up
(e) Connect Q_0 to CP_1 and clock signal to CP_0; ground MR_1, MR_2, MS_1, and MS_2.
(f) Connect clock signal to CP_1; connect Q_3 to CP_0; ground all MR and MS inputs.

7-22. Changes are shown below.

7-25.

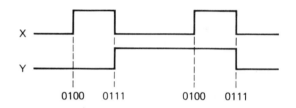

7-27. (a) when the counter goes from 0111 to 1000
7-28. Twelve FFs
7-29. Four; sixteen; thirteen
7-30. (a) $J_A = B\overline{C}$, $K_A = 1$, $J_B = CA + \overline{C}\overline{A}$, $K_B = 1$, $J_C = \overline{A}B$, $K_C = B + \overline{A}$
(b) $J_A = B\overline{C}$, $K_A = 1$, $J_B = K_B = 1$, $J_C = K_C = B$
7-31. $J_A = K_A = 1$, $J_B = K_B = \overline{A}$, $J_C = K_C = \overline{A}\,\overline{B}$, $J_D = K_D = \overline{A}\,\overline{B}\,\overline{C}$
7-33.

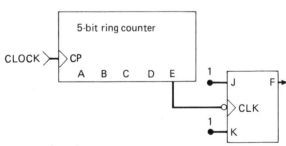

Counting sequence

A	B	C	D	E	F	
1	0	0	0	0	0	
0	1	0	0	0	0	
0	0	1	0	0	0	
0	0	0	1	0	0	
0	0	0	0	1	0	Repeats
1	0	0	0	0	1	
0	1	0	0	0	1	
0	0	1	0	0	1	
0	0	0	1	0	1	
0	0	0	0	1	1	

7-34. Counting sequence is: 00000, 10000, 11000, 11100, 11110, 11111, 01111, 00111, 00011, 00001, and repeat.

7-35. Frequency at z is 5 Hz.

7-36. (b) 257 (c) 323

7-37. (a) 22 (b) 450 (c) 0 or 1

7-38. Connect Q_3 of the 74290 to $\overline{CP}_1$ of the 74293.

7-40. Add a second J-K flip-flop.

7-42. Connect Q_3 of one to D_S of the other.

7-43. The 74178 is loaded with 1101 on the NGT of the sixth clock pulse.

7-44. 64 μs

7-45. (b) $Q_0 \cdot \overline{Q}_1 + Q_0 \cdot Q_7$ (c) $Q_3 \cdot \overline{Q}_6$

7-47. (a) asynchronous (b) true (c) left shift (d) right shift (e) parallel load (f) hold data (g) recirculating left

7-48. MR_1 to Q_1 is open, or Q_3 or MR_2 is shorted to ground.

7-49. DC CLR of flip-flop X is open or stuck HIGH or not responding.

7-50. MSB of MOD-6 is stuck LOW.

7-51. (a) Display = 001 (b) Display = 000

7-52. Counters are not being cleared before each SAMPLE interval.

7-53. (a) Flip-flop C will toggle on every NGT of the input. (b) Will not count up properly.

7-54. P_1 open or S3 faulty.

7-56. (a) parallel (b) binary (c) down counter (d) MOD-10 (e) asynchronous (f) ring (g) Johnson (h) all (i) presettable (j) up/down (k) asynchronous (l) BCD (m) synchronous

CHAPTER EIGHT

8-1. (a) A, B (b) A (c) A

8-2. (a) 728.3 pJ (b) 459.2 pJ (c) 58.95 pJ (d) 27.41 pJ (e) 67.26 pJ

8-3. (a) 0.9 V (b) 1.4 V

8-4. (a) I_{IH} (b) I_{CCL} (c) t_{PHL} (d) speed-power product (e) V_{NH} (f) surface-mount (g) current sinking (h) fan-out (i) totem-pole (j) pull-down transistor (k) 4.75 to 5.25 V (l) 2.4 V; 2.0 V (m) 0.8 V; 0.4 V

8-5. (a) 0.7 V; 0.3 V (b) 0.5 V; 0.4 V (c) 0.5 V; 0.3 V

8-6. (a) 40 μA; 1.6 mA (b) AND, NAND (c) Unconnected inputs

8-7. (c) five

8-8. (a) 30/15 UL (b) 24 mA

8-10. Fan-out not exceeded.

8-11. Fan-out not exceeded.

8-12. 67 ns

8-13. (a) 500 Ω (b) 2 kΩ

8-14. (b) 4.7-kΩ resistor is too large

8-15. (a) amplitude too low (b) t_P is too small (c) $t_W(L)$ too low

8-18. six AND gates

8-19. AB + CD + FG

8-20. $R_P > 359$ Ω

8-21. (a) 5 V (b) $R_S = 110$ Ω for LED current of 20 mA

8-23. (a)

x	y	Data on Bus
0	0	C
0	1	B
1	0	B
1	1	A

8-24. R_P (max) = 5.77 kΩ

8-25. AND

8-29. $V_{DD} = 5$ V, $f_{max} = 10$ kHz.

8-30. 3.6 V for both states

8-31. Using values at $V_{CC} = 6$ V, the speed-power product is 1.44 pJ per gate.

8-32. 0 V, 1.22 V

8-33.

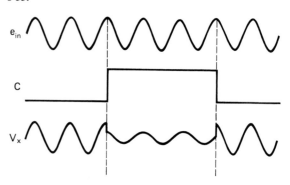

8-34. −1 and −2

8-35. (a) 74HCT (b) converts logic voltages (c) CMOS cannot sink TTL current (d) false

8-36. The choice in (c) is a possible reason.

8-37. (a) none (b) two

8-42. The choice in (b) is a possible fault.

8-43. NAND CMOS gate is shorted to +5 V.

8-44. 0 V to −11.25 V and back to −6 V.

CHAPTER NINE

9-1. (a) all HIGH (b) $\overline{O}_0$ = LOW (c) $\overline{O}_7$ = LOW (d) all HIGH

9-2. six inputs, 64 outputs

9-3. $[A]$ = 110, $E_3 = 1$, $\overline{E}_1 = \overline{E}_2 = 0$

9-4.

9-5.

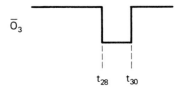

9-6. Connect $\overline{E}_2$ to ground and take output from $\overline{O}_2$.

9-7. Enabled when $D = 0$.

9-8.

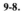

9-10. Resistors are 250 Ω.

9-12.

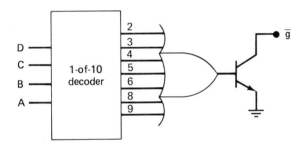

9-13. (a) encoder (b) encoder (c), (d), (e) decoder
9-14. 0111
9-17. The fourth key actuation would be entered into the MSD register.
9-18. Choice (b)
9-19. Choice (a) or (d)
9-20. (a) yes (b) no (c) no
9-21. A_2 bus line is open between Z2 and Z3.
9-22. E_2 of Z4 is stuck LOW.
9-23. g segment or decoder output transistor would burn out.
9-24. Decoder D and C inputs are reversed.
9-25. Decoder outputs: a and b are shorted together.
9-27. A 4-to-1 MUX
9-28. Use nine 74151s.
9-29.

9-30. (b) The total number of connections in the circuit using MUXes is 63, not including V_{CC} and GND, and not including the connections to counter clock inputs. The total number for the circuit using separate decoder/drivers is 66.

9-31.

9-32.

9.22

9-33.

A	B	C	
0	0	0	$0 \rightarrow I_0$
0	0	1	$0 \rightarrow I_1$
0	1	0	$0 \rightarrow I_2$
0	1	1	$1 \rightarrow I_3$
1	0	0	$0 \rightarrow I_4$
1	0	1	$1 \rightarrow I_5$
1	1	0	$1 \rightarrow I_6$
1	1	1	$1 \rightarrow I_7$

9-34. Connect I_1, I_5, I_8, I_{11}, I_{14}, I_{15} to V_{CC}. All other inputs connect to ground.
9-35. Z = HIGH for $DCBA$ = 0010, 0100, 1000, 1010.
9-36. (b) $Z = \overline{A}B\overline{C} + AB\overline{C} + \overline{A}\,\overline{B}C + A\overline{B}C + \overline{A}\,\overline{B}\,\overline{C}\,\overline{D} + ABCD$
9-37. (a) encoder, MUX (b) MUX, DEMUX (c) MUX (d) encoder (e) decoder, DEMUX (f) DEMUX (g) MUX
9-38.

9-39. Each DEMUX output goes LOW, one at a time in sequence.
9-40. (a) All LEDs off. (b) Each LED will flash. (c) LEDs 2 and 6 will flash.
9-41. Five lines
9-43. (a) Sequencing stops after actuator 3 is activated. (b) Same as (a).
9-44. Probable fault is short to ground at MSB of tens MUX.
9-45. Q_0 and Q_1 are probably reversed.
9-46. Inputs 6 and 7 of MUX are probably shorted together.
9-48. (a) no (b) yes
9-51. Final binary output is 1000101.
9-52. A_0 and B_0 probably reversed.
9-53. $\overline{OE}_C = 0$, $\overline{IE}_C = 1$; $\overline{OE}_B = \overline{OE}_A = 1$; $\overline{IE}_B = \overline{IE}_A = 0$; apply a clock pulse
9-54. (a) At t_4, each register will have contents of 1011.

9-55.

9-56. (b) ON, ON, ON, OFF (left to right).
9-57. (a) At t_3, each register holds 1001.
9-58. (a) 57FA (b) 5000 to 57FF (c) 9000 to 97FF (d) no

CHAPTER TEN

10-1. (f, g) false
10-2. 3.58 V
10-3. LSB = 20 mV
10-4. 20 mV; approx. 0.4%
10-5. Approx. 5 mV
10-6. eight
10-7. 14.3%, 0.286 V
10-9. 250.06 RPM
10-10. 10 mV, 0.1%, 6.95 V
10-11.

Binary	BCD
3.88 mV	10 mV
0.392%	1.01%

10-13. 800 ohms; no
10-14. (a) −0.3125 V, −4.6875 V (b) 4.27 kilohms
(c) −0.0267
10-15. Uses fewer different R values.
10-16. 20 μA; yes
10-17. (a) seven
10-19. 242.5 mV is not within specifications.
10-20. Offset out of spec.
10-21. Bit 1 of DAC is open or stuck HIGH.
10-22. Bits 0 and 1 reversed.
10-24. (a) 10010111 (b) 10010111 (c) 102 μs, 51 μs
10-27. (a) 1.2 mV (b) 2.7 mV
10-28. (a) 0111110110 (b) 0111110111
10-29. 2.834 to 2.844 V
10-31. Digital ramp: a, d, e, f, h,   SAC: b, c, d, e, g, h
10-34. 80 μs
10-35. 100101
10-36. 2.96 V
10-37. 2.37 V
10-39. Nine comparators, ten resistors
10-40. Flash: a, c, d, f, g SAC: b, f, g
Digital-ramp: b, e, f
10-41. (a) up/down (b) flash (c) voltage-frequency
(d) voltage-frequency (e) dual-slope (f) dual-slope
10-42. 386 steps
10-43. The MSB of the MSD is not affecting V_{AX}.
10-45. Switch is stuck closed; switch is stuck open, or capacitor is shorted.
10-47. Address is EAxx.

CHAPTER ELEVEN

11-1. 16,384; 32; 524,288
11-2. 16,384
11-3. 64K × 4
11-4. 16; 16; 13; 16,384

11-6. (a) data, address, control (b) address (c) data
(d) CPU
11-7. (a) Hi-Z (b) 11101101
11-8. (a) register 11 (b) 0100
11-9. (a) 16,384 (b) four (c) two 1-of-128 decoders
11-11. 120 ns
11-12. 180 ns
11-13. (a) PROM (b) MROM (c) EPROM, EEPROM
(d) EPROM (e) EEPROM (f) PROM (g) EEPROM
11-14. Q_{13}, Q_{14}, Q_{15}
11-15. The following transistors will have open base connections:
Q_0, Q_2, Q_5, Q_6, Q_7, Q_9, Q_{15}.
11-17. The waveform at D_0 is:

11-18. Hex data: 5E, BA, 05, 2F, 99, FB, 00, ED, 3C, FF, B8, C7, 27, EA, 52, 5B
11-19.

A_5	A_4	A_3	A_2	A_1	A_0	D_2	D_1	D_0	C_3	C_2	C_1	C_0
1	0	0	1	0	1	0	1	1	0	1	1	1
0	0	1	1	1	0	0	0	1	0	1	0	0
1	1	1	1	1	0	1	1	0	0	0	1	0
0	1	1	1	0	0	0	1	0	0	1	1	0
1	1	0	1	0	1	1	0	1	0	0	1	1
0	0	0	1	1	0	0	0	0	0	1	1	0
0	1	0	0	1	1	0	0	1	1	0	0	1
1	1	0	0	0	0	1	0	0	1	0	0	0

11-20. $O_3 = D$; $O_2 = C\overline{D} + \overline{C}D$; $O_1 = B\overline{C} + \overline{B}C$;
$O_0 = A\overline{B} + \overline{A}B$
11-21. $O_3 = D + 0 + 0 + 0$
$O_2 = C\overline{D} + \overline{C}D + 0 + 0$
$O_1 = B\overline{C} + \overline{B}C + 0 + 0$
$O_0 = A\overline{B} + \overline{A}B + 0 + 0$
11-23. (a) 100 ns (b) 30 ns (c) ten million (d) 20 ns
(e) 30 ns (f) 40 ns (g) ten million
11-24. eight

11-25.

11-26.

11-29. *MUX* of Figure 11-55 is inverted and connected to ROW ENABLE of Figure 11-35.

11-32. Add four more PROMs (PROM-4 through PROM-7) to the circuit. Connect their data outputs and address inputs to data and address bus respectively. Connect AB_{13} to *C* input of decoder, and connect decoder outputs 4 through 7 to *CS* inputs of PROMs 4 through 7 respectively.

11-33. Connect AB_{13}, AB_{14}, and AB_{15} to a 3-input OR gate, and connect the OR gate output to *C* input of the decoder.

11-34. (a) 32×8 (b) RAMs 2 and 3 (c) 00–0F; 10–1F

11-37. * PEACE

11-39. F000–F3FF; F400–F7FF; F800–FBFF; FC00–FFFF

11-41. *B* input of decoder is open or stuck HIGH.

11-42. The OR gate output/*C* input node is stuck LOW.

11-43. Only RAM modules 1 and 3 are getting tested.

11-44. The RAM chip with data outputs 4 through 7 shows a fault at each bit.

11-45. RAM module 3 output 7 is open or stuck HIGH.

11-47. Checksum = 11101010.

INDEX